Climate Development and History of the North Atlantic Realm

Springer

Berlin
Heidelberg
New York
Barcelona
Hong Kong
London
Milan
Paris
Tokyo

Gerold Wefer • Wolfgang H. Berger
Karl-Ernst Behre • Eystein Jansen
Editors

Climate Development and History of the North Atlantic Realm

With 189 Figures, 18 in Color, and 16 Tables

Springer

Editors:

Professor Dr. Gerold Wefer
Universität Bremen, Fachbereich Geowissenschaften,
Klagenfurter Straße 2, 28359 Bremen, Germany

Professor Dr. Wolfgang H. Berger
Scripps Institution of Oceanography, Geological Research Division,
University of California, San Diego, La Jolla, CA 92093-0215, USA

Professor Dr. Karl-Ernst Behre
Niedersächsisches Institut für historische Küstenforschung,
Viktoriastraße 26/28, 26483 Wilhelmshaven, Germany

Professor Dr. Eystein Jansen
University of Bergen, Department of Geology,
Allégaten 41, 5007 Bergen, Norway

Cover motif:
Landsat 5 Thematic Mapper Data,
Brockmann Consult, Geesthacht, Germany

ISBN 3-540-43201-9 Springer-Verlag Berlin Heidelberg New York

Library of Congress Cataloging-in-Publication Data

Hanse Conference on Climate History.
Climate development and history of the North Atlantic realm / Gerold Wefer... [et al.], editors.
p. cm. Includes bibliographical references.
ISBN 3540432019 (alk. paper)
1. Paleoclimatology--North Atlantic Ocean Region--Congresses. 2. Paleoclimatology--Holocene--Congresses. 3. North Atlantic Ocean Region--Climate--History--Congresses. 4. Nature Effect of human beings on--North Atlantic Ocean Region--Congresses. I. Wefer, G. (Gerold) II. Title.
QC884.5.N67 H36 2002 551.69163´1--dc21 2002020799

Springer-Verlag Berlin Heidelberg New York
a member of BertelsmannSpringer Science+Business Media GmbH

http://www.springer.de

Printed in Germany

Production: PRO EDIT GmbH, Heidelberg
Cover Design: Erich Kirchner, Heidelberg
Typesetting: Camera-ready by Volker Diekamp

Printed on acid-free paper SPIN: 10865363 32/3130/Di 5 4 3 2 1 0

Preface

The global environment is changing rapidly under the impact of human activities, and an important element of this change is related to global climate modification.

Can the study of climate and history help in devising strategies for coping with this change? What might be the type of information most useful in this context? What are the pitfalls awaiting the unwary? These are the kinds of questions that led us to bring together experts from the natural and social sciences with a strong interest in history, to promote discussion between workers in different disciplines by focussing on a common topic of great interest to society.

The meeting was arranged in the framework of a "Hanse Conference" within the interdisciplinary program of the Hanse-Wissenschaftskolleg, a foundation set up to promote interdisciplinary studies in collaboration between the universities of Bremen and Oldenburg. The aim of the Hanse Conferences in general is to provide opportunities for experts from different fields of the sciences and humanities to come together and explore the larger framework of topics of common interest. What unites the participants is their desire to look over the fence to neighboring disciplines. Young colleagues who wish to build an interdisciplinary career are particularly welcome.

In the Hanse Conference on Climate and History, we have endeavoured to build bridges between the climate sciences and the sociological sciences concerned with environmental impacts on human activities. The geological sciences, we felt, are especially well suited to the purpose because they already comprise historical aspects.

The theme of the conference is extremely broad. We decided to narrow the scope by putting the North Atlantic realm at the center of the discussion, and by emphasizing evidence for change during the Holocene. In many ways, the North Atlantic realm is central to the global climate, through its role as a hub for deep convection of the ocean, which is an important factor in the heat budget of our planet. For the northern hemisphere, and especially for Europe, the storms originating in the regions where the cold air masses over the Greenland ice meet the warm air masses brought by the Gulf Stream system and its extension are a crucial part of the weather machine. In addition to its status as a climate amplifier on a global scale, the North Atlantic realm has many of the most detailed and most interesting climate records, and the countries of this realm have historical records that are well kept and accessible.

A strong scientific tradition in Europe, in relating climate to history, is exemplified by the fundamental works of H.H. Lamb in England and E. Le Roy Ladurie in France and documented in a long list of relevant publications (most recently in the monumental series "European Climate and Man" edited by B. Frenzel). This tradition provides another motivation to concentrate on the North Atlantic realm with emphasis on Europe.

In conducting the conference, we have attempted to avoid the disadvantages common to many large scientific meetings characterized by information-overload and lack of time for discussion. Instead, we have loosely followed the model of the "Dahlem Konferenzen", introduced by the late Dr. Silke Bernhard. An advisory committee of six scientists from different disciplines (W.H. Berger, J. Grove, E. Jansen, J. Pätzold, C. Pfister, G. Wefer) met well before the conference to formulate the overall goal and the themes of four discussion groups. This committee also was responsible for producing an initial list of invited participants, a list subsequently expanded through the recommendations of invitees. We aimed for about 40 scientists, complemented by selected graduate students and postdoctoral researchers. The conference was set for four days. Within each of the four theme sections, several participants were asked to provide background papers in their fields, as a basis for discussion. The aim is to have these papers sent as drafts to all participants one month before the conference, to stimulate the formulation of questions and critical comments.

The focus of activity within the Hanse Conferences is discussion, and not presentation of talks. The participants come with the background knowledge acquired through study of the overview papers prepared for the conference. On the first day of the conference each of the four discussion groups agree on a list of topics derived from the questions and comments that arise from the study of the background papers. The following two days are dedicated to debating these topics, within the four discussion groups. On the fourth day, each group considers a summary prepared by its rapporteur, who presents the most important results of the discussions. Suggestions for modifications to the summary are incorporated into the final summary, which is given at the end of the conference to the entire assembly, by each of the rapporteurs. At this meeting, comments are invited by all participants on any of the points raised.

The final proceedings, which are published in this book, begin with a general section containing an overall introduction and overview papers of broad interest. The thematic sections then follow, ordered from physical process to historical narrative and from long-term to short-term change. The final report from each group introduces each of the thematic sections, which contain the revised background papers and any additional material on pertinent subjects solicited by the editors. All papers benefited from peer review. It is hoped that they will be useful in informing the ongoing discussions on climate change and potential societal impacts, wherever such debate may take place. We especially hope that high-school and college teachers find much in these proceedings to enrich their courses in environmental sciences. In the educational realm, a marriage between physical understanding of Earth's life support systems and an appreciation of history leading to responsibility will be necessary to provide the basis for political action which can deal with the challenge of man-made climate change.

We are grateful to the following colleagues for their efforts in reviewing and offering many valuable suggestions for improving the manuscript: J. Alheit, J. Baten, T. Bickert, R. Brázdil, L. Dupont, K. Emeis, B.W. Flemming, J. Grove, C. Hammer, D. Hebbeln, E. Jansen, F. Lamy, H.J. Lindeboom, Luterbacher, W. Karlen, J. Meincke, J. Negendank, A. Nesje, A. Paul, R. Schlitzer, M. Schulz, V. Smetacek, T. van Weering, H. von Storch.

For their support we would like to thank V. Diekamp, T. Klein, A. Grimm-Geils, W. Hale, I. Mehser, M. Wachholz-Logemann, S. Middendorf, B. Schilder-Lübben and Antje Spitzner. Publication of this book was supported by the Hanse-Wissenschaftskolleg, and by the University of Bremen (Research Center Ocean Margins of German Research Foundation).

Gerold Wefer and Wolfgang H. Berger
(Convenors)

Contents

Climate History and the Great Geophysical Experiment

Wolfgang H. Berger

University of California, California Space Institute, La Jolla, CA 92093-0524, USA
e-mail: wberger@ucsd.edu

Abstract: Climate history is an important resource in the study of the Great Geophysical Experiment being performed on the planet, that is, the large-scale release of greenhouse gases (carbon dioxide and methane, and also others). Climate history is crucial in establishing that man-made global warming is an historic fact. The probability that recent warming has occurred independently of greenhouse gases derived from human activity is very small. The task is to make the best possible assessment of potential effects of global warming on regional agriculture and public health. The Great Experiment is a test of the response of the climate system to disturbance. It is also a test of the ability of human society to deal with environmental problems on a global scale.

In history there is a rich store of experience concerning climate change, which we need to mobilize to assess the risks of the future. Information can be derived on many different time scales: from the last 1000 years (containing the "Little Ice Age"), from deglaciation (containing the enigmatic "Younger Dryas" cold spell) and even from periods millions of years ago. Climate history is not supposed to deliver analogs, but to expand our thinking and make us ask pertinent questions. Perhaps the worst of the bad-case scenarios associated with global warming is the notion that ocean warming can lead to large-scale release of methane from the sea floor. Apparently it did in the distant past, at the end of the Paleocene, 55 million years ago. Venting of methane is proceeding right now. The question is, will it accelerate.

Traditional historians are reluctant to admit a strong role for climate in human history. However, as climate is becoming a product of human activities, it automatically becomes part of history; climate has entered the narrative of what humans do to each other.

Introduction

The oceanographer and geologist Roger Revelle (1909-1991), director of Scripps Institution of Oceanography in the 1950s, coined the widely adopted phrase about the "great geophysical experiment" that humankind is now performing with the entire planet as its laboratory. Revelle was among the first to realize that the ocean's reaction to the large-scale release of carbon dioxide would be of crucial importance in the course of this global experiment (Revelle and Suess 1957). Also, well before global warming became an issue, he was aware that the most pressing need, decades later, would be for baseline studies to provide the background for the developments to come. Thus, Revelle was a strong supporter of the International Geophysical Year (in 1957) and of starting systematic measurements of the chemistry of the atmosphere. That same year Charles D. Keeling, who joined Scripps Institution of Oceanography for the purpose, began such measurements. The constantly updated "Keeling Curve", showing the inexorable rise of carbon dioxide superimposed on a surprisingly strong annual fluctuation, has since become the central symbol of the study of global change (Fig. 1). The continuous measurements started by Keeling on Mauna Loa in Hawaii mark the beginning of a systematic study of the chemistry of the atmosphere in the context of global change.

It has since become clear that the instrumental record is much too short to contain the full range of

From WEFER G, BERGER W, BEHRE K-E, JANSEN E (eds), 2002, *Climate Development and History of the North Atlantic Realm.* Springer-Verlag Berlin Heidelberg, pp 1-16

natural variation of the system. Thus, the necessity to obtain more comprehensive information on natural background fluctuation has made climate history an important topic in geophysics. In the process, the distance from climate history to human history has become less daunting, with a growing realization that changes in climate have impacted human affairs in profound ways. The collapse of the Classic Maya civilization (Hodell et al. 1995), the pre-history of Egypt (Sandweiss et al. 1999), the demise of the Akkadian empire (Weiss et al. 1993; Cullen et al. 2000), and the pre-history of Peru with its El Niño events (Sandweiss et al. 1996; DeVries et al. 1997; Wells and Noller 1997) are topics arising in this context, engendering much discussion.

Also, there is a proliferation of books and symposia on climate history, beginning with the pioneers (Ladurie 1971; Lamb 1972, 1977, 1982; Wigley et al. 1981) and greatly expanding within the last several years (Flohn and Fantechi 1984; Berger and Labeyrie 1987; Grove 1988; Bradley and Jones 1992; Frenzel et al. 1992; Hughes and Diaz 1994; Glantz 1996; Jones et al. 1996; Alverson et al. 1999; Pfister et al. 1999). Climate is usually the focus in these works, rather than history, but boundaries become blurred as history is mined for geophysical information.

Geophysicists, used to talking mainly to each other, are learning to cope with societal reactions to their findings. A few years ago, the Intergovernmental Panel on Climatic Change (IPCC), an organization which advises governments on the progress of global warming and its implications, stated that "the balance of evidence suggests that there is a discernible human influence on global climate" (Houghton et al. 1996). This cautiously phrased assessment brought forth a surprisingly vigorous (and at times vicious) attack from a number of quarters.

Some critics even suggested that man-made climate change is an invention of scientists with a political agenda. This proposition ranks the assessments of professional geophysicists (and of trained ecologists) with the various pronouncements by commentators whose opinions may be of some interest in the context of politics and sociology, but are of little import in the context of the physics and ecology of Earth's climate.

As shown below, climate history can help clarify the issues and settle the disputes arising.

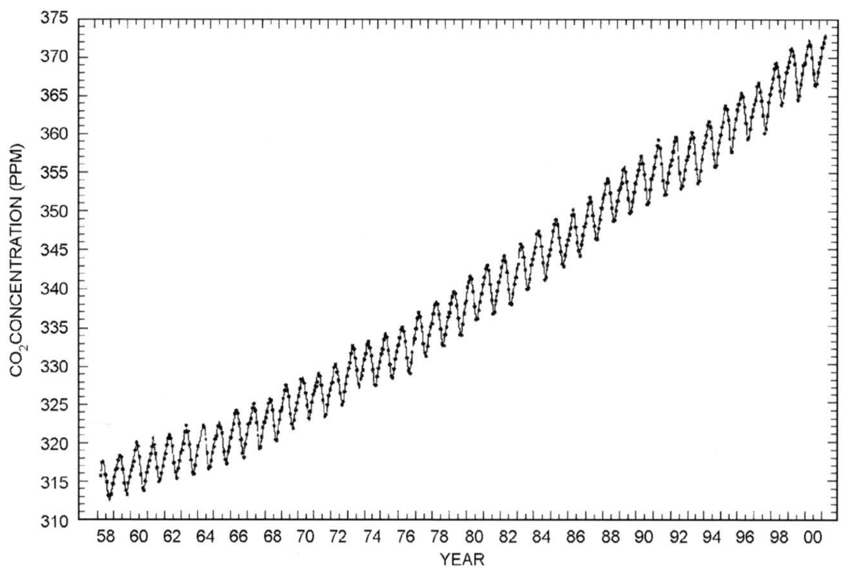

Fig. 1. The Keeling Curve, showing the rise of carbon dioxide in the atmosphere, since the first systematic measurements were made, in 1957. The large amplitude of the annual variation of the atmospheric carbon dioxide content, reflecting mainly growth and decay of plant matter in the northern hemisphere, was a major discovery. Data courtesy C.D. Keeling, Oct. 2000.

Most importantly, climate history has tremendous potential for expanding our horizons regarding the range of possible responses of the Earth's climate system to the Great Experiment now being performed on the planet, by the combustion of fossil fuels. History has comparatively little to offer about how society responds to major challenges arising from human environmental impact — the Great Experiment is unprecedented in human history for vastness of scale and rate of change.

The Reality of Climate Change and Its Denial

A surprisingly common response to the geophysical findings regarding the Great Experiment is to ex-press denial that anything is wrong and an unwillingness to discuss options for remedial action. The main strategy of denial is to raise doubts about scientific judgment.

Setting aside the various sociological and psychological aspects of denial, how can such criticism be answered?

Can we demonstrate that the Great Experiment has indeed resulted in global warming?

Computer models are one possible approach. In fact, it has been suggested, quite reasonably, that global warming first became an issue when computer models showed that it should occur. When we turn to computer models of the ocean-atmosphere system to test the climate system's response to the introduction of greenhouse gases (mainly carbon dioxide and methane), we are likely to obtain interesting results concerning the possible course of events. For example, we might see a certain rise in temperature, greater in high latitudes, within the troposphere, and a certain response of the ocean, with subsequent feedback to further climate modification. However, such models do not agree (as recently pointed out by a prominent member of the U.S. Senate). Thus, in the view of many a reasonable lay person, all but one of the models must be wrong. It takes no great logical leap to conclude that probably all of them are wrong, on some level.

Should we then trust that part of the model output in which all of the models agree, that is, that substantial high-latitude warming will accompany the continued introduction of greenhouse gases?

What to scientists are simply numerical experiments designed to deepen the understanding of climate mechanisms may be seen by the intelligent lay observer as (failed) attempts to predict the future.

Not so with climate history. That the past decade was the warmest since instrumental records began (Bradley 2000) is a verifiable proposition. Whether a computer model can generate output within the range of past experience can be tested. If it is unable to do so the model may not be suitable for simulation of future developments. Finally, whether or not the climate of the last decade, or the last two decades, or the last three decades, has been highly unusual can be ascertained by a statistical exploration of the more distant past and a comparison with the recent past. The word "unusual" means the same as "out of the ordinary". If we are to employ the concept "unusual" we need to define the range of conditions that are to be counted as "ordinary". Only history can tell us what is "ordinary" and what is not. Computer models cannot, although they can let us run experiments without risking damage to the environment.

Climate history has become an integral part of the study of the geophysics of climate. As the great climate experiment proceeds, we shall increasingly turn to its store of information to find clues to how the system works outside of the framework of ordinary experience.

As noted in the story of the Great Flood (Genesis 6:13) there have been times in human history when people felt they were to blame for bad weather. In fact, there is an ancient and deeply rooted belief that human action can influence the changing course of climate (see von Storch and Stehr, this volume). The burning of witches during the Little Ice Age apparently is one outgrowth of such a belief (Behringer 1999). The rain dances among Native Americans, in the southwestern U.S., are part of a long tradition reflecting the feeling that human needs and wishes are important in climate history. Presumably, the geophysical implications of such attitudes and activities are nil. (On the other hand, large-scale burning of forests, bush and grasslands may indeed have had climatic effects of some importance, on several continents, within the Holocene.)

It now appears, perhaps for the first time in human history, that feelings of responsibility toward climate change are appropriate and scientifically justifiable. We, the human species, are rapidly changing the balance of radiation of the planet, and we are changing the Earth's climate by doing so.

The course of recent temperature history within the fluctuations of the last ten centuries leaves little doubt about the reality of unprecedented climate change (Fig. 2). Recent average northern hemisphere temperatures are well outside of three standard deviations from the mean of the last 1000 years. The probability of finding a run of unusually high values such as seen, centered at a specified time within the last millennium (that is, the very end) may be taken as somewhere between 1/100 and 1/1000, depending on how expectations are calculated. In other words, given the temperature distribution through history, the probability that the warming since the late 1980s is a chance event within the general background of fluctuations of the last millennium is very small.

Faced with such evidence, what are we to make of a statement, offered by "skeptics", that man-made climate change cannot be "demonstrated"? It is difficult to deal with this question when "demonstrate" is interpreted to mean "prove beyond any possible doubt."

"Proof beyond reasonable doubt" is a concept appropriate to court action, where we wish to avoid the error of punishing the innocent. It is irrelevant to a large portion of human affairs, however, wherever probable cause and effect is involved. We do not wait for such certainty to prepare for action in the face of a potential enemy. We do not wait till disease strikes, but administer the vaccine beforehand, even before we can be sure that there will (or would) be an epidemic. And we buy insurance for our homes even though we are quite sure that they will *not* burn down.

What then are we to make of the vigorous attacks on the IPCC statement quoted above? What does the reaction tell us about the difficulties ahead concerning acceptance of responsibility by society and its leaders?

The reaction is the more puzzling as there is actually nothing alarmist or alarming at all about the IPCC statement. The postulated "influence" could be either beneficial or detrimental or irrelevant to human affairs. The distinguished Russian hydrologist Mikhail Budyko, for example, takes the position that emitting carbon dioxide will allow the Si-

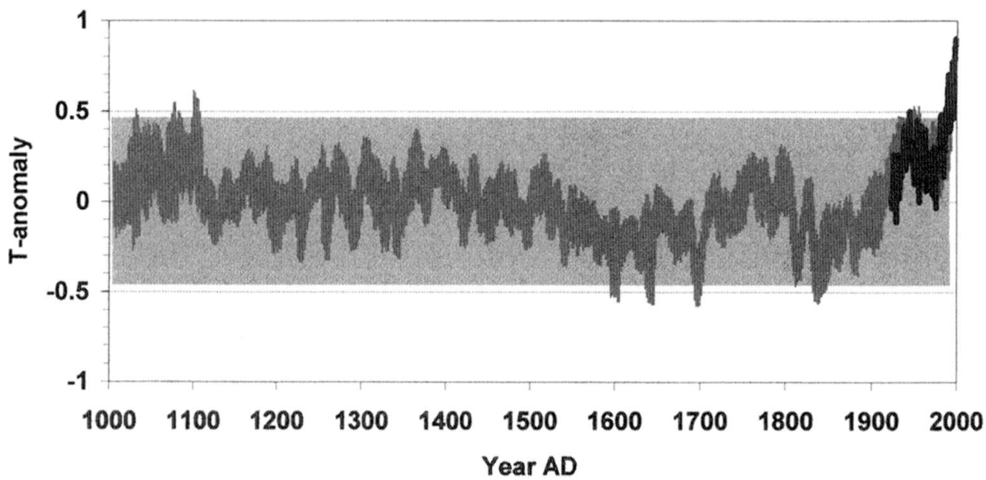

Fig. 2. Reconstruction of northern hemisphere temperature fluctuations for the last millennium, and the special position of recent temperature history. Grey zone: two standard deviations around the annual mean (set to zero). Dark band: one standard deviation around a ten-year gliding mean. Heavy line: actual measurements. Data from Jones et al. (1998) and from the Climate Research Unit at the University of East Anglia (www.cru.uea.ac.uk; compilation by Phil Jones).

berian steppe to become a bread-basket for a hungry world (dinner speech at the annual meeting of the American Geophysical Union, San Francisco 1995). A "neither-good-nor-bad" position is taken by von Storch and Stehr (2000) who have studied societal response to climate change (see Stehr and von Storch, this volume). They "strongly believe that near-surface temperatures are rising in response [to the accumulation of greenhouse gases]", but "are not convinced that present and future climate change will have a significant impact on society and global ecosystems." Others (for example people living at sea level on atolls in the South Pacific) have expressed a less sanguine view of the future.

In any case, given the neutrality of the IPCC statement, the attacks published in such newspapers as the Wall Street Jounal seem out of line. They betray an underlying agenda quite independent from scientific inquiry. The use of scientific uncertainty as a tool to attack scientific findings in public debate is familiar from the discussion of the link between health and tobacco use. In that context, residual scientific uncertainties were used to deny the obvious, in order to protect the economic interests of the tobacco industry.

The Great Experiment Proceeds

We should not be surprised that warming is occurring.

The content of greenhouse gases has risen substantially in the past century — roughly by the equivalent of a factor of 1.5 over background, as concerns the effect on radiative balance. (This is higher than the factor of increase of 1.3 for carbon dioxide, because of the addition of methane and other greenhouse gases.) Thus a rise in temper-ature must occur. This is not some tentative theory, but plain physics, the principles of which have been well understood for some time. The investigation of the issue began with the study of radiation balance by the famous Swedish chemist Svante Arrhenius (1859-1927), and continued through the discoveries regarding the role of the ocean in absorbing a part of the industrial carbon dioxide, by Revelle and Suess (1957) and their contemporaries. Since then, tremendous strides have been made in understanding (see Hansen et al. 1984; Houghton et al. 1990, 1996; Peixoto and Oort 1992), and are still being made, both conceptually and through the use of rapidly expanding computing power and large-scale observation from satellites.

Actually, for lay persons, it is not necessary to understand the physics. When trees topple and houses sink into the melting permafrost in Alaska, when Arctic sea ice shrinks as never before observed, and when tropical glaciers in the Andes melt for the first time in 10,000 years, we must assume that changes are afoot.

As mentioned, recent historical-geophysical studies have now delivered incontrovertible evidence that warming has arrived (Fig. 2). Jones et al. (1998) used information from tree-rings, ice-cores, corals and historical documents to reconstruct the global temperature history of the last six centuries and before. The warming in the 20th century is unprecedented over the time interval considered. Mann et al. (1998) used a similar set of records, including long instrumental records, to reconstruct the temperatures of the northern hemi-sphere for the last six centuries. In a careful consideration of possible forcing factors, they find that in the 20th century the release of greenhouse gases by human activities were responsible for much of the warming observed.

Briffa et al. (1998) also have generated a 600-year reconstruction for northern hemisphere summer temperature, from tree-ring information. They found that before 1950, tree-ring data were well matched to instrumental temperature data. After that time, however, tree growth apparently increasingly reacted to other environmental factors, responding as though summers were getting colder when the opposite is true. Presumably, the detri-mental factors are linked to human activities such as the introduction of acidic aerosols resulting in acid deposition ("acid rain"). It is a reminder that global warming is not the only large-scale ex-periment in progress. The global biosphere is being stressed in many other ways as well.

As concerns the warming in the 20th century, it is now clear that it is in large part a response to an ever-increasing input of trace gases to the atmosphere, from an ever-increasing population with ever-increasing energy demands. While population growth is skewed toward the less developed countries, energy demands are highly skewed toward the industrialized nations. The trace gases being added are carbon dioxide, methane, chlorofluorocarbons and nitrogen oxides. The molecules of these gases intercept a portion of the outgoing infrared radiation, from the Earth's surface and from the lower atmosphere, and re-radiate in all directions. In essence, they constitute a road-block to achieving balance between the incoming sunlight and the outgoing heat radiation (a balance that keeps the Earth habitable). Positive feedback from albedo changes (e.g. replacing snow and ice with dark ground and water surface) and from vapor (a greenhouse gas that becomes more abundant as the atmosphere warms) amplifies this effect.

Two effects retarding warming have prevented temperature to rise distinctly above background fluctuations until recently: (1) uptake of heat by the ocean, and (2) an increased amount of dust in the atmosphere, presumably mainly from soot caused by both internal combustion and open fires. Sulfurous particles have been proposed as playing a crucial role in shading the Earth (Mitchell and Johns 1997), in analogy to volcanic winter effects (LaMarche and Hirschboeck 1984; Briffa et al. 1998).

The heat uptake by the ocean, and the shading from aerosol particles have reduced the rate of warming well below that expected from a simple greenhouse model. This has given rise to unwarranted optimism, by extrapolation into the future. Unfortunately, much of the feedback in a warming climate is positive; that is, re-enforcing. Water vapor increases with warming, and it itself adds to the greenhouse effect. As the ocean warms, it can hold less carbon dioxide, making it more difficult to get rid of much of the man-made carbon dioxide through uptake in the ocean. As the ground warms in high latitudes reflective snow is replaced with dark soil, which absorbs more sunlight. Even increased winter snow, in places, results in trapping heat in the ground from the previous summer. In addition, methane production appears to be increasing, on the whole, as the planet warms.

Should we rely on negative feedback to keep things in balance?

In the tropics, increased cloud formation can increase the reflectivity of the planet, sending sunlight back to space, and this is indeed negative feedback. For negative feedback to work, of course, warming first has to occur. While such negative feedback may prevail in the wet tropics (at least for a while) it is highly unlikely that it will prevent warming elsewhere.

In any case, the Great Experiment itself, and not our computers, will decide the questions concerning the various feedback mechanisms and their interactions. We shall have more accurate answers at the end of the century.

Computer modeling is difficult because the climate system is extremely complex (Fig. 3). Many of the physical processes of the climate machine cannot be resolved. Examples are clouds in the atmosphere, eddies in the ocean, individual tree response to rain on the ground. We neither have the data nor the theory for detailed modeling of all the relevant processes involved. Thus, models are inherently flawed when used as a tool for prediction. The main source of uncertainty is the role of cloud formation. Clouds can act to trap heat attempting to escape to space, or they can provide shade, by reflecting sunlight. Whether they act as blankets or umbrellas in a given situation is not always readily resolved in climate simulation.

Problems also arise from the many nonlinear feedback mechanisms that make up the climate system. Warming in high latitudes, as mentioned, melts snow and ice, which darkens the ground (or the sea), which allows absorption of more sunlight and hence produces additional warming. The transition is not gradual but more like a switch (depending on the elevation of the ground, for example). This type of feedback introduces large uncertainties into the computations.

Quite generally, the various feedback mechanisms are nonlinear, that is, they show different strengths depending on the range of conditions, and they may have thresholds, where the response changes drastically. For example, when permafrost melts and the arctic forest on it dies, this changes albedo irreversibly in ways difficult to guess before the event.

The path, then, from climate model to prediction (and detection of human-induced change) is

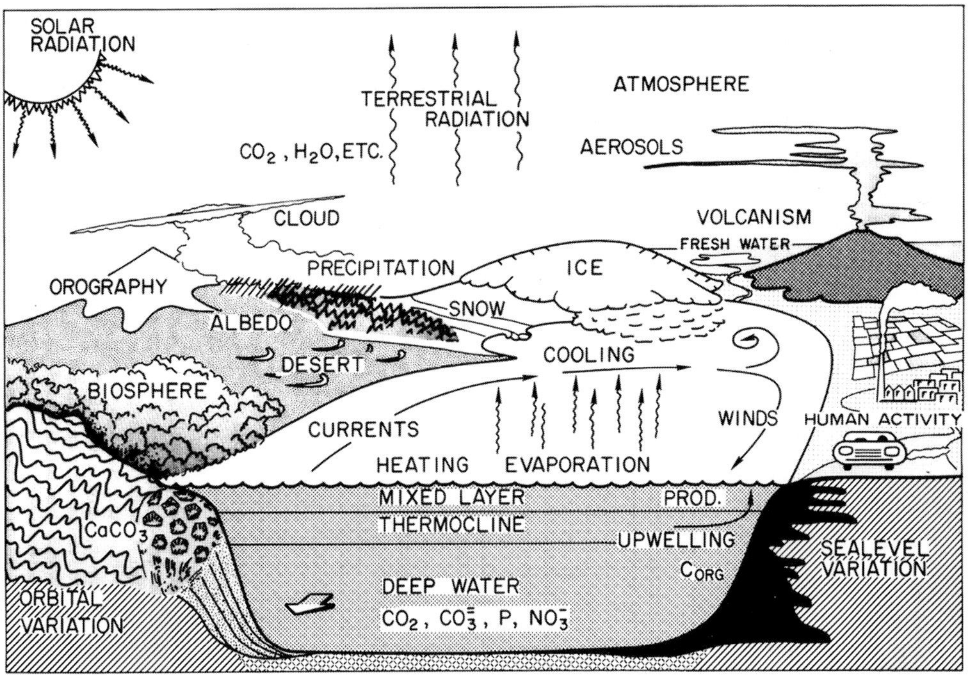

Fig. 3. The major elements of the climate system, greatly simplified (from Berger and Labeyrie 1987.) Major sources of instability are: ice mass, volcanism, variable deep-water production, upwelling, gases within upwelling sediments, and human activities.

fraught with difficulties. It starts with the uncertainties inherent in the climate models, which are literally crammed with different types of feedback mechanisms. We can be quite sure we have not found them all, and the ones we have found we have not comprehended precisely. The result is that climate models cannot tell us, with high accuracy, what happens when doubling the carbon dioxide content of the atmosphere (or the effective trace gas content). Combining this uncertainty with the amplitude of natural background fluctuations in climate, we remain unsure just what is the human contribution to change. It could be less, or it could be more, than the calculated amount.

Paths to Policy

In order to be useful, predictions of climate conditions have to lead into statements about possible environmental and economic consequences (Watson et al. 1998). As a step in this direction, Hulme and collaborators from various climate centers in the UK have made an effort to assess the relative impacts of human-induced climate change and natural climate variability, on a regional scale, using both climate simulations and environmental response models (Hulme et al. 1999). They compared the effects of natural climate variability (30-year time-scales) and of the projected range of human-induced climate change on river runoff and agricultural potential in Europe. They find that the human-induced impact by 2050 will still be undetectable for many regions when set against historic experience. (Thus, in those regions, any adaptive or preventive action by society would presumably be perceived as unnecessary and ill-informed, especially by those bearing the economic burden of such action.)

The most significant human-induced impacts, according to Hulme et al., will be experienced throughout Scandinavia, especially in the far North. Northern Scotland (and the Faroes and Orkneys) also belong to this regime. Another zone of expected high impact surrounds the Mediterranean Sea, from eastern Spain to southern Italy, to Greece and southern Turkey, and back along the northern coast of Africa to northern Algeria. The third zone of high projected impact includes the mountain ranges from the Pyrenees to the Alps, from the high Tatra Mountains to the Carpathian on to the Dinarids and the Balkan Mountains, and Mount Olympus itself.

The irony of the picture outlined by Hulme et al. is that the very regions that have the least to contribute to change, in terms of issuing trace gases, are the ones where impact is potentially greatest. Why should this be so?

The answer may not be so difficult: In the far North, albedo feedback is especially strong — warm summers melt snow and ice. Mountains, with their high latitude environments at high elevations, are outposts of the far North. Also, they are unusually sensitive to changes in wind fields or moisture content in the air impinging. Neither the far North nor mountains are particularly hospitable to people, so population density remains low. Hence, these regions suffer (and amplify) but do not generate human-induced change. In the dry regions around the Mediterranean, even a small change in average or seasonal precipitation can make much difference in terms of impact. On the whole, projected changes are favorable to agri-culture in the Mediterranean countries, according to Hulme et al. (1999), implying an increase in precipitation as the crucial factor.

Of course, even the kind of advanced geophysical-economic modeling exemplified by the study of Hulme et al. cannot deal with highly irregular risk-bearing phenomena such as result from the spread of insects. Cutting back on winter frosts and increasing wetness has all kinds of consequences for the range extension of agricultural pests and vectors of disease, where chance events play an important role (e.g. see Dowlatabadi 1997). Prediction of possible consequences of this sort will strongly depend on a sound ecologic data base combined with a deep understanding of the complicated relationships between ecology, economics and epidemics. The problems associated with influenza outbreaks illustrate the type of conundrum that arises in this context.

A number of common diseases may be closely linked to climate. Prominent suspects are the various forms of malaria, which are transmitted by certain mosquitoes ("vectors"). These are controlled by temperature and availability of moisture. Malaria outbreaks were once not uncommon in northern Europe, including England. Presumably, many of them were reported as poorly defined maladies associated with recurring fever. Occurrence depends on mean summer temperatures staying above 15°C. Thus, a general warming should be favorable for re-establishment of the disease. Nevertheless, according to Knottnerus (this volume) there is little relationship between the rise and fall of malaria and the fluctuations of global climate. Thus, apparently, regional or local conditions were of paramount importance.

Appleby (in Rotberg and Rabb 1981) suggests that the eventual decline of malaria in England and most of northwestern Europe was linked to changes in agricultural practices. Likewise, according to Appleby, a connection between plague and climate is difficult to recognize. He contends that while plague appeared during a cool and unstable period in the mid-fourteenth century, it also recurred regularly through relatively warm and stable periods, for example between 1500 and 1550. It disappeared in the mid-seventeenth century, during a cool and unstable period. The most widely accepted explanation for the disappearance, according to Appleby, is not linked to climate change but to improvements in quarantine methods. Likewise, he rules out dominance of climatic forcing for both smallpox epidemics and childhood diarrhea, although admitting climate as a possible factor.

The fact is that malaria is on the rise in Europe (according to the WHO, cases in the European Union went from 2882 in 1981 to 12,328 in 1997). Civil disorder and irrigation is seen as the culprit (MacKenzie 1999). Chilly winters, it is thought, will prevent malaria from re-establishing itself in northern Europe.

When discussing climate change, much of the disputes turn on the uncertainties in the geophysical models. If both the risk of doing nothing and the benefit of doing something are poorly known, is it not better to do nothing, and save the expense of attempting to influence the course of events? Why make a sacrifice if we don't know whether it will do any good? These are surely reasonable questions. But they have no ready answers.

Answers will have to found within the following framework: (1) There is not now, nor will there be in the future, certainty in scientific climate prediction. All we can do is improve on probabilities with more or less vague error limits. (2) In an economic context, those showing greater constraint in adding greenhouse gases to the common atmosphere, voluntarily, will be rewarded with the better conscience, but might bear, conceivably, a greater economic burden. (3) Those pre-adapting their economies to a new world-order of climate responsibility will be rewarded (for example, by selling "green" technology), but only if such a new order does come about in a reasonable time frame and if it includes all major players.

In coping with risk, doing nothing may be much better than panic. Nevertheless, doing nothing is rarely the best strategy. Problems arising can be solved more easily in the early stages of development, before a full-scale crisis dominates the scene. This is one lesson that history has taught over and over.

In history is the store of human experience that we need to mobilize to assess the risks of the future. History is a bountiful resource. We need to use it because physics is inadequate for prediction. The bugbear of prediction by physics is the elusive threshold, the crossing of which throws the system into a different mode of operation. The probability of such unrecognized thresholds, leading to runaway situations or to abrupt climatic change, can only be gauged through the study of history.

The "Little Ice Age" Conundrum

Of course, history is not just a bountiful resource, it is also a source of confusion. Specifically, in assessing global warming and the human role in it, we have to deal with the fact that the middle of the 19th century was unusually cold (Fig. 2). Naturally, this raises the question of whether the warming that Earth has experienced since was to be expected anyway; that is, whether the system simply started moving back toward some more normal condition.

Indeed, the standard answer to this question is that the warming right after 1850 is mostly natural. The weather in the mid- and late 1830s was highly unusual and highly stressful, with severe winters and bad harvests. (The great Irish famine falls into this period.) These years are embedded within a larger period with an unusual number of cold winters, the so-called "Little Ice Age", which lasted roughly from 1400 to 1850 or so (see Grove, this volume). The "Little Ice Age" is known as a time when mountain glaciers advanced in many regions of the northern hemisphere and when extensive permanent snowfields developed in Canada (as is seen in the age distributions of lichens). Among the possible reasons given for its existence are low solar activity and increased volcanism. (See Lean, this volume, concerning solar forcing, and Hammer, regarding the link between volcanism and climate.)

In the conventional view, then, the "Little Ice Age" is an anomaly (indeed it was the coldest period in the last several thousand years) and the warming after 1850 simply gets us back on track. This concept also supports the questionable but attractive idea that warming is good for humankind (and plants and animals, too) because it re-installs the previous regime of a more benign climate.

It is difficult to argue with the conventional view that the end of the Little Ice Age was natural — but we must be aware that it is only one of several possible views. Our present understanding of climate is insufficient to categorically forbid a continuation or worsening of the "Little Ice Age" conditions after 1850 and right into the present. On the contrary, the long-term view of climate change — the one that includes the coming and going of ice ages — actually demands increased cooling for the last 3000 years or so. (Indeed, this is seen in the re-expansion of glaciers in Scandinavia and in the expansion of the Sahara, for the last several millennia.) In this perspective, the "Little Ice Age" was right on target. There is no particularly pressing reason why it should be treated as an anomaly to be explained, with an expectation that things will soon revert to "normal".

Coming out of the "Little Ice Age" is as much in need of explanation as getting into it.

There are at least three plausible reasons for the warming after 1850: (1) the system was rebounding from an anomaly (caused by low solar input) due to increase in solar activity, (2) the system was rebounding from an anomaly (caused by high volcanic activity) due to cessation of vol-canic dust production, and (3) the system was not rebounding at all, but was forced toward a warmer climate by the increase in carbon dioxide.

The simplest explanation of the warming after 1850 is that the system was responding to the addition of trace gases to the atmosphere, mainly carbon dioxide and methane. Using only the known rise in carbon dioxide, and scaling to the statement "a doubling of carbon dioxide produces a global temperature increase of 2°C", we actually obtain a remarkable fit to the overall trend of the temperature rise (Fig. 4). The coefficient of regression is 0.75 for this fit. (It is a striking 0.89 if we use a gliding decadal average for describing the warming). The assumed response to doubling of carbon dioxide is on the conservative side of the range of estimates in the literature (about 1.5°C to 4.5°C). This is prudent, because these estimates refer to equilibrium conditions, while one would expect that the real warming is retarded due to the heat uptake by the ocean (Levitus et al. 2000).

Similar (but more complicated and computer-assisted) exercises have been carried out by a number of investigators, using different proportions of human (trace gas) impact, plus forcing from a brightening sun (see Lean, this volume), shading by particles in the air, and heat uptake by the ocean (e.g. Free and Robock 1999; Crowley 2000, and references therein). Typically, these investigators have concluded that there is an important portion of the warming which has to be ascribed to man-made forcing, that is, to the large-scale release of carbon dioxide and other trace gases to the atmosphere.

Lessons from the Distant Past

What is distant for historians is not so distant for geologists. In any case, going back 12,000 y certainly moves the action into prehistory, into a time when the world was very different from today. The main reason to go back so far is to study events that are extremely rare, and which involved the global environment in abrupt climatic change.

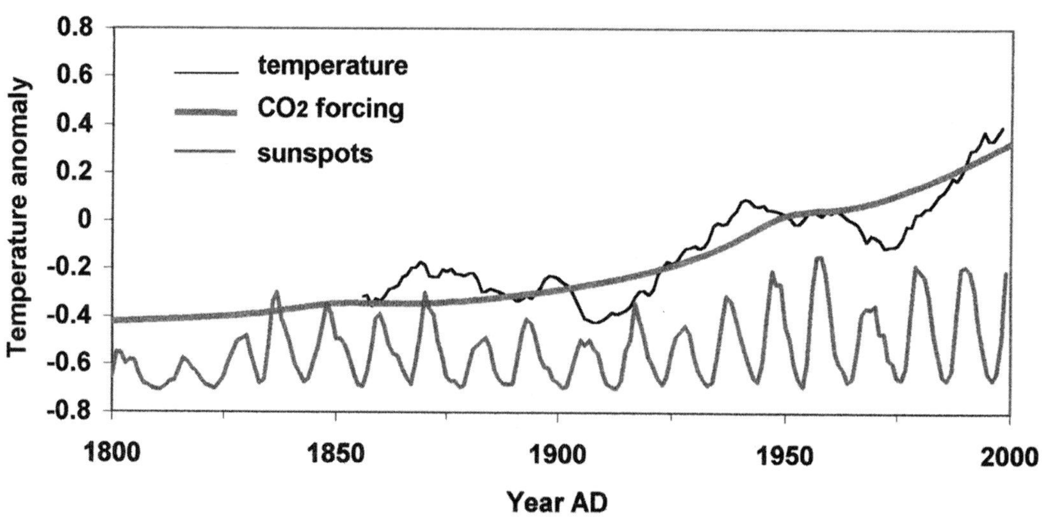

Fig. 4. The history of average annual temperature as measured in the northern hemisphere (compiled by P.D. Jones and colleagues; see Fig. 2) and the warming expected if doubling of carbon dioxide produces a 2°C rise in average global temperature. Estimates of sunspot activity (Hoyt and Schatten 1997, and Judith Lean, pers. comm.) are plotted for comparison (see text).

The Younger Dryas event, a cold spell during the general warming which brought about the end of the last ice age, is such an event (see Troelstra et al. 1995). A number of questions arise when studying this event, which has an abrupt change from warm (Alleroed period) to cold (Younger Dryas) and an abrupt change from cold (Younger Dryas) to warm (Preboreal period). These were among the largest climate changes experienced in the last 100,000 years.

One important issue is that of the stability of the world's thermohaline circulation (see Meincke, this volume; Paul and Schulz, this volume; Schulz and Paul, this volume). The thermohaline circulation involves deep convection in high latitudes, supply of cold deep water to all of the world's ocean basins, and overall poleward transport of warm waters. The result is re-distribution of heat from the tropics to the frigid regions. One major source of deep water is in the regions around Greenland, in the Norwegian-Greenland Sea (or Nordic Sea) and in the Labrador Sea. In principle, warm water moves northward in the Gulf Stream and its extensions, gives off heat on its way north while being cooled, and sinks to depth when cold (and heavy). The cold water then moves south across the equator to join the Antarctic Circum-polar Ocean, the great mixing ring of the world ocean. The heat delivered to NW Europe by this conveyor-like circulation is enormous. It is responsible for the great difference in living conditions in Scandinavia versus Greenland.

As the details of the Boelling-Alleroed-Younger Dryas story emerged, from ice cores on Greenland and from periglacial deposits and lake sediments in western Europe, the question arose, naturally, what caused this severe one-thousand year cold spell (see Troelstra et al. 1995). Major warming and melting was already well underway — why did the system revert back into the glacial mode?

One possibility that has received much attention and discussion is that a shutdown of North Atlantic Deepwater production greatly decreased the heat flux from the tropics to the high northern latitudes during the crucial period (see Broecker 1997, and refs. therein). A shutdown of deepwater production accompanies sufficiently large meltwater input, since an admixture of freshwater at the surface of the ocean builds up a water layer of low density that cannot sink below underlying waters, when cooled, but stays on top. This is the case in today's Arctic Ocean, for example, where a low-salinity water layer on the surface provides for stable stratification. When the water cannot sink, the influx of warm surface waters from the south is then correspondingly decreased, making the Nordic heat pump much less efficient. (This heat pump apparently started to work more actively at the start of the deglaciation period, as seen in deep-sea sediments; reviews in Berger and Labeyrie 1987.)

Rapid warming of the ocean surface is analogous to adding freshwater. Since warm water is less dense than cold water, it can (and does) generate stable stratification. Thus, by analogy, the meltwater theory of the origin of the Younger Dryas has been used to support the concept that global warming (which affects high northern latitudes especially seriously) could lead to substantial re-ductions in North Atlantic Deep Water (NADW) production, presumably with dire effects for the proper workings of the ocean-atmosphere system (Broecker 1997). In various tests of the basic theory (NADW shutdown from meltwater influx) using coupled ocean-atmosphere models, it is readily verified that the expectation holds true, and that surface water temperatures drop when the conveyor is slowed or switched off. The change in temperature accompanying deep convection changes is very rapid (e.g. Manabe and Stouffer 1995; Rahmstorf 1995; see Paul and Schulz, this volume).

As concerns the future, warming of the seas surrounding Greenland (where deepwater formation takes place) and increased precipitation (from a higher vapor content of the air arriving at the polar front) should indeed result in a slowing of NADW production. The process may be expected to have positive feedback, since once deep convection slows, stable stratification can build up more easily.

Of course, the origin of the Younger Dryas episode could well be visualized as a consequence of ice and melting dynamics, rather than ocean dynamics. In the view of Berger and Jansen (in Troelstra et al. 1995) the Younger Dryas simply represents a pause between the melting of the more vulnerable ice masses (the ones grounded below sea level) and the more stable ones (the ones which had to be warmed before they could melt). Also, in their view, the conveyor had already stopped working when melting was going on, in the Alleroed. Since the Alleroed was very warm, even in the more northern parts of Europe, the heat had to be pumped in by some other means. Thus, the history of the Younger Dryas, as an analog for the future of thermohaline circulation, is probably of limited relevancy. Deglaciation history, however, does remind us that freshwater input and heating interferes with deepwater formation, and encourages us to try to understand the geophysics of oscillations for different boundary conditions. The main task of historical study is not to provide analog cases, but to help us expand our horizons so we can ask the right questions.

How far back should we look?

However far we can, while we learn about climate dynamics. The further back we look, the more unusual events we are confronted with, challenging our understanding of climate dynamics in situations well out of bounds of experience or imagination.

A case in point are the recent findings regarding events near the Paleocene-Eocene boundary, about 55 million years ago, involving large excursions in the ratio of the two stable isotopes of carbon, as recorded in shell-bearing foraminifers in deep-sea sediments. What has been observed (Kennett and Stott 1991) is that the lighter of the two isotopes (^{12}C and ^{13}C) is greatly enriched in deep waters for a rather brief period, embedded within a time of general warming. During the time of the carbon isotope spike, there also is a brief distinct warming over background. The event is known as the "Latest Paleocene Thermal Maximum", or LPTM. It is marked by widespread extinction in the benthic fauna. Significantly, it is the deep ocean that is warming, along with the high latitudes; tropical temperatures apparently were less affected. Extinctions, it is suggested, are the result of lack of oxygen at depth.

A plausible explanation for the sudden warming (and the only one that can explain both the warming and the strange change in carbon isotopes) is major injection of methane gas into the ocean and from there into the atmosphere (Dickens et al. 1995). Two-thirds of the carbon isotope anomaly occurred within no more than a few thousand years (Norris and Röhl 1999), suggesting that the methane was released catastrophically, presumably upon reaching a threshold in the general warming. Once the methane was being injected, its greenhouse effect would have resulted in much additional warming, generating release of more methane. By all accounts, it was a classical run-away situation.

The obvious candidate for such release is a breakdown of methane ice on the sea floor. Such ice (clathrate, with space to accommodate the gas) is formed in cold water under high pressure, in the presence of methane (and also of CO_2 in cases) (see Sloan et al. 1994). Conditions favorable (at present) are continental slopes in regions of high organic matter supply (to produce the gas), high-latitude shelves (wherever gas is present and it is cold enough), and in certain permafrost regions. The total amount of methane in gas hydrates on the sea floor is not known. A general ballpark estimate is 10,000 Gt of carbon (1gigaton is 10^{15}g), more than ten times all the carbon in atmosphere, and more than the readily recoverable coal and petroleum (K. A. Kvenvolden in Sloan et al. 1994).

Even at one tenth the estimate, the amount of methane now available in the sea floor presumably exceeds all the carbon now in the atmosphere (as carbon dioxide). Sudden release of even a modest fraction of this methane, from general warming, would constitute a major disturbance on the climate system, because of the great effectiveness of methane as a greenhouse gas. The threat of methane release from clathrates in high latitude shelves and permafrost, in fact, is probably one of the more serious among the various worst-case scenarios that one might generate, regarding possible effects of global warming (Kvenvolden 1988; Haq 1998).

Climate in History

What can students of climate and of history learn from each other, if anything?

There is no question that climate historians can benefit greatly from the archives kept by historians, regarding the time of the grape harvest, for example, or the frequency of flooding of great rivers, or the incidence of drought. Such clues are enormously useful in the attempt to reassemble the course of climatic change through past centuries and millennia, as so brilliantly shown by Eduard Brückner in the 19th century (see Stehr and von Storch 2000), and more recently by Emmanuel Le Roy Ladurie (1971) and by H. H. Lamb (1972). In turn, there is little doubt that important elements of the history of nations, nations whose economies depend on agriculture, have been influenced by climate, starting with the origin of agriculture itself and its spreading across Eurasia (Harris 1996). Yet, as a cursory review of history textbooks will easily demonstrate, climate change is not generally accepted as an important part of the historical narrative. On the contrary, reference to climate is commonly missing altogether, with the notable exception of archeologists, especially those working in regions at the edge of the great deserts, where the impact of drought is paramount, and at the edge of the Arctic zones, where harsh winters limit existence. Where historians have urged climatic forcing of history they have found considerable skepti-cism among their peers (see DeVries, in Rotberg and Rabb 1981).

There is good reason why many historians are disinclined to tackle climate as an important factor in history. One is that effects of climate fluctuations on the course of history are difficult to show, because societies and their economies have many internal mechanisms compensating for adverse climatic effects. In fact, much of the social and economic history of mankind can be interpreted as an effort to diminish dependency on nature and particularly on the vicissitudes of climate. Long-term trends in climatic conditions, as well, are difficult to relate to the evolution of societies, because societies evolve in any case, with or without outside forcing, according to their own dynamics and those of their neighbors.

However, to deny the impact of climate on human affairs is to assert that the human species is somehow exempt from being limited by natural resources, especially food. There may be a bias in the reporting structure: "The hungry rarely write history, and historians rarely are hungry", write Millman and Kates (1990). Given that hundreds of millions of people on the planet may be considered undernourished, food shortage would indeed seem to be a factor at least in modern history. In the past, when trade was less efficient than today and agriculture was still the dominant economic activity determining much of the wealth of European countries, availability of food presumably was quite important in peace and war. During the height of the "Little Ice Age", between about 1550-1750, famines were not uncommon in western Europe, and some of these famines were associated with epidemics that took a high toll in mortality. The settling of North America by Europeans may have owed as much to poor harvests in the 18th and 19th centuries as it did to the oft-quoted desire for religious freedom and freedom from conscription.

The occurrence of climate-induced famine would seem to provide a test case for the economic resiliency of a society, including government response. Such a test can provide considerable insight into the resources and workings of a social system, and the social philosophy reigning at the time and in the region. The same is true for other externally forced calamities such as flood, water shortage, or earthquake damage. An entirely new type of test of this sort has arisen from industry-related distress such as oil spills, mercury pollution, river poisoning, and regional contamination with radioactive substances. The remedies offered by a government in such cases to the injured are a ready-made standard experiment on the type of social contract in force between the government and the people.

The potential of crisis as a repeatable experiment may not be fully appreciated. Thus, Jan de Vries (in Rotberg and Rabb 1981) criticizes the focus on extreme crisis in the contributions by his colleagues, commenting as follows (p. 23): "Unless these crises can be shown to be something other than unique, exogenous shocks, a skeptic might feel justified in concluding that short-term climatic crises stand in relation to economic history as bank robberies to the history of banking." That may be so, but a history of bank robberies would surely provide interesting insights into the police system of the time, which is in fact an important part of a society's history.

"The challenge now", according to Jared Diamond (1999, p. 408), "is to develop human history as a science, on a par with acknowledged historical sciences such as astronomy, geology, and evolutionary biology."

To outline the approach, Diamond proceeds from resources such as plants and animals suitable for domestication. He stresses the importance of knowledge transmission, weaponry and other technology, the evolutionary balance between human populations and their germs, and the social organization that derives from government and religion. His is an anthropologic and ecologic view of history, unabashedly adopting those elements of "geographic determinism" that are useful in ex-plaining why the world of humans became what it is. Inasmuch as climate is tied to geography, it is implicitly recognized as a strong force of history.

This is not a generally accepted view of what history is about. (Perhaps there is no such view.) In many a narrative on the history of Europe, or the World, climate and other natural constraints and resources are rarely mentioned (except when weather impacted military operations). To many traditional historians, history is about what Man does to Man. A respectable textbook on World Civilization, such as that by F. Roy Willis (1982), for example, has no index entries for climate, the Little Ice Age, earthquakes, volcanism or catastrophe. A lengthy entry for Jericho makes no mention that it was destroyed many times by earthquakes. An even longer entry for Minoan-Mycenaean civilization makes no mention of the possibility of volcanic catastrophe as a cause of its demise, but only refers to war. It is as though geophysics had no relevance to human affairs. "A master key to the understanding of history is not so easily, nor so early, available," writes historian Hugh Thomas (1979), in criticizing those who would seek a strong role for climate in history. Toynbee (1961), inclined to occasionally allow considerations of geographic conditions into the study of history, is roundly trounced by Ortega (1973). The philosopher opines (p.97): "Fortunately the phrase 'natural history' is becoming obsolete, and the word 'history' refers exclusively to the history of human things." On the other hand, he also insists that "...physics is the organ of human happiness and the renewal of this science has, in human affairs, been the most important event in universal history." Here Ortega focuses only on the desire to remake the world to fit human needs. But he forgets that the world being remade may have a dynamics of its own.

While climate change may not rank high as a driving force in human history among traditional historians, it is unlikely to have been unimportant. In any case, it is very unlikely to be unimportant in the future, as the Great Geophysical Experiment proceeds. We humans are making climate history, and in so doing, we make climate a part of the history of what Man does to Man.

Acknowledgments

I am indebted to many colleagues, too numerous to mention, for educating me in matters of climate and history. Discussions with Richard Somerville have been especially helpful over the years. I thank Jürgen Pätzold and Gerold Wefer for reading the draft and making useful suggestions for improvement.

References

Alverson KD, Oldfield F, Bradley RS (eds) (1999) Past Global Changes and Their Significance for the Future. Pergamon, Elsevier Science, pp 1-479

Behringer W (1999) Climatic change and witch-hunting: The impact of the Little Ice Age on mentalities. Climate Change 43:335-351

Berger WH, Labeyrie LD (eds) (1987) Abrupt climatic change: Evidence and implications. Reidel, Dordrecht, pp 1-425

Bradley R (2000) 1000 years of climate change. Science 288:1353-1355

Bradley RS, Jones PD (1992) Climate Since A. D. 1500. Routledge, London, pp 1-679

Briffa KR, Jones PD, Schweingruber FH, Osborn TJ (1998) Influence of volcanic eruptions on northern hemisphere summer temperature over the past 600 years. Nature 393:450-455

Broecker WS (1997) Thermohaline circulation, the Achilles heel of our climate system: Will man-made CO_2 upset the current balance? Science 278:1582-1588

Crowley TJ (2000) Causes of climate change over the past 1000 years. Science 289:270-277

Cullen HM, deMenocal PB, Hemming S, Hemming G, Brown FH, Guilderson T, Sirocko F (2000) Climate change and the collapse of the Akkadian empire: Evidence from the deep sea. Geology 28:379-382

DeVries TJ, Ortlieb L, Diaz A, Wells L, Hillaire-Marcel C

(1997) Determining the early history of El Niño [discussion]. Science 276:965-966
Diamond J (1999) Guns, Germs and Steel. The Fates of Human Societies. W.W. Norton, New York, pp 1-480
Dickens GR, O'Neil RR, Rea DK, Owen RM (1995) Dissociation of oceanic methane hydrate as a cause of the carbon isotope excursion at the end of the Paleocene. Paleoceanography 10:965-971
Dowlatabadi H (1997) Assessing the health impacts of climatic change. Clim Change 35:137-144
Flohn H, Fantechi R (eds) (1984) The Climate of Europe: Past, Present and Future, D. Reidel, Dordrecht, pp 1-356
Free M, Robock A (1999) Global warming in the context of the Little Ice Age. J Geophys Res 104:19,057-19,070
Frenzel B, Pfister C, Gläser B (eds) (1992) European Climate Reconstructed from Documentary Data: Methods and Results. Gustav Fischer, Stuttgart, pp 1-265
Glantz MH (1996) Currents of Change: El Niño's Impact on Climate and Society. Cambridge University Press, Cambridge, pp 1-194
Grove JM (1988) The Little Ice Age. Routledge, London, pp 1-498
Haq BU (1998) Gas Hydrates: Greenhouse nightmare? Energy panacea or pipe dream? GSA Today 8:1-6
Hansen J, Lacis A, Rind D, Russell G, Stone P, Fung L, Ruedy R, Lerner J (1984) Climate sensitivity: analysis of feedback effects. AGU Geophysical Monogr 29:130-163
Harris DR (ed) (1996) The Origin and Spread of Agriculture and Pastoralism in Eurasia. UCL Press, London, pp 1-594
Hodell DA, Curtis JH, Brenner M (1995) Possible role of climate in the collapse of Classic Maya civilization. Nature 375:391-394
Houghton JT, Jenkins GJ, Ephraums JJ (1990) Climate Change The IPCC Scientific Assessment. Cambridge Univ. Press, New York, pp 1-365
Houghton JT, Meira Filho LG, Callander BA, Harriss N, Kattenberg A, Maskell K (eds) (1996) Climate Change: The 1995 IPCC Assessment. Cambridge Univ. Press, pp 1-572
Hoyt DV, Schatten KH (1997) The Role of the Sun in Climate Change. Oxford University Press, pp 1-304
Hughes MK, Diaz HF (eds) (1994) The Medieval Warm Period. Kluwer Academic, Dordrecht, pp 1-342
Hulme M, Barrow EM, Arnell NW, Harrison PA, Johns TC, Downing TE (1999) Relative impacts of human-induced climate change and natural climate variability. Nature 397:688-691
Jones PD, Bradley RS, Jouzel J (eds) (1996) Climatic Variations and Forcing Mechanisms of the Last 2000 Years. Springer-Verlag, Berlin Heidelberg, pp 1-649
Jones PD, Briffa KR, Barnett TP, Tett SFB (1998) High-resolution palaeoclimatic records for the last millennium: Interpretation, Integration and Comparison with general circulation model control-run temperatures. The Holocene 8:455-471
Kennett JP, Stott LD (1991) Abrupt deep-sea warming, paleoceanographic changes and benthic extinctions at the end of the Paleocene. Nature 353:225-229
Kvenvolden KA (1988) Methane hydrates and global climate. Global Biogeochemical Cycles 2:221-230
Ladurie EL (1971) Times of Feast, Times of Famine: A History of Climate since the Year 1000. (Transl. B. Bray.) Doubleday, Garden City, New York, pp 1-426
LaMarche VC, Jr, Hirschboeck KK (1984) Frost rings in trees as records of major volcanic eruptions. Nature 307:121-126
Lamb HH (1972) Climate: Present, Past and Future - Vol. 1: Fundamentals and Climate Now. Methuen, London, pp 1-613
Lamb HH (1977) Climate: Present, Past and Future - Vol. 2: Climatic History and the Future. Methuen, London, pp 1-835
Lamb HH (1982) Climate History and the Modern World. Methuen, London, pp 1-387
Levitus S, Antonov JI, Boyer TP, Stephens C (2000) Warming of the world ocean. Science 287:2225-2229
MacKenzie D (1999) The comeback killer: malaria is taking hold again in Europe's mosquitoes. New Scientist, September 18, p 13
Manabe S, Stouffer RJ (1995) Simulation of abrupt climate change induced by freshwater input to the North Atlantic Ocean. Nature 378:165-167
Mann ME, Bradley RS, Hughes MK (1998) Global-scale temperature patterns and climate forcing over the past six centuries. Nature 392:779-787
Millman S, Kates RW, (1990) Toward understanding hunger. In Newman LF (ed) Hunger in History. Basil Blackwell, Cambridge, Mass, pp 3-24
Mitchell JFB, Johns TC (1997) On the modification of global warming by sulphate aerosols. J Clim 10:245-267
Norris RD, Röhl U (1999) Carbon cycling and chronology of climate warming during the Palaeocene-Eocene transition. Nature 401:775-778
Ortega JG (1973) An Interpretation of Universal History. W.W. Norton & Co. New York, pp 1-302 (Transl. fr. Spanish)
Peixoto JP, Oort AH (1992) Physics of Climate. American Inst. Phys., New York, pp 1-520
Pfister C, Brazdil R, Glaser R (eds) (1999) Climatic Variability in Sixteenth-Century Europe and Its Social

Dimension. Kluwer Academic, Dordrecht, pp 1-351
Rahmstorf S (1995) Bifurcations of the Atlantic thermohaline circulation in response to changes in the hydrological cycle. Nature 378:145-149
Revelle R, Suess HE (1957) Carbon dioxide exchange between atmosphere and ocean and the question of an increase of atmospheric CO_2 during the past decades, Tellus 9:18-27
Rotberg RI, Rabb TK (eds) (1981) Climate and History. Princeton University Press, Princeton New Jersey, pp 1-280
Sandweiss DH, Richardson JB, Reitz EJ, Rollins HB, Maasch KA (1996) Geoarchaeological evidence from Peru for a 5000 years b.p. onset of El Niño. Science 273:1531-1533
Sandweiss DH, Maasch KA, Anderson DG (1999) Transitions in the Mid-Holocene. Science 283:499-450
Sloan ED, Happel J, Hnatow MA(1994) Natural Gas Hydrates. Annals of the N.Y. Academy of Sciences, 715, pp 1-579
Stehr N, von Storch H (eds) (2000) Eduard Brückner — The Sources and Consequences of Climate Change and Climate Variability in Historical Times. Kluwer Academic, Dordrecht, pp 1-338
Thomas H (1979) A History of the World. Harper and Row, New York, pp 1-700
Toynbee AJ (1961) A Study of History. Vol.XII. Reconsiderations. Oxford Univ Press, pp 1-740
Troelstra AR, van Hinte JE, Ganssen GM (eds) (1995) The Younger Dryas. North Holland, Amsterdam, pp 1-224
Von Storch H, Stehr N (2000) Climate change in perspective. Nature 405:615
Watson RT, Zinyowera MC, Moss, RH Dokken DJ (eds) (1998) The Regional Impacts of Climate Change: an Assessment of Vulnerability. Intergov. Panel Clim. Change, Cambridge Univ Press, pp 1-517
Weiss H, Courty M-A, Wetterstrom W, Guischard F, Senior L, Meadow R, Curnow A (1993) The genesis and collapse of third millennium North Mesopotamian civilization. Science 261:995-1004
Wells LE, Noller JS (1997) Determining the Early History of El Niño [discussion]. Science 276:966
Wigley TML, Ingram MJ, Farmer G (eds) (1981) Climate and History. Studies in past climates and their impact on Man. Cambridge University Press, Cambridge, UK, pp 1-530
Willis FR (1982) World Civilizations. D.C. Heath & Co, Lexington Mass, pp 1-1485

Towards a History of Ideas on Anthropogenic Climate Change

Hans von Storch[1] and Nico Stehr[2]

[1]Institute for Coastal Research, GKSS Research Center, Max-Planck-Strasse 1, D-21502 Geesthacht, Germany
[2]Sustainable Development Research Institute, The University of British Columbia, 6201 Cecil Green Park Road, Vancouver, BC V6T 1Z1, Canada
**corresponding author (e-mail): storch@gkss.de*

Abstract: In this essay we show that the notion of anthropogenic climate change is not novel. Concerns about transformations of the Earth's climate by human activities have been expressed since the 18th century Enlightenment and earlier. It Is reasonable, therefore, to speak of a "history of anthropogenic climate changes". Most of the instances were not "real". But all cases were associated with the perception of significant discontinuities and in most instances the perceived change was seen as a threat to society. We briefly discuss the possible implications of this "history" for the functioning of the scientific community in the present debate.

Introduction

Most observers take it for granted that the concept of anthropogenic climate change is of relatively recent origin. The present notion primarily refers to expected climate changes associated with ongoing emissions of greenhouse gases and aerosols from burning fossil fuels. The scientific literature as well as the media in many countries are teeming with analyses related to this issue. The world's governments established the *Intergovernmental Panel of Climate Change* (IPCC) to deal scientifically with the problem of anthropogenic climate change (Houghton et al. 1990, 1992, 1996).

When we speak in the following of "climate change" we do not mean changes of the local climate by the expansion of cities, clearing of forests and other local modifications of land use. Instead we are referring to changes on a regional, continental and global scale.

We will demonstrate that the notion of anthropogenic climate change is by no means novel. Concerns about extensive transformations of the Earth's climate have been expressed since the 18th century enlightenment and earlier. It is reasonable to speak of a "history of concepts of anthropogenic climate changes". Most of the cases were not real; as a matter of fact, none of them proved to be associated with significant impacts related to the suggested dynamical link. But all cases were associated with the perception of significant discontinuities; in many of the most recent instances the perceived change was seen as a threat; only in the past they were sometimes welcomed as an improvement of natural conditions.

The following list of cases is not exhaustive. Many of the examples included were encountered almost fortuitously. A more systematic analysis by historians could lead to more illustrative cases.

From WEFER G, BERGER W, BEHRE K-E, JANSEN E (eds), 2002, *Climate Development and History of the North Atlantic Realm.* Springer-Verlag Berlin Heidelberg, pp 17-23

List of Cases

1. Religious interpretations of climate anomalies, such as the prolonged wet period in England in the early 14th century, explained the adverse climatic conditions as the divine response to people's lifestyle (Stehr and Von Storch 1995). In medieval times, for instance, it was proposed that climatic anomalies, or extreme events, were a punishment for parishes which were too tolerant of witches. Of course, witches were believed to be able to directly cause adverse weather (Behringer 1998). It would be interesting to learn what people thought about the cold spells within the Maunder Minimum at the end of the 17th century (Wanner 1995). This practice is also used nowadays. *Berlingske Tidende* (11 April 1998) has reported a recent example about the religious interpretations of violent tornadoes in Alabama at Easter in 1998.

2. Our oldest case documented by contemporary scientific writing refers to the climate of the North American colonies (Williamson 1771). The physician Williamson analyzed the changes of climate, and related them to the clearing of the landscape by the settlers. This is a case in which human action was perceived as having a beneficial impact on climate. Glacken (1967) describes more cases during the medieval times which were related to the colonization by monks.

3. In many parts of Europe, the summer of 1816 was unusually wet, presumably because of the eruption of the volcano Tambora. However, people ascribed the adverse conditions to the new practice of using lightning conductors. The case is documented in two articles published in the newspaper *Neue Züricher Zeitung* (21 June and 9 July 1816). The authorities called the concerns unsubstantiated and issued grave warnings concerning violent and illegal acts against the conductors. Interestingly, it is mentioned that some years earlier people in Germany blamed the conductors for being responsible for a drought.

4. In the 19th century scientists in Europe and North America were confronted with the concept that the climate would be constant on historical time scales; however, scientists found significant differences between mean precipitation and temperature when averaged over various multi-year periods (e.g. Brückner 1890). Also, scientists claimed that the water levels of rivers would fall continuously. This led to questioning the assumption of constant climatic conditions — in modern terms: interdecadal natural variability — and, alternatively, to the hypothesis that the observed changes are caused by human activities, mainly deforestation or reforestation. It seems that the majority adopted the concept of man-made causes over the natural variability hypothesis (Brückner 1890; Stehr et al. 1996).

5. There are reports that blame both the extensive gun-fire during World War I and the initiation of short wave trans-Atlantic radio communication for the wet summers during the 1910s and 20s (Kempton et al. 1995; Hinzpeter, pers. comm.).

6. In the first half of the 20th century a remarkable warming took place in large parts of the world. In 1933, this warming was documented, and the uneasy question "Is the climate changing?" was put forward in *Monthly Weather Review* (Kincer 1933). Some years later, Callendar (1938) related the warming to anthropogenic emissions of carbon dioxide into the atmosphere, a mechanism described some 40 years earlier by Arrhenius (1896). Interestingly, Arrhenius himself stated that anthropogenic emissions of CO_2 would cause a significant climate change only after several hundred years (Arrhenius 1903). Flohn (1941) also brought this line of reasoning into the scientific debate. In the 1940s global mean temperatures began to decline – which eventually led to claims that the Earth was heading towards a new Ice Age. Today, the risk associated with increasing concentrations of greenhouse gases such as carbon dioxide and methane is the overriding concern. There is no doubt about the increasing atmospheric concentrations. In a survey including 412 North American and German climate scientists, a majority of 70% out of 412 respondents were convinced that "*global warming is a process already underway*". Yet the respondents had some difficulty in specifying what the impacts of expected climate change would be and how they would be harmful. Only 34% responded positively to the question "*To what degree can we explicitly state the detrimental*

effects that climate change will have on society?", and only 1% answered with *"a great deal"* (Bray and Von Storch 1999).

7. After World War II scientists noticed a cooling and some speculated about whether this cooling was the first indication of a new Ice Age, possibly brought on by human actions, mostly emissions of dust and industrial pollution. It was speculated that human pollution would increase by a factor of as much as 8, which meant that the opacity of the atmosphere could increase within one hundred years by 400%. This in turn would significantly reduce incoming sun light causing the global mean temperature to sink by 3.5 °C. Such a cooling would almost certainly be enough to force the Earth into a new Ice Age (Rasool and Schneider 1971). The prospect was illustrated with the words: "Between 1880 and 1950, Earth's climate was the warmest it has been in five thousand years. ... It was a time of optimism. ... The optimism has shriveled in the first chill of the cooling. Since the 1940s winters have become subtly longer, rains less dependable, storms more frequent throughout the world." (Ponte 1976: 89).

8. After World War II, the new practice of exploding nuclear devices in the atmosphere caused a widespread concern about the climatic implications of these experiments. According to Kempton's analysis, even nowadays many lay-people are concerned about this link (Kempton et al. 1995; see also Kimble in *New York Times* 8 July 1962).

9. In Russia, plans for re-routing Siberian rivers southward have been discussed since the beginning of this century. The plans visualize benefits in supplying semi-arid regions with water, and an improved regional climate. A byproduct was thought to be an ice-free Arctic ocean because of the reduced fresh water input from the rivers. This would shorten the winters and extend the growth season; the increase of evaporation from the open water would transform the Arctic climate into a maritime climate with moderate temperatures and busy harbors along the Soviet Union's northern coast (Ponte 1976: 136). Such plans were formally adopted in 1976 at the 25th Assembly of the Soviet Communist Party. Scientists from the west as well as from the Soviet Union opposed to these plans and warned that the formation of an ice-free Arctic could significantly affect the circulation of the global ocean and thus global climate. Eventually, the plans were abandoned although more careful analyses indicated that the probability of melting the Arctic sea ice associated with a re-routing of the rivers was overestimated (Lemke 1987; Micklin 1981; Semtner 1984; Aagard and Coachman 1975; Cattle 1985).

10. The concept of engineering or manipulating the climate system became popular in the first half of the 20th century. Re-routing Siberian rivers was such a scheme; another was put forward by the New York engineer Riker, who in 1912 suggested changing the Gulf Stream with the purpose of improving the climate not only in North America, but also in the Arctic and Europe. Riker's idea was: "The Gulf Stream travels up along the American coast without any problem, ... but when it turns east to cross the Atlantic Ocean it collides with the Icy Labrador Current coming down from the Arctic. This collision in relatively shallow water weakens the Gulf Stream ... But this would change ... if a simple jetty 200 miles long could be built from Cape Race on Newfoundland to a point just beyond the underwater Grand Banks. The jetty would keep the two currents apart ... Off the tip of Greenland ...the more powerful Gulf Stream would divide. Half would throw increased warmth against Northern Europe, and half would thrust into the Arctic... The benefits of this would be enormous .... Fog would disappear, ... all ice in the Arctic would melt. The melting of the Arctic would improve the world climate in two ways. ... Europe and North America would be freed of chilling storms and icy ocean currents... And without the North Polar ice, the surviving ice pack at the South Pole would become the heaviest part of our planet. Centrifugal force would then tip the Earth ... With the Northern hemisphere tipped more towards the sun, Europe and North America could expect warmer climate." (Ponte 1976: 138). It is interesting to note that Riker thought of warming as an improvement of climate. The same view is put forward by Lamb (1982). Furthermore, the idea of modifying ocean currents was later pursued by scientists from the USA,

USSR and other nations. In most cases, these schemes revolved around the building of a dam, which would, for instance, block the flow through the Bering Strait.

11. Close to the idea of climate engineering is the military use of climate modifications. The idea to change the course of the Gulf Stream had been already put forward in the 18th century by Benjamin Franklin, who envisaged a northward diversion of the Gulf Stream as a powerful weapon against the British Empire (Ponte 1976:137). A perceived attack using climate as a weapon is a purported Soviet plan in the 1950s to build a "jetty 50 miles or more long out from near the eastern tip of Siberia. The jetty would contain several atomic powered pumping stations that would push cold Arctic waters down through the Bering Strait. This would ... inject increasing amounts of icy waters into the ocean current that flows down the West Coast of Canada and the United States. The result would be colder, more stormy weather throughout North America and enormous losses to the American economy in agriculture, work days and storm damage." (Ponte 1976: 169-170). Concern about the development of climate weapons lead to a series of diplomatic discussions. During a summit meeting 1974 the United States and the Soviet Union issued a Joint Draft Treaty: "Each State Party to this Convention undertakes not to engage in military or other hostile use of environmental modification techniques having widespread, long-lasting or severe effects as the means of destruction, damage or injury ... the term 'environmental modification techniques' refers to any technique for changing – through the deliberate manipulation of natural processes – the dynamics, composition of the Earth, including its biota, lithosphere, hydrosphere and atmosphere ... so as to cause such effects as ... changes in weather pattern, ...in climate patterns, or in ocean currents." (Ponte 1976: 259-263)

12. In the 1960s and 70s aircraft industries in the USA, Europe and Soviet Union designed supersonic civil air planes. These plans provoked substantial criticism. Scientist argued that the exhaust from such planes would damage the ozone layer in the stratosphere and the climate in general. In the USA the plans were stopped, but in Europe the Concorde was built and in the Soviet Union the TU 144. Of course, numerous military supersonic aircraft nowadays cruise the lower stratosphere. For many years, the discussion about the impact of air traffic on the climate ceased. But in the early 1990s the topic re-entered the public debate, this time regarding high-flying conventional jetliners. The focus of concern is the effect of contrails and exhaust gases on the radiative balance of Earth. Scientists (e.g. Sausen and Schumann 1998) regard present effects from these sources as minor, compared to other effects. However, some argue that with present projections of future passenger numbers and technology the effect maybe or will be significant.

13. A popular, but for natural scientists somewhat surprising mechanism links space traffic to the deterioration of global climate. In interviews made by Kempton et al. (1995) with lay people, this mechanism is mentioned several times. 43% of the respondents in Kempton's survey considered the statement *"there may be a link between the changes in the weather and all the rockets they have fired into outer space"* as plausible.

14. The ongoing deforestation of tropical forests is of great concern to many people, who are afraid not only of reduction in the variety of species but also of changes in global climate (Kempton et al. 1995; Dunlap et al. 1993). Model calculations indicate that these land use modifications cause significant local and regional changes whereas in most model calculations global effects are marginal. Interestingly, similar results were obtained for the climatic implications of the transformation of the North American wilderness into agricultural land (Copeland et al. 1996).

15. Anthropogenic aerosols are considered powerful agents for changing the global climate. One scenario deals with the emission of aerosols mainly from burning forests and fossil fuels. A dramatic version is that of "nuclear winter". It was assumed that the explosion of a multitude of nuclear bombs in a future war would create a high flying veil of soot particles which would effectively shield off solar radiation and cause a collapse of the biosphere (Cotton and Pielke 1995). Support came from a number of computer simulations. The ignition of the Kuwait oil wells in the aftermath of the Gulf War in 1991 led some scientists to expect a minor

nuclear winter, particularly with respect to the Indian Monsoon. It turned out that the effect was severe locally but insignificant on the larger scales (Cahalan 1992; Bakan et al. 1991).

16. A new issue of concern, especially in Europe, refers to the stability of the Gulf Stream in the Atlantic Ocean. Ocean models exhibit a markedly nonlinear behavior of the Atlantic circulation with two stable states, one with an active Gulf stream and another with a weakened northward transport moderating the European climate. Both states are stable within a certain range of conditions, but when the system is brought to the margins of these ranges, it can switch abruptly to the other state (Rahmstorf 1995). Paleoclimatic reconstructions using evidence from ice cores and other indirect sources support the existence of such stable states and frequent rapid changes from one state to another. During the present interglacial period from about 10,000 years ago to the present, such rapid climate changes have not been detected and quasi-realistic climate models featuring interactive ocean and atmosphere models have not exhibited such behavior (Manabe and Stouffer 1996; Von Storch et al. 1997). In the global warming debate the risk of a "collapse" of the Gulf Stream is put forward (Rahmstorf 1997). While the globe is becoming warmer, Europe and Northeast America would experience colder conditions with the possibility of a new ice age. However, most scientists regard such an evolution as unlikely, at least as long as carbon dioxide concentrations are not increased by a factor of four and more (Morgan and Keith 1995). Not only global warming is presented as a human lever for terminating the Gulf Stream. Another hypothesis was published by Johnson (1997) with a significant reaction in the public media. This time the culprit was the Assuan Dam in Egypt which would reduce the flow of fresh water into the Mediterranean Sea. This reduction along with an enhanced evaporation caused by global warming would result in a saltier outflow from the Mediterranean Sea into the Atlantic eventually causing the Gulf Stream to flip over. Subsequent quantitative analysis revealed that the impact of the Assuan Dam would be much too small to have this effect on the Gulf Stream (Rahmstorf 1998).

Conclusions

It is interesting to speculate about the social and cultural processes which made and make the concept of anthropogenic climate change not merely an episodic but an almost permanent issue that challenges scientists and alarms non-experts. It would also be interesting to find out why the notion of an anthropogenic challenge to large-scale climate has been repeatedly forgotten and reinvented.

First, scientists are part of society, and not immune to traditional, non-scientific knowledge. Secondly, more specific considerations apply, as documented in a recent survey among North American and European scientists (Bray and Von Storch 1999). The question *"How often do you think experts frame problems so that the solution fits his or her area of expertise?"* was answered positively by more than 70% of the 412 scientists who responded to the questionnaire. 80% answered positively to *"To what degree do you think there is growing pressure for climate research to be justified in terms of policy relevance?"*. The question *"To what degree does exposure to the media have the potential to change the attitude of a scientist?"* was answered positively by about 50% of the respondents. More than 30% (in Germany even a majority) of the respondents agreed to *"Some scientists present the extremes of the climate debate in a popular format with the claim that it is their task to alert the public. How much do you agree with this practice?"* This means that scientists perceive the need for framing their research problems, so that the solution will fit their area of expertise. Some members of the scientific community are willing to engage in public agenda setting with a certain bias toward more dramatic scenarios. Others are lead by ide-ological elements such as genuine general concerns about the state of the natural environment. A minority of scientists seems to be driven by the desire to have a public presence in the media. Thus, there is a distinct possibility that the scientists for entirely mundane reasons tend to exaggerate the dangers associated with anticipated climate change (see also the concept of rainmaking in the 1960s and 70s; Cotton and Pielke 1995). In that sense climate change is a social construction, consistent with the traditional knowledge and professional needs.

On the other hand, the concept of anthropogenic climate change is hardly a purely scientific construction like that of ether in the 19th century. In past decades a range of solid information has been assembled. The key processes of climate formation, like clouds and oceanic circulation, are better understood. The detailed process-based climate models now describe many components, like the ocean and sea ice and sometimes even vegetation and cycles of carbon, and have demonstrated their utility in several applications. A multitude of proxy data on paleoclimatic conditions support the concept that varying greenhouse gas concentrations are associated with different climatic regimes. Quality controlled collections of instrumental data with global coverage have identified the latest changes of globally distributed temperature as likely being not within the range of natural variations. After examining the evidence in great detail (Houghton et al. 1990, 1992, 1996), the IPCC made in 1995 its famous statement that "the balance of evidence suggests that there is a discernible human influence on global climate. "In fact, one could argue that "there has been a shift from absurd statements to uncertainty at the edge of knowledge of complicated systems" (Berger, pers. comm.). There are good reasons to consider the present prospect as more real than the earlier cases, and our list should not be mistaken as an attempt to negate the at present voiced concerns about anthropogenic climate change.

In summary, we should view the prospect of anthropogenic climate change as both, scientifically and socially constructed as well as underpinned by strong evidence. As common in post-normal science (Funtowicz and Ravetz 1985; Bray and Von Storch 1999), solid scientific knowledge is blended with lay understanding, and political con-cepts. In fact, now as in the previous cases, political bodies act upon scientific warning. Contemporary examples are not only the partly political IPCC process but the Enquete Commission of the German Parliament (1988). Historical examples are committees set up one hundred years ago by parliaments and governments in Europe (e.g. Prussia, Italy, and Russia) to deal with the reality of anthropogenic climate change related to deforestation (Brückner 1890; Grove 1975).

Acknowledgements

We are grateful to Wolfgang Berger and an anonymous reviewer for helpful comments.

References

Aagard, K., Coachman LK (1975) Toward an ice-free Arctic Ocean. EOS 56:484-486

Arrhenius SA (1903) Lehrbuch der kosmischen Physik. Volume 2, Leipzig: S. Hirzel

Arrhenius SA (1896) On the influence of carbonic acid in the air upon the temperature of the ground. Philosophical Magazine and Journal of Science 41:237-276

Bakan S, Chlond A, Cubasch U, Feichter J, Graf H, Grassl H, Hasselmann K, Kirchner I, Latif M, Roeckner E, Sausen R, Schlese U, Schriever D, Schult I, Schumann U, Sielmann F, Welke W (1991) Climate response to smoke from the burning oil wells in Kuwait. Nature 351:367-371

Behringer W (1998) Hexen. Glaube - Verfolgung - Vermarktung. C.H. Beck Wissen, pp 1-115

Bray D, Von Storch H (1999) Climate Science and the transfer of knowledge to public and political realms. In: Von Storch H, Flöser G (eds) Anthropogenic Climate Change, Springer Verlag, pp 287-328

Bray D, Von Storch H (1999) Climate Science. An empirical example of postnormal science. Bull Amer Met Soc 80: 439-456

Brückner E (1890) Klimaschwankungen seit 1700 nebst Bemerkungen über die Klimaschwankungen der Diluvialzeit.Geographische Abhandlungen herausgegeben von Prof. Dr. Albrecht Penck in Wien; Wien and Olmütz, Hölzel ED, pp 1-325. (engl. Translation in: Stehr N, Von Storch H (eds) (2000) Eduard Brückner - The Sources and Consequences of Climate Change and Climate Variability in Historical Times. Kluwer Publisher, pp 1-338)

Cahalan R (1992) Kuwait Oil Fires as seen by Landsat. J Geophys Res 97:14,565-14,57

Callendar GS (1938) The artificial production of carbon dioxide and its influence on temperature. Q J Roy Met Soc 64:223-239

Cattle H (1985) Diverting Soviet rivers: some possible repercussions for the Artic Ocean. Polar Record 22:485-98

Copeland JH, Pielke RA, Kittel TGF (1996) Potential climatic impacts of vegetation change: a regional modeling study. J Geophys Res 101, D3, 7409-7418

Cotton W, Pielke RA (1995) Human impacts on weather and climate, Cambridge University Press, pp 1-288

Deutscher Bundestag (1988) Schutz der Erdatmosphäre: Eine internationale Herausforderung. Deutscher Bundestag, Referat Öffentlichkeitarbeit, Bonn

Dunlap RE, Gallup GH Jr., Gallup AM (1993) Health of the Planet: A George H. Gallup Memorial Survey Gallup International Institute Princeton, New Jersey, USA

Flohn H (1941) Die Tätigkeit des Menschen als Klimafaktor. Z Erdkunde 9:13-22

Funtowicz SO and JR Ravetz (1985) Three types of risk assessment: a methodological analysis. In: Whipple C, Covello VT (eds) Risk Analysis in the Private Sector, New York, Plenum, pp 217-231

Glacken CJ (1967) Traces on the Rhodian Shore. University of California Press, pp 1-763

Grove RH (1975) Green Imperialism. Expansion, Tropical Islands Edens and the Origins of Environmentalism 1600 – 1860. Cambridge University Press, pp 1-540

Houghton JL, Jenkins GJ, Ephraums JJ (eds) (1990) Climate Change. The IPCC scientific assessment. Cambridge University Press, pp 1-365

Houghton JT, Callander BA, Varney SK (eds) (1992) Climate Change 1992. Cambridge University Press, pp 1-200

Houghton JT, Meira Filho LG, Callander BA, Harris N, Kattenberg A, Maskell K (eds) (1996) Climate Change 1995. The Science of Climate Change. Cambridge University Press, pp 1-572

Johnson RG (1997) Climate control requires a dam at the Strait of Gibraltar, EOS 27:277, 280-281

Kempton W, Boster JS, Hartley JA (1995) Environmental values in American Culture. MIT Press, Cambridge MA and London, pp 1-320

Kincer JB (1933) Is our climate changing? A study of long-term temperature trends. Mon Wea Rev 61:51-259

Lamb HH (1982) Climate, history and the modern world. Methuen Co. London, pp 1-433

Lemke P (1987) A coupled one-dimensional sea ice-ocean model. J Geophys Res 92 (C12), 13,164-13,172

Manabe S, Stouffer RJ (1996) Low frequency variability of surface air temperature in a 1000 year integration of a coupled ocean-atmosphere model. J Climate 9:376-393

Micklin PP (1981) A preliminary system analysis of impacts of proposed Soviet river diversions on Artic sea ice. EOS 62:89-493

Morgan MG, Keith DW (1995) Subjective judgements by climate experts. Environmental Science and Technology 29:68A-476A

Ponte L (1976) The Cooling. Prentice-Hall Inc. Englewood Cliffs., NY, pp 1-306

Rahmstorf S (1995) Multiple convection patterns and thermohaline flow in an idealised OGCM. J Climate 8:028-3039

Rahmstorf S (1997) Risk of sea-change in the Atlantic. Nature 388:25-826

Rahmstorf S (1998) Influence of Mediterranean outflow on climate. EOS 79, 281-282

Rasool SI, Schneider SH (1971) Atmospheric carbon dioxide and aerosols: Effects of large increases on global climate. Science 173:138-141

Sausen R, Schumann U (1998) Estimates of the climate response to aircraft emission scenarios. Institut für Physik der Atmosphäre 95, DLR, pp 1-26

Semtner AJ (1984) The climatic response of the Artic Ocean to Soviet river diversions. Climatic Change 6:109-130

Stehr N, Von Storch H (1995) The social construct of climate and climate change. Clim Res 5:99-105

Stehr N, Von Storch H, Flügel M (1996) The 19th century discussion of climate variability and climate change: analogies for present day debate? World Res Rev 7:589-604

Von Storch J-S, Kharin V, Cubasch U, Hegerl G, Schnur R, Schriever D, Von Storch H, Zorita E, 1997: A 1260 year control integration with a coupled general circulation model. J Climate 10, 1526-1544

Wanner H, Pfister C, Brádzil R, Frich P, Frydendahl K, Jonsson T, Kington J, Lamb HH, Rosenørn S, Wishman E (1995) Wintertime European circulation patterns du ring the late Maunder Minimum cooling period (1675-1704). Theor Appl Climatol 51:167-175

Williamson H (1770) An attempt to account for the change of climate, which has been observed in the Middle Colonies in North America. Trans Amer Phil Soc 1, 272

Climate Dynamics of the North Atlantic and NW-Europe: An Observation-Based Overview

Jens Meincke

University of Hamburg, Institute of Oceanography, Troplowitzstraße 7, D-22529 Hamburg, Germany
e-mail: meincke@ifm.uni-hamburg.de

Abstract: The North Atlantic is the site for major ocean-atmosphere interaction because its wide meridional connection with the Arctic allows for extreme poleward propagation of warm subtropical waters and hence for extreme heat release to the high-latitude atmosphere. The cooled surface waters sink and lead to the formation of North Atlantic Deep Water, which is spreading southward in the lower limb of the Atlantic meridional overturning circulation and the global thermohaline conveyor. The ocean/atmosphere system experiences considerable variability on interannual to multi-decadal timescales. On five examples of observed North Atlantic ocean variability and possible links with low-frequency atmospheric driving related to the North Atlantic Oscillation the range of climate-variability in the past decades is described.

Introduction

Within the global climate system the North Atlantic realm is a key-site for studies on climate-change. The wide meridional connection between the Atlantic and the Arctic Oceans allows for extreme poleward advection of warm and saline subtropical waters which causes one of the globes's largest surface temperature anomaly. It is centered just offshore NW-Europe, providing the well known climate advantage of northern Europe against other regions within the same latitude belt. (Fig.1a). It is understood that changes of climate conditions will show the largest effects in areas of strong anomalies. Fig.1b exemplifies two types of changes: There is the global-scale change of the atmosphere's radiative properties due to increasing CO_2 and there is the regional scale change in ocean-atmosphere interaction due to an extreme trend in the north-south atmospheric pressure difference over the North Atlantic since the 1960's. Superimposed to the trend is a strong decadal variability. Taken the relatively short period of direct ocean observations (max. 100 years), the North Atlantic time series offer to look for links between observed ocean variability and observed changes in low frequency atmospheric driving.

The following overview will be given in chapters on the North Atlantic current system, the NAO and the Atlantic/European variability.

The North Atlantic Circulation System

Reference is made to comprehensive reviews of Reid (1994), Schmitz and McCartney (1993) and Talley (1996) . The basic structure of the large scale horizontal circulation is forced by the mean wind pattern and by thermohaline (buoyancy) processes. The wind-induced transports in the upper layers (Ekman transports) cause a system of mass-convergences/divergences which establish pressure gradients in the ocean's interior. They set up the system of gyre circulation (the Sverdrup-regime) in the hemispheric ocean basins. The buoyancy forcing is by cooling, evaporation and freezing, resulting in the formation of dense waters which mix downward and create an interior field of density/pressure differences. These differences are of a global scale and drive the thermohaline circulation.

From WEFER G, BERGER W, BEHRE K-E, JANSEN E (eds), 2002, *Climate Development and History of the North Atlantic Realm.* Springer-Verlag Berlin Heidelberg, pp 25-40

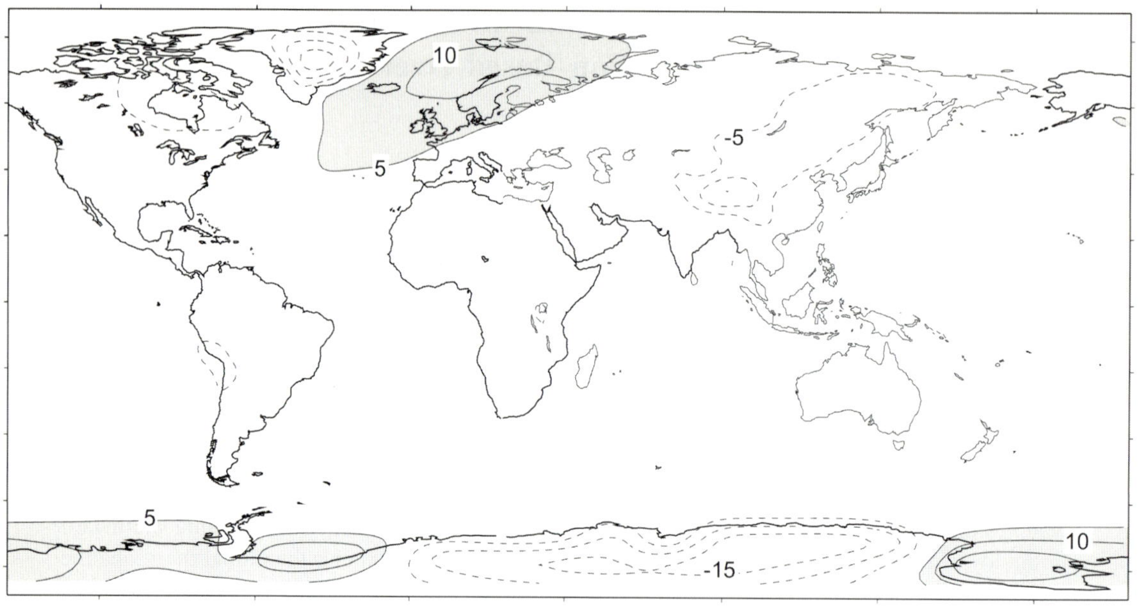

Fig. 1a. Anomalies of the zonal mean surface temperatures given in Δ T=5°C. Shaded areas, **full lines:** Positive anomalies, **broken lines:** Negative anomalies. The positive anomaly of the Northern Hemisphere is mainly due to heat advection by the North Atlantic Current (after Rahmstorf and Ganopolski 1999).

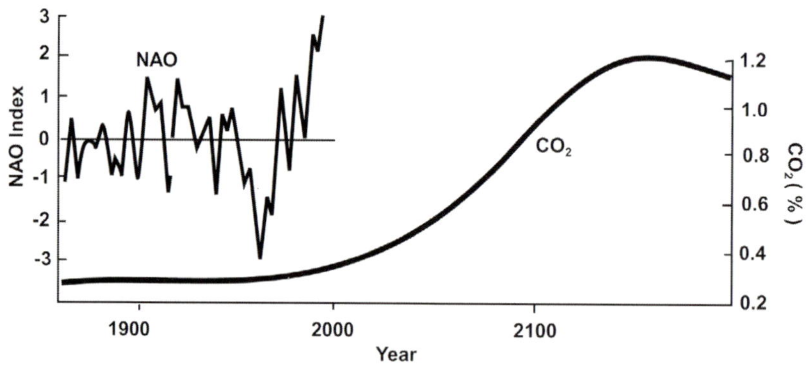

Fig. 1b. Two reasons for changes to be expected in our present climate: **(a)** For the North Atlantic sector the North Atlantic Oscillation (NAO) is undergoing the most prominent change since the 1960's. **(b)** The modelled CO_2-increase within the present century will cause a pronounced change of the earth's radiation budget (after Rahmstorf and Ganopolski 1999).

With respect to the North Atlantic, we focus on those components of the circulation that are relevant to the examples of ocean variability presented in chapter 4. Fig. 2a shows the basic structure of the Meridional Overturning Circulation (MOC) in the North Atlantic. Its elements are the warm to cold transformation of the northward flowing upper waters, the downward mixing of dense surface water in the convective areas of the Labrador and Nordic Seas and the Mediterranean and southward spreading of North Atlantic Deep Water, made up of its differently dense components from the Labrador Sea and the overflows across the Greenland-Scotland ridge and the Gibraltar sill. The currents forming the upper limb of the MOC are mapped in Fig. 2b, the pattern of water mass

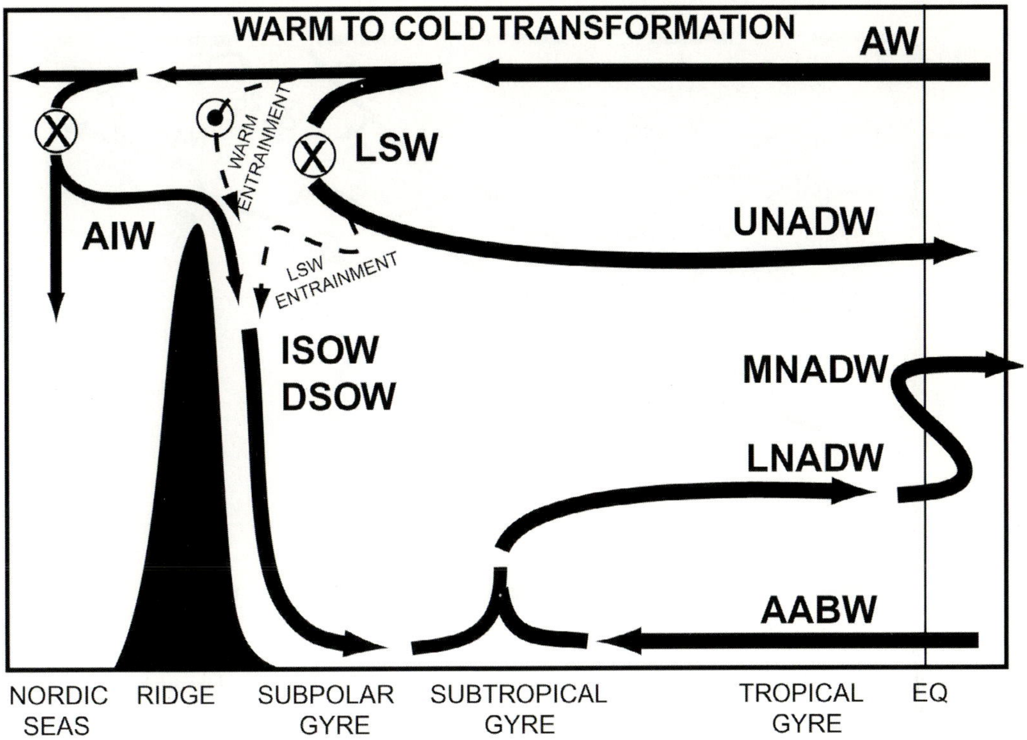

Fig. 2a. Schematic of the meridional overturning circulation (MOC) in the North Atlantic. Poleward flowing Atlantic Water (AW) forms the upper limb of the MOC. It is cooled and sinks in the convection regions of the Labrador Sea (forming Labrador Sea Water, LSW) and the Nordic Seas (forming Arctic Intermediate, AIW, and Deep Waters). The AIW leaving the Nordic Seas as Iceland Scotland Overflow Water (ISOW) and as Denmark Strait Overflow Water (DSOW) form the Lower and the Middle North Atlantic Deep Water (LNADW + MNADW). The LSW flowing southward forms the Upper North Atlantic Deep Water (UNADW). The NADW's together form the lower limb of the MOC, which compensates both the northward flux of AW and of Antarctic Bottom Water (AABW). The circled crosses denote the process of convection, the circled dots denote entrainment of warm AW into colder waters (after Schmitz 1996).

spreading making up the deep limb are shown in Fig. 2c. The connection between the limbs is provided through the processes of convection (vertical mixing due to surface cooling and salt input from evaporation or freezing) and entrainment (mixing in the ocean's interior due to strongly sheared currents). They are depicted in Fig. 2b and 2c by specific symbols.

For the study of climate variability the following features from the graphical presentation of the North Atlantic circulation are of importance: The paths of the major warm surface currents of the Subtropical and the Subpolar gyres are the primary regions of effective air-sea interaction. For the variability in the vertical and deep components of the MOC it is the convective water mass transformation within the Subpolar gyre and the subsequent export from the convective centers into the deep boundary currents, including water mass changes by entrainment.

The North Atlantic Oscillation (NAO) – Interannual to Multi-Decadal Variability in Atmospheric Forcing

The distribution of air flows and water movements as well as the related distribution of atmospheric and oceanic masses are by no means stationary. Next to energetic short term variability and pronounced seasonal changes they are subject to larger-

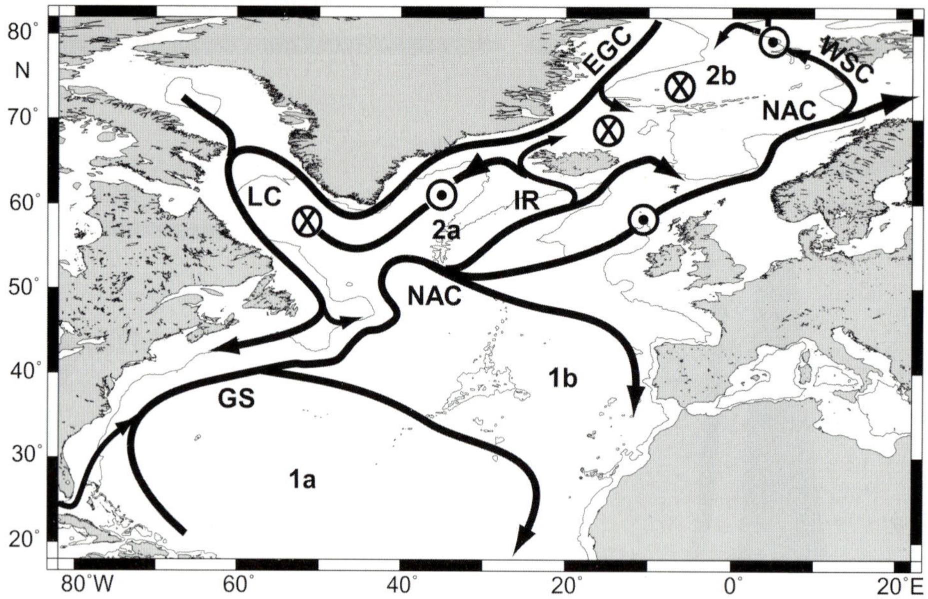

Fig. 2b. Schematic surface circulation of the North Atlantic Ocean. The northward components form the upper limb of the MOC. Numbers denote the subtropical gyre (1a) and its northern extension (1b) and the subpolar gyre in the Northwestern North Atlantic (2a) and the Nordic Seas (2b).

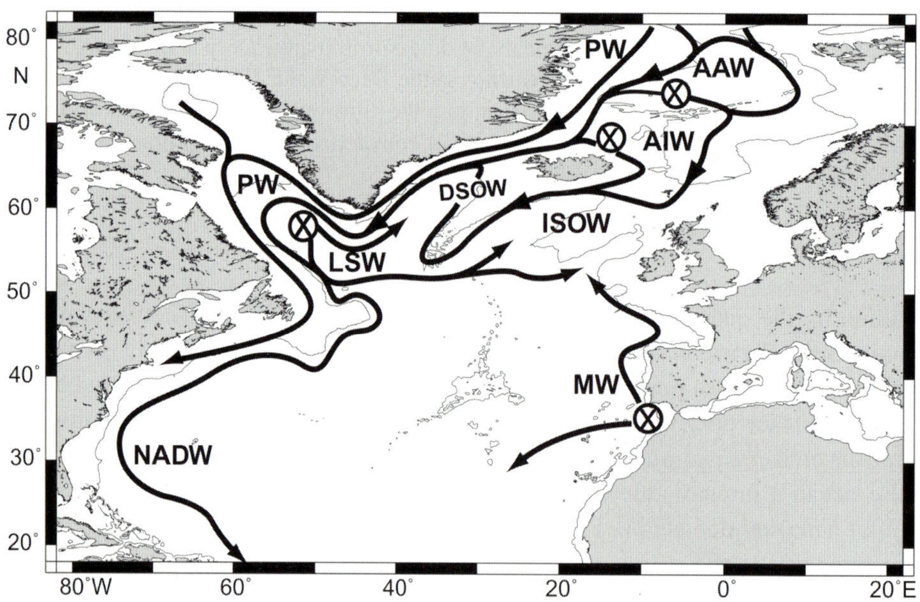

Fig. 2c. Schematic map of intermediate and deep water flows. The southward components form the lower limb of the MOC. NADW North Atlantic Deep Water, MW Mediterranean Water, LSW Labrador Sea Water, DOW Denmark Strait Overflow Water, ISOW Iceland Scotland Overflow Water, AIW Arctic Intermediate Water, AAW Arctic Atlantic Water (Atlantic Water after recirculation in the Arctic), PW Polar (Surface) Water. Circled crosses denote centers of convection. The center off Gibraltar stands for convection within the Mediterranean.

scale variability pattern (modes) of transient periodicity, ranging from interannual to decadal. It is this low frequency variability that needs to be understood in order to advance the predictability of the ocean-atmosphere system from "weather-" to "climate-"time scales.

Based on the operational meteorological network of global synoptic measurements there is detailed evidence for some 10 to 15 recurrent modes of the atmospheric circulation. Due to the much poorer situation for long term ocean observations there are only a few indications of oceanic modes. Therefore only limited studies are available on coupled climate modes in the ocean-atmosphere system. The best known is ENSO, i.e. the coupling between the oceanic El Niño phenomenon and the Southern Oscillation (SO) of the tropical Pacific atmosphere. Its intense investigation over the past two decades has revealed the physics of this large scale ocean-atmosphere interaction and has lead for the first time to a prediction of upwelling intensity off the Peruvian coast and of anomalous precipitation pattern in areas bordering the Eastern Pacific several months in advance.

In contrast to the tropical ENSO, the research on coupled variability in higher latitudes is more complex and consequently less advanced: The wintertime deep convection in the high latitude requires to include the deep ocean with its lagged response. For the atmosphere the energetic high-frequency variability in the belt of the westerlies causes problems in observing the lower-energy, low-frequency changes. The most dominant recurrent atmospheric mode in higher latitudes is the North-Atlantic Oscillation (NAO). It is related to the variability of the system composed of the Azores High and Icelandic Low, which is controlling the band of westerlies across the North Atlantic (Fig. 3).

Rogers (1984) defined a station-based index of the NAO using sea level pressure anomalies from Ponta Delgada, Azores and Akureyri, Iceand. Hurrell (1995, 1996) extended the index further back in time to 1864 by using data from Lisbon, Portugal and Stykkisholmur, Iceland. Jones et al. (1997) finally extended the instrumental record back to 1821 by using data from Gibraltar and SW Iceland (Fig. 4). A "high-index" pattern indicating strong midlatitude westerlies, is characterised by

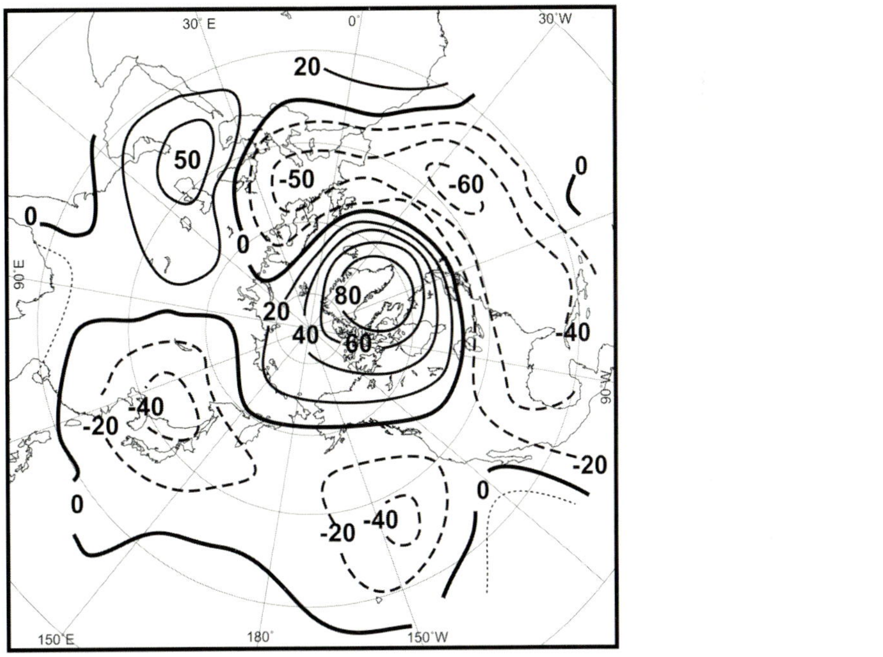

Fig. 3. Correlation map of atmospheric pressure variance in winter for the level of 700 hectopascal. Reference location is 65°N, 45°W. CalCOFI Atlas, 29,1981.

a strong Iceland Low and a strong Azores High to its south, while in the "low-index" case the signs of the anomaly cells are reversed.

Considerable effort has been put into extending the time series of the NAO index further back by using proxy data. Cook et al. (1998) constructed the index back to 1701 from tree ring analysis. Appenzeller et al. (1998) used ice accumulation rates from cores in W-Greenland as proxy for NAO-related precipitation and extended the time series back to 1650. Stockton and Glueck (1999) used tree-ring and ice-core data for the reconstruction back to 1429 and Luterbacher et al. (1999) used both instrumental station pressure, temperature and precipitation data and proxy data in a statistical reconstruction back to 1675. One example of a spectral analysis of the time series is taken from Appenzeller et al. (1998) and presented as Fig. 5. It is fair to summarize the various analyses by stating that the NAO appears to be an intermittent climate oscillation with no sign of a single, persistent, multiannual frequency. There have been several phases, when most of the energy is located in the period band below 15 years. There is also variability on time scales of 70 to 90 years, in particular since the mid-19th century.

Here a critical remark on the use of an index for the NAO is in place. A fixed two-point pressure difference for description of an atmospheric pattern is bound to fail in the case of pattern displacement. This point is important when correlating NAO-type atmospheric forcing with regional oceanic response. It is also important in interpreting the mentioned apparent intermittency of the NAO from time series analysis of indices from fixed stations.

Nevertheless, of the many recurrent pattern of atmospheric behaviour the NAO is the most robust from 13 atmospheric circulation modes worldwide. The NAO is the only mode that is present in every month of the year (Barnston and Livizey 1987). It was shown, that the NAO accounts for the largest amount of interannual variability in monthly North

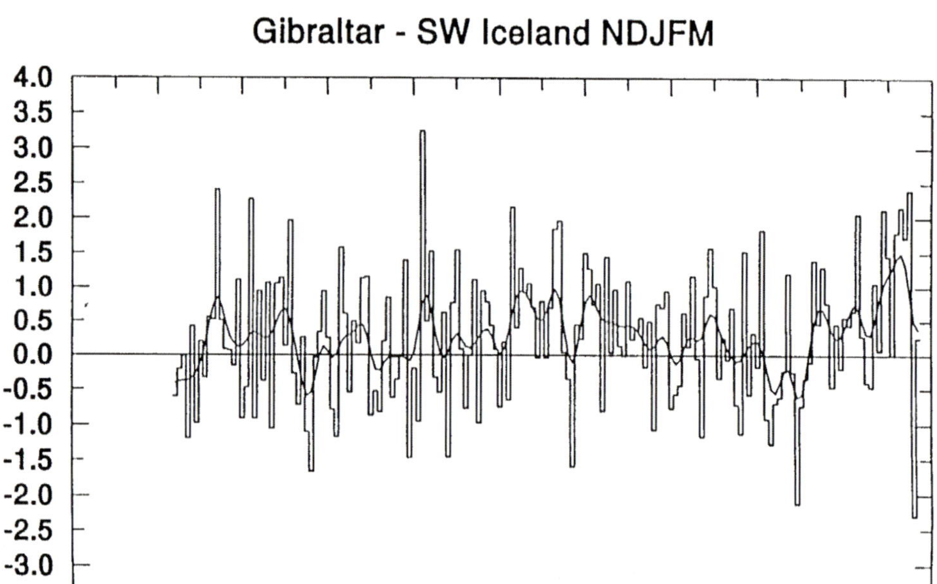

Fig. 4. The normalised NAO-index after Jones et al. (1997). It is a 2-point pressure difference between the long-term observations of sea level pressure at Gibraltar and SW-Iceland . The data cover the winter months from November to March only, since during the season of strongest forcing the signal to noise ratios is greater than for annual averages.

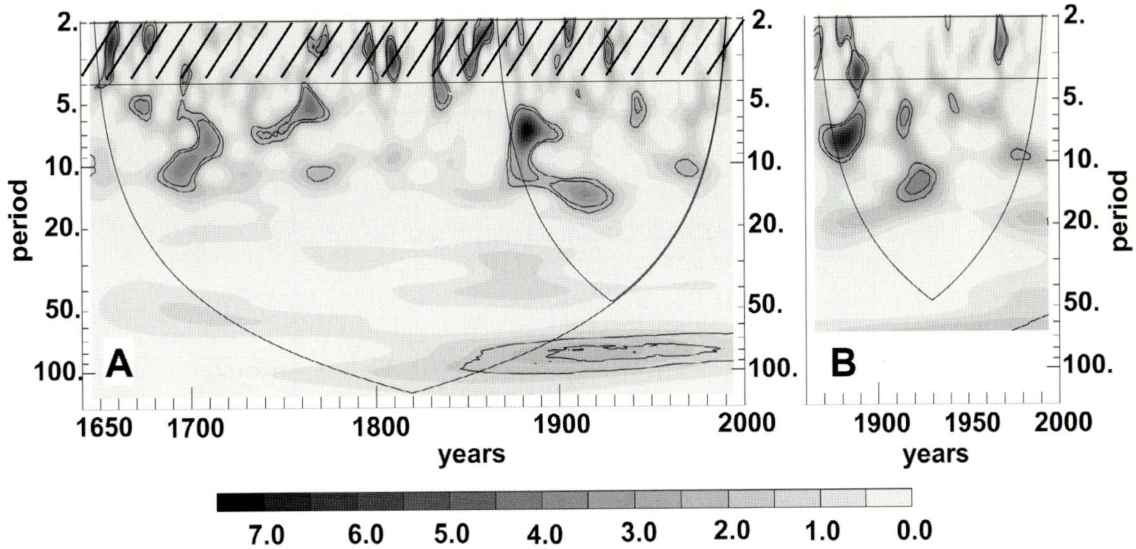

Fig. 5. Wavelet power spectrum of a reconstructed NAO index **(A)** and the instrumental NAO-index **(B)**. The 90% and 95% confidence limits are shown as thin lines, the cone of influence marks regions where edge effects might underestimate amplitudes. Hatched region indicates periods, for which the proxy should not be interpreted. After Appenzeller et al. (1998).

Atlantic sea level pressure (Rogers 1990) and the NAO proved to be the only circulation mode tested that was coherent with the periodic behaviour of the 700 year stable isotope record from the GISP-2 ice core from Central Greenland (Barlow et al. 1993).

Recently it was suggested (Thompson and Wallace 1998) that the NAO may be the regional manifestation of an annular hemispheric mode of variability characterised by a seesaw of atmospheric mass between the polar cap and the middle latitudes in both the Atlantic and the Pacific sector. This was named the Arctic Oscillation (AO). The NAO and the AO are nearly indistinguishable in the time domain and Deser (2000) has found that the correlation between the two relevant indices is as high as 0.95. This comes about because of the dominance of the Atlantic-Arctic sector in the AO-pattern (Dickson et al. 2000).

With respect to the North Atlantic/European importance of the NAO the analysis by Hurrell (1996) on global warming is very clear: Using multivariate linear regression, the effect of observed changes in the NAO and the SO on northern hemisphere extratropical (20°N-90°N) temperatures during winters since 1935 was analysed. The individual and combined response to these circulation changes is shown in Fig. 6. Together the two circulation modes explain 44% of the observed interannual variance in hemispheric extratropical temperatures, with the NAO accounting for 31% and the SO for 16%. Moreover, when the linear effects of the NAO and the SO are removed, the residual time series is found to exhibit no significant trend. From this it can be concluded, that the mechanisms of hemispheric warming - whatever its cause - is acting through an amplification of the NAO in the Atlantic sector, and of the SO signal in the Pacific. Either way, the NAO is revealed as a major source of interannual variability in Atlantic/European weather and climate.

On the more regional scale the interannual variability of the NAO has been found responsible for generating systematic, large amplitude pattern in the anomalies of wind speed, latent and sensible heat fluxes. The extreme states of the NAO have been suggested to determine deep temperatures and salinities on the West Greenland Banks (Buch

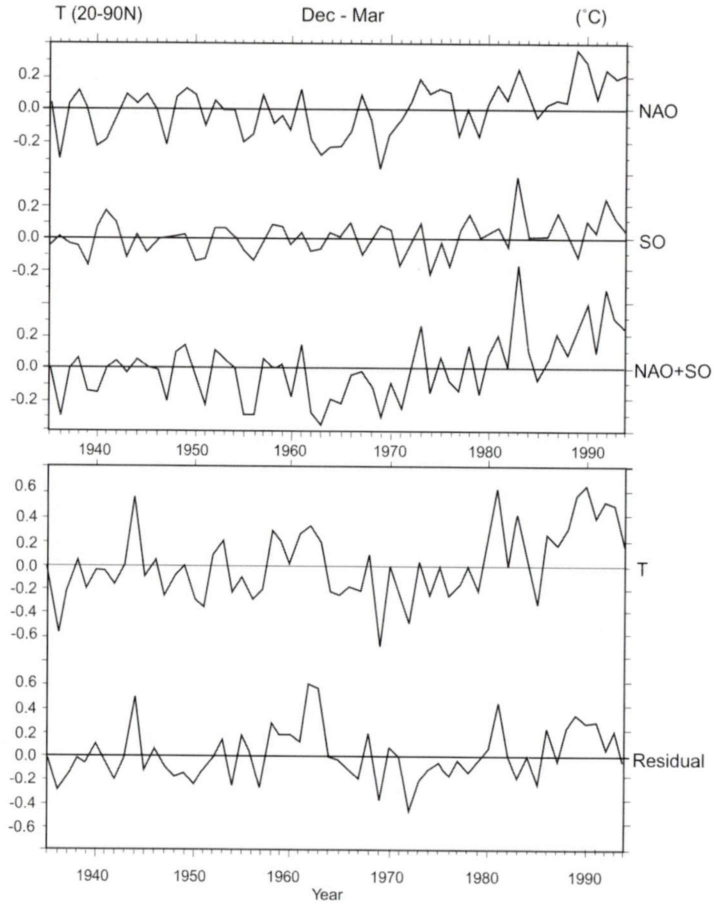

Fig. 6. Temperature changes associated with NAO, the SO and their sum averaged from 20°N to 90°N. The Northern Hemishpere extratropical temperature anomalies relative to the 1935 to 1994 mean, and the residual after removing the linear effects of the NAO and SO are shown in the lower panel (From Hurrell 1996).

1995), heat and salt transports in the currents of Greenland-Labrador (Reverdin et al. 1997), the storm-track in the North Atlantic (Rogers 1990), the midlatitude westerly wind strength and significant wave height (Kushnir et al. 1997), evaporation and precipitation pattern (Hurrell 1995), Arctic sea ice extent (Fang and Wallace 1994) and the iceberg flux past Newfoundland (Drinkwater 1994). It has recently been held responsible for modulating and coordinating the intensity of deep convection in the three main Atlantic sites, thus driving interannual changes to considerable depth in the ocean (Dickson et al. 1996) Figs. 7a and b summarize significant NAO-effects for the North Atlantic.

The fact that the year to year changes in the NAO index are exceeding the range of past experience suggests that mere "persistence" is likely to be a poor estimate of the future behaviour of the NAO (see Fig. 1). Instead, prediction of the NAO will require an adequate understanding of the factors which provide its long-term behaviour, be it from pure atmospheric dynamics or from intensive ocean-atmosphere coupling, both with and without the effects of anthropogenic activities.

Atlantic/European Sector Variability

As mentioned in chapter 3, there are ample examples of correlations between oceanic variability and

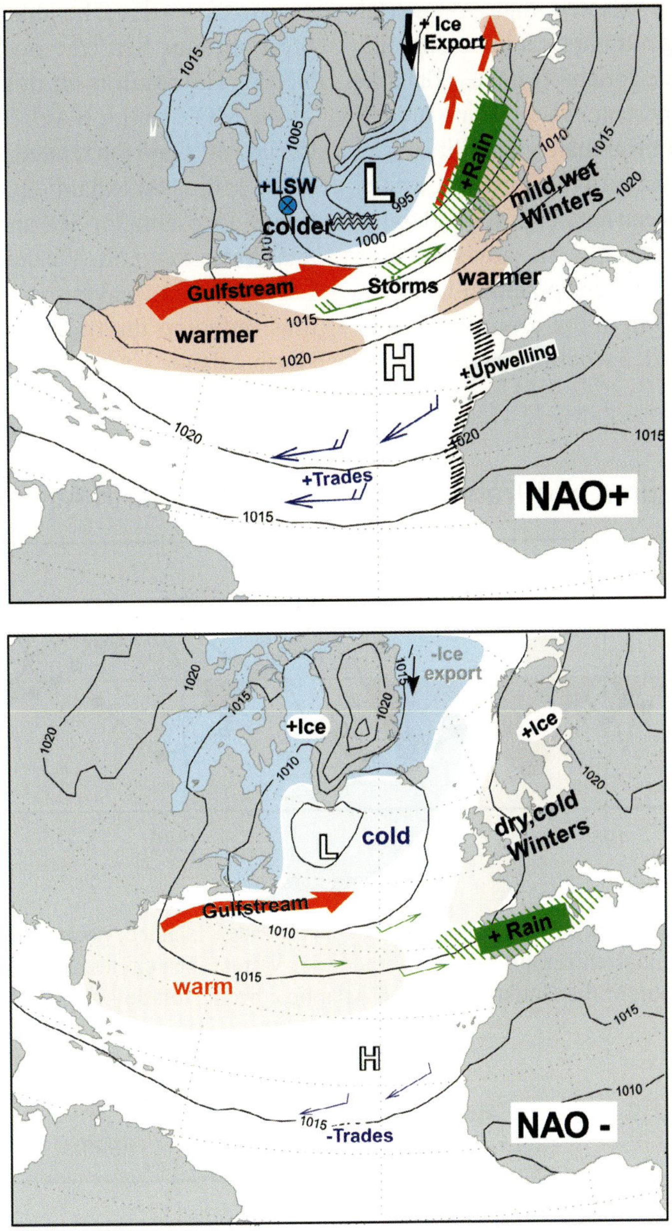

Fig. 7. (a) Composite of effects of an NAO-extreme positive state. The enlarged pressure difference between the Azores High (H) and the Iceland Low (L) causes an intensified band of westerly winds stretching from southwest to northeast across the northern North Atlantic. Warm and humid air masses reach northwestern Europe and the Arctic, the North Atlantic Current and the ice flux from the Arctic are intensified. Massive outbreaks of cold arctic air affect the Labrador Sea and intensify convection. Enhanced trade winds cause strong upwelling off Northwest Africa. **(b)** Composite of effects of an NAO-extreme negative state. A diminished pressure difference between the Azores High and the Iceland Low reduces the intensity of the band of westerly winds, which is more zonally oriented and reaches southeastern Europe. Northern Europe experiences dry and cold winters. The oceanic circulation is reduced, convection in the Labrador Sea is ceased.

NAO-related atmospheric variability. For this chapter five examples were selected, that have a high potential in improving our physical understanding of the ocean-atmosphere variability on decadal time scales and that are important in supporting modelling of the North Atlantic climate variability.

The first example was provided by McCartney et al. (pers. com.) on ocean temperatures and transports related to the MOC. Fig. 8a shows the time series of the NAO, the average potential temperature of the upper 2000 m in the central Labrador Sea and the eastward baroclinic transport of the Gulf Stream/NAC as measured by the difference in the sea level heights between Bermuda and the Labrador Sea (the oceanic analogue to the NAO). The result is an indication of the reaction of the MOC to the NAO: The low NAO-state of the late 60's and the related low flux of cold Arctic air over the NW Atlantic have decreased convection in the Labrador Sea and caused the Labrador Sea Water to warm. Weakened convection in the Labrador Sea weakens the driving of the MOC, thus reducing the transport of the Gulf Stream/North Atlantic Current (see Fig. 2). The long-term

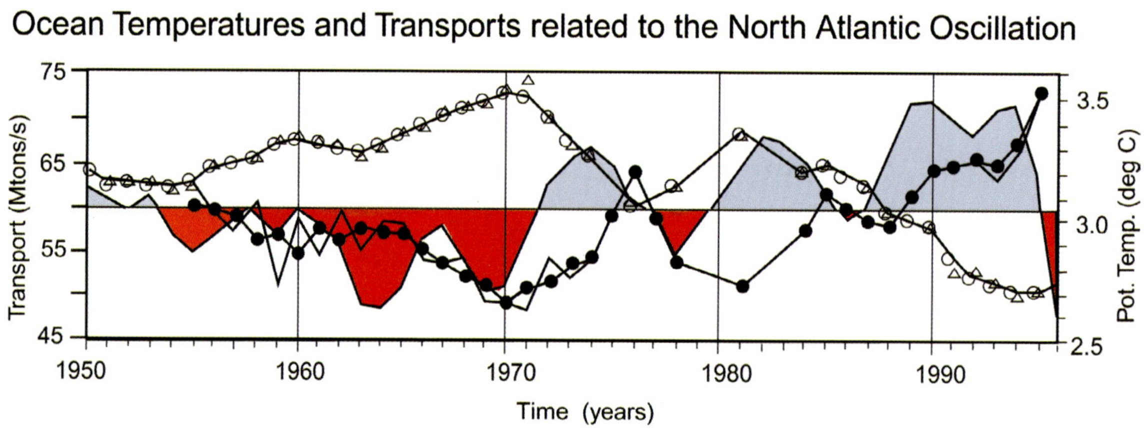

Fig. 8a. Demonstration of the correlation between the NAO-Index (shaded, positive above the reference line negative below), the potential temperature of the Labrador Sea Water (open circles, average over the upper 2000m) and the baroclinic transport of the Gulfstream/North Atlantic current (full dots) (After McCartney, pers. com.).

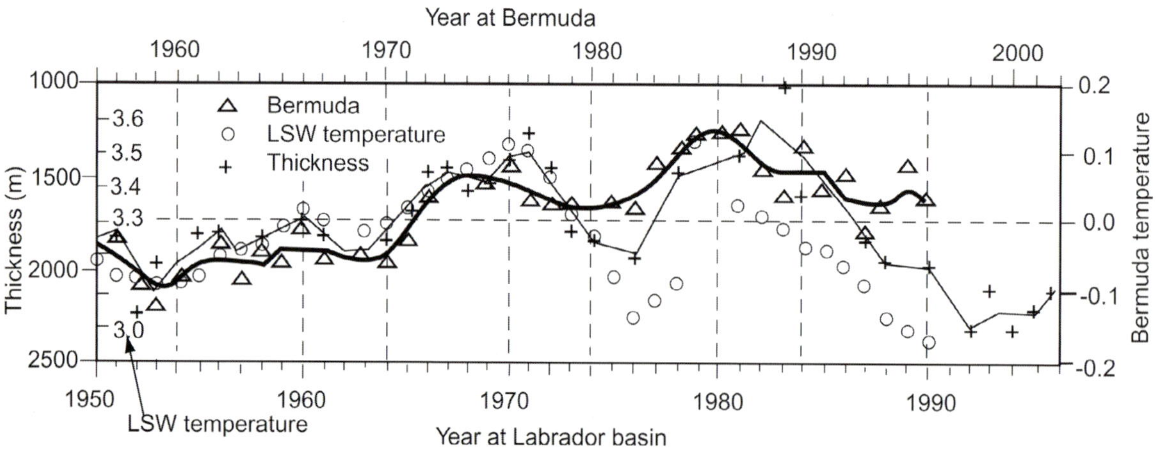

Fig. 8b. Demonstration of the 6year-lagged correlation of the temperature anomaly at 1500 to 2500m near Bermuda with the temperature and the thickness of the Labrador Sea Water (Modified after Curry et al. 1998).

increase in the NAO towards the 90's, the corresponding cooling of the Labrador Sea Water temperature by enhanced cold air outbreaks over the Labrador Sea and subsequent intensified convection increase the Gulf Stream/North Atlantic Current transports accordingly. Note the amplitudes: A 0.8°C temperature range of the large pool of subpolar water in the Labrador Sea goes together with more than 30% change in transports within the upper limb of the MOC.

Whereas Fig. 8a shows the relation between NAO-driven changes of the vertical component of the MOC and its upper limb, Fig. 8b completes the picture by showing the relation between changes of the vertical component and the deep limb of the MOC (Curry et al. 1998). According to Fig. 2c, the convected water from the Labrador Sea should spread southward as upper North Atlantic Deep Water. This is shown by comparing the temperature changes of the Labrador Sea Water to the deep temperature changes near Bermuda. The maximum correlation occurs at a phase lag of 6 years, i.e. it takes 6 years for temperature signals to propagate from the Labrador to the Sargasso Sea. Then Figs. 8a and b together nicely demonstrate the coherence between NAO-driven changes of convection in the Labrador Sea and related variability in the upper and lower limb of the MOC.

The second example is on northward oceanic heat transport across 48°N. This quantity describes the oceanic heat advection from the subtropics to the northern North Atlantic. Lorbacher (2000) has plotted heat transport estimates based on repeat trans-Atlantic hydrographic sections along 48°N on top of NAO-time series (Fig. 9). Periods of a high NAO-index correspond to high heat transports and vice versa (see Fig. 7a and b). The heat transport has a phase lag of about one year. The few realisations show a correlation from the decadal to the event-time scale (see example number 5) and result in changes of up to 60% of the meridional heat transport. This is in contrast to present modelling, which can only yield changes lower by one order of magnitude for maximum possible parameter variation.

The third example was published by Dickson et al. (1996) on the NAO-control of the North Atlantic convective activity: The convective renewal of intermediate and deep waters in the Labrador Sea and Greenland/Iceland Sea both contribute to the production and export of North Atlantic Deep water thus driving the global MOC. For as long as

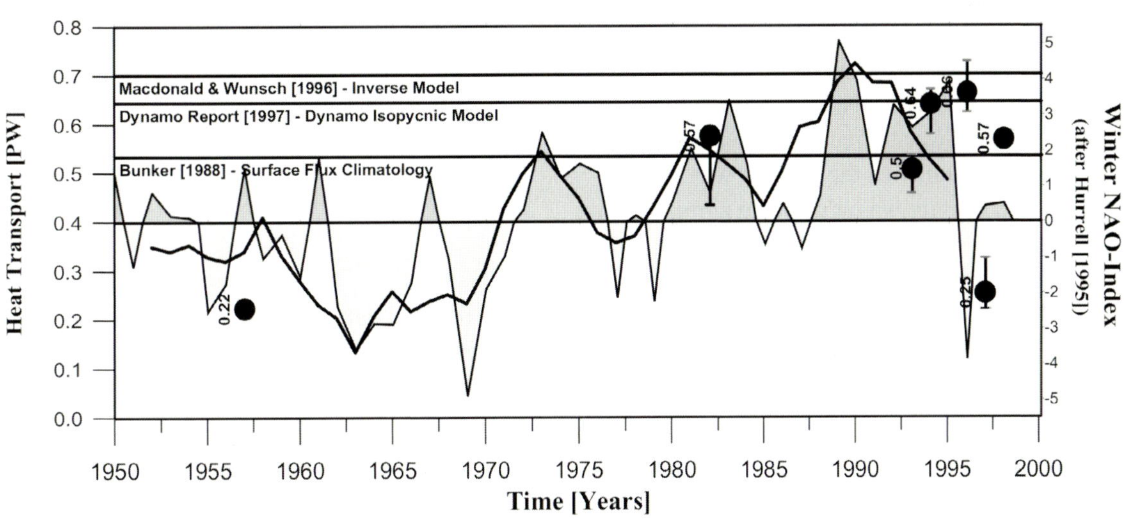

Fig. 9. Northward oceanic heat transport (PW=10^{15} Watt) across 48°N in the Atlantic. Dots are estimates based on hydrographic transects. Winter NAO-index after Hurrell (1995) and 5 year running means (full line). (From Lorbacher 2000)

time series of adequate precision have been available (Fig. 10), it has been apparent that the intensity of convection at each of these sites, and the hydrographic character of their products have been subject to major interannual change. The recent history of these changes shows that the major convective centres of the Greenland and Labrador Seas have been at opposite convective extrema until 1995, with vertical exchange at the former site limited to 1000 m or so, but with Labrador Sea convection reaching deeper than previously observed, to over 2300 m. As a result, the deep water of the Greenland Sea has become progressively warmer and more saline since the early 70's because of increased horizontal exchange with the Arctic Ocean through Fram Strait, while the Labrador Sea Water has become progressively colder and fresher over the same period through increased vertical exchange; most recently, convection has become deep enough there to reach into the more saline North Atlantic Deep water which underlies it, so that cooler, but now saltier and denser Labrador Sea Water has resulted. The theory is advanced that the scales of atmospheric forcing have imposed a degree of synchrony on convective behaviour at the two sites over the present century, with ventilation of the Greenland Sea undergoing a multi-decadal evolution to reach a long term maximum in the 1960's, driven by the NAO. During the NAO minimum of the 1960's, an extreme Greenland ridge was feeding record amounts of fresh water into the subpolar gyre in the form of the Great Salinity Anomaly, and storm activity retracted in a south-ward direction . Therefore in the area of the Labrador Sea the surface was freshening and storm activity was at its postwar minimum. This brought a progressive reduction, and ultimately a cessation of wintertime convection there.

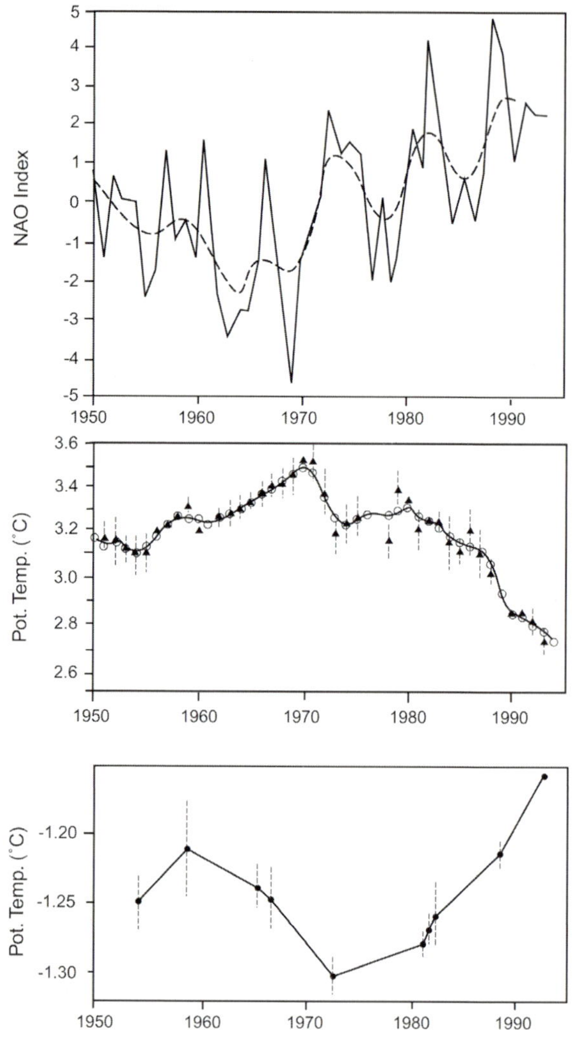

Fig. 10. Coordinated change of North Atlantic convective activity in relation to the NAO. (From Dickson et al. 1996). Shown are the NAO-Index, the potential temperature of the upper 2000m in the central Labrador Sea and the potential temperature of the waters below 2000m in the central Greenland Sea.

In the fourth example Dickson et al. (2000) present the Arctic Ocean response to the NAO: The climatically sensitive zone of the Arctic Ocean lies squarely within the domain of the NAO. However, the specific response of the Arctic to annual and longer-period changes in the NAO is not well understood. The authors investigate that response using a wide range of data-sets, but concentrating on the winter season when the forcing is maximal and on the postwar period which includes the most comprehensive instrumental record. This period also contains the largest recorded low-frequency change in NAO activity – from its most persistent and extreme low index phase in the 1960's to its most persistent and extreme high index phase in the late 1980's/early 1990's. This long-period shift between contrasting NAO extrema was accompanied, amongst other changes, by an intensifying storm track through the Nordic Seas, a radical increase in the atmospheric moisture flux convergence and winter precipitation in this sector (Fig.11), an increase in the amount and temperature of the Atlantic water inflow to the Arctic Ocean via both inflow branches (Barents Sea Throughflow and West Spitsbergen Current), a decrease in the late-winter extent of sea-ice throughout the European subarctic, and (temporarily at least) an increase in the annual volume flux of ice from Fram Strait. In the context of this paper it has become evident, that the strong amplification of the northern cell of the NAO dipole to record levels in the 1960's was the key factor in generating the largest known dislocation of the freshwater balance of the northern subpolar gyre, the so-called Great Salinity Anomaly (Dickson et al. 1988).

The last example is to demonstrate how effective the NAO-related atmospheric driving can affect the North Atlantic/European sector on a large spatial scale, but on an event time-scale: Dickson and Meincke (pers.com.) have considered the anomalous drop of the NAO-index from its second highest value over the instrumental record in winter 1994/95 to the observed absolute minimum in winter 1995/96 and its subsequent recovery in the period '96 to '99 (see Fig.4). Five parameters were correlated with the drop of the NAO-index:

(a) The northward oceanic heat transport across 48°N dropped with a lag of 1 year (Fig. 9),

(b) the zonal location of the boundary between the subtropical and the subpolar gyre at 55°N (Fig. 12) shifted westward,

(c) the deep convection in the Labrador Sea stopped,

(d) the sea level height of the northern North Atlantic increased by 8cm and

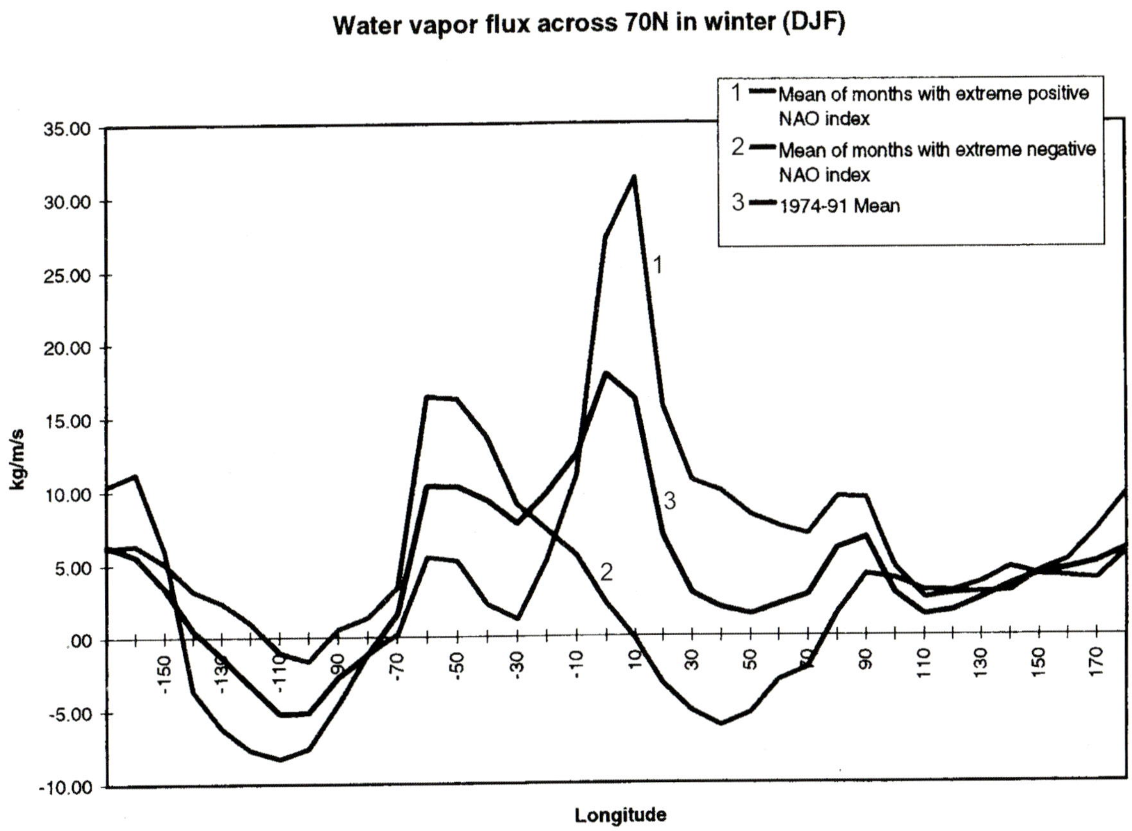

Fig. 11. Vertically-integrated meridional moisture flux crossing 70°N in winter expressed as a function of longitude for composites of winter (DJF) months representing high-index and low-index NAO extrema. The 1974-91 winter mean is shown. (From Dickson et al. 2000)

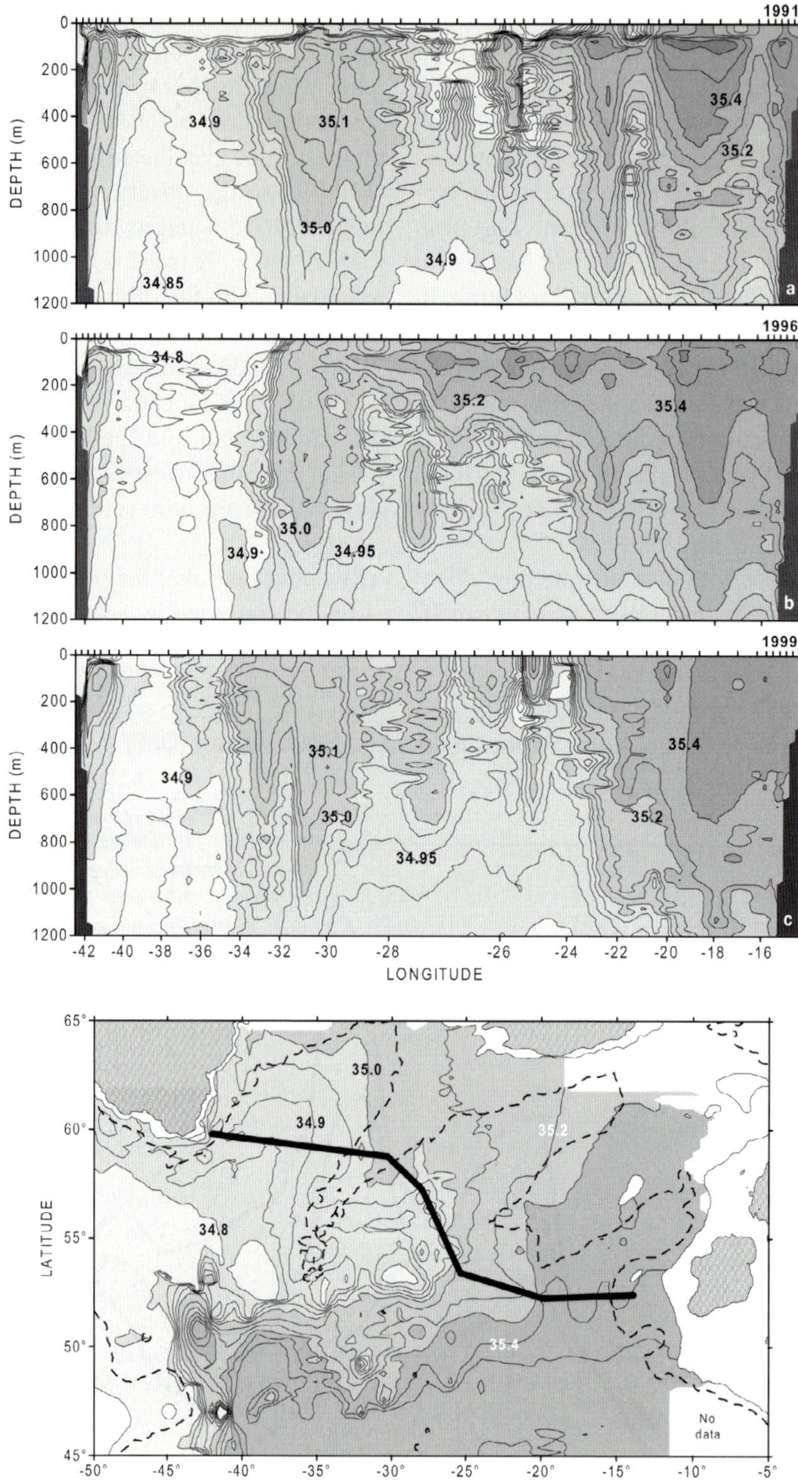

Fig. 12. The shift of the boundary between the subtropical gyre/subpolar gyre in reaction to the large NAO-drop in winter 1995/1996 (see Fig. 4). Shown is the salinity distribution along a repeat hydrographic section between Greenland and Ireland. (Modified after Bersch et al. 1999).

(e) the precipitation pattern from northern Africa to northern Norway and Spitsbergen showed a drastic shift from wet to dry in NW-Europe.

Their response to the present recovery of the NAO-index remains to be seen. First indications are, that the North Atlantic does return to its high-index-state in the eastern part (see Fig. 7a), but not in the western part. It is suspected, that the NAO-pattern has shifted eastward, so that there is no NAO-type atmospheric driving over the western North Atlantic. Here the critical remark from the foregoing chapter seems to apply.

Conclusions

This overview has been solely based on observational evidence for interannual to decadal changes in the North Atlantic region. To understand and eventually predict the complex dynamical ocean-atmosphere-land system in the North Atlantic and its decadal-centennial interaction, future emphasis will be placed on numerical modelling. The modelling efforts must determine whether variability simulated in models is fundamentally coupled, whether one component of the climate system drives variability in another, whether the simulated variability modes are related to observed variations, and whether modes of variability once identified are predictable. The modelling component is also particularly crucial because of the present lack of observational records of sufficient length, combined with the need to synthesise sparse time-series data from both the ocean and the atmosphere. This hints to the need, in addition to modelling, to continue with the efforts of constructing long time series of sufficient resolution from atmospheric and oceanic proxy-data. It is expected, that the Climate Variability and Predictability (CLIVAR) Study, launched as a long-term component of the World Climate Research Programme, will contribute to these ends.

References

Appenzeller C, Stocker TF, Anklin M (1998) North Atlantic Oscillation Dynamics recorded in Greenland Ice Cores. Science 282:446-449

Barlow LK, White JWC, Barry RG, Rogers JC, Grootes PM (1993) The North Atlantic Oscillation Signature in Deuterium and Deuterium Excess Signals in the Greenland Ice Sheet Project-2 Ice Core, 1840-1970. Geophys Res Lett 20:2901-2904

Barnston AG, Livezey RE (1987) Classification, seasonality and persistence of low-frequency at-mospheric circulation patterns. Mon Wea Rev 115:1083-1126

Bersch M, Meincke J, Sy A (1999) Interannual thermohaline changes in the North Atlantic 1991-1996. Deep Sea Res II (46):55-75

Buch E (1995) A Monograph on the Physical Oceanography of the Greenland Waters. Roy Dan Admin of Nav and Hydrogr, Copenhagen, pp1-405

Cook ER, D'Arrigo RD, Briffa KR (1998) The North Atlantic Oscillation and its expression in circum-Atlantic tree-ring chronologies from North America and Europe. The Holocene 8:9-17

Curry RG, McCartney M, Joyce T (1998) Oceanic transports of subpolar climate signals to mid-depth subtropical waters. Nature 391:575-577

Deser C (2000) On the teleconnectivity of the "Arctic Oscillation". Geophys Res Lett Vol. 27, 6:775

Dickson RR, Meincke J, Malmberg S-A, Lee A (1988) The "Great Salinity Anomaly" in the northern North Atlantic 1968-1982. Prog Oceanogr 20:103-151

Dickson RR, Lazier J, Meincke J, Rhines P, Swift J (1996) Long-term coordinated changes in the convective activity of the North Atlantic. Prog Oceanogr 38:241-295

Dickson RR, Osborn TJ, Hurrell JW, Meincke J, Blindheim J, Ådlandsvik B, Vinje T, Alekseev G, Maslowski W (2000) The Arctic Ocean response to the North Atlantic Oscillation. J Clim 13 (15):2671-2696

Drinkwater,KF (1994) Environmental Changes in the Labrador Sea and some Effects on Fish Stocks. ICES CM 1994/MINI:4, pp 1-19 (mimeo)

Fang Z, Wallace JM (1994) Arctic sea-ice variability on a time scale of weeks and its relation to atmospheric forcing. J Clim 7:1897-1914

Hurrell JW (1995) Decadal trends in the North Atlantic Oscillation: regional temperatures and precipitation. Science 269:676-679

Hurrell JW (1996) Influence of variations in extratropical wintertime teleconnections on Northern Hemisphere temperature. Geophys Res Lett 23:665-668

Hurrell JW, Von Loon H (1997) Decadal variations in Climate associated with the North Atlantic Oscillation. Climate Change 36:301-326

Jones PD, Jonsson T, Wheeler D (1997) Extension to the North Atlantic Oscillation using early instrumental pressure observations from Gibraltar and SW Iceland. Climatol 17:1433-1450

Kushnir Y, Cardone VJ, Greenwood JG, Cane M (1997) The recent increase in North Atlantic wave heights. J Climate 10, 8:2107-2113

Lorbacher K (2000) Niederfrequente Variabilität meridionaler Transporte in der Divergenzzone des nordatlantischen Subtropen- und Subpolarwirbels. Ber. BSH 22, pp 1-156

Luterbacher J, Schmutz C, Gyalistras D, Xoplaki E, Wanner H (1999) Reconstruction of monthly NAO and EU indices back to AD 1675. Geophys Res Lett 26:2745-2748

Rahmstorf S, Ganopolski A (1999) Long-term global warming scenarios computed with an efficient coupled climate model. Climate Change 43:353-367

Reid JI (1994) On the total geostrophic circulation of the North Atlantic Ocean: flow patterns, tracers and transports. ProgOceanogr 33:1-92

Reverdin G, Cayan DR, Kushnir Y (1997) Decadal variability of hydrography in the upper northern North Atlantic, 1948-1990. J Geophys Res 102, CC4, 8505-8532

Rogers JC (1984) The Association between the North Atlantic Oscillation and the Southern Oscillation in the Northern Hemisphere. MonWea Rev 112:1999-2015

Rogers JC (1990) Patterns of low-frequency monthly sealevel pressure variability (1899-1986) and associated wave cyclone frequencies. J Clim 3:1364-1379

Schmitz WI (1996) On the World Ocean circulation: Vol. I, WHOI, Tech Rep, 96-03, pp 1-150

Schmitz WI, McCartney MS (1993) On the North Atlantic circulation. Rev Geophys 31:29-49

Stockton CW, Glueck MF (1999) Long-term variability of the North Atlantic Oscillation (NAO). In Preprint of the Am. Met. Soc. 10th Symposium on Global Change Studies, 10-15 Jan 1999, Dallas, Texas, 290-293

Talley LD (1996) North Atlantic circulation and variability. Physica, D, Elsevier, 625-646

Thompson DW, Wallace JM (1998) The arctic oscillation signature in the wintertime geopotential height and temperature fields. Geophys Res Lett 25:1217-1300

Holocene Climate Variability on Centennial-to-Millennial Time Scales: 1. Climate Records from the North-Atlantic Realm

Michael Schulz[1] and André Paul[2, 3]

[1] *University of Kiel, Institute for Geosciences, Universität Kiel, Olshausenstr. 40, D-24118 Kiel, Germany*
e-mail: mschulz@email.uni-kiel.de
[2]*Scripps Institution of Oceanography, La Jolla, 9500 Gilman Drive, CA 92093–0244, USA*
[3]*Now at: University of Bremen, Department of Geosciences, Postfach 33 04 40, D-28334 Bremen, Germany*
e-mail: apau@palmod.uni-bremen.de

Abstract: Holocene oxygen isotope data from the GISP2 ice core reveal temperature oscillations in Greenland with a periodicity of ~900 y, which can be correlated to climate perturbations in northern and central Europe. We suggest that the 900-y climate fluctuations are generated within the climate system, and are probably triggered by negative salinity anomalies in the North Atlantic. A simple template is used to show that two such triggering events centered at ~8.3 and 4.7 ky BP are required to explain temporal evolution of 900-y climate cycles between ~3.5–8.5 ky BP as sequence of damped oscillations. Although pacing of the 900-y cycles by changes in the Earth's orbit cannot be ruled out, we regard this scenario as unlikely. We show that the existing paleoceanographic evidence for ~1400–1500-y climate oscillations during the Holocene is questionable. Instead we suggest that deep-sea records from the North Atlantic may be reconciled with 900-y climate oscillations during this period.

Introduction: The Seemingly Stable Holocene

Recent results from ice-cores, drilled at the summit of the Greenland ice sheet, boosted paleoclimate research since they provide clear evidence for recurrent and abrupt climatic changes during the last glacial period. In particular oxygen isotope (δ^{18}O) data, which are assumed to monitor air temperature, reveal that large and rapid temperature fluctuations (warming by up to 10°C within a few decades) dominated climate in Greenland between ~11–74 thousand years before present (ky BP; all ages are reported as calendar years) (Dansgaard et al. 1993; Grootes and Stuiver 1997; Fig. 1). Grootes and Stuiver (1997) demonstrated that warm peaks (so called Dansgaard-Oeschger inter-stadials) occurred with a fundamental period of 1470 years (y). The significance of this millennial-scale climate variability is corroborated by similar climate variations on a global scale which can be correlated to the Dansgaard-Oeschger interstadials (e.g. Sirocko et al. 1993; Behl and Kennett 1996; Schulz et al. 1998; Sarnthein et al. 2001). Accordingly, the ice-core δ^{18}O series at least recorded climate in the North Atlantic region (Grootes and Stuiver 1997) and to some extent global climate variability (Blunier et al. 1998) during the last glacial period.

A salient feature of the Greenland δ^{18}O record is its relative small variation during the Holocene, compared to the preceding time interval, implying that no large temperature variations occurred in Greenland during the last 10 ky (Fig. 1). However, this finding contrasts with data from various Euro-

From WEFER G, BERGER W, BEHRE K-E, JANSEN E (eds), 2002, *Climate Development and History of the North Atlantic Realm.* Springer-Verlag Berlin Heidelberg, pp 41-54

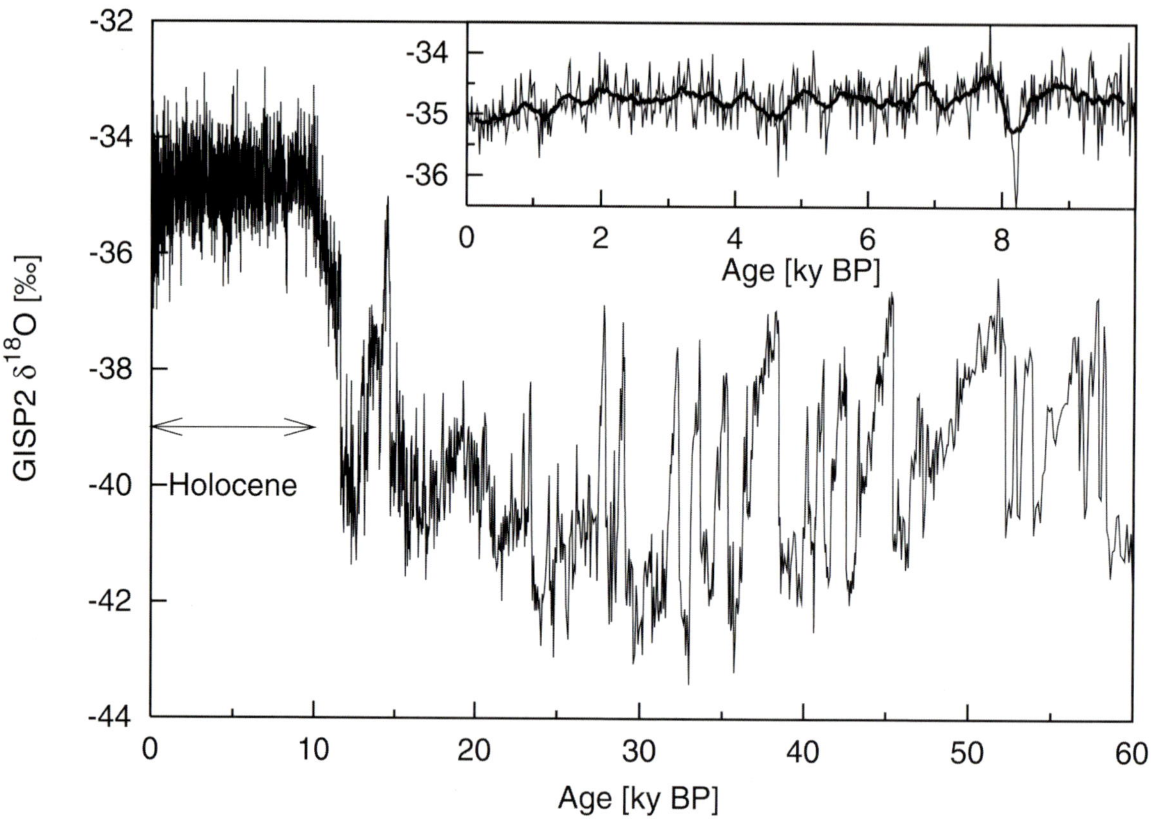

Fig. 1. Oxygen isotope ($\delta^{18}O$) time series from Greenland (GISP2 ice core; Grootes and Stuiver 1997). The inset shows expanded Holocene portion (thin line) and 300-y running mean (bold line).

pean paleoclimatic archives, which reveal marked climate variations during the Holocene, as indicated for example by fluctuations of glaciers and tree-lines (e.g. Karlén 1993; Dahl and Nesje 1996) as well as net precipitation (Harrison and Digerfeldt 1993; Schettler et al. 1999). Although the dating of these events is not without problems, it appears that European climate varied on multicenntennial-to-millennial time scales during the Holocene. These natural climate fluctuations were probably strong enough to affect early human societies in Europe. Insights into the temporal evolution of Holocene multicentennial-to-millennial scale climate variability are not only important for interpreting archeological data, but also because of the potential interference with today's man-made climate perturbation.

Moreover, recent paleoceanographic proxy data from the North Atlantic (Bond et al. 1997; Bianchi and McCave 1999) suggest millennial-scale climate variability with a periodicity at ~1400–1500 y during the Holocene, implying that Dansgaard-Oeschger events, although attenuated, may have occurred during the last 10 ky, too. The lack of similar events in the $\delta^{18}O$ data from Greenland (Fig. 1, inset) raises the question if the ice-core data can be regarded as climate indicator for the *entire* North-Atlantic realm during the Holocene.

A Holocene 900-y Temperature Cycle in Greenland

Despite its random appearance, the Holocene bidecadal $\delta^{18}O$ record from the Greenland Ice Sheet Project Two (GISP2; Grootes and Stuiver 1997) contains a distinct low-frequency component, which becomes clearly visible after applying a 300-y running mean filter to the data (Figs. 1, 2). The running mean varies between approximately

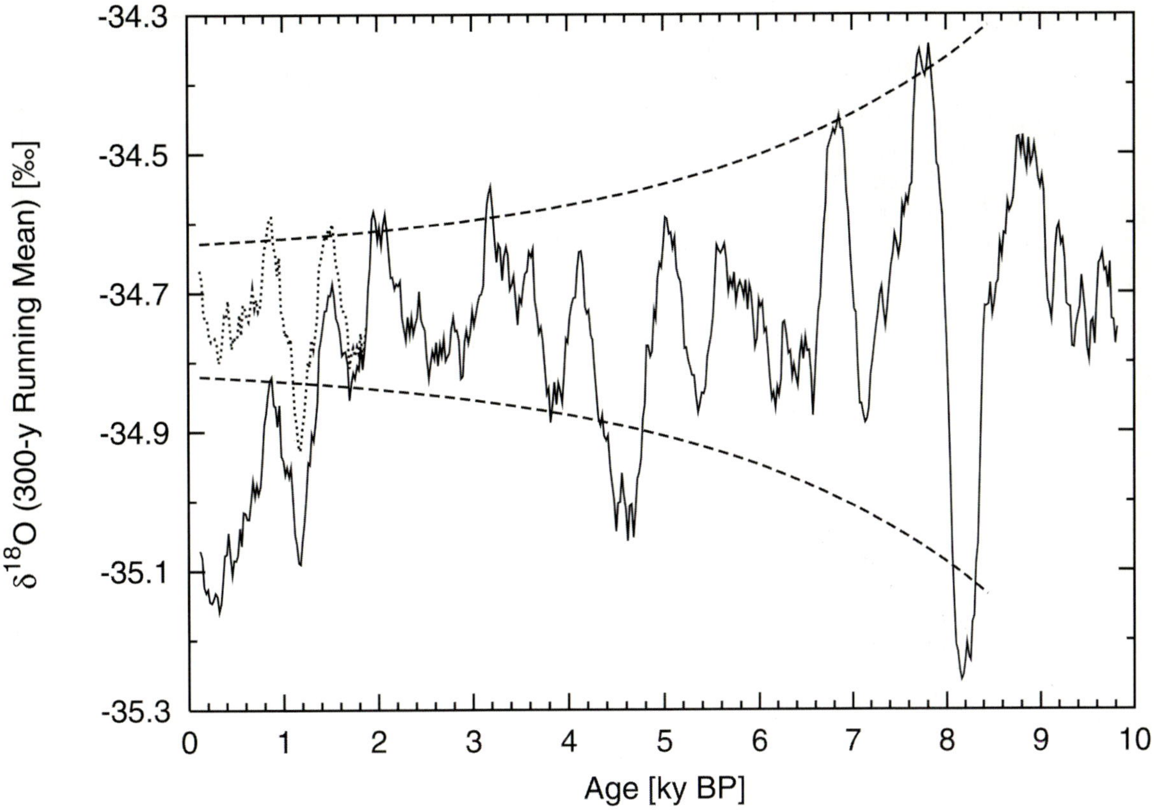

Fig. 2. Smoothed $\delta^{18}O$ time series (300-y running mean [solid line]; as in Fig. 1, inset). For subsequent analyses, linear trend between 0-2 ky BP was subtracted (dotted line). Exponential envelopes (dashed lines) depict general temporal evolution of $\delta^{18}O$ minima and maxima.

± 0.5‰ to ± 0.1‰ around its average value (Fig. 2). It appears that the amplitude of the running-mean fluctuations decreases exponentionally from a maximum around 8.2 ky BP. This tendency is, however, somewhat obstructed by a trend in the youngest part of the record (<2 ky BP). Most extreme $\delta^{18}O$ minima occur approximately every 3.4 ky (at 1.2, 4.7 and 8.2 ky BP; Fig. 2). Between 3–8.5 ky BP the fluctuations of the running mean appear to be almost periodic, with a recurrence time of ~900 y. Harmonic analysis of the unsmoothed $\delta^{18}O$ data (Fig. 3) indicates that the oscillations are indeed contained in the time series and are not an artefact of the running-mean filter (cf. similar periodicity in Grootes and Stuiver 1997).

To test the robustness of the above inference (i.e. a damped ~900 y oscillation), one can use the $\delta^{18}O$ record from the North GRIP ice core (Hammer this volume). Between 5–9 ky BP the temporal evolution of the smoothed $\delta^{18}O$ time series from North GRIP (not shown) is similar to the GISP2 data (Fig. 2). Therefore, the North GRIP data corroborate the existence of damped ~900-y temperature oscillations in Greenland in the early Holocene. However, the North GRIP record shows no evidence for a cooling at ~4.7 ky BP. Instead a trend towards lighter $\delta^{18}O$ values dominates the last 5 ky in this data set. Possible causes for this unexpected discrepancy between the two ice-core records remain open for debate (cf. Hammer, this volume).

With a $\delta^{18}O$-surface temperature gradient of 0.5‰ per °C (Jouzel et al. 1997 and refs. therein), the estimated changes in the running mean imply temperature fluctuations, ranging from ±1°C in the early Holocene to only ±0.2 °C in the youngest part of the record. Compared with temperature variations associated with glacial Dansgaard-Oeschger

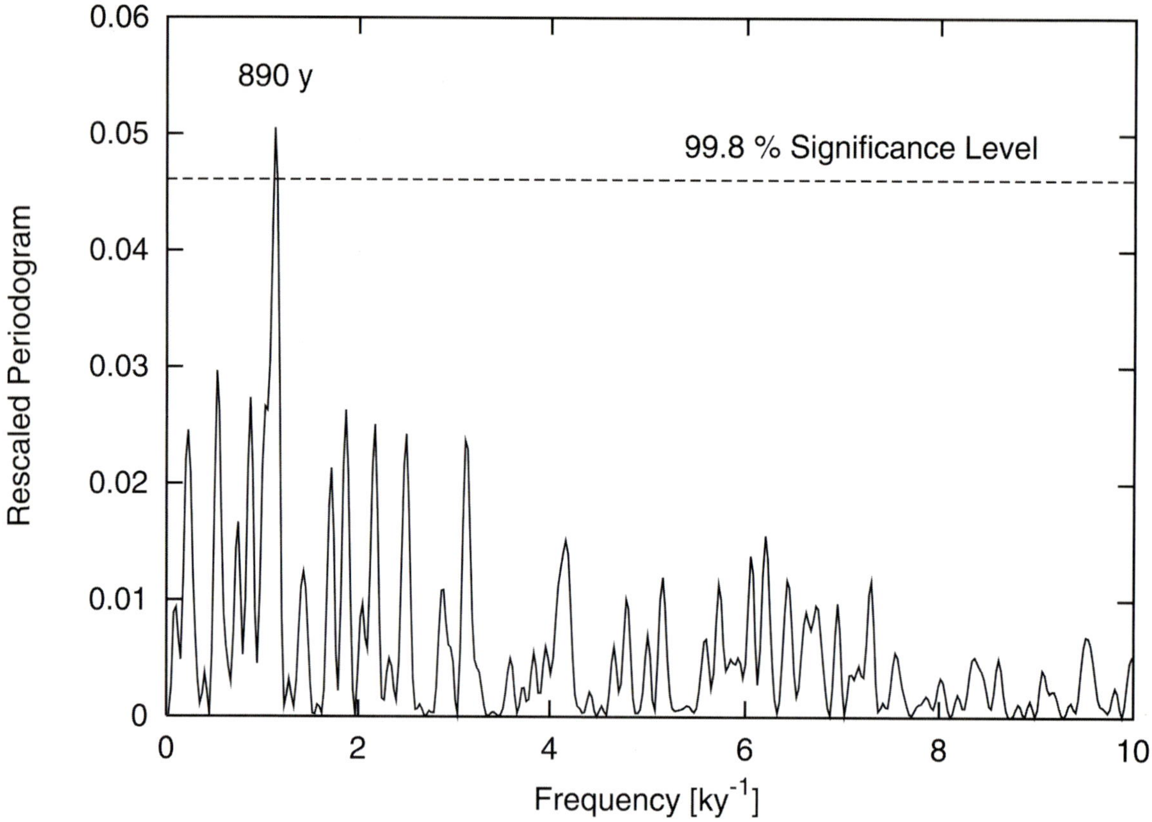

Fig. 3. Harmonic analysis (Fisher 1929; Schulz and Stattegger 1997) of unsmoothed $\delta^{18}O$ time series between 0-10 ky BP. The time series can be distinguished from white noise at the (1-1/*N*) × 100% = 99.8% significance level (cf. Thomson 1990; *N* = number of data points). The sharp peak at periodicity of 890 y is highly significant and indicates the presence of a harmonic signal component.

interstadials (up to ~10°C; cf. Fig. 1), the multi-centennial temperature perturbations during the Holocene are rather small. Based on the premise that $\delta^{18}O$ fluctuations preserved in Greenland ice-cores are at least significant for the North Atlantic region (Grootes and Stuiver 1997), one has to ask if these seemingly small temperature oscillations are relevant to climate variations in Europe.

Climate Variations in Northern and Central Europe

Dahl and Nesje (1996) estimated Holocene summer temperature in south-central Scandinavia based on variations of the altitude limiting pine-tree growth (Fig. 4). Furthermore, these authors reconstructed changes of the equilibrium-line altitude (ELA) of glaciers in southern Norway (Fig. 4) during the Holocene. ELA variations provide a measure for shifts in net glacier growth.

A comparison of the Scandinavian summer temperature with the smoothed $\delta^{18}O$ record from Greenland (Fig. 4) reveals a good correspondence between extremes of these two climate proxies. (It should be noted that the stratigraphies of the records are *not* tuned to each other.) Before 6.6 ky BP cold events in Greenland (negative $\delta^{18}O$ anomalies) coincide with cold summers in Scandinavia, whereas between 4–6.6 ky BP cold conditions in Greenland correlate with warm summers in northern Europe. This distinct change of the phase relationship between the temperature re-

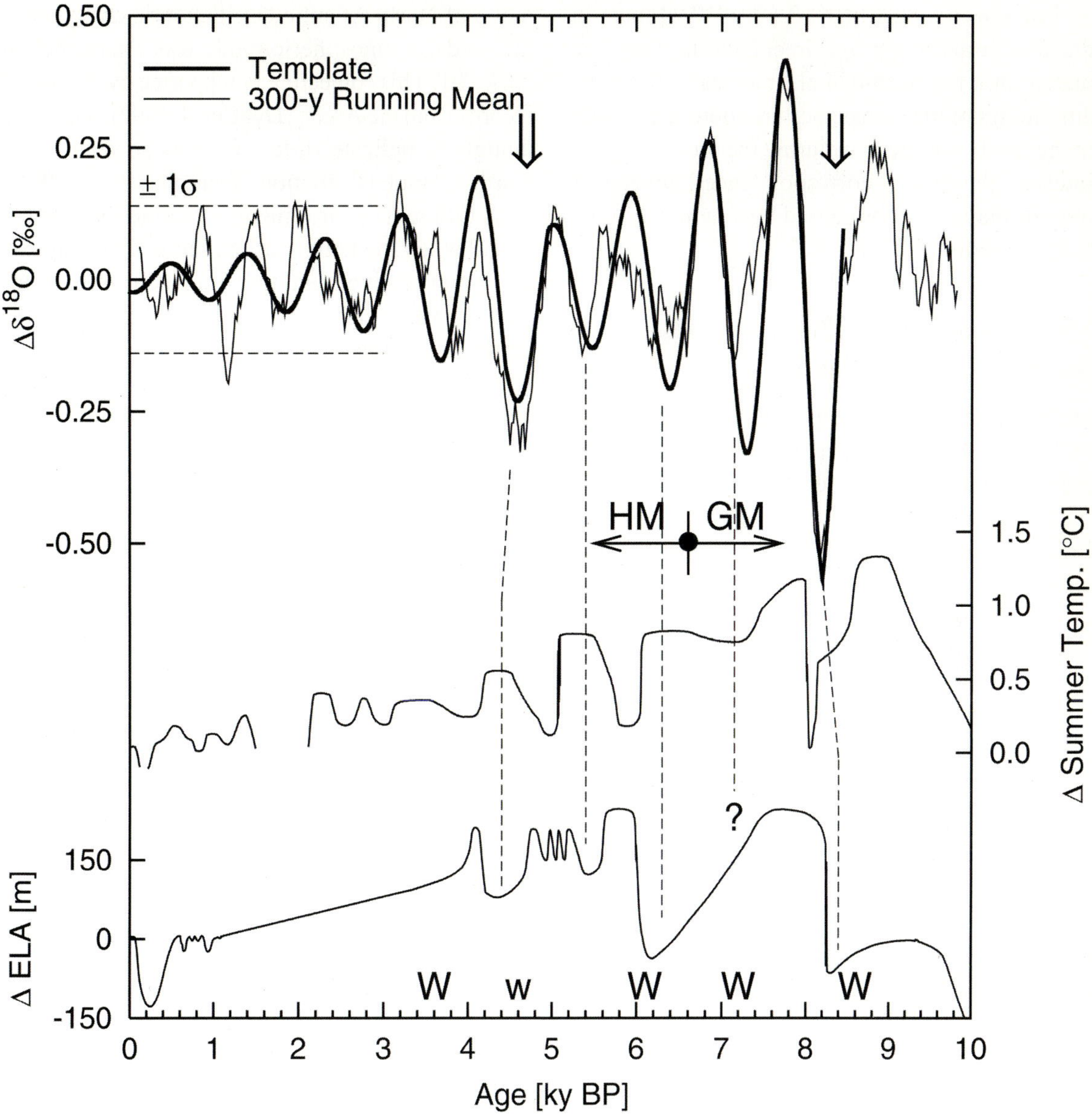

Fig. 4. Top: Smoothed $\delta^{18}O$ record (as in Fig. 2, but after subtraction of mean value) and output of the template. Arrows indicate forcing of the template by cooling phases. Dashed horizontal lines between 0-3 ky BP depict uncertainty of $\delta^{18}O$ data (1σ confidence interval). Center: Deviation of summer temperature from modern value in Scandinavia (Dahl and Nesje 1996). Bottom: Departure of equilibrium line altitude from modern position in Scandinavia (Dahl and Nesje 1996). Negative values indicate glacier advance. 'W' denote wet periods in central Europe (Schettler et al. 1999). Dashed vertical lines depict inferred correlation between climate records. The inferred transition from glacial (GM) to Holocene mode (HM) is indicated. See text for further details.

cords around 6.6 ky BP is, however, not seen between $\delta^{18}O$ and ELA. Instead, glacier advances (ELA minima) go along with cold temperatures in Greenland between 4–8.5 ky BP. Moreover, between ~3.6–8.6 ky BP cold spells in Greenland occur together with periods of enhanced rainfall in central Germany (Schettler et al. 1999), except for the cold period at 5.4 ky BP (Fig. 4).

For ages younger than ~3.6 ky BP, climate indicators from Europe and Greenland show no clear match on a multicentennial time scale. Since the amplitudes of the climate perturbations are smaller in the late Holocene than during the preceding time interval, the lack of correspondence between the records may in part be caused by a reduced signal-to-noise ratio.

The Icelandic Low: Climate Interface Across the North Atlantic

Observations, the earliest going back to the second half of the eighteenth century, indicate that mild (severe) winters in western Greenland go along with severe (mild) winter conditions in northern Europe (Barlow et al. 1997 and refs. therein). This temperature „seesaw" across the North Atlantic is closely linked to the variability of the Icelandic Low. Warm winters in Europe coincide with a strong Icelandic Low, located between Iceland and Greenland (Fig. 5b). This pressure distribution steers warm and humid air masses from mid-latitudes eastward into northern and central Europe, whereas polar air masses reach Greenland. In contrast, when the Icelandic Low is weak and located near the southern tip of Greenland (Fig. 5a) the flow of cold polar air to Greenland is reduced, whereas more cold polar air reaches Europe.

The anti-phase relationship of temperature anomalies between Greenland and Scandinavia between 4–6.6 ky BP (Fig. 4) can thus be explained in terms of the temperature „seesaw" mediated by the Icelandic Low. This inference is corroborated by the simultaneous occurrence of warm and humid periods in Europe; the latter is thought to lower the ELA by increased snowfall (Dahl and Nesje 1996). Accordingly, we surmise that the 900-y climate oscillations between 4–6.6 ky BP may be linked to multicentennial variations in strength and location of the Icelandic Low.

Today, extremes of the temperature „seesaw" between Greenland and northern Europe are linked to positive and negative phases of the decadal North Atlantic Oscillation (e.g. Kushnir 1999; Fig. 5). Accordingly, minima and maxima of the 900-y temperature cycles may be an expression of a modulation of North Atlantic Oscillation intensity, once the modern atmospheric mode was established at 6.6 ky BP. This conjecture is supported by driftwood records from the Arctic (Dyke et al. 1997), which are thought to indicate shifts in the mean state of the North Atlantic Oscillation (Tremblay et al. 1997).

A different picture emerges for times prior to 6.6 ky BP when the temperature „seesaw" was apparently inoperative, resulting in synchronous temperature variations in Greenland and Europe. This situation is reminiscent of a glacial climate mode, with the Icelandic Low located further south (e.g. Harrison et al. 1992), allowing an almost unrestricted flow of cold polar air into the entire North-Atlantic realm. Based on the phase relationship between temperature variations in Greenland and Europe, we can only speculate that the glacial climate mode operated in this region until 6.6 ky BP. The transition to today's climate mode at ~6.6 ky BP is consistent with a marked change in the prevailing wind direction from northeast to west in south-western Sweden between 6–7 ky BP (Mörner 1980). However, this timing of the mode transition contrasts results of Stager and Mayewski (1997), who suggested a global transition from glacial to postglacial mode between 7.8–8.2 ky BP, that is, simultaneous to the final decay of the North American ice sheet (Licciardi et al. 1998).

Making 900-y Climate Oscillations: Internal Feedbacks *vs.* Orbital Control

The existence of 900-y cycles in climate records does not provide direct information about the origin of these cycles. We will consider two mechanisms, which are not mutually exclusive: (1) Internal oscillations of the climate system and (2) external forcing by secular changes of the Earth's orbital parameters.

Internal origin: Based on the finding that the amplitude of the 900-y temperature oscillations decays almost exponentionally (Fig. 2), one can build simple mathematical templates to mimic this be-havior. We use the term template in order to distinguish the following excercise from modeling efforts which treat the involved physical processes explicitly. One possibility for such a template is to

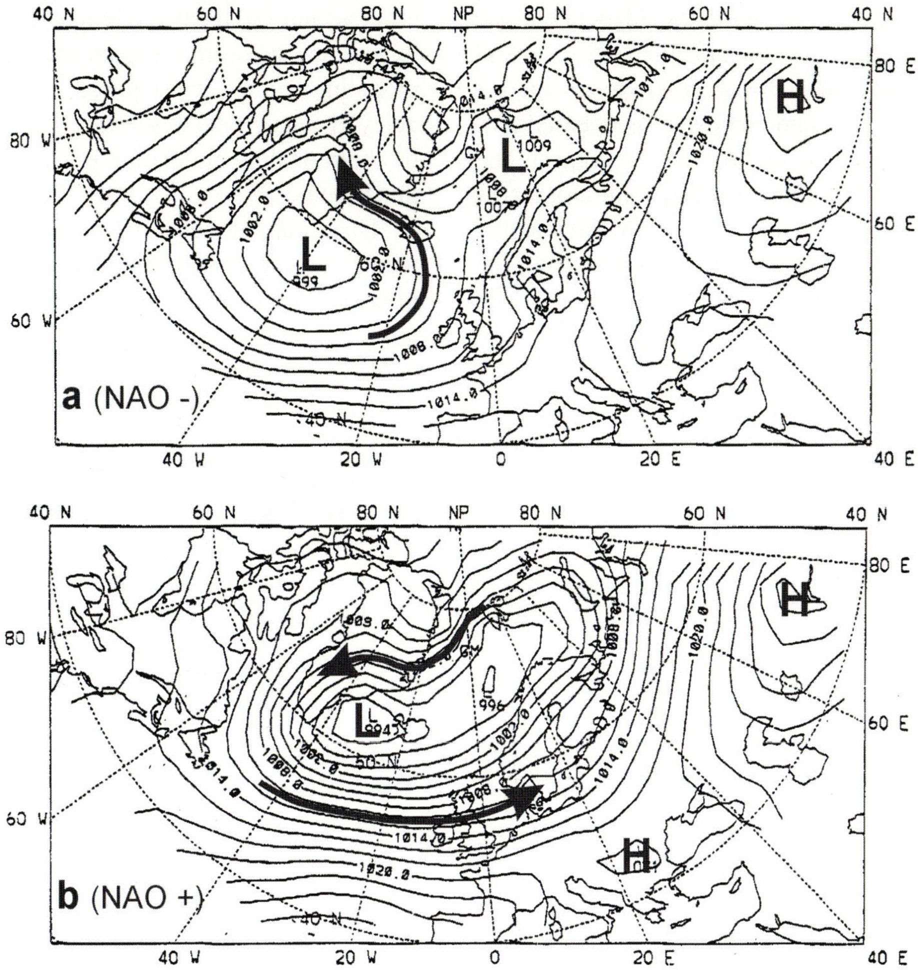

Fig. 5. Mean sea-level pressure in the North Atlantic realm for opposing modes of the temperature „seesaw" (modified from Barlow et al., 1997). **(a)** Southwesterly position of the Icelandic Low causes mild Greenland winters and severe temperatures in Europe and corresponds to the negative phase of the North Atlantic Oscillation (NAO). **(b)** Opposite situation with cold Greenland winters (positive NAO phase). Main centers of high (H) and low (L) pressure are indicated.

express the temperature anomaly in Greenland (δT in units of $\delta^{18}O$) as function of the average value of δT during the preceding τ years, denoted by $\langle \delta T \rangle_\tau$:

$$\frac{d\delta T}{dt} = a \, \langle \delta T \rangle_\tau + c(t), \qquad (1)$$

with t being time, a = -0.009 ‰/y, τ = 415 y and

$$c(t) \; [‰/\mathrm{y}] = \begin{cases} -0.0025 \text{ between 8.20-8.45 ky BP} \\ -0.0007 \text{ between 4.45-4.95 ky BP} \\ 0.0 \text{ else} \end{cases}$$

With a less than zero, the „memory" of the system (provided by $\langle \delta T \rangle_\tau$) determines the current evolution of δT; a positive (negative) value of the average temperature anomaly during the past 415 y will cause a decrease (increase) of the current temperature. The combination of $a < 0$ and „memory" for past states results in a delayed negative feedback, which in principle allows the template in eq. (1) to produce temperature oscillations. However, without an initial perturbation the template cannot produce any deviations from its equilibrium ($\delta T = 0$). Such external forcing is therefore included into eq. (1) by the time-dependent parameter c; setting $c < 0$ for some time interval results in a cooling event. The observation that the $\delta^{18}O$ anomalies go beyond the lower exponential envelope at ~ 4.7 and ~8.2 ky BP (Fig. 2) suggests cooling events at these times. Actual values for $c(t)$ and a were found after some numerical experiments by matching the template to the observed $\delta^{18}O$ anomalies („fit by eye").

The template agrees well with the observed $\delta^{18}O$ anomalies between 3–8.5 ky BP (Fig. 4). Between 0–3 ky BP, the match degrades, possibly due to the low signal-to-noise ratio of the data. The template describes the 900-y cycles as damped oscillations, triggered by two cooling events at ~8.3 and 4.7 ky BP, with the magnitude of the first cooling event being ~4-times larger than the mid-Holocene cooling. The periodicity of the predicted temperature cycles is set by the length of the system „memory" (τ). To assess the appropriateness of the template to real world processes, it is important to identify the possible origin of the delayed negative feedback, in particular, to locate the „memory" of the system.

Today, the large-scale circulation of the Atlantic Ocean is dominated by deep-water formation in the northern North Atlantic, where surface water gains density, through heat and freshwater exchange with atmosphere, and sinks to greater depth. The southward flow of newly formed deep water is compensated by a northward return flow of near-surface and intermediate water masses (e.g. Schmitz 1995). Experiments with ocean general circulation models (England 1995) indicate that ventilation ages (i.e. the time since the last contact of a water parcel with the atmosphere) of intermediate and deep waters in the Atlantic Ocean measures up to several hundred years. Hence, it should be possible to propagate a hydrographic anomaly from the sea surface near the deep-water formation sites in the North Atlantic southward, with the deep-water flow. Subsequent upwelling of the anomaly and its entrainment into northward flowing water masses, which feed deep-water formation, returns the (attenuated) anomaly to its origin. The reappearance of the anomaly at its source after several hundred years constitutes a delayed feedback loop. A smorgasbord of processes within the feedback loop can be envisioned to produce an overall negative feedback. Careful modeling studies are required to assess possible delayed negative feedback loops within the Atlantic thermohaline circulation (cf. Paul and Schulz, this volume).

The timing of the inferred cooling events in the template coincide with negative salinity anomalies in the North Atlantic Ocean, suggesting a potential physical mechanism: Deep-water formation in the North Atlantic results in the transport of heat from low to high latitudes (e.g. Peixoto and Oort 1992). Modeling studies (e.g. Bryan 1986) suggest that freshening of surface-waters in the North Atlantic could attenuate or even suppress deep-water formation, and as a consequence, reduce northward heat transport. Since the final disintegration of the North American ice sheet delivered meltwater into the Labrador Sea around 8.2 ky BP (Barber et al. 1999), it is likely that this freshwater influx suppressed deep-water formation in this area and thereby caused the inferred cooling in the North-Atlantic realm. However, the lack of deglacial meltwater influx makes this scenario inappropriate for the mid-Holocene cooling at ~4.7 ky BP. A change in the freshwater budget of the North Atlantic seems nevertheless a reasonable assumption, since a major decrease in sea-surface salinity occurred at ~5 ky BP in the northeastern Atlantic Ocean (Duplessey et al. 1992). Since the cause of this salinity perturbation is not yet known, we can only surmise that it may have weakened deep-water formation in the North Atlantic and induced the mid-Holocene cooling event in the North-Atlantic realm. This inference is supported by benthic isotope data which suggest indeed a weakening of deep-water production at this time (Maslin et al. 1996; Knaack 1997).

Orbital forcing: Loutre et al. (1992) identified a ~900-y periodicity in eccentricity variations of the Earth's orbit, which modulates incoming solar radiation. Accordingly, it might be possible that the observed 900-y climate fluctuations are linked to this external forcing. The amplitude of the ~900-y component of summer insolation at 65°N is less than 1 mW m^{-2} (Fig. 6), that is, two orders of magnitude smaller than the estimated radiative forcing resulting from solar output variability

(~300 mW m^{-2}; Schimel et al. 1996). The effect of the latter on Earth's climate is probably rather small (Schimel et al. 1996), making it unlikely that the extremely small 900-y insolation variations should affect climate directly. Nevertheless, one could envision that internal oscillations of the climate system are paced by the orbital variations, resulting in a phase-lock between forcing and climate variations.

Between ~3.5–8 ky BP the 900-y component of northern summer insolation shows indeed a good correlation with the inferred long-term temperature variations in Greenland (Fig. 6). However, close inspection of the phase relationship between the two time series reveals that the phasing varies in time: While there is no discernable phase-lag between 7.6–8.4 ky BP, warming trends seen in the $\delta^{18}O$ running mean show an increasing delay to changes in insolation from ~0 y at 8 ky BP to 300 y at 4.4 ky BP. This suggests that the frequency of the insolation variation slightly exceeds the frequency of the climate oscillations; a finding which is difficult to reconcile with a phase-lock between the two signals. Dating uncertainties of the ice-core $\delta^{18}O$ record amount to 1–2% during the Holocene (Meese et al. 1997) and are thus too small to account for the increasing phase lag. Furthermore, between 1–3.5 ky BP and prior to 8.4 ky BP we observe more mismatches than matches between peaks of the insolation and $\delta^{18}O$ curve. Finally, we note that the exponential envelope of the $\delta^{18}O$ data (Fig. 2) contrasts the more linear decrease of the amplitude of the 900-y insolation anomaly (Fig. 6). Although we cannot rule out orbital pacing of the 900-y climate oscillations, the above findings neither support direct orbital forcing nor a simple phase-locked behavior. Whether the good correlation between insolation forcing and climate change

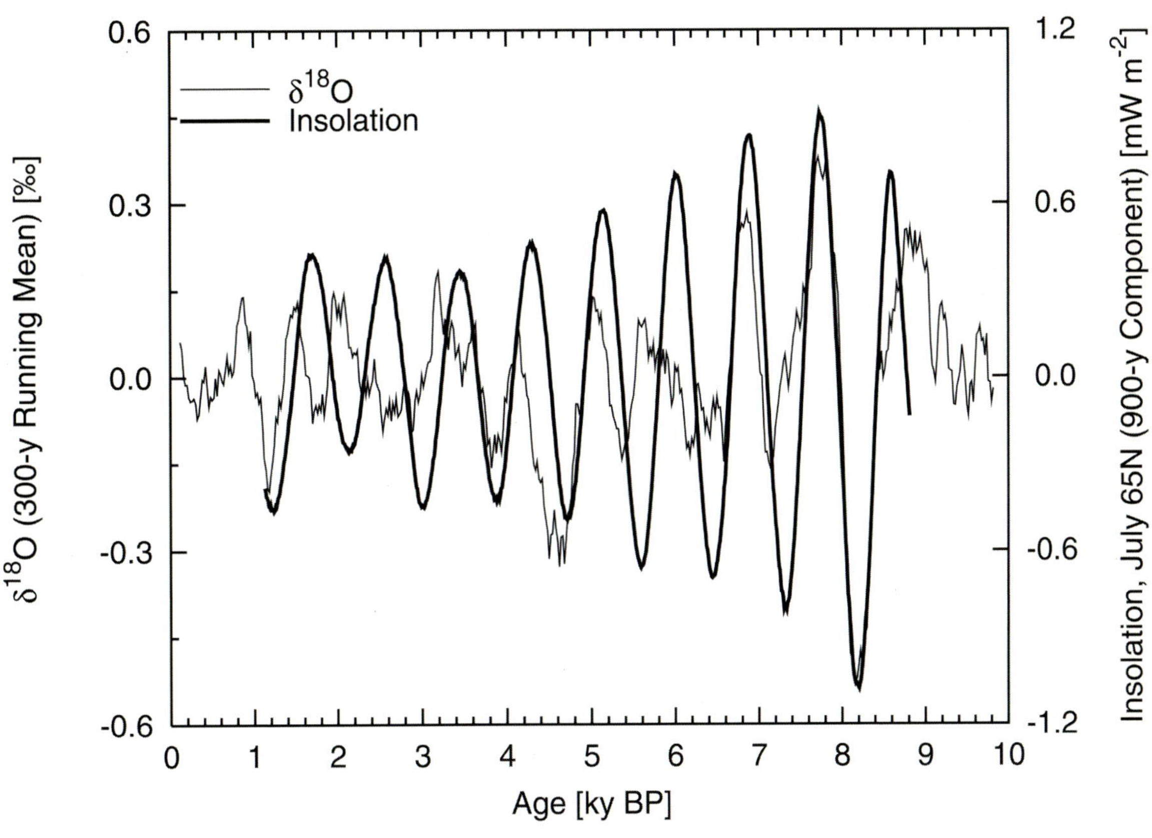

Fig. 6. Smoothed $\delta^{18}O$ time series (as in Fig. 4) together with 900-y signal component in summer insolation at 65°N (Loutre et al. 1992), which was estimated by a harmonic-filtering algorithm (after Ferraz-Mello 1981). Prior to filtering a polynomial trend was subtracted from the insolation time series.

between 7.6–8.4 ky BP points to a causal mechanism or occurred by happenstance remains an open question.

The Deep-Sea Record of Holocene Climate Oscillations

The deep-sea record of continental rock fragments transported by icebergs are thought to be a sensitive indicator of past climate change. Bond et al. (1997) identified periods in the Holocene during which the input of ice-rafted debris (IRD) from Greenland and Iceland to a location west of Ireland increased (Fig. 7). Bond and coworkers consider ocean-surface coolings as immediate cause of these IRD events, which occur with a recurrence time of 1370 ± 500 y (1-σ confidence interval) during the Holocene. We note that the large uncertainty of the estimated recurrence time does not exclude a 900-y pacing. Considering chronostratigraphic uncertainties, the average sampling interval of the two IRD time series (~220 y in the Holocene), possible smoothing due to bioturbation and a non-linear response of IRD abundance to climatic cooling in the North-Atlantic realm, the IRD events may be tentatively correlated to $\delta^{18}O$ minima in the ice-core record, that is, to air-temperature minima in Greenland (Fig. 7). Given the above limitations of the IRD records, it cannot be ruled out that the 900-y climate cyclicity, observed in the atmospheric records, is also contained in the IRD time series but was conceived by Bond et al. (1997) as longer cycle. Hence, the important question regarding the cyclicity of North Atlantic IRD

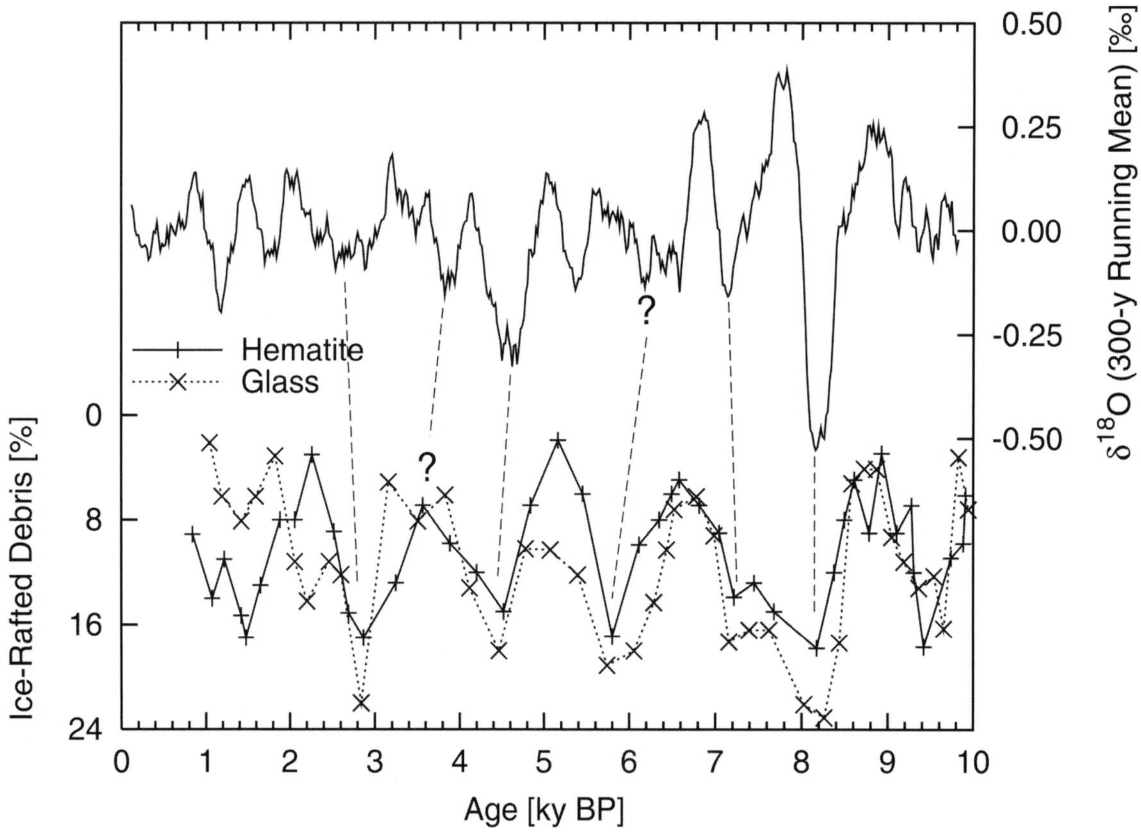

Fig. 7. Smoothed $\delta^{18}O$ time series (as in Fig. 4) together with two ice-rated debris records from the North Atlantic (Bond et al. 1997). Hematite-coated sediment grains stem from East Greenland or Svalbard whereas glass fragments originate from Iceland. Note inverted scale for ice-rafted debris; high values are thought to be caused by cool time intervals (Bond et al. 1997). Dashed vertical lines depict possible correlation between the records.

events in the Holocene cannot be answered unequivocally until deep-sea records with a better temporal resolution become available.

Bianchi and McCave (1999) inferred changes in deep-water flow in the North Atlantic from sedimentary grain-size data. According to their results, velocity of deep-water flow varied with a periodicity of ~1500 y during the Holocene, implying concomitant changes in oceanic heat transport. This finding contrasts the existence of 900-y climate oscillations and seems to be more in line with the 1370 ± 500 y IRD cycle. However, times of reduced deep-water flow do not coincide with the IRD events described by Bond et al. (1997; not shown). Furthermore, harmonic analysis reveals that the time series cannot be distinguished from white noise at a 95% significance level (Fig. 8). This finding most likely explains why the deep-water flow record shows no consistent correlation to the IRD or smoothed $\delta^{18}O$ time series.

Conclusions

Proxies of atmospheric temperature and humidity from Greenland and northern/central Europe show evidence for 900-y climate oscillations between 3-8.5 ky BP. The magnitude of the climate perturbations in Europe was probably large enough to affect human societies, especially since they occurred during the important transition from hunting-gathering life style to sedentary agriculture (e.g. Willis and Bennett 1994).

Based on the temporal evolution of the temperature proxy from Greenland, we regard it as more

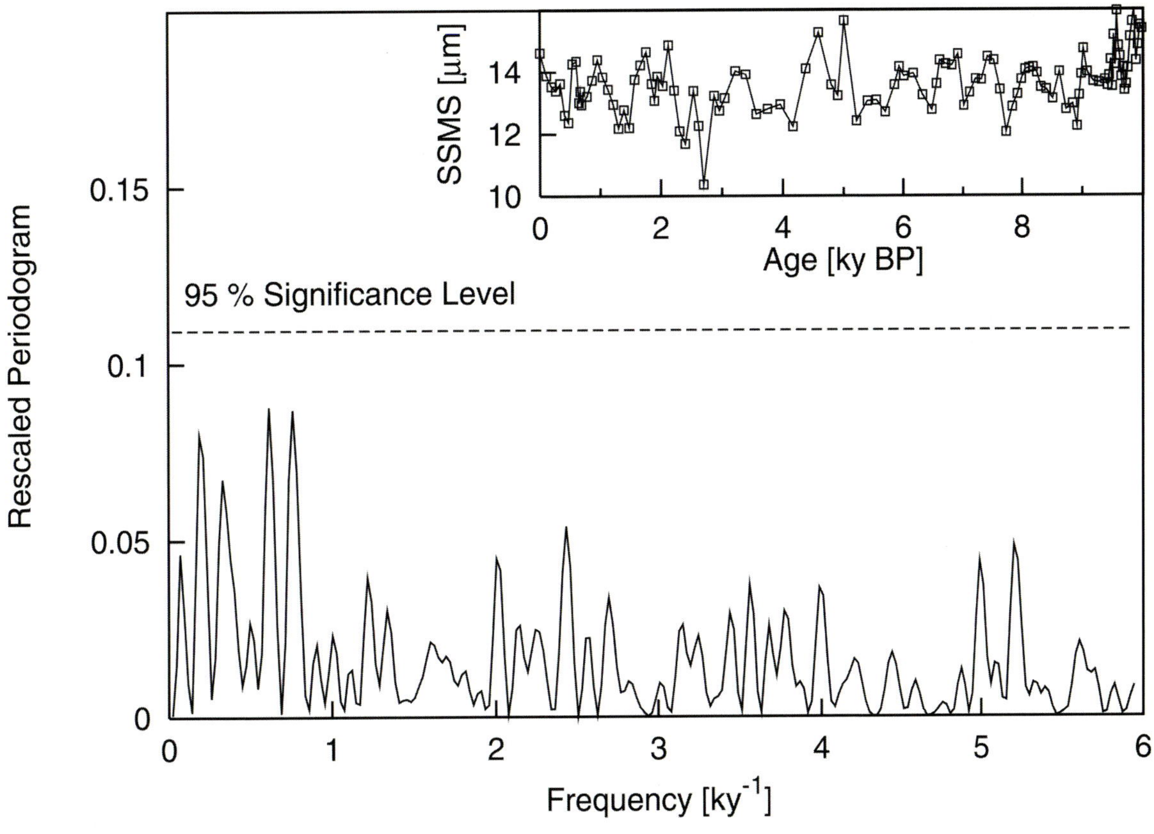

Fig. 8. Harmonic analysis (Fisher 1929; Schulz and Stattegger 1997) of sedimentary grain-size data from the North Atlantic (Bianchi and McCave 1999). The time series (inset; SSMS = sortable silt mean size) cannot be distinguished from white noise at the $(1-1/N) \times 100\% = 99.2\%$ significance level (cf. Thomson 1990; N = number of data points) and not even at the 95% significance level.

likely that the climate oscillations were generated within the climate system than by external orbital forcing. Using a simple template, the climate cycles can be described as damped oscillations, triggered by cooling events. We suggest that negative salinity anomalies in the North Atlantic are the immediate cause of these cold spells, by means of reduced northward heat transport via thermohaline circulation. However, uncertainty remains regarding the ultimate cause of the salinity anomalies. While the first event at ~8.2 ky BP may be linked to the final decay of glacial ice sheets, the cause of the younger event remains unknown. Claussen et al. (1999) simulated an abrupt transition in the hydrological cycle as highly nonlinear response to slow variations in orbital forcing in the mid-Holocene. Although the timing of this transition occurs at ~5.4 ky BP in the model, it may nevertheless point to a potential mechanism, capable of perturbing the salinity field in the north Atlantic.

So far, evidence for the existence of Dansgaard-Oeschger cycles in the Holocene depends largely on the identification of the characteristic *glacial* Dansgaard-Oeschger periodicity (~1470 y) in Holocene climate records (cf. Bond et al. 1997; Bianchi and McCave 1999). Our analysis shows that the evidence for ~1470 cycles in deep-sea records from the North-Atlantic is, however, contestable and that these time series do not contradict 900-y climate oscillations during the Holocene. Considering the current lack of a generally accepted, comprehensive theory for Dansgaard-Oeschger cycles, we cannot specify the possible relation between the Holocene 900-y cycles and the glacial 1470-y Dansgaard-Oeschger cycles. However, the persistence of the 900-y climate oscillations, across the inferred transition from glacial to Holocene mode of atmospheric circulation at ~6.6 ky BP, suggests that the underlying physical mechanism might also work under glacial boundary conditions. Accordingly, the oscillations might be linked to recurrent glacier surges from Greenland, which are in turn generated by feedbacks in the ocean-atmosphere-cryosphere system as proposed by Sarnthein et al. (2001) for glacial Dansgaard-Oeschger cycles.

Until further paleoclimatic data become available, the 900-y cycles should be considered as a regional signal of the North Atlantic realm. Moreover, it is conceivable that different components of the climate system varied on different centennial-to-millennial time scales during the Holocene. For example Wang et al. (1999) identified a ~780-y periodicity in the East Asian monsoon regime. Episodes of enhanced dust deposition occurred in Greenland approximately every 2.6 ky, starting at ~8.3 ky BP (O'Brien et al. 1995). The lack of an obvious correlation between these dust events and inferred temperature perturbations in Greenland (not shown) suggests a different origin of the two signals. Analysis of modern air-mass trajectories (Kahl et al. 1997) suggest that dust sources are mainly located in northern Eurasia and North America. It is beyond the scope of this study to investigate possible links between climate records from these source regions and the dust accumulation in the Greenland ice cores. We note however that times of increased dust deposition seem to coincide with glacier advances in North America (Denton and Karlén 1975) and increased East Asian summer monsoon (Wang et al. 1999).

The smoothed $\delta^{18}O$ data as well as the template suggest only small-amplitude multicentennial climate variability between 0–3 ky BP, making it difficult to predict any future interference with a possible man-made climate perturbation. However, it is possible that the climate system is still capable of generating 900-y climate oscillations. A decrease in North Atlantic sea-surface salinity, predicted by climate models for a future greenhouse climate (e.g. Wood et al. 1999), may therefore restart the climate oscillator – a potential hazard for human societies in Europe.

Acknowledgments

We thank W. Berger for focusing our attention to the variable Holocene climate. We appreciate constructive comments on the manuscript by C. Hammer and an anonymous reviewer. MS is grateful to M. Sarnthein for many discussions about the role of the oceans in climate change. Thanks are also due to M.-F. Loutre for providing Earth's orbital data. MS is supported by the German Klimaforschungsprogramm and AP by the Alexander v. Humboldt Stiftung (through a Feodor Lynen Research Fellowship) and the California Space Institute.

References

Barber DC, Dyke A, Hillaire-Marcel C, Jennings AE, Andrews JT, Kerwin MW, Bilodeau G, McNeely R, Southon J, Morehead MD, Gagnon J-M (1999) Forcing of the cold event of 8,200 years ago by catastrophic drainage of Laurentide lakes. Nature 400:344-348

Barlow LK, Rogers JC, Serreze MC, Barry RG (1997) Aspects of climate variability in the North Atlantic sector: Discussion and relation to the Greenland Ice Sheet Project 2 high-resolution isotopic signal. Geophys Res C102:26333-26344

Behl RJ, Kennett JP (1996) Brief interstadial events in the Santa Barbara basin, NE Pacific, during the past 60 kyr. Nature 379:243-246

Bianchi GG, McCave IN (1999) Holocene periodicity in North Atlantic climate and deep-ocean flow south of Iceland. Nature 397:515-517

Blunier T, Chappellaz J, Schwander J, Dällenbach A, Stauffer B, Stocker TF, Raynaud D, Jouzel J, Clausen HB, Hammer CU, Johnsen SJ (1998) Asynchrony of Antarctic and Greenland climate change during the last glacial period. Nature 394:739-743

Bond G, Showers W, Cheseby M, Lotti R, Almasi P, deMenocal P, Priore P, Cullen H, Hajdas I, Bonani G (1997) A pervasive millennial-scale cycle in North Atlantic Holocene and glacial climates. Science 278:1257-1266

Bryan F (1986) High-latitude salinity effects and interhemispheric thermohaline circulation. Nature 323:301-304

Claussen M, Kubatzki C, Brovkin V, Ganopolski A, Hoelzmann P, Pachur HJ (1999) Simulation of an abrupt change in Saharan vegetation in the mid-Holocene. Geophys Res Lett 26:2037-2040

Dahl SO, Nesje A (1996) A new approach to calculating Holocene winter precipitation by combining glacier equilibrium-line altitudes and pine-tree limits: a case study from Hardangerjokulen, central southern Norway. Holocene 6:381-398

Dansgaard W, Johnsen SJ, Clausen HB, Dahl-Jensen D, Gundestrup NS, Hammer CU, Hvidberg CS, Steffensen JP, Sveinbjörnsdottir AE, Jouzel J, Bond G (1993) Evidence for general instability of past climate from a 250-kyr ice-core record. Nature 364:218-220

Denton GH, Karlén W (1975) Holocene climatic variations - their pattern and possible cause. Quat Res 3:155-205

Duplessy JC, Labeyrie L, Arnold M, Paterne M, Duprat J, van Weering TCE (1992) Changes in surface salinity of the North Atlantic Ocean during the last deglaciation. Nature 358:485-488

Dyke AS, England J, Reimnitz E, Jetté H (1997) Changes in driftwood delivery to the Canadian Arctic Archipelago: the hypothesis of postglacial oscillations of the transpolar drift. Arctic 50:1-16

England MH (1995) The age of water and ventilation timescales in a global ocean model. J Phys Oceanogr 25:2756-2777

Ferraz-Mello S (1981) Estimation of periods from unequally spaced observations. Astron J 86:619-624

Fisher RA (1929) Tests of significance in harmonic analysis. Proc R Soc Lond, Ser A 125:54-59

Grootes PM, Stuiver M (1997) Oxygen 18/16 variability in Greenland snow and ice with 10^{-3}- to 10^{5}-year time resolution. J Geophys Res C102:26455-26470

Harrison SP, Digerfeldt G (1993) European Lakes as Palaeohydrological and Palaeoclimatic Indicators. Quat Sci Rev 12:233-248

Harrison SP, Prentice IC, Bartlein PJ (1992) Influence of insolation and glaciation on atmospheric circulation in the north Atlantic sector: implications of general circulation model experiments for the late Quaternary climatology of Europe. Quat Sci Rev 11:283-299

Jouzel J, Alley RB, Cuffey KM, Dansgaard W, Grootes P, Hoffmann G, Johnsen SJ, Koster RD, Peel D, Shuman CA, Stievenard M, Stuiver M, White J (1997) Validity of the temperature reconstruction from water isotopes in ice cores. Geophys Res C102:26471-26487

Kahl JDW, Martinez DA, Kuhns H, Davidson CI, Jaffrezo JL, Harris JM (1997) Air mass trajectories to Summit, Greenland: A 44-year climatology and some episodic events. Geophys Res C102:26861-26875

Karlén W (1993) Glaciological, sedimentological and paleobotanical data indicating Holocene climatic change in Northern Fennoscandia. In: Frenzel B, Eronen M, Vorren K-D, Gläser B (eds) Oscillations of the alpine and polar tree limits in the Holocene. Gustav Fischer Verlag, Stuttgart, pp 69-83

Knaack J (1997) Eine neue Transferfunktion zur Rekonstruktion der Paläoproduktivität aus Gemeinschaften mariner Diatomeen. Berichte - Reports, Geol-Paläont Inst Univ Kiel 83:1-119

Kushnir Y (1999) Europe's winter prospects. Nature 398:289-290

Licciardi JM, Clark PU, Jenson JW, Macayeal DR (1998) Deglaciation of a soft-bedded Laurentide Ice Sheet. Quat Sci Rev 17:427-448

Loutre MF, Berger A, Bretagnon P, Blanc P-L (1992) Astronomical frequencies for climate research at the decadal to century time scale. Clim Dyn 7:181-194

Maslin M, Sarnthein M, Knaack J-J (1996) Subtropical eastern Atlantic climate during the Eemian. Naturwissenschaften 83:122-126

Meese DA, Gow AJ, Alley RB, Zielinski GA, Grootes PM, Ram M, Taylor KC, Mayewski PA, Bolzan JF (1997) The Greenland Ice Sheet Project 2 depth-age scale: Methods and results. Geophys Res C102:26411-26423

Mörner N-A (1980) The northwest European "sea-level laboratory" and regional Holocene eustacy. Palaeogeogr, Palaeoclimatol, Palaeoecol 29:281-300

O'Brien SR, Mayewski PA, Meeker LD, Meese DA, Twickler MS, Whitlow SI (1995) Complexity of Holocene climate as reconstructed from a Greenland ice core. Science 270:1962-1964

Peixoto JP, Oort AH (1992) Physics of climate. American Inst. of Physcis, New York

Sarnthein M, Stattegger K, Dreger D, Erlenkeuser H, Grootes P, Haupt B, Jung S, Kiefer T, Kuhnt W, Pflaumann U, Schäfer-Neth C, Schulz H, Schulz M, Seidov D, Simstich J, van Kreveld S, Vogelsang E, Völker A, Weinelt M (2001) Fundamental modes and abrupt changes in North Atlantic circulation and climate over the last 60 ky - Concepts, reconstructions and numerical modeling. In: Schäfer P, Ritzrau W, Schlüter M, Thiede J (eds) The Northern North Atlantic: A changing environment. Springer-Verlag, Berlin Heidelberg New York, pp 365-410

Schettler G, Rein B, Negendank JFW (1999) Geochemical evidence for Holocene palaeodischarge variations in lacustrine records from the Westeifel Volcanic Field, Germany: Schalkenmehrener Maar and Meer-felder Maar. Holocene 9:381-400

Schimel D and 26 others (1996) Radiative forcing of climate change. In: Houghton JT et al (ed) Climate Change 1995. Cambridge Univ Press, Cambridge, pp 65-131

Schmitz WJ (1995) On the interbasin-scale thermohaline circulation. Rev Geophys 33:151-173

Schulz H, von Rad U, Erlenkeuser H (1998) Correlation between Arabian Sea and Greenland climate oscillations of the past 110,000 years. Nature 393:54-57

Schulz M, Stattegger K (1997) SPECTRUM: Spectral analysis of unevenly spaced paleoclimatic time series. Comput Geosci 23:929-945

Sirocko F, Sarnthein M, Erlenkeuser H, Lange H, Arnold M, Duplessy JC (1993) Century-scale events in monsoonal climate over the past 24,000 years. Nature 364:322-324

Stager JC, Mayewski PA (1997) Abrupt early to mid-Holocene climatic transition registered at the equator and the poles. Science 276:1834-1836

Thomson DJ (1990) Time series analysis of Holocene climate data. Phil Trans R Soc Lond A 330:601-616

Tremblay L-B, Mysak LA, Dyke AS (1997) Evidence from driftwood records for century-to-millennial scale variations in the high latitude atmospheric circulation during the Holocene. Geophys Res Lett 24:2027-2030

Wang L, Sarnthein M, Erlenkeuser H, Grimalt J, Grootes P, Heilig S, Ivanova E, Kienast M, Pelejero C, Pflaumann U (1999) East Asian monsoon climate during the late Pleistocene: High-resolution sediment records from the South China Sea. Mar Geol 156:245-284

Willis KJ, Bennett KD (1994) The Neolithic transition - fact or fiction? Palaeoecological evidence from the Balkans. Holocene 4:326-330

Wood RA, Keen AB, Mitchell JFB, Gregory JM (1999) Changing spatial structure of the thermohaline circulation in response to atmospheric CO_2 forcing in a climate model. Nature 399:572-575

Holocene Climate Variability on Centennial-to-Millennial Time Scales: 2. Internal and Forced Oscillations as Possible Causes

André Paul[1,2] and Michael Schulz[3]

[1]Scripps Institution of Oceanography, La Jolla, 9500 Gilman Drive, CA 92093–0244, USA
[2]Now at: University of Bremen, Department of Geosciences, Postfach 33 04 40, D-28334 Bremen, Germany
e-mail: apau@palmod.uni-bremen.de
[3]University of Kiel, Institute for Geosciences, Olshausenstr. 40, D-24118 Kiel, Germany
e-mail: mschulz@email.uni-kiel.de

Abstract: The climate of Western and Northern Europe is closely linked to the meridional circulation of the Atlantic Ocean, and climate variability on centennial-to-millennial time scales in this realm may well be caused by fluctuations in the associated heat transport. The sensitivity of the meridional circulation is determined by a number of feedbacks in the climate system, which can also interact to sustain internal oscillations. Another source of climate variability is (quasi-) periodic external forcing of the climate system, which gives rise to forced oscillations. A climate model is used to generate internal oscillations of one sort (i.e. deep-decoupling oscillations) with periods ranging approximately from 1600-2000 years. During all phases of these oscillations, there is active deep water formation and export from the North Atlantic to the Southern Ocean, and the maximum of the meridional overturning does never drop below 6×10^6 $m^3 s^{-1}$.

Introduction

Compared to other regions at similar geographic latitudes, the climate of Western and Northern Europe has been relatively mild for most of the Holocene. This is the effect of a fairly remote cause: The net meridional water transport in the upper layer of the *South* Atlantic Ocean is directed equatorwards rather than polewards as in the other oceans. This net meridional water transport is a continuous source of heat and salt for the *North* Atlantic Ocean that mainly benefits the Gulf Stream System. Consequently, the North Atlantic Ocean is about 3°C warmer and 1 unit saltier than the North Pacific Ocean, and up to 9°C warmer and 1.5 units saltier than a completely water-covered Earth, at similar geographic latitudes (Dietrich 1950). Upon forming North Atlantic Deep Water (NADW), the heat gained by the North Atlantic Ocean is released to the atmosphere.

Historically, the first to envision a meridional circulation in the Atlantic Ocean, with currents of cold water at the bottom and the warm Gulf Stream as its manifestation at the surface, was Thompson, Count Rumford, by the end of the 18th century. He called the ocean an "equalizer of Heat" and already noted its moderating effect on climate (see Warren 1981; Weaver et al. 1999 for an account of his work). In the second half of the 19th century, the Scottish geologist Croll (1875) related the ice ages in Europe to changes in the heat conveyed by the Gulf Stream and released to the atmosphere. It then was the analysis of the unique data of the 1925/27 "Meteor" expedition that revealed the surprising phenomenon that in the South Atlantic Ocean the net meridional water transport in the upper layer (0 to 1400 m depth) is directed equatorwards, i.e. against the temperature gradient. Wüst attempted to calculate the associated heat transport, but he never published his results, apparently because they

From WEFER G, BERGER W, BEHRE K-E, JANSEN E (eds), 2002, *Climate Development and History of the North Atlantic Realm.* Springer-Verlag Berlin Heidelberg, pp 55-73

indicated a direction contradicting his intuition (Willebrand 1993). It was left to Model (1950) to present the first estimate of the net meridional heat transport implied by the "Meteor" data and point out the role of the South Atlantic Ocean as a supplier of heat to the North Atlantic Ocean.

The situation has not always been as favorable as it is at present. At the last glacial maximum, the meridional circulation of the Atlantic Ocean was probably weaker by some ten percent (cf. Schäfer-Neth and Paul 2001), which would add to a climate on the surrounding continents much colder than today. During the entire last glacial period, the temperature of the North Atlantic Ocean experienced rapid changes and strong fluctuations on millennial time scales (Bond et al. 1997; Sarnthein et al. 2001). These were obviously linked to the very similar behavior of the air temperature at the summit of the Greenland ice sheet and in Europe (e.g. Lowe and Walker 1997). The Little Ice Age could well be the most recent cold phase of this millennial-scale climate variability (Bond et al. 1997).

While the climate fluctuations in the North Atlantic Ocean and Europe could be traced into the early and middle Holocene, the climate of Greenland as archived in the ice became seemingly stable. However, harmonic analysis of the Holocene $\delta^{18}O$ record from the Greenland Ice Sheet Project Two (GISP2, Grootes and Stuiver1997) shows that some variability remained, most of which occurred in a general band centered on about 900 a, as opposed to 1470 a during the last glacial period (Schulz and Paul, this volume).

If climate fluctuations in the North Atlantic realm are indeed linked to variations in the meridional water, heat and salt transports, the question is what in turn caused the variations in the underlying meridional circulation. To answer this question, the complex interplay of feedbacks and external forcings in the climate system needs to be studied. Feedbacks give rise to the unforced variability of the climate system, such as internal oscillations. They also determine its sensitivity to changes in external forcings (Schulz et al., this volume), which, if (quasi-) periodic, produce forced oscillations. The insight gained from such studies should be useful in predicting future climate change, e.g. the fate of the meridional circulation of the Atlantic Ocean during global warming, and could be an aid in separating natural from human influences on climate.

Feedbacks in the Climate System

The climate system forms a complex net of interactions that couple the atmosphere, the ocean and the land and involve physical, chemical and biological processes. The relationship between the climate-change response to a climate forcing of a defined strength is called the climate sensitivity, and a process that alters the climate sensitivity is called a feedback. A positive feedback amplifies the climate change response, while a negative feedback damp-ens it (Appendix in Schulz et al., this volume). A recipe for constructing a model of climate change is to order the feedbacks according to importance, and then consider the most important feedbacks first (Hartmann 1994).

In the following, we describe a number of feedbacks (F1-F12) that affect the sensitivity of a meridional circulation with sinking at high latitudes, such as in the present-day North Atlantic Ocean. The climate forcing we consider is a small freshening of the high northern latitudes. This can result from an external perturbation of the surface freshwater flux (through excess precipitation or ice melting), or from the internal dynamics of the ocean itself. In either case, such a freshening leads to a decrease in high-latitude salinity and hence high-latitude density, which diminishes the deep-water formation and eventually weakens the meridional circulation.

F1 The mean meridional circulation advects the negative salinity anomaly away and restores the initial salinity gradient *(advective feedback, negative).*

F2 As the meridional circulation weakens, the poleward transport of salty water is reduced, which in turn further decreases the high-latitude salinity *(haline feedback, positive).*

F3 As the meridional circulation weakens, the poleward transport of warm water is reduced, which leads to a decrease in high-latitude tem-perature and hence an increase in high-latitude density. This strengthens the deep water formation and eventually the meridional circulation *(thermal feedback, negative).*

F4 A lower high-latitude temperature causes less evaporation, which in turn further decreases the high-latitude salinity *(temperature-evaporation feedback, positive).*

F5 If the high-latitude density falls below a certain threshold, convection will stop, deep water formation cease and the meridional circulation collapse *(convective feedback, positive).*

While feedbacks F1-F3 are purely oceanic, feedback F4 involves the atmosphere. Due to its dependence on a density threshold, feedback F5 is highly nonlinear.

The reduced poleward transport of warm water leads to a decrease in sea surface temperature at high latitudes and an increase in sea-surface temperature at low high latitudes. Consequently, the atmospheric temperature gradient increases. This yields two additional feedbacks F6-F7 that involve the atmospheric heat and moisture transports:

F6 The poleward transport of moisture by the atmosphere is enhanced, which implies increase in high-latitude precipitation and an additional decrease in high-latitude salinity and hence high-latitude density *(atmospheric moisture transport feedback, positive).*

F7 The poleward transport of heat by the atmosphere is enhanced (in part coupled to the transport of moisture through the latent heat), which causes a cooling of the low-latitude ocean and a warming of the high-latitude ocean. This decreases the density of the high-latitude surface water even further *(atmospheric heat transport feedback, positive).*

The following three feedbacks F8-F12 are linked to the formation and melting of sea ice in high latitudes and its equatorward advection:

F8 During sea-ice formation, brine is rejected, the salinity below the ice is increased and the meridional circulation is strengthened *(brine rejection feedback, negative).* This feedback has no effect in the steady state, but may play an important role in a changing climate.

F9 As the meridional circulation weakens, the high-latitude temperature is reduced. This leads to the growth of sea ice and an increase in high-latitude albedo, which cools the high latitudes further and hence helps to restore the strength of the meridional circulation *(ice albedo feedback, negative).*

F10 As the sea ice grows, the heat loss from the ocean to the atmosphere is reduced due to the insulating effect of the sea ice. This keeps the high-latitude surface water warmer than in the absence of sea ice, reducing its density even further *(ice insulation feedback, positive).*

F11 The equatorward transport of negative heat (i.e. cooling induced by the atmosphere) stored in the sea ice keeps the high-latitude surface water warmer than in the absence of sea-ice advection, reducing its density even further *(ice heat transport feedback, positive).*

F12 The equatorward transport of freshwater stored in the sea ice keeps the high-latitude surface water saltier than in the absence of sea-ice advection, increasing its density *(ice freshwater transport feedback, negative).*

Many of these feedbacks have been studied intensively by Willebrand (1993), Nakamura (1994), Marotzke (1996), Lohmann and Gerdes (1998) and Jayne and Marotzke (1999) in simplified climate models. In addition, the water vapor, cloud and wind feedbacks are important, although their study re-quires more complex climate models (Hartmann 1994; Mikolajewicz 1996).

Internal Oscillations

Our interest is in the climate variability in the North Atlantic realm on centennial-to-millennial timescales (Schulz and Paul, this volume). In particular, we look for a mechanism that can produce internal oscillations of the meridional circulation of the Atlantic Ocean with periods from centuries to millennia. For internal oscillations to occur in a dynamical system, the effect of an initial anomaly (the deviation from an arbitrary initial state) must be counteracted by a negative feedback, with a certain time delay. Equivalently, the dynamical system must possess some memory or inertia. Internal oscillations do not rely on any external forcing. Nevertheless, their phase can be set by external forcing, or they can draw energy from it.

Here we take as an initial anomaly a small freshening of the high northern latitudes (cf. Section 2). The immediate effect of this initial perturbation depends on its size. Below a certain threshold, deep water formation is diminished, but still active.

Above this threshold, convection is interrupted, a halocline develops and deep water formation ceases. In both cases, the haline feedback F2 amplifies the per-turbation, while the thermal feedback F3 dampens it. The negative salinity anomaly must be large enough for the haline feedback to dominate the thermal feedback.

There are at least two mechanisms that can produce internal oscillations of the ocean with centennial-to-millennial periods. If a halocline forms, it will quickly spread southward and largely sup-press deep-water formation (this is the convective feedback F5). The Atlantic Ocean will enter a "deep-decoupled phase", during which the meridional circulation and heat transport are weak. To terminate the deep-decoupled phase, processes are called for that eventually break down the polar halocline and start a "deep-coupled phase", during which deep water is formed and the meridional circulation and heat transport are large. Such processes are the import of heat and salt by ocean advection and diffusion and the export of freshwater by sea-ice advection. They form a delayed negative feedback, which is the prerequisite for internal oscillations. If these processes are neither too weak nor too strong, a "deep-decoupling oscillation" will emerge. This type of oscillation has been studied in detail by Winton (1997 and references therein). Apparently, it has been observed first in the coupled ocean-energy balance climate-sea ice model of Harvey (1992).

The second mechanism applies if no halocline is formed and the deep-water formation remains active. Then the negative salinity anomaly will be advected with the mean meridional circulation: it will be mixed down by convection, flow southward at depth, upwell in the Southern Hemisphere and return to the Northern Hemisphere in the upper layer of the Atlantic Ocean. The meridional circulation is retarded, when the negative salinity anomaly reappears in the high northern latitudes, and recovers, when it is elsewhere, due to the action of the advective feedback F1. The oscillation that ensues is often called a "loop oscillation". Its period is determined by the time it takes to travel around the mean meridional circulation loop, which is given by ratio of the volume of the ocean basin and the rate at which new deep water is formed. This "flushing timescale" is about 500 a for the Atlantic Ocean. A loop oscillation with a period of 320 a has been noted by Mikolajewicz and Maier-Reimer (1990) in the response of an ocean model to stochastic forcing. The underlying mechanism has been clarified with the help of simple models of a fluid contained in a circular loop (e.g. Welander 1986).

The amplitude of deep-decoupling oscillations (several °C) that have been simulated so far is larger than that of loop oscillations, and their period (1500 a and up) is longer because they involve the slow vertical diffusion in the ocean.

Forced Oscillations

Another possible cause for climate variability is (quasi-) periodic external forcing, which produces forced oscillations. A prominent example are changes in solar insolation caused by variations of the Earth's orbital elements on timescales from 20,000 to 400,000 a (Milankovitch Theory; Berger 1992; Paul and Berger 1999). There are also a number of cycles with much shorter periods such as a 900 a-eccentricity cycle, which gives rise to changes in summer insolation at 65°N of less than 1 mW m^{-2}, probably too small to induce noticeable effects on the climate system (see Schulz and Paul, this volume). A huge literature exists on solar activity cycles and their relevance to recurrent climate change (e.g. Hoyt and Schatten 1997; van Geel 1999; Lean, this volume). More recently, Keeling and Whorf (2000) proposed that a 1800-a oceanic tidal cycle exists and stimulates episodic cooling at the sea surface by modulating the intensity of vertical mixing in the ocean.

According to the current paradigm, the meridional circulation is driven by surface heat and freshwater fluxes. This is possibly not the case. Rather the strength of the meridional circulation and associated heat transport may be primarily determined by the power available for returning the deep water to the surface (Munk and Wunsch 1998). There are actually two pathways for deep water removal: Vertical mixing, as it is instrumental in the mechanism put forward by Keeling and Whorf (2000), and wind-induced upwelling, as advocated by Toggweiler and Samuels (1995). In both cases, the Southern Ocean winds could play a decisive role in sustaining the meridional circulation. First, the winds are one possible source of energy for mixing, with about 70% of the global wind energy input occurring south of 40°S (Wunsch 1998). Unlike tidal dissipation and geothermal heating, they are clearly subject to climate change (cf. Huang 1999). Second, the latitude band containing the Drake Passage imposes a unique dynamic constraint on the large-scale ocean circulation (the Drake Passage effect). The northward, wind-induced Ekman drift can only be balanced by a southward, density-driven geostrophic flow at depths where there are oceanic ridges to support an east-west pressure gradient. This facilitates the upwelling of deep water from below the tops of the oceanic ridges.

The possible sources for the power required to return the deep water to the surface are the winds (0.88 TW - Wunsch 1998; 1 TW = 10^{12} W), the tidal dissipation in the open ocean (0.9 TW - Munk and Wunsch 1998) and the geothermal heat flux (0.46 TW - Huang 1999). Thus the total power available is about 2.2 TW.

Internal Oscillations in a Climate Model

In order to illustrate the mechanisms that can give rise to internal oscillations on century-to-millennial time scales, we developed a zonally averaged numerical model of the atmosphere-sea ice-ocean system (the "Hanse" model — see Appendix). The atmospheric component of the "Hanse" model is based on the formulation by Winton (1997), supplemented by a fixed latitudinal profile of the zonally averaged annual mean wind stress. Its oceanic component is closely related to that of Stocker and Wright (1996). Furthermore, we included a simple sea-ice component, which alleviates the problem of subfreezing temperatures. We consider a single ocean basin that represents the Atlantic Ocean. It is 60° wide, 5000 m deep and extends from 70°S to 90°N. Between 50° and 60°S there is a channel of about 2200 m depth, which simulates the effect of the Drake Passage on the meridional circulation.

In the reference state (Experiment A; Tab. 1, Fig. 1), the temperature-insensitive part of the outgoing longwave radiation is set to A_0 = 206.40 W m^{-2}, corresponding to an approximate mean atmospheric temperature of T_{am} = 17°C. The meridional heat transport (Fig. 1a) is southward but small south

Experiment	A_0 (W m^{-2})	ΔA_0 (W m^{-2})	T_{am} (°C)	
A	206.40	—	17	reference state
B	208.63	2.23	16	
C1	210.88	4.48	15	
C2	210.88-208.63	4.48-2.23	15-16	linear increase of T_{am}
C3	210.88-208.63	4.48-2.23	15-16	linear increase of T_{am}, meltwater input at t = 42 ka
D	210.88	4.48	15	insulating effect of sea ice included
E	210.88	4.48	n.a.	sea ice albedo feedback included

Table 1 List of experiments. Note that for Experiment E, the mean air temperature is not constant.

of 20°S and northward elsewhere. In particular, there is a northward heat transport across the Equator, as in the present-day Atlantic Ocean. The meridional overturning streamfunction (Fig. 1b) indicates the formation of deep water ("NADW") at a rate of 12 Sv, most of which upwells in the Equatorial and South Atlantic Oceans. Only 2 Sv of "NADW" flow southward below the sill depth of the Drake Passage to partly balance the northward Ekman flow at the surface. The formation of bottom water in the Southern Ocean is suppressed on purpose by excessive freshening of the southern high latitudes (Fig. A1b). The temperature distribution (Fig. 1c) shows that the whole basin is very cold. The basin mean temperature is 1.4°C, more than 2°C colder than in the present-day Atlantic Ocean, and the warm water sphere (defined as water warmer than 8°C) extends to depths of 300-400 m, about 200 m less than observed. According to the salinity distribution (Fig 1d), the deep water salinity is close to the prescribed basin mean value of 34.7 everywhere.

An increase of the outgoing longwave radiation by $\Delta A_0 = 4.4$ W m^{-2} (Experiment C1), corresponding to a cooling of the approximate mean atmospheric temperature by $\Delta T_{am} = 2$ °C , gives rise to deep-decoupling oscillations (Fig. 2). The maximum of the meridional streamfunction varies between 8 and 15 Sv (Fig. 2a). The temperature of the surface water (layers 1 and 2, see Tab. A1) north of 60°N indicates alternating phases of cold, ice-covered and warm, ice-free conditions (Fig. 2b). The temperature variations of the intermediate water (layers 3 to 10) are equally large, while those

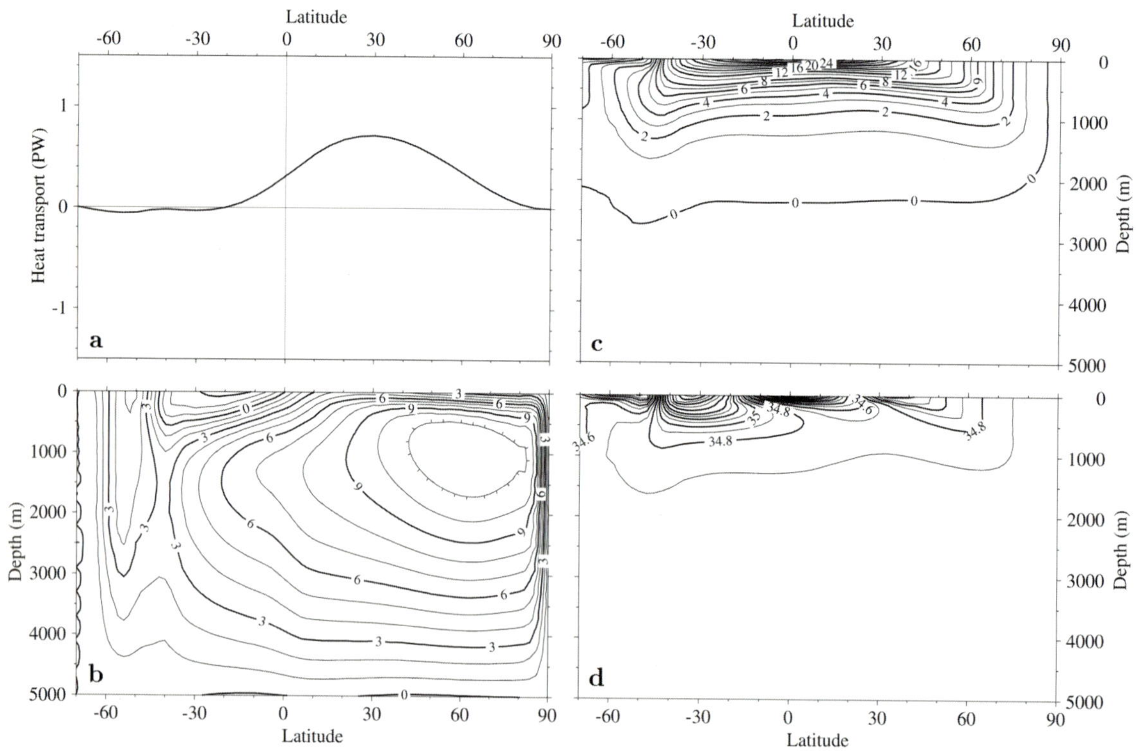

Fig. 1. Reference state of the "Hanse" model (Experiment A, $T_{am} = 17$°C) . **(a)** Net meridional heat transport (continuous line). Positive (negative) values indicate northward (southward) direction. **(b)** Meridional overturning stream-function.The flow is directed along the streamlines. Positive (negative) contour levels indicate clockwise (anti-clockwise) overturning. The flow rate through any section between two adjacent streamlines is given by the contour interval, i.e. 1 Sv. **(c)** Potential temperature distribution. The contour interval is 1°C below a value of 8°C and 2°C above. **(d)** Salinity distribution. The contour interval is 0.1 units below a value of 35.0 and 0.2 units above.

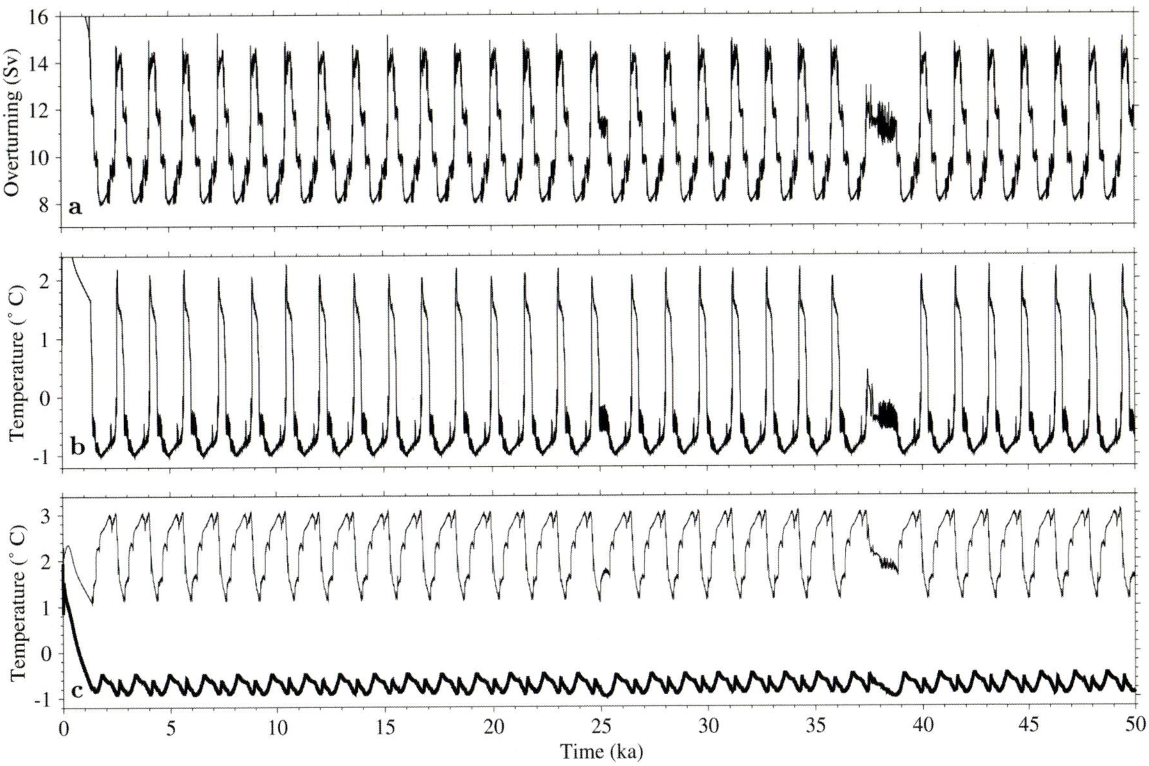

Fig. 2. Long sequence of deep-decoupling oscillations in the "Hanse" model (Experiment C1, T_{am} = 15°C). **(a)** Maximum of meridional streamfunction north of the Equator (Sv). **(b)** Average temperature north of 60°N (°C) of surface water. **(c)** Average temperature north of 60°N (°C) of intermediate water (thin line) and deep water (thick line).

of the deep water (layers 11 to 17) have an amplitude of only 0.5°C. Note that length and shape of the individual cycles are not constant; near time t = 37 ka, the oscillations even "crumble", but recover afterwards. This is a first indication that predictability is limited.

Fig. 3 shows a series of snapshots of a particular deep-decoupling cycle. At the end of the coupled phase, the maximum of the meridional streamfunction is 2-3 Sv higher than in the reference state (Fig. 3a), and the deep ocean is even colder than in the reference state (Figs. 1c and 3b). As a result of deep convection, temperatures in the upper 1500 m of the northern North Atlantic Ocean are nearly homogeneous.

During the decoupling phase, the meridional circulation is weakening (Fig. 3c) and the convection receding, as indicated by the non-vertical isotherms (Fig. 3d). Everywhere below 2500-3000 m depth temperatures are now lower than -1°C. A halocline is forming near the northern boundary, indicated by an inverted thermocline down to about 500 m depth.

At the midpoint of the decoupled phase, the meridional circulation has detached from the bottom (Figs. 3e and f), i.e. the southward outflow from the North Atlantic Ocean is now limited to depths above 3000 m. The halocline has deepened and spread southward, but there is already a tongue of warm intermediate water underneath it that destabilizes the water column. The intermediate water has also warmed the deep water by vertical mixing (cf. the downward shift of -1°C isotherm in Fig. 3f vs. Fig. 3d). Finally, Figs. 3g and h show the onset of deep convection at the beginning of the next coupled phase. The halocline is largely eliminated, and very cold deep water is formed.

To better understand the mechanism underlying the oscillations, we focus on one particular deep-decoupling cycle, starting from a coupled phase

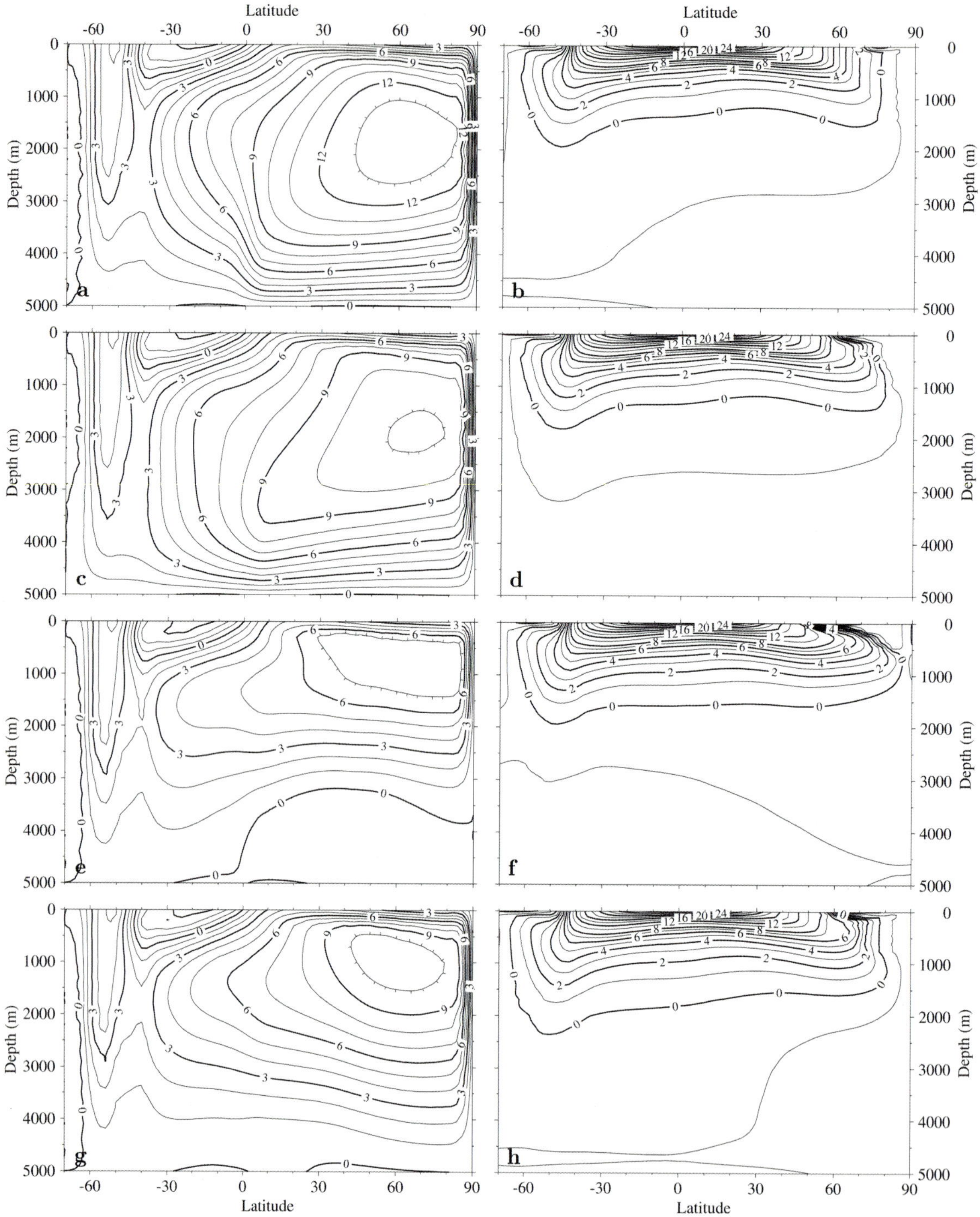

Fig. 3. Representative deep-decoupling oscillation in the "Hanse" model (Experiment C1, T_{am} = 15°C). Meridional streamfunction (Sv) and temperature distribution (°C) for **(a), (b)** year 40,200 (coupled phase), **(c), (d)** year 40,400 (decoupling phase), **(e), (f)** year 40,800 (decoupled phase), **(g), (h)** year 41,520 (coupling phase). Contour intervals as in Fig. 1(b) and (c).

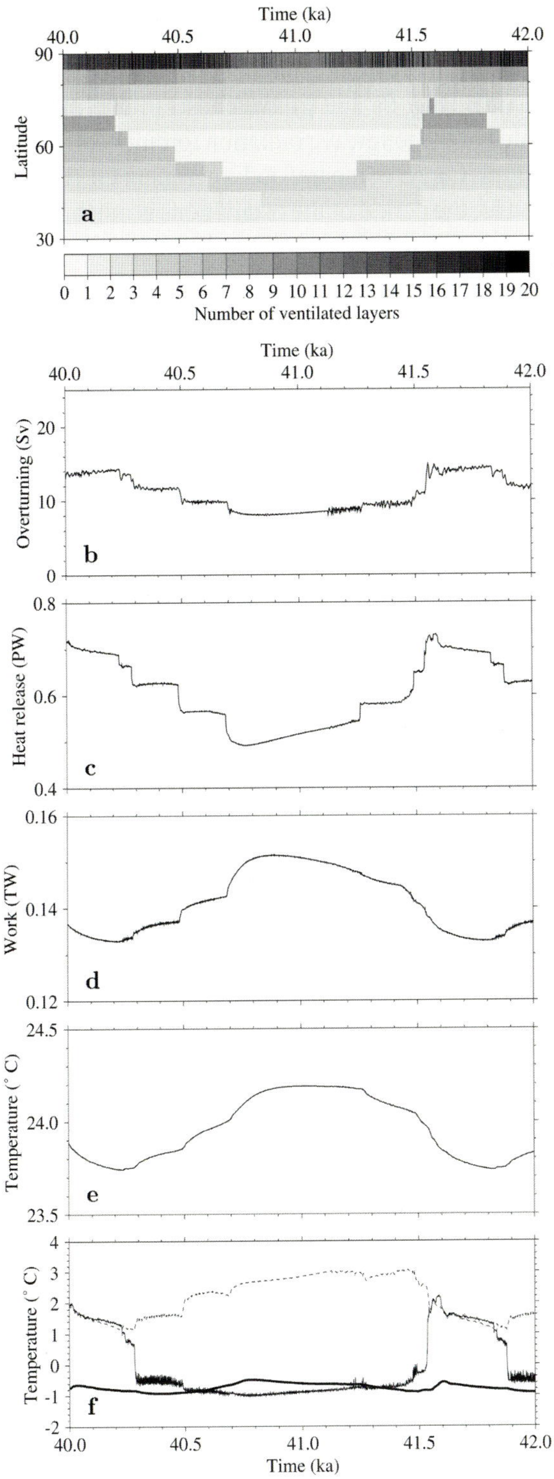

Fig. 4. (left) Representative deep-decoupling oscillation in the "Hanse" model (Experiment C1, $T_{am} = 15°C$). **(a)** Depth of convection (number of layers ventilated from the surface) as a function of time and latitude. **(b)** Maximum of meridional streamfunction north of the Equator (Sv). **(c)** Total heat released north of 35°N (PW). **(d)** Total rate of work performed against buoyancy forces (TW). **(e)** Average sea-surface temperature between 10°S and 30°N (°C). **(f)** Average temperature north of 60°N (°C) of surface water (thin line), intermediate water (dashed line) and deep water (thick line).

(Fig. 4). Convection never ceases completely, but as the decoupling progresses, it is shifting southward and is becoming shallower (Fig. 4a). It does so in steps, which leave their imprint on the strength of the meridional circulation (Fig. 4b) and related quantities (Figs. 4c to f). The heat released to the atmosphere north of 35°N (Fig. 4c) is closely correlated to the maximum of the meridional streamfunction and varies between 0.5 and 0.7 PW (1 PW = 10^{15} W).

Fig. 4d illustrates that a deep-decoupling cycle at a fixed value of the vertical diffusivity coefficient implies large changes in the total rate of work required to return deep water to the surface. Since the wind forcing is fixed and the tidal forcing is implicitly assumed constant, this violates the conservation of energy (Section 2).

When the meridional circulation slows down and less heat is transported northward, the tropical surface temperature rises by 0.5°C (Fig. 4e). At the same time the temperature of the surface water north of 60°N is low (Fig. 4f), indicating the presence of sea ice. Thus, during a deep-decoupling cycle surface temperatures at low and high latitudes are in anti-phase. In contrast, the tropical surface temperature and the temperature of the intermediate water north of 60°N are largely in phase. The intermediate water warms gradually until deep convection resumes. Apparently, part of the heat not released to the atmosphere is stored in the intermediate water (cf. Fig. 3f). The deep water temperature slightly increases during the first part of the decoupled phase, then decreases as the heat release to the atmosphere slowly recovers, and finally it shows a small overshoot at the be-ginning of the next coupled phase (Fig. 4c and f).

The power spectra of the strength of the meridional circulation and the temperature of the high-latitude surface, intermediate and deep water in Experiment C1 show a red noise shape (Fig. 5). The continuous part of a power spectrum indicates a nonperiodic, irregular and even chaotic behavior that limits predictability. On top of it, there are a number of discrete peaks, which are all harmonics of the main peak at a frequency of 1/(1640 a), with a half-width at half-height of about 8% of this frequency. (The harmonics owe their existence to the Fourier analysis of square-wave type time series.)

To explore the sensitivity of the "Hanse" model, we compare Experiment C1 with further experiments (Table 1). We again focus on the time-series of the maximum of the meridional streamfunction and the temperature of the high-latitude surface, intermediate and deep water. For Experiment B (Fig. 6a and g), we chose an intermediate value of the longwave radiation forcing (ΔA_0=2.23, ΔT_a=1°C). As compared to Experiment C1 (Fig. 6b and h), the period of the oscillations is much longer and their amplitudes are much smaller. Only the variations of the high-latitude surface water temperature, with an amplitude of about 1°C, are clearly discernible. Thus, Experiment B marks the transition to deep-decoupling oscillations, while the stable reference state (Experiment A) is already quite close it.

In Experiment C2 (Fig. 6c and i), we linearly increased the longwave radiation forcing ΔA_0 from 4.48 to 2.23 Wm^{-2}, corresponding to a linear rise

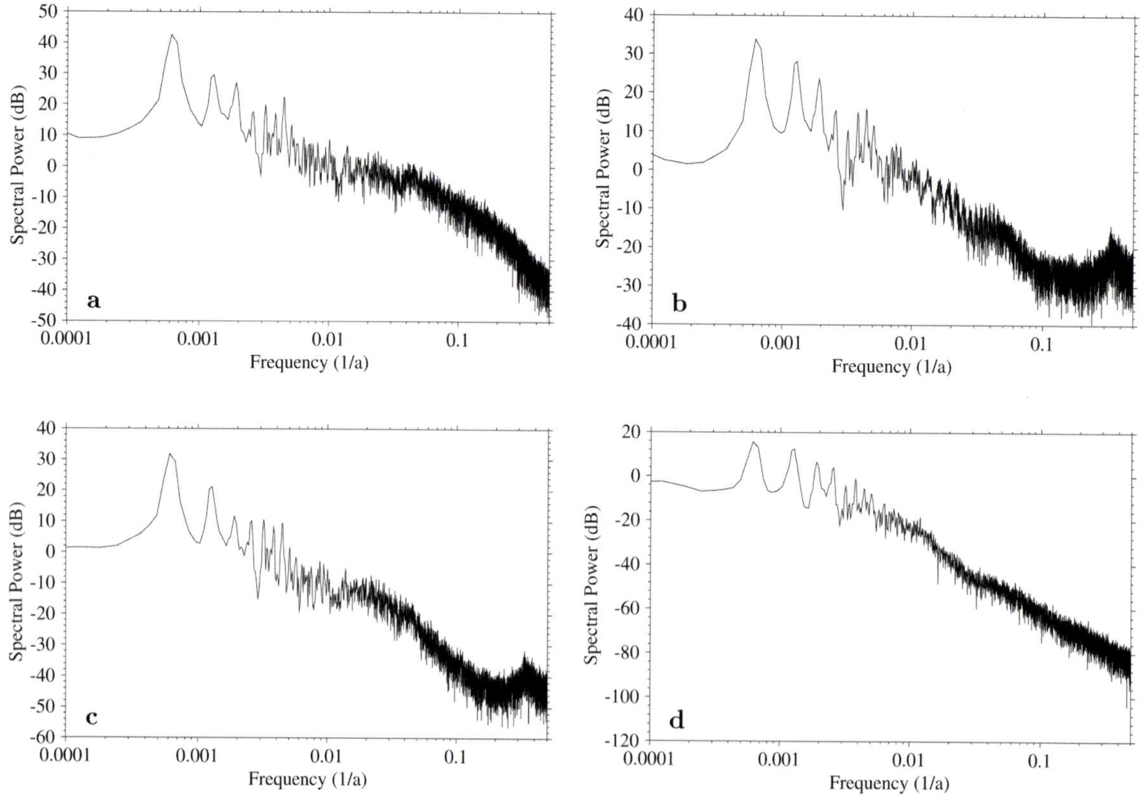

Fig. 5. Spectra of a long sequence of deep-decoupling oscillations in the "Hanse" model (Experiment C1, T_{am} = 15°C, years 2,000 to 37,000). Logarithmic ordinate (frequency) and abscissa (spectral power density). Cf. Fig. 4(b) and (f). **(a)** Meridional overturning. **(b)** Surface-water temperature. **(c)** Intermediate-water temperature. **(d)** Deep-water temperature.

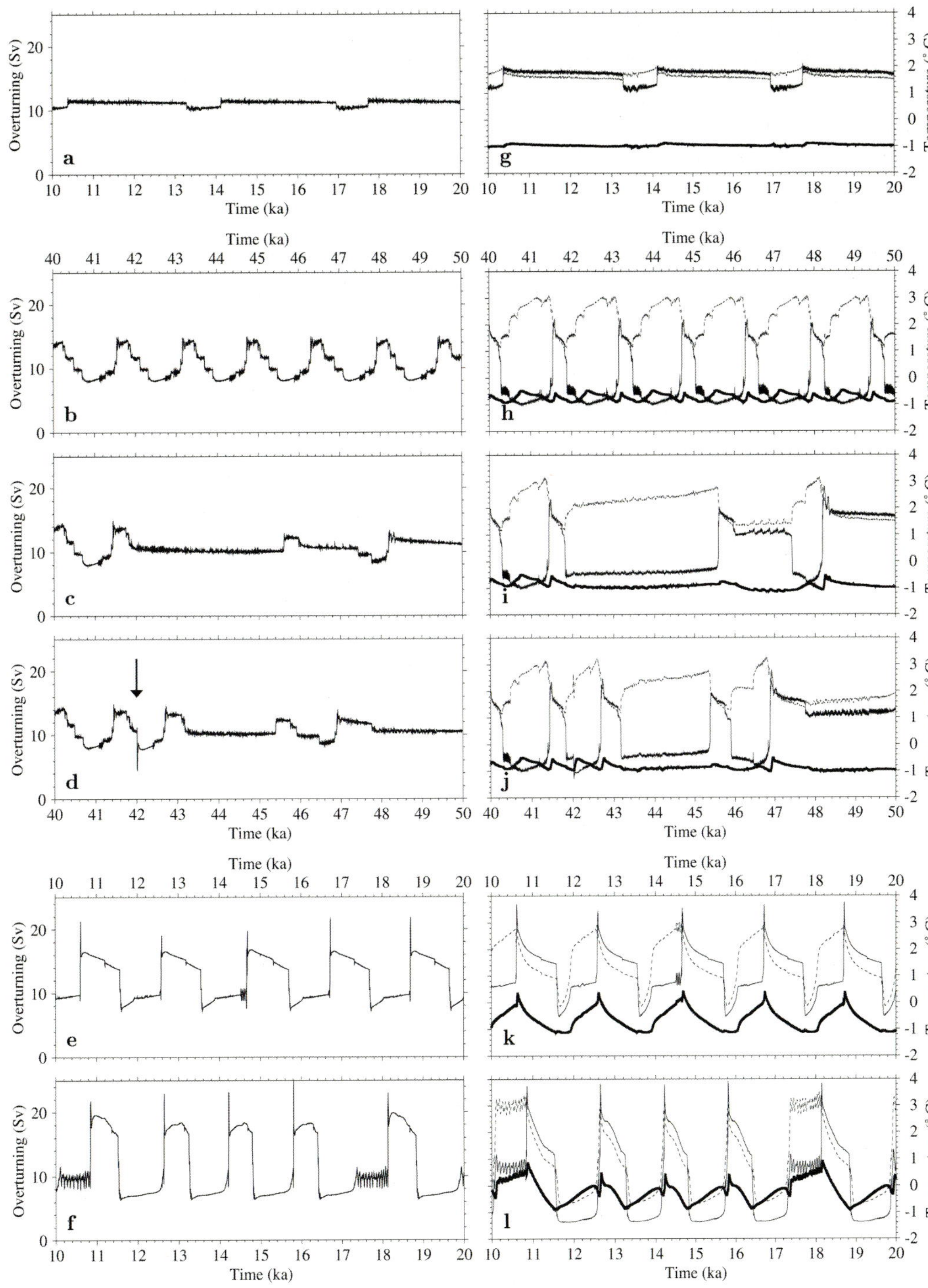

Fig. 6. Selected sequences of deep-decoupling oscillations in the "Hanse" model. **(a)-(f)** Maximum of meridional streamfunction north of the Equator (Sv) and **(g)-(l)** average temperature north of 60°N (°C) of surface water (thin line), intermediate water (dashed line) and deep water (thick line). Cf. Fig. 4b) and f). **(a), (g)** Experiment B (T_{am} = 16°C). **(b), (h)** Experiment C1 (T_{am} = 15°C). **(c), (i)** Experiment C2 (T_{am} linearly increasing from 15°C to 16°C). **(d), (j)** Experiment C3 (T_{am} linearly increasing from 15°C to 16°C) with additional freshwater input in year 42,000 (arrow). **(e), (k)** Experiment D (T_{am} = 15°C), including ice-thickness effect. **(f), (l)** Experiment E (T_{am} = 15°C), including ice-thickness and ice-albedo effect.

in the mean atmospheric temperature T_{am} from 15 to 16°C. Hence at the end of the deep-decoupling cycle at about t=42 ka the atmosphere has already warmed to a level that is less favorable for further deep-decoupling oscillations to occur. In fact, the next deep-decoupling cycle only follows 4 ka later, slightly deformed and with a reduced amplitude. Thereafter the "Hanse" model approaches a state similar to that of Experiment B. Experiment C3 (Figs. 6c and i) was identical to Experiment C2 except for an additional meltwater input at t=42 ka, mimicking the catastrophic drainage of North American proglacial lakes at 8.2 ka BP (Barber et al. 1999). The meltwater was directly added to the surface of the Atlantic Ocean between 45 and 60°N, at a constant rate of 0.6 Sv for a duration of 10 a. During the meltwater input, deep-water formation is suppressed, but recovers quickly afterwards. A sudden cooling of the high-latitude surface water ensues, accompanied by a sudden warming and subsequent cooling of the high-latitude intermediate water, then followed by an additional deep-decoupling cycle. The next deep-decoupling cycle occurs slightly earlier than in Experiment C2, and within the final 2 ka the "Hanse" model has not arrived yet at the state of Experiment B.

So far, the role of the sea ice component of the "Hanse" model was merely to prevent the occurrence of subfreezing temperatures in an energy-conserving way. Because of its equatorward advection, sea ice was not accumulated continously as in the original formulation by Large et al. (1997), but it still piled up to the unrealistic height of several tens of meters (not shown). Hence, in Experiment D (Figs. 6e and k) we included as a remedy the insulating effect of sea ice to reduce the heat loss of the underlying ocean to the atmosphere. The deep-decoupling oscillations persist, but they change in character: The period increases to 2000 a. The pervasive high-frequency variability of Experiment C1 is restricted to few intermittent episodes. The maximum of the meridional streamfunction now shows sharp spikes at the onset of each coupled phase, and there are no distinct steps as in Experiment C1. Furthermore, the temperature variations of the deep water reaches 1.3°C, and the surface water is significantly warmer than the intermediate water during the coupled phases.

Finally, in Experiment E (Figs. 6f and l) we not only included the insulating effect of sea ice, but also accounted for the sea-ice albedo effect. The deep-decoupling oscillations again change in character, their period is about 1700 a, shorter than the one in Experiment D, but similar to the one in Experiment C1. The intermittent episodes of high-frequency variability become longer and can prolong the onset of the next coupled phase by almost 1 ka, which severely reduces its predictability. Any such episode is marked by an intermediate value of the temperature of the high-latitude surface water. During an uncoupled phase, the temperature of the high-latitude intermediate water is more than 3°C colder than either in Experiment C1 or Experiment D.

As compared to Experiment C1 (Fig. 4a), the insulating effect of sea ice in Experiment D goes along with deeper convection during the coupled phase (Fig. 7a), which is due to the much reduced sea ice cover: In Experiment C1, the four northernmost grid boxes are permanently covered by sea ice (not shown), whereas in Experiment D, the northernmost grid box is only temporarily covered by sea ice (Fig. 7b). During the decoupled phase, convection under the sea ice never ceases completely. In Experiment E, the sea ice albedo effect leads to an increase in sea ice relative to Experiment D, both in extent and thickness (Figs. 6c and d).

Discussion

The "Hanse" model is an extension of the climate model used by Winton (1997). It is reassuring that given the same atmospheric heat transport divergence and surface fresh water flux, it also produces deep-decoupling oscillations with millennial periods. Convection never ceases and the meridional circulation never breaks down completely in any of the Experiments B-E. During the deep-decoupled phase of, e.g. Experiment C1 (Fig. 3e), deep water formation does not reach all the way to the bottom, but still to 3000-4000 m depth. In contrast to Winton (1997) it is not only diffusive heating from below that destroys the halocline, but also the ongoing

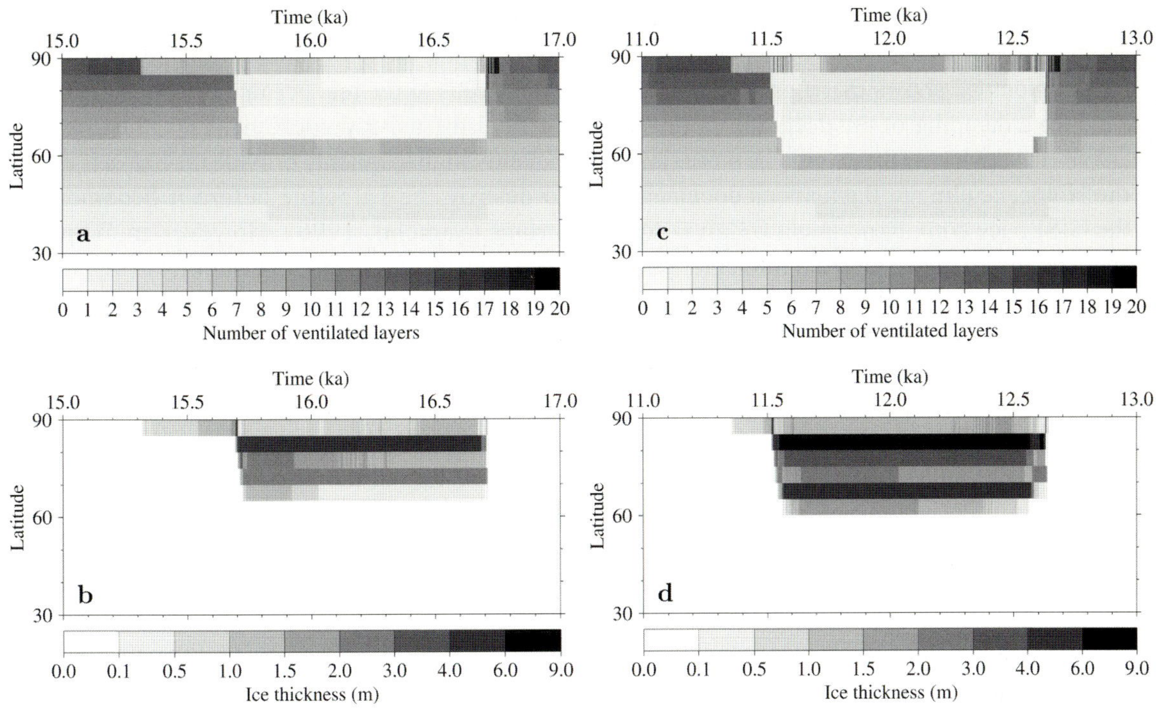

Fig. 7. Depth of convection (number of layers ventilated from the surface) and sea-ice thickness (m) as a function of time and latitude in the "Hanse" model. **(a), (b)** Experiment D (T_{am} = 15°C, including ice-thickness effect), cf. Figs. 5(e) and (k). **(c), (d)** Experiment E (T_{am} = 15°C, including ice-thickness and ice-albedo effect), cf. Figs. 6(f) and (l).

poleward advection of warm water by the meridional circulation. Both shutdown and reinitiation of deep convection are rapid processes that occur within a few years. The timescale of the deep-decoupling oscillations is set by the slow buildup of the halocline during the deep-coupled phase and its equally slow breakdown during the deep-decoupled phase. The time span required for buildup is largely determined by the prescribed surface freshwater flux (Fig. A1), whereas the duration of the breakdown are set by the combined effect of diffusive and advective heat transports, that is, by the negative feedback F1. This is supported by experiments without the sea-ice component, in which temperatures below freezing are simply reset to the freezing point. The ensuing deep-decoupling oscillations differ from Experiment C1 in that the convection stops completely, the meridional circulation does not penetrate the halocline and the period increases up to about 4000 a (not shown).

During all phases of the deep-decoupling oscillations, there is the same export of 2-3 Sv of deep water from the North Atlantic Ocean to the Southern Ocean across 50 S (Fig. 3). This is a manifestation of the Drake Passage effect (Section 4): As long as there is convection in the North Atlantic Ocean, "NADW" contributes to the southward geostrophic flow that balances the northward Ekman drift, which is constant due to the imposed wind stress forcing. Moreover, even during the deep-decoupled phase (i.e., without deep convection in the North Atlantic Ocean) the maximum of the meridional overturning does not drop below 8 Sv with an associated poleward heat transport of 0.4 PW (Experiment C1, Figs. 4b and c). These two finding supports the notion that, given surface heat and freshwater fluxes to allow for deep convection in the North Atlantic Ocean, the strength of the meridional circulation is mainly controlled by the wind forcing in the Southern Ocean and vertical mixing (Toggweiler and Samuels 1995; Munk and Wunsch 1998).

Experiment C2 shows that a gradual warming can induce a damping of the deep-decoupling oscillations, in which case they would be more similar to the Holocene climate fluctuations documented in the GISP2 ice core. Experiment C3 gives some further support to the conceptual model of Schulz and Paul (this volume), in that an additional meltwater input can indeed trigger another deep-decoupling cycle of the meridional circulation. In comparison to the conceptual model, the damping in Experiments C2 and C3 is too strong to admit further deep-decoupling cycles.

Including the insulating effect of sea-ice (Experiments D and E) reduces the surface heat flux and its sensitivity to changes in sea-surface temperature. This dampens the high-frequency variability of the convection, and hence that of meridional circulation and high-latitude temperatures (Figs. 6 and 7). Additional consideration of the sea-ice albedo feedback (Experiment E) increases the amplitude of the meridional overturning during the deep-decoupling oscillations by about 4 Sv and similarly of the high-latitude temperature oscillations (Fig. 6). Since the higher albedo of sea ice amplifies sea-ice growth, convection is more readily suppressed and the deep-coupled phase is shortened.

The temperature evolution of the intermediate water differs greatly between Experiments C1, D and E. The most obvious differences exist during the deep-decoupled phase: In Experiment C1 the intermediate-water temperature rises continuously (except for small steps; cf. Section 5), while in Experiment D first a brief temperature drop occurs. In Experiment E, the warming is slow until the start of the deep-coupling phase (Fig. 6). In both Experiments D and E intermediate water temperature shows a rapid rise. However, while the increase precedes the onset of the deep-coupled phase by about 700 a in Experiment E, the offset is only about 50 a in Experiment E. These contrasting predictions on the phasing of surface- and intermediate-water temperatures prompt a comparison with high-resolution paleoceanographic proxy data, which should become available in the near future.

In this context, there are three important limitations upon the applicability of deep-decoupling oscillations to the real ocean: (1) The value of the vertical diffusivity coefficient D_v must be sufficiently high to quickly provide the heat necessary to destroy the halocline before it becomes too strong (Sarachik 1996). However in Winton's (1997) as well as in the "Hanse" model D_v is set to 0.5 $cm^2 s^{-1}$. Although this value is 5 times higher than observed in the interior of the ocean, it is probably a reasonable value for the zonal mean (Sjöberg and Stigebrandt 1992). Furthermore, in the "Hanse" model, advection of heat also contributes to the breakup of the halocline. Hence, the deep-decoupling oscillations could possibly persist at an even lower value of D_v. (2) A fixed value of D_v implies large changes in the total rate of work performed against buoyancy forces although the mechanical (i.e., wind and tidal) forcing does not change. In a version of the "Hanse" model in which we replaced the fixed D_v by a fixed rate of work against buoyancy forces, deep-decoupling oscillations still occur but their period increased by about 20% (not shown). (3) Any deep water formed in the Southern Hemisphere competes with the polar region of the Northern Hemisphere for the heat supplied by the intermediate water and therefore might prevent the breakup of the halocline. Both, Winton's (1997, Fig. 9b) as well as our reference state (Experiment A) resemble the real Atlantic Ocean mainly in that they are equilibrium states with sinking in the Northern Hemisphere only. Since both models lack deep-water formation in the Southern Hemisphere, they do not allow to test if deep-decoupling oscillations could persist in the presence of deep-water of a southern origin.

The main peaks in the power spectra of Experiment C1 (Fig. 5) are almost as narrow as that of a pure sinusoidal wave. In contrast, from the internal variability of a real world fluid-dynamical system, one would expect a broader peak (Wunsch 2000). The reason for this discrepancy lies probably in the limited complexity of the "Hanse" model and its low number of internal degrees-of-freedom.

The second mechanism that can produce oscillations on centennial-to-millennial time scales is the loop oscillations (Section 3). They are an attractive mechanism because the shortening of the period of the reconstructed oscillations during the glacial-to-Holocene transition (Schulz and Paul, this volume) would find an elegant explanation in terms of a strengthening of the meridional circulation. From the ratio of the glacial and Holocene periods 1500 a/900 a, the formation of new NADW would have increased by 40%, which is a reasonable estimate according to a number of ocean model results (e.g. Schäfer-Neth and Paul 2001). Taking the Holocene rate of deep water formation to be equal to the present-day rate, the glacial rate would have been 900 a/1500 a $\times$ 14 Sv = 8.4 Sv.

In the "Hanse" model, the distinction between deep-decoupling and loop oscillations appears to be partly lost. Because the meridional circulation never ceases completely, the mean flow takes a negative salinity anomaly around a closed path, as in the case of pure loop oscillations.

Conclusions

(1) In our experiments, internal oscillations with periods ranging approximately from 1600-2000 a are a persistent feature. Although the influence of dynamic atmospheric heat and moisture transports is still to be included, it is tempting to translate this result to the real climate system, in which case millennial-scale internal variability could be superimposed on and might even mask the human influence on climate.

(2) During all phases of the internal oscillations, there is active deep-water formation and export from the North Atlantic to the Southern Ocean. In all experiments, the maximum of the meridional overturning does never drop below 6 Sv.

(3) Our experiments also underline the importance of the sea-ice component, which prevents a complete breakdown of the meridional circulation by always maintaining convection. However, the effects of, e.g. fractional sea-ice coverage and snowfall on ice need to be studied further.

(4) Future studies of the sensitivity of the meridional circulation of the Atlantic Ocean should include further feedbacks of the climate system, e.g. the water vapor feedback. Moreover, they should address important effects that have been largely neglected so far, e.g. the Drake Passage dynamic constraint and tidal mixing.

Acknowledgments

We greatly appreciate many lively discussions with W. Berger on the role of internal and forced oscillations in the climate system. AP also thanks R. Keeling who drew his attention to the question what actually drives the meridional circulation of the Atlantic Ocean. We acknowledge the benefical comments on the manuscript by an anonymous reviewer. AP is supported by the Alexander v. Humboldt Stiftung (through a Feodor Lynen Research Fellowship) as well as the California Space Institute and MS by the German Klimaforschungsprogramm.

References

Barber DC, Dyke A, Hillaire-Marcel C, Jennings AE, Andrews JT, Kerwin MW, Bilodeau G, McNeely R, Southon J, Morehead MD, Gagnon J-M (1999) Forcing of the cold event of 8,200 years ago by catastrophich drainage of Laurentide lakes. Nature 400:344-348

Berger AL (1992) Astronomical theory of paleoclimates and the last glacial-interglacial cycle. Quat Sci Rev 11:571-581

Bond G, Showers W, Cheseby M, Lotti R, Almasi P, deMenocal P, Priore P, Cullen H, Hajdas I, Bonani G (1997) A pervasive millennial-scale cycle in North Atlantic Holocene and glacial climates. Science 278:1257-1266

Croll J (1875) Climate and time in their geological relations: A theory of secular changes of the Earth's climate. Daldy, Isbister & CO., London, pp 1-517

Dietrich G (1950) Kontinentale Einflüsse auf Temperatur und Salzgehalt des Ozeanwassers. Dt hydrograph Z 3:33-39

Farrow DE, Stevens DP (1995) A new tracer advection scheme for Bryan and Cox type ocean general circulation models. J Phys Oceanogr 25:1731-1741

Graves CE, Lee W-H, North GR (1993) New parametrizations and sensitivities for simple climate models. J Geophys Res 98 (D3):5025-5036

Grootes PM, Stuiver M (1997) Oxygen 18/16 variability in Greenland snow and ice with 10^{-3}- to 10^{5}-year time resolution. J Geophys Res 102 (C12):26455-26470

Hartmann DL (1994) Global Physical Climatology. Academic Press, San Diego, pp x + 411

Harvey LDD (1988) Development of a sea ice model for use in zonally averaged energy balance climate models. J Clim 1:1221-1238

Harvey LDD (1992) A two-dimensional model for long-term climate simulations: Stability and coupling to atmospheric and sea ice models. J Geophys Res 97 (C6):9435-9453

Hoyt DV, Schatten KH (1997) The role of the sun in climate change. Oxford University Press, New York, pp viii + 279

Huang RX (1993) Real freshwater flux as a natural boundary condition for the salinity balance and thermohaline circulation forced by evaporation and precipitation. J Phys Oceanogr 23:2428-2447

Huang RX (1999) Mixing and energetics of the oceanic thermohaline circulation. J Phys Oceanogr 29:727-746

Jayne SR, Marotzke J (1999) A destabilizing thermohaline circulation-atmosphere-sea ice feedback. J Clim 12:642-651

Kalnay E, Kanamitsu M, Kistler R, Collins W, Deaven D, Gandin L, Iredell M, Saha S, White G, Woolen J, Zhu Y, Chelliah M, Ebisuzaki W, Higgins W, Janowiak J, Mo KC, Ropelewski C, Leetmaa A, Reynolds R, Jenne R (1996) The NCEP/NCAR reanalysis project. Bull Am Meteorol Soc 77:437-471

Keeling CD, Whorf TP (2000) The 1,800-year oceanic tidal cycle: A possible cause of rapid climate change. Proc Natl Acad Sci USA 97:3814-3819

Large WG, Danabasoglu G, Doney SC, McWilliams JC (1997) Sensitivity to surface forcing and boundary layer mixing in a global ocean model: Annual-mean climatology. J Phys Oceanogr 27:2418-2447

Lohmann G, Gerdes R (1998) Sea ice effects on the sensitivity of the thermohaline circulation. J Clim 11:2789-2803

Lowe JJ, Walker MJC (1997) Reconstructing quaternary environments. Addison Wesley Longman, Harlow, Essex, pp xxii + 446

Marotzke J (1996) Analysis of thermohaline feedbacks. In: Anderson DLT, Willebrand J (eds) Decadal Climate Variability: Dynamics and Predictability. NATO Advanced Science Institutes Series I: Global Environmental Change 44. Springer-Verlag, Berlin, Heidelberg, New York, pp 333-378

Mellor GL (1991) An equation of state for numerical models of oceans and estuaries. J Atmos Oceanic Technol 8:609-611

Mikolajewicz U (1996) A meltwater induced collapse of the 'conveyor belt' thermohaline circulation and its influence on the distribution of $\Delta^{14}C$ and $\delta^{18}O$ in the oceans. Report 189, Max-Planck-Institut für Meteorologie, Hamburg, pp 1-25

Mikolajewicz U, Maier-Reimer E (1990) Internal secular variability in an ocean general circulation model. Clim Dyn 4:145-156

Model F (1950) Warmwasserheizung Europas. Ber Dt Wetterdienst US-Zone 12

Munk W, Wunsch C (1998) Abyssal recipes II: energetics of tidal and wind mixing. Deep-Sea Res I 45:1977-2010

Nakamura M, Stone PH, Marotzke J (1994) Destabilization of the thermohaline circulation by atmospheric eddy transports. J Clim 7:1870-1882

Paul A, Berger WH (1999) Climate cycles and climate transitions as a response to astronomical and CO_2 forcings. In: Harff J, Lemke W, Stattegger K (eds) Computerized Modeling of Sedimentary Systems. Springer-Verlag, Berlin, Heidelberg, pp 223-245

Rahmstorf S (1996) A fast and complete convection scheme for ocean models. In: Pacanowski RC (ed) MOM 2. Documentation, User's Guide and Reference Manual. Technical Report 3.2, Geophysical Fluid Dynamics Laboratory, Princeton, New Jersey, pp 1-329

Sarachik ES, Winton M, Yin FL (1996) Mechanisms for decadal-to-centennial climate variability. In: Anderson DLT, Willebrand J (eds) Decadal Climate Variability: Dynamics and Predictability. NATO Advanced Science Institutes Series I: Global Environmental Change 44. Springer-Verlag, Berlin, Heidelberg, New York, pp 157-210

Sarnthein M, Stattegger K, Dreger D, Erlenkeuser H, Grootes P, Haupt B, Jung S, Kiefer T, Kuhnt W, Pflaumann U, Schäfer-Neth C, Schulz H, Schulz M, Seidov D, Simstich J, van Kreveld S, Vogelsang E, Völker A, Weinelt M (2001) Fundamental modes and abrupt changes in North Atlantic circulation and climate over the last 60 ky - Concepts, reconstructions and numerical modeling. In: Schäfer P, Ritzrau W, Schlüter M, Thiede J (eds) The Northern North Atlantic: A changing environment. Springer-Verlag, Berlin, pp 365-410

Schäfer-Neth C, Paul A (2001) Circulation of the glacial Atlantic: A synthesis of global and regional modeling. In: Schäfer P, Ritzrau W, Schlüter M, Thiede J (eds) The Northern North Atlantic: A changing environment. Springer-Verlag, Berlin, pp 441-462

Sjöberg B, Stigebrandt A (1992) Computations of the geographical distribution of the energy flux to mixing processes via internal tides and the associated vertical circulation in the ocean. Deep-Sea Res 39:269-291

Stocker TF, Wright DG (1996) Rapid changes in ocean circulation and atmospheric radiocarbon. Paleoceanography 11:773-795

Toggweiler JR, Samuels B (1995) Effect of Drake Passage on the global thermohaline circulation. Deep-Sea Res 42:477-500

van Geel B, Raspopov OM, Renssen H, van der Plicht J, Dergachev VA, Meijer HAJ (1999) The role of solar forcing upon climate change. Quat Sci Rev 18:331-338

Warren BA (1981) Deep circulation of the world ocean. In: Warren BA, Wunsch C (eds) Evolution of Physical Oceanography. Scientific surveys in honor of Henry Stommel. MIT Press, Cambridge, Massachusetts, pp 6-41

Weaver AJ, Bitz CM, Fanning A, Holland AM (1999) Thermohaline circulation: High-latitude phenomena and the difference between the Pacific and Atlantic. Ann Rev Earth Planet Sci 27:231-281

Welander P (1986) Thermohaline effects in the ocean circulation and related simple models. In: Anderson DLT, Willebrand J (eds) Large-Scale Transport Processes in Oceans and Atmospheres. NATO Advanced Science Institutes Series C: Mathematical and Physical Sciences 190. D Reidel Pub Co, Norwell, Massachusetts, pp 163-200

Willebrand J (1993) Forcing the ocean with heat and freshwater fluxes. In: Raschke E, Jacob D (eds) Energy and Water Cycles in the Climate System. NATO Advanced Science Institutes Series I: Global Environmental Change 5. Springer-Verlag, Berlin, New York, pp 215-233

Winton M (1997) The effect of cold climate upon North Atlantic Deep Water formation in a simple ocean-atmosphere model. J Clim 10:37-51

Wunsch C (1998) The work done by the wind on the oceanic general circulation. J Phys Oceanogr 28:2332-2340

Wunsch C (2000) On sharp spectral lines in the climate record and the millennial peak. Paleoceanography 15:417-424

Wright DG, Vreugdenhil CB, Hughes TMC (1995) Vorticity dynamics and zonally averaged ocean circulation models. J Phys Oceanogr 25:2141-2154

Appendix: Description of the "Hanse" Model

The "Hanse" model consists of three components: atmosphere, sea-ice and ocean.

Atmosphere: There is no explicit representation of the atmospheric momentum balance. Instead, the zonal component of the zonally averaged global annual mean wind stress is prescribed as a function of latitude according to the NCEP/NCAR reanalysis data (Kalnay et al. 1996) between 70°S and 45°S. The atmospheric heat and moisture capacities as well as the land heat capacity are assumed to be zero. The atmospheric zonal heat transport is taken to be infinitely rapid. Thus, the atmospheric heat balance at latitude φ reads

$$0 = -(1 - f_l)\, K\, (T_a - T_s) - (A + B\, T_a)\,,$$

where f_l is the land fraction of the latitude belt, $F = K\,(T_a - T_s)$ is the surface heat flux (positive downward),

$$A = \operatorname{div} H_a - R_a^{SW} + A_0\,,$$

is the sum of the heat transport divergence div H_a, minus the shortwave radiation balance R_a^{SW}, and the temperature-insensitive part of the longwave radiation balance A_0. Furthermore, B = 2.23 W m^{-2}K^{-1} is the coefficient of the temperature-sensitive part of the longwave radiation balance, $K = 40$ W m^{-2} K^{-1} the air-sea heat exchange coefficient, T_a atmospheric temperature and T_s sea-surface temperature. Our formulation reduces to the choice of Winton (1997) for A if we set

$$A_0 = R_{am}^{SW} - B\, T_{am}$$

and

$$\operatorname{div} H_a = A_1 P_2 + R_a^{SW} - R_{am}^{SW}\,.$$

Here $R_{am}^{SW} = 244.31$ Wm^{-2} is the global mean shortwave radiation balance, and T_{am} the global mean atmospheric temperature in equilibrium. Finally, $P_2(\varphi) = 1/2\,(3\sin^2\varphi - 1)$ is the second Legendre polynomial and $A_1 = 80$ W m^{-2} an adjustable amplitude. The atmospheric energy balance can be solved for the atmospheric temperature,

$$T_a = \frac{(1 - f_l)\, K T_s - A}{(1 - f_l)\, K + B}\,;$$

which in turn can be inserted into the surface heat flux equation

$$F = \frac{KB}{(1-f_l)\,K + B}\,(-A/B - T_s)\,.$$

The term $KB / [(1-f_l)\, K + B]$ can be thought of as an effective thermal damping, and $-A/B$ as an effective restoring temperature. The effective thermal damping is weak and varies between 5 W m^{-2} K^{-1} in the Northern Hemisphere and 2 W m^{-2} K^{-1} south of 30°S, which is much smaller and more realistic than in models using traditional restoring boundary conditions. The land fraction f_l is 1/2 in the Northern Hemisphere, 1/3 between 30°S and the Equator, and 0 south of 30°S.

The surface freshwater flux is given by 0.8 times the "shape-1" freshwater forcing curve defined by Winton (1997). The atmospheric heat transport divergence and the surface freshwater flux are fixed in all experiments and shown in Fig. A1.

In computing the shortwave radiation balance R_a^{SW}, the planetary albedo parameterization of Graves et al. (1993) is used. If the sea ice-albedo feedback is included (Experiment E), the planetary albedo is enhanced by 7% in the presence of sea ice.

Sea ice: Any subfreezing water in the uppermost model layer is immediately heated to the local freezing point with a corresponding increase in salinity (Large et al. 1997). The associated sea ice (more precisely, the negative heat stored in sea ice Q_{ice}) is accumulated locally and advected equatorward. The sea ice advection velocities are based on Harvey (1988). In the Northern Hemisphere, a southward velocity of 0.5 cm s^{-1} is prescribed south of 70°N, with a linear decrease to 0.05 cm s^{-1} at 80°N. Similarly, in the Southern Hemisphere, a northward velocity of 3.0 cm s^{-1} is prescribed north of 60°S, with a linear decrease to 1.0 cm s^{-1} at 70°S. If sufficient heat is supplied, the sea ice is molten, and a corresponding decrease in salinity occurs.

If the insulating effect of sea ice is included (Experiments D and E), the sea-ice surface temperature T_s is calculated from the balance of the conductive heat flux through the ice and the heat flux at the air-sea ice interface

$$\frac{k_{ice}}{h_{ice}}\,(T_s - T_f) = K\,(T_a - T_s)\,,$$

where k_{ice} = 2.16 W m^{-1} K^{-1} is the sea ice conductivity, h_{ice} the sea-ice thickness and T_f the local freezing point. The resulting formula is similar to Eq. 13 of Lohmann and Gerdes (1998):

$$T_s = \frac{k_{ice}\,T_f + K\,h_{ice}\,T_a}{k_{ice} + K\,h_{ice}}\,.$$

The sea ice thickness h_{ice} is diagnosed from the negative heat stored in sea ice Q_{ice} as follows:

$$h_{ice} = \frac{-Q_{ice}}{\rho_{ice}\,L_f}\,,$$

where ρ_{ice} = 905 kg m^{-3} is the sea-ice density and L_f=0.334 × 10^6 J kg^{-1} the latent heat of fusion. The surface freshwater flux is directly added to the uppermost model layer and does not influence sea-ice growth.

Ocean: The oceanic component is closely related to the two-dimensional ocean model of Stocker and Wright (1996). Only a single basin is considered, which is 60° wide, 5000 m deep and extends from 70°S to 90°N.

The parameterization of the zonal density gradient (which is required to compute the zonal pressure gradient) is based on vorticity dynamics (Wright et al. 1995). In the Southern Ocean, between 60°S and 50°S, a 2157 m deep channel is included in which the zonal pressure gradient is set to zero. A fully nonlinear equation of state is used (Mellor 1991) and a convection scheme that completely stabilizes the water column after every time step (Rahmstorf 1996). Furthermore a real freshwater flux boundary condition is imposed, as opposed to a virtual salt flux boundary condition (Huang 1993). For the transport of tracers (here temperature and salinity) an explicit discretization in time is chosen. The time step is 9.125 days. The spatial discretization of the advection terms follows the QUICK scheme, which has low numerical diffusion and dispersion (Farrow and Stevens 1995). The meridional resolution is 5°, and there are 20 layers in the vertical (Tab. A1).

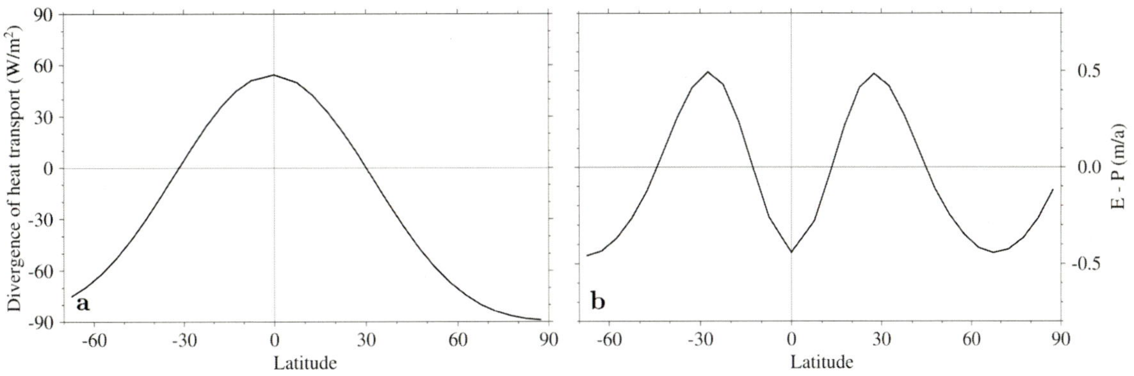

Fig. A1. Components of atmospheric surface fluxes in the "Hanse" model. **(a)** Divergence of atmospheric heat transport (W m^{-2}). **(b)** Surface freshwater flux (m a^{-1}). Positive (negative) values indicate net evaporation (precipitation).

No. of layer	Depth of mid-point (m)	Depth of bottom (m)
1	25.00	50.00
2	75.00	111.97
3	148.94	187.93
4	226.92	280.02
5	333.12	390.44
6	447.76	521.39
7	595.02	675.00
8	754.98	853.21
9	951.44	1057.69
10	1163.94	1289.76
11	1415.58	1550.25
12	1684.92	1839.45
13	1993.98	2157.00
14	2320.02	2501.87
15	2683.72	2872.31
16	3060.90	3265.86
17	3470.82	3679.39
18	3887.96	4109.16
19	4330.36	4550.91
20	4771.46	5000.01

Table A1. Vertical resolution of the "Hanse" model.

Solar Forcing of Climate Change in Recent Millennia

Judith Lean

E.O. Hulburt Center for Space Research, Naval Research Laboratory, Washington, D.C., USA
email: lean@demeter.nrl.navy.mil

Abstract: Solar radiation varies continuously; so too does Earth's global environment. Numerous empirical relations exist among assorted solar and climate parameters. Reliable specification of the true extent of Sun-climate connections requires continuous solar monitoring with sufficiently high stability to specify long-term solar irradiance variability and to characterize spectral irradiance variability at all wavelengths. A suite of cavity-type electrical substitution radiometers on various spacecraft have tracked day-to-day and solar cycle fluctuations in the Sun's total irradiance during the past two decades. Space-based spectroradiometers have also measured concomitant changes in the shorter wavelength ultraviolet spectrum. Primary irradiance variability mechanisms - dark sunspots and bright faculae - have been identified, and models developed that replicate much of the observed variance in the extant irradiance databases. When extended to centennial time scales, the models predict irradiance variability in the range 0.2 to 0.5%. General circulation model simulations are beginning to explore the physical pathways through which the climate system responds to changes in total solar irradiance. These models suggest that surface warming of order 0.5°C may be associated with increasing solar activity since the 17th Century Maunder minimum. Improved knowledge of Sun-climate connections requires future simulations that include an adequate representation of the spectral nature of the Sun's radiative output changes, and of the indirect pathways that solar UV irradiance variability may influence climate via its control of ozone concentrations and the stratosphere. Since the increase in solar activity from anomalously low Maunder Minimum levels to presently high Modern Maximum levels is the most recent of similar episodes apparent throughout the Holocene, improved understanding of both solar forcing and climate response in the past 400 years, will likely contribute new understanding of climate change in past millennia.

Climate Change

Climate change, manifest in the surface temperature of the Earth, occurs on time scales of years, decades, centuries and millennia. Although surface temperatures during the Holocene, recorded by $\delta^{18}O$ in ice-cores (Fig. 1, upper panel) fluctuate less than in the preceding Ice Age, real fluctuations are nevertheless apparent. The Sun's activity also varies over multiple time scales, indicated in Fig. 1 (lower panel) by $\delta^{14}C$ cosmogenic isotopes in tree-rings (Stuiver et al. 1997). Statistical analysis of the $\delta^{14}C$ data in Fig. 1, whose results Fig. 2 shows, indicate that fluctuations in solar activity during past millennia are non-stationary and occur at many periods. Whereas the surface temperature record in Fig. 1 is from a single polar site, Fig. 3 compares reconstructions of global and hemispheric surface temperatures during the last millennium, compiled from combinations of archives in tree-rings, ice-cores, bore-holes and other paleoclimate proxies. Prior to the industrial epoch (i.e. before 1850) fluctuations of a few tenths °C are evident. For comparison in Fig. 4 is the record of solar activity over the same time period. In this figure, the $\delta^{10}Be$ cosmogenic isotopes in ice-cores (Bard et al. 1997) are seen to be consistent with the $\delta^{14}C$ tree-ring data about the nature of solar activity during the past millennium, and recorded by direct sunspot observations more recently.

From WEFER G, BERGER W, BEHRE K-E, JANSEN E (eds), 2002, *Climate Development and History of the North Atlantic Realm.* Springer-Verlag Berlin Heidelberg, pp 75-88

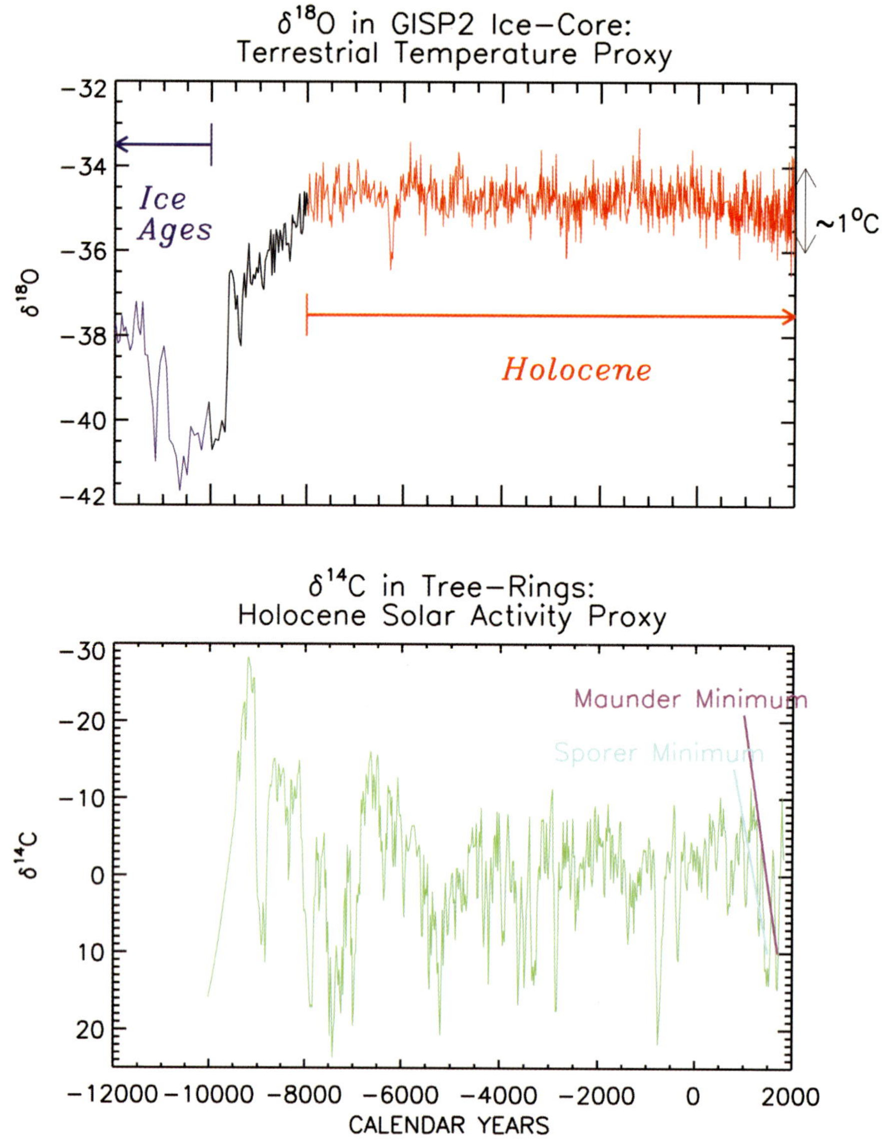

Fig. 1. Compared are variations in $\delta^{18}O$ in ice-cores, a proxy for Earth's surface temperature (upper panel) and in $\delta^{14}C$ in tree-rings, a proxy for solar activity (lower panel).

Potential causes of climate change are manifold. They include anthropogenic forcings by atmospheric gases (CO_2, O_3, N_2O, CH_4, CFCs), aerosols and landuse changes, natural forcings by solar variability and volcanic aerosols, and unforced variability within the climate system itself (ENSO, NAO). The extent of surface temperature change ΔT (°C) resulting from total climate forcing ΔF (Wm^{-2}) depends on the sensitivity ϑ (°C per Wm^{-2}) of the climate system. Climate sensitivity is estimated to be in the range 0.3 to 1°C per Wm^{-2}. When multiple forcings are present, $\Delta T = \Sigma \vartheta_i \Delta F_i$ where ϑ_i for an individual forcing depends on the amplitude, spatial and altitudinal distribution and temporal history of that forcing. Reliable attribution of climate change both prior to and during the

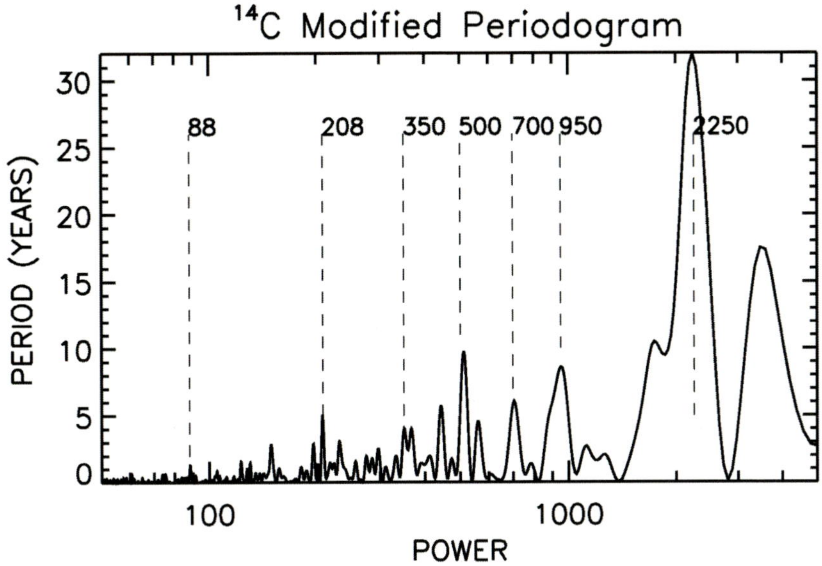

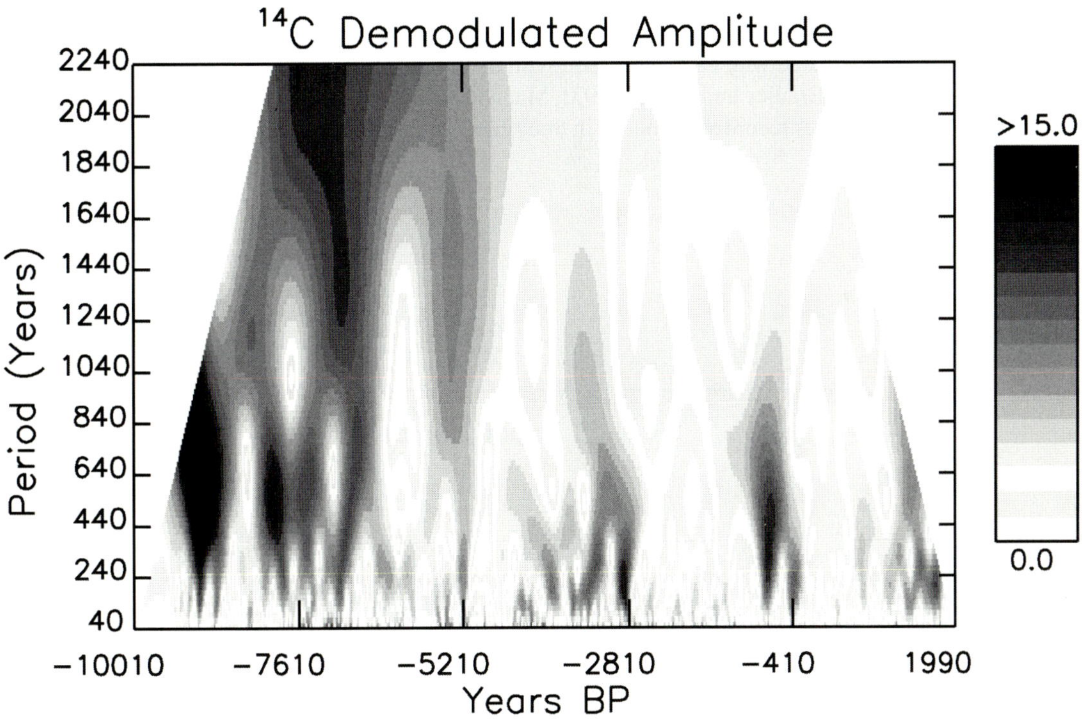

Fig 2. Shown (upper panel) is the spectral density of centennial to millennial scale power in the $\delta^{14}C$ time series in Fig. 1, and the evolution with time during past millennia of power in the range of 40 to 2240 years, determined by complex demodulation of the time series.

industrial era requires reliable specification of all climate forcings, and of the climate sensitivity to each individual forcing. It should be kept in mind that sensitivity to one type of forcing can change as a function of total forcing.

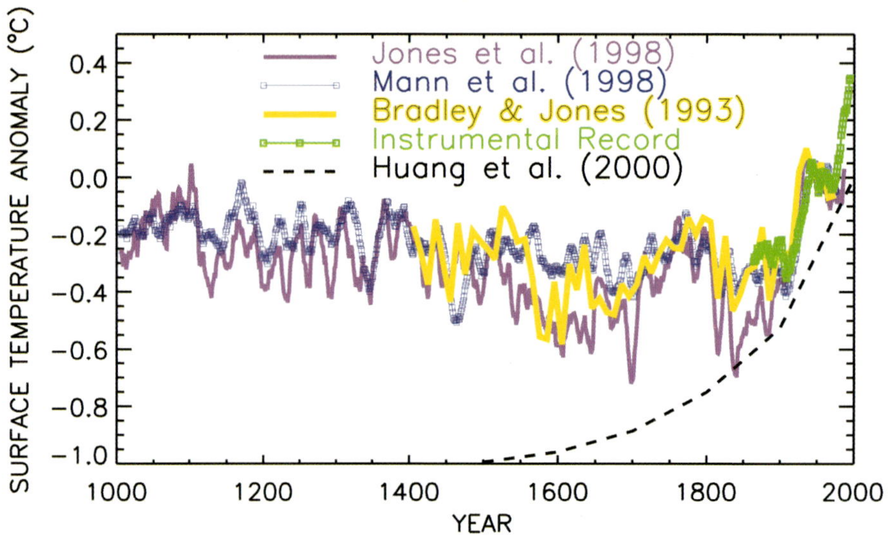

Fig. 3. Reconstructions of Earth's surface temperature anomalies (departures from the 1961-1990 averages) during the past 1000 years are compared with each other, and with the instrumental temperature record (thin line). The reconstructions compiled by Bradley and Jones (1993), Mann et al. (1998) and Jones et al. (1998) utilize various combinations of climate proxies recorded primarily in tree rings, ice-cores and corals. That of Huang et al. (2000) is from borehole records.

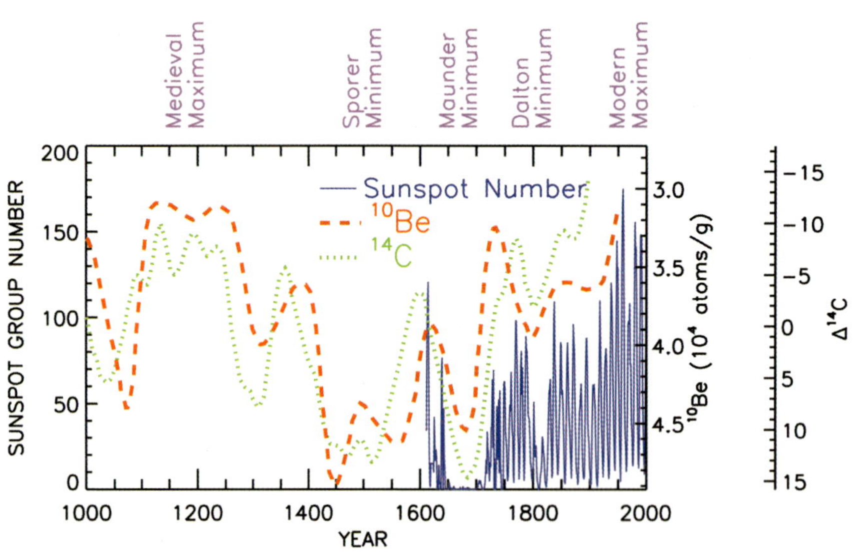

Fig. 4. Shown are decadal to centennial time scale variations in solar activity over the past 1000 years seen in sunspot numbers (thin solid line) and in the δ^{10}Be (dashed line) and δ^{14}C (dotted line) cosmogenic isotopes, which track the envelope of the sunspot numbers during more recent times. Also identified are epochs of overall high solar activity - the Medieval and Modern maxima - and anomalously low solar activity - the Spörer, Maunder and Dalton minima.

Solar Activity and Radiative Climate Forcing

Solar Activity

The Sun has the potential to influence climate because the energy it provides to the Earth (primarily in the form of radiation) varies continuously (Lean 1997). This variability occurs because the Sun, like other stars of comparable age, mass and evolution, is a magnetically active star. The most familiar evidence of the Sun's magnetic activity is the fluctuation in the number of small, compact dark features — called sunspots — on its surface. Sunspots occur frequently during certain epochs (e.g. 1980 and 1990) but are much reduced at other times (e.g. 1986 and 1996). The approximate 11-year cycle in sunspot numbers, shown in Fig. 4, is only one manifestation of solar magnetism. Accompanying the sunspot cycle are fluctuations in a wide range of other solar parameters, including the eruption of flares and coronal mass ejections, radiant energy from X-rays to radio fluxes, and particle energy outputs.

The Sun's activity cycle is constant neither in amplitude nor phase. In particular, the amplitudes of recent cycles of sunspot numbers are seen in Fig. 4 to be some of the largest of the past 400 years. Solar activity is presently at a high overall level relative to the anomalously low levels from 1645 to 1715, when sunspots were absent entirely from the Sun's disk for years at a time. During this period of low solar activity, the Maunder Minimum, levels of $\delta^{14}C$ and $\delta^{10}Be$ cosmogenic isotopes in, respectively, tree rings and ice cores are seen in Fig. 4 to be elevated. Low solar activity weakens the magnetic coupling of the Sun and the Earth, permitting enhanced fluxes of galactic cosmic rays to enter the atmosphere where they produce isotopes such as ^{14}C and ^{10}Be. Contemporary solar activity is also at high overall levels in comparison with other stars of similar mass and age. This may suggest that the overall level of solar activity is more likely to decrease, rather than increase, in the future.

Observed Solar Radiative Forcing

The Sun's radiation varies in concert with the 11-year "sunspot" activity cycle. Solar irradiance increases when solar activity is high, relative to times of solar minima. Evident in the composite space based record of the total (i.e. spectrally integrated) solar irradiance, shown in Fig. 5, is an increase of about 0.1% during activity maxima in 1980 and 1990 relative to the 1986 and 1996 minima. This observed total irradiance change of 1.3 Wm^{-2} at the top of the Earth's atmosphere is equivalent to an equilibrium radiative climate forcing change of 0.22 Wm^{-2}. Prior to the space era, dependency of the Sun's total irradiance on solar activity was not well documented hence the term "solar constant" - because a century of ground based monitoring had failed to detect its true variability.

Fluctuations in the spectrum of solar radiation, shown in Fig. 6, have yet to be defined observationally at most wavelengths. The exception is in the UV spectrum where changes of 1.1% and 0.25%, respectively, in the 200-300 nm and 300-400 nm wavelength bands are estimated to contribute 13% and 18% (a total of 31%) of solar cycle total irradiance variations. Radiation in the visible and near infrared spectrum (400 to 1000 nm) is estimated to vary in the range 0.07% to 0.11%, which is comparable to the variability of total irradiance.

The epoch of direct measurements of solar irradiance from space is too short, by far, to permit unambiguous detection of solar forcing changes on climatically relevant multi-decadal and centennial time scales. Inaccuracies and sensitivity drifts in the solar radiometers preclude reliable detection of multi-decadal scale irradiance fluctuations in the space-based observations, which cover less than two 11-year cycles. Depending on the interpretation of solar versus instrumental components in the extant data records, long-term irradiance fluctu-ations may be negligible over the decade from 1986 to 1996, or as much as 0.04%. That models of irradiance variability predict similar irradiance levels during the 1986 and 1996 minima suggests that claims of a 0.04% change may be spurious. The variability models, developed from

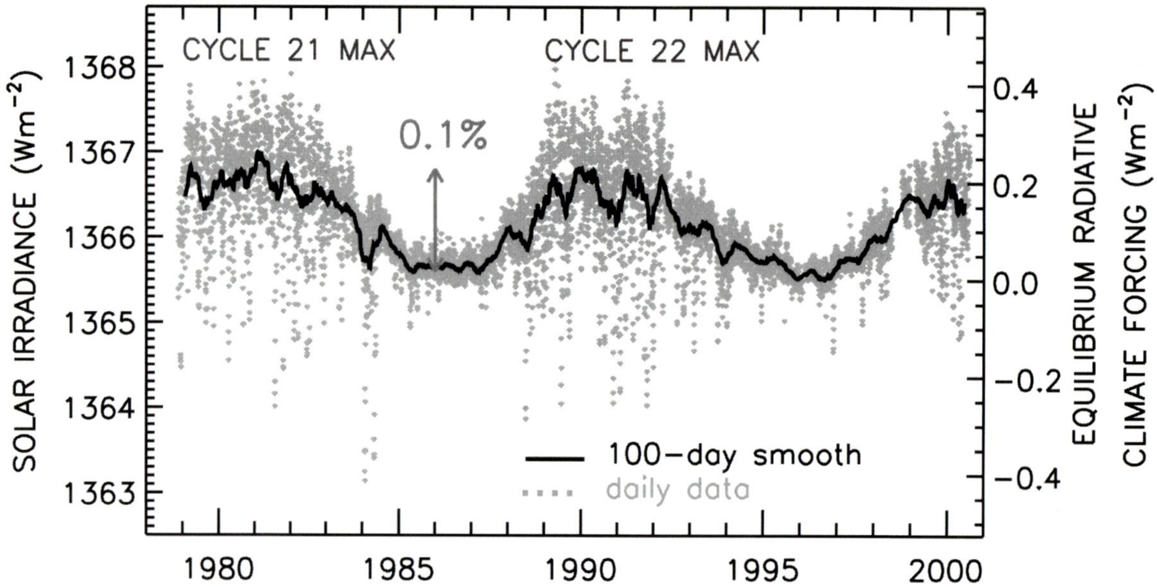

Fig. 5. Shown is a composite record of total solar irradiance compiled by Fröhlich and Lean (1998) from detailed cross-calibrations of various space-based radiometric measurements from the end of 1978 to the present, and adjusted to the absolute scale of the Space Absolute Radiometric Reference.

magnetic features in the solar atmosphere — dark sunspots and bright faculae and plages — account for 88% of the variance in the composite total irradiance record in Fig. 5.

Speculated Historical Solar Radiative Forcing

In lieu of direct space-based solar observations, historical data such as sunspots, geomagnetism and cosmogenic isotopes that pertain to various aspects of solar magnetism are utilized to infer the extent of long-term solar irradiance variability. Relating these activity proxies to observed solar irradiance variability, and within the broader context afforded by monitoring of Sun-like stars, yields long-term solar irradiance reconstructions such as depicted in Fig. 7. Although explicit physical connections between solar irradiance and activity proxies have yet to be adequately deduced over longer time scales, the empirical historical reconstructions suggest that solar irradiance was reduced in the mid 1600s relative to the present period, analogous to the reduced overall flux levels present in non-cycling stars (like the Maunder Minimum Sun) relative to their cycling counterparts (the Modern Sun?). Estimates of the total irradiance decrease are uncertain, and range from 0.2 to 0.5%. The temporal footprint of solar forcing change over this 350 year period is also uncertain because solar activity proxies exhibit somewhat different temporal fluctuations. For example, using the length of the 11-year sunspot cycle as an irradiance proxy predicts solar forcing fluctuations that are at times out of phase by as much as 20 years with a proxy based on sunspot cycle amplitude, as comparison of two different total irradiance reconstructions in Fig. 7 shows. The present lack of understanding of the mechanisms for long-term solar irradiance variability inhibits improved definition of historical solar forcing changes.

Empirical Sun-Climate Connections

Numerous empirical and statistical similarities exist between historical solar and climate data, on time scales that include 11-, 22-, 80- and 200- year periodicities, and longer (Hoyt and Schatten 1997). Land and ocean surface temperatures, US drought, rainfall, forest fires, cyclones, cloud cover, tropo-

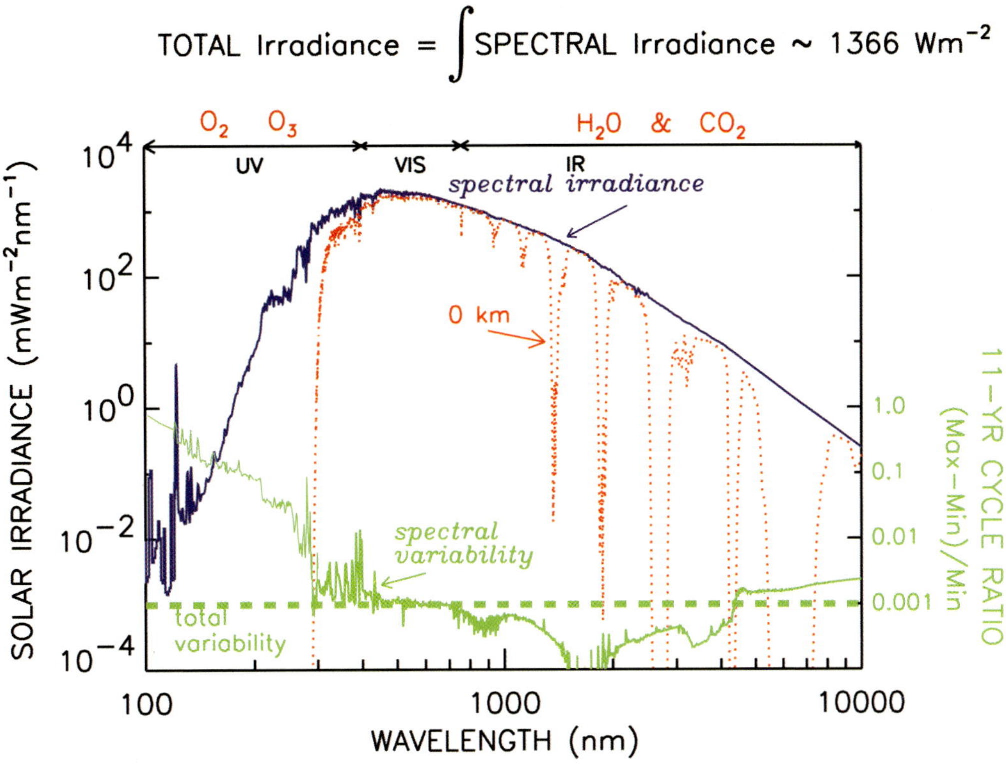

Fig. 6. Shown is the spectrum of the Sun's radiation incident on top of the Earth's atmosphere (dark solid line), and at the Earth's surface (0 km, dotted line). Various atmospheric constituents absorb part of the Sun's radiation, especially in the ultraviolet spectrum, and in some near infrared bands. The total solar irradiance is the integral over all wavelengths of the spectrum at the top of the atmosphere. It has a nominal value of 1366 Wm^{-2}. Also shown are variations of the solar spectrum estimated for a recent 11-year activity cycle (light solid line). At wavelengths longer than 400 nm, these estimates, by Solanki and Unruh (1998), are speculative because no observational data exist. The dashed line indicates the variation in the spectrally integrated (i.e. total) solar irradiance solar cycle during recent 11-year activity cycles.

spheric temperatures, ice-core $\delta^{18}O$ and dust layer thickness, and tropical corals all exhibit various of these cycles during some portion of the recent past. Centennial changes in both the instrumental and proxy climate records have been correlated with long-term solar activity changes during the recent Holocene. Many reported Sun-climate similarities occur in local geographical regions but, importantly, connections encompass global and NH surface temperatures as well.

The most recent of the centennial Sun-climate associations is the parallel increase in surface temperatures since the coldest portion of the Little Ice Age with the increase in solar activity since the end of the Dalton Minimum. From the mid seventeenth century (end of Maunder Minimum) to the present (Lean and Rind 1998, 1999), Earth's surface temperature increased in the range of 0.6 to 1°C, coincident with a solar irradiance increase estimated to be in the range of 0.2 to 0.5%. This apparent Sun-climate connection exists separately in each of the three different ocean basins. However, the empirical relationship that connects common solar and climate variations in the pre industrial era differs from that in the twentieth century. From 1650 to 1790, Fig. 8 shows a northern hemisphere surface temperature increase of ΔT=0.3°C, corresponding to ΔF_{sun}=0.35 Wm^{-2}.

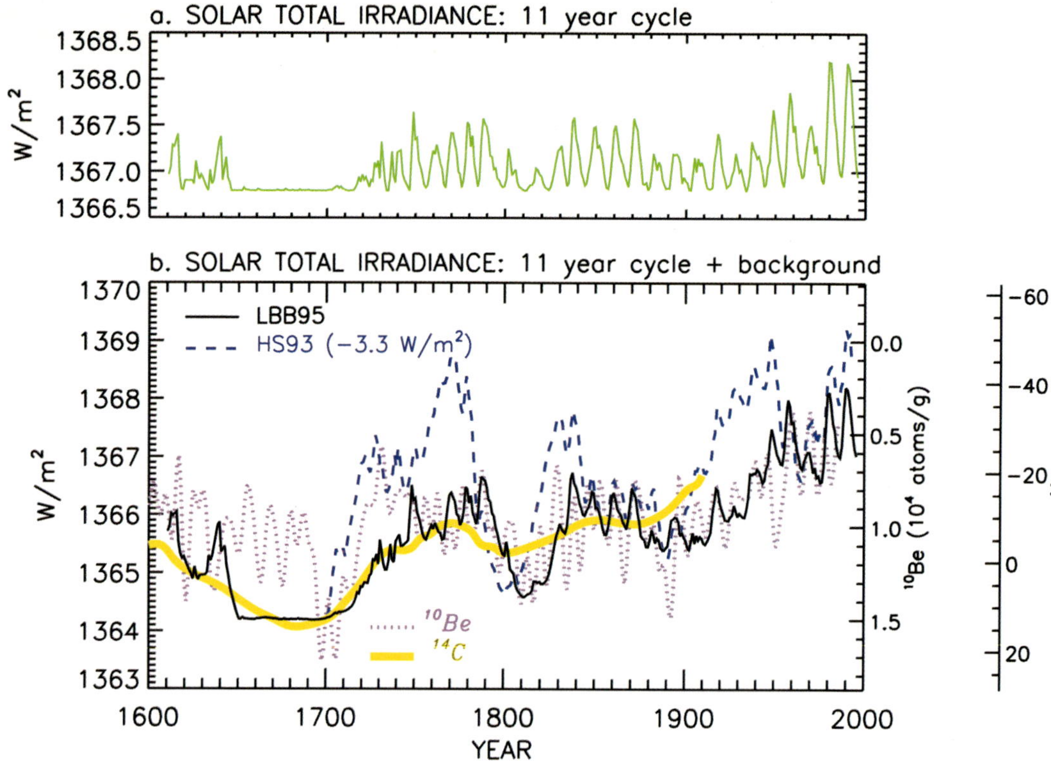

Fig. 7. Shown are reconstructions the Sun's total irradiance, assuming in **(a)** an 11-year cycle alone and in **(b)** the 11-year activity cycle and a longer term component. In the LBB95 reconstruction (dark solid line) by Lean et al. (1995) the longer term component is based on the average amplitude of each sunspot cycle. It is compared with δ^{10}Be (small symbols) and δ^{14}C (thick line) cosmogenic isotope records and with another irradiance reconstruction (dashed line), by Hoyt and Schatten (1993). The HS93 model assumes that longer term irradiance changes track the length of the 11-year solar activity cycle (rather than its average amplitude).

Since 1850, however, ΔT=0.6°C and ΔF_{sun}=0.38 Wm^{-2}. Although the overall twentieth century increase in both surface warming and solar activity has lead to claims that solar forcing can, alone, explain recent climate change, the pre-industrial Sun-climate relationship of 0.85°C per Wm^{-2} con-strains this to only about half of the industrial warming. As Fig. 8 depicts, a plausible interpretation of climate change in the past four hundred years includes a notable solar influence from 1600 to 1800, a period of cool temperatures as a result of signif-icant volcanic activity in the nineteen century, followed by twentieth century warming attributable in part to solar activity, but primarily to anthro-pogenic influences, especially during the past three decades. From 1976 to 1996, annual mean surface temperatures are seen in Fig. 9 to increase 0.2°C even though the levels of solar irradiance were similar in 1976 and 1996, according to the space-based record of total solar irradiance in Fig. 5, whose annual means are shown in Fig. 9.

While the overall climate warming from 1976 to 1996 is apparently not the result of solar variability (based on the comparisons in Fig. 9), adjusting the surface temperatures for volcanic cooling in 1982 and 1992, does reveal a decadal signal that may be solar-related. More generally, climate parameters frequently exhibit decadal variability. An example, in Fig. 10, is the association of a decadal mode of surface temperature with solar activity during the three strong 11-year cycles since the 1950s. The peak-to-peak amplitude of the decadal climate component is roughly 0.1°C, in phase with an 11-year solar cycle radiative forcing of

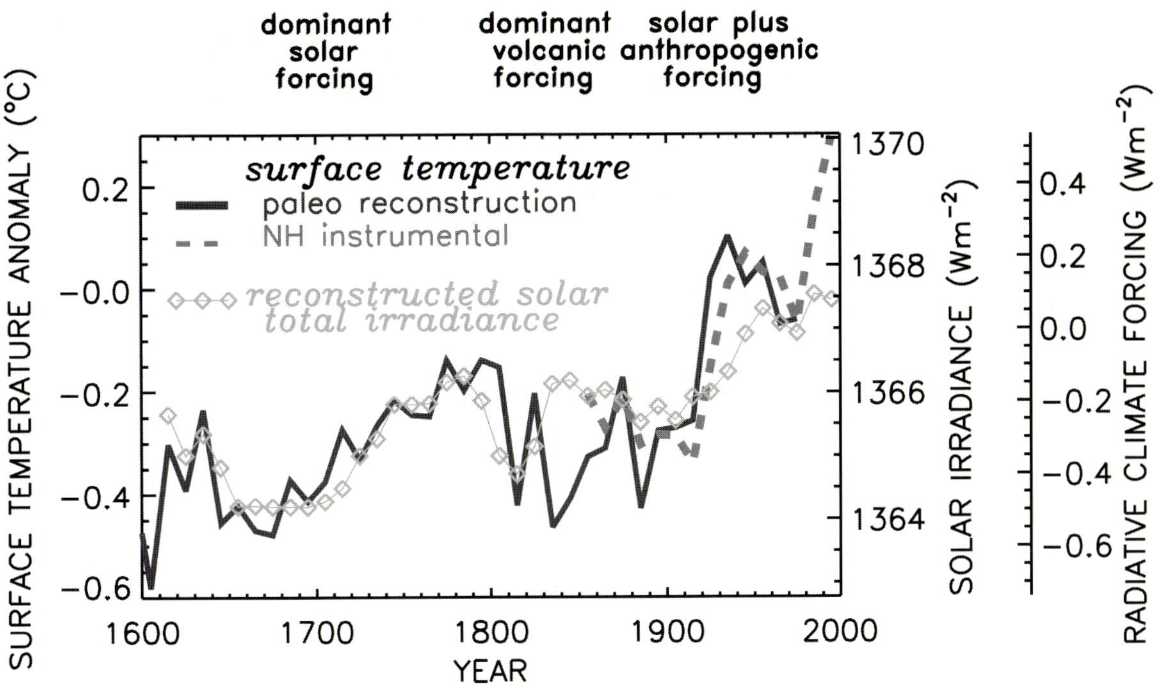

Fig. 8. Compared are decadally averaged values of reconstructed total solar irradiance (LBB95, in Fig. 7) and NH summer temperature anomalies from 1610 to the present. The solid line is the Bradley and Jones NH summer surface temperature reconstruction from paleoclimate data (primarily tree rings), as shown in Fig. 3 scaled to match the NH instrumental data (IPCC 1992) (dashed line) during the overlap period.

0.2 Wm^{-2}. Like the centennial scale variability, this decadal climate change is present not only globally but separately in each of the three different ocean basins, and is detected, as well, from independent wavelet analysis of the surface temperature record.

The extent of climate change that is actually attributable to solar forcing remains uncertain in part because the common solar and climate cycles are typically neither stationary, nor deterministic, nor evident in all climate parameters at all times, nor always global, nor phase locked with each other. This lack of rigorous statistical associations leads climate researchers to often reject empirical evidence for solar effects on climate. Yet climate cycles in general (including the now ubiquitous ENSO), are neither stationary nor phase stable nor even linearly related to the assumed forcings, whatever their origins. The abundance of solar-related cycles and correlations in numerous climate records cautions against premature dismissal of this large body of empirical evidence as coincidental instead of a real Sun-climate connection, and motivates better specification of the possible physical processes.

Simulated Climate Response to Solar Radiative Forcing

Predicting the climate system's response to natural and anthropogenic influences requires simulations by models of physical processes that involve water vapor, sea ice and clouds. The models range in complexity from simple energy balance models to coupled atmosphere ocean general circulation models (AO-GCM), that have varying degrees of spatial and altitudinal resolution (Cubasch et al. 1997; Rind et al. 1999).

Time dependent simulations with the Goddard Institute for Space Studies (GISS) atmospheric GCM, shown in Fig. 11, predict a global surface warming of 0.45°C in response to an adopted 0.25%

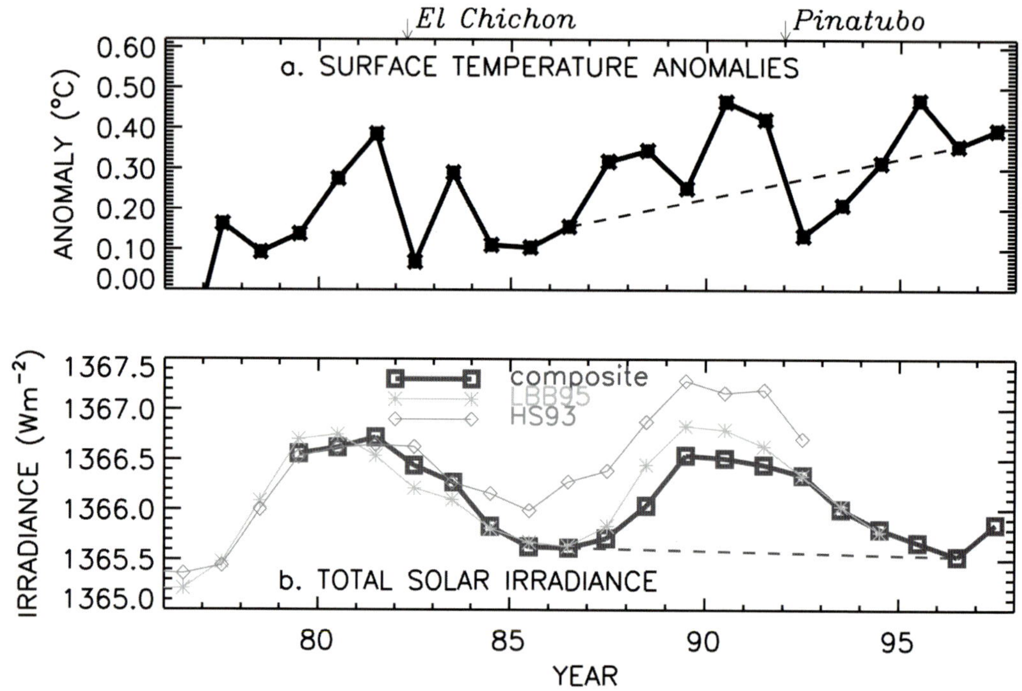

Fig. 9. Globally averaged annual mean surface temperatures compiled by the Goddard Institute for Space Studies (GISS) and shown in the upper panel, exhibit an overall increase of more than 0.2°C from 1986 to 1996, indicated by the dashed line. In contrast, the annual mean values of the composite total irradiance record shown in the lower panel, from Fig. 5, indicate similar levels over this decade. Compared in the lower panel with the annual mean composite irradiance of record are two different irradiance reconstructions, LBB95 and HS93 (from Fig. 7), that have been used elsewhere to investigate historical solar forcing of climate change.

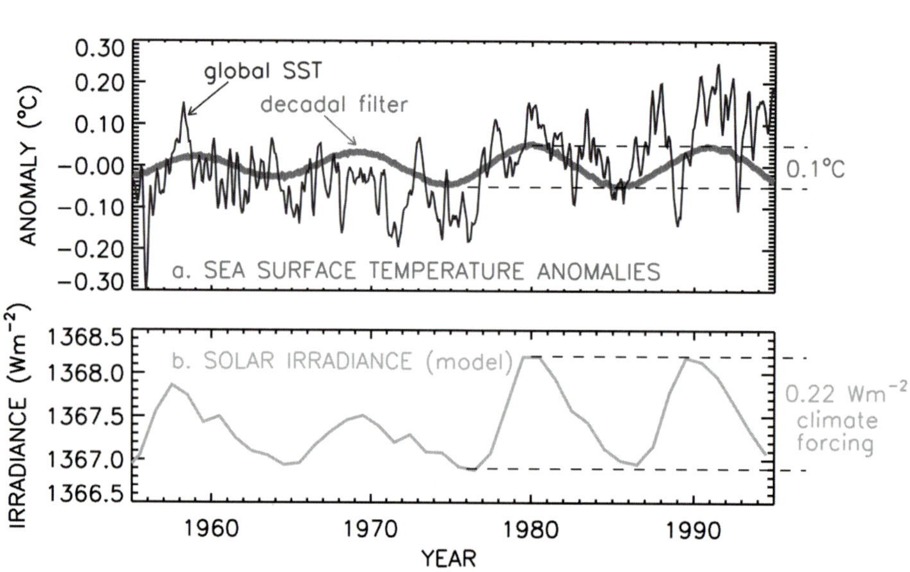

Fig. 10. White et al. (1997) identified in global averaged sea surface temperature anomalies compiled from bathythermographs (BT), shown in **(a)**, significant annual and interannual variability as well as a decadal component that tracks the LBB95 (from Fig. 7) solar irradiance reconstruction shown in **(b)**.

increase in the Sun's total irradiance since the Maunder Minimum (as indicated in Fig. 7). The GISS model has 9 vertical atmospheric layers, 8°x10° horizontal resolution, and a mixed layer ocean of variable depth with heat transport by specified "Q" fluxes. Its sensitivity is in the range 0.8 to 1°C per Wm^{-2}, as a result of known climate feedbacks, whose response to solar irradiance forcing are shown in Fig. 12. Feedbacks by water vapor, cloud cover and sea-ice/snow achieve their full equilibrium response on centennial time scales, and account for a total of 65% (respectively, 35%, 20% and 10%) of the 0.45°C surface warming simulated in response to solar radiative forcing since 1650. The average of four model simulations shown in Fig. 11 supports the inferences from the pre-industrial empirical relationship in Fig. 8 that solar variability potentially accounts for most of the surface temperature changes in the pre-industrial epoch from 1600-1800, about half of the warming from 1850 to 1970, but very little of the long-term warming since then.

On decadal time scales climate model feedbacks are unable to achieve their full equi-librium response, in part because of the large thermal inertia of the oceans. As a result, the GCM surface temperature simulations do not show a clear response to solar forcing during the 11-year solar cycle, and they cannot replicate the empirical relationships between decadal solar and global ocean surface temperature fluctuations evident in Fig. 10. One explanation for this disagreement is that an internal oscillation within the ocean-atmosphere system, rather than solar forcing, is the source of decadal climate variability. Another explanation may lie in the inability of present climate models to simulate rapid climate response to solar variability because of a lack of processes that amplify internal decadal oscillations by forcing of a similar period.

Climate models thus far have simulated only those processes that respond to direct forcing by changes in the bulk radiation of the Sun, without proper spectral differentiation of the variability.

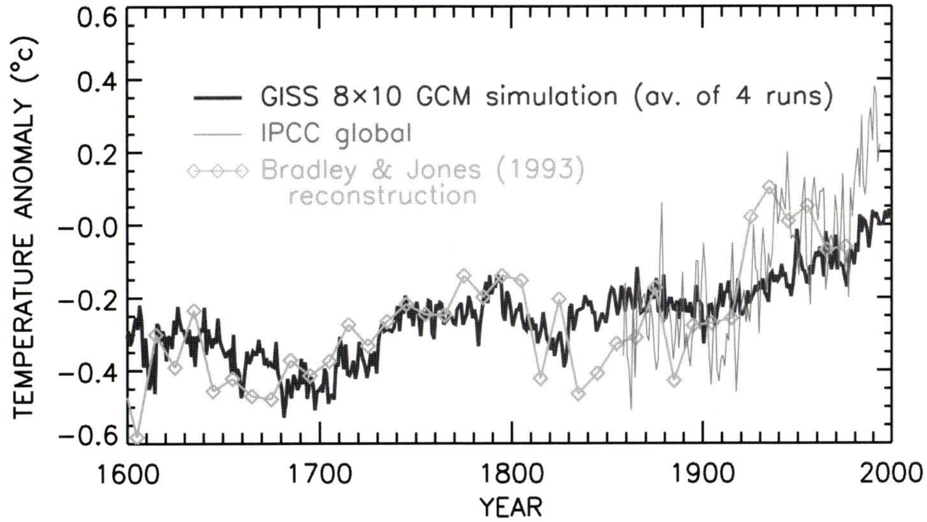

Fig. 11. Shown is the average of four GISS GCM simulations of the global surface temperature response to the solar radiative forcing in Fig. 7 (dark solid line), compared with the IPCC observed surface temperatures and reconstructed surface temperatures from Fig. 3. Rind et al. (1999) provide further details.

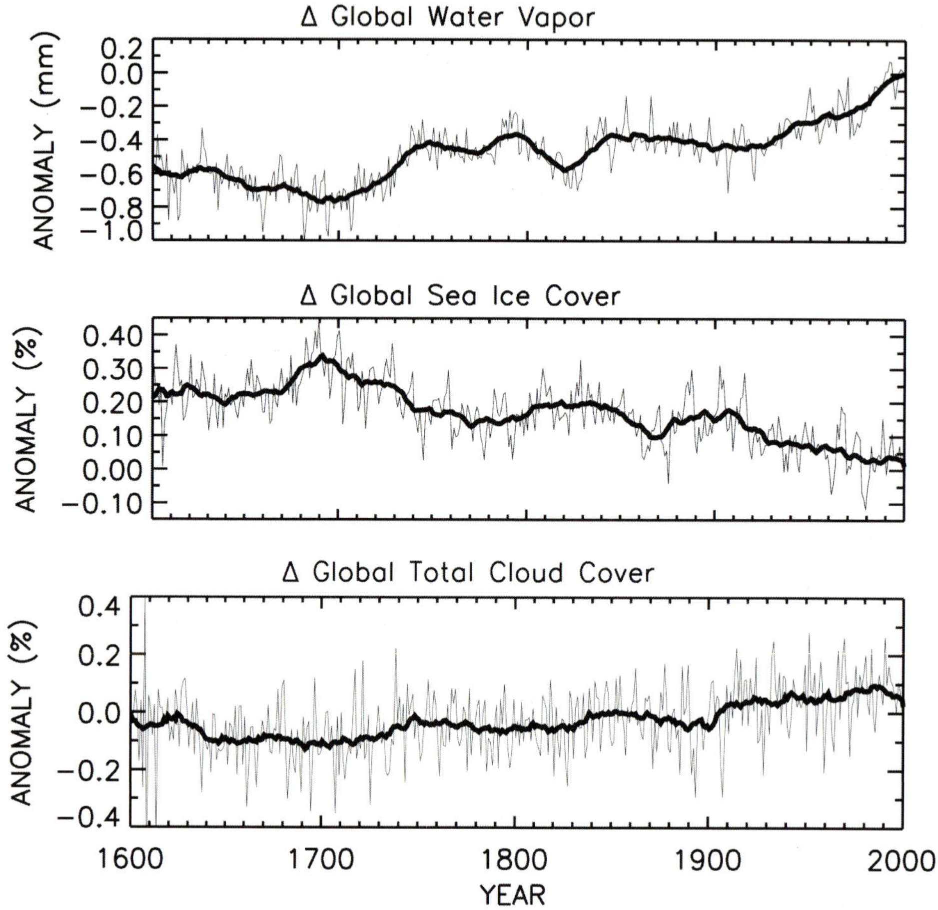

Fig. 12. Climate system feedbacks are shown in response to a 0.25% change in solar radiative forcing. Changes in global water vapor (upper panel), sea ice cover (middle panel) and total cloud cover (lower panel) account for 65% of the corresponding 0.45°C global surface temperature change simulated by the GISS GCM, shown in Fig. 11. Rind et al. (1999) provide further details.

The models also lack interactive pathways through which solar variability may influence climate indirectly, such as by changing the middle atmosphere and ozone concentrations, and by modulation of energetic particles in the atmosphere. Solar cycle changes of a few percent occur in the total ozone column and significant 11-year cycles occur in upper troposphere and lower stratosphere temperatures and pressures. In modifying the heating and dynamics of the atmosphere, these solar-induced changes affect the propagation of planetary waves, radiative forcing of the troposphere, and strength of the Hadley cell circulation (Shindell et al. 1999). The extent of the indirect solar-induced climate change depends strongly on the altitudinal profile of the ozone changes, which are as yet poorly specified. Simulations with middle atmosphere models have identified plausible pathways for this indirect climate effect of solar UV modulation of the atmosphere, although sensitivities in the models are smaller than the observational relationships infer.

Summary and Future Research

Changes in the Sun's radiation, clearly evident in 20 years of space based solar monitoring, are a plausible mechanism of climate change. Model simulations indicate that a Sun-climate connection can account for a significant fraction (0.45°C) of global surface warming since 1650 by adopting magnitudes of solar forcing (0.25%), climate sensitivity (0.8°C per Wm^{-2}) and surface temperature changes (0.7°C) that are all compatible with present understanding of these values. But the attribution of about 0.25°C of the observed 0.6°C surface temperature increases since 1900 to direct solar radiative forcing requires that other forcings of order 2 Wm^{-2} - such as by direct and indirect aerosol cooling and albedo changes, for example - be invoked to offset the warming by greenhouse gases. Generalizing the association of the Spörer and Maunder minima and the coldest portion of the Little Ice Age as a paradigm for similar Sun-climate connections during the past millennium suggests that surface temperature changes associated with solar variability are of order 0.5°C.

Because the duration of space-based solar monitoring is only two decades, it is not yet known whether long term changes actually occur in solar forcing, such as the speculated increase since the seventeenth century. Detecting such changes, if they exist, is a future challenge requiring long term solar monitoring with exceedingly high radiometric precision and stability, by, for example, the programs identified in Fig. 13. Since the processes by which solar forcing impacts terrestrial global change are highly wavelength dependent, equally challenging will be the determination of solar spectrum variability, which is essentially unknown

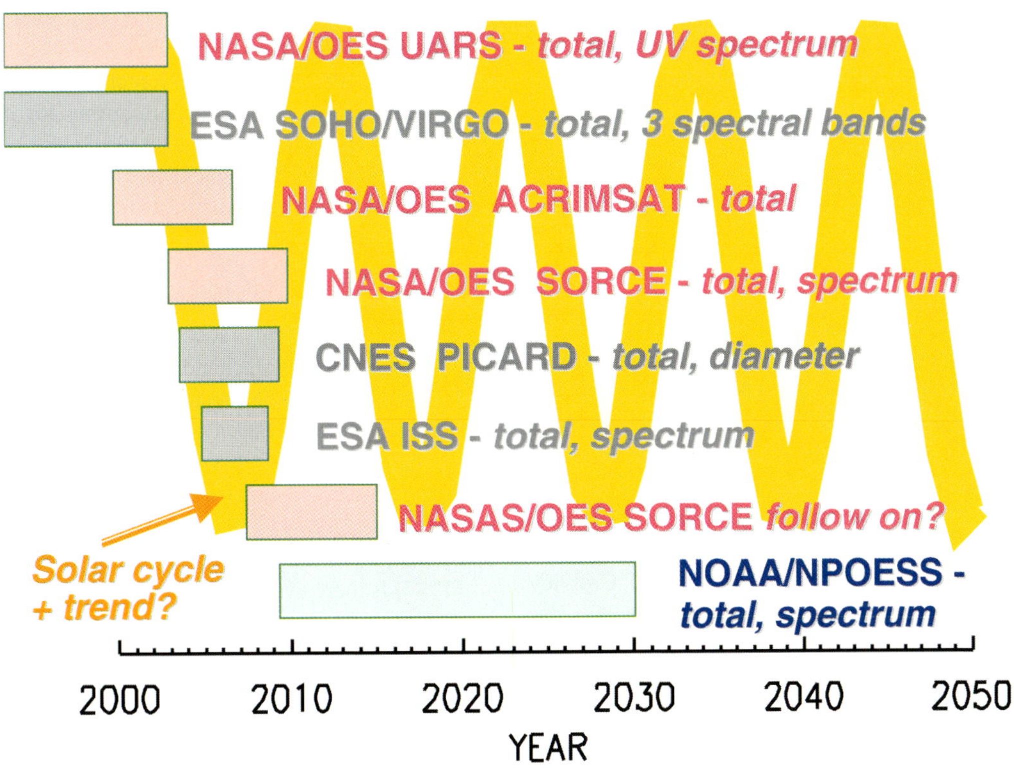

Fig. 13. A suite of radiometers on a variety of space platforms has monitored the Sun's radiation from space since the late 1970s. A critical aspect of the space-based data is the overlap of successive missions. Sufficient overlap is essential to provide the cross-calibration of the radiometers necessary to construct a composite record of solar irradiance, such as that shown in Fig. 5. Presently, radiometers on the Upper Atmosphere Research Satellite (UARS) and the Solar and Heliospheric Observatory (SOHO) are maintaining the total solar irradiance database. NASA, ESA and France (CNES) all plan future solar irradiance monitoring at the beginning of the next century. Not yet identified are missions in the 2006 to 2010 time frame that will be needed to connect the extant solar monitoring with operational monitoring by NPOESS.

because of the lack of monitoring at all but the shortest UV wavelengths.

Not known at all, yet, is the mutual impact of the various forcings on the ability of the climate to respond to each - so that, for example, might solar forcing in the pre-industrial epoch have impacted climate differently than in the present era of large greenhouse gas and aerosol concentrations? The non-linear and chaotic character of the climate system suggest that such a scenario is possible. Also lacking is adequate specification of indirect solar forcings via UV modulation of ozone and the middle atmosphere and by modulation of energetic particles in the atmosphere that affect cloud formation processes. Pathways that facilitate indirect solar UV radiative impacts on climate are known but sensitivities in existing models are smaller than empirical evidence suggests. In the third case, both the physical pathways and the sensitivities need better specification.

Acknowledgments

NASA funded this work.

References

Bard E, Raisbeck GM, Yiou F, Jouzel J (1997) Solar modulation of cosmogenic nuclide production over the last millennium: Comparison between ^{14}C and ^{10}Be records. Earth and Planet Sci Lett 150:453-462

Bradley RS, Jones PD (1993) "Little Ice Age" summer temperature variations: their nature and relevence to recent global warming trends. The Holocene 3(4):367-376

Cubasch U, Voss R, Hegerl GC, Waszkewitz J, Crowley T (1997) Simulation of the influence of solar radiation variations on the global climate with an ocean-atmosphere general circulation model. Clim Dyn 13:757-767

Fröhlich C, Lean J (1998) The Sun's total irradiance: Cycles, trends and climate change uncertainties since 1976. Geophys Res Lett 25:4377-4380

Hoyt DV, Schatten KH (1993) A discussion of plausible solar irradiance variations, 1700-1992. J. Geophys. Res. 98:18895-18906

Hoyt DV, Schatten KH (1997) The Role of the Sun in Climate Change. Oxford Univ. Press NY

Huang S, Pollack HN, Shen P-Y (2000) Temperature trends over the past five centuries reconstructed from borehole temperatures. Nature 403:756-758

Intergovernmental Panel on Climate Change, Climate Change 1992 (1992) The supplementary report to the IPCC scientific assessment. Houghton JT, Callander BA, Varney SK (eds), Cambridge University Press

Jones PD, Briffa KR, Barnett TP, Tett SFB (1998) High-resolution paleoclimatic records for the last millennium: interpretation, integration and comparison with General Circulation Model control-run temperatures. The Holocene 8(4): 455-471

Lean J (1997) The sun's radiation and its relevance for Earth. Annu Rev Astron Astrophys 35:33-67

Lean J, Rind D (1998) Climate forcing by changing solar radiation. J Climate 11:3069-3094

Lean J, Rind D (1999) Evaluating Sun-climate relationships since the Little Ice Age. J Atmos Solar Terrest Phys 61:25-36

Lean J, Beer J, Bradley R (1995) Reconstruction of solar irradiance since 1610: Implications for climate change. Geophys Res Lett 22:3195-3198

Mann ME, Bradley RS, Hughes MK (1998) Global-scale temperature patterns and climate forcing over the past six centuries (1998) Nature 392:779-787

Rind D, Lean J, Healy R (1999) Simulated time-dependent climate response to solar radiative forcing since 1600. J Geophys Res 104:1973-1990

Shindell D, Rind D, Balachandran N, Lean J, Lonergan P (1999) Solar cycle variability, ozone and climate. Science 284:305-308

Solanki SK, Unruh YC (1998) A model of the wavelength dependence of solar irradiance variations. Astron Astrophys 329:747-753

Stuiver M, Braziunas TF, Grootes PM, Zielinski GA (1997) Is there evidence for solar forcing of climate in the GISP2 Oxygen isotope record? Quaternary Res 48:259-266

White WB, Lean J, Cayan DR, Dettinger MD (1997) Response of global upper ocean temperature to changing solar irradiance. J Geophys Res 102:3255-3266

Times of Quiet, Times of Agitation: Sverdrup's Conjecture and the Bermuda Coral Record

Wolfgang H. Berger*[1], Jürgen Pätzold[2] and Gerold Wefer[2]

[1]*University of California, California Space Institute, La Jolla, CA 92093-0524, USA*
[2]*University of Bremen, Department of Geosciences, Postfach 33 04 40, D-28334 Bremen, Germany*
**corresponding author (e-mail): wberger@ucsd.edu*

Abstract: The distinguished Norwegian oceanographer Harald U. Sverdrup (1888-1957) proposed, in 1918, that an increased amount of energy in the system, during periods of high sunspot activity, results in increased strength of winds in the North Atlantic. Changes in wind strength and mixing in the North Atlantic should affect coral growth on Bermuda. We test this proposition with an 800-year record of *Montastrea cavernosa* from this island, comparing the cycles found within the series of growth bands (that is, rates of growth) with the cycles found in the sunspot record. The main lines in the solar record, for the last 300 years, are as follows: 11.2, 9.9, 10.6, 12.0, and 9.3 (in the order of strength). In the coral record, strong periods appear near 37.5, 30.8, 27.0, 21.7, 7.5 and 6.0 years. There is no peak near 11 years. However, when testing for predictability (that is, deviation from expectation by forward extrapolation of thickness of growth bands), periods appear (in the order of amplitude) at 30, 10.5, 21, 9.1, 33.6, 55, 7.2, 80 and 6.2 years. All can conceivably be interpreted as whole-number sunspot multiples (3 x 9.9; 1 x 10.6; 2 x 10.6; 1 x 9.3; 3 x 11.2; 5 x 11.2; 2/3 x 11.2; 7 x 11.2 or 8 x 9.9; 2/3 x 9.3) thus giving support to Sverdrup's conjecture. However, other factors also must be considered, including internal oscillations and tidal cycles.

Introduction

The idea that solar cycles influence climate has met with varying degrees of acceptance over the last century, strong endorsement alternating with stern rejection (see reviews in Brooks 1949; Eddy 1981). More recently, the possibility of a solar component in climate variability has (again) emerged in a serious manner (Anderson 1992; Labitzke and Van Loon 1993; Frenzel et al. 1995; Hoyt and Schatten 1997; Lean 1997). Early in the 20th century, the distinguished Norwegian meteorologist and oceanographer Harald U. Sverdup (1888-1957) (lead author of the first comprehensive textbook in oceanography), stated that climatic changes in the North Atlantic (and elsewhere) are periodic, and that an eleven-year period had been recognized and successfully correlated with the corresponding sunspot cycle (Sverdrup 1918). He suggested that large-scale air-pressure gradients are the main factor whose variation can be tied to the sunspot cycle: gradients are large during sunspot maxima and small during sunspot minima.

Thus, Sverdrup focused on the energy available for moving air masses (parallel to ideas expressed somewhat earlier by Sir Gilbert Walker, and by Bjoern Helland-Hansen and Fridtjof Nansen). According to Sverdrup, “the newest measurements show that the solar constant is relatively large during sunspot maxima and small during sunspot minima, that is, the energy received by the sun varies as the number of sunspots”. While such a direct relationship was subsequently denied by some, recent satellite data support the correlation (see Roederer 1995). Sverdrup argued that relatively modest periodic pulsing in the delivery of energy from the sun to the “climate machine” would introduce periodic changes in the rates of heat transport and circula-

From WEFER G, BERGER W, BEHRE K-E, JANSEN E (eds), 2002, *Climate Development and History of the North Atlantic Realm.* Springer-Verlag Berlin Heidelberg, pp 89-99

tion, which would in turn introduce periodic changes into all meteorologic phenomena. He called for analysis of periods both shorter and longer than the eleven-year cycle, surmising that it would "not be improbable", "that most of these periods would also be evident in the solar constant and thus be explainable in the same fashion".

We define "Sverdrup's conjecture" to mean that solar output is tied to sunspots and that it periodically changes the energy in the climate system (especially the strength of winds). Corals in Bermuda are at the limit of distribution of Caribbean tropical species; only a few species survive here. The growth record of such corals, therefore, should be highly sensitive to modest change in conditions involving temperature and nutrient supply. An 800-year record of annual growth of the massive coral *Montastrea cavernosa* is available for analysis (Pätzold et al. 1999). In the following, we search for evidence that this growth record is influenced by solar activity cycles. We show that there is indeed evidence for the presence of solar cycles in the change-of-growth record (rather than in the rate of growth itself). This suggests that, periodically, the environment of growth is changed suddenly by subtle changes in solar output tied to sunspot activity. On a long time scale, extremes in growth rate are tied to extremes in abundance of carbon-14 and beryllium-10 in the atmosphere. The abundance of these elements has been tied to solar activity (although other explanations are conceivable).

The Coral Record

Coral material was recovered from the Bermuda Platform, which is located at 32°20'N and 64°50'W in the North Atlantic. It has an area of approximately 55 square km consisting of about 120 individual islands. Positioned in the northwestern region of the Sargasso Sea, Bermuda is bathed by warm water brought with the Gulf Stream system.

The climate record presented in Pätzold et al. (1999) and analyzed here is based on a three-meter high colony of *Montastrea cavernosa*, a massive coral exhibiting a columnar growth habit. The material was collected on the Bermuda Platform at the North East Breakers in August 1987 from a water depth of 11 meters. Slices of the cores 5-10 cm wide and 7-9 mm thick were sawed out parallel to the growth direction of colony members, and subsequently X-rayed. The X-radiographs show annual growth bands in the calcareous skeletons, produced by density differences between summer and winter. Each pair of growth bands, one dense and the other less dense, represents a one-year period.

A precise time scale (sclerochronology) can be produced by counting the annual layers. To verify the uniformity of deposition of the individual layers and to determine an accurate age for the various density layers, samples for oxygen isotope composition analyses (for determination of the ratio between ^{18}O and ^{16}O) were taken using a precision drill. Up to 20 samples per annual layer were retrieved, representing a temporal resolution of two to three weeks. Thus it was possible to assign seasons of growth to the dense and less dense layers. As indicated by relatively lower ^{18}O-to-^{16}O ratios, dense layers were formed during times of warmer temperature, and vice versa. Detailed investigation shows that the corals secrete the denser layers from June/July to February/March. The densest material is formed in October/November, a few months after the highest water temperatures are observed (August). This observation, that the densest bands are produced late in summer, agrees with previous investigations of Bermuda corals (Fairbanks and Dodge 1979).

The annual growth of the skeleton is determined by the distance between the most dense growth bands, which are formed in October/November in Bermuda. Therefore, the yearly growth of these corals is referred to, not in calendar years, but in terms of the period from October/November to autumn of the following year. For clarity in presenting the interannual differences, the annual growth rates of the corals are normalized. The growth rate of Montastrea cavernosa ranges from 1.78 to 5.0 mm per year, with an average of 3.1 mm per year.

The record of growth rates is shown as a normalized growth index centered on unity, and with a standard deviation of 0.25 (Fig. 1a). One notes a slight asymmetry: there are only a few cases when rates dropped below two standard deviations, but there are many when rates exceeded two standard deviations beyond the mean. There are five time spans containing unusually high growth rates: in the earliest part of the 13th century, around AD 1200 and AD 1500, in the middle of the 16th century, in the early 17th century, and around AD 1900. Rather low growth rates cluster around AD 1300, AD 1450, in the early 18th century, and in the most recent years. There seems to be no obvious relationship to general cooling and warming in the North Atlantic realm.

A comparison of growth rates with sunspot cycles back to AD 1700 (Fig. 1 b) shows but little predictive power of solar activity, regarding the growth of the coral. There is some indication that high sunspot activity leads to lowered rates, from visual comparison of the records. However, regression analysis yields no significant correlation.

Spectra of Coral Record and Sunspot Activity

If the growth record of the coral is influenced by solar activity, in the manner expected from Sverdrup's conjecture, it should show the periodicity of the sunspot cycles. This expectation is not fulfilled (Fig. 2). The solar cycles have periods near 11 years, 10 years, 56 years, and 100 years. The coral record of the last 300 years (for which period we have good control on the sunspot cycles) shows periods of 39 years, 30 years, 27 years, and 22 years (Fig. 2). Of these, only the 22-year period qualifies directly as a candidate for a solar cycle (the Hale magnetic cycle, which is twice the 11-year cycle). For comparison, the oxygen isotope record in GISP2 (Stuiver et al. 1995; Grootes and Stuiver 1997) shows periods at 29.5 and 20.3 years, and also some power near 83 years.

The main lines in the sunspot record (Hoyt and Schatten 1997), found by Fourier analysis of autocorrelation results, are as follows: 11.2, 9.9, 10.55, 12.0, and 9.3 (in the order of strength). The uncertainty in these values is of the order of 1 percent. The main beat periods of these lines are as follows: 83, 182, 153, and 174. Thus, the period near 83 years (Fig. 2a, upper) could well be a solar cycle beat. The period near 30 years, which is seen in both the growth record and the GISP2 record, is ubiquitous in the northern hemisphere. It has been observed in other long-term records, in varves from marine sediments off Pakistan (Von Rad et al. 1999) and in similar varves from off Santa Barbara, California (C.B. Lange and W. H. Berger, unpublished). It may represent variations in solar output (3 times 9.9) or it may represent internal oscillation (Timmermann et al. 1998), perhaps stimulated by sunspot cycles. Also, conceivably, it could be related to tidal forcing, since the difference tone between the solar cycle (11 y) and the tidal cycle (17.7 y) is 29 years (for an 11.2 y solar cycle it is 30.5 y). In any case, this ca.-30-year cycle is a rather prominent feature in several records.

Subjecting the entire coral record to Fourier analysis (with and without previous application of autocorrelation) yields the spectrum shown in Fig. 2b. Strong periods appear near 37.5 years and near 31 years. Solar cycles (near 11 and 22 years) are weak or absent. There is some indication of a cycle near 83 years, but it is not strong. In the short-cycle realm, there are cycles near 7.5 years (North Atlantic Oscillation, Hurrell 1995) and near 6 years and shorter (perhaps related to El Niño, Kaplan et al. 1998). A solar influence is not evident from these results.

Spectral Landscape of the Coral Record

If solar cycles stimulate an oscillatory climate system (analogous to Milankovitch input producing ice-age fluctuations), we should expect that the cycles hold true to their period and are persistent through time. This expectation can be tested using evolutionary spectra, that is, spectra for a moving window of data. (Wavelet analysis is a special case of a moving-window analysis, whereby the window is constantly readjusted to the wavelength of the probe.) The result of the spectral analysis of consecutive or overlapping sections is a spectral landscape (Fig. 3). Each profile represents a spectrum centered on the year shown to the left. The height variation of each profile indicates the amplitude for the period shown on top of the graph (ranging from 100 years to 3.16 years).

The expectation of persistent cycles is not fulfilled. Although there is a steep drop in amplitude near the 50 year period (corresponding to the lack of power left of 40 years, in Fig. 2b), there is no crest for a period near 37.5 years running from the

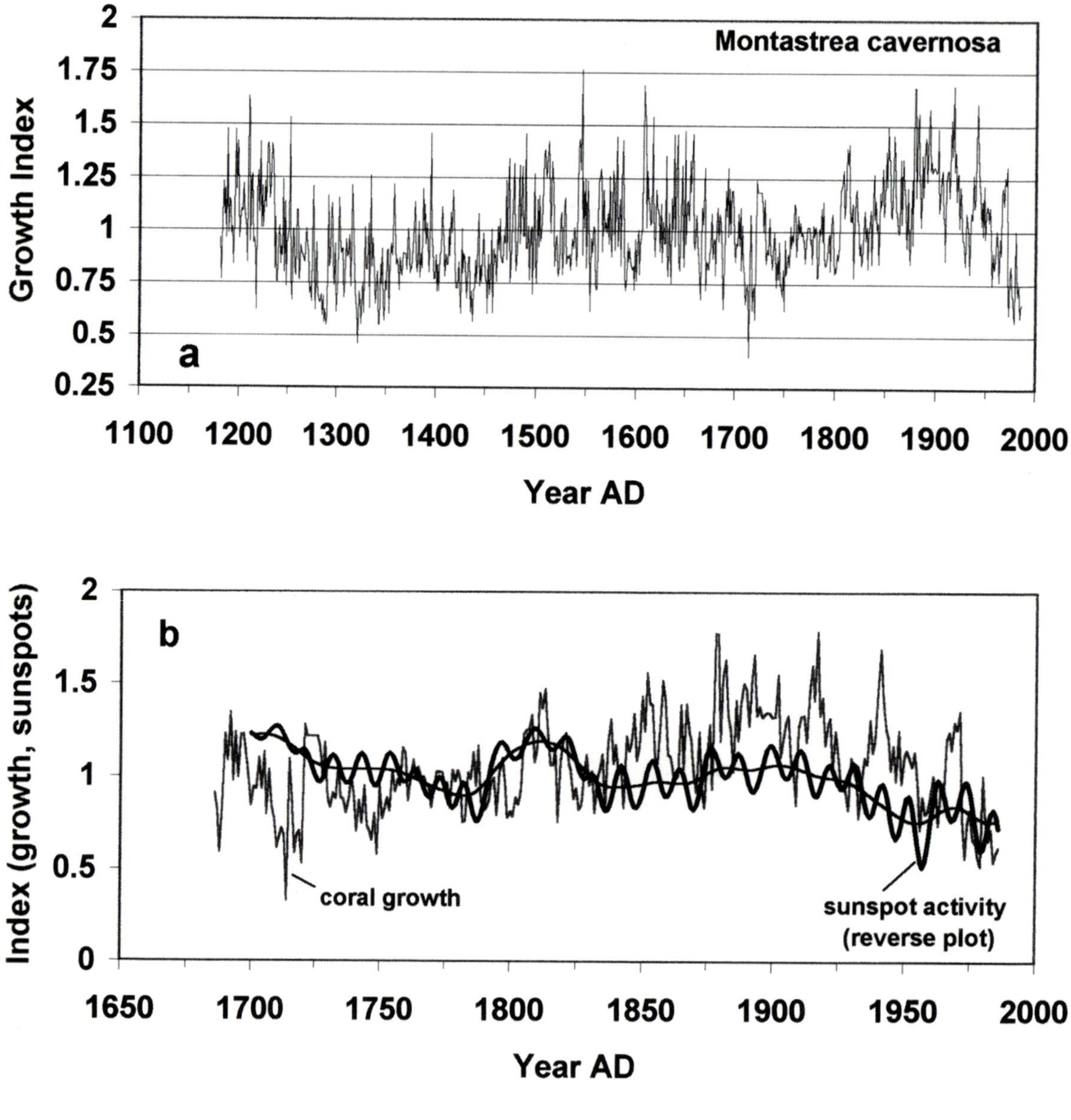

Fig. 1. Growth record of *Montastrea cavernosa* from Bermuda compared with sunspot record. **(a)** Growth values normalized to average of unity and standard deviation of 0.25. **(b)** Attempt to fit smoothed sunspot record for the last 300 years. Sunspots inverted for best fit. The fit is not statistically significant. For sunspot data see Hoyt and Schatten (1997) and Lean (1997).

top to the bottom of the graph. Instead, the peak amplitude jumps back and forth around this value, yielding a rather irregular sequence of hills for periods between 50 years and 25 years (Fig. 3). A most interesting feature of the landscape, besides the precipice near 40 years, is the occurrence of low-energy valleys running left to right, that is, low energy profiles tied to certain portions of historic time. The earliest of these begins near AD 1350 and lasts for 110 years or so, with a hill for a 12.5-year

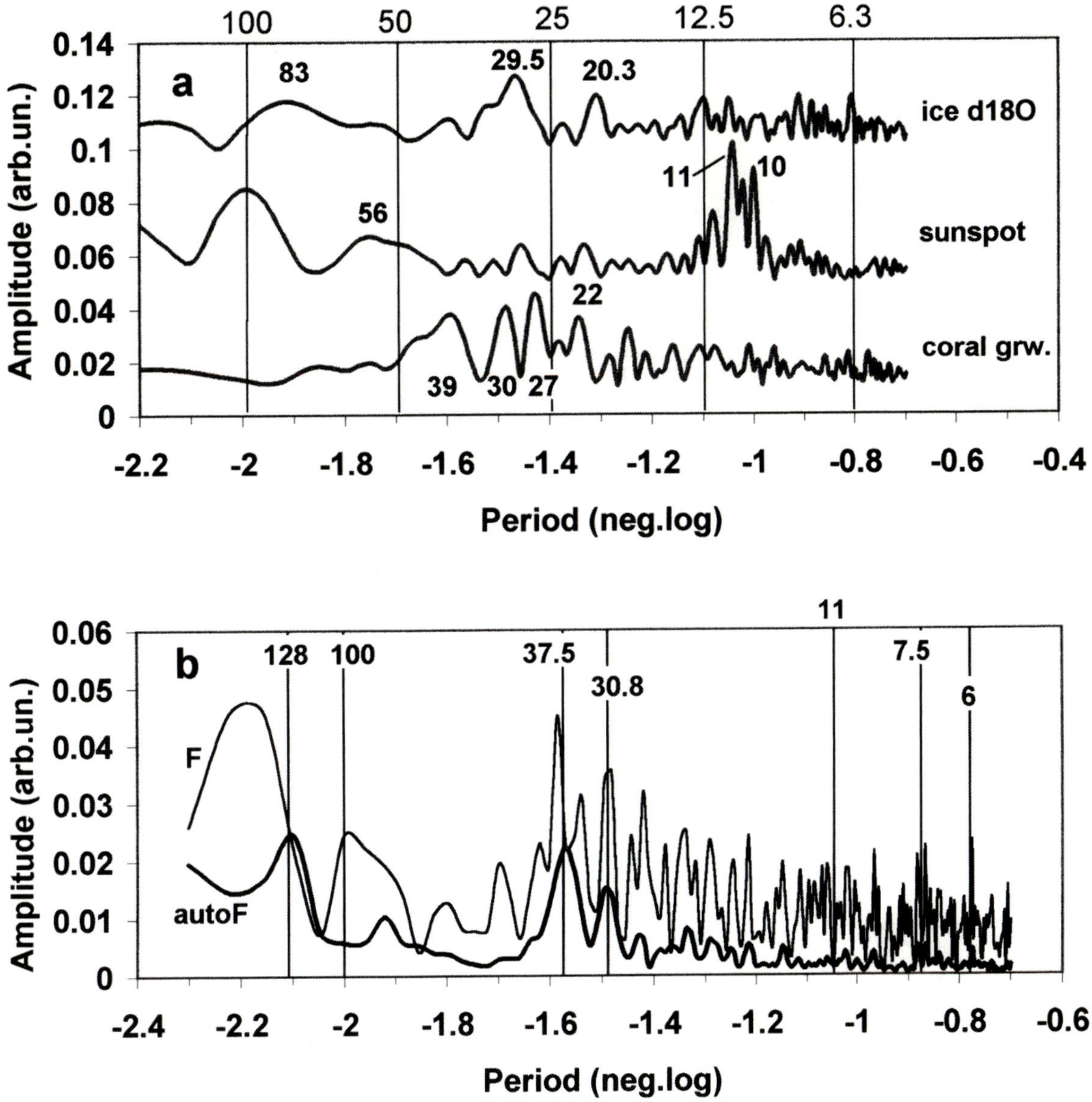

Fig. 2. Spectra of coral growth record and sunspot record. Ice-$\delta^{18}O$: oxygen isotope record from GISP2 (Stuiver et al.1995; Grootes and Stuiver 1997), for comparison. **(a)** Last 300 years. **(b)** Entire coral record. Thick line: autocorrelation and Fourier analysis. Other: Fourier analysis of untreated series.

period in the middle. Another such "time of quiet" is centered near AD 1780 and lasts about 60 years. In contrast, "times of agitation" are located between AD 1500 and AD 1700, and again from AD 1900 to the present. No obvious relationship to solar cycles emerges from this distribution, which does not match the reconstructed minima and maxima of solar activity (max. AD 1372; "Spoerer" min. AD 1450-1540; "Maunder" min. AD 1645-1715).

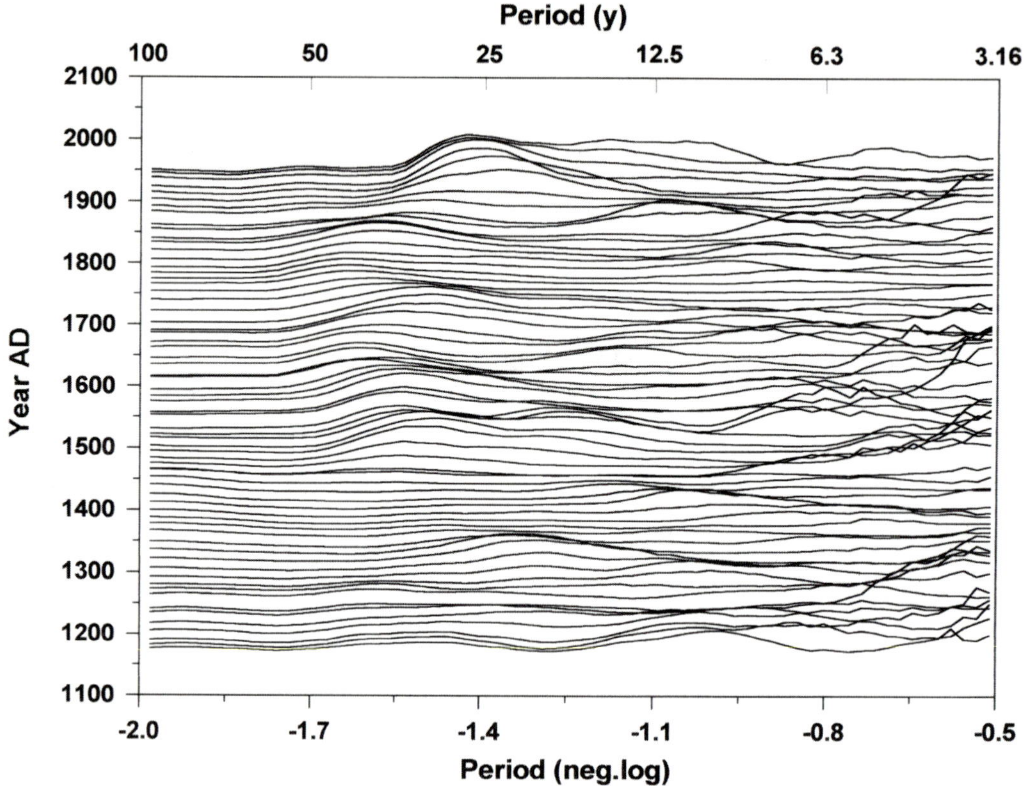

Fig. 3. Spectral landscape of coral growth record. Each profile is a spectrum, plotted over log (frequency) and centered on the mid-year of the window. Periods are given at top of graph (cycles beyond 100-year period not shown).

The Record of Sudden Change

The spectral landscape of the coral record shows that, on the whole, predictability will be poor, except for limited time spans, when times are uneventful (that is, when the prediction that changes will be small is true).

During "times of quiet", prediction will be rather simple: the average conditions of the past few years will predict the growth for the year to come. For all other times, especially for "times of agitation", prediction will be poor. Rapid change - characteristic for agitation - will stand out. The predictability of the record, defined as the square of the deviation of any one value from the average of 6 (or 8) previous values, is plotted in Fig. 4a. One notes that the "times of quiet" and the "times of agitation" are quite well defined, by low and high density of bad predictions, respectively. There is a tendency, for the "times of quiet", to show moderately slow growth, over some length of time. In contrast, during "times of agitation", the growth record jumps from one extreme to another.

We do not know what factor or factors precondition the system to enter the quiet or the agitated state. But we can surmise that a small disturbance (for example a change in solar output) will be much more effective in triggering a change if the condition is "agitated" than if it is "quiet". We next test, therefore, whether the record of sudden change contains information about solar forcing.

To this end, we analyze the change record (Fig. 4a, bottom), using autocorrelation and Fourier expansion. The result is shown in Fig. 4b. The strongest line next to the familiar 30-year cycle is a 10.5-year period, clearly an excellent candidate for solar periodicity. A doubling yields 21 years, also strongly represented. An 11.2 cycle may be hidden

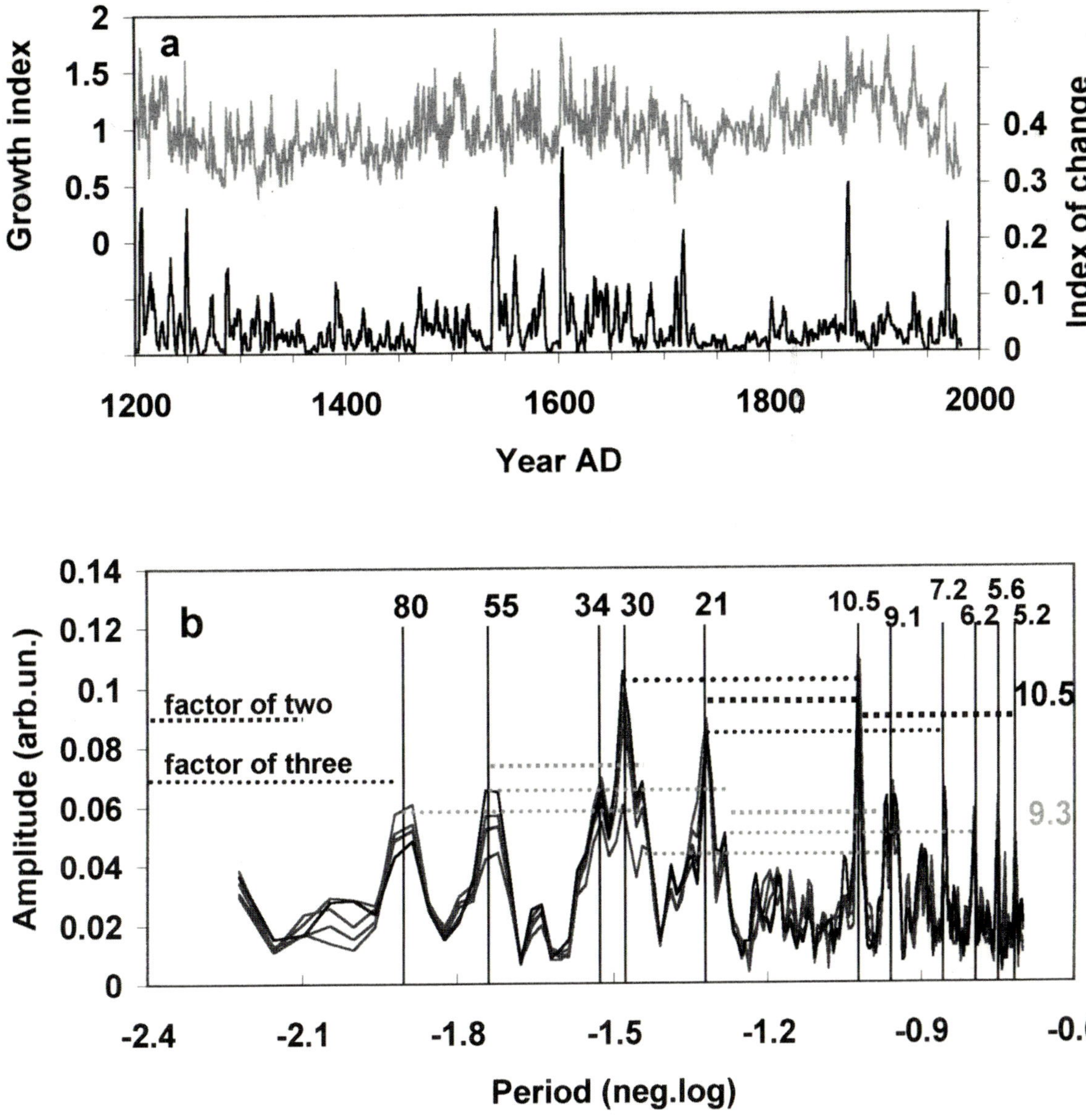

Fig. 4. The record of rapid change. **(a)** Original record (upper) compared with predictability (lower). The index of change is defined as the square of the difference between prediction (average of previous several years) and actual value. **(b)** Spectra of indices of change, based on using 4, 5, 6, 7, and 8 previous years as predictor; index is square of difference between data point and prediction. The horizontal bars, dotted, denote factor-of-two and and factor-of-three distances, suggesting the presence of harmonies in two groups, the "10.5" group (black) and the "9.3" group (gray).

within the 33.6-year line. The 55-year period recalls the 56-year cycle in the sunspot record (Fig. 2a), the 80-year cycle looks like the so-called "Gleissberg" modulation of the 11-year sunspot cycles. The 9.1 -year line (split) seems close enough to the sunspot line at 9.3 years to qualify. (It also happens to be close to one half the lunar node cycle, however, which also is 9.3) Thus, four or five of six of the prominent peaks in the spectrogram are arguably related to solar cycles. However, at least one of them also has lunar affinity. In exploring such affinities it will be useful to map doubles and triples of the basic periods with the most energy (e.g. 10.5 and 9.3). The application of this strategy (illustrated by the horizontal bars in Fig. 4b) suggest that the 7.3 NAO cycle is one third of the strong 21 period which in turn is double the 10.5 cycle. The 5.6 cycle is one half of the sunspot-cycle. Thus, it appears that these cycles can be grouped according to "familiy", which should facilitate detection of origin.

The Record of Extremes

We have argued that sudden change relies on several factors crucially collaborating, so that our recording system is sensitive even to the weaker factors. If those weaker factors are periodic (while others are not) the record of change will show periodicities. A similar argument can now be made regarding extremes of growth. Whenever growth is extremely fast, or extremely slow, all factors are at work, including the one we are searching for. Is there a special pattern to the occurrence of extremes?

Indeed, the extremes fall into several groups (Table 1). On the whole, maximum growth groups occur whenever sunspot activity is high (or proposed to be high) and minimum growth years cluster within sunspot minima (Fig. 5). The only clear exception to this pattern is the most recent group of extremely low growth rates, which characterizes the 1970s and 1980s. Apparently, the overall patterns of coral growth for the last 1000 years are now no longer valid, presumably because conditions have moved outside the usual framework. Regional changes in Bermuda (e.g. pollution, overfishing) may be at fault. Alternatively, we are seeing the effects of global change on oceanographic conditions.

The sequences in the various clusters shown in Table 1 are compatible with an 11-year sunspot cycle, as follows:

Group I, high rate (Medieval Maximum):
1188, 1196, 1198, 1209, 1210, 1221, 1229, 1233, 1252

Group II, low rate (post-MM minimum):
1278, 1282, 1283, 1285, 1287, 1288, 1289, 1290, 1299,1320, 1321, 1323, 1324, 1327, 1333, 1339, 1342, 1343, 1347, 1348, 1353

Group III, low rate (Spörer Minimum):
1425, 1435, 1437, 1452, 1457

Group IV, high rate (pre-Maunder maximum):
1580, 1586, 1607, 1608, 1609, 1616, 1638, 1641, 1649, 1657

Group V, low rate (Maunder Minimum):
1688, 1710, 1714, 1717, 1720

Group VI, high rate (modern maximum):
1852, 1858, 1878, 1879, 1881, 1882, 1889, 1890, 1892, 1893, 1902, 1915, 1917

Group VII, low rate (post-modern):
1973, 1978, 1979, 1984, 1985, 1986, ...?

Table 1. Clusters of high and low growth rates.

Group I: 1187, 1198, 1209, 1220, 1231, 1242, 1253.
Group II: 1277, 1288, 1299, 1310, 1321, 1332, 1343.
Group III: 1425, 1436, 1447, 1458.
Group IV: 1586, 1597, 1608, 1619, 1630, 1641, 1652.
Group V: 1688, 1699, 1710, 1721 (close to sunspot minima).
Group VI: 1859, 1870, 1881, 1892, 1903, 1914 (ca. 3 years past sunspot minima).

Discussion and Conclusions

There is evidence, in the coral growth record from Bermuda, that Sverdrup was right when he claimed that solar activity introduces periodicity into the climate system, and that sunspot cycles are an expression of relevant solar activity.

However, as Sverdrup pointed out, even without any variation in solar input, the climate would be characterized by many aperiodic changes that would be very difficult to explain. The reason is that a host of different factors interact in complex ways to make the system unpredictable. In fact, inter-decadal variability (Bjerknes 1964; Kushnir 1994) has both "noise" and cycles (Mann et al. 1995), and while many of the latter may be oscillations of some kind stemming from the interaction of various elements of the climate machine (Delworth et al. 1993; Weisse et al. 1994; Dickson et al. 1996), such oscillations are expected to emerge most clearly when stimulated by outside forcing, on comparable time scales. The solar input makes itself felt in subtle ways, and it takes special conditions for it to come forth. Such special conditions are reached during "times of agitation", when the system is highly sensitive (for whatever reason) toward disturbance. During such times, cycles of different periods all seem to gain in amplitude simultaneously, suggesting transfer of energy from the shorter cycles (El Niño, North Atlantic Oscillation) to the longer ones.

There is no evidence in the coral record for the presence of persistent cycles. If there are oscillations (for example at periods between 25 and 50 years), they drift in the length of the period. What is interesting is the drop-off of power between 40

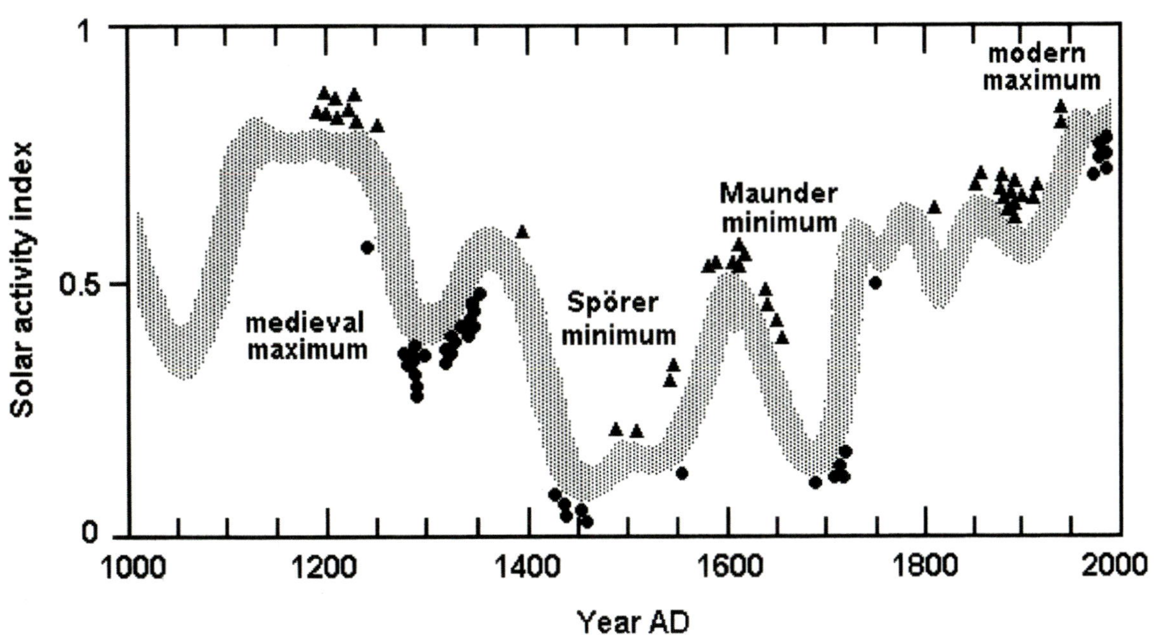

Fig. 5. Extremes of coral growth (highest 5% and lowest 5%) compared with inferred solar activity, based on ^{10}Be and ^{14}C. (For last 300 years, the sunspot activity was used also, to draw the solar activity index.) For data sources and discussion, see Damon and Linick (1986), Stuiver and Braziunas (1993), Stuiver and Pearson (1993), and Beer et al. (1988, 1990). High growth: triangles. Low growth: circles.

and 80 years, a kind of "quiet" zone in variability. While cycles are linked in the periods shorter than 50 years, there is an abrupt drop in linkage to longer cycles. Different mechanisms of cycling may dominate in this long-period realm. The drop-off is well expressed since 1450 AD, but not before. Thus, it may be related to the development of severe winters (and sea ice formation).

The puzzle of the ca.-30-year cycle (which seems pervasive, albeit fuzzy) calls for some kind of internal oscillation, perhaps stimulated by a number of interfering forcings from solar cycles (3 times 11), North Atlantic Oscillation (4 times 7.7) and tides (combination tone 11 and 17.7). For mechanisms of amplification one might look to motions of the polar front, extent of sea ice cover, and salinity changes in the Nordic Sea and the Labrador Sea, influencing deepwater production (among other things) (see Dickson et al. 1988; Lazier 1988; Mysak et al. 1990; Wohlleben and Weaver 1995).

The coral record we studied may be unusually sensitive to climatic variations, because it is situated at the limits of coral distribution, and because growth here is apparently controlled as much by nutrient supply (that is, wind strength and stirring) as it is by temperature (Pätzold et al. 1999). The method we have used (searching for unusual deviation from prediction) will tend to double inherent frequencies. Thus, our 10.6-y cycle could be closely related to the 22-y forcing postulated by Baranyi et al. (1998). A sorting out of oscillations from solar (and lunar) forcing will come from comparing the spectra of records from several ocean basins, which may be expected to have different oscillatory patterns. In this context, it is of interest that the tree-ring record of Tasmania (!) shows "regular oscillations on the order of 30, 56, 80 and 180 years in length" (Cook et al. 1992). These are almost precisely the main cycles, in this range, seen in our coral growth (change) record (30, 55, 80, ca.172). If they are to be regarded as "arising from the internal dynamics of the ocean-atmosphere-cryosphere system" (as suggested by Cook et al.) we should have to postulate rather well-timed oscillations involving both hemispheres. It seems to us that the question of oscillation versus outside forcing is not an either-or proposition and that Sverdrup's conjecture regarding fast climate response to variable solar input has merit. What needs to be clarified is why the system seems to move from quiet states to agitated ones, and back. Transitions between different climatic regimes, on long time scales, may play a role in introducing disequilibrium (Schulz et al. 1999) and thereby increasing sensitivity. The role of tidal action needs to be investigated in the context (see companion paper in this volume).

References

Anderson RY (1992) Possible connection between surface winds, solar activity and the Earth's magnetic field. Nature 358:51-53

Baranyi T, Ludmany A, Coffey H (1998) 22 Year solar modulation of Earth's northern hemisphere temperatures. Geophys Res Letters 25:2269-2272

Beer J, Siegenthaler U, Bonani G, Finkel RC, Oeschger H, Suter M, Woelfli W (1988) Information on past solar activity and geomagnetism from ^{10}Be in the Camp Century ice core. Nature 331:675-679

Beer J, Blinov A, Bonani G, Finkel RC, Hofmann HJ, Lehmann B, Oeschger H, Sigg A, Schwander J, Staffelbach T, Stauffer B, Suter M, Woelfli W (1990) Use of ^{10}Be in polar ice to trace the 11-year cycle of solar activity. Nature 347:164-166

Bjerknes J (1964) Atlantic air-sea interaction. Advances in Geophysics 10:1-82

Brooks CEP (1949) Climate through the Ages. 2nd ed. Ernest Benn. Ltd. (Republished by Dover, New York, in 1970)

Cook E, Bird T, Peterson M, Barbetti M, Buckley B, D'Arrigo R, Francey R (1992) Climatic change over the last millennium in Tasmania reconstructed from tree-rings. The Holocene 2(3):205-217

Damon PE, Linick TW (1986) Geomagnetic-heliomagnetic modulation of atmospheric radiocarbon production. Radiocarbon 28:266-278

Delworth TL, Manabe S, Stouffer RJ (1993) Interdecadal variations of the thermohaline circulation in a coupled ocean-atmosphere model. J Climate 6:1993-2011

Dickson RR, Meincke J, Malmberg S-A, Lee AJ (1988) The "great salinity anomaly" in the northern Atlantic 1968-1982. Progr Oceanogr 20:103-151

Eddy JA (1981) Climate and the role of the Sun. In: Rotberg RI, Rabb TK (eds) Climate and History. Studies in Interdisciplinary History. Princeton U Press, Princeton, New Jersey, pp 145-167

Fairbanks RG, Dodge RE (1979) Annual periodicity of the $^{18}O/^{16}O$ and $^{13}C/^{12}C$ ratios in the coral *Montastrea annularis*. Geochim Cosmochim Acta 43:1009-1020

Frenzel B. (ed) (1995) Solar Output and Climate during the Holocene. Gustav Fischer Verlag, Stuttgart, pp 1-186

Grootes PM, Stuiver M (1997). Oxygen 18/16 variability in Greenland snow and ice with 10^{-3}- to 10^{5}-year time resolution. J Geophysical Res C102:26455-26470

Hoyt DV, Schatten KH (1997) The Role of the Sun in Climate Change. Oxford University Press, pp 1-304

Hurrell J (1995) Decadal trends in the North Atlantic Oscillation: regional temperatures and precipitation. Science 269:676-679

Kaplan A, Cane M, Kushnir Y, Clement A, Blumethal M, Rajagopalan B (1998) Analyses of global sea surface temperature 1856-1991. J Geophys Res 103:18567-18589

Kushnir Y (1994). Interdecadal variations in the North Atlantic sea surface temperature and associated atmospheric conditions. J Climate 7:141-157

Labitzke K, Van Loon H (1993) Some recent studies of probable connections between solar and atmospheric variability. Ann Geophysicae 11:1084-1094

Lazier JRN (1988) Temperature and salinity changes in the deep Labrador Sea, 1962-1986. Deep-Sea Res 35:1247-1253

Lean J (1997) The Sun's variable radiation and its relevance for Earth. Ann Rev Astron Astrophys 35:33-67

Mann ME, Park J, Bradley RS (1995) Global interdecadal and century-scale climate oscillations during the past five centuries. Nature 378:266-270

Mysak LA, Manak DK, Marsden RF (1990) Sea-ice anomalies observed in the Greenland and Labrador Seas during 1901-1984 and their relation to an interdecadal Arctic climate cycle. Clim Dyn 5:111-133

Pätzold J, Bickert T, Flemming B, Grobe H, Wefer G (1999) Holozänes Klima des Nordatlantiks rekonstruiert aus massiven Korallen von Bermuda. Natur und Museum 129:165-177

Roederer J.G (1995) Solar variability effects on climate. In: Frenzel B (ed) Solar Output and Climate during the Holocene. Gustav Fischer Verlag, Stuttgart, pp 1-21

Schulz M, Berger WH, Sarnthein M, Grootes PM (1999) Amplitude variations of 1470-yr climate oscillations during the last 100,000 years linked to fluctuations of continental ice mass. Geophys Res Lett 26:3385-3388

Stuiver M, Braziunas TF (1993) Sun, ocean, climate and atmospheric $^{14}CO_2$: an evaluation of causal and spectral relationships. The Holocene 3:289-304

Stuiver M, Pearson GW (1993) High-precision bidecadal calibration of the radiocarbon time scale, A.D. 1950-500 B.C. and 2500-6000 B.C. Radiocarbon 35:1-23

Stuiver M, Grootes PM, Braziunas TF (1995) The GISP2 18O record of the past 16,500 years and the role of the sun, ocean and volcanoes. Quat Res 44:341-354

Sverdrup HU (1918) Die Beziehung der elfjährigen Klima-schwankungen zur Sonnentätigkeit. Annalen der Hydrographie und Maritimen Meteorologie. Mai/Juni 1918, pp 191-193

Timmermann A, Latif M, Voss R, Groetzner A (1998) Northern hemisphere interdecadal variability: a coupled air-sea mode. J Climate 11:1906-1931

Von Rad U, Schaaf M, Michels KH, Schulz H, Berger WH, Sirocko F (1999) A 5000-yr record of climate change in varved sediments from the oxygen minimum zone off Pakistan, Northeastern Arabian Sea. Quat Res 51:39-53

Weisse R, Mikolajewicz U, Maier-Reimer E (1994) Decadal variability of the North Atlantic in an ocean general circulation model. J. Geophysical Res 89:12411-12421

Wohlleben TMH, Weaver AJ (1995) Interdecadal variability in the subpolar North Atlantic. Clim Dyn 11:459-467

A Case for Climate Cycles: Orbit, Sun and Moon

Wolfgang H. Berger*[1], Jürgen Pätzold[2] and Gerold Wefer[2]

[1]*University of California, California Space Institute, La Jolla, CA 92093-0524, USA*
[2]*University of Bremen, Department of Geosciences, Postfach 33 04 40, D-28334 Bremen, Germany*
**corresponding author (e-mail): wberger@ucsd.edu*

Abstract: The success of Milankovitch Theory – which states that ice age cycles are real and are a result of orbital forcing – has made climate cycle research thoroughly respectable. This was not the case when E. Brückner (1862-1927) pioneered climate cycles on the scale of decade-to-century, for the last 1000 years, publishing a century ago. In the last 20 years or so, Brückner's claim that climate cycles exist at this scale has received much support. Work on solar-forced climate cycles suggests that such cycles interfere with the assessment of human-induced global warming. A portion of the general warming observed over the last century may be owing to solar activity. If so, this shows considerable sensitivity of climate to modest changes in forcing. The original Brückner cycle of 35 years is here suggested to be of tidal origin (twice the lunar perigee period). Likewise, the 1470-y Greenland cycle is proposed to be tidal in nature (as a beat between nodal and perigee cycles, effective during the summer season).

Introduction

Most geologists know Eduard Brückner (1862-1927) as the co-author of the monumental opus on the ice age geology of the Alps published early in the century in collaboration with Albrecht Penck (Penck and Brückner 1901-1909). This volume, among other things, contains the then available evidence for repeated expansion of Alpine glaciers during the ice ages, at a time when ice age cycles were considered unlikely by many experts (including the geologist A. Penck). However, much of the research carried out by Brückner (a meteorologist) was not concerned with Alpine geology at all, but with the evidence of climate change on historical time scales, a field that had no tradition at the time. Fortunately, Brückner's writings on the subject have now been rescued from their rather obscure repositories and made readily accessible (Stehr and Von Storch 2000). His research documented that climate has changed substantially over the last thousand years or so, as reflected in such recorded items as lake levels, floods, ice on rivers and lakes, and the timing of harvests.

Perhaps the most striking proposition emerging from Brückner's work was his insistence that many of his records are characterized by the presence of a cycle near a 35-y period. He thought that the accuracy of estimate for the length of this period was better than 1 year, based on long historic records (see Stehr and Von Storch 2000, p. 144). Brückner's 35-y cycle has been forgotten since, but recently it was found that the Brückner cycle appears in certain finely laminated sediments recovered in the Arabian Sea off Pakistan and it was suggested that the cycle is identical to the 35.4-y tidal period (Berger et al. 2000; Berger and Von Rad, submitted).

Of Brückner's claims, the historian Emmanuel Le Roy Ladurie (who pioneered the interpretation of history in terms of climate change in France) had this to say (Ladurie 1971, p. 11):

"Such researches are to the real history of climate what the philosopher's stone was to oxygen.

From WEFER G, BERGER W, BEHRE K-E, JANSEN E (eds), 2002, *Climate Development and History of the North Atlantic Realm.* Springer-Verlag Berlin Heidelberg, pp 101-123

Yet they have been carried on ceaselessly by generations of researchers. But now they have received a deathblow. ..."

While the circumstances of the "deathblow" are not elaborated, it is clear that Ladurie judged the search for cycles in climate history a complete waste of time. This opinion still has many adherents, but it is rapidly making room for a more pragmatic approach. Cycles have become respectable again. There is growing interest in the subject of climate cycles and the literature on the topic is expanding rapidly (e.g. see references in Hoyt and Schatten 1997).

Several reasons are responsible for this development:

(1) The cyclicity of the ice ages is now well established, as is the fact that changes in the contrast of summer and winter insolation in high latitudes are driving the cycles. These changes are cyclic and are tied to the wobbles of Earth's axis and the changing shape of its orbit, as first correctly described by Milutin Milankovitch (in 1930). The advice of the then-famous Penck to the young Milutin Milankovitch that his pursuit of cycles was doomed (in Berger et al. 1984, related by V. Milankovitch, p. XXXII) proved quite wrong. In the end, Penck's ice age scheme survived in the nomenclature, but the concepts introduced by Milankovitch provided the path to geophysical understanding.

(2) The question of which portion of the global warming of the last hundred years is man-made, if any, can only be answered if one has a notion what the climate state would be without human interference. In this context, changes in solar irradiance have been identified as an important factor in controlling Earth's global temperature (Frenzel 1995; Lean et al. 1995; Lean 1997; Bard et al. 2000; Beer et al. 2000; Crowley 2000). The fluctuating intensity of solar radiation is closely tied to sunspot activity, which in turn is definitely cyclic. Thus, if the variability of the sun is important in climate change, such change will have a cyclic component.

(3) The recent great expansion of efforts in finding the patterns of climate change on a decadal scale for the last century or so, have led to the discovery of a number of internal oscillations in the ocean-atmosphere system. Among the best studied of these are the El Niño - Southern Oscillation (ENSO) phenomenon (e.g. Diaz and Markgraf 1992) and the North Atlantic Oscillation (NAO) (e.g. Hurrell 1995). The ENSO involves large changes in air pressure patterns in the tropical Pacific, associated with drastic redistribution of warm water at the sea surface. A commonly used index is the difference in air pressure between Darwin, Australia, and Papeete, Tahiti. During strong El Niño conditions, when the eastern Pacific becomes abnormally warm (and air pressure drops in the east), unusual temperature and rainfall patterns are observed all over the world. The NAO likewise involves changes in air pressure and the redirecting of ocean currents. In this case, a commonly used index is the fluctuating pressure difference between Iceland and the Azores, which has implications for the strength of the Gulf Stream extension and the routing of warm air masses associated with it. A system that oscillates is a candidate for amplification of periodic stimulation.

(4) Long and detailed proxy series of climate change have become available in the last two decades or so, based on the study of tree rings, marine varves, polar ice cores, coral skeletons and lake sediments (e.g. Jones et al. 1996). They document climatic change for thousands of years, in cases with year-to-year resolution. Given the rapid rise of computing power, and the ready availability of statistical techniques for time series analysis, such records have invited a scanning for cycles on scales from years to centuries (for examples see Stuiver and Braziunas 1993; Stuiver et al. 1995; Plaut et al. 1995; Jones et al. 1996; Vos et al. 1997).

(5) There have been a number of reports about oscillations developing within climate-simulating computer models containing algorithms describing internal circulation and interactions between the ocean and atmosphere. While little can be said about the precise meaning of such oscillations (in terms of their presumed analogs in nature) they do demonstrate that systems that combine positive feedback (leading to enhancement of a disturbance) with lagged negative feedback (leading to a delayed return to the initial condition) tend to oscillate. The climate machine is such a system.

Geologists have long been interested in whether one finds cycles in the climate record or not, whether such cycles, if present, are seen in different types of proxies (representing different parts of the climate system), and whether they persist or appear only under certain circumstances (Schwarzacher 1993). Here we make a case that Milankovitch theory – the study of orbital cycles in the record of the ice ages – has greatly advanced geophysical understanding of the workings of the climate machine. We argue that a search for cyclic forcing within the Holocene record might have similarly beneficial effects. While outside forcing of climate in the Holocene is likely to be weak, compared with Milankovitch forcing in the Quaternary, it may still be strong enough to detect, and to employ in the study of the climate system.

Milankovitch Cycles: Astronomic Forcing and Internal Oscillation

Why should we care about ice age cycles when discussing climate change in the Holocene?

One reason is, ironically, that the large climatic excursions over the last 100,000 years are better understood than the much lesser fluctuations of the last 10,000 years, mainly because of the dominance of astronomical forcing in driving the ice-age climate. If we now realize that the warm period of the early Holocene had strong monsoon rains generating lakes and vegetation in northern Africa (Gasse 2000), it is because such monsoon periods occurred many times before, at predictable intervals. They are easily recognized, for example, in the sapropel deposits of the eastern Mediterranean (Rossignol-Strick 1985). Every time the Sun appeared large during summer solstice (that is, when Earth happened to be closest to the Sun during June) the Nile flooded, the Mediterranean Sea became less saline and more productive, and the oxygen at depth in the Levantine Basin disappeared, allowing black mud to accumulate. At those favored times, as beautifully recorded in the stone-age drawings in the caves of Tassili n'Ajjer and elsewhere in the northern Sahara, game was plentiful in regions where rocks and sand dunes now dominate the landscape. The times of sapropel development (and of a greening Sahara) can be precisely predicted from celestial mechanics. Computer models confirm that the monsoon was stronger during the early Holocene (Kutzbach and Liu 1997; see Fig. 1) and that the change toward a dry climate was accelerated by the positive feedback from a disappearing vegetation.

The reason why the Sahara became dry is because Earth's closest approach to the Sun moved from summer to spring to winter, as a result of precession of the Earth's axis (which has the effect of equinox occurring a little before completion of each orbit). The closest approach (perigee) now occurs in early January. This weakens the monsoon. It is unlikely that we would have found the reason for the great change in Holocene climate in northern Africa without the clues from earlier cycles. Repeated experiments are more instructive than single events.

In analogy, we can state quite confidently that the initial agricultural expansion in Scandinavia about 4000 y.b.p. benefited from the fact that the solar disk was still large when snow melt began (as well as from a sun that moved slightly farther north in summer than today). These same circumstances also account for the higher elevation of the tree line for *Pinus silvestris* between 8000 and 4000 years ago in Fennoscandia and Scotland (see Lowe 1991). The decline in climate which marks the transition from the Atlantic period to the Sub-Boreal (and which required considerable adaptation in terms of farming culture in the circum-Baltic region; Rowley-Conwy 1985; Zvelebil 1996) may be ascribed to perigee becoming ineffective as an aid in melting snow, as the time of maximum solar disk size moved into February. As snow now persisted longer into the year, the growing season became shorter and summer temperatures dropped. Another important factor in the expansion of agriculture, of course, was the fact that the land rose in response to removal of the ice load. In due course, this provided areas that could be settled (and drained) and, where once covered by the sea, soils that are buffered by calcium carbonate. Loess-based soils elsewhere in Europe likewise proved excellent for agriculture, another inheritance from the last ice age.

A second reason for bringing up ice-age climate cycles is to remind ourselves that the concept of climate cycles has merit. The realization that this is so only took hold in the 1970s and 1980s, in the

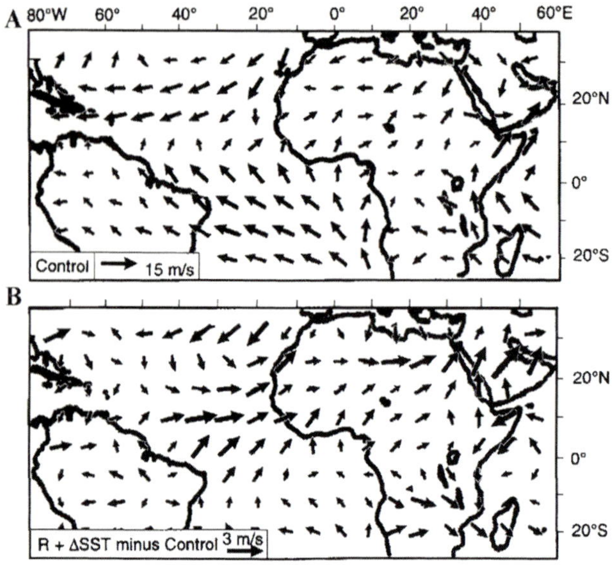

Fig. 1. Simulated surface winds for summer months for (A) Control and (B) R+DSST minus Control. Control = present conditions. R+DSST = departure of the situation 6000 years ago from the present, based on differences in seasonality due to changes in the annual distribution of heat received by the Sun. Note the change in wind speed scale, going from A to B. "R" denotes the effects of summer insolation, with SST fixed. DSST denotes an adjustment, in the model, for heating of the sea surface. Source: Kutzbach and Liu (1997), as simplified by Berger et al. (1998).

midst of much debate. The ice-age climatic cycles had been predicted a century earlier, by James Croll, and were discovered by Cesare Emiliani in the middle of the 20th century, within deep-sea sediments. They were subsequently dated by Shackleton and Opdyke using magnetic stratigraphy, and tied into astronomic forcing by the British science writer Nigel Calder (using a conceptual model and simple arithmetic) and by the paleoceanographers James Hays, John Imbrie and Nicklas J. Shackleton (using spectral analysis) (see Imbrie and Imbrie 1979). What we now call Milankovitch theory made its public debut in the writings of Köppen and Wegener (in 1924), and of Milankovitch (1930). It took more than half a century to achieve textbook status.

Milutin Milankovitch (who was greatly encouraged by Wladimir Köppen) formulated the core of the ice age theory that is now the reigning paradigm. He emphasized seasonal contrast in the solar energy received in high northern latitudes, with the proviso that cool summers allow snow and ice to accumulate, and that warm summers remove snow and ice. At the same time, increasing or decreasing the cover with snow and ice greatly changes the amount of sunlight reflected into space from the ground, providing for strong feedback. Thus, the regions where snow and ice have a marginal existence, where snow can stay or melt, are the ones that contribute most to the sensitivity of the climate system. This principle holds even today, of course, except that the remnants of snow and ice on the northern hemisphere are of pitiful proportions when compared with the last ice age. We must assume that changing snow fields and sea ice still play that role today, in high latitudes.

The reason why seasonal contrast changes in high latitudes, over long time spans, is rather straightforward. The amount of energy received from the Sun depends on how high it is over the horizon, and how large its disk appears. Its maximum height over the horizon is determined by the obliquity of the Earth's axis. This parameter changes on a cycle of 41,000 years. Thus, for several thousand years, the Sun is unusually high above the horizon (this was the case in the very early Holocene), and 20,500 years later in the cycle its maximum ascent is at a minimum. At high latitudes, the relatively small changes (within a range of about three degrees) make a lot of difference. In low latitudes, they do not. (Here the Sun is nearly overhead anyway, in mid-summer.)

As mentioned, the Sun changes its apparent size as seen on Earth, as our planet moves around its star. The reason is the eccentricity of the orbit. Right now, we are closest to the Sun early in January. This has the effect of diminishing the seasonal contrast (cool summers, mild winters). In 10,000 years, the Sun will be closest in June. This will greatly increase seasonal contrast in high northern latitudes (warm summers, cold winters). The change is a result of the precession of Earth's axis (it does not stay fixed on the North Star, but describes circles in the sky), and also of changes in the shape of Earth's orbit. A full (climatic) preces-sional cycle takes about 23,000 years.

Since the 1980s it is established that the changing orbital and rotational conditions of planet Earth, which lead to changes in seasonal insolation, are entirely sufficient to force the changes in climate seen on time-scales between 10,000 and 100,000 years. The only condition is that internal oscillation must be admitted as well, a condition not clearly recognized by the pioneers but well understood since the 1980s (Fig. 2).

The conceptual model ("mt414" in Fig. 2a) consists of an equation relating the input of solar energy ("Milankovitch forcing") to the change in ice mass. The equation says that the change in ice mass results from constant ice growth and variable melting (which is set to depend on summer insolation and existing and past ice mass). The melting term is of the form "IF summers are unusually warm, and IF there is ice to melt, and IF the ice is ready to melt because a lot of it has been around for a long time, THEN ice shall melt." (Details in Berger 1999.)

The resulting output is compared with the oxygen isotope record from the deep sea. In the figure shown, the record is not tuned (that is, it is taken as found, without adjustment for changes in sedimentation rate) to show the power of the Milankovitch-based algorithm. For the last 700,000 years, the climate system has displayed a strong cycle near 100,000 years, closely corresponding to one of the cycles of changing eccentricity (96,000 y). This is the time span (650,000 y) for which Milankovitch generated his insolation curves, and which he attempted to date on the basis of his theory. (He was quite close to the truth, in hindsight, but there was no way to verify his age scale, in his time.)

The Milankovitch chron is probably the best-understood period in geologic history. It is also the period of human evolution. We can now say, with great confidence, that at no time during that period did we have a trace gas content in the atmosphere that was as high as today, not even during the warmest of the interglacials, Stage 11. We know this from the analysis of Antarctic ice cores from the Vostok station (Petit et al. 1999), which gets us back 400,000 years, and from extrapolating these results into the past, using indirect indicators from the deep-sea record (Berger et al. 1996).

Solar Cycles and Implications for Global Warming

An important issue in the context of the global warming within the last century (Fig. 3) is the question to what extent a variable Sun may be held responsible (e.g. Wigley and Raper 1990). The warming observed is quite striking, against the background of previous centuries, and especially against the background of the so-called Little Ice Age (Grove 1988; Bradley and Jones 1992) which lasted from around AD 1350 or 1450 to about 1850 or 1880 (with the beginning being more difficult to define than the end). Already 25 years ago the solar physicist and astronomer Eddy (1976) suggested that a less energetic sun was responsible for this cold spell, correlating a minimum in sunspot activity (the "Maunder Minimum") with the Little Ice Age. According to this hypothesis, much of the recent warming would be attributable to the Sun becoming more active again. The implication is that strong sunspot activity is tied to higher energy output - an implication long suspected but only recently proved, using satellite data.

The main arguments for an effect from solar forcing are laid out in the article by Judith Lean (this volume). Briefly, she not only reconstructs the solar activity cycles (10 to 11-year period, typically), but also a longer term variability component, to obtain an estimate for the history of solar energy output. Reconstruction of the longer-term background series (which is crucial for the argument) is based on ingenious (but somewhat

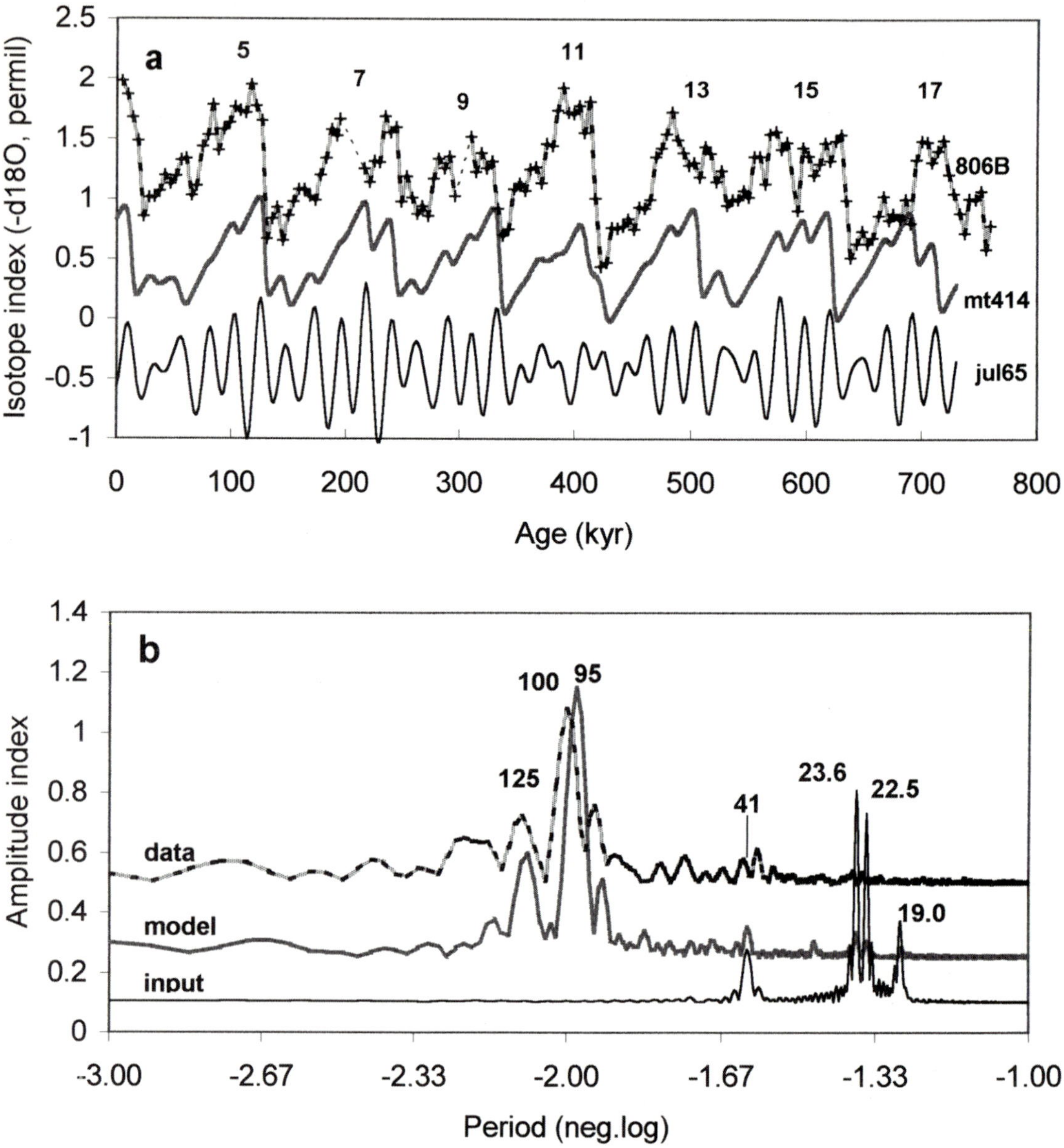

Fig. 2. Conceptual model of climate change compared with observed variations in the oxygen isotope record from ODP Site 806B, in the western equatorial Pacific. Target series consists of untuned data (unvarying sedimentation rate assumed) and model output is produced by Milankovitch-based algorithm (mt414, with oscillation near 100 ky; see Berger et al. 1996). **(a)** Time domain. **(b)** Fourier domain.

tenuous) arguments about solar physics, using nearby sun-like stars. Some of these stars are in a cyclic phase, some are not. The ones in the cyclic phase are brighter than the others. From this, Lean and colleagues conclude that the Sun was 0.24 percent dimmer during the Maunder Minimum than now. A change of 0.24 percent goes well beyond the 0.1 percent change ascribed to the 11-year cycle itself, so that a climate impact becomes much more probable. In addition, Lean assumes a change in UV output that is 6 times larger than that of visible light (with interesting implications for the history of the ozone layer).

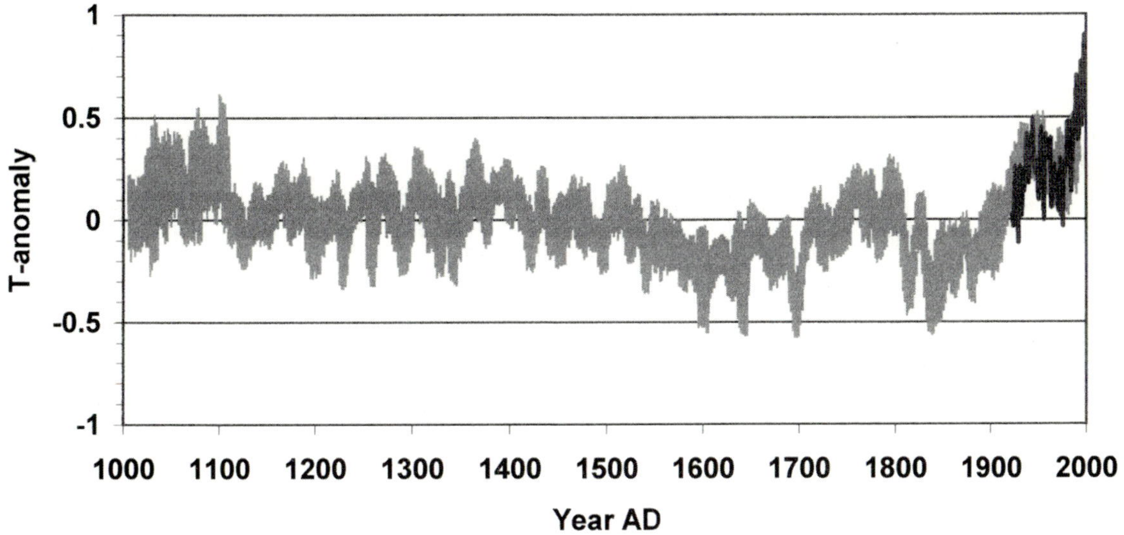

Fig. 3. Northern hemisphere paleotemperature reconstruction of Jones et al. 1998. Note that the last decade is unprecedented in the temperature record of the last 1000 years. (Data are set so that one standard deviation is 0.25 and the mean is zero.)

Lean's reconstruction procedure is equivalent to taking a running mean of the cyclic solar activity (as in Fig. 4a) and reading the result as "extra" energy output from the sun. Peak times are seen, roughly, around 1750, 1850 and 1950, with amplitudes higher at each peak. Low energy periods are centered near 1700 (at the edge of the graph), near 1810 and, less clearly, near 1900. Again, the levels of the minima rise at each successive instance. Thus, we see (in addition to the well-known eleven-year cycle) a centennial cycle and a portion of a longterm change of unknown proportions but possibly with a half-length between 250 and 300 years (distance between lowest and highest points in the graph of Fig. 4a).

In documenting their reconstruction of solar output, Lean et al. (1995) show that it tracks well with the cosmogenic isotope ^{10}Be, whose abundance follows solar activity (mainly), providing strong support for the reconstruction. The correlation coefficient between paleotemperature and solar output for the period between 1610 and 1800 was found as 0.86 (75%), suggesting a strong participation of the Sun's variable output in climatic change during that part of the "Little Ice Age". Extrapolating forward using the same regression, Lean et al. find that "solar forcing may have contributed about half of the observed 0.55°C surface warming since 1860 and one third of the warming since 1970." Crowley (2000) using a number of different estimates of solar activity as well as different paleotemperature reconstructions for the period AD 1000-1850, obtains a range of results centered on r=0.45 (20%). Beer et al. (2000) suggest a solar contribution to global warming since AD 1850 of about 40%.

If the Sun helps warm the climate since 1850, a simple Arrhenius-type warming model (that is, a warming by 3°C for a doubling of carbon dioxide in the atmosphere) cannot be correct. The trace-gas effect must be less than anticipated. Heat storage by the ocean has to be considered. Also, there is much evidence that minute particles created by human (industry and firewood) block sunlight. When this is considered, the sum of all anthropogenic effects on change in temperature are well below that anticipated from trace-gas release alone. We then actually *need* the change in solar output to explain the warming since 1850.

Some "climate skeptics" (a term loosely applied to those doubting that anthropogenic warming has taken place or can be demonstrated) may take

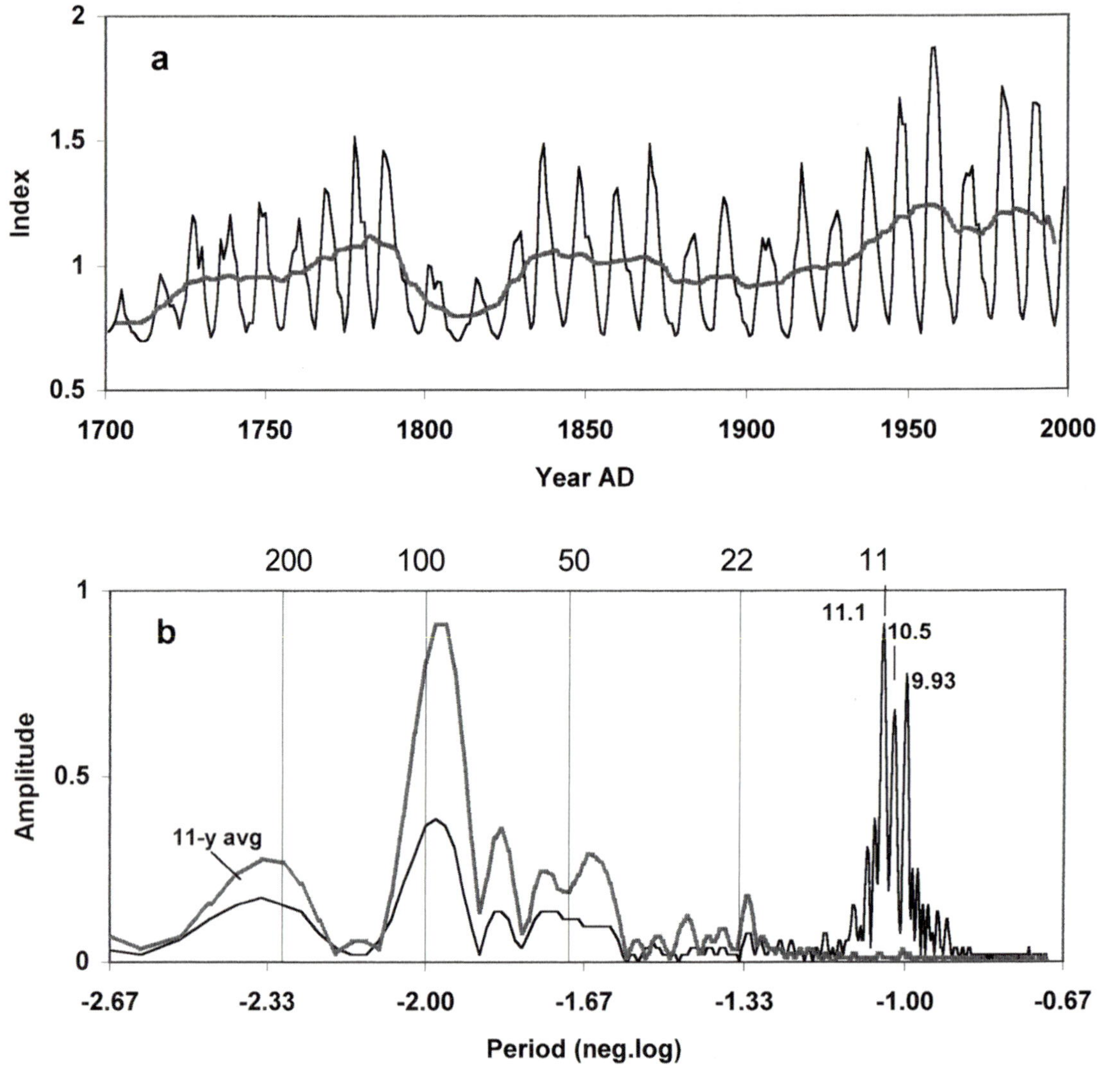

Fig. 4. Sunspot cycles of the last 300 years. Data courtesy J. Lean (see Lean et al. 1995; Hoyt and Schatten 1997). **(a)** Arbitrary index of Wolf numbers, centered on unity, and running mean. **(b)** Spectral properties of the sunspot series. Amplitude: Fourier power of autocorrelation series, arbitrary units.

comfort from the fact that a brightening Sun may be to blame for a substantial portion of 20th-century warming. Such a stance would be quite unwarranted. Two things should be kept in mind: (1) If the Sun controls substantial climate fluctuations by changing its brightness by only 0.25 percent, a change in virtual brightness of more than 1 percent (from present trace gas forcing) should have considerably greater impact than generally admitted, and (2) it cannot be assumed that the dust in the sky will keep on increasing in the measure necessary for balancing the effect from the excess greenhouse effect. On the contrary, clean burning (which is desirable for various reasons) will presumably diminish that effect from air pollution. Finally, (3) the reconstruction of sunspots, their relationship to solar energy output, and the link to overall background brightness are all quite in flux, as yet (Hoyt and Schatten 1997). Thus, solar contributions may be much less (or also somewhat more) than those estimated.

Solar Cycles in the Coming Century

Seen against the history of the last 400 years (since Galileo first described spots on the face of the Sun), the second half of the 20th century has been entirely unusual in terms of intensity of solar activity, having all four of the maximum amplitude sunspot cycles, as far as this can be ascertained. As mentioned, this fact is now recognized as an important ingredient in the global warming debate. Naturally, it would be very interesting to know whether the Sun will stay as hot as it has been since its overall background peak in 1950 (Fig. 4a) or whether it is on the way to a cooler state. Several years ago, Sir John Maddox (then editor of the science magazine Nature) speculated that the ca.-100-y cycle in sunspot activity would assert itself and take the Sun back to a quieter state, "saving our bacon" as far as global warming (at least for awhile) (Maddox 1995).

Can we predict how the Sun will change in the coming century? Will the Sun get dimmer soon? How soon?

To estimate the future behavior of the Sun, we make use of its cyclic behavior seen in the sunspots. As mentioned, based on inspection, one recognizes two cycles, one close to a decade (28.5 cycles in 300 years, Fig. 4a) and one close to a century. Both cycles emerge nicely in the spectral analysis (Fig. 4b). For this transform (from time series to spectral representation) we made two autocorrelation series, one with the first third of the time series as a moving template, and one with the last third. We then merged these two series such that they share the unity correlation for zero lag at the point of joining. The resulting series is 4/3 of the length of the original series, and needs no taper. We then used standard Fourier analysis on the merged autocorrelation series.

The most prominent sunspot cycle is 11.1 years long, precisely as indicated by Rudolf Wolf (1816-1893), erstwhile director of the Zurich Observatory, and the person who started a systematic collection of historic sunspot observations. (Hence the term "Wolf number" for the sunspot index.) Also, there are two other precisely defined cycles: 10.5 years and 9.93 years, with the cycle at 10.5 years being the weakest of the three. The relative strength is in the ratio 100 to 74.7 to 84.6.

One notes that nine cycles of the main period (11.1) make a century, while 10 cycles of the second most important period (9.93) just falls short of a century by one year. In fact, the beat cycle between 11.11 and 9.93 is 93.5 years, which is also close to a whole number multiple of 10.5. The beat cycles between 11.11 and 10.53 and between 10.53 and 9.93 are 201.7 and 174.3, respectively. We should expect, therefore, to see power in the spectrogram at periods just below 100 y and around 200 y, when analyzing the running average, precisely as noted (Fig. 4). (For a record only 300 y long, a period of 200 y cannot be defined with confidence.)

We next investigate the possibility that the length of the sunspot cycles changes as a function of overall activity of the Sun. To this end, we use the running mean to modify the amplitude of the cycles in the manner shown in Fig. 5 (a and b). First, we enhance the amplitude for those parts of the record for which the Sun shows a high level of activity (Fig. 5a). Then we do the opposite enhancement (Fig. 5b). After subtracting the running mean for each case, we calculate the spectrogram of the residual series by the above-described procedure. The result shows that all three main cycles are strong when the Sun is active (black line) but only the longest cycle (11.1) is important when the Sun is in a quiet phase, within the last 300 y (Fig. 5c). The result may be interpreted as showing that the cycle at 11.1 y persists, whereas the other two provide for an interference pattern that produces high energy states.

Our analysis shows that the cyclicity of the Sun is affected by the overall state of activity (cf. Hoyt and Schatten 1997). Nevertheless, for this demonstration, we use the simplest possible procedure for making a prediction of future activity of the Sun: extending the three basic decadal cycles beyond the present. We use sinusoidal cycles, although the sunspot cycles have aspects of a certain sawtooth format, with rapid rise and slow decay and with a fixed minimum and a variable maximum. For the exercise, we arrange the following sine curves in three series: periods of 11.11, 10.53, and 9.93, with amplitudes in the ratio of 0.91 to 0.68 to 0.77 (as

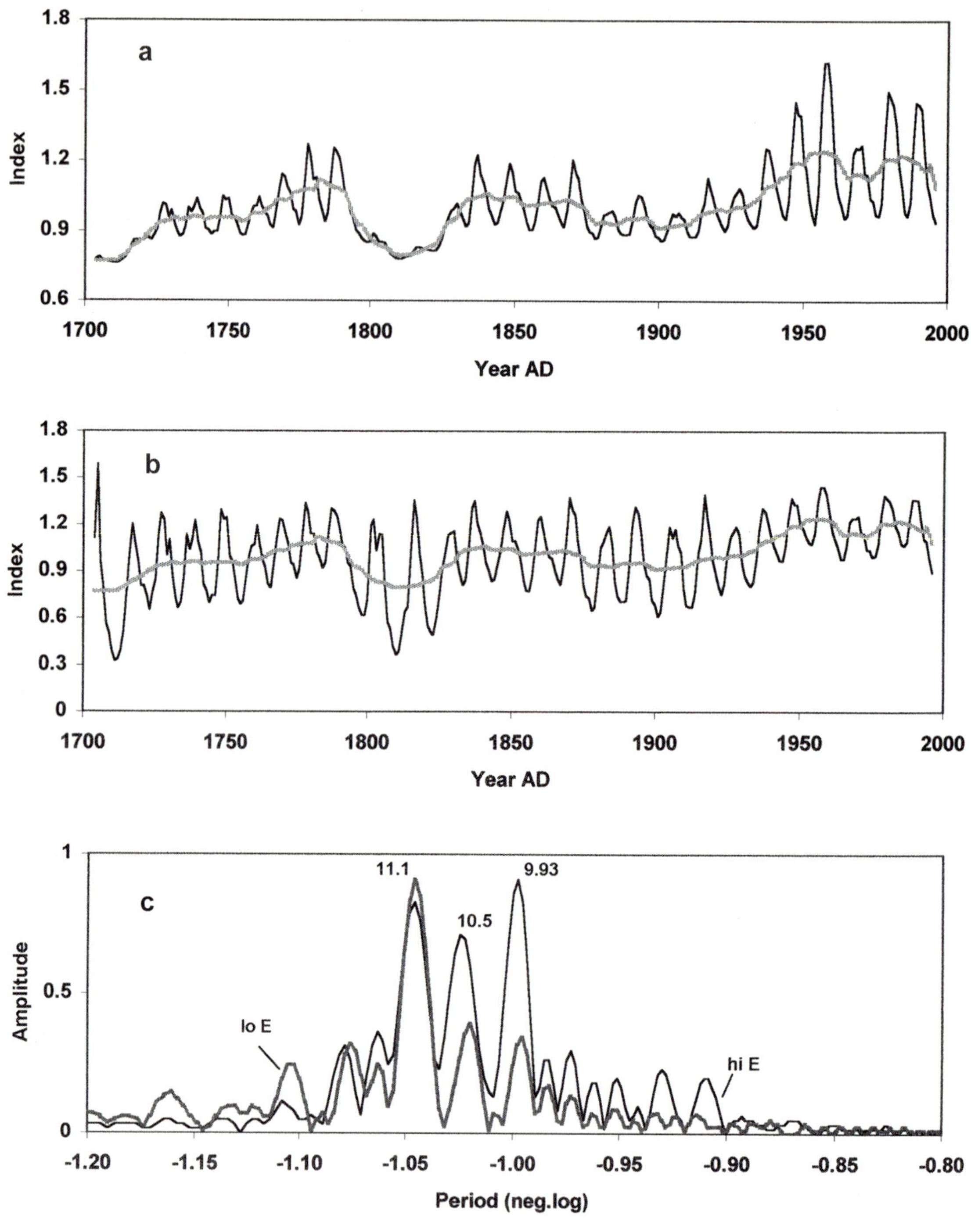

Fig. 5. Test for change in the periodicity of sunspots as a function of running mean. A high level of sunspot activity is taken as denoting high-energy output. **(a)** Enhancing the amplitude of cycles during high energy output, using the running mean for guidance. **(b)** Enhancing the amplitude of cycles during low energy output, using the running mean for guidance. **(c)** Spectrogram of the residuals of **(a)** and **(b)** after subtracting the running mean. "lo E", low energy state of the Sun. "hi E", high energy state of the Sun.

obtained from spectral analysis shown in Fig. 4b). We assume that the sum of the cycles yields an estimate of the maximum sunspot activity for any given cycle. Noting that the maximum activity was centered in the year 1958, we set the maximum for each cycle for 1958. This procedure defines the phase of all three cycles (Fig. 6a). Next, we simply add the values for the cycles for each year, into the past and into the future. We also set values below a certain threshold (chosen arbitrarily) to an overall minimum value. The resulting simulation has a remarkable similarity with the history of sunspot cycles back to 1830, that is, looking back 130 years from the time of maximum activity (Fig. 6b). The estimates do not match the data well whenever the Sun is in low-energy mode ("lo E" in the graph). During this condition the rules here used do not work well.

Looking forward, the interaction of the three main decadal cycles suggests we should see a reduction in solar activity, with a low-energy period centered on around 2020. Thus, the suggestion of Maddox (1995) that a decrease in solar activity will mitigate the global warming (at least temporarily) in the coming decades must be taken seriously. Warming from trace-gas emissions will continue,

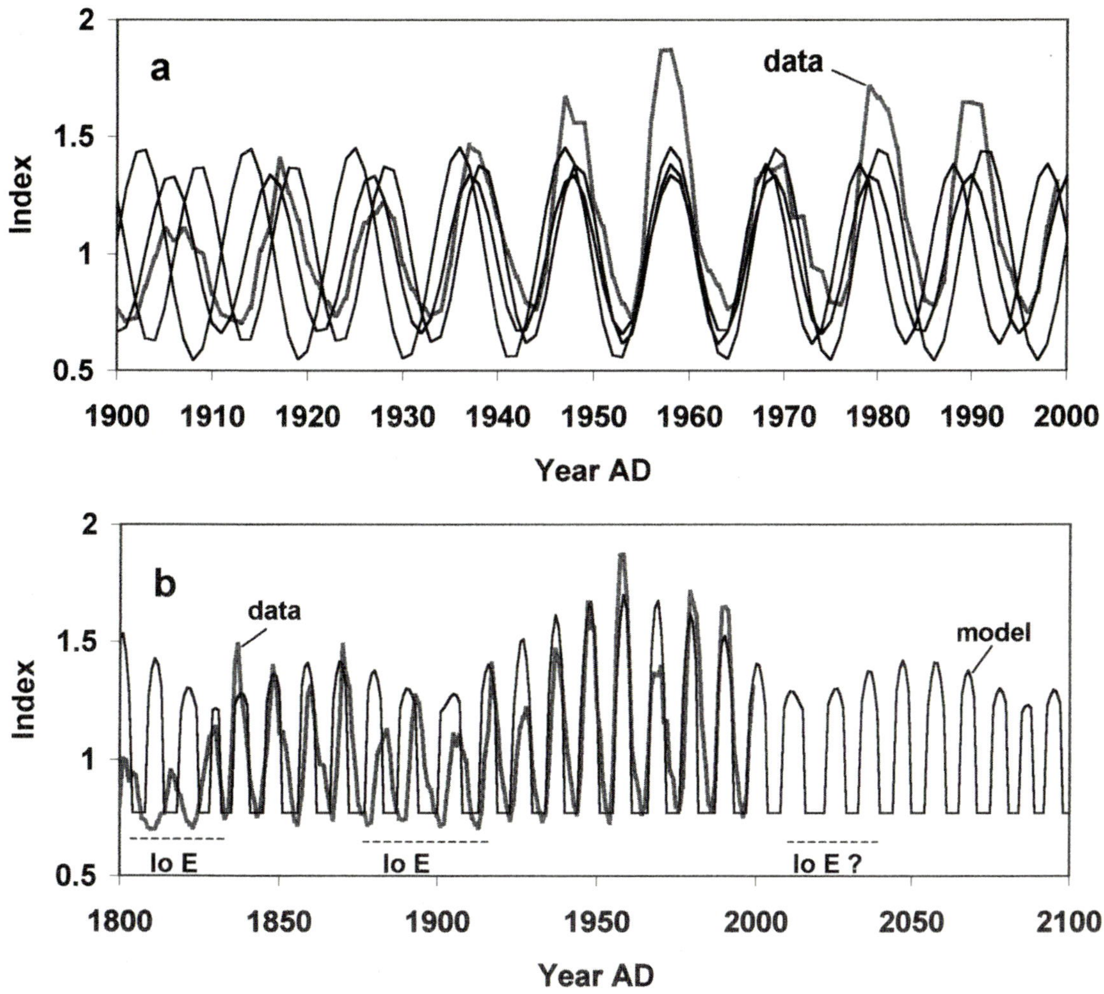

Fig. 6. Qualitative prediction of solar activity, using only three quasi-decadal cycles (11.11, 10.53, 9.93) and assuming they were precisely in phase in 1958. **(a)** Relationship of the individual cycles to solar sunspot activity. **(b)** Sum of the cycles compared with sunspot activity for the last 200 years and extrapolated into the future for 100 years. (Values are set to a constant minimum below a certain (arbitrary) threshold.)

but will be slowed by a cooling sun. Our analysis also suggests that by the middle of the coming century this temporary relief from a cool sun will be reversed again, at least partially.

We offer this exercise in prediction to demonstrate the usefulness of cycles, not as a state-of-the-art concerning forward extrapolation of solar activity. When extrapolating longterm solar cycles into the future (based largely on radiocarbon and beryllium-10) quite different results are obtained. Combining the 11-y cycle with a 88-y (Gleissberg) and a 208-y (Suess) cycle, Jirikowic and Damon (1994) suggest that the solar activity has not peaked, but will keep increasing toward a maximum centered near AD 2035, and will drop during the second part of the beginning century. They suggest that the present solar maximum is analogous to the one which presumably helped fuel the Medieval Warm Period (the time of colonization of Iceland and Greenland). Our basic argument is comparatively simple: the three quasi-decadal cycles are nearly in phase within the second half of the 20th century, re-enforcing each other, and they will be maximally out of phase within 30 years, given the basic 94-y cycle. The unusually strong sunspot activity in 1991 suggests that we should center common phase in 1969 (rather than 1958), which would delay the predicted drop in activity by one cycle.

Sunspots and the Weather in Bergen

The comparison of long-term solar brightness variation with long-term trends in temperature must cope with two difficulties: (1) the long-term solar brightness is not known within a factor of two (say), and (2) the actual effect on global warming from a given change in brightness also is not known. What we do have in hand, then, is a correlation of trends, with plausible arguments regarding possible forcing.

If the sun indeed forces climate, perhaps we should see solar cycles imprinted in the weather record of sensitive regions. Where might such a sensitive region be? An obvious candidate is any region that is much too warm for its latitude (or too cold). Presumably, such an area is kept in an anomalous condition by unusually strong import or export of heat, and such import or export would be vulnerable to disturbance. If the Sun is a source of such disturbance, its variation in brightness, over a sunspot cycle, should leave its mark in such an anomalous area.

The sea off Norway is the region with the greatest warm-water temperature anomaly in an otherwise frigid realm, and the rainfall in Bergen is tied to the flux of heat into the region. Thus, the rain (and snow) in Bergen is a good time series with which to test the possibility of solar influence (Fig. 7).

Two things are immediately obvious from the raw Bergen record and its spectrum (Fig. 7a, black line in 7b). Firstly, winter precipitation has increased within the last two decades, suggesting a link to a warming ocean offshore. Secondly, there is no indication of a weather cycle in the vicinity of the sunspot cycle of 11 years (the so-called Schwabe cycle), within the precipitation record. The test fails to produce the evidence we seek, and hence puts into question the entire argument relative to solar forcing.

Let us modify our expectations. Let us tentatively assume that the anomalous situation around Bergen is vulnerable to change in both directions, that is, the pulsed solar forcing can move the system into either warmer or colder conditions with equal probability, depending on circumstances. Differences in circumstances might arise from switches set elsewhere, for example the intensity of the high pressure region around the Azores. If we take this approach, it is not the heating that counts, but the fact that the system is being agitated by solar input (as proposed by Sverdrup in 1918; see Berger et al., this volume). The only interesting thing then is the size of deviations from average conditions, no matter in which direction. To capture this effect, we standardize and square the residual after subtracting the smoothed record from the original series. This procedure emphasizes large excursions from typical conditions. (It also produces exclusively positive values, and tends to double any cycles present.)

The periodicity of unusual excursions does show support for the concept that the Sun's cycle in-fluences the weather in Bergen: the period with the highest amplitude is 22 years, precisely that of

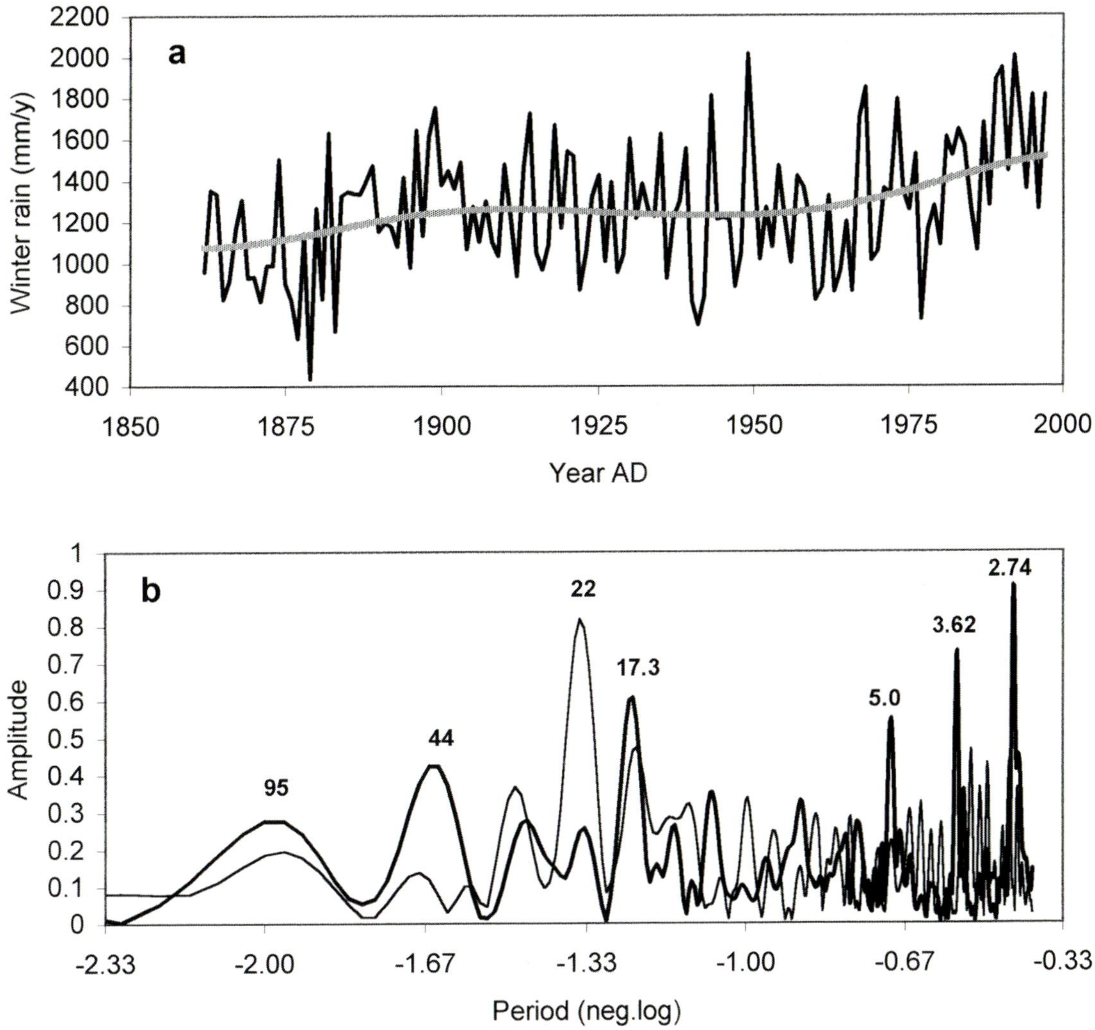

Fig. 7. Precipitation in Bergen and a test for solar forcing. **(a)** Variations in the amount of winter precipitation. Data University of Bergen. **(b)** Spectrum of winter precipitation (heavy line), and a derivation based on subtracting the trend, standardizing the residual, and squaring all the terms (thin line).

the well-known Hale cycle. This cycle is twice the length of the Schwabe cycle. It is in fact the "true" sunspot cycle, because the Sun's magnetic polarity changes from one 11-year cycle to the next, so that the repeat of conditions on the Sun occurs over two Schwabe cycles.

Why should we see the 22-year Hale cycle, but not the 11-year Schwabe cycle? The answer is not known. One possibility is that the Sun's effects on the upper atmosphere differ depending on the magnetic polarity of the sunspots, and that effects in the upper atmosphere somehow translate into climate change. Another possibility is that the system including the Norwegian Sea avoids oscillating in cycles near 11 years. Instead, the energy goes into a doubling of the Schwabe cycle (22 y) or into whole-number fractions. A one-third Hale cycle (at 7.3) may help produce the "North Atlantic Oscillation" (see Meincke, this volume). A third possibility is that the 22-y cycle has nothing to do with the sun at all, but reflects different forcing or internal oscillation. The significance of the cycles at

5.0, 3.62 and 2.74 is not clear. It may be noted that twice 5 is close to the second most powerful solar cycle (at 9.93), while both 3.62 and 2.74 have as their closest whole-year multiple the period of 11 years (10.9 and 10.96). (Whole-year multiples are important because, ultimately, climate cycles have to be tied to seasons.) Thus, these cycles could have a simple relationship to solar forcing. The cycle at 17.3 is close to the tidal period of 17.7 (the perigee cycle).

Tidal Cycles: Lunatic Weather Patterns?

Testing for the reality of solar forcing led us into a controversial field, but testing for lunar cycles is likely to strain credulity even further, among many climate scientists. In fact, as Walter Munk has pointed out on occasion, there is a name for people who become strongly attached to theories involving the Moon.

Suggestions that tidal forcing may influence weather patterns are nothing new, actually (see authors cited in Hoyt and Schatten 1997), and there is evidence bearing on the subject. Recently, for example, a satellite-derived 17-year dataset of daily temperatures in polar regions was found to contain a strong lunar cycle (Shaffer et al. 1997). Among tree-ring experts, Currie (1984, 1992) has long urged a lunar component in precipitation patterns, and impressive support for this position has been recently provided by Cook et al. (1997). They find cycles at 20 to 23 y, 7.8 y, 4.1 y and 2.5 y at the 95% confidence level, for a comprehensive data set for the USA, comprising the time span AD 1700-1978. A maximum entropy spectrum of the series shows narrow peaks at 22.2 y and 19.2 y, which they link, tentatively, to solar and lunar forcing, respectively.

Conceivably, bi-decadal cycles arise from internal oscillation, from feedback mechanisms between ocean and atmosphere (e.g. Latif and Barnett 1996). Of course, the existence of such oscillations (whether observed or computed) do not preclude outside forcing of similar periodicity. (In fact, as we have pointed out above, an interaction between periodic forcing and an oscillatory response is widely accepted as the solution to the ice age riddle.) The concept of lunar forcing has also entered the climate debate through a postulated effect of tides on deep ocean mixing (Sjöberg and Stigebrandt 1992; Munk and Wunsch 1998; Egbert and Ray 2000; Wunsch 2000) and a direct effect on the temperature of surface waters, through tidal mixing in shallow waters (Keeling and Whorf 1997).

Let us examine the Bergen record with the prospect of lunar forcing in mind.

The lunar orbit has a number of different elements. The well-known "synodic month" reckons time from one New Moon to the next (29.530588 days). Equally interesting are the following measures: the Moon's orbit relative to the intersection with Earth's orbital plane, the Moon's return to the same longitude in the sky, and the Moon's return to closest approach between Earth and Moon. The "draconic month" has a length of 27.21222 days, the "sidereal month" of 27.321661 days, and the "anomalistic month" of 27.554551 days. The difference beats between these three latter periods are 18.6 years, 8.85 years, and 6 years. These three periods together produce one more additional beat, 16.9 years. These then are periods that are relevant to tidal forcing (that is, to the range between the tidal maxima and minima).

The Bergen record (Fig. 7b) has cycles near 95, 44, 22, 17.3, 11.8, 9.9, 7.7, 5.0 and 3.62. These numbers are close to the following multiples and fractions of tidal periods (in the same sequence): 5 times 18.6, 5/2 times 17.7, one half previous, halfway between 17.7 and 16.9, 2/3 times 17.7, 5/3 times 6, with the rest unassigned. Thus, the main periods observed in the Bergen record could conceivably be related to tidal action. What is interesting is that we do not see much power in the base periods (18.6, 17.7, 6, 16.9). This is, of course, the reason that a tidal connection is not readily recognized, even if it were present. Two strategies in finding tidal connections may be important: searching for whole-number fractions and for smallest whole-number multiples. For example, the GISP record (see Berger et al., this volume) is seen to have a strong line near 29.5. This is 5/3 times 17.7. Similar peaks in the Bermuda coral record (30.8, 30, 27; see Berger et al., this volume) could be ascribed to analogous multiples (5/3 times 18.6, 5 times 6, 3/2 times 17.7). The main driver for moving energy off the base periods presumably is a lock-in to seasons, that is, whole-year multiples.

Additional examples for a possible tidal connection are readily found. The radiocarbon record of tree-rings in the Pacific northwest (AD 1514 to 1954) shows a prominent cycle at a period of 16.9 (Stuiver and Braziunas 1993; their Fig. 2). The cycle, as mentioned, is a lunar beat cycle. The second most prominent cycle (13.2) may be taken as 3/4 times 17.7.

Admittedly, the list of tidal cycles generated by using multiples and whole-number fractions offers much opportunity for finding suitable matches, given some arbitrary list of periods found in a series of observations. To keep it simple, one might prefer to strictly stay with the main periods seen in the actual gravitational effect on sea level, in mid-latitudes (Fig. 8). The spectrum shown (Fig. 8b) is based on expected maximum sea level stands between 1938 to 1994, off San Francisco, as calculated for a water-covered planet, based on Newtonian mechanics (output series courtesy D. Agnew and R. Flick, S.I.O.). The main lines are at 18.6, 17.7 and 4.425. The smallest whole-number multiple of 18.6 that is in phase with seasons is 93 for 18.6, and 177 years for 17.7 and 4.425 (although 53 is close). Beating (by difference tone) 18.6 with 17.7 yields 365.8, 18.6 with 8.85 yields 16.9, and the beat of 4.425 with the quasi-bidecadal cycles yields 5.8 to 5.9. The sum cycles are 9.3, 9.1, 6.0, 3.57, 5.9, 3.54, 2.95 and 2.2. From this reasoning, the main cycles to look for when testing for lunar forcing are 366, 177, 93, 53, 18.6, 17.7, 16.9, 9.3, 9.1, 6.0, 5.87, 4.425, 3.57, 3.54, 2.95 and 2.2, with a requirement that a "match" should be very close to the tidal period.

Yet more simple: the shortest period for which syzygy (full or new moon), perigee, eclipse potential, and perihelion nearly coincide is 18.03 years (Keeling and Whorf 1997) and this period, and its whole-number multiples and fractions, is central therefore (in analogy to the 11-y sunspot cycle) (see Fig. 8b). On long time-scales, a 93-y cycle is dominant, although over many centuries a 90-y cycle takes precedence (Wood 1986). Working back from a strong tidal event in 1973/1974, we obtain 1880 and 1787 as years of unusually high tides. Working forward from AD 1247, we have 1340, 1433, 1526, and 1619 as years of high tides (Cartwright 1974). Wood (1986) has urged that storm floods occurring during such times of high tides are especially damaging, although the most important factor of a storm tide, of course, is the storm itself (Dietrich et al. 1975, p.436).

The longest well-expressed cycle that may well be of lunar origin is the 1470-year cycle described from the GRIP2 ice core (Grootes and Stuiver 1997; Schulz et al. 1999). It will be noted that the main interference cycle between the two lunar master periods (18.6 and 17.7) is 366, almost precisely one fourth of the Greenland cycle (1464 vs. 1470). This is the strongest of the beats for periods of 100 y or more. We suggest it is the underlying cause of the prominent 1470-y cycle seen in the oxygen isotope record of the Greenland ice (Fig. 9). This cycle apparently is modulated both by the amount of ice present on the globe (as seen in the oxygen isotope record of Ontong Java Plateau), and by the rate of change in the ice mass (Schulz et al. 1999). In a scenario of lunar forcing of this cycle, unusually large tides during the height of summer, when acting on marine-grounded unstable ice shelves, would tend to free large sheets of ice from the shelves, which would then drift south and change the climate, periodically. The factor of 4 over the 17.7||18.6 beat would reflect the requirement that the maximum tidal action occur a relatively narrow window during the summer season. Similar cycles, presumably, could resume whenever the West-Antarctic ice sheet becomes responsive to tidal action, as the sea level rises.

But why should the Moon influence the weather in Bergen?

A remarkable passage in a standard oceanography text (Dietrich et al. 1975, p. 440) contains hints for a possible answer. The text explains how the turbulence generated by tidal action in shelf seas sets up seasonal temperature anomalies (with consequences for the occurrence of fog). The seasonal temperature contrast is diminished, and the timing of the temperature maximum is delayed, in the case of the North Sea and Irish Sea, for example. Thus, we may safely assume that a change in intensity of mixing on the shelf of the North Sea, in line with the great lunar rhythms of maximum tidal range, would alter the seasonal course of the surface temperatures of that sea and hence modify the atmospheric pressure in the region. This could conceiv-

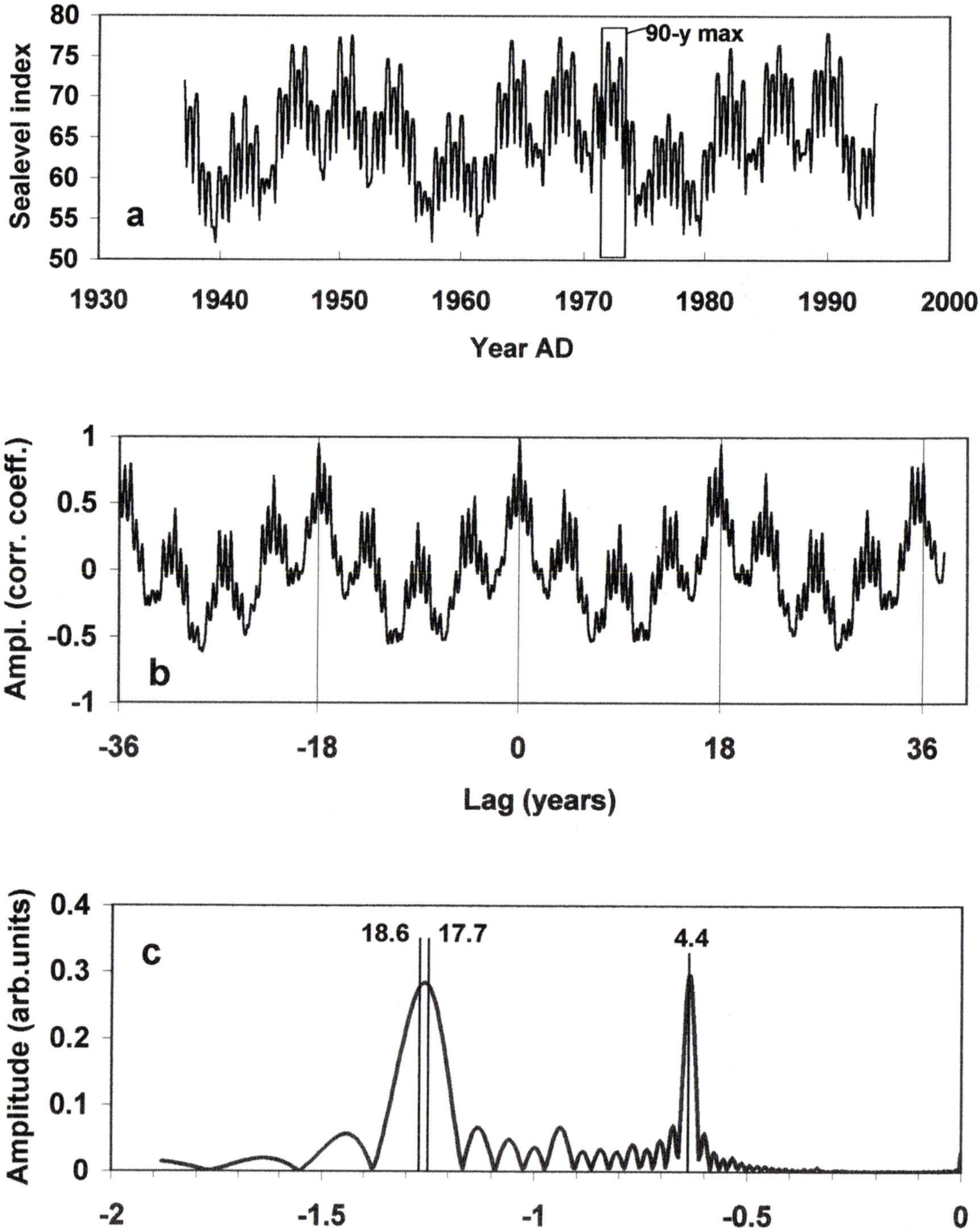

Fig. 8. Periodicity of mid-latitude tides. **(a)** Sealevel indices for high-stands, based on squaring calculated monthly maxima for sea level off San Francisco, on a water-covered planet. Sealevel output series courtesy of R. Flick, S.I.O. (program by A. Duncan). The rectangle indicates the timing of a 90-year maximum expected from perigee and nodal position, according to Keeling and Whorf (1997). **(b)** Autocorrelation series of series in 8a. **(c)** Spectrum (Fourier scan with 1 percent offset) of 8b.

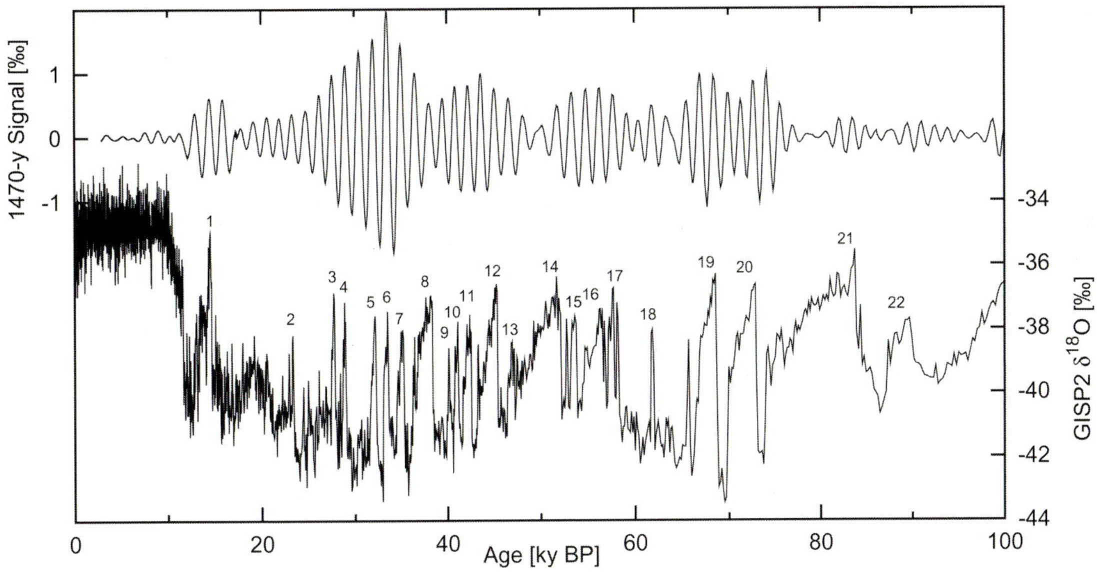

Fig. 9. Amplitude variations in the 1470-year cycle in Greenland ice. From Schulz et al. (1999).

ably set up a flickering switch, influencing the path of westerly marine winds. Analogous mixing processes on the Barents Sea shelf also could be very important in admitting or blocking southerly winds.

Of course, these speculations are by no means "explanations", but rather thought experiments to demonstrate that it is possible to conceive of ways to link the Moon to the weather in the Nordic Sea. It should be noted that even the 22-year cycle (here addressed as "Hale cycle") could be related to tidal forcing, since five times 4.425 (the shortest of the tidal range cycles in mid-latitudes) comes out as 22.1 years. (The multiple could be rationalized by saying that there is a 10-week window within the year, when maximum forcing has to be applied, to be effective.)

In discussing solar and tidal cycles, it must not be forgotten that the presence of cycles in the Bergen rain- and snowfall (and in the related North Atlantic Oscillation; Hurrell 1995) is not generally accepted. Windowed spectral analysis, resulting in spectral landscapes, show that such cycles come and go and are usually ill-defined. While Sun and Moon surely influence the weather, the climate system reacts to the forcing in unpredictable ways and captures the pulsed energy into its own oscillations (e.g. Mysak and Venegas 1998). (In the case of Bermuda coral, increased tidal mixing is readily linked to increased growth, through nutrient supply, so there may be no strong climate connection for the cycle.)

Nature's Experiments: Volcanic Winter

It is no surprise that a major deglaciation will trigger volcanism. The changes in elevation, through isostatic response to redistribution of ice and water masses, as well as the corresponding changes in the rotation of the planet, set up stresses in the crust and upper mantle which can result in the release of pent-up volcanic energy. Volcanogenic particles in the stratosphere can cause noticeable cooling of the climate, by blocking incoming sunlight over an entire hemisphere (Schneider and Londer 1984). Embedded within a major cold spell, such as the Younger Dryas, the effects of a "volcanic winter" on animals and people living in the tundra would surely be catastrophic.

We can study details of effects on climate from volcanic eruptions using information from the last several centuries. Briffa et al. (1998) summarized the relationship between volcanic eruptions and extremely low tree-ring density averaged around the northern hemisphere. The densities are taken as indicators for summer warmth. The timing of lowest densities closely follows major eruptions in almost all cases. The following is a list of low-density years, with the immediately preceding eruption (-1 yr) indicated in parentheses (derived from their Tables 1 and 2):

1601 (Huaynaputina, Peru), 1816 (Tambora, Lesser Sunda Isl.), 1641 (Parker, Philippines), 1453 (Kuwae, Vanuatu, SW Pacific), 1817 (Tambora, 1815), 1695 (n.d.), 1912 (Novarupta, Katmai Alaska, same year) 1675 (Gamkonora, Halmahera, 1673), 1698 (n.d.), 1643 (Parker, Philippines, 1641), 1699 (n.d.), 1666 (possible eruptions in Japan), 1884 (Krakatoa, west of Java), 1978 (n.d.), 1837 (Csiguina, Nicaragua, 1835), 1669 (Shikotsu, Japan, 1667), 1587 (Kelut, Java), 1740 (Shikotsu, Japan), 1448 (possible eruptions in Alaska), 1992 (Pinatubo, Philippines), 1836 (Coseguina, Nicaragua), 1818 (Tambora, 1815), 1495 (n.d.), 1968 (n.d.), 1742 (Shikotsu 1739), 1783 (n.d.), 1667 (Shikotsu, same year), 1642 (Parker, Philippines), 1819 (Tambora 1815), 1446 (n.d.). Acid layers in ice cores from Greenland suggest that the events of 1666, 1695 and 1698 likewise were linked to volcanic eruptions, according to Briffa et al. (1998).

Uncertainties of dating disallow unique assignment in many cases. Multiple eruptions (not necessarily from the same volcano) may be assumed for long-lasting spells of unusually low temperatures (see also the discussion of the 1601 eruption of Huaynaputina, by De Silva and Zielinski 1998).

It appears, from the compilation of Briffa et al. (1998) that volcanic activity was an important factor in the cold spells of the Little Ice Age. Presumably such spells were the more severe, since they were superimposed on the background of a generally cooler climate. The largest of the eruptions, in Peru in 1601, is thought to have caused severe economic damage in Peru and its neighbors, such that it took 150 years to recover (De Silva and Zielinski 1998). The second largest eruption (Tambora 1815) was earlier identified as having emitted enormous amounts of particles into the stratosphere (Lamb 1972), the most effective shading being provided by sulfurous aerosol. Severe winters, late frost, and cool summers tend to follow such eruptions, with attendant poor harvests (Lamb 1982). One notorious year in this regard, in western Europe, was 1739 (Appleby, in Rotberg and Rabb 1981). It greatly stimulated emigration, being embedded in a series of poor harvests. Similarly, the Tambora and Cosiguina eruptions resulted in much economic stress. (Ladurie says that all of Europe spent the summer around the fire, in 1816; there was frost in July, and famine followed.)

It appears evident that some of these occurrences of tree-recorded stress are clumped (most prominently in the 1690s) and that periods between events or clumps of events are of the order of 25 to 40 years. To check on this idea, we performed spectral analysis of the stress record of Briffa et al. (1998), assigning weights from 6 to 2 to the 30 low-density-ring years, pro-rated according to rank. We allowed these values to diffuse into neighboring years, before analysis (in part to take account of uncertainties in dating, and also to allow for small deviations from any underlying cycles). We then obtained three autocorrelation series, one using the top third as template, the second using the middle third, and the third using the bottom third of the input. The resulting autocorrelation series were expressed in terms of Fourier matrices. The average amplitude of the three series shows peaks near 200, 75 (4 times 18.6), 46.5 (5/2 times 18.6), 36.1 (twice 18.03), 29.9, 24.5 (4/3 times 18.6) and 19.2 years (Fig. 10). (Expressions in parentheses identify close-by tidal multiples.) These periods may be compared to those found in the Bermuda coral record, which we report elsewhere in this volume (80, 37.5 (twice 18.6), 33.6 (twice 16.9), 30.8, 30, 27, 21.7, 21). In particular, the dominant period in this range, at 30 years (in the disturbance record, Fig. 4b of Berger et al., this volume) is identical to the central volcanism cycle at 29.9. The main period in the GRIP2 ice record, in this general range, is 29.5 years (Berger et al., this volume, Fig. 2a). As mentioned, Currie (1992) found a 19.2-y cycle in the tree-growth record of western Europe, which he assigns to lunar forcing.

If we follow the arguments of Briffa et al. (1998) that the tree stress reflects volcanic erup-tions, we must seriously consider the possibility that the periods seen in the coral-growth record also largely reflect the climatic effects of volcanic activity. (The elevated growth rate after the great Peruvian eruption of AD 1601 suggests favorable effects

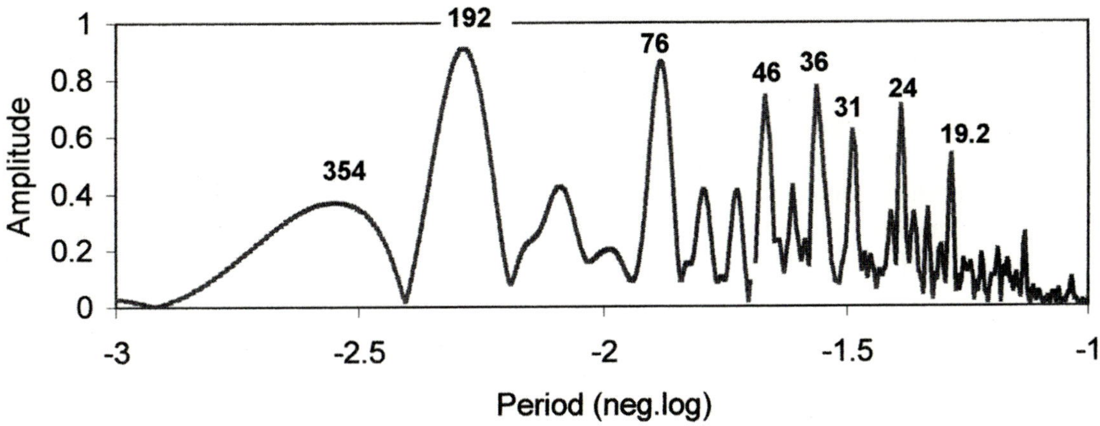

Fig. 10. Spectrum of extreme stress events in trees from the northern hemisphere, as listed by Briffa et al. (1998). These are related to volcanic eruptions, according to Briffa et al. Analysis is by application of autocorrelation to the treated series, and Fourier analysis of the autocorrelation series (see text). The numbers labeling the peaks are the lengths, in years, of the cycles contained in the major volcanic eruption series (as seen in the response of trees).

from increased mixing during winter storms, in this nutrient-starved region.)

That volcanic eruptions may show periodicity is not a new idea. V. A. Alekseyev (1990) reports on previous studies indicating the possible existence of cycles with periods of 5-6, 11, 19, 23, 55, 60-90 and 180 y. He analyzed a rather large data set (4244 eruptions in 1760-1980) using both Fourier and maximum-entropy statistics. He divided the set into two groups, the first (1437 eruptions) containing active volcanoes (no less than 10 eruptions each for the interval studied), the second comprising the remainder (2751 cases). The active volcanoes (thought to be the more sensitive) showed periods of 17 and 16.7, depending on the analysis (Fourier or maximum entropy). Also, maximum entropy analysis yielded a modest peak near 9 years. The remainder of the eruptions showed peaks near 23 and 21, respectively, for an average of 22 years.

Alekseyev (1990) suggests that the underlying forcing period is the nodal cycle of the Moon, at 18.6 years, arguing that the overall average of 17, 16.7, 21 and 23 is close to that value. Whether this be accepted or not, the periods of 17 and 22 years (which are the ones he actually found) are the two most prominent in the Bergen winter precipitation record (Fig. 7b) — implying yet another cause for these periods, that of volcanic activity, conceivably forced by Earth tides (or the pressure effects from the great Kelvin waves generated by the tides). If eruptions occur in cycles, at least in part, and if the periods of these cycles are linked to tides, this would support Baillie's suggestion that bolide impacts have triggered certain major volcanic eruptions over the last several thousand years. If volcanoes listen to the tides, they can surely react to the seismic waves generated by large impacts.

Circular conclusions

Are climate cycles real? If yes, are they useful?

Some climate cycles surely are real. In this category we should put, without hesitation, the Milankovitch cycles of the ice ages. Not only do they exist, but a plausible mechanism for forcing has been found and is generally accepted (albeit with some discussion; see Muller and MacDonald 2000). Another cycle, much shorter than the ice-age cycles, is likely real: the 1470-y Greenland cycle. We suggest it is a fourfold multiple of the 366-y beat of the two main tidal periods (18.6 and 17.7). If it tests for availability of marine-based ice (as we suggest) it should be a useful tool in paleoglaciology and paleoclimatology in general. Solar brightness cycles and lunar and solar gravitational cycles also are real, of course, but their effects on climate, if any, are as yet in need of documentation. It is likely that such effects exist, but that they are easily masked by stochastic behavior of the system and perhaps by internal oscillations. Cyclicity in volcanic activity, at this point, is entirely in the realm of speculation.

At the beginning of this chapter we quoted Le Roy Ladurie (1971), citing his emphatic dismissal of the work of Brückner and of other students of climate cycles. Ladurie, to his credit, offers both wine harvest data and oak-ring data, which makes it possible to test his assessment. In lieu of a final conclusion on the subject, we offer an analysis of these data (Fig. 11). It appears that both time series contain cycles familiar from other environments. In any case, both contain the Brückner cycle (at 35 plus or minus a year).

The wine harvest has a strong cycle at 112 (precisely six times the lunar nodal cycle) and at 58 (close to three times the nodal cycle) and a weak expression of both the nodal and the perigee cycle (two peaks close to 36; expected is 37.2 and 35.4).

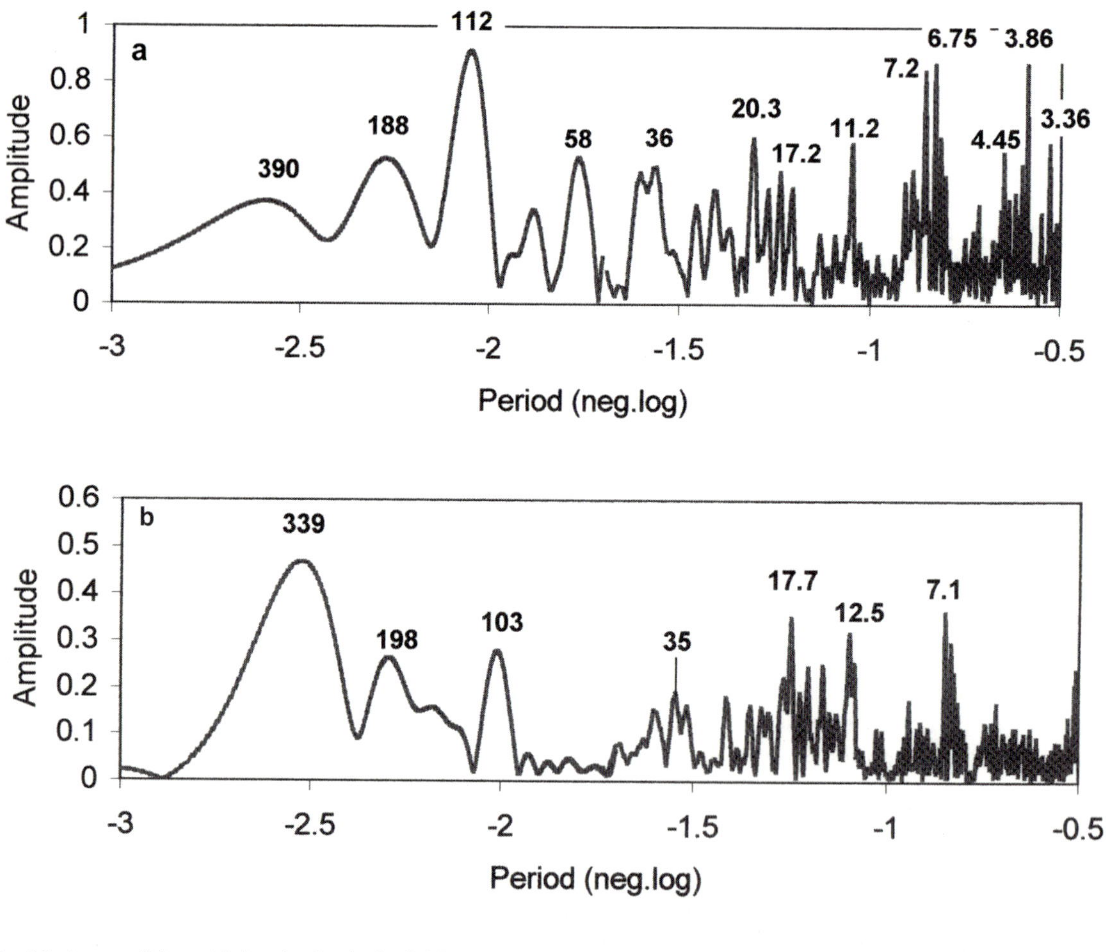

Fig. 11. A test of the criticism by Ladurie (1971) regarding the study of climate cycles, using the data provided in the appendix of his book "Times of Feast, Times of Famine, a History of Climate since the Year 1000". **(a)** Spectrum of wine harvest data. **(b)** Spectrum of oak-ring data.

Also, at 17.2, there is a line half-way between the fundamental perigee cycle (17.7) and the beat at 16.9 (one half perigee with the nodal cycle). Unaccounted for, in the range between 10 and 100, are cycles at 20.3 and 11.2 y, which could be of solar origin. (It would be fitting for the wine to reflect the Sun's activity and not just the Moon's.) The strong lines at 7.2 and 6.75 are reminiscent of the North Atlantic Oscillation. It will be noted that the smallest whole-number multiple of 7.2 is 36. For 6.75 it is 27 (3/2 times 17.7). The cycle at 4.45 (although weak) belongs to perigee (4.425). The cycles at 3.86 and 3.36 are of unknown origin; their smallest whole-number multiples are 27 and 10.

The oak tree data (for which only the series since 1401 was used) have an important cycle near 3-and-a-half centuries, reminiscent of the beat between the nodal and the perigee periods (366). A cycle near 103 seems linked to the Brückner cycle at 35 (factor of three), which in turn is linked to the perigee cycle (at 17.7, and well expressed). The cycle at 12.5 is of unknown origin (although familiar from elsewhere, see Biondi et al. 1997). The period of 7.1 again recalls the North Atlantic Oscillation.

One might criticize this type of analysis as rather arbitrary, especially when no reference is made to the phase between forcing and presumed climate response, and to possible mechanisms. After all, Fourier analysis is meant to produce cycles; it should not come as a surprise that one can find them when converting time series to spectra. In reply, we simply note that (a) the Brückner cycle is present in the data of Ladurie, (b) periods close to perigee and nodal cycles of the Moon, and simple multiples of these, seem over-represented in the spectra of climate proxies, and (c) the search for cycles in climate change and the search for explanation of such cycles is likely to result in some interesting questions and insights. The ad hoc dismissal of climate cycle research, as reflected by Penck's remarks to Milankovitch and by Ladurie's comments on Brückner's work, is now largely a thing of the past. Whatever prejudice lingers has found more subtle expression: It tends to focus on the details of time series analysis.

References

Alekseyev VA (1990) Periodicity in current terrestrial volcanism. Geochem. Int. 27:124-127 (Transl. from Geokhimiya, 4, 599-603, 1990)

Bard E, Raisbeck G, Yiou F, Jouzel J(2000) Solar irradiance during the last 1200 years based on cosmogenic nuclides. Tellus 52B:985-992

Beer J, Mende W, R. Stellmacher R (2000) The role of the sun in climate forcing. Quaternary Science Reviews 19, 403-415.Beer, J., W. Mende, and R. Stellmacher, 2000. The role of the sun in climate forcing. Quaternary Science Reviews 19:403-415

Berger AL, Imbrie J, Hays J, Kukla G, Saltzman B (eds) (1984) Milankovitch and Climate (Part 1). Hingham, Mass. (Reidel), pp 1-510

Berger WH (1999) The 100-kyr ice-age cycle: internal oscillation or inclinational forcing? Int J Earth Sciences 88:305-316

Berger WH, Von Rad U (submitted) Decadal to millennial cyclicity in varves and turbitides from the Arabian Sea: Hypothesis of Tidal Origin. Global and Planetary Change.

Berger WH, Bickert T, Yasuda MK, Wefer G (1996) Reconstruction of atmospheric CO_2 from ice-core data and the deep-sea record of Ontong Java plateau: the Milankovitch chron. Geol Rundsch 85:466-495

Berger WH, Wefer G, Richter C, Lange CB, Giraudeau J, Hermelin O, and Shipboard Scientific Party (1998) The Angola-Benguela upwelling system: paleoceanographic synthesis of shipboard results from Leg 175. In: Wefer G, Berger WH, Richter C et al., Proc. ODP, Init Repts 175:505-532

Berger WH, Lange CB, Von Rad U (2000) Brückner cycles in marine varves - detection and possible significance. European Geophys. Society, Nice, March 2000 (abs.)

Biondi F, Lange CB, Hughes MK, Berger WH (1997) Inter-decadal signals during the last millennium (AD 1117-1992) in the varve record of Santa Barbara basin, California. Geophys Res Letters 24(2):193-196

Bradley RS, Jones PD (1992) Climate Since A. D. 1500. Routledge, London, pp 1-679

Briffa KR, Jones PD, Schweingruber FH, Osborn TJ (1998) Influence of volcanic eruptions on northern hemisphere summer temperature over the past 600 years. Nature 393:450-455

Cartwright DE (1974) Years of peak astronomical tides. Nature 248:656-657

Cook ER, Meko DM, Stockton CW (1997) A new assessment of possible solar and lunar forcing of the bidecadal drought rhythm in the western United States.

J Climate 10:1343-1356

Crowley TJ (2000) Causes of climate change over the past 1000 years. Science, 89:270-277

Currie RG (1984) Periodic (18.6-year) and cyclic (11-year) induced drought and flood in western North America. J Geophys Res 89:7215-7230

Currie RG (1992) Deterministic signals in tree-rings from Europe. Ann Geophysicae 10:241-253

De Silva SL, Zielinski GA (1998) Global influence of the AD 1600 eruption of Huaynaputina, Peru. Nature 393:455-458

Diaz H F, Markgraf V (eds) (1992) El Niño, Historical and Paleoclimatic Aspects of the Southern Oscillation. Cambridge Univ. Press, Cambridge UK, pp 1-476

Dietrich G, Kalle K, Krauss W, Siedler G (1975) Allgemeine Meereskunde, Eine Einführung in die Ozeanographie. Gebrüder Bornträger, Stuttgart, pp 1-593

Eddy JA (1976) The Maunder Minimum. Science 192:1189-1202

Egbert GD, Ray RD (2000) Significant dissipation of tidal energy in the deep ocean inferred from satellite altimeter data. Nature 405:775-778

Frenzel B (ed) (1995) Solar output and climate during the Holocene. Gustav Fischer Verlag, Stuttgart, pp 1-186

Gasse F (2000) Hydrological changes in the African tropics since the last Glacial Maximum. Quaternary Science Reviews 19:189-211

Grootes PM, Stuiver M (1997) Oxygen 18/16 variability in Greenland snow and ice with 10^{-3} - to 10^{5}-year time resolution. J Geophys Res C102:26455-26470

Grove JM (1988) The Little Ice Age. Routledge, London, pp 1-498

Hays JD, Imbrie J, Shackleton NJ (1976) Variations in the Earth's orbit: pacemaker of the Ice Ages. Science 194:1121-1132

Hoyt DV, Schatten KH (1997) The Role of the Sun in Climate Change. Oxford Univ. Press, New York Oxford, pp 1-279

Hurrell JW (1995) Decadal trends in the North Atlantic Oscillation, regional temperatures and precipitation. Science 269:676-679

Imbrie J, Imbrie KP (1979) Ice ages: solving the mystery. Enslow, Short Hills, New Jersey, pp 1-224

Jirikowic JL, Damon PE (1994) The medieval solar activity maximum. In: Hughes MK and Diaz HF (eds) The Medieval Warm Period. Kluwer Academic, Dordrecht, pp 309-316

Jones PD, Bradley RS, Jouzel J (1996) Climatic Variations and Forcing Mechanisms of the Last 2000 Years. Springer-Verlag Berlin Heidelberg, pp 1-649

Jones PD, Briffa KR, Barnett TP, Tett SFB (1998) High-resolution palaeoclimatic records for the last millennium: Interpretation, Integration and Comparison with general circulation model control-run temperatures. The Holocene 8:455-471

Keeling CD, Whorf T (1997) Possible forcing of global temperature by the oceanic tides. Proc Natl Acad Sci USA 94:8321-8328

Köppen V, Wegener A (1924) Die Klimate der geologischen Vorzeit. Berlin

Kutzbach JE, Liu Z (1997) Response of the African monsoon to orbital forcing and ocean feedbacks in the middle Holocene. Science 278:440-443

Ladurie EL (1971) Times of Feast, Times of Famine: A History of Climate since the Year 1000. (Transl. B. Bray.) Doubleday, Garden City, New York

Lamb HH (1972) Climate: Present, Past and Future - Vol. 1: Fundamentals and Climate Now. Methuen, London, pp 1-613

Lamb HH (1982) Climate History and the Modern World. Methuen, London, pp 1-387

Latif M, Barnett TP (1996) Decadal climate variability over the North Pacific and North America: Dynamics and predictability. J Climate 9:2407-2423

Lean J (1997) The Sun's variable radiation and its relevance for Earth. Ann Rev Astron Astrophys 35:33-67

Lean J, Beer J, Bradley R (1995) Reconstruction of solar irradiance since 1610: Implications for climate change. Geophysical Research Letters 22:3195-3198

Lowe JJ (1991) Integration of pollen and other proxy data: The Holocene paleoclimate of the British Isles and adjacent parts of Europe. In: Frenzel B, Pons A, Gläser B (eds) Evaluation of climate proxy data in relation to the European Holocene. Gustav Fischer Verlag, Stuttgart, pp 37-50

Maddox J (1995) Natural antidote to global warming? Nature 377:193

Mann ME, Bradley RS, Hughes MK (1998) Global-scale temperature patterns and climate forcing over the past six centuries. Nature 392:779-787

Milankovitch M (1930) Mathematische Klimalehre und astronomische Theorie der Klimaschwankungen. Handbuch der Klimatologie, Bd 1, Teil A. Bornträger, Berlin, pp 1-176

Muller RD, MacDonald GJ (2000) Ice Ages and Astronomical causes. Springer Praxis, Chichester, pp 1-318

Munk WH, Wunsch C (1998) Abyssal recipes II: Energetics of tidal and wind mixing. Deep-Sea Res 45:1977-2010

Mysak LA, Venegas SA (1998) Decadal climate oscillations in the Arctic: A new feedback loop for atmosphere-ice-ocean interactions. Geophys Res Letters 25:3607-3610

Penck A, Brückner E (1901-1909) Die Alpen im Eiszeitalter. 3 vols, Tauchnitz, Leipzig, pp 1-1199

Petit JR, Jouzel J, Raynaud D and 16 others (1999) Climate and atmospheric history of the past 420,000 years from the Vostok ice core, Antarctica. Nature 399:429-436

Plaut G, Ghil M, Vautard R (1995) Interannual and interdecadal variability in 335 years of central England temperature. Science 268:710-713

Rotberg RI, Rabb TK (eds) (1981) Climate and History. Princeton University Press, Princeton New Jersey, pp 1-280

Rowley-Conwy P (1985) The origin of agriculture in Denmark: a review of some theories. J Danish Archaelogy 4:188-195

Rossignol-Strick M (1985) Mediterranean Quaternary sapropels, an immediate response of the African monsoon to variation of insolation. Palaeogeogr, Palaeoclimat, Palaeoecol 49:237-263

Schneider SH, Londer R (1984) The Coevolution of Climate and Life. Sierra Club Books, San Francisco, pp 1-563

Schulz M, Berger WH, Sarnthein M, Grootes PM (1999) Amplitude variations of 1470-year climate oscillations during the last 100,000 years linked to fluctuations of continental ice mass. Geophys Res Letters 26:3385-3388

Schwarzacher W (1993) Cylostratigraphy and the Milankovitch Theory. Elsevier, Amsterdam London, pp 1-225

Shaffer JA, Cerveny RS, Balling RC (1997) Polar temperature sensitivity to lunar forcing? Geophys Res Letters 24:29-32

Sjöberg B, Stigebrand A (1992) Computations of the geographical distribution of the energy flux to mixing processes via internal tides and the associated vertical circulation in the ocean. Deep Sea Research 39:269-291

Stehr N, Von Storch H (eds) (2000) Eduard Brückner — The Sources and Consequences of Climate Change and Climate Variability in Historical Times. Kluwer Academic, Dordrecht, pp 1-338

Stuiver M, Braziunas TF (1993) Sun, ocean, climate and atmospheric $^{14}CO_2$: an evaluation of causal and spectral relationships. The Holocene 3:289-305

Stuiver M, Grootes PM, Braziunas TF (1995) The GISP2 $\delta^{18}O$ climate record of the past 16,500 years and the role of the sun, ocean and volcanoes. Quaternary Res 44:341-354

Vos H, Sanchez A, Zolitschka B, Brauer A, Negendank JFW (1997) Solar activity variations recorded in varved sediments from the Crater Lake of Holzmaar - a maar lake in the Westeifel volcanic field, Germany. Surveys in Geophysics 18:163-182

Wigley TML, Raper SCB (1990) Natural variability of the climate system and detection of the greenhouse effect. Nature 344:324-327

Wood FJ (1986) Tidal Dynamics. Reidel, Dordrecht

Wunsch C (2000) Moon, tides and climate. Nature 405:743-744

Yiou P, Jouzel J, Johnsen S, Rognvaldsson OE (1995) Rapid oscillations in Vostok and GRIP ice cores. Geophys Res Letters 22:2179-2182

Zvelebil M (1996) The agricultural frontier and the transition to farming in the circum-Baltic region. In: Harris DR (ed) The origins and spread of agriculture and pastoralism in Eurasia. Univ. College London Press, London, pp 323-345

Tracing Climate-Variability: The Search for Climate Dynamics on Decadal to Millennial Time Scales

Michael Schulz[1], Wolfgang H. Berger[2], Michael Baillie[3], Jürg Luterbacher[4], Jens Meincke[5], Jörg F. W. Negendank[6], Andre Paul[2,7] and René O. Ramseier[8]

[1]*University of Kiel, Institute for Geosciences, Olshausenstr. 40, D-24118 Kiel, Germany*
[2] *Scripps Institution of Oceanography, La Jolla, 9500 Gilman Drive, CA 92093–0244, USA*
[3]*Queen's University of Belfast, School of Archaeology and Palaeoecology, University Road, UK-Belfast BT7 1NN, Northern Ireland*
[4]*University of Berne, Institute of Geography, Hallerstrasse 12, CH-3012 Bern, Switzerland*
[5]*University of Hamburg, Institute of Oceanography, Troplowitzstraße 7, D-22529 Hamburg, Germany*
[6] *GeoForschungsZentrum Potsdam, Telegrafenberg, D-14473 Potsdam, Germany*
[7]*Now at: University of Bremen, Department of Geosciences, Postfach 33 04 40, D-28334 Bremen, Germany*
[8]*Microwave Group-Ottawa River, Inc., 3954 Armitage Rd., Ontario, K0A 1T0, Canada*
[*] *corresponding author (e-mail): Mschulz@email.uni-kiel.de*

Abstract: Prognoses regarding future climate scenarios hinge on the predictive skills of climate models. They must produce reliable estimates of the future mean climate state as well as future climate variability about the mean, resulting from the interference between natural climate variability and anthropogenic perturbations of the climate system. Natural climate variability originates from interactions between components of the climate systems and from perturbations external to the climate system. To validate model-generated climate variability against the whole range of natural variability, it is necessary to gather information on the temporal progression of Holocene climate variability. Distinct spatial climate regimes (climate-variability patterns) can be extracted from climate observations and facilitate the comparison between paleoclimate reconstructions and model results. Examples of instrumental observations and proxy data demonstrate the potential of such records to trace climate regimes through time. The dominant mode of today's climate variability in Europe - the North Atlantic Oscillation - constitutes a climate regime, that can serve to provide an expectation in assessing Holocene climate variability. First attempts to reconstruct the history of the North Atlantic Oscillation during the last 300 years or so are summarized. Volcanic eruptions and solar variability constitute important external perturbations of the climate system that account for a significant portion of natural variability.

Introduction

Given that human activities influence climate in increasing ways (e.g. Houghton et al. 1996), how can we make useful scenarios for the future, which can inform policy decisions? Predictions of the future climate state can be based either on experience, that is, awareness of past climate states and the likelihood of their occurrence in the future, or detailed application of the physical principles which determine differences between climate states, that is, by means of climate modeling. A useful time scale for considering climate impact on today's human societies is that of societal change itself, that is, the scale of a decade to a century.

From WEFER G, BERGER W, BEHRE K-E, JANSEN E (eds), 2002, *Climate Development and History of the North Atlantic Realm.* Springer-Verlag Berlin Heidelberg, pp 125-148

To capture different climate states on these time scales, we turn to instrumental and proxy observations of climate variability on time scales up to millennia for different intervals of the Holocene.

Mean and variability about the mean of the future climate state is of utmost importance. To describe climate variability associated with any given climate state as well as shifts between different climate states, it is helpful, to identify characteristic climate "regimes". Such regimes delineate regions of coherent climate fluctuations and can be considered as climate-variability patterns. We present the North Atlantic Oscillation (NAO) as example for a climate regime of the present climate state. The NAO constitutes a decadal variation of the atmospheric pressure pattern in the North Atlantic realm. It controls the intensity and location of the band of the westerly winds, which is a dominant component for European climate. Using statistical tools, the history of the NAO can be reconstructed from historical climate observations as well as paleoclimate records. In conjunction with climate modeling, important insights into the generation of natural climate variability can be gained from such reconstructions. Accordingly, it appears that the NAO originates from interactions between ocean and atmosphere, that is, its generating mechanism is located entirely within the climate system. External perturbations of the climate system (e.g. volcanic eruptions, solar variations) provide an additional source for climate variability. The history of this externally forced climate variability can be reconstructed from a variety of geological archives. A comprehensive understanding of the whole range of natural climate variability is indispensable for risk analysis with respect to future climate change.

Climate Models As Prediction Machines

The task of climate modeling is to derive subsequent states of climate from given states, using well-defined rules. Models built to perform this task are sets of equations describing processes within the climate system. These equations contain information about physical mechanisms governing the rates of change of the elements of the system. Alternatively, the mechanisms are represented by statistical relationships derived from present conditions as based on short-term observations (the last 150 years or so). Ideally, the information contained in the physical and statistical rules are sufficient to provide reliable predictions of future climate states, given the present state (or arbitrarily defined initial states). The necessity for prediction arises from the fact that humankind is now engaged in a large-scale (and one-time) geophysical experiment testing the response of the climate system to substantial perturbation (such as emission of greenhouse gases). Reliable climate prediction will be one element in setting climate policy.

Difficulties in building climate models arise from insufficient knowledge of physical mechanisms involved in the climate system and from insufficient resolution of known processes. An inherent problem of all such models (independent of resolution and accuracy) is that internal interactions between climate processes lead to some degree of unpre-dictability, that is, predictions can only be made regarding the ranges of possible outcomes. Thus, although any one sequence of calculations might result in one particular prediction of climatic state at the end of the experiment, such a prediction is just one possibility among many. Lorenz (1963) drew attention to this fundamental limitation of the power of prediction; it is sometimes referred to as the "butterfly effect", with the implication that even extremely small disturbances of the system can eventually lead to noticeable differences in the outcome.

The range of uncertainty in possible outcomes can be explored by making a number of "repeat" runs, that is, model experiments which are identical to each other except for small changes in initial state (so called ensemble predictions). Other methods focus not so much on the inherent internal variability, but on the uncertainty contained in the rules that govern change. Such sensitivity experiments can be extremely expensive in terms of time requirements on large computers. Another check on the reliability of predictions is a comparison between different models. If they disagree markedly, they cannot all be correct, and a search for the reason of disagreement is indicated. (Agreement is not a guarantee of correctness, of course.)

Special demands on predictive capability result from the requirement to calculate future variability about the future mean state, rather than the mean state alone. In the same fashion, a requirement for regional prediction greatly increases the computational task, and also the scope of uncertainty. Both these items, variability about the mean and a regional focus, are of great interest in a planning context, much more so than mean state. As concerns future variability, the time scales of interest range from decades to centuries.

The sources of variability are the subject of intense discussion. Variability is generated both within the climate subsystems - ocean, atmosphere, cryosphere and biosphere - and, more importantly, through their interaction. Historically, there are two principal ways of representing these interactions and these have led to two types of climate models: "stochastic" models (Hasselmann 1976) and "dynamic" models (Bjerknes 1964). In the stochastic representation of the interaction of ocean and atmosphere, the ocean is responding passively to a highly variable atmosphere. Atmospheric variability is characterized by a "white spectrum", that is, its energy is evenly distributed over a considerable range of time scales, typically from minutes to days. The ocean responds with a "red spectrum", that is, its variability moves preferentially into long time-scales, much like a swing in the wind. In contrast, dynamic climate models are based on two-sided interactions between climate subsystems. Thus, the ocean's long-period variability feeds back into the atmosphere, generating long-term variability here as well.

The potential reliability of climate models can be tested by comparing simulated with observed climate variability on various time scales. Observations on climatic fluctuations of the time scales of interest include both direct measurements and proxy records. Instrumental measurements of meteorological variables on land (e.g. temperature, air pressure) go back 300 years at a maximum, but generally no more than about 150 years. At sea, series are somewhat shorter (about a century). Proxy time series (e.g. observations based on geological or botanical records) can provide information on climate variability for the entire Holocene (and even further back). One important problem of model verification is the lack of a sufficient number of instrumental observations. In lieu of these, proxy time series are very useful.

Modeling of climate variability, using general circulation models, at this juncture can only be done on centennial time scales due to present limitations imposed by computing power. One can anticipate that computing power will continue to increase substantially, and that prediction of multiple sce-narios will progress along with this trend. Precisely for this purpose, we shall need ways to validate model-generated variability against the whole range of reconstructed variability.

Climate Variability: Instrumental Observations

To gain insight into the dynamics of the climate system, we need to study its variability as expressed in time series of various types of observations. Observations range from direct measurements, to historical observations, to paleoclimatic reconstruction from growth patterns of organisms, ice layers and sediments (Table 1). In this section we discuss time series based on instrumental observations.

The Central-England temperature record (CET), first compiled by Manley (1974) and continued to the present, is the longest homogenous series for any region of the world (Fig. 1a). The series represent monthly (starting January AD 1659) and daily (starting AD 1772; Parker et al. 1992) mean temperature of the English Midlands (three-station average). Several studies show a close relationship between CET and atmospheric circulation. Trends in this record can be used, with caution, as indicator for trends in northern-hemisphere land temperature (Jones and Hulme 1997). According to this study, temperature over the British Isles has risen by about 0.7°C since AD 1700 and by about 0.5°C since the turn of the 20th century, with warming occurring in all four seasons.

Paris air pressure. Station pressure data for Paris (Fig. 1b) are available for the period AD 1675–1713 (Legrand and Le Goff 1992). After 50 years with no observations, monthly pressure data are available from AD 1764 on.

Proxy	Time Scale [y]	Seasonal Resolution	Transfer Function	Human Effect on TF	Example
Lacustrine varve thickness	$1\text{-}10^4$	+	$f_{bio}(T, E, \text{lake dyn})$	+	Figure 4 (Negendank, this volume)
Tree-ring chronologies	$10\text{-}10^3$	+	$f_{bio}(T, P)$	+	Figure 7 (Baillie, this volume)
Ice-core $\delta^{18}O$	$1\text{-}10^5$	(+)	$f_{phy}(T, P_{seas}, \delta^{18}O_{src})$	--	Figure 8
Marine flora/fauna	$10\text{-}10^5$	(+)	$f_{bio}(T, S)$	+ (?)	Koç and Jansen (this volume)
Coral $\delta^{18}O$	$1\text{-}10^3$	+	$f_{phy}(T, \delta^{18}O_w)$	--	Berger et al. (this volume)
Coral growth rate	$1\text{-}10^3$	+	$f_{bio}(T, L, N)$	+ (?)	Berger et al. (this volume)

Table 1. Representative terrestrial and marine climate proxies discussed in the text. Time scale indicates the typical range on which climate variability can be reconstructed from a proxy. $\delta^{18}O$ denotes oxygen-isotope ratio. Biological (bio) and physical (phy) transfer functions (TF) for each proxy are expressed as function of temperature (T), evaporation rate (E), precipitation (P), runoff (R), seasonality of precipitation (P_{seas}), $\delta^{18}O$ of moisture source ($\delta^{18}O_{src}$), $\delta^{18}O$ of ambient water ($\delta^{18}O_w$), light (L) and nutrient availability (N). Potential bias of the transfer functions by human interference with the environment are indicated.

Kew precipitation. The longest measured precipitation series is available from Kew near London. Monthly values start AD 1697 and continue to the present (Peterson and Vose 1997 and references therein; Fig. 1c).

Lund wind direction. The time series consists of frequencies of relative wind directions at Lund in southern Sweden from AD 1740 until today with monthly resolution (Jönsson and Fortuniak 1995). Records of wind direction are the most reliable and consistent set of historical meteorological observations. Wind-direction frequencies reflect atmospheric circulation pattern on a regional scale and may serve as important record for reconstructing past climate variations. The time series seem to be reliable throughout the whole period. Trend analysis shows significant decreasing trends for north and north-east winds during all seasons between AD 1741–1990.

North Atlantic hydrographic time series going back to about 1900 (Fig. 2) or earlier are listed in Table 2. In contrast to land-based observations, no instrumental data (i.e. temperature or salinity) exist prior to AD 1876. This lack of data makes the need for paleoceanographic reconstructions for earlier time intervals especially urgent.

Sea-ice observations. Coupled ocean-atmosphere models show significant increase in winter temperatures in the Arctic (between 3-4°C) during the next 50 years (Mitchell et al. 1995). This predicted warming would enhance the melting of sea ice, thus providing an early indication of global change. In fact, the changes in Northern Hemisphere sea-ice extent show a trend of −2.8% per decade for the period 1978-1996 (Parkinson et al. 1999). Regionally the largest decreases are observed in the Kara and Barents Seas, followed by the seas of Okhotsk, the Arctic Ocean, Greenland Sea, Hudson Bay and the Canadian Archipelago. Increasing ice extent has been observed in Baffin Bay/Labrador Sea, Gulf of St. Lawrence and Bering Sea.

The history of ice extent has been reconstructed for the Baltic, for the period between AD 1720 and 1940. Although these data are not strictly instrumental in origin, they are based on direct observations (Fig. 3a). Another sea-ice reconstruction, back to AD 1501, is the *"Western Baltic Sea-Ice Index"* (Koslowski and Glaser 1999; Fig. 3b). Ice-winter-severity is strongly negative correlated with the NAO: Strong westerlies over the eastern North Atlantic are characterized by weak ice production and vice versa. Identification of variations in ice-winter-severity therefore indicates the probable state of the NAO during the last five centuries. The ice-severity index is regarded as suitable for de-

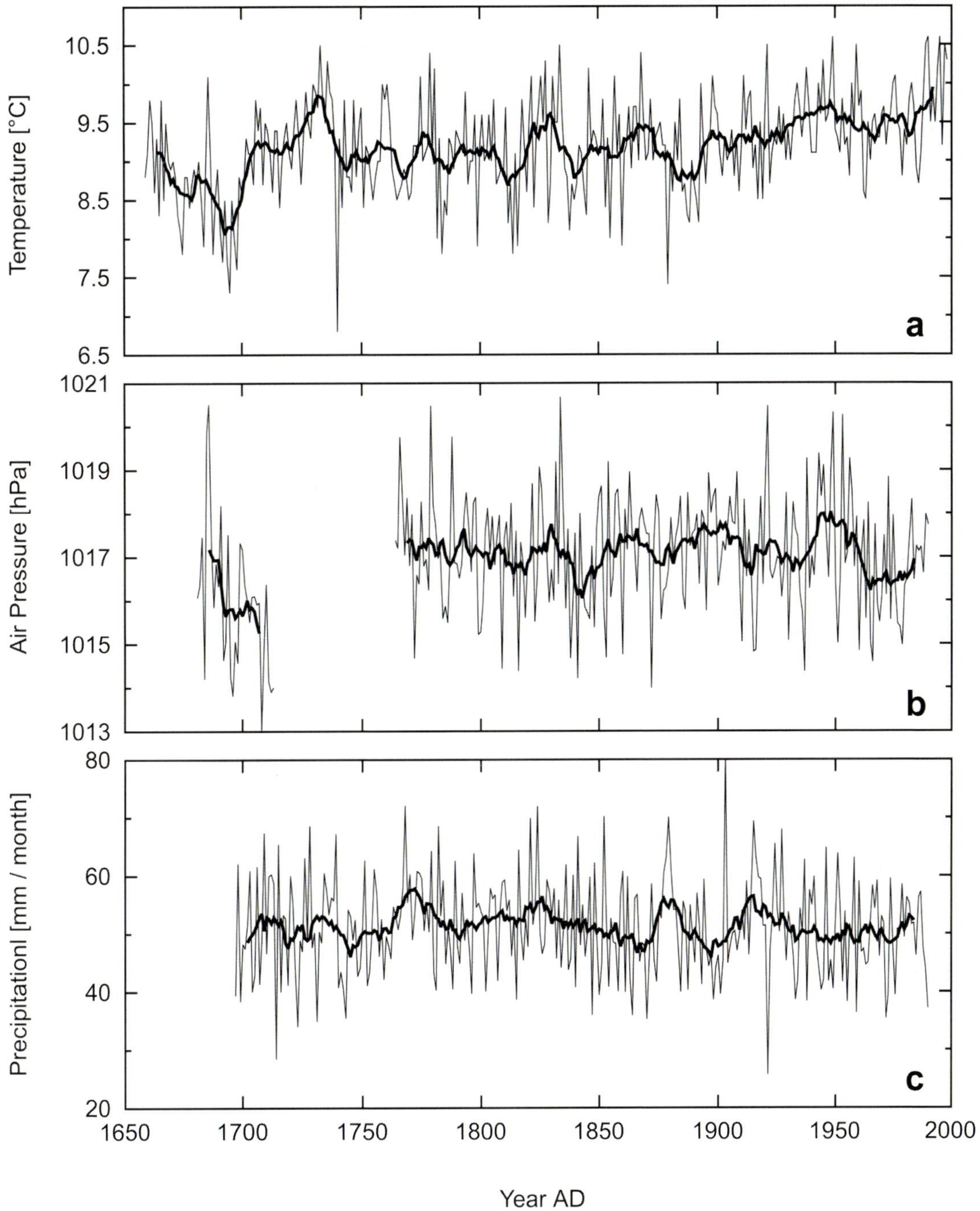

Fig. 1. Example time series for long-term meteorological observations. **(a)** Air temperature (annual mean) in central England (Manley 1974; Parker et al. 1992). Uncertainty of the data is of the order of some tenths of a degree; data prior to AD 1721 are considered less reliable. **(b)** Air pressure in Paris (Legrand and Le Goff 1992). Uncertainties for the early period (AD 1675–1713) are ± 0.8 hPa, and for periods from AD 1764 on even less than that. **(c)** Precipitation (annual mean) in Kew near London (Peterson and Vose 1997; and references therein). Data from the first few decades are considered less trustworthy than the remaining part. Heavy lines indicate 10-yr running means.

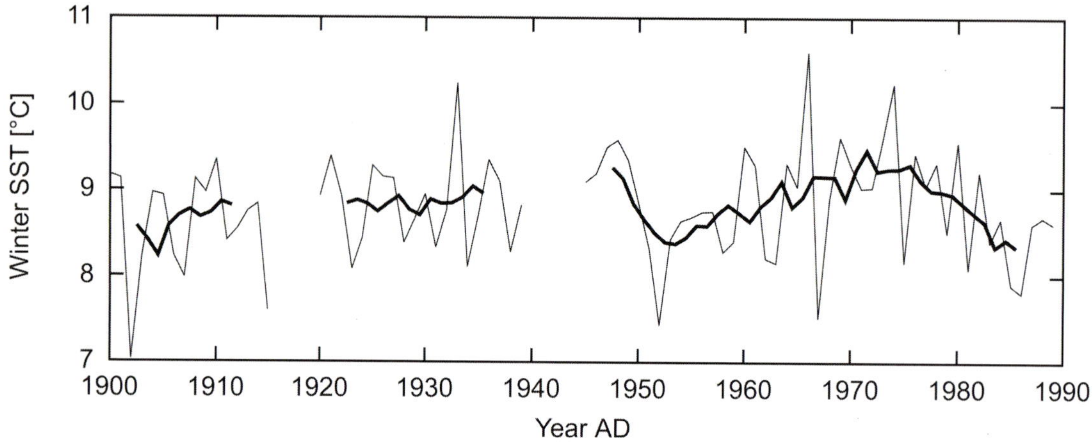

Fig. 2. Average winter sea-surface temperature in the North Atlantic (Faeroe-Shetland Channel area; 60–65°N, 5–10°W). Heavy line depicts 5-yr running mean.

Location/Area	Type of Measurements	Parameters	Start/Stop of Observation	Temporal Resolution	Source
Bermuda Panulirus	Fixed station, surf.-bottom	T, S	1954-ongoing	monthly-seasonal	NODC
Faeroe-Shetland Channel	Sections, surf.-bottom	T, S	1902-ongoing	seasonal-annual	ICES Figure 2
Barents sea Kola-section	Sections, surf.-bottom	T, S	1899-ongoing	seasonal	ICES
Norwegian waters	Sections, surf.-bottom	T, S	1902-ongoing	seasonal-annual	ICES
North Atlantic	Merchant vessels, surface	T	1876-ongoing	monthly	ICES/ COADS
Coastal North Atlantic	Light-vessels, light houses, surface	T	1876-ongoing	daily-seasonal	ICES/ COADS
Baltic Sea	Visual observations, satellite imagery	sea-ice cover	1720-ongoing	ice season	FIMR

Table 2. Overview of hydrographic time series covering more than about one century. Fixed stations: Regular hydrographic profiles from surface to bottom by weather ships or research vessels. Sections are regular repeat-sections, station distance 10–50 km and hydrographic profiles from surface to bottom. Parameters: T = temperature; S = salinity. Sources: COADS = Comprehensive Ocean-Atmosphere Data Set; ICES = International Council fort the Exploration of the Sea; NODC = National Oceanographic Data Center; FIMR = Finish Institute of Marine Research.

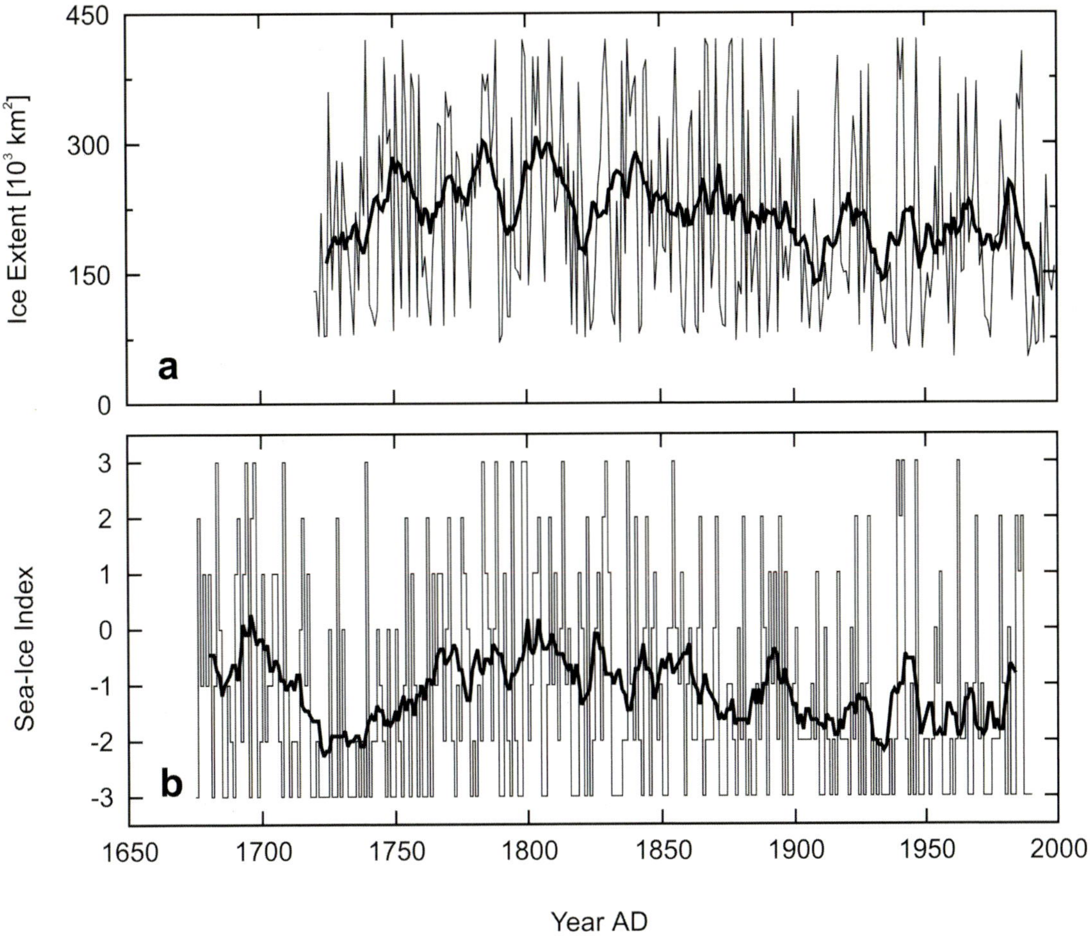

Fig. 3. Sea-ice indices for the Baltic Sea. **(a)** Reconstructed (AD 1720-1940) and observed maximum sea-ice extent during the winter season to the line Grimstad-Skagen in the Skagerrak region (Seina and Palosuo, 1996). Maximum aerial extent is 420 × 10^3 km^2. **(b)** Western Baltic ice winter severity index (Koslowski and Glaser, 1999) is a dimensionless measure for sea-ice cover (+3: very strong and extreme winter sea ice distribution; 0: normal conditions; –3: weaker than normal sea-ice conditions). Reconstruction for the pre-1850 period is based on the duration of ice occurrence and ice cover in the Belts, Öresund and Baltic, in particular around Bornholm. After AD 1878 ice-winter-severity types were derived from accumulated areal ice volume along the German Baltic coast. The pre-1871 data might be less trustworthy. Heavy lines indicate 10-yr running mean.

scribing the degree of severity of meteorological winters in central Europe with an approximately decadal resolution (Koslowski and Glaser 1999).

Climate Variability: Proxy Observations

Laminated lake sediments are potentially a rich source of climate information, especially if an annual calendar-year ("varve") chronology can be established. Various types of varves (clastic, organic, evaporitic and mixtures of these) are known from the geologic record (Anderson and Dean 1988; and references therein). Mixed varves are common and reflect biological and geological processes within a lake and its drainage basin (Negendank et al. 1999). These various processes are linked to regional climate, topography and vegetation, and their precise relationships to varve properties have to be established on a case-by-case basis (Zolitschka 1998). In some lakes varved sediments hold the potential for paleoclimatic reconstructions with seasonal resolution (Brathauer et al. 2000; Fig. 4). In Lake Holzmaar (Eifel Region; Germany), for example, diatom concentration decreases in the varves during cold periods, because of a decrease in production or an increase in dilution by sediments brought in from the outside, or both. Also, anthropogenic influence may lead to a significant increase of erosion products. For example, the sedimentation rate in lake Holzmaar shows a sudden increase around AD 1000 which coincides with intensified forest clearance, which led to enhanced runoff (Zolitschka 1998; Fig. 5).

Tree-ring chronologies. Variations in tree-ring width (Fig. 6) and wood density reflect the response of a given tree to changes in local temperature and local precipitation (e.g. Fritts 1976). In most cases a mixture of the two factors determines tree-ring growth. Careful consideration of a local tree habitat is therefore required to infer paleoclimatic information from tree-ring data. A "tree-ring chronology" is a time series of tree-ring properties such as an index representing annual departures of growth from the average. Examples for tree-ring chronologies for Irish oak, Fennoscandian pine and Finnish pine are shown in Figs. 7 and 8. Irish oak is a temperate deciduous species while the pines

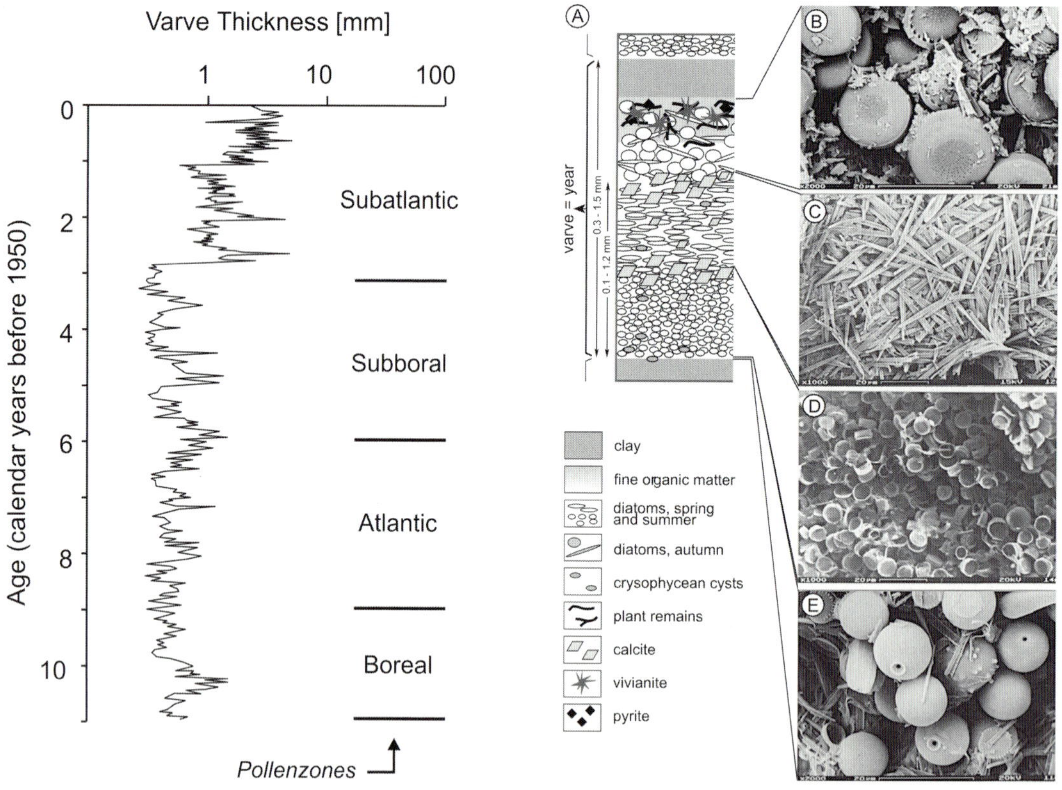

Fig. 4. Left: Holocene varve-thickness record from Lake Holzmaar (Eifel Region; Germany). Note logarithmic abscissa scale. **Right:** A: composition of a typical varve of the Boreal-Atlantic pollenzone with seasonal sublayers deposited during spring (D, E), summer (C), autumn and winter (B). Photographs show the seasonal succession of diatoms *Cyclotella radiosa* (B); *Nitzschia sp.* (C); *Stephanodiscus sp.* (D) and *Chrysophycean cysts* (E). Scale bars are depicted at the bottom of each photograph.

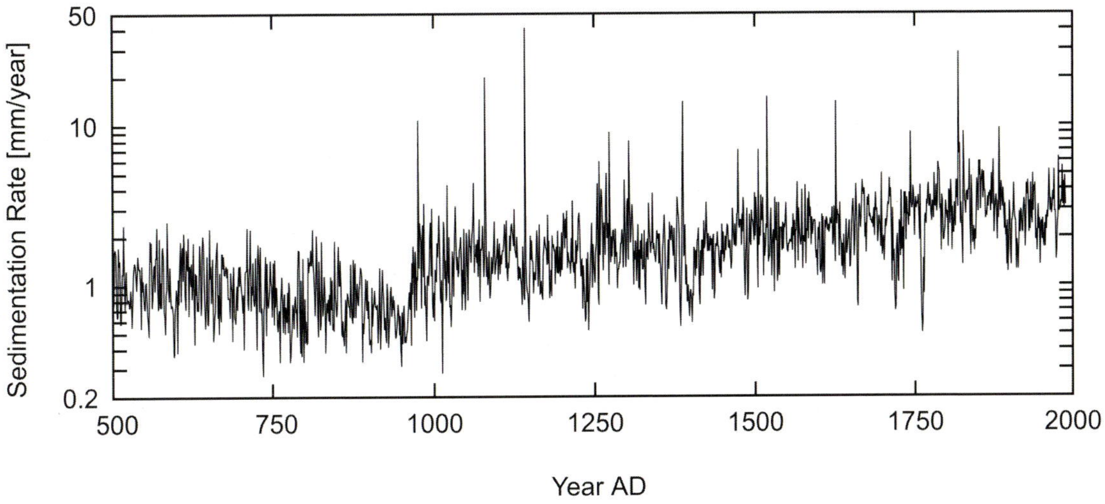

Fig. 5. Sedimentation rate from varved sediments in Lake Holzmaar (central Germany) (Zolitschka 1998). Starting around AD 1000, sedimentation rate increased due to anthropogenic wood clearance. Note logarithmic abscissa scale.

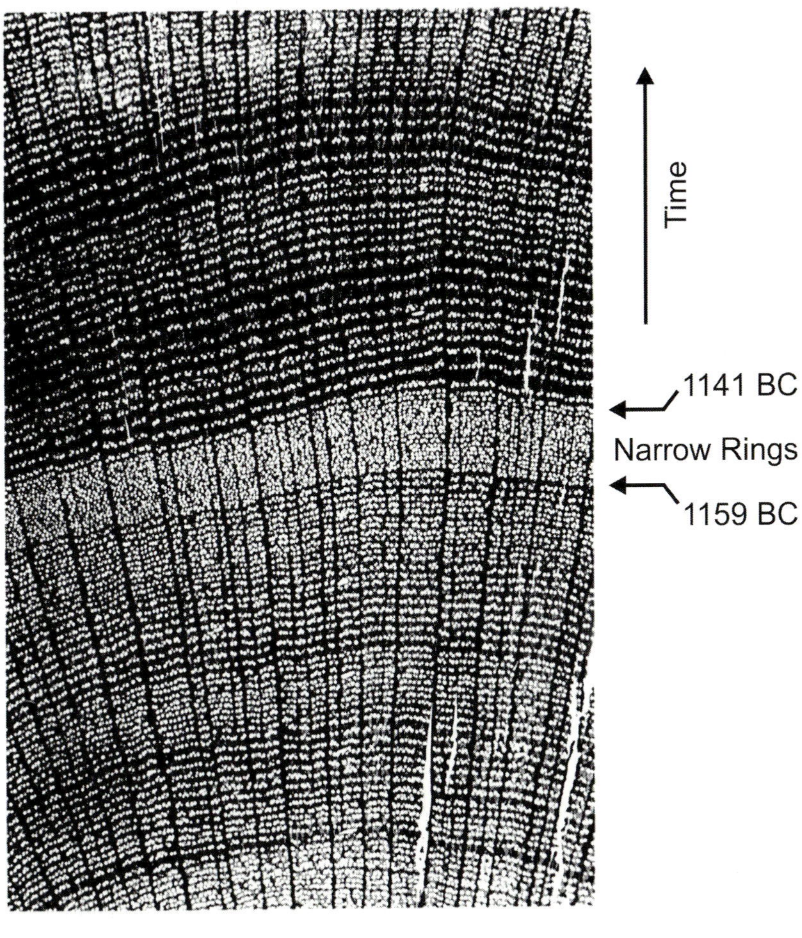

Fig. 6. Bog oak ring patterns spanning the 12th century BC. Sample Q8001 from Garry Bog, Ireland. The band of extremely narrow rings in the center of the figure spans 1159–1141 BC and indicates a climatic deterioration (Baillie 1998).

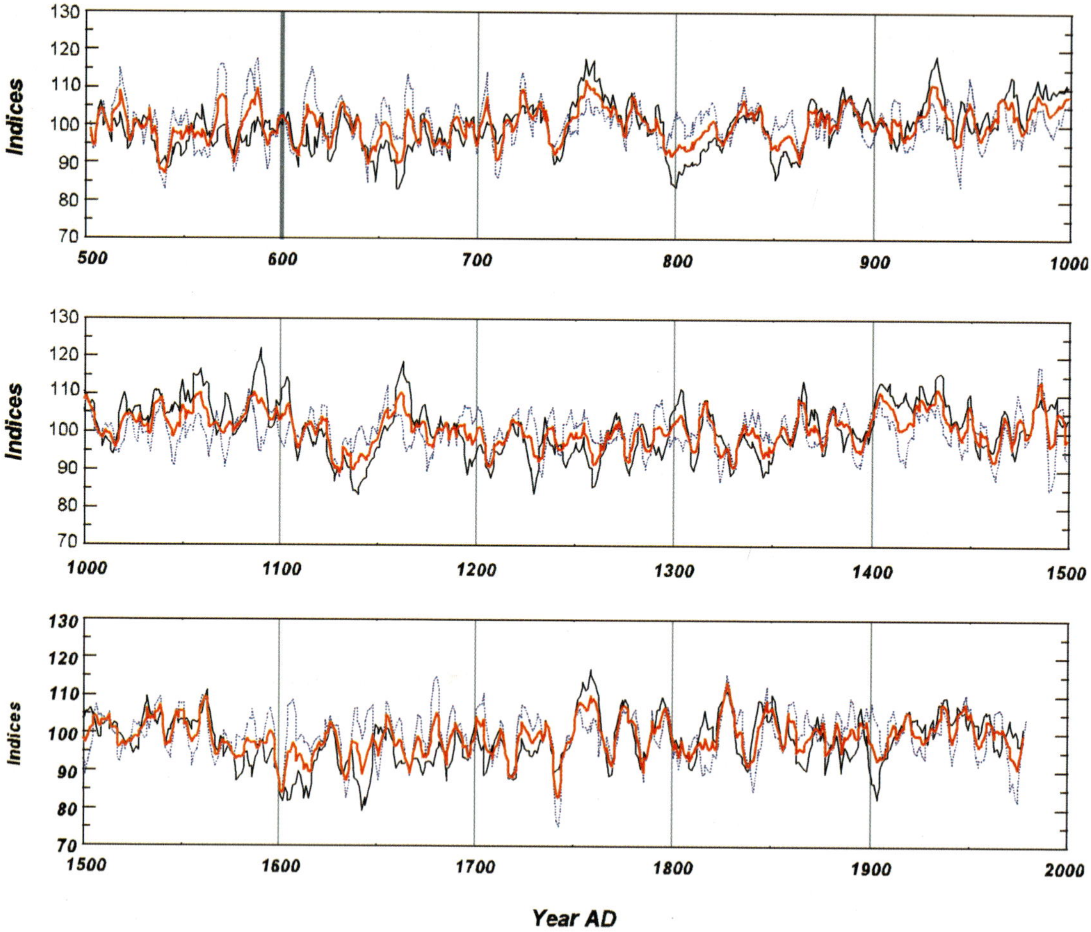

Fig. 7. Tree-ring based climate proxies. Smoothed records of growth indices from European oak (dotted line) and Fennoscandian pines (Briffa et al. 1992; thin solid line). Growth indices are positively correlated with temperature. Mean of the two curves is indicated (bold solid line) to show the degree of common response.

used in the two other chronologies are mainly sensitive to summer temperatures. The oak chronology records some of the notable cold events in the last 600 years, many of which are linked to volcanic eruptions (Briffa et al. 1998). For a combined tree-ring chronology of the northern hemisphere (Briffa et al. 1998), the most extreme negative departure in the period is the one connected to the eruption of the volcano Huaynaputina in southern Peru (AD 1601) (de Silva and Zielinski 1998). The Irish oaks likewise show a strong reaction, but they record their most significant negative growth departure in the early 1740s following the extremely cold year AD 1740 recorded in the Central England temperature record (Fig. 1a) and other European proxy data.

Ice-core isotope records. Time series of the isotopic composition of ice provide important information on past climate. In particular, oxygen and hydrogen isotope ratios (expressed as $\delta^{18}O$ and δD, respectively) are assumed to be largely controlled by the air temperature at the altitude of snow formation (Dansgaard 1964). As a rule-of-thumb for Holocene oxygen isotope records from Greenland, a 1‰ increase in $\delta^{18}O$ indicates a 2°C temperature rise (Stuiver et al. 1995; Jouzel et al. 1997).

However, a variety of other factors such as origin and seasonality of precipitation (as well as cloud physics) affect the isotopic composition of ice, so that the calibration may be time-dependent (e.g. Jouzel et al. 1997; Hammer this volume). Ice cores can be dated by counting annual or even seasonal ice layers. For the last 1200 years or so, $\delta^{18}O$ time series from the GISP2 ice core (Stuiver and Grootes 2000) has a resolution of about 0.1 year, whereas the remaining part of the Holocene is resolved with an approximately 2-year resolution (Fig. 9). The high time resolution of the isotopic stratigraphy affords the opportunity for precise dating of various climate-relevant parameters recorded in ice cores, such as volcanic eruption tracers (Hammer, this volume). Comparison of paleovolcanic records with other climate records provide important insights into the climate forcing potential of large volcanic eruptions (Briffa et al. 1998). Based on model experiments, Crowley and Kim (1999) suggest that approximately 20% of reconstructed climate variability between AD 1400–1850 can be explained by volcanic forcing.

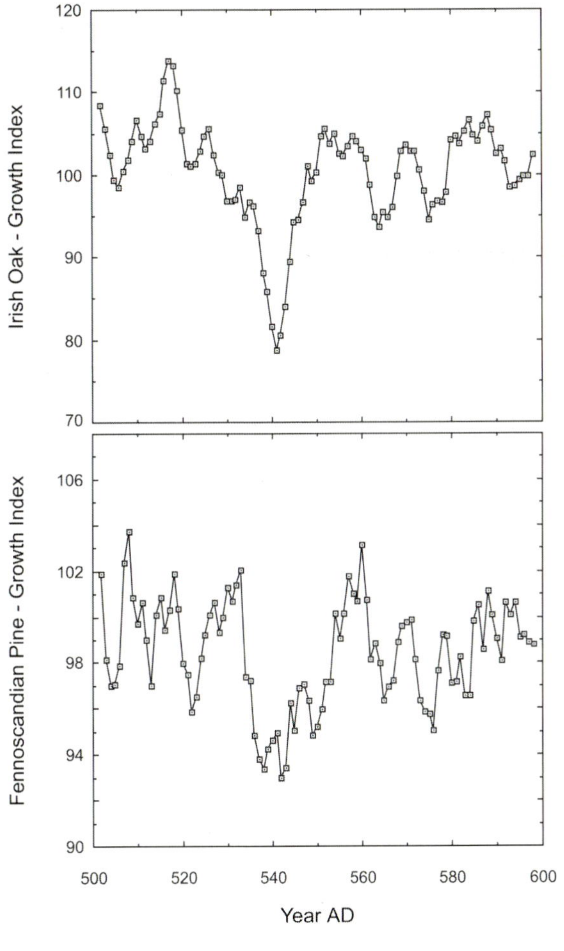

Fig. 8. Detail of the growth-index chronologies from Figure 7 for the 6th century AD: European oak (top) and Fennoscandian pine - temperature indices (bottom; Briffa et al. 1992). Trends of the two records show a considerable degree of coherence across this century. Pronounced decrease around AD 540 in both records indicates a marked cooling event (see text for further details).

Sea-surface conditions from fossils of marine organisms. Variations of species assem-blages, and $\delta^{18}O$ in calcareous fossils in marine sediment records, can be used to reconstruct past changes in sea-surface conditions. High-resolution studies are restricted to regions of extremely high sedimentation rate (normal rates are near 2 cm per thousand years, with a resolution of 1000 to 2000 years at best). An unusually detailed record was recently recovered from the Cariaco Basin off Venezuela, where anaerobic conditions prevent the stirring of sediment by bottom-living organ-isms. The record provides information on the history of upwelling and trade-wind variability over the last 825 years (Black et al. 1999). High sedimentation rates on the margins of formerly glaciated regions also contain information relevant to centennial-scale climatic fluctuations (Keigwin and Pickart 1999; Koç and Jansen, this volume). The various methods for estimating sea-surface temperature from paleontological and geochemical data are summarized in Fischer and Wefer (1999).

Coral $\delta^{18}O$ and coral growth rate. Corals contain a valuable record of climate-related information about the tropical oceans. In the Atlantic, corals on the Bermuda Islands provide an outpost for monitoring a number of growth-linked factors such as temperature, light and nutrient supply (which is tied to wind-stirring of the upper ocean) (Nozaki et al. 1978; see Berger et al., this volume).

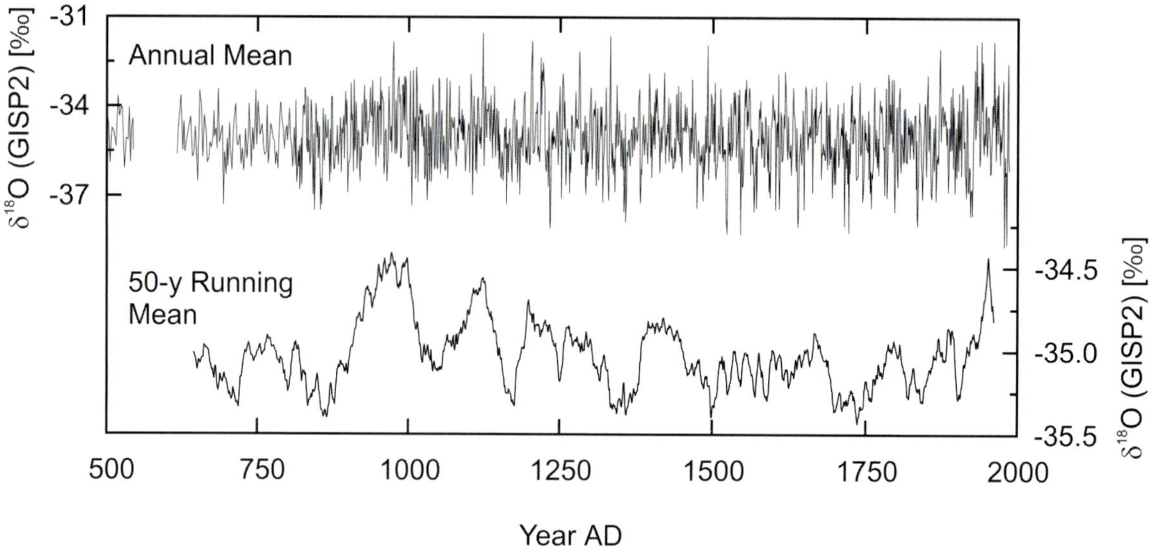

Fig. 9: Oxygen isotope ($\delta^{18}O$) time series from the GISP2 ice core in central Greenland (Stuiver and Grootes, 2000). Data are assumed to be largely controlled by temperature above Greenland. 50-y running mean record shows decadal variability.

Climate Regimes: Modern Climate-Variability Patterns in the North-Atlantic Realm

The various observations of climate fluctuations - whether instrumental or by proxy - suggest that the mean state of the climate as well as its variability change considerably through time (cf. Huntley et al. Group 2 report, this volume). The different climatic states can be characterized by a set of descriptive statistics. On a temporal scale, such statistics include mean, variability around the mean, and the contribution of the different time scales to the overall variability (spectral features). In addition, the identification of regions of coherent climate fluctuations, delineating climate regimes, greatly simplifies the task of describing fluctuations of climate states on a spatial scale. In contrast to local information on climate variability, the larger spatial scale of climate regimes make them a prime target for comparing results from climate models with paleoclimatic reconstructions.

At present, climate variability in the North Atlantic realm is dominated by the North Atlantic Oscillation (NAO). The "oscillation" denotes large-scale shifts in air masses over the North Atlantic (as seen in air pressure), with centers of action near Iceland and the Azores. The seesaw fluctuations between the Iceland Low and the Azores High comprise a large part of the atmospheric variability in the North-Atlantic sector throughout the year, accounting for more than one-third of the total variance in sea-level pressure (Cayan 1992; Fig. 10). The state of the seesaw oscillation is based on the difference in sea-level pressure anomalies between Iceland on one side and either the Azores (Rogers 1989) or Portugal (Hurrell 1995) or Gibraltar (Jones et al. 1997) on the other, depending on availability (Fig. 11). A "high-index" pattern indicates strong mid-latitude westerlies and is characterized by a strong Iceland Low combined with a strong Azores High and *vice versa* (see Meincke, this volume).

The NAO dominates climate variability in NW-Europe. It determines the storm track over the North Atlantic and into Europe as well as the precipitation patterns. In the NAO-high regime the storm track crosses the North Atlantic from southwest to northeast, that is, atmospheric depressions and excess precipitation are steered preferentially into Scandinavia. The NAO-low regime has a west-to-east ("zonal") storm track, that is, depressions

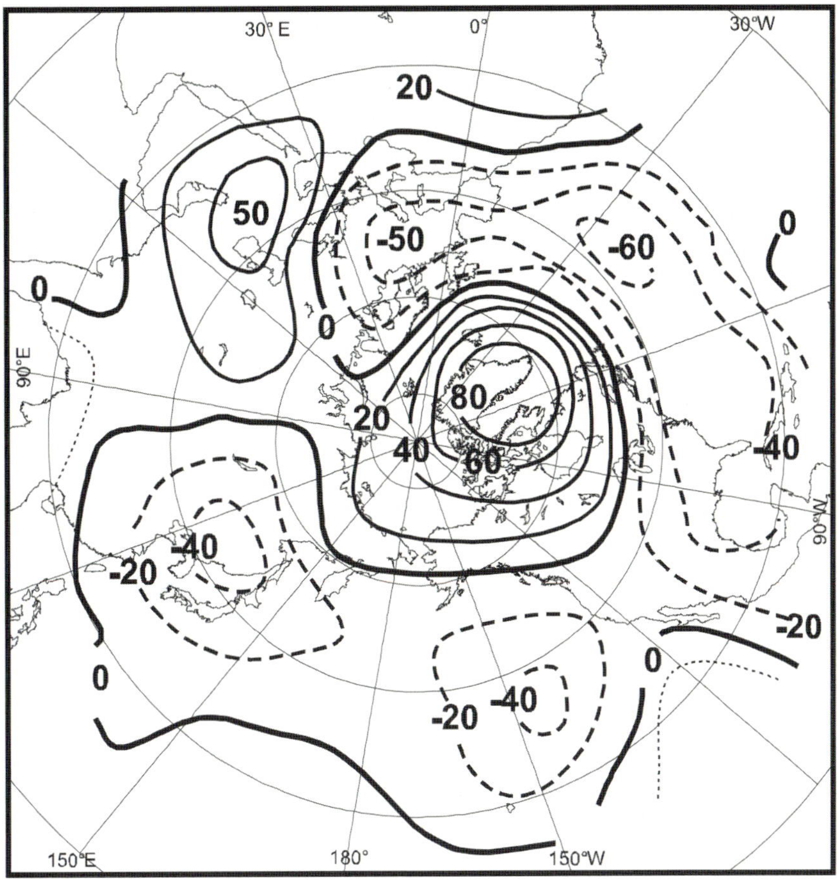

Fig. 10: Correlation map (correlation coefficients × 100) of atmospheric pressure variance for the level of 700 hPa. Reference location is 65°N, 45°W (after Namias 1981).

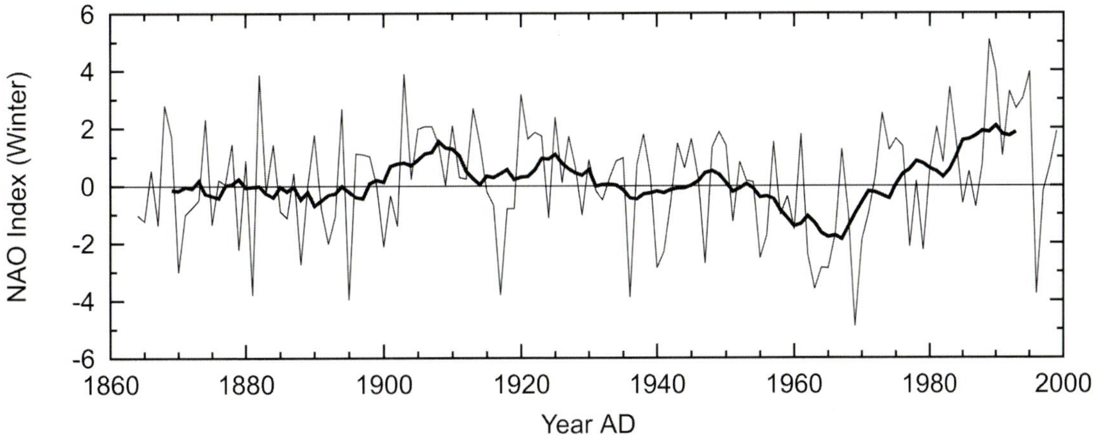

Fig. 11. NAO index for winter (December through March) based on the difference of normalized sea-level pressures between Lisbon, Portugal and Reykjavik, Iceland (Hurrell 1995). Heavy line indicates 10-yr running mean.

and their precipitation reach Europe much farther south, including the Mediterranean realm. Analysis of climate fluctuations on a global scale shows that the NAO is a major source of inter-annual variability in global weather and climate (Hurrell 1996; Meincke this volume). The central importance of NAO for climate prediction emerges from the fact that the hemispheric warming of recent decades - whatever the cause or causes - appears to go parallel with an amplification of the NAO.

The present NAO regime can usefully serve as a working hypothesis for the operation of the North Atlantic climate system in the past. The tools to be used, therefore, in reconstructing the climate system, must be appropriate for extracting "regimes" and fluctuations within and between these regimes.

Climate Regimes: Tools for Detection and Reconstruction through Time

A large number of methods are available for the quantitative analysis of shifts between climate states and regimes (von Storch and Zwiers 1999). Techniques for detecting shifts between climate states must be capable to estimate regime statistics as a function of time. A simple approach is to calculate mean and variance of a data set within a "sliding window", which is moved along the time series. Selection of the window width is always a compromise between minimizing the uncertainty of an estimated parameter (which decreases with a wider window) and maximizing the temporal resolution (which requires a narrow window).

Processes acting on different time scales contribute differing amounts of variability to the overall variability of the climate system. Various types of spectral analysis allow to decompose a time series into the various contributions of variability from the frequencies associated with the different time scales. Standard spectral analysis techniques are based on the assumption that a time series is stationary (i.e. the "spectral character" is time invariant). Traditionally, in order to estimate temporal changes in the spectrum of a time series (i.e. time-frequency analysis), "windowed spectral analysis" has been used. In this procedure, a time series is split into segments of equal length and spectra are estimated for each segment. Regarding long-period variation, the disadvantage of this method is that information may be distorted as the window-length approaches and falls below the length of the period. Obversely, a window adjusted for long periods will fail to resolve temporal variations in amplitude in the high frequency range.

Wavelet analysis (e.g. Lau and Weng 1995; Torrence and Compo 1998) comes without this disadvantage, by adjusting the width of the window to each period analyzed. Wavelet analysis is therefore well suited for time-frequency analysis of non-stationary climatic time series. In particular, this techniques can be used to identify short-lived high-frequency signal components in time series, which may be caused by abrupt changes of the observed variable. Torrence and Compo (1998) offer public-domain wavelet software (<paos.colorado.edu/research/wavelets/>), especially designed for the analysis of climatic time series.

Most algorithms for time-series analysis require evenly spaced data. A problem arises from the fact that paleoclimatic time series commonly are unevenly spaced. Even spacing is generated by interpolation, which causes a significant bias by overestimating the amplitude of low-frequency signal components at the expense of high-frequency components (Horowitz 1974). Programs for time-frequency analysis which deal with this problem are available (Foster 1996; Schulz et al. 1999).

The mapping of climate regimes and their evolution through time usually proceeds on the basis of rather large data sets which are difficult to comprehend. For the purpose of data reduction there are a number of multivariate analysis techniques. One of the most widely used of these is empirical orthogonal function (EOF) analysis, also known as principal component analysis. EOF analysis extracts patterns, that is, main signals based on a few artificial variables (eigenvectors, principal components), which reflect most of the information of interest within the original multivariate data set (Barnett and Preisendorfer 1987; von Storch and Zwiers 1999). Canonical correlation analysis (CCA) extends EOF analysis to two multivariate data sets. Much as in ordinary linear regression, CCA identifies the transform function converting one set (in this case multivariate) into another (also multivariate). For instance, CCA expresses the large-scale patterns of the atmospheric circulation over the North Atlantic-European area ("predictands") as a function of proxy or measured station data ("predictors") (Barnett and Preisendorfer 1987; von Storch and Zwiers 1999). A comprehensive public-domain software package for statistical computation and graphics (including EOF and CCA analysis) is "R", which is available from <www.r-project.org/>.

Recurrent patterns in single time series can be detected by singular-spectrum analysis (SSA, Vautard and Ghil 1989). SSA is closely related to EOF analysis. However, despite its name, it is not a spectral-analysis technique in the classical sense (i.e. frequency-dependent decomposition of a time series). SSA breaks down a time series into independent "typical sequences of events" (von Storch and Zwiers 1999). Reconstruction of a time series based only on those typical sequences which together explain a significant portion of the total variance (e.g. say approximately 90%), results in a filtered time series with enhanced signal-to-noise ratio (e.g. Ghil and Yiou 1996). Public-domain software for SSA analysis (for Unix operating systems) is available from the SSA-MTM Group (Dept. Atm. Sci., Univ. of California, Los Angeles; <www.atmos.ucla.edu/tcd/ssa/>).

Climatic Regimes: Hindcasting NAO

To reconstruct ("hindcast") the history of the NAO beyond the period of direct observations we turn to proxy observation. Given the good correlation between the NAO regime and local climate fluctuations in western Europe, we can attempt to reconstruct the NAO regime through time, to capture the essence of climate change. When reconstructing past climate from proxies and historical archives, one uses multivariate statistical techniques and 20th-century instrumental data (for calibration and verification). The aim is to achieve an understanding of the extent to which the local surface climate (pressure, temperature, precipitation, proxy data) reflects the large-scale patterns of atmospheric circulation over the North Atlantic-European area. In other words, regional patterns are estimated from local variables ("upscaling procedure"). The transfer function serving this purpose is derived for the 20th century and is then applied to the data in the period to be reconstructed. The 20th-century patterns, as expressed in the transfer function, are assumed to be applicable to earlier time intervals.

Among its rich blend of frequencies, the NAO index has exhibited considerable long-term variability. Cook et al. (1998) analyze tree-ring chronologies from eastern North America and northern Fennoscandia (back to AD 1701) and identify concentrations of variance around periods of 24, 8 and 2 years, and also at 70 years. Appenzeller et al. (1998) use ice-accumulation rates from W-Greenland as proxy for NAO-related precipitation (back to AD 1650). They summarize the NAO as an intermittent climate oscillation with temporally active (coherent) and passive (incoherent) phases. No indication for a single, persistent and multiannual NAO frequency is found. In active phases, most of the variability is located in the frequency band with periods of less than 15 years. Variability on time scales of 80 to 90 years has been evident since the mid-19th century.

These reconstructions only apply to the winter season or refer to annual to multi-annual mean values. To resolve variability of the NAO on a much finer level, Luterbacher et al. (1999) used early instrumental and proxy data for monthly NAO reconstructions back to AD 1675. In addition, these authors reconstruct a Eurasian climate index, which represents the history of an important climate regime across central and eastern Europe (Fig. 12). Reconstructions are based on canonical correlation analysis. Similarly, Luterbacher et al. (2000) reconstructed a record for the monthly mean surface pressure field for the eastern North Atlantic-European region for the Late Maunder Minimum (AD 1675–1715). The reconstructions were based on 20 proxy and instrumental records (predictors) from several European sites. The station pressure from Paris is the most important predictor for the surface pressure field for all seasons accounting for more than 50% of the total variance. Unimportant variables are generally indexed rainfall data from the Mediterranean area. Thus incorrect temperature or pressure observations for Paris could potentially cause a large error in the reconstructed surface pressure distribution, although the reconstructions

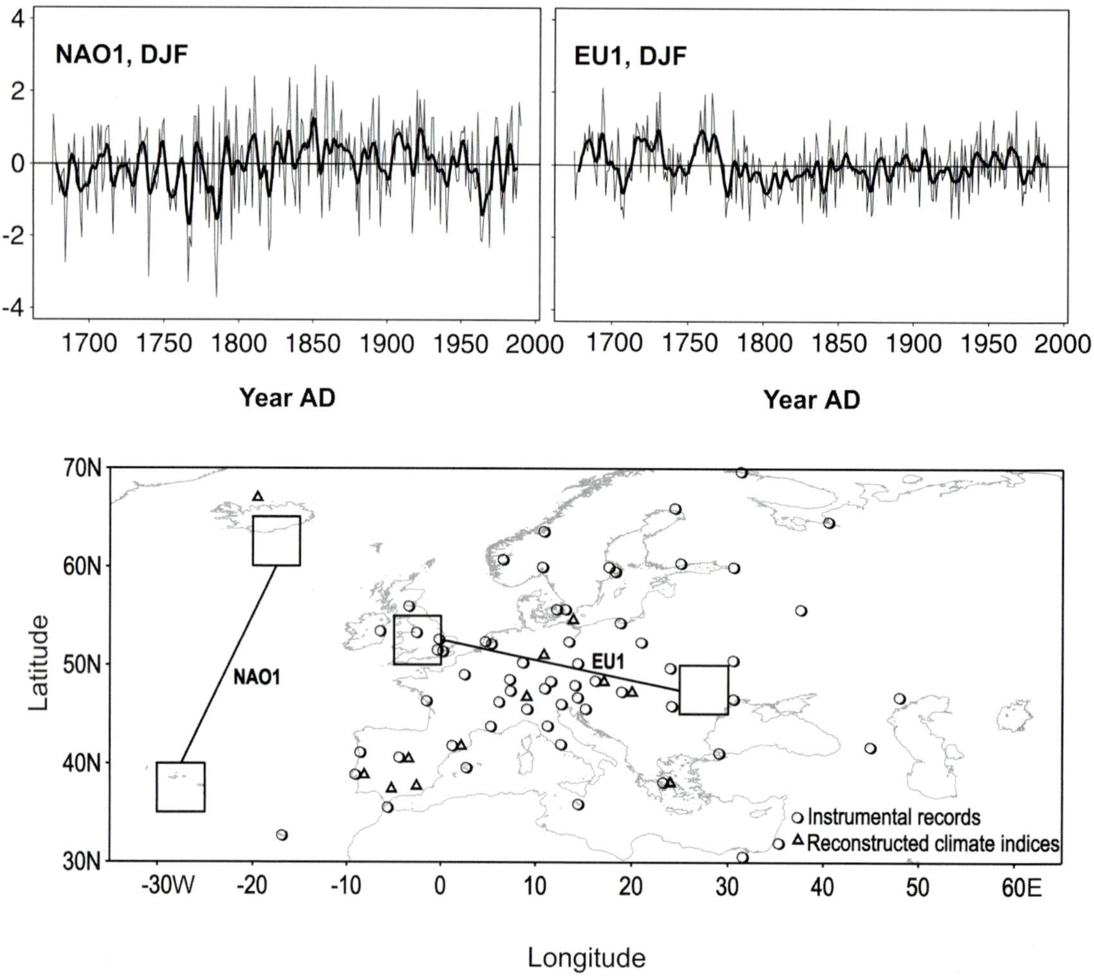

Fig. 12. Top: Reconstructed NAO1 and EU1 indices back to AD 1675 for winter (mean of Dec., Jan. and Feb.; DJF); thick lines are 7-point low-pass filtered time series (Luterbacher et al. 1999). **Bottom:** Geographical distribution of the data locations and location of the two reconstructed circulation indices (modified from Luterbacher et al. 1999). All time series exhibit strong decadal to interdecadal variations.

are robust against errors in a few data points (Luterbacher et al. 2000).

Climate Dynamics: Internal Aspects

The reconstruction of climate states through time and of regimes (regions of coherent fluctuations) by themselves do not necessarily result in greatly improved understanding of dynamics. However, in conjunction with climate modeling, such constraints can be very helpful. One important problem is the existence of multiple quasi-stable states of the climate (see Appendix). The simplest steady states are fixed or oscillatory (repetitive). Some steady states may be associated with high variability ("agitated" times), others with low variability ("quiet" times). The detection of climate states (through time) and their associated climate regimes (in space) would enable us to discover the likelihood of transitions from one state to another. An interesting problem in this context is the stability of the Gulf Stream circulation and associated production of North Atlantic Deep Water (see Meincke, this volume; Paul and Schulz, this volume; and references therein).

Free oscillations can be understood in terms of interplay between positive and negative feedbacks within the system. Positive feedback reinforces the direction of current change in the climate system, while negative feedback counteracts the change. Positive feedback destabilizes the system, favoring an increase in amplitudes of climate excursions, while negative feedback tends to decrease amplitudes. Whenever positive and negative feedbacks are shifted in time with respect to each other oscillations will result from perturbations.

These concepts of positive and negative feedback are crucial in understanding the quasi-periodic climate oscillations in the North-Atlantic realm (that is, the NAO). Two general notions about generating the observed variability provide a relevant framework for discussion, that is, "uncoupled" and "coupled" modes.

Uncoupled variability. Atmospheric general circulation models, forced with a given distribution of constant sea surface temperatures (SST), display NAO-like fluctuations. Thus, it would seem that the origin of the NAO lies within the dynamics of the atmosphere. In the same vein, Frankignoul et al. (1997) have shown that substantial decadal SST anomalies can indeed be generated by stochastic wind stress forcing. Accordingly, there is a possibility that the decadal variability of the North Atlantic hydrography represents the response of the ocean to chaotic, high- frequency forcing by the atmosphere. However, a feedback of the oceanic changes produced by the atmospheric forcing to the atmospheric circulation itself is most likely, leading into coupled modes of variability. Alternatively, uncoupled variability could also be viewed as the passive response of the atmosphere to internally-generated oceanic variations.

Coupled variability. There is growing observational and modeling evidence to suggest that decadal variability over the northern ocean basins is governed by the coupled ocean-atmosphere system, including processes which might determine longer period fluctuations in the NAO. In such scenarios, low-frequency responses of the ocean to atmospheric forcing, and their feedback on the atmosphere, result in oscillatory behavior. Arguments for this type of dynamics were made by Bjerknes (1964), and these were supported by Kushnir (1994), who concluded that pronounced decadal changes in SST and sea-level pressure over the subpolar and subtropical Atlantic gyres were indicative of coupled ocean-atmosphere interactions. The advantage of the coupled-mode concept is that long-term memory is available in the heat-storage capacity of the ocean and the ocean's ability to transport heat anomalies to sensitive regions where they can act as perturbation. For example, model results of Latif and Barnett (1994) and Groetzner et al. (1998) suggest that positive SST and sub-surface heat content anomalies in the central North Atlantic are created by an enhanced subtropical ocean-gyre circulation which transports warm tropical waters poleward by the western boundary current and its extension. The response of the atmosphere is an anticyclonic circulation and a weakening of the storm tracks, which enhances the SST anomalies through anomalous surface heat fluxes and wind-induced upwelling. The atmospheric response, however, also consists of a wind stresses which oppose the spin of the subtropical gyre, thereby reducing the northward transport of heat and creating negative SST and sub-surface heat-content anomalies.

More recently, Timmerman et al. (1998) used a coupled model to show that positive SST anom-alies occur in the North Atlantic if the thermo-haline circulation is anomalously strong. The atmospheric response to the SST anomalies is a strengthened NAO, which leads to anomalous fresh water fluxes off Newfoundland and the Greenland Sea. The resulting negative sea surface salinity anomalies are advected by the subpolar gyre, eventually reaching the convectively active region south of Greenland. The convection and subsequently the strength of the thermohaline circulation are weakened, leading to reduced poleward oceanic heat transport and the formation of negative SST anomalies. The completion of this phase reversal takes about 35 years.

Another example for the generation of a quasi-decadal climate cycle in the coupled atmosphere-ice-ocean system was put forward by Mysak and Venegas (1998), based on observations from the Arctic Ocean. The proposed feedback loop involves Arctic sea-ice concentration and winter sea-level pressure: The cycle starts during a NAO-low phase with a positive sea-ice concentration anomaly in the Greenland and Barents Sea, linked to a combination of anomalous northerly winds and relatively small northward sensible-heat transport

(Greenland Sea) and weaker-than-normal advection of warm water by the northward flowing North Atlantic Current (Barents Sea). The positive sea-ice concentration anomalies are then advected into the Labrador Sea by the local mean-ocean circulation over period of 3–4 years. As soon as the southern part of the Greenland Sea becomes relatively ice free, strong heating of the atmosphere during winter sets in, which is thought to cause a deepening of the Icelandic Low and may thereby help to change the phase of the NAO. During the subsequent NAO-high phase (deep Icelandic Low), wind anomalies create positive sea-ice concentration anomalies in the Beaufort Sea, which are slowly advected out of the Arctic Ocean via the Beaufort Gyre and Transpolar Drift Stream over an interval of another 3–4 years. As a consequence, the Greenland Sea becomes increasingly ice covered, cutting off the heat flux to the atmosphere during winter. It is envisioned that the interaction with the atmosphere causes the Icelandic Low to weaken and, again, to contribute to the following change in the NAO "polarity" which, finally, brings the system back to the beginning of the cycle after about 10 years (Mysak and Venegas 1998).

Climate Dynamics: External Forcing

The stage of the climate system is the surface of the Earth. Perturbations that arrive from space or from inside the Earth may be taken to be "external" to the system. Events such as volcanic eruptions or the penetration of cometary debris into the atmosphere are experiments on the climate system offering an opportunity to study transient response to short-term disturbance. In both types of experiment, the radiation balance of the atmosphere may be affected, resulting in climate cooling (e.g. Covey et al. 1994; Crowley and Kim 1999). Volcanic sulphur dioxide injected into the stratosphere oxides to form a sulphate aerosol layer that coarsens slowly, and drifts around the globe (Pyle 1998). This aerosol layer intercepts incoming solar radiation, disturbing Earth's surface temperature. Events are important in their own right, of course, because short-term climate perturbations are expected to have noticeable impact on human activities and well-being. The reason is that human societies will not readily adapt to very rapid change and neither will the biological systems they depend on.

The impact of a volcanic eruption (or similar disturbance) depends on the sensitivity of the system to the short-term forcing experienced. Formally, climate sensitivity is defined as the ratio of a change in the climate variable of interest (e.g. the mean global temperature, or the mean winter temperature) and the change in forcing (e.g. in the sunlight arriving on the surface of the Earth) (see Appendix for details). A process that dynamically changes the climate sensitivity is called a feedback. The magnitude of a response following a perturbation may depend on the state of the system. For example, if it is cold to begin with, a "volcanic winter" should bring more damage than when it is warm.

Response to volcanic eruptions. Based on oxygen-isotope and volcanic proxy data from the GISP2 ice core (Greenland), Stuiver et al. (1997) showed that for the ten largest volcanic eruptions since AD 818 the annual cooling amounts up to 2°C in Greenland immediately after the deposition of the volcanic aerosols in the ice core. For the northern hemisphere, Briffa et al. (1998) estimated summer cooling by 0.2-0.4°C for most volcanic eruptions during the past 600 years. An extreme cooling by 0.8°C occurred in AD 1601. The mid-latitudes of the Northern Hemisphere seem to be particularly sensitive to the effects of volcanic eruptions, ex-periencing slight winter warming and marked summer cooling (Briffa et al. 1998; Pyle 1998). Based on the proxy records it appears that effects from volcanic climate perturbations decay within a few years (Lindzen 1995; Stuiver et al. 1997). This implies that a cumulative effect may result from multiple small eruptions within a few years (Lindzen 1995; Briffa et al. 1998). Overall, impacts and recovery times after volcanic eruptions show that the Holocene climate system is commonly quite resilient to disturbance, at least to short-term cooling events.

Recently attention has been drawn to the possibility that abrupt climate effects may have been induced during the Holocene by cometary debris (Baillie 1998, 1999). Although the case is circumstantial, the occurrence of one widespread environmental downturn centered on AD 540 (Baillie 1994; Fig. 8), does conform to a prior hypothesis that the Earth was at increased risk of bombardment in the time window AD 400-600 (Bailey et al. 1990). The Bailey et al. hypothesis sees the major hazard being cometary debris orbiting the sun and giving rise to well known meteor showers each year. Bailey et al. (1990) suggest that the major hazard is collision with a large number of such objects in a short period of time (minutes to years). These considerations raise interesting questions regarding the visibility of such impacts in ancient records. Airbursts leave no obvious craters. Impacts over oceans leaving no traces, while many impacts over land may destroy little but vegetation. For example, the impact which hit Tunguska, Siberia in 1909 devastated approximately 2000 km^2 of forest which are now in the course of growing back (see Baillie, this volume).

Response to solar forcing. The Sun's energy output plays a central role in determining Earth's climate, because variations in solar irradiance constitute a direct climate forcing mechanism. Direct observations of solar output changes are only available for the past two decades, whereas reconstructions of solar irradiance go back to about AD 1600 (Lean, this volume). These reconstructions are based on transfer functions, which relate solar output to observed solar features (e.g. sun-spot number, solar-cycle length). (Crucial information for these transfer functions is derived from studying sun-like stars.) According to these reconstructions, solar output has varied by 0.2– 0.5% since approximately AD 1600 on multi-decadal to centennial time scales (Lean, this volume). Because of planetary albedo and geometrical considerations, actual radiative climate forcing is however only approximately 17.5% of these values (Cubasch et al. 1997), that is, 0.5–1.2 Wm^{-2}, which is approximately 20–50% of the anthropogenic radiative forcing (Schimel et al. 1996). With a climate sensitivity of 0.5°C /(Wm^{-2}) (Hartmann 1994), this yields an equilibrium climate response of less than 0.6°C.

Because the multi-decadal to centennial time scale of the forcing is of the same order than the response time of the upper ocean, the transient response of the climate system will be less than the equilibrium response. A recent study with a coupled ocean-atmosphere model (Cubasch et al. 1997) suggests a maximum response of approximately ± 0.25°C in global mean temperature, upon forcing by solar irradiance variations for the last 300 years. The finding that modeled temperature response is smaller than the observed temperature variations for the last 300 years suggests that additional climate forcing factors or feedback mechanisms have to be considered. Changes in solar forcing may be underestimated, because the empirically derived transfer function are not valid throughout the entire low-frequency solar cycles. Additional feedback mechanisms may exist, which amplify the impact of solar forcing on climate. A potential candidate for such mechanism is the feedback between solar forcing and stratospheric ozone concentration (Haigh 1996; Shindell et al. 1999). Additional factors, such as volcanic aerosols (Crowley and Kim 1999; Free and Robock 1999) or internal climate variability (Free and Robock 1999) also have to be considered.

In addition to changes in the Sun's radiative energy output, variations in Earth's orbit on decadal to multi-centennial time scales lead to changes in seasonal and latitudinal distribution of incoming solar energy, as well as variations annual-mean insolation (Loutre et al. 1992). According to these authors, insolation since AD 1700 at 65° N includes short-term variations with a magnitude of approximately 0.1 Wm^{-2} which are superimposed on a long-term decline of about 1 Wm^{-2}. These magnitudes of the same order as changes in solar activity, hence, they should be considered in future modeling studies of multi-decadal to centennial climate variability.

Conclusions

The various observations of Holocene climate fluctuations - whether instrumental or by proxy - suggest that both, mean state of the climate system and its variability change considerably through time. Variability arises naturally within the climate system from interactions between ocean, atmosphere, cryosphere and biosphere as well as from external perturbations such as volcanic eruptions or solar variations. The interference of natural climate variability with the anthropogenic climate perturbations makes a comprehensive understanding of the range of natural climate variability indispensable for risk analysis with respect to future climate change. Time scales of key interest with respect to future climate change range from decades to about a century. Climate models are essential for predicting future climate states and their predictive capability requires to calculate future climate variability about the future mean state.

The reliability of climate models can be tested by comparing simulated with observed climate variability on various time scales. Accordingly, paleoclimate data need to supply information on past climate variability, rather than the mean state alone. For the next stage of development, this provides a challenging task for climate reconstructions, which are hitherto largely restricted to "time slices" (i.e. mean states). The longest European instrumental climate observations span approximately the last 300 years. In contrast, hydrographic observations in the North Atlantic rarely cover the last century. Given the increasingly better understanding of how climate signals are transformed into geological proxy records, we anticipate that the rich blend of geological climate proxies available, is indeed sufficient to reconstruct past climate variability on regional scales.

In contrast to local information on climate variability, the large spatial scale associated with climate regimes (i.e. regions of coherent climate fluctuations) makes them a prime target for comparing results from climate models with paleoclimatic reconstructions. Today, the North Atlantic Oscillation (NAO) regime accounts for approximately 30% of the interannual climate variability in extratropical temperature of the Northern Hemisphere (Hurrell 1996) and is the dominant mode of climate variability in Europe. However, the amount of explained variance changes considerably through time and may even drop to zero (e.g. between 1895-1920; Osborn et al. 1999). The central importance of NAO for climate prediction emerges from the fact that the hemispheric warming of recent decades seems to go parallel with an intensification of the NAO (Hurrell 1996). An appealing conjecture is that the NAO regime can serve as a working hypothesis for the operation of the North Atlantic climate system in the past. First successful attempts to trace the NAO back through time indicate that the NAO has varied on decadal and perhaps even longer time scales during the last few centuries. Whenever the expected NAO regime do not hold at all, we can still use descriptive statistics to define former climate regimes. It will be interesting to see if the regimes thus identified repeat in time and can be simulated in climate models.

External perturbations of the climate system (e.g. volcanic eruptions and solar-activity variations) are another source of natural climate variability. Model experiments suggest that for the last few centuries approximately 20–35% of the reconstructed climate variability on (multi)-decadal time scales can be explained by volcanic forcing and solar variability (Crowley and Kim 1999). The mid-latitudes of the Northern Hemisphere seem to be particularly sensitive to the effects of volcanic eruptions (Briffa et al. 1998; Pyle 1998). Accordingly, the potential effects of volcanic eruptions and solar variations have to be considered in predictions of future climate change.

References

Anderson RY, Dean WE (1988) Lacustrine varve formation through time. Palaeogeogr, Palaeoclimatol, Palaeoecol 62:215-235

Appenzeller C, Stocker TF, Anklin M (1998) North Atlantic oscillation dynamics recorded in Greenland ice cores. Science 282:446-449

Bailey ME, Clube SVM, Napier WM (1990) The Origin of Comets. Pergamon Press, London

Baillie MGL (1994) Dendrochronology raise questions about the nature of the AD 536 dust-veil event. Holocene 4:212-217

Baillie MGL (1998) Evidence for climatic deterioration in the 12th and 17th centuries BC. In: Hänsel B (ed) The Bronze Age: the first golden age of Europe. Oetker-Voges Verlag, Kiel, 49-56

Baillie MGL (1999) A view from outside: Recognising the big picture. Quat Proc 7:625-635

Barnett T, Preisendorfer R (1987) Origins and levels of monthly and seasonal forecasts skill for United States surface air temperature determined by canonical correlation analysis. Mon Wea Rev 115:1825-1850

Bhattacharya K, Ghil M, Vulis IL (1982) Internal variability of an energy-balance model with delayed albedo effects. J Atmos Sci 39:1747-1773

Bjerknes J (1964) Atlantic air-sea interaction. Adv Geophys 10:1-82

Black DE, Peterson LC, Overpeck JT, Kaplan A, Evans MN, Kashgarian M (1999) Eight centuries of North Atlantic ocean atmosphere variability. Science 286:1709-1713

Brathauer U, Brauer A, Negendank JFW, Zolitschka B (2000) Rasche Klimaänderungen am Beginn der heutigen Warmzeit. Zweijahresbericht GeoForschungsZentrum Potsdam 1998/1999, pp 29-33

Briffa KR, Jones PD, Bartholin TS, Eckstein D, Schweingruber FH, Karlén W, Zetterberg P, Eronen M (1992) Fennoscandian summers from AD 500: temperature changes on short and long timescales. Clim Dyn 7:111-119

Briffa KR, Jones PD, Schweingruber FH, Osborn TJ (1998) Influence of volcanic eruptions on Northern Hemisphere summer temperature over the past 6000 years. Nature 393:450-455

Cayan DR (1992) Latent and sensible heat flux anomalies over the northern oceans: the connection to monthly atmospheric circulation. J Climate 5:354-369

Cook ER, D'Arrigo RD, Briffa KR (1998) A reconstruction of the North Atlantic Oscillation using tree-ring chronologies from North America and Europe. The Holocene 8:9-17

Covey C, Thompson SL, Weissman PR, MacCracken MC (1994) Global climatic effects of atmospheric dust from an asteroid or comet impact on Earth. Glob Planet Change 9:263-273

Crowley TJ, Kim KY (1999) Modeling the temperature response to forced climate change over the last six centuries. Geophys Res Lett 26:1901-1904

Cubasch U, Voss R, Hegerl GC, Waszkewitz J, Crowley TJ (1997) Simulation of the influence of solar radiation variations on the global climate with an ocean-atmosphere general circulation model. Clim Dyn 13:757-767

Dansgaard W (1964) Stable isotopes in precipitation. Tellus 16:436-468

De Silva SL, Zielinski GA (1998) Global influence of the AD 1600 eruption of Huaynaputina. Peru. Nature 393:455-458

Fischer G, Wefer G (eds) (1999) Use of proxies in paleoceanography. Springer Verlag, Berlin, pp 1-735

Foster G (1996) Wavelets for period analysis of unevenly sampled time series. Astron J 112:1709-1729

Frankignoul C, Müller P, Zorita E (1997) A simple model of the decadal response of the ocean to stochastic wind forcing. J Phys Oceanogr 27:1533-1546

Free M, Robock A (1999) Global warming in the context of the Little Ice Age. J Geophys Res D104:19057-19070

Fritts HC (1976) Tree rings and climate. Academic Press, London

Ghil M (1994) Cryothermodynamics: The chaotic dynamics of paleoclimate. Physica D 77:130-159

Ghil M, Yiou P (1996) Spectral methods: what they can and cannot do for climatic time series. In: Anderson DLT, Willebrand J (eds) Decadal climate variability - dynamics and predictability. Springer Verlag, Berlin, 445-482

Groetzner A, Latif M, Barnett TP (1998) A decadal cycle in the North Atlantic as simulated by the ECHO coupled GCM. J Climate 11: 831-847

Haigh JD (1996) The impact of solar variability on climate. Science 272:981-984

Hartmann DL (1994) Global physical climatology. Academic Press, San Diego

Hasselmann K (1976) Stochastic climate models: Part I. Theory. Tellus 28:473-485

Horowitz LL (1974) The effects of spline interpolation on power spectral density. IEEE Trans Acoust, Speech, Signal Processing ASSP-22:22-27

Houghton JT, Meira Filho LG, Callander BA, Harris N, Kattenberg A, Maskell K (eds) (1996) Climate Change 1995 - The science of climate change. Cambridge University Press, Cambridge

Hurrell JW (1995) Decadal trends in the North Atlantic oscillation: regional temperatures and precipitation. Science 269:676-679

Hurrell JW (1996) Influence of variations in extratropical wintertime teleconnections on Northern Hemisphere temperature. Geophys Res Lett 23:665-668

Jönsson P, Fortuniak K (1995) Interdecadal variations of surface wind direction in Lund, Southern Sweden. Int J Clim 15:447-461

Jones PD, Hulme M (1997) The changing temperature of 'Central England'. In: Hulme M, Barrow E (eds) Climates of the British Isles, present, past and future. Routledge, London, 453

Jones PD, Jónsson T, Wheeler D (1997) Extension to the North Atlantic Oscillation using early instrumental pressure observations from Gibraltar and south-west Iceland. Int J Clim 17:1433-1450

Jouzel J, Alley RB, Cuffey KM, Dansgaard W, Grootes P, Hoffmann G, Johnsen SJ, Koster RD, Peel D, Shuman CA, Stievenard M, Stuiver M, White J (1997) Validity of the temperature reconstruction from water isotopes in ice cores. J Geophys Res C102:26471-26487

Keigwin LD, Pickart RS (1999) Slope water current over the Laurentian fan on interannual to millennial time scales. Science 286:520-523

Koslowski G, Glaser R (1999) Variations in reconstructed ice winter severity in the Western Baltic from 1501 to 1995, and their implications for the North Atlantic Oscillation. Climatic Change 41:175-191

Kushnir Y (1994) Interdecadal variations in North Atlantic sea surface temperature and associated atmospheric conditions. J Climate 7:142-157

Latif M, Barnett TP (1994) Causes of decadal climate variability over the North Pacific and North America. Science 266:634-637

Lau K-M, Weng H (1995) Climate signal detection using wavelet transform: how to make a time series sing. Bull Amer Met Soc 76:2391-2402

Legrand JP, Le Goff M (1992) Direction de la Météorologie Nationale, Monographie Nr. 6: Les observations météorologiques de Louis Morin entre 1670 et 1713'. Météo-France, Trappes

Le Treut H, Ghil M (1983) Orbital forcing, climatic interactions, and glaciation cycles. J Geophys Res C88:5167-5190

Lindzen RS (1995) Constraining possibilities versus signal detection. In: National Research Council (ed) Natural climate variability on decade-to-centennial time scales. National Academy Press, Washington, D.C., 182-186

Lorenz EN (1963) Deterministic nonperiodic flow. J Atmos Sci 20:130-141

Loutre MF, Berger A, Bretagnon P, Blanc P-L (1992) Astronomical frequencies for climate research at the decadal to century time scale. Clim Dyn 7:181-194

Luterbacher J, Schmutz C, Gyalistras D, Xoplaki E, Wanner E (1999) Reconstruction of monthly NAO and EU indices back to AD 1675. Geophys Res Lett 26:2745-2748

Luterbacher J and 33 others (2000) Reconstruction of monthly mean sea level pressure over Europe for the Late Maunder Minimum period (1675-1715) based on canonical correlation analysis. Int J Clim

Manley G (1974) Central England temperatures: monthly means 1659 to 1973. Quart J Roy Met Soc 100:389-405

Mitchell JFB, Johns TC, Gregory JM, Tett SFB (1995) Climate response to increasing levels of greenhouse gases and sulphate arrosols. Nature 376:501-504

Mysak LA, Venegas SA (1998) Decadal climate oscillations in the Arctic: A new feedback loop for atmosphere-ice-ocean interactions. Geophys Res Lett 25:3607-3610

Negendank JFW, Zolitschka B, Rein B, Brauer A, Brüchmann C, Sanchez A, Vos H (1999) Varves and solar variability (Lake Holzmaar, Eifel, Germany). Bulletin de la Société belge de Géologie 106 (1997): 53-61

Namias J (1981) Teleconnections of 700-mb height anomalies for the Northern Hemisphere. In: Fleminger, A (ed) CalCOFI Atlas No. 29, Marine Life Research Program, Scripps Institution of Oceanography

Nozaki Y, Rye DM, Turekian KK, Dodge RE (1978) ^{13}C and ^{14}C variations in a Bermuda coral. Geophys Res Lett 5:825-828

Osborn, TJ, Briffa KR, Tett SFB, Jones PD, Trigo RM (1999) Evaluation of the North Atlantic Oscillation as simulated by a coupled climate model. Clim Dyn 15:685-702

Parker DE, Legg TP, Folland CK (1992) A new daily Central England Temperature series, 1772-1991. Int J Clim 12:317-342

Parkonson CL, Cavalieri DJ, Gloersen P, Zwally HJ, Comiso JC (1999) Arctic sea ice extents, areas, and trends, 1978-1996. J Geophys Res 104:20837-20856

Peterson TC, Vose RS (1997) An overview of the global historical climatology network temperature database. Bull Amer Met Soc 78:2837-2850

Pyle DM (1998) How did the summer go? Nature 393:415-416

Rahmstorf S (1996) On the freshwater forcing and transport of the Atlantic thermohaline circulation. Clim Dyn 12:799-811

Rogers JC (1989) Patterns of low-frequency monthly sealevel pressure variability (1899-1986) and associated wave cyclone frequencies. J Climate 3:1364-1379

Schimel D and 26 others (1996) Radiative forcing of climate change. In: Houghton JT et al (ed) Climate Change 1995. Cambridge Univ. Press, Cambridge, 65-131

Schulz M, Berger WHB, Sarnthein M, Grootes PM (1999) Amplitude variations of 1470-year climate oscil-lations during the last 100,000 years linked to fluctuations of continental ice mass. Geophys Res Lett 26:3385-3388

Seina A, Palosuo E (1996) The classification of the maximum annual extent of sea ice cover in the Baltic Sea 1720-1995. Meri-Report series of the Finnish Institute of Marine Resarch 27:79-81

Shindell D, Rind D, Balachandran N, Lean J, Loonergan P (1999) Solar cycle variability, ozone, and climate. Science 284:305-308

Stuiver M, Grootes PM (2000) GISP2 oxygen isotope ratios. Quat Res 53: 277-284

Stuiver M, Grootes PM, Braziunas TF (1995) The GISP2 δ^{18}O climate record of the past 16,500 years and the role of the sun, ocean and volcanoes. Quat Res 44:341-354

Stuiver M, Braziunas TF, Grootes PM, Zielinski GA (1997) Is there evidence for solar forcing of climate in the GISP2 oxygen isotope record? Quat Res 48:259-266

Timmermann A, Latif M, Voss R, Groetzner A (1998) Northern Hemisphere interdecadal variability: a coupled air-sea mode. J Climate 11:1906-1931

Torrence C, Compo GP (1998) A practical guide to wavelets. Bull Amer Met Soc 79:61-78

Vautard R, Ghil M (1989) Singular spectrum analysis in nonlinear dynamics, with applications to paleoclimatic time series. Physica D 35:395-424

Von Storch H, Zwiers FW (1999) Statistical analysis in climate research. Cambridge Univ. Press, Cambridge

Zolitschka B (1998) Paläoklimatische Bedeutung laminierter Sedimente. Gebr. Borntraeger, Berlin

Appendix: Mathematical Excursion on Climate Dynamics and Sensitivity

Nonlinearity of the climate system allows for **multiple steady states**, that is, the possibility of different climate states for, in the simplest case, the same value of one so-called bifurcation parameter. This bifurcation parameter can be external as well as internal, that is, determined by the internal dynamics of the climate system itself. An examples for such steady states is given by Rahmstorf (1996) for the thermohaline circulation of the ocean, where the bifurcation parameter is excess freshwater input into the North Atlantic.

According to the theory of dynamical systems, the simplest steady states are stationary (fixed points) or oscillatory (limit cycles or free oscillations). Some steady states may be associated with climatic extremes, others with low variability ("quiet times"). The detection of climate states (through time) and their associated climate regimes (in space) would enable us to draw the bifurcation diagram of the climate system (or a subsystem thereof). Hence we would learn how close the present climate state is to any bifurcation point and thus, upon some external perturbation, how probable a transition to a different climate state is.

Limit cycles or free **oscillations** result from a delayed negative feedback on one state variable z or from the coupling of two state variables z_1, z_2. They are the solutions to either one delayed differential equation (with delay τ) or two first-order ordinary differential equations. Simple linear examples are (t denotes time)

$$\frac{dz}{dt} = -z(t-\tau)$$

or

$$\frac{dz_1}{dt} = z_2 \quad \text{and} \quad \frac{dz_2}{dt} = -z_1.$$

The set of two first-order ordinary differential equations can also be written as one second-order ordinary differential equation,

$$\frac{d^2 z_1}{dt^2} = -z_1.$$

All three representations are, in a sense, equivalent (Bhattacharya et al. 1982). They show that the effect of delay or "memory" is to introduce "inertia" into the climate system and allow for the possibility of an internal (or free or self-sustained) oscillation with frequency $f \sim 1/(2\,\tau)$.

If the climate system is nonlinear, there will be additional free oscillations associated with the harmonics of the fundamental internal frequency. If, furthermore, the climate system is forced (e.g. by variations in solar activity), there will be harmonics and combination tones of the external frequencies, too (Le Treut and Ghil 1983). In the nonlinear, forced climate system, part or all of the variability occurs at the forcing frequencies, instead of the internal frequencies only. That is, energy is transferred from the internal to the forcing frequencies. This phenomenon is called nonlinear resonance, and can even occur for large differences between the external and internal frequencies (Ghil 1994). However, if the internal frequency is lower than the forcing frequencies, then the amplitude of high-frequency harmonics and combination tones can be small or insignificant.

If the number of active state variables of a system is three or larger (as in the climate system), it can undergo a transition to chaos, with a corresponding change in the character of the oscillations and the spectrum (the behavior becomes more nonperiodic [quasi-periodic] and irregular, and the background noise rises). The interaction of external and internal frequencies can cause an intermittent character of the oscillations and hence transitions between climate states. These transitions occur without explicitly changing a bifurcation parameter, and the climate system may spend more time in one state than in others (Bhattacharya et al. 1982).

The relationship between the size of the external perturbation (e.g. change in radiative forcing dQ) and the magnitude of its impact (e.g. global-mean surface temperature response dT) is called the **climate sensitivity** λ_R,

$$\frac{dT}{dQ} = \lambda_R.$$

The climate sensitivity is a simple constant only in a special case. In the general case, it is a function of a subset Ψ' of a set of variables $\Psi = (\psi_1,...,\psi_n)$ (e.g. water-vapor content of the atmosphere, snow and ice cover, etc.) that describe the state of the climate system:

$$\lambda_R = \lambda_R(\Psi').$$

A process that dynamically changes the climate sensitivity is called a **feedback**. In our example, it is the global-mean surface-temperature response dT that would "feed back on itself". Thus T would beco me part of the subset Ψ', and we could write symbolically

$$\lambda_R = \lambda_R(T, \Psi''),$$

where Ψ'' denotes the subset of Ψ' without the element T, that is, all state variables that would influence the climate sensitivity besides global-mean surface temperature.

Even if the climate sensitivity depends only linerarly on T, an initial temperature perturbation will be increasingly amplified ($\lambda_R(T, \Psi'') > 0$) or increasingly dampened ($\lambda_R(T, \Psi'') < 0$). Hence, a feedback makes the climate system nonlinear. Moreover, the magnitude of a change in T upon a perturbation depends on the state of the system (e.g. T). It is the combined effect of positive (amplifying) and negative (dampening) feedbacks that prevents run-away scenarios, that is, which help to stabilize the climate system.

Holocene Climate and Past Volcanism: Greenland - Northern Europe

Claus U. Hammer

University of Copenhagen, Niels Bohr Institute of Astronomy, Physics and Geophysics, Department of Geophysics, Juliane Maries Vej 30, DK-2100 Copenhagen O, Denmark
e-mail: cmhammer@get2net.dk

Abstract: Holocene climate data from Greenland ice cores, European tree ring widths and their relation is discussed. The problems related to such proxy data as well as to a volcanic paleo index for stratospheric volcanic aerosol loading are presented. The stacking of data from especially the Greenland and Canadian ice cores and the problems involved are treated. Possible solutions to the problems are presented.

Background: Two Problems

The First Problem

The title of the review paper which I have been asked to write for this conference is rather precise but very long: So I have shortened it a bit as I also want the reader to know the job I have been given, I here repeat the working title i.e. *"The Greenland Ice Core Record of Climatic Changes During the Holocene with Relevance for Northern Europe" (Emphasis to be placed both on the regional climate change and possible coupling to the volcanic activity).*

How to approach this task? Let me first deal with it in a hypothetical way i.e. let us assume, that our paleoclimatic and paleovolcanic information was on the following form: A paleotemperature record representing the Greenland-Canadian region in a realistic way existed for the Holocene period and a similar one for Northern Europe plus a fairly good idea of the northern hemispheric stratospheric volcanic aerosol loading over the same period. In this case a reasonable approach to "live up to the title" would be to analyse the two climate records by various mathematical techniques in order to identify e.g. correlation coefficients, eigen functions, special features etc. The next step would be more complicated, because now one would have to argue why the similarities in the records are as found and why the differences should be as found. The optimal outcome would be to present a climate model, which actually concurred with all 3 records! Would this be possible if all 3 records were true?

In my opinion the answer is both yes and no. Yes, because it is possible today to run an advanced climate model in such a way that it could encompass a fair percentage of the variability in the various records. No, because the time development of two regional temperature records over the Holo-cene can hardly be fully described in a climate model unless the model includes in a realistic way internal couplings in the climate system as well as external forcings. Even if the volcanic influence could be calculated it would only be under a number of assumptions (some of these assumptions might be based on other paleoclimatic and environmental information) but we would evidently be far from synthezing a comprehensive set of such "true" data.

Even when we concentrate our efforts only on the 20th century, a period offering a multitude of climatic-environmental information, we have great difficulties in agreeing on the influence of the "anthropogenic greenhouse effect"; at least concerning its quantification.

So we do need to improve both the number and quality of relevant paleodata as well as the incoorperation of "realistic" submodels into a general climate model.

From WEFER G, BERGER W, BEHRE K-E, JANSEN E (eds), 2002, *Climate Development and History of the North Atlantic Realm.* Springer-Verlag Berlin Heidelberg, pp 149-163

The above describes a general problem in climate research which will not be solved at this conference. I just want to remind you that more paleodata are fine, but not sufficient to understand climate change. The second problem has to do with the paleorecords themselves.

The Second Problem

While the first problem is the concern of all climate researchers and has been an area of research showing great progress over the past decades (modern computers have played an essential role in this development (see e.g. Bengtsson 1999), less attention has been given to the problems concerning the paleoclimatic records themselves. I do of course not mean that paleoresearchers do not know about the problems related to the paleorecords, but in my opinion there has been a tendency to accept records as e.g. tree-ring widths, ice core isotopic composition, lake sediment data, peat bog data, and sediment records from the ocean as more or less true proxies of climate change; this lead me to one of the tasks I was given for this conference - to compare the Greenland ice core data to Northern Europe climate data.

In order to compare such data over such a long period as the Holocene, it is nescessary to secure that the records are comparable. In some cases we are on "safe ground" e.g. the dramatic drop in the isotopic values around 8200 BP in all Greenland and Canadian ice core records (Hammer et al. 1986; Dansgaard et al. 1993; Fisher et al. 1996) is no doubt real and signifies a major climate event, but what about the general accepted high temperatures in the climatic optimum? The information is somewhat confusing as some records clearly show the optimum, while others do not. Evidently there are problems with the records! Problems which are not due to regional or local climatic conditions, but problems related to the records per se.

All this may be considered as a quite pesimistic point of view, but how can one compare paleorecords from the Greenland region to the records from Northern Europe if the records are influenced by factors which distort them? This distortion of the records is caused by changes in the climatic system; so in some way there is a relation to climate change. The question is how is this relation?

Below I shall especially discuss two parameters which are used as indicative of temperature changes in the past i.e. ice core isotopic composition and tree-ring widths, but before doing that let me recapitulate some of the findings relating Greenland-Canadadian climate data to those of Northern Europe.

Holocene Climate Changes - Greenland Ice Cores and Northern Europe

The Regional Character

Paleoclimatic proxy data are obtained from certain sites i.e. usually they are assumed to reflect the climatic conditions at a local site. The validity of the record can usually be extended to a larger region by reasonable arguments based on meteorological considerations e.g. data from the Netherlands, Northern Germany, part of Scandinavia and the British Isles must generally show a fair degree of correlation. The situation may, however, change if such data are compared to data from Southern Europe, the West of Scandinavia, Ireland or Iceland. The regional character is of course even more pronounced if correlation studies are made between Northern Europe and e.g. India, Antarctica, South America, or the United States. An easy way to demonstrate this is to compare annual temperature records covering the past 100 years from all over the Globe: They do not always correlate well even though several regions show similar trends i.e. either warming over the past century or no trend at all. Some regions - far apart - show the same trends, which is to be expected due to the interconnections of the atmospheres dynamic behaviour e.g. a rule-of-thumb used for North European weather compared to Greenland weather has been that "when it is cold in Greenland it is warm in Northern Europe". It is a rule which concur fairly well with the existing meteorological data over the past 20 years, but it is certainly not a rule which concur with the entire 20th century. To use it for the climatic changes over the Holocene would be a great mistake.

Another aspect of regional paleoclimatic time-series over the Holocene is their variability. It is not surprising that regional proxy climate records will show much more variability than stacked paleoclimatic series from all over the Earth. Regional changes in climate is, however, of outmost importance as it is these changes which are experienced by societies, towns, and the people living in the regions: The change of the average Global climate is important in order to understand the Earth Climate System in general, but the regional (or local) climate is what people experience. In principle the global climate could change very little, while dramatic changes could happen on a regional scale.

Three Older Records

H.H. Lamb tried in 1965 and 1970 to reconstruct a Central England Temperature record reaching back to 800 AD (Lamb 1965, 1970). The record was not on an annual basis and it was based on actual temperature data mixed with proxy data and historical observations. This record was often quoted during the 1970-ies and early 1980-ies and it offered two advantages i.e. the chronology was correct and it was based on a rather comprehensive study.

In 1974 La Marche presented the White Mountains, California, tree ring widths as a well founded regional proxy climate record; in this case the chronology was as good as the counting of tree rings offer (± a few years). The record showed good agreement with Lamb's Central England temperatures: In both records the "Little Ice Age" was clearly seen as well as the "Warm Middle Ages". The records also showed a slight warming between 1500-1550 AD.

It should be remembered that in the 1970-ies very few well dated records existed, which could be considered a representative for "temperature". The "Little Ice Age" and "Medieval Warmth" therefore became fairly widespread concepts in the climate litterature. European historical sources on the 16th to 17th centuries had many examples of extreemely cold winters e.g. the frozen Danish Belts which allowed the Swedish king Karl Gustav to bring his army over the frozen belts between the Danish islands in January-February 1658, the freezing of the Thames, etc. Of course such historical events do not make a scientific climate record, but they add to the verification of a concept like the "Little Ice Age".

I would like to add yet another historical information which indicate special climate conditions 400 years ago: On May the 7th 1598 AD people in Oldenburg (Holstein) saw 3 suns on the sky and the event was apparently so unusual that a woodcut was made (Fig. 1). At that time the superstition was strong and much nonsense has survived in the written sources of the 16-17th centuries, but as a glaciologist I have experienced the 3-suns phenomenon on the sky over the Greenland Ice Sheet several times: Today we would call the phenomenon "Sun dogs" and it is related to the reflection in special snow crystals. Now these crystal configurations are associated with cold conditions and they are certainly rare phenomena in to-day's Oldenburg not to say Northern Europe.

Apparently atmospheric layers of an "Arctic like" snow crystal figuration penetrated to Northern Germany in 1598 AD - in May: Yet another indication that the weather was cold during "The Little Ice Age"! Such extreme events are not confined to the 16-17th century, but have also occured in more recent times. In 1816 AD the year after the major eruption of Tambora in 1815, New England experienced "The year without a summer" (Stothers 1984); snow was falling even in July. 15 years later several major eruptions overlapped in time between 1830-1840 AD (Hammer et al. 1980): In Copenhagen the Danish painter Christen Købke noted in a letter, that snow was still present in the streets at the shadow-side of the town houses in the month of June! It sounds almost impossible to me living in Copenhagen since 1945. I know of no one in the 20th century having experienced "permanent" snow cover in Copenhagen during June.

So, even though we can not construct climatic time series from such events, we should not forget that they are related to human experience in a direct way, and is it not this relation which actually is the important goal of climate research?

The records of climate change of Lamb and La Marche was published at a time, when the dating of Greenland ice cores was strongly improved and the isotopic composition of the ice and its relation to the local condensation temperature (Dansgaard

Fig. 1. On May the 7th 1598 the inhabitants of Oldenburg observed 3 suns on the clear sky (From the "Chronicle of Oldenburg" 1599). To-day the phenomenon would be called "Sundogs" and would indicate cold air with special snow crystals. The sky appears blue even though crystals are present. (From The Royal Library, Copenhagen).

1964) came into focus as a proxy for past temperature changes in Greenland (Dansgaard et al. 1975).

The subject of paleoclimatic ice core studies was extended to the study of climatic forcing parameters by means of ice cores, when greenhouse gas concentration in the ice bubbles (Stauffer et al. 1985) and past volcanism was found to be inscribed in the ice (Hammer 1977; Hammer et al. 1980). Actually Hammer et al. (1980) attempted to compare a Northern Hemisphere Temperature Index to past volcanism. Perhaps the name "Northern Hemisphere Temperature Index" was a little bold in 1980, but by using the data of Lamb, La Marche and the data from the Greenland Ice Sheet we did the best we could i.e. the state of the art in 1980 (Fig. 2).

The resulting conclusion was that past volcanism could explain a large fraction of the climatic varia-bility in the Northern Hemisphere over the period 550-1950 AD; though other causes of climatic change must also have been in action.

To-day we know that major volcanic eruptions do indeed change the climate on a global or hemispherical scale, but we are also much more aware of the variability of the climate (without external forcing like volcanism, changes in the sun and the atmospheric greenhouse gas concentrations). The tools for analysis has been "sharpened", the requirements for a larger coverage by proxy climatic data series and a tremendous increase in computer power and the internet access to data sources have changed the research area dramatically.

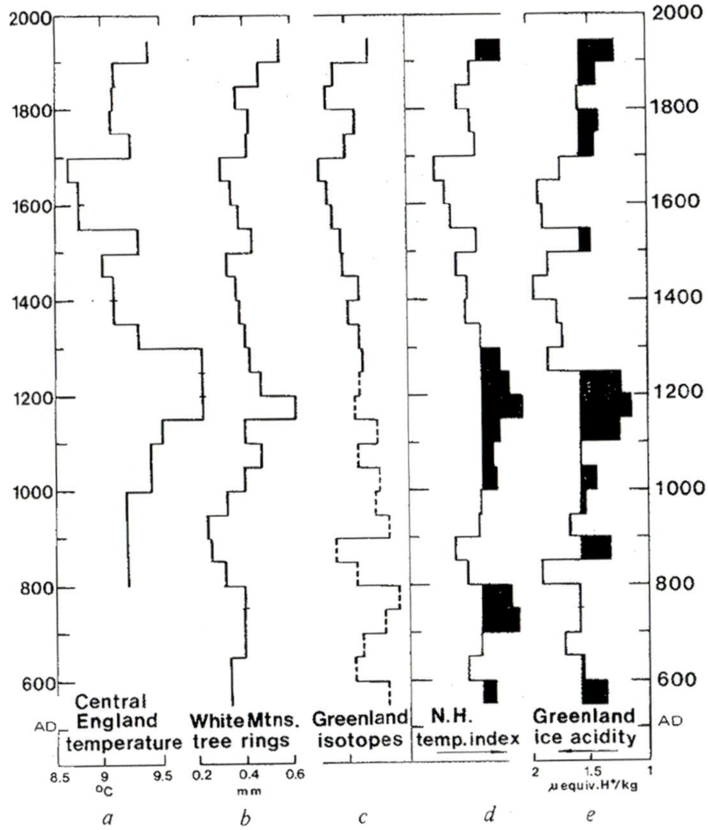

Fig. 2. Correlation between climatic temperature variations(curve d, composed of the three indirectly determined records a through c; warm periods are set out in black) and volcanic activity (curve e, showing an acidity profile along the Crête ice core; lower than normal volcanism corresponds to low acidities set out in black). (From Hammer et al. 1980).

Phenomena like ENSO, NAO, SO have to be brought into the analysis. Also the "noise" in the records have called for more data in order to validate the older "single" records from a certain local site; it be in Greenland, in Europe, or in any locality on Earth.

The number of proxy climatic data from the Greenland-Canadian sites have grown over the past 20 years and so have the tree ring data from Europe and around the world. What influence has this on the older conclusions and how much do we know to-day?

Stacking Data

Ice Cores

Two attempts have been made in recent times to stack ice core data from Greenland/Canadian (Fisher et al. 1996) records and from Central Greenland (White et al. 1997). The time perspective in the two attempts were quite different: Fisher et al. (1996) extended the analysis to the entire Holo-cene, while White et al. concentrated on the GRIP- and GISP2 records as well as on the past century. Their approach was different due to the different periods involved. As the studies of Fisher et al. (1996) covered the entire Holocene they also had to accept that the "stacked" records could be influenced by glacio dynamic problems, while the stacking of data by White et al. only covered the last few centuries and reduced the "noise" caused by snow deposition. The latter study (White et al. 1997) could con-centrate on simple mathematical correlation studies, while Fisher et al. had to conclude with a question i.e. "One is left with the question of why the GRIP/GISP2 records show virtually no Holocene trend at least since about 8000 (BP) while other records do".

Let us first look on the conclusions of the White et al. (1997) paper, which deals with recent times. By stacking five GRIP and one GISP2 well dated ice cores the glaciological noise could be strongly diminished. This Central Greenland stacked record was then used as a paleoclimatic record covering the time period 1770-1989 AD The record was compared to Greenland coastal temperatures, SST, NAO, and the sunspot record. In their test the period 1900-1987 was used due to the limited data from meteorological records. Their conclusion can be summarized in the following way: The Central Greenland isotope record contains information on the NAO, the Sea Saw, SST, and Greenland Coastal temperatures. The reader is advised to seek details in White et al. (1997).

The important point is, however, that the ice record of Greenland over 1900-1987 AD apparently is the result of several features of the North Atlantic climate system: This indicates the importance of modelling the climate changes as well as to avoid overinterpreting the Greenland ice core climate data.

Fisher et al. (1996) included Canadian ice cores in their study over longer time intervals, which added to the regional validity of the stacked records, but also indicated the problems associated with the differences between the records. Let it be that glaci-ological problems entered the interpretation, but the various ice core records over the Holocene did also suggest that the records were not easy to inter-prete! While Fisher et al. (1996) by using the EOF technique (Empirical Orthogonal Functions) got reasonable results for the past thousand years (Fig. 3) the comparison over the entire Holocene left several questions to be answered.

In my opinion the problem is caused by an overinterpretation of the isotopic data: Glaciological noise has little to do with this, but the dynamics of ice sheets and the break down of the assumptions of isotopic "Paleothermometer" are the real causes of the inconsistent results. It should, however, be noted that the cold climate spell of 8200 B.P. seems to be a consistent feature in all the Greenland/Canadian ice records (the chronological accuracy may differ, but the strong cooling around 8200 BP seems to be a fact).

Apparently the climate records from the Greenland Ice Sheet conserve important information on the past climatic changes over and around Greenland, but their validity seems to be restricted to the past few thousand years. Actually the relation between the isotopic composition of the snow and the local temperature (Dansgaard 1964) seems only to hold for the past few thousand years. This is not surprising as the δ^{18}O-T relation has only been verified for this period.

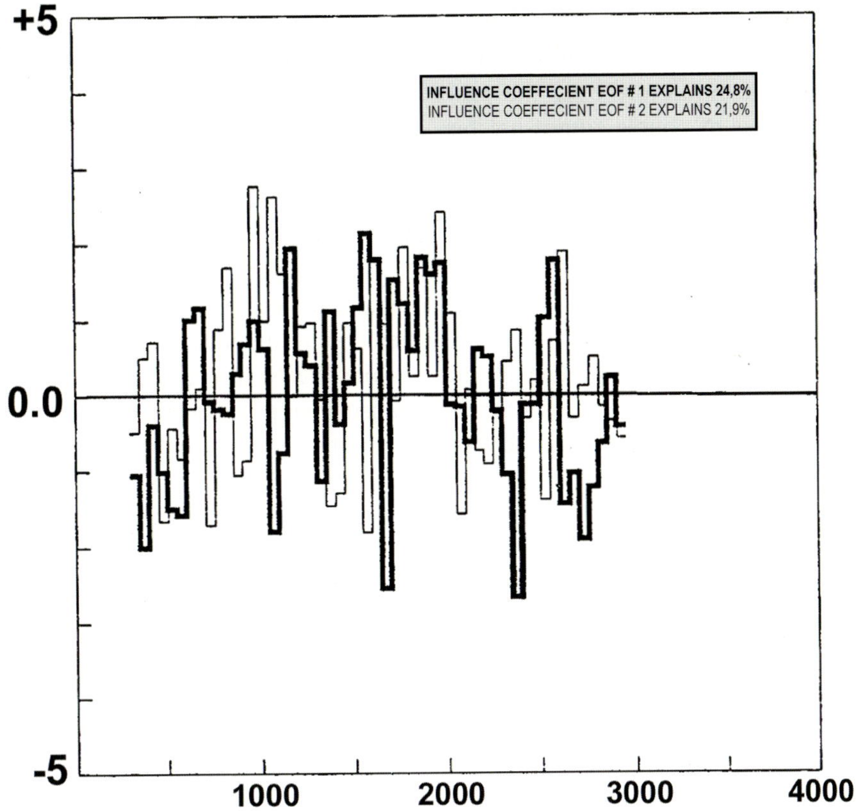

Fig. 3. The influence cofficient of the first EOF (thick) line explains 24.8% of the variance and represents the paleotemperature in the analysis of the stacked ice core data. (From Fisher et al. 1996).

Tree Ring Widths

Tree ring widths and density data are of course available from much wider regions than the corresponding ice core data, but as climatic proxy data, most of them contain a mixed atmospheric signal i.e. the thicknesses depend not only on temperature and precipitation, but also on various environmental and botanical parameters.

Stacking of data sets may help to smooth out local "noise" in the records, but even for Europe there are only a few time series data sets which can be representative for air temperature over the Holocene.

Some of the European tree ring data have played a crucial role in chronological studies and recently Kromer and Becker (1993) (plus later corrections) have extended the Holocene dating into the Younger Dryas: In this way Greenland ice core data and European data of the entire Holocene can be linked closely in a chronological way.

The question remaining is, however, can we stack North European data and in this way get something like a correspondence to stacked Greenland ice core records over the Holocene? If we can, one cannot help raising another question: "What does this help if, as we have seen, the Greenland data already leaves us with several unanswered questions?"

I believe we have to work our way backwards in time, scrutinizing the data millenium by millenium, and to find new methods, which can help us to perhaps correct the records in such a way that they become comparable. In order to be more specific I have chosen to discuss two problematic features in the reconstruction of Holocene climatic changes: The differences observed in the climatic records of two Central Greenland deep ice cores covering the Holocene and the recent attempt by Mann et al. (1999), to reconstruct the Northern Hemisphere average temperature over the past millenium. The latter is of course somewhat far from elucidating Northern Europe's Holocene climate, but it illus-trates the problems involved.

The GRIP and NORTH GRIP Holocene Climate Records

When the European GRIP drilling in Greenland and the adjacent US GISP2 drilling could present their isotopic analysis of the cores in a comparative way (Dansgaard et al. 1993; Grootes et al. 1993) it was quite clear that the records were similar (at least for the past 90 kyrs) Both records showed a fairly "stable" climate over the Holocene apart from an expected increased warming over the Pre-Boreal, the abrupt climate change around 8200 years BP and short term fluctuations in climate over the period. The stability was, however, not apparent, when e.g. methane concentrations were analyzed (Blunier et al. 1998). Evidence from North Atlantic deep sea cores have revealed a Holocene cyclicity of 1470 ± 500 years (Bond et al. 1997) and one would have expected this to show up in a fairly strong way in the Summit ice cores.

Later the analysis of the borehole temperatures suggested that the isotopic records underestimated the temperature changes from glacial to past glacial times and a climatic optimum was clearly indicated in the Holocene (Dahl-Jensen et al. 1998). Even though ice bore hole temperatures have to be transformed into atmospheric temperatures by glaciodynamic modelling and only general trends of temperatures can be derived, the results showed that the δ^{18}O-T relation did not hold for the older ice layers.

This became even more evident, when the NORTH GRIP core (315 km north west of the GRIP drill-site) was analyzed. Both cores were retrieved from Central Greenland and it was expected that they would show similar climatic trends over the Holocene. As can be seen in Fig. 4 this was not the case.

The explanation can not be due to the ice flow as the Holocene flow regime has hardly changed much after the Pre-Boreal. It is possible that changes in the elevations of the drill site locations over time could cause some minor changes over the Holocene. It is, however, hard to imagine how this could suppress the rather warm "temperatures" before 4 kyrs BP as indicated by bore hole measurements. This is particular pronounced in the GRIP record.

What is particular annoying in Fig. 4 is the difference between the isotopic changes, over the past century in the NORTH GRIP and GRIP records: While the NORTH GRIP record clearly shows unusual warming in the 20th century, as compared to the last Millenium, this is not the case for the GRIP core. As mentioned earlier stacked data can remove depositional noise, but that the differences over a whole century should be due to this is very unlikely.

We know from temperature data from the Greenland coastal stations that the NORTH GRIP 20th century profile concur with these. Also the NORTH GRIP profile over this period resembles the paleotemperatures from the GRIP borehole (Dahl-Jensen et al. 1998). In my opinion the most likely explanation of the differences is due to changes in the seasonal snow deposition pattern. The linear δ^{18}O-T relation presently used for Greenland ice cores is (Johnsen et al. 1989):

$$\delta^{18}O = 0.67\ T - 13.7‰ \text{ (T in degrees Celsius)}$$

It is, however, a prerequisite that the seasonal snow deposition is unchanged; the absolute amount of annual snow accumulation (over e.g. 30 years) may change as long as the seasonality is maintained.

How serious could this be if, say, the amount of winter to summer snow ratio did change? (Such a change should a priori be expected if the climate changes). It could be quite serious as the isotopic oxygen composition of the Greenland Ice Sheets' winter and summer snow differs by some 10 to 15‰! Let me give a simple example: The temperatures prior to 4 kyrs BP were 4°C warmer than the last millenium; if this warming preferentially made the winters in Greenland more moist (a not unlikely consequence) the ratio between winter and summer snow deposition would change. A 20% increase in the winter to summer snow ratio would then decrease the annual δ^{18}O value by some 1 to

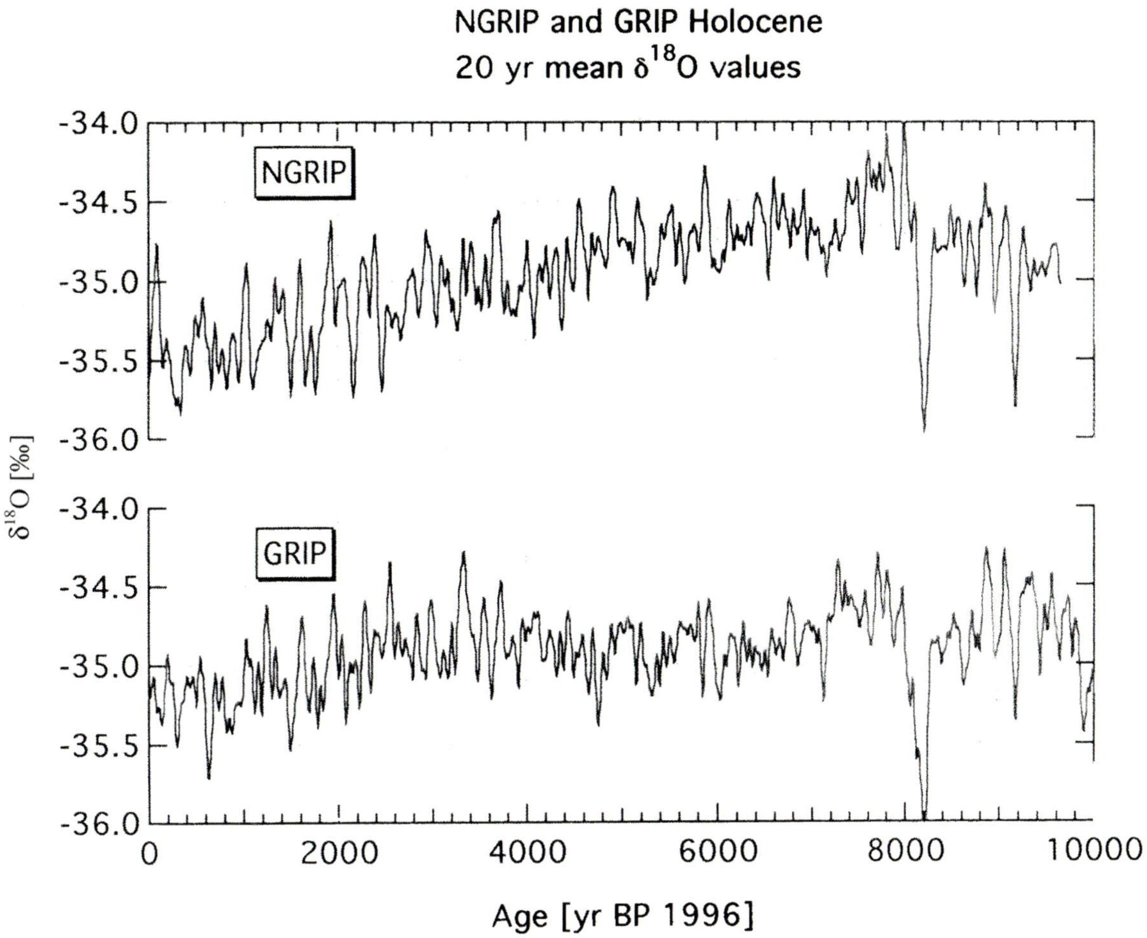

Fig. 4. Isotopic profiles over the Holocene past of the NORTH GRIP and GRIP ice cores from Central Greenland. Note the difference in trends, the "lack" of a clear climatic optimum, and the difference in the 20th century.

1.5‰, which corresponds to an apparent change in annual temperature of some 1.5 - 2.2°C . The result would be that even though a warmer climate should increase the $\delta^{18}O$ values the increased amount of winter snow substantially decrease the $\delta^{18}O$ values. This is clearly a serious problem, but has escaped sufficient attention because the $\delta^{18}O$-T linear relation has been empirical verified for the ice sheet. The empirical relation has been established by using a fair amount of isotopic measurements and average annual temperature measurements (10 m temperatures in the upper part of the ice sheet). The problem is that the $\delta^{18}O$-T relation is established by using recent data i.e. over periods assumed to represent a reasonable stable climate. Hence it is not valid for the climatic optimum and the high temperatures in Greenland between the late 1920-ies to early 1960-ies!

Is there any way out of this dilemma? The obvious answer is to measure some parameters along the ice cores which can lead to an indication of the seasonal snow deposition characteristics, but such parameters are not so easy to identify. One possibility is to make highly resolved measurements of nitrate and dust concentrations along the cores: The nitrate concentration is believed, in general, to peak during the summer, while the dust concentration peaks in late winter-early spring (Hammer 1977).

Technically this is possible to do but it requires a substantial amount of work. Also there is no guarantee that such a project will solve the problem and remove the inconsistencies.

Northern Hemisphere Temperatures - Reconstruction Over the Last Millenium

Now, let me turn to the attempt of Mann et al. (1999) to improve the NH record of temperatures for the past millenium. Mann et al. have made a very meticulous and careful study of the proxy records covering the last millenium. They have selected the time series in a restrictive way and treated them in a mathematical and careful way. Their work is depending on many other researchers' work and they clearly state the problems of uncertainties as they go from recent temperature data to tree ring- and ice core data for the times before 1700 AD Their main result is shown in Fig. 5.

From Fig. 5 it can be seen, that the "Medieval warm period" and the "Little Ice Age" are not the most evident features. There is a somewhat higher level of temperatures prior to 1400 AD and slightly lower temperatures between 1400 - 1900 AD, but nothing resembling e.g. the older records mentioned earlier and shown in Fig. 2. The most evident features in Fig. 5 are a general decrease of temperature over the past millenium consistent with long-term astronomical forcing (according to Mann et al. 1999) and an anomalous warming during the 20th century.

Should we trust this NH temperature construction? Mann et al. are quite clearly pointing out the problems involved in the reconstruction; especially before 1400 AD, but their main purpose is to illustrate the exceptional high temperatures in the 20th century compared to the last millenium.

A really stricking feature in Fig. 5 is the low variability of the NH climate before 1900 AD. Also

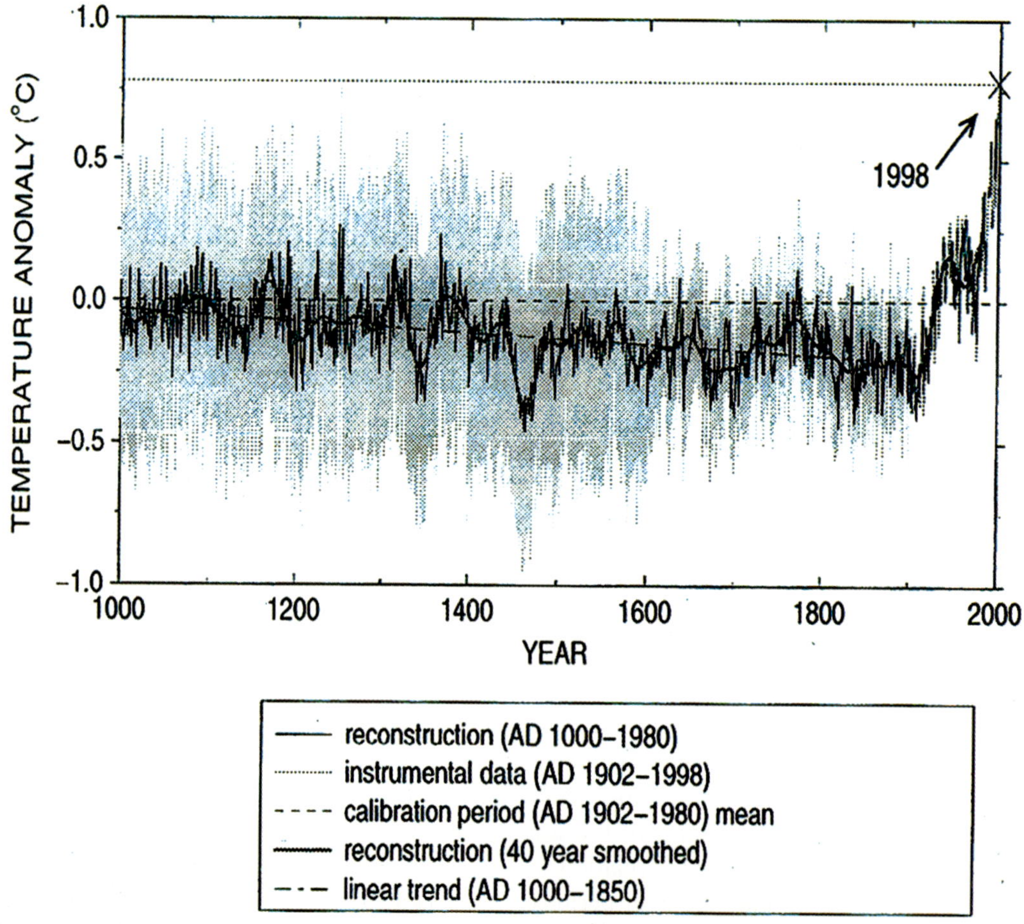

Fig. 5. Northern Hemisphere temperature reconstruction by Mann et al. (1999).

the uncertainty of the temperature estimate is increasing strongly for the data representing the time before 1600 AD; especially for the data prior to 1400 AD. The reason for this is due to both the fewer data covering the early parts of the millenium and their proxy character.

Recently this temperature reconstruction was discussed at an international workshop in Stockholm and support of the technique and criticism of the chosen data set were aired (W. Karlén et al. 1999).

Is the absence of the Medieval warm period and the Little Ice Age due to the inclusion of more proxy time series than the ones used in Fig. 2. The answer is again both yes and no! The Central England temperature is not used in the analysis of Mann et al. Instead proxy indicators from tree ring data of Fennoscandia, the Polar Urals, Marocco, France, USA, and Greenland stacked ice core data were used. This is, of course, an improvement compared to the older reconstructions, but is it enough and do these data really represent the NH temperature development over the past millenium? We do not know and as shown earlier even stacked ice core data contains unknown biases.

Recently the exceptional warm characteristics of the 20th century was demonstrated by the high Arctic Hans Tausen Iskappe (ice cap) ice cores' $\delta^{18}O$ profile (Hammer and Højmark 1998; Hammer et al. 2001): This ice cap is situated in the extreme north of Greenland. The $\delta^{18}O$ profile did, however, not confirm the recent 20 years of warming as in Mann et al. (1999) N.H. construction; in this respect the Hans Tausen $\delta^{18}O$ profile is typical for the Greenland/Canadian region.

To me it seems evident that in lack of sufficient data there is a tendency to *get all you can out of the existing data while you have the chance* and a little more than is justified. In principle this is OK from a scientific point of view, but it should be remembered that in the public area such scientific findings are appearing as "the truth" without any knowledge about all the scientific reservations.

The Aspect of Volcanism

Above I have only slightly discussed the influence of volcanism on the Holocene climate apart from the discussion related to Fig. 2. A number of volcanic eruptions do of course influence climate and the way they do this or have done this is well known. For the Holocene we have a fairly good idea about how much and when the stratosphere was loaded with climatic relevant volcanic products. So in principle we could just enter this information into the models of climate change. There are, however, still several unsolved problems related to establishing a realistic volcanic index parameter to be used in the models.

The problem concerning the effect of volcanism on the past climate is not related to the major climatic relevant eruptions; these have left clear "signals" in both the Arctic and Antarctic ice cores, so we are on "solid ground" i.e. we can estimate both the stratospheric volcanic debris load and how long the climatic relevant volcanic load was present.

The real problem is to estimate the stratospheric volcanic aerosol load from volcanic eruptions of intermediate size to background volcanic activity i.e. how do we from ice core analysis distinguish between the volcanic sulfate concentrations over time from biogenic or continental sulfate concentrations? Only the stratospheric load has a major direct effect on climate while the tropospheric concentrations deriving from the sea and the continents have only indirect effects on the climate. This problem has not yet been solved, but the situation is not hopeless.

Comparison between Arctic and Antarctic ice cores, which have a well established chronology, can help: Actually the EU project MilEClim has as one of its main goal to improve the volcanic index over the last millenium. So even though the present index of past volcanism is used as an input parameter to climate models, this index have several deficiencies: First of all it does only partly rely on data which are linked to the actual aerosol loading of the stratosphere, i.e. one might say: "If there are no climatic relevant atmospheric debris found in polar ice cores it is simply because there was nothing *up there!*"

Some scientists are still using information on past volcanism which only to a minor extent has been verified by the real facts inscribed in the ice sheets. It is rather frustrating as estimates of the influence of past volcanism on climate is still to a large extent based on data which have not been confirmed by the real world!

If there are no climatic relevant volcanic signals in the ice cores (especially in the Greenland and the Antarctic Ice Sheets) there is no reason to use an index of past volcanism, which is based on subjective evidence. This is the reason why we still need to achieve a better index for past volcanism of relevance to climatic research.

Robock and Free (1995) used stacked ice core acidity to construct a volcanic index, but we still need to address the latitudinal aspect i.e. to relate the volcanic time series in ice cores to latitudinal zones of the eruptions.

A first approach and perhaps the most realistic one would be to divide eruption sites into 5 zones: 60°-90°N, 30°-60°N, 30°N-30°S, 30°-60°S. This subdivision is important because the influence of a major eruption on climate depends on how the volcanic veil is spread in the Earth stratosphere. This zone division might seem somewhat in contradiction to pyrheliometric data covering the period 1881-1960 (Stothers 1996), because e.g. the stratospheric loading from the 1912 Katmai eruption in Alaska could be clearly observed down to 47°N and with a weaker change of optical depth down to 30°N. The volcanic impurity deposition as observed in Greenland ice cores do, however, not a priori have to reflect the stratospheric loading above the ice sheet location, but is also depending on tropospheric volcanic loading and the average weather in a region around the drill-site. The suggested division is therefore based on actual ice cores observations.

The deposition over time of the volcanic aerosol in Greenland ice cores is actually depending of the zone of eruption (Fig. 6). In the case of a single major eruption this helps to identify the zone of the eruption; if the eruption took place in the zone 30°N-30°S one would expect a volcanic signal in ice cores from both Greenland and Antarctica; Langway et al. 1995 compared volcanic signals in ice cores from Greenland and Antarctica covering the last millenium and identified many common volcanic signals. Eruptions north of 30°N would only in extreeme cases be detectable in Antarctic ice cores.

The use of the deposition over time and a bipolar approach is therefore a strong tool for identifying the eruption zone. The requirements for doing this is, however, not always present: The ice core must be well dated, the annual precipitation must not be too small and the volcanic aerosol concentration must clearly exceed the ice core non-volcanic background concentration. Further depositional noise calls for stacking data from several cores either from the same site or from areas, which offers similar depositional characteristics. The GRIP, GISP2, and Crête cores are all from Central Greenland and apart from the depositional noise, they are all representative for Central Greenland - but not for Southern Greenland, where even major Arctic eruptions like Katmai 1912 can hardly be detected.

A special, but important, challenge is the entanglement of several intermediate eruptions mixed with many smaller eruptions over a short period like in the period 1815 - 1850 AD; how do we separate these "in time" overlapping volcanic signals? It is a particular important problem, because overlapping volcanic signals intensify and strengthen the influence of volcanism on climate (Crowley and Yul Kim 1999).

Several scientific projects are just now trying to obtain more and better data as well as analyzing the already existing data e.g. the European MilEClim project is working to improve the situation and the US' ITASE traverses in Antarctica plans to enlarge the geographical coverage of Antarctica, while the deep drillings of EPICA and the Japanese Dome Fujii core expand the coverage back in time.

It will take time and a lot of data analysis, but in the horizon we can at least see the first contures of a realistic volcanic climate index. In the meantime the climate modellers will of course play with what they can get.

Some Final Remarks

Trying to relate Greenland or Canadian ice core data to the climate of Northern Europe will only be succesful if a sufficient number of paleoclimatic proxy data from the Northern Hemisphere are analyzed. Also it is nescessary that the data series are treated mathematically in a way, which tell us something about the geographical pattern of the

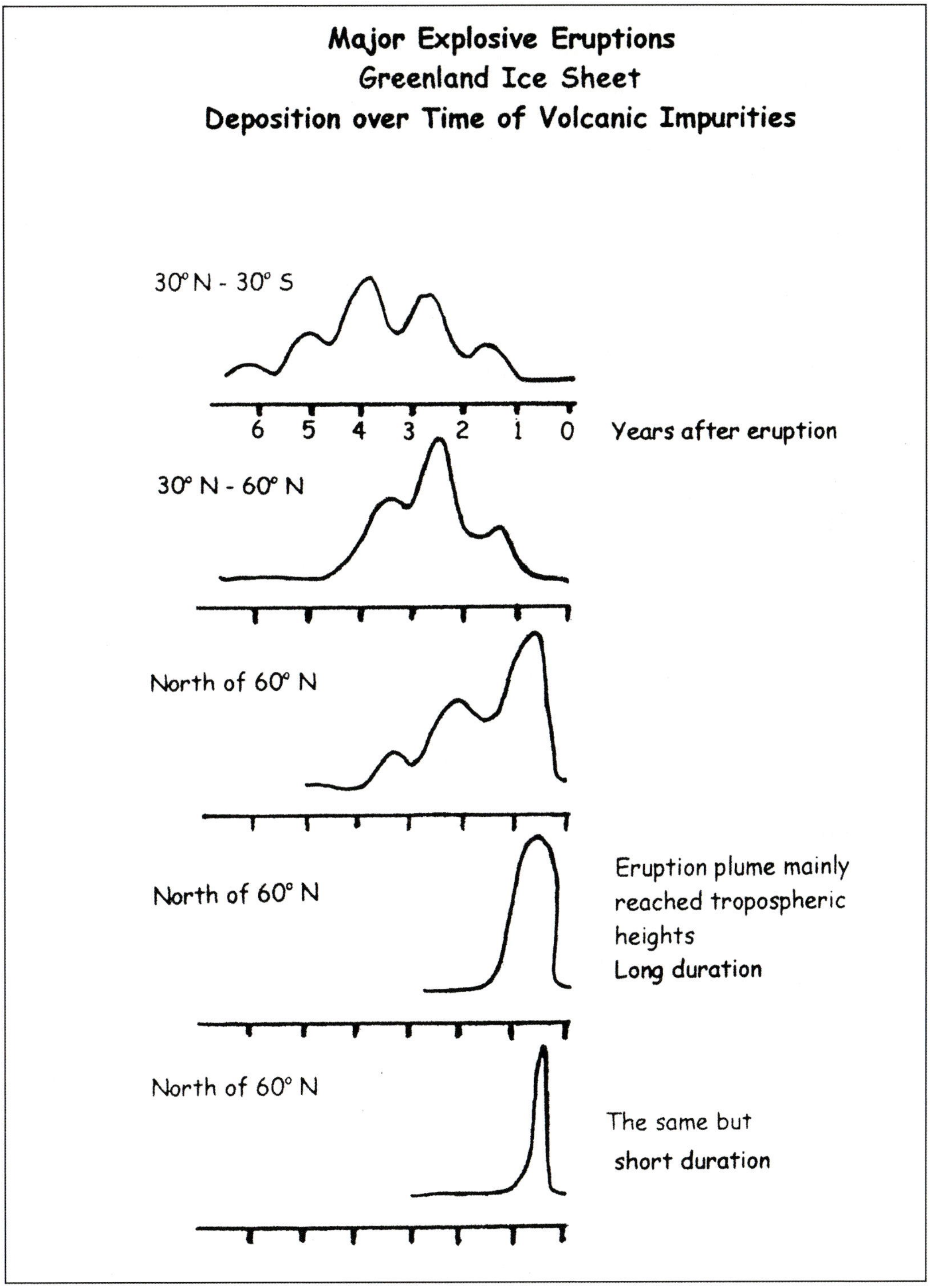

Fig. 6. Scematic illustration of how the volcanic aerosol is deposited on the Greenland Ice Sheet.

signal as well as how much common variability can be explained e.g. by EOF techniques.

Fisher (Abstract 1997 and personal communication 1999) attempted to use this technique on 50 sites north of 40°N and included tree rings, ice cores, and historical annual time series from 1761-1970 AD The results agreed fairly well with the results of Bradley and Jones (1993). Overpeck et al. (1997) and Mann et al. (1999), however, the geographical coverage was extended and the high frequency changes could be included. This study also indicated where the geographical similarities should be expected e.g. the 1 EOF vector shows that Northern China, Northern Canada, and Greenland have much in common with respect to climate variations. Northern Europe, East Scandinavia, Eastern England, Northern Russia, Greenland, and most of USA (apart from the South East USA) all experienced a high degree of common variability.

What is really lacking is annual proxy data over the oceans, but that is an entirely different problem.

Acknowledgments

The author wants to thank the Hanse Institute for Advanced Study, Delmenhorst for inviting me to join the Conference and for giving me the opportunity to write this paper. Also I would like to thank the Danish Natural Sciences Research Council and the European Commission, DG 12 (Project MilEClim and EPICA) for making it possible for me to scrutinize ice core data on past climate and volcanism.

References

Bengtsson L (1999) From short-range barotropic modelling to extended-range global weather prediction: a 40-year perspective. Tellus vol 51A-B 1:13-22

Blunier T, Chappelaz J, Schwander J, Dällenbach A, Stauffer B, Stocker TF, Raynaud D, Jouzel J, Clausen HB, Hammer CU, Johnsen SJ (1998) Asynchrony of Antarctic and Greenland climate change during the last glacial period. Nature 394:739-743

Bond G, Showers W, Cheseby M, Lotti R, Almasi P, deMenocal P, Priore P, Cullen H, Hajdas I, Bonani G (1997) A pervasive millenial-scale cycle in North Atlantic Holocene and glacial climates. Science 278:1257-1266

Bradley RS, Jones PD (1993) 'Little Ice Age' summer temperature variations: their nature and relevance to recent global warming trends. Holocene 3:367-376

Crowley TJ, Kwang-Yul Kim (1999) Modeling the temperature response to forced climate change over the last six centuries. J Geophs Res Lett 26 13:1901-1904

Dahl-Jensen D, Mosegaard N, Gundestrup N, Clow GD, Johnsen SJ, Hansen AW, Balling N (1998) Past Temperatures Directly from the Greenland Ice Sheet. Science 282:268-271

Dansgaard W (1964) Stable isotopes in precipitation. Tellus 16:436-468

Dansgaard W, Johnsen SJ, Reeh N, Gundestrup N, Clausen HB, Hammer CU (1975) Climatic changes, Norsemen and modern man. Nature vol 255 5503:24-28

Dansgaard W, Johnsen SJ, Clausen HB, Dahl-Jensen D, Gundestrup NS, Hammer CU, Hvidberg CS, Steffensen JP, Sveinbjörnsdottir AE, Jouzel J, Bond G (1993) Evidence for general instability of past cimate from a 250 ky ice-core record. Nature 364:218-220

Fisher DA (1997) High resolution Northern Hemisphere temperatures for the last centuries; using regional average tree ring, ice core and historical annual time series. Suppl. EOS, Transact, AGU, vol 78, no 46

Fisher DA, Koerner RM, Kuivinen K, Clausen HB, Johnsen SJ, Steffensen JP, Gundestrup N, Hammer CU (1996) Intercomparison of ice core (0-18) and precipitation records from sites in Canada and Greenland over the last 3500 years and over the last few centuries in detail using EOF techniques. In: Jones PD, Bradley RS, Jouzel J (eds) Climate Variations and Forcing Mechanisms of the Last 2000 Years. NATO ASI Series I 41:297-328

Grootes PM, Stuiver M, White JWC, Johnsen S, Jouzel J (1993) Comparison of oxygen isotope records from the GISP 2 and GRIP Greenland ice cores. Nature 366:552-554

Hammer CU (1977) Past volcanism revealed by Greenland Ice Sheet impurities. Nature 270:482-486

Hammer CU, Højmark Thomsen H (1998) Fælles nordisk indsats - tæt på verdens top. Geografforlaget Geologi-Nyt fra GEUS 2:1-14

Hammer CU, Clausen HB, Dansgaard W (1980) Greenland Ice Sheet Evidence of Post-Glacial Volcanism and Its Climatic Impact. Nature 288:230-235

Hammer CU, Clausen HB, Tauber H (1986) Ice-core dating of the Pleistocene/Holocene boundary applied to a calibration of the 14C time scale. Radiocarbon 28 (2A):284-291

Hammer CU, Johnsen SJ, Clausen HB, Dahl-Jensen D, Gundestrup N, Steffensen JP (2001) The Paleoclimatic

Record from a 345 m long Ice Core from the Hans Tausen Iskappe. Meddelser Om Groenland. Special issue on the Hans Tausen Iskappe project. Geoscience 39:87-95

Johnsen SJ, Dansgaard W, White JW (1989) The origin of Arctic precipitation under present and glacial conditions. Tellus 41:452-469

Karlén W, Källen E, Rodhe H, Backman J (1999) Man-made versus natural climate change. AMBIO 28:376-377

Kromer B, Becker B (1993) German oak and pine 14C calibration, 7200-9439 BC. Radiocarbon 35:125-135

La Marche VC Jr (1974) Paleoclimatic inferences from long tree-ring records. Science 183:1043-1048

Lamb HH (1965) The early Medieval warm epoch and its sequel. Paleogeogr Paleoclimatol Paleoecol 1:13-37

Lamb HH (1970) Volcanic dust in the atmosphere; with a chronology and assessment of its meteorological significance. Phil Trans R Soc A 266:425-533

Langway CC Jr, Osada K, Clausen HB, Hammer CU, Shoji H (1995) A 10-century comparison of prominent bipolar volcanic events in ice cores. J Geophys Res 100 (D8):16,241-16,247

Mann ME, Bradley RS, Hughes MK (1999) Northern Hemisphere Temperatures During the Past Millenium: Inferences, Uncertainties and Limitations. Geophys Res Lett 26 6:759-762

Overpeck J, Hughen K, Hardy D, Bradley R, Case R, Douglas M, Finnet B, Gajewski K, Jacobi G, Jennings A, Lamoureux S, Lasca A, MacDonald G, More J, Retelle M, Smith S, Wolfe A, Zielinski G (1997) Arctic environmental change of the last four centuries. Science 278:1251-1256

Robock A, Free MP (1995) Ice cores as an index of global volcanism from 1850 to the present. J Geophys Res 100:11,549-11,567

Stauffer B, Neftel A, Oeschger H (1985) CO_2 concentration in air extracted from Greenland ice samples. In: Langway CC, Oeschger H, Dansgaard W (eds) Greenland Ice Core: Geophysics, Geochemistry, and the Environment. AGU Monogr Washington DC 33:85-90

Stothers RB (1984) The Great Tambora Eruption in 1815 and Its Aftermath. Science 224:1191-1198

Stothers RB (1996) Major Optical Depth Perturbations to the Statosphere from Volcanic Eruptions: Pyrheliometric Period, 1881-1960. J Geophys Res 101:3901-3920

White JWC, Barlow LK, Fisher D, Grootes P, Jouzel J, Johnsen SJ, Stuiver M, Clausen H (1997) The climate signal in the stable isotopes of snow from Summit, Greenland: Results of comparisons with modern climate observations. In: Greenland Summit Ice Cores. AGU Monogr Washington DC 102 C12:26,425-26,439

Holocene Climate Evolution of the North Atlantic Ocean and the Nordic Seas - a Synthesis of New Results

Nalân Koç[1] and Eystein Jansen[2]

[1]*Norwegian Polar Institute, Polar Environmental Center, N-9296 Tromsø, Norway*
email: Nalan.Koc@npolar.no
[2]*University of Bergen, Department of Geology, , Allégt 41, N-5007 Bergen, Norway*
email: Eystein.Jansen@geol.uib.no

Abstract: Synthesis of oceanic records shows that the surface waters of the North Atlantic and the Nordic Seas have been modulated by the strength of the Northern Hemisphere insolation during the Holocene. This is manifested by the early Holocene optimum temperatures and the general cooling trend of the surface ocean since then. However, the duration of this optimum decreased both towards the north and towards the margins of the Nordic Seas suggesting that some regions of the oceans being more sensitive to climatic forcings than others. There is increasing evidence that superimposed on the late Holocene cooling trend there is also a higher frequency (albeit subtle) variability of multi-decadal and centennial time scales in the ocean. The magnitude of this variability is in the order of a few degrees. At the moment the underlying mechanisms for this variability remains as a topic of debate.

Introduction

The documentation of rapid and high amplitude fluctuations both in atmospheric and oceanic temperatures of the last glacial period has raised questions about the extent to which interglacial periods are stable, and whether switches in climatic conditions akin to those well documented in the last glacial have occurred in this or previous interglacial periods. Also the possibility that greenhouse gas forcing may drive the system into a mode of more instable conditions is of considerable concern. At the heart of these discussions lies possible changes in the strength of the thermohaline circulation of the North Atlantic/ Nordic Seas which is thought to be a main driver for rapid climatic change. As a result of this there is a need to document the Holocene natural climate variability of the North Atlantic/Nordic Seas, both to understand the causes for changes whether they be natural or man made, or both and to in order to assess the climate change due to anthropogenic forcing and distinguish these from the background natural variability. Here we summarise present knowledge of the Holocene climate evolution for the North Atlantic Ocean and the Nordic Seas areas.

Long Term Evolution

The Beginning

The most accurate datings for the beginning of the Holocene period is established from the Greenland ice cores and Swedish varv chronology by layer counting methods. According to the Greenland ice cores it started 11,640 (GISP2) or 11,550 (GRIP) calender years BP, and according to results from the Swedish varv chronology it started 11,525 calender yrs BP (Johnsen et al. 1992; Alley et al. 1993; Andrén et al. 1999). It is not possible to date the beginning of the Holocene as precisely in marine sediments since it falls within a radiocarbon age plateau. Still, both the Greenland ice core records and the oceanic records indicate that the tempera-

From WEFER G, BERGER W, BEHRE K-E, JANSEN E (eds), 2002, *Climate Development and History of the North Atlantic Realm.* Springer-Verlag Berlin Heidelberg, pp 165-173

tures rose very rapidly at the inception of the Holocene (~7°C and ~9°C within 50 yrs, respectively) (Dansgaard et al. 1989; Koç, Karpuz, and Jansen 1992; Alley et al. 1993; Haflidason et al. 1995) (Fig. 1). A drastic change of climatic conditions occurred in the Nordic Seas as summer insolation reached its maximum between 11 and 9 ka (Koç et al. 1993; Hald and Aspeli 1997; Andruleit and Baumann 1998). Within <50 yr the sea-surface temperature rose ~9°C from the cold Younger Dryas values to Preboreal values. The seasonal maximum sea-ice margin and the polar front, which had been positioned in the eastern Nordic Seas through the deglaciation period, suddenly retreated to a north-western position off Greenland, documenting a strong influx of warm Atlantic waters (Koç et al. 1993) (Fig. 2). In mid-latitude North Atlantic the transition into the Holocene was not as dramatic, since the Younger Dryas cooling was only about 1-2°C (Koç et al. 1996).

The warming into the Holocene climate optimum is regarded as a "2 step warming". After the initial warming the sea surface temperatures cooled (~2°C off southeastern Norwegian Sea and ~0.5°C in mid-latitude North Atlantic) between 9800-9500 yrs BP (The Preboreal oscillation). However, the surface ocean warmed again ~5°C more (off southeastern Norwegian Sea) around 9500 yrs BP (Fig. 1).

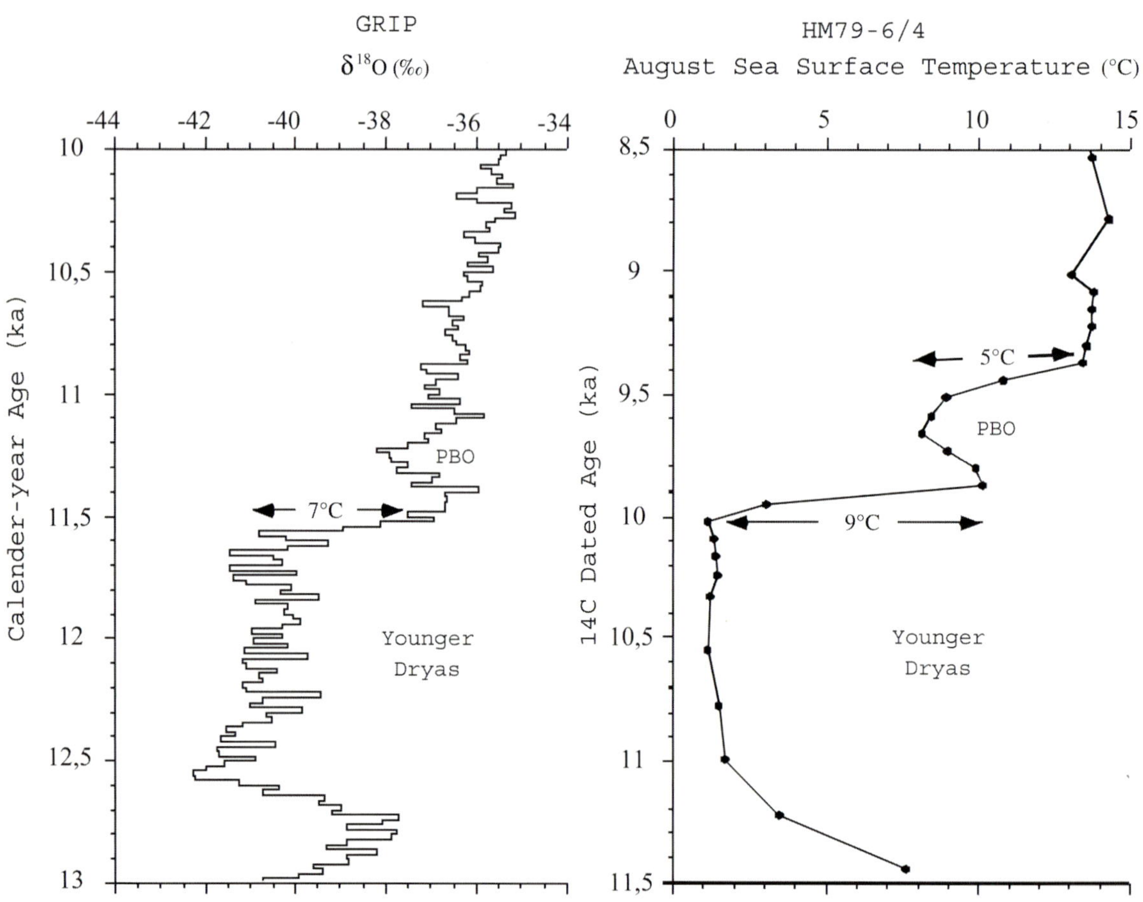

Fig. 1. Rapid warming of the Northern hemisphere atmosphere and the surface waters at the Younger Dryas-Holocene transition as depicted by the GRIP ice core $\delta^{18}O$ (‰) (modified from Johnsen et al. 1992) and the SE Nordic Seas sediment core HM79-6/4 (62°58'N 02°42'E) summer sea surface temperatures (SST) reconstructed from diatom assemblages (modified from Koç Karpuz and Jansen 1992). Age scale of HM79-6/4 is modified by assigning an age of 10,300 yrs BP to the Vedde ash. PBO is Preboreal Oscillation.

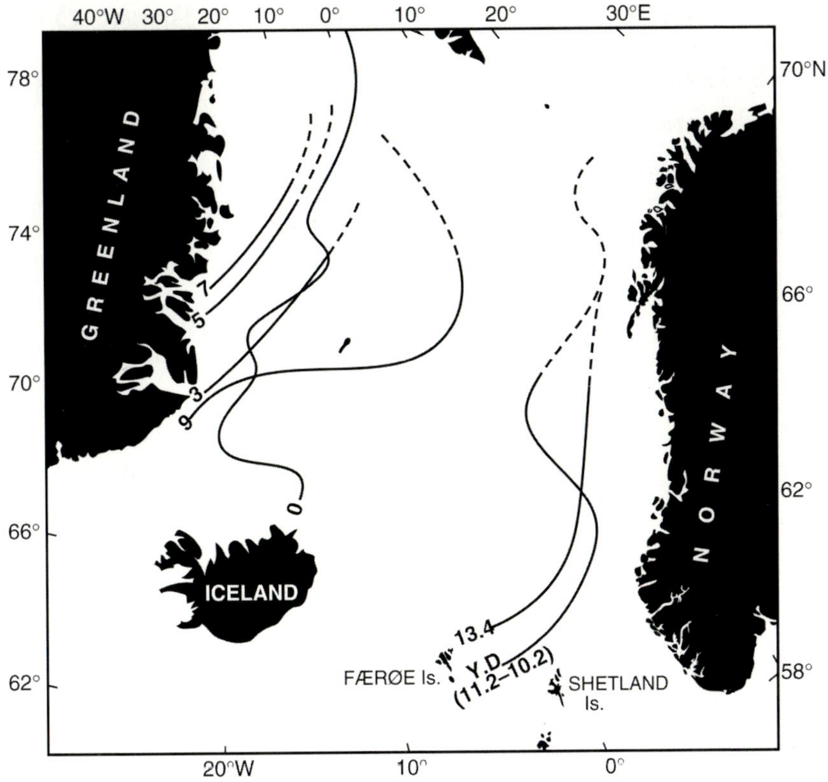

Fig. 2. Spatial variation of polar front and accompanying sea-ice edge in Nordic seas through last 13.4 ka (as compiled from Koç et al. 1993, 1996; Koç and Jansen 1994). Numbers indicate ages in ka and dashed lines indicate assumed position of polar front. YD is Younger Dryas.

Holocene Climatic Optimum

The first half of the Holocene is recorded as the warmest period during the last 13,400 yrs BP in the North Atlantic and the Nordic Seas (Koç et al. 1993 and 1996; Fronval and Jansen 1997; Hald and Hagen 1998). In step with the high insolation values, the area underwent warmest climatic conditions between 8 and 5 ka (Fig. 3). This is only 1-2 ka after the insolation maximum, and the delay was probably caused by the influence of the remnants of the Laurentide ice sheet. Summer SST's were around 15°C and winter SST's around 10°C in the southeastern Norwegian Sea. This climatic optimum lasted until about 5000 yr BP in both the southeastern Norwegian Sea and in the eastern North Atlantic (Koç et al. 1996). Over on the western Iceland Plateau, the SSTs started to decrease already after 7000 yr BP. The duration of this optimum decreased both towards the north and towards the margins of the Nordic Seas.

Late Holocene Cooling

The second half of the Holocene is characterized by a cooling trend which is in step with decreasing summer insolation in the northern hemisphere. During this period an increase in the extent of seasonal sea-ice cover of the western Nordic Seas is observed (Koç and Jansen 1994; Koç et al. 1996) (Fig. 3). This happened as early as 7 ka, documenting that the western Nordic Seas responded very early to the falling summer insolation, in the manner foreseen as a "preglacial state" by a spread of sea-ice cover that increased the albedo of the region (Imbrie et al. 1992). In the eastern part of these seas, which today is still influenced by the advection of warm Atlantic water, the climatic

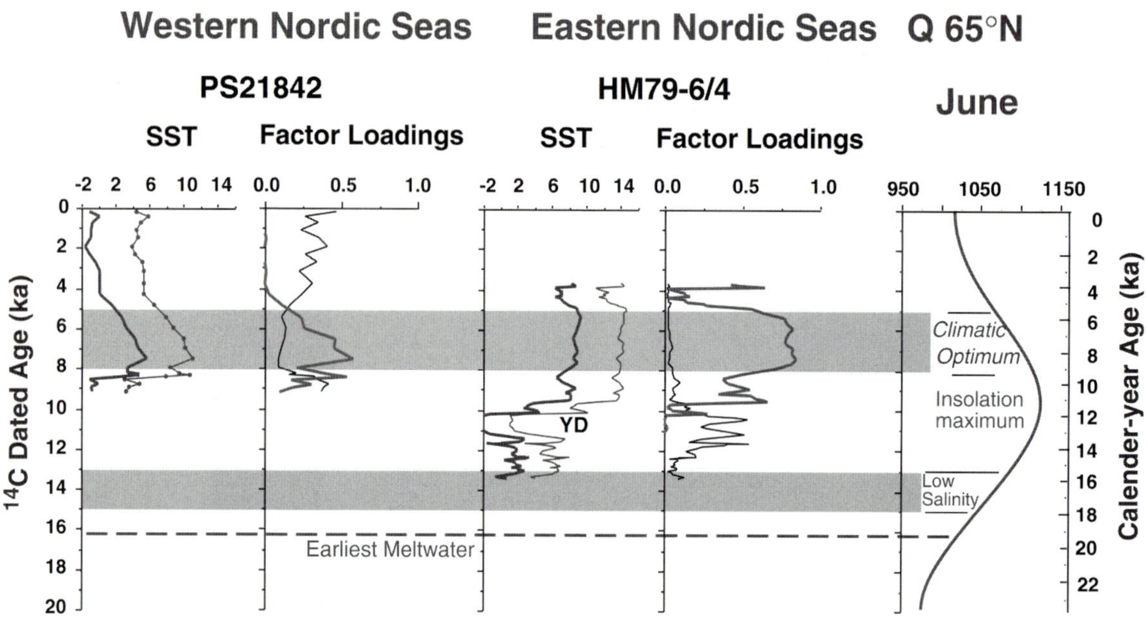

Fig. 3. Diatom-based winter (light blue lines) and summer (pink lines) sea-surface temperature estimates (SST), factor loadings of diatom factors typical of warm Atlantic waters (red lines), and waters with seasonal sea-ice cover (blue lines) in Nordic seas, compared with June insolation at latitude 65°N (from Koç and Jansen 1994). The higher the factor loadings, the higher is influence of specific water masses associated with factors. YD = Younger Dryas. Shaded area indicates the period of marine climatic optimum and hatched area interval of major deglacial signal in Nordic seas. Data for core HM79-6/4 is from Koç Karpuz and Jansen (1992).

optimum lasted until ~5 ka. Since then the surface waters have cooled to modern values. Also in the eastern North Atlantic, a cooling since the climate optimum is observed.

N-S SST Gradients/ Regional Differences

There is emerging evidence for marked regional differences in the development of the Holocene climate in the area. As reported above there are E-W and N-S differences in the duration of the early Holocene optimum in the Nordic Seas, which suggest a time-transgressive evolution of climate. It is also possible to interprete this as some regions of the oceans being more sensitive to climatic forcings than others, and therefore reacting faster. There also seems to be differences between the east and west basins of the North Atlantic. In the eastern basin, optimum SSTs are recorded between 6-8 ka. However, in the western basin stable and present high SSTs are recorded between 10-8.5 ka and a progressive cooling since then (L. Labeyrie unpublished). These results may suggest a possible antiphase relation between the eastern and western basins of the North Atlantic .

Holocene Events

Superimposed on this half sinusoidal curve with a cooling trend, there is evidence for some climate events. The most prominent of these events are the Preboreal oscillation, the so-called "8.2 event" and the Little Ice Age. However, evidence for more events are accumulating as higher resolution climate records are being generated.

The Preboreal Oscillation

The Preboreal oscillation is part of the "2 step warming" at the inception of the Holocene between 9800-9500 yrs BP (Fig. 1). It is recorded as a cooling in other ways a warming trend in the Nordic

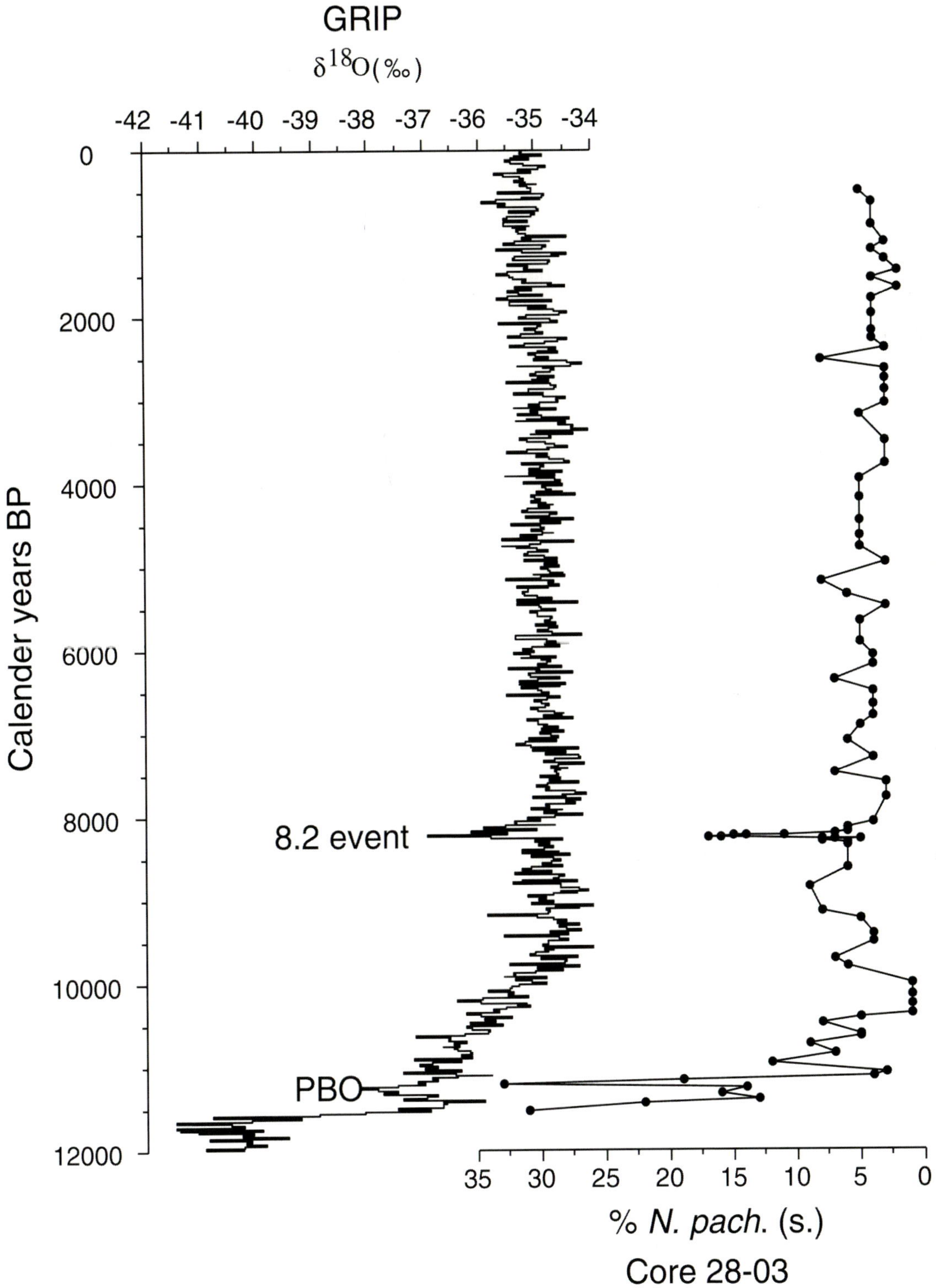

Fig. 4. Holocene variability of $\delta^{18}O$ (‰) of the GRIP ice core (from Dansgaard et al. 1993) and the percentage variability of the polar planktonic species *N. pachyderma* (s.) in the marine core 28-03 (60°52'N, 03°44'E) from the northern North Sea (Klitgaard Kristensen et al. 1998). Two early Holocene cooling events (the Preboreal oscillation - PBO - and the 8.2 event) are easily detectable in both records.

Seas and the North Atlantic Ocean (Koç, Karpuz, and Jansen 1992; Koç et al. 1996; Hald and Hagen 1998). In the tropical Atlantic Ocean, an increase in the trade-wind strength, reflecting higher north-south temperature gradients during this time interval, is recorded (Hughen et al. 1996). The Preboreal oscillation corresponds in time to the second global meltwater pulse, MWP 1b. Even though MWP 1b has been invoked as a possible cause for this cooling event, no clear direct evidence has been found yet (Koç Karpuz, and Jansen 1992; Hald and Hagen 1998).

The 8.2 Event

This event is most obvious in the $\delta^{18}O$ records of the GRIP and GISP ice cores as a short but distinct decline of the $\delta^{18}O$ values corresponding to a cooling of 6±2°C (Johnsen et al. 1992; Alley et al. 1997). The event occurred ~8000 to 8400 yr ago, peaking at ca. 8.25 ka or ca. 7500 ^{14}C yr BP. This event has also been recorded as a 15% increase in the abundance of the polar *Neogloboquadrina pachyderma* (s), which corresponds to a minimum of 2°C SST decrease from a core off southeastern Norway (Klitgaard Kristensen et al. 1998) (Fig. 4), and in planktonic oxygen isotope records in the Eastern Norwegian Sea, indicating a 2-3 degree cooling (B. Risebrobakken pers. comm. 1999). A high-resolution grey-scale record from the Cariaco basin (tropical Atlantic Ocean) indicates increased trade-wind strength, reflecting higher north-south temperature gradients during this time interval (Hughen et al. 1996). These evidences indicate that both the atmosphere and the surface ocean has been perturbed during the 8.2 event. Recently Barber et al. (1999) published evidence for massive outflow of freshwater from the Hudson Strait as the glacial lakes Agassiz and Ojibway drained catastrophically about 8470 calender years ago.

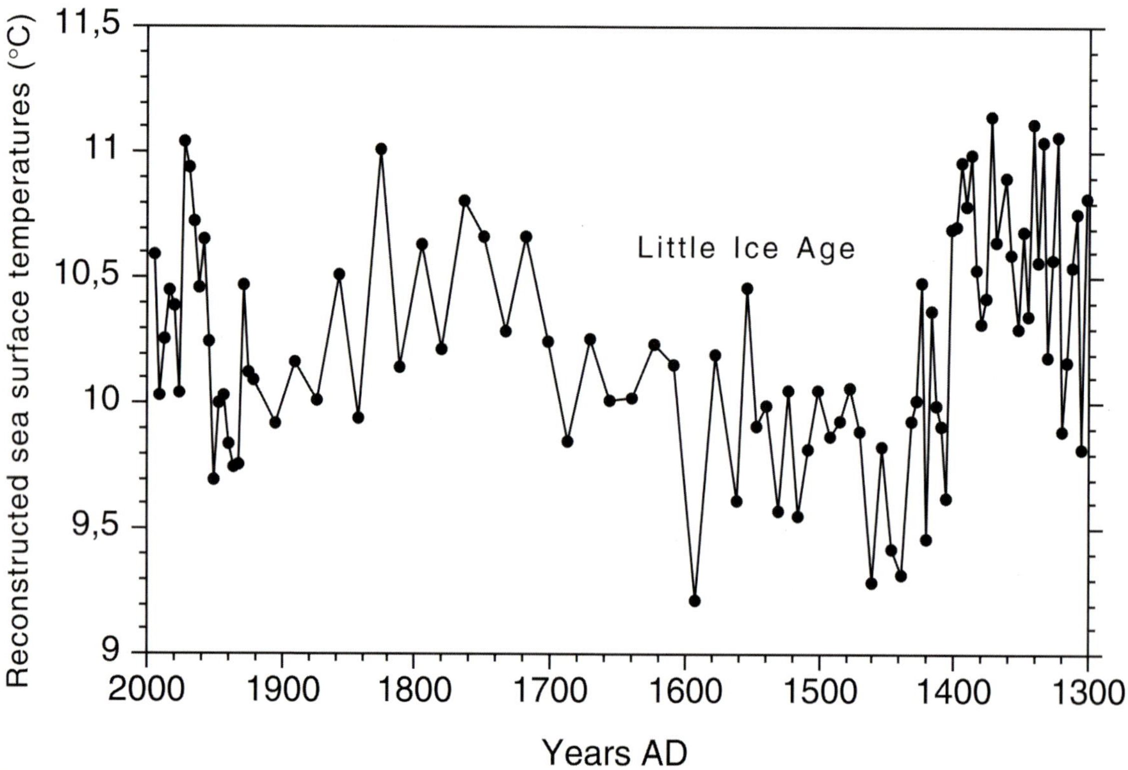

Fig. 5. Summer sea surface temperatures reconstructed from diatom based transfer functions in cores JM97948/MD952011 (66°41'N, 04°34'E) from the Vøring Plateau, eastern Nordic Seas (N. Koç and C. Andersen unpublished data). The cooling between 1400-1700 AD is correlated to the Little Ice Age.

The authors suggest that this massive freshwater flux perturbed the North Atlantic thermohaline circulation, and therefore is the probable cause of the 8.2 event. Yet, the corollary to this, i.e. the freshwater forcing on the North Atlantic remains to be identified and quantified. Hence, this causal mechanism is not a "proven fact".

The Little Ice Age

There is ample historical and glaciological evidence that climate conditions about 500 yrs ago were often quite different than the present (Bradley and Jones 1995). During this time period the most recent neoglaciation occurred, where glaciers grew and advanced down-valley in most glaciated mountain regions of the world. There are, however, very few investigations of the paleoceanographic conditions during this time span. In the existing records, there is a signifant cooling associated with periods of "The Little Ice Age". Core MD 2011/JM97 from the Vøring Plateau documents a general cooling of 1-2 degrees compared with modern temperatures during the Little Ice Age (Fig.5). Higher temperatures are recorded for the Medieval period in the same record. However, none of these periods were monotonously warm or cold. Cooler than present conditions are recorded also off Bermuda (Keigwin 1996) and in Nansen Fjord, eastern Greenland (Jennings and Weiner 1996) during parts of the Little Ice Age. Off Bermuda planktonic foraminiferal oxygen isotopes documents ~1°C cooling. Whereas, variable climatic conditions with frequent intervals of severe cold characterize a Little Ice Age type interval in Nansen Fjord, eastern Greenland (Jennings and Weiner 1996).

Recent studies give evidence for regional differences in the manifestation of climatic signals, which may provide clues to the mechanisms behind century to millenial scale climate variability. Keigwin and Pickart (1999) document a warm Little Ice Age period from the Laurentian Fan. The authors suggest that the observed antiphase relation documented between the Laurentian Fan and the Bermuda Rise during the Little Ice Age period is similar to the North Atlantic Oscillation (NAO), and that the NAO might be a useful model for millenial-scale ocean variability during interglacial climate states.

Is There a Cyclicity?

As increasingly higher resolution Holocene climate proxy records are being generated, some periodicities/quasi-periodicities are also being observed. However, the significance and the underlying causes of these cycles still remain more or less an enigma. A pervasive millenial-scale cyclicity close to ~1500 yrs (though not strictly periodic) in both the glacial and the Holocene North Atlantic sedi-ments have been suggested (Bond et al. 1997). In the Holocene, the variability has been interpreted as southward shifts of cool ice-bearing surface waters deep into the subpolar North Atlantic. Recently, Bianchi and McCave (1999) by studying the variability in sortable silt size in Holocene sediments from the Gardar drift south of Iceland (core NEAP-15K), also documented a quasi-periodicity of ~1500 yrs in the flow speed of the Iceland-Scotland Overflow Water. These two records suggest a possible existence of a quasi-periodic variability of ~1500 yrs in both the surface and the deep-ocean of the North Atlantic during the Holocene.

New studies apparently reveal even higher frequency periodicities. Spectral analysis of sediment lightness from core NEAP-15K, which indicates variations in the delivery of terrigeneous sediment to the oceans as a result of iceberg rafting and/or changes in surface productivity, reveals cyclicities at 1650, 1000 and 450 yrs. through the Holocene (Chapman et al. 1999). Reconstructed high-resolution sea surface temperatures from the Vøring Plateau (core MD952011) indicate variability in the order of 1°-2°C during the last 2500 years. Spectral analysis of this record reveals the presence of ~220 yrs periodicity during the latest Holocene. Oxygen isotope records of benthic foraminifers of a core from a fjord in western Norway show cyclic changes with strong perio-dicities of 380 and 206 years, which is interpreted to reflect bottom water temperature changes of up to 3-4°C through the last 2500 years (Mikalsen and Sejrup submitted). As has been previously suggested from instrumental data (Reid 1987; Lean et al. 1992; 1995) and from modelling experiments (Tett et al. 1999; Wood et al. 1999), these authors also suggest that the solar radiation most likely has been an important forcing mechanism for the oceanic surface temperature changes and thermo-haline circulation variability through the last 2500 years based on the good correlation between the 206 year cycle in their record and in the atmospheric $\Delta^{14}C$ record.

Conclusions

There is good evidence that the surface waters of the North Atlantic and the Nordic Seas have been modulated by the strength of the Northern Hemisphere insolation during the Holocene. In the ocean this is manifested by the early Holocene optimum temperatures and the general cooling trend of the surface ocean since then. However, there is increasing evidence that superimposed on this cooling trend there is also a higher frequency (albeit subtle) variability in the ocean. The magnitude of this variability is in the order of a few degrees. This variability is manifested in both multi-decadal time scales and centennial time scales. While it is possible to invoke feedback mechanisms related to melting of remnant ice sheets for climate variability documented during the deglaciation and the early Holocene, it is not possible to invoke these mechanisms for the late Holocene climate variability. At the moment it is not easy to understand the underlying mechanisms for this variability. However, better documentation of the variability and mapping of its areal distribution and documentation of possible anomaly patterns will no doubt provide clues to its source(s).

Acknowledgments

This study was supported by grants from the Research Council of Norway and from the EU-Environment and Climate Programme.

References

Alley RB and 10 others (1993) Abrupt increase in Greenland snow accumulation at the end of the Younger Dryas event. Nature 362:527-529

Alley RB, Mayewski PA, Sowers T, Stuiver M, Taylor KC, Clark PU (1997) Holocene climatic instability: A prominent, widespread event 8200 yr ago. Geology 25:483-486

Andrén T, Björck J, Johnsen S (1999) Correlation of Swedish glacial varves with the Greenland (GRIP) oxygen isotope record. Journal of Quaternary Science 14:361-371

Andruleit HA, Baumann KH (1998) History of the last deglaciation and Holocene in the Nordic seas as revealed by coccolithophore assemblages. Marine Micropaleontology 35:179-201

Barber DC and 10 others (1999) Forcing of the cold event of 8,200 years ago by catastrophic drainage of Laurentide lakes. Nature 400:344-348

Bianchi GG, McCave N (1999) Holocene periodicity in North Atlantic climate and deep-ocean flow south of Iceland. Nature 397:515-517

Bond G, and 9 others (1997) A pervasive millennial-scale cycle in North Atlantic Holocene and glacial climates. Science 278:1257-1266

Bradley RS, Jones PD (1995) Climate Since A.D. 1500. London, Routledge

Chapman MR, Shackleton NJ, Bianchi GB, McCave IN (1999) Millenial-scale climate fluctuations in the subpolar North Atlantic during the Holocene. Terra Nova 11 (4):169

Dansgaard W, White JWC, Johnsen SJ (1989) The abrupt termination of the Younger Dryas climate event. Nature 339:532-533

Dansgaard W and 10 others (1993) Evidence for general instability of past climate from a 250-kyr ice-core record. Nature 364:218-220

Fronval T, Jansen E (1997) Eemian and early Weichselian (140-60 ka) paleoceanography and paleoclimate in the Nordic seas with comparisons to Holocene conditions. Paleoceanography 12:443-462

Hald M, Aspeli R (1997) Rapid climatic shifts of the northern Norwegian Sea during the last deglaciation and the Holocene. Boreas 26:15-28

Hald M, Hagen S (1998) Early preboreal cooling in the Nordic seas region triggered by meltwater. Geology 26:615-618

Haflidason H, Sejrup HP, Klitgaard Kristensen D, Johnsen S (1995) Coupled response of the late glacial climatic shifts of northwest Europe reflected in Greenland ice cores: Evidence from the northern North Sea. Geology 23:1059-1062

Hughen KA, Overpeck JT, Pewterson LC, Trumbore S (1996) Rapid climate changes in the tropical North Atlantic region during the last deglaciation. Nature 380:51-54

Imbrie J and 17 others (1992) On the structure and origin of major glaciation cycles 1. Linear responses to Milankovitch forcing. Paleoceanography 7: 701-738

Jennings AE, Weiner NJ (1996) Environmental change in eastern Greenland during the last 1300 years: evidence from foraminifera and lithofacies in Nansen Fjord, 68°N. The Holocene 6:179-191

Johnsen SJ and nine others (1992) Irregular glacial interstadials recorded in a new Greenland ice core. Nature 359:311-313

Keigwin LD (1996) The Little Ice Age and Medieval Warm Period in the Sargasso Sea. Science 274:1504-1508

Keigwin LD, Pickart RS (1996) Slope Water Current over the Laurentian Fan on interannual to millenial time scales. Science 286:520-523

Klitgaard Kristensen D, Sejrup HP, Haflidason H, Johnsen SJ, Spurk MA (1998) A regional 8200 cal. yr BP cooling event in NW Europe. J Quaternary Sci 13:165-169

Koç Karpuz N, Jansen E (1992) A high resolution diatom record of the last deglaciation from the SE Norwegian Sea: Documentation of rapid climatic changes. Paleoceanography 7:499-520

Koç N, Jansen E (1994) Response of the high latitude Northern Hemisphere to orbital climate forcing: evidence from the GIN-Seas. Geology 22:523-526

Koç N, Jansen E, Haflidason H (1993) Paleoceanographic reconstructions of surface ocean conditions in the Greenland, Iceland and Norwegian seas through the last 14 ka based on diatoms. Quaternary Science Reviews 12:115-140

Koç N, Jansen E, Hald M, Labeyrie L (1996) Lateglacial-Holocene sea-surface temperatures and gradients between the North Atlantic and the Norwegian Sea: Implications for the Nordic heat pump. In: Andrews JT, Austin WEN, Bergsten HE (eds) The Late Glacial Paleoceanography of the North Atlantic Margins. Geological Society Special Publication No. 111, pp 177-185

Lean J, Beer J, Bradley R (1995) Reconstruction of solar irradience since 1610: Implications for climate change. Geophysical Research Letters 22:3195-3198

Lean J, Skumanich A, White O (1992) Estimating the sun's radiative output suring the Maunder Minimum. Geophysical Research Letters 19:1591-1594

Mikalsen G, Sejrup HP A high-resolution 2500 year long temperature record from Sognesjøen, western Norway; correlations to atmospheric Δ14C cycles. Submitted to Geology

Reid GC (1987) Influence of solar variability on global sea surface temperatures. Nature 329:142-143

Tett SFB, Stott PA, Allen MR, Ingram WJ, Mitchell JFB (1999) Causes of twentieth-century temperature change near the Earth's surface. Nature 399:569-572

Wood RA, Keen AB, Mitchell JFB, Gregory JM (1999) Changing spatial structure of the thermohaline circulation in response to atmospheric CO_2 forcing in a climate model. Nature 399:572-575

Holocene Climatic History of Northern Europe – the Evidence from Lake Deposits

Jörg F. W. Negendank

GeoForschungsZentrum Potsdam, Telegrafenberg, D-14473 Potsdam, Germany
e-mail: neg@gfz-potsdam.de

Abstract: Lacustrine sediments are ideal archives for deciphering terrestrial climatic changes. Since sediment records from lakes often reflect a mixture of global, regional and local signals, it is necessary to combine multiproxy data sets from lakes in different regions to exhibit the larger-scale climatic evolution including the recognition of regional differences. Such an approach is presented in this paper for the reconstruction of Holocene climate in Europe. Difficulties arise from correlation problems and dating uncertainties as well as from the fact that lake records from key regions like Siberia are rare and that not all sediment profiles cover the whole time window. A basic requirement for climate reconstruction from lake sediments is a reliable time scale in calendar years derived from varve counting. Therefore, the network of lake sediment archives presented here is grouped around two of the best dated records, Lake Gosciaz, Poland, and Holzmaar/Meerfelder Maar in the Eifel, Germany. Tie-point for these records is the Late Glacial/Holocene boundary with an almost perfect agreement in dating. The main characteristic of the Holocene climate evolution, a temperature maximum during the Atlanticum, is well documented in both records and most of the other regions described. In contrast, the warmest part of the Holocene in Siberia appears to be slightly earlier during the Boreal. Despite a general trend of climatic deterioration interpreted from all records, changing patterns of maritime influences and continentality have been detected besides an increasing influence of man during the last ~3000 years.

Introduction

European lake deposits provide insights into the climatic history of Europe because sedimentation in large or small lakes is controlled by autochthonous production (e.g. phyto- and zooplankton, precipitation of calcite) and /or allochthonous (fluvial, aeolian, detrital and organic) input. Therefore, lacustrine records reflect local, regional or global environmental influences.

Most promising are facies patterns from the deepest part of the lakes recording time and environment on an annual scale. These varved (annually laminated) sequences are important records for establishing independent time scales for correlation, even with marine records and ice cores (Brauer et al. 1999a).

Originally, palynostratigraphy was the tool to develop a stratigraphic subdivision of Late Glacial and Holocene records. This is now complemented by AMS^{14}C-dating (calibrated by dendrochronology), varvechronology, tephro-chronology, magnetostratigraphy and oxygen isotope stratigraphy. As most lake records are non-varved, different procedures have been proposed for a stratigraphic scale, for example the INTIMATE proposal (Björck et al. 1998; Walker et al. 1999) which is based on the GRIP Greenland ice core as a stratotype for the last Termination. However, it seems to be more rational in terms of stratigraphical procedures to use annually laminated sediments with biozones (palynologically defined), tephra layers and stable isotopes. One candidate for a regional strato-type for the Late Glacial – Holocene in central Europe (Eifel region – NW-Germany – Poland) is the record from the Eifel maars (Meerfelder Maar and Holzmaar).

This could be the nucleus for an absolute synthetic, consistent time scale for the Late Glacial and the Holocene and be the base for correlation of all European lake records and for environmental and climate reconstruction.

From WEFER G, BERGER W, BEHRE K-E, JANSEN E (eds), 2002, *Climate Development and History of the North Atlantic Realm.* Springer-Verlag Berlin Heidelberg, pp 175-190

Independent precise dating is a prerequisite for correlation of lake sediments besides the widely used method of "wiggle" matching with biological, geochemical or geophysical parameters. A multiproxy approach is used because no direct temperature/precipitation indicator such as $\delta^{18}O$ in marine records or ice cores is available.

Until today the basic climatic information is provided by palynology. However, the multiproxy approach applying biological (cladocera, chironomids, ostracods, diatoms, chrysophycean cysts, pollen, plant macrorelicts) and geochemical parameters (total organic carbon (TOC), dry density, water content, biogenic silicia, inorganic carbon content, isotopes, major and trace elements, etc.) is slowly but steadily increasing. These proxies can be used in combination to derive average temperatures or other climate variables (transfer functions).

After the last glaciation the Scandinavian ice shield retreated step by step from S to N so that complete sequences documenting the transition from Late Glacial to the Holocene are located in southern, i.e. periglacial or non-glacial areas in Europe, where lakes survived up to several glacial cycles (e.g. Lac du Bouchet: ~250,000 years). Some of these lakes provide the additional advantage of annual layering (varves), allowing the establishment of a varvechronology (Eifel region, Germany: 23,000 years (Zolitschka et al. 1999), Monticchio, Italy: 102,000 years (Allen et al. 1999)).

A precise dating is the basic prerequisite to understand the beginning of the Holocene warming and to determine climatic changes in Europe as well as gradients from S to N and from W to E (maritime – continental). However, high-resolution lake records are sparse. Therefore, the reconstruction is based only on a few key locations (Fig. 1) and on two stratigraphic units of Late Glacial and the Holocene.

The Holocene (11,590 Varve Years until Today)

The Holocene stratigraphy (Preboreal, Boreal, Atlanticum, Subboreal, Subatlanticum) is based on well known pollen zones. Climatic oscillations are of minor importance in comparison to the Late Glacial time. The beginning of the Holocene, characterized by a distinct warming, is dated to 11,590 varve years BP (Table 1, Figs. 2-4). The transition from the YD (Younger Dryas) cold spell to the warming took around 50 years due to the varved record in the Eifel region (Brauer et al. 1999b, 2000).

Climate Change During the Holocene

Discussing the climate history of the Holocene starts mostly with the Late Glacial because tying and dating of the Holocene sequences can often be done only by determining the rapid shifts of the Late Glacial and especially the Younger Dryas (Tables 1, 2, Figs. 2-4).

For the purposes of this discussion, the available records from locations in Northern Europe (Germany, Poland, Scandinavia, Russia) (Fig. 1) are selected, showing the pronounced climatic deterioration during the YD (Figs. 2-4) before the beginning of the Holocene. But this is based on the pollen zonation and only in few locations independent chronologies have been developed. Therefore, these biozones are not chrono-stratigraphic markers in most of these sequences.

However, two independent high resolution records have been selected as examples for the climate history of the Holocene achieved by using multiproxy networks based on varve counting and detailed microstratigraphy.

Firstly, one of the best investigated and documented varved sequences, Lake Gosciaz (Poland), and, secondly, varved sequences from the Eifel region, Germany.

Lake Gosciaz (Poland)

Ralska-Jasiewiczowa et al. (1998) have published an extraordinarily detailed volume about Lake Gosciaz, central Poland. This laminated record has an excellent time control (Table 1, Fig. 4a) and the multiproxy network is documented in Tab. 2a and 2b. The Younger Dryas/Holocene transition is dated to 11,510 cal. BP (11,480 – 11,550) and marked by a rapid rise of $\delta^{18}O$ and blooms of the cladoceran *Tetraedron minimum* indicating a rise of mean annual temperature by ~5°C within 70 years. This warming led to forest expansion.

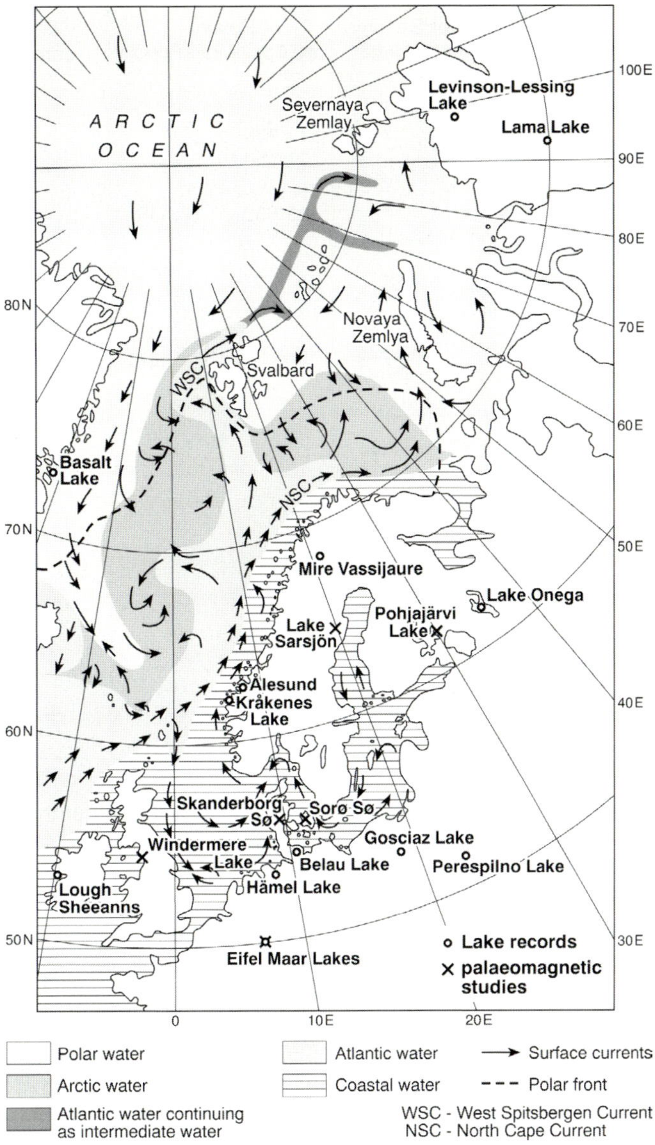

Fig. 1. Locations of selected lake sediments for the Holocene climate history.

The Preboreal chronozone (11.5 – 10.0 kyr BP) demonstrates the early Holocene warming with maximum carbonate precipitation in the record. Two stages of forest evolution can be traced: 11.5 – 10.5 kyr BP: *Betula*, *Populus* stage; 10.5 – 10.0 kyr BP: *Pinus* dominance.

Within the Boreal chronozone (10.0 – 9.0 kyr BP) deciduous forests of different types developed after 9850 cal. BP and progressive warming is demonstrated by the appearance of *Viscum* (mean July temperature >17°C) at ~9200 cal. BP.

Hedera (January temperature >2°C) pollen at the beginning of the Atlantic chronozone (9.0 – 5.8 kyr BP) at ~8950 cal. BP document milder winter seasons. Additionally, the increase of temperature caused the closing of deciduous forests and the increase of lake trophy. The whole chronozone was a period of maximum development of mixed deciduous forests (dominance of *Ulmus*, *Fraxinus*, and *Quercus*). First human influences can be traced between 8.0 and 6.6 kyr BP locally affecting the forests (Mesolithic tribes). From 6.4 cal. BP

Biozones NW Germany		Eifelmaar region Meerfelder Maar		Lower Saxony Hämelsee		Poland Lake Gosziaz	Poland Lake Perespilno	Dendro-chronology	GRIP (INTIMATE)
HOLOCENE									HOLOCENE
			11.590		11.560	11.480-11.550	→	11.570	11.550
LATEGLACIAL	Younger Dryas	4		III	(no varves)				Greenland Stadial GS-1
			12.680			12.650	12.630		12.650
	Allerød	3c 3	(200a above LST)	IIc	(200a above LST)				Greenland Interstadial GI-1a
			LST: 12.880		LST				12.900
		3c 2	(Gerzensee)	IIb					GI-1b
			entire duration: 670a		entire duration: 625a				13.150
		3c 1		IIa					GI-1c1
			13.350						
	Older Dryas	3b		Ic	(no varves)				GI-1c2
			13.540						
	Bølling	3a		Ib	(ca. 200 varves)				G-1c3
			13.670						13.900
	Oldest Dryas	2		Ia	(no varves)				GI-1d
			13.800						14.050
	Meiendorf	1		Mei	(no varves)				GI-1e
			ca. 14.450						14.700
PLENIGLACIAL									Greenland Stadial GS-2

Table 1. Synchronisation of Late Glacial continental sequences in northern central Europe based on varved lacustrine sediments (Litt et al. 2000).

Neolithic populations are indicated by charcoal occurrences in the records.

The Atlantic/Subboreal transition is marked by an abrupt extinction of *Ulmus* trees (Subboreal chronozone 5.8-2.6 kyr BP). This *Ulmus* decline and forest openings emphasize the intensification of human settlements, dominating the Subboreal and Subatlantic chronozones. So human impact masks the natural climate influence in the 150 years after the *Ulmus* fall leading to open woodlands with high abundance of *Corylus* and intensively flowering *Quercus*.

Bronze age cultures appeared around 4100 cal. BP and tribes of the Lusitanian culture destroyed the *Carpinus betulus* forests around 3400 cal. BP

The Subatlantic chronozone after 2.6 kyr BP is clearly overprinted by human activities and this human impact intensified during the "Roman period" with a local culture (Przeworska) causing large- scale deforestation accompanied by the cultivation of *Secale cereale* and *Cannabis sativa*.

The reduction of human settlement during the migration period (AD 400 – 1000) can be traced again by expansion of *Betula* copses and *Carpinus* woods. Later on, in the 10th century the modern settlement started leading to the gradual extinction of deciduous forests with the remaining *Pinus* and mixed woods lasting till today.

Maar Lakes of the Eifel Region (Germany)

In the West Eifel volcanic field maar lakes have been used as archives for reconstructing the palaeoclimatic evolution during the last 23,000 calendar years. These small lakes have advantages in comparison to large lakes, being sediment traps without or with minor inlets and outlets. According to size and depth, sedimentation in the centre of these maar lakes guarantees a complete sequence and favours the formation of annual laminations (varves).

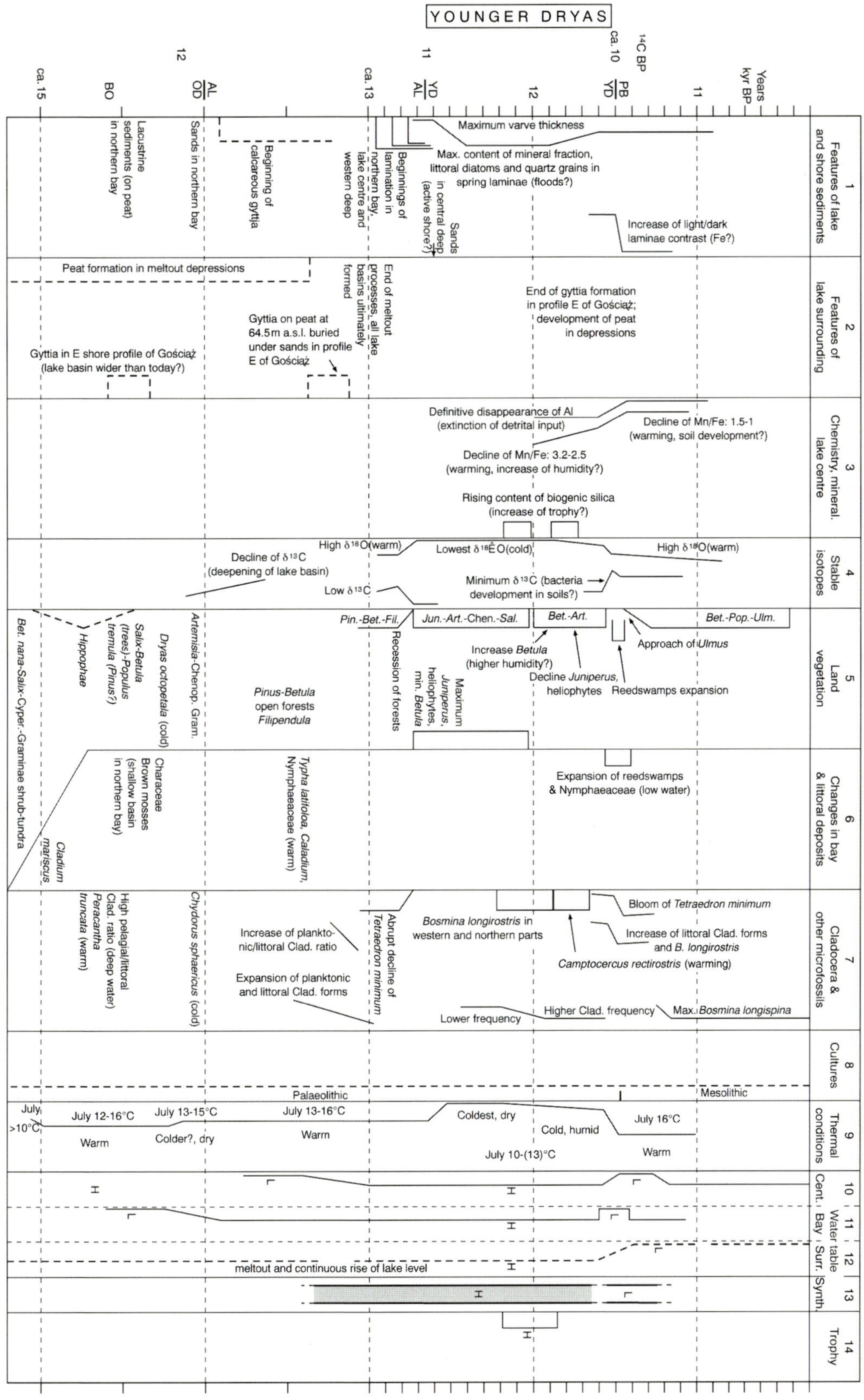

Table 2. Multiproxy network for Late Glacial and Holocene of Lake Gosciaz (Poland) (Ralska-Jasiewiczowa et al. 1998) **a)** Late Glacial.

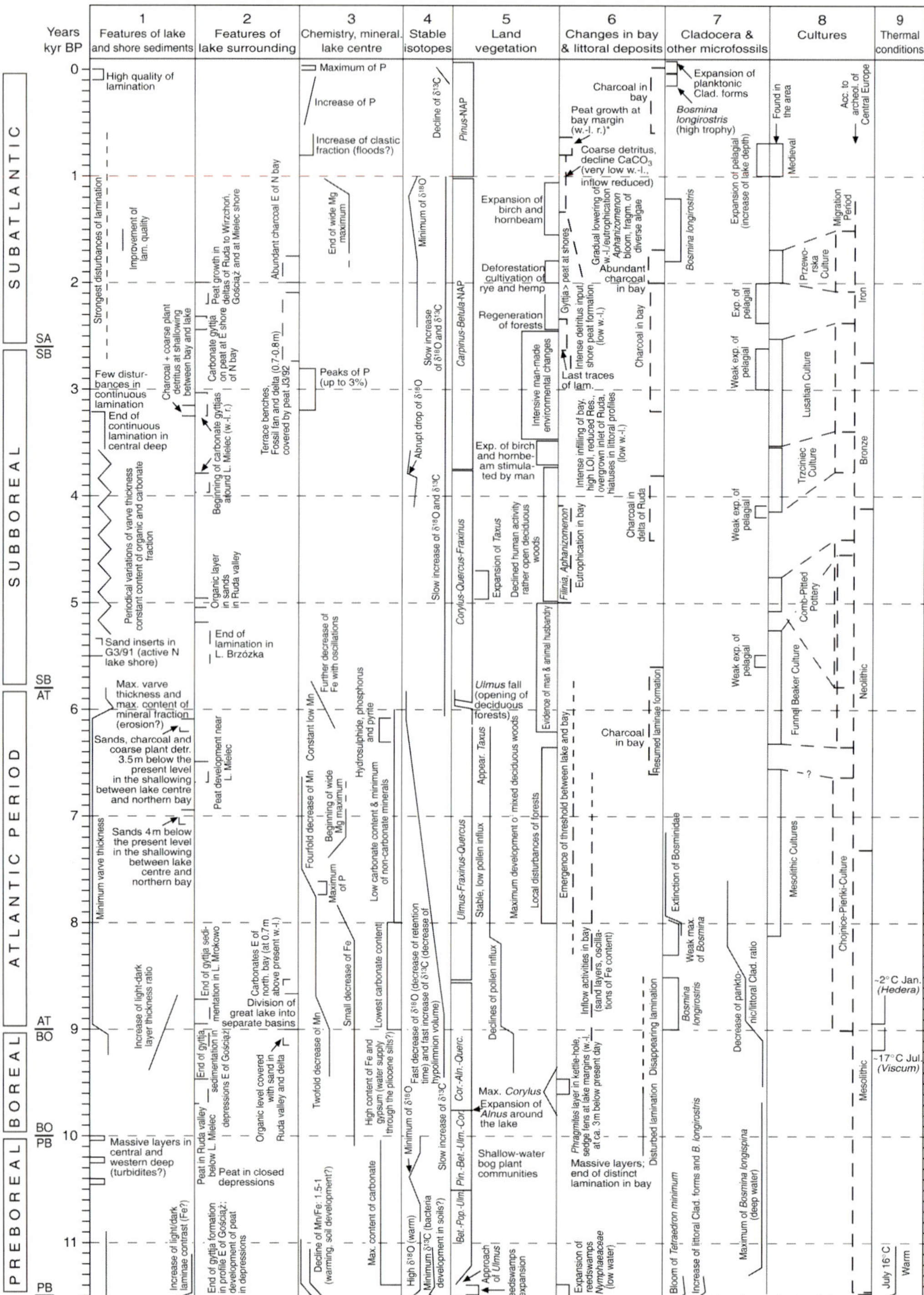

Table 2. (b) Holocene.

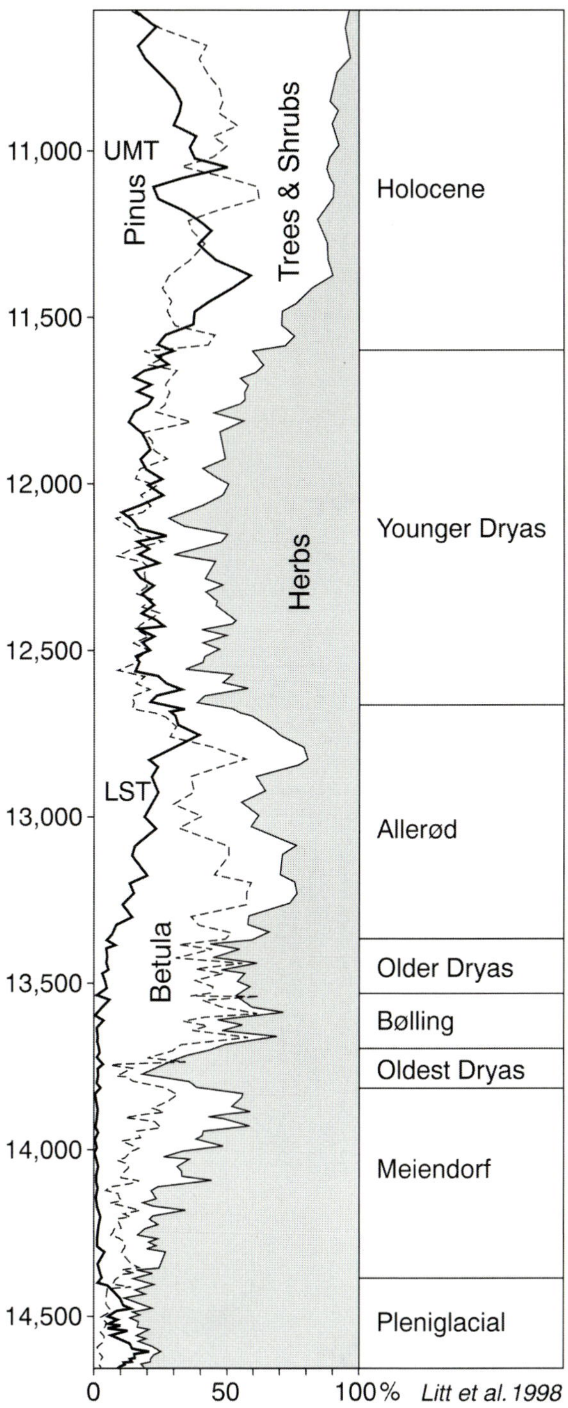

Fig. 2. The Meerfelder Maar, Eifel region pollen record (Litt et al. 1998).

The varvechronology of the Eifel region is based on records from Meerfelder Maar [MFM] and Holzmaar [HZM]. Several cores from each lake and multiple counting of several investigators provide a Pleni–Late Glacial sequence of predominantly clastic laminites and a Holocene unit of varved diatomaceous organic sediments, interrupted by two tephra layers: the European-wide distributed Laacher See tephra (LST; 12,880 varve years BP) and the regional tephra of the Ulmener Maar eruption dated to 11,000 varve years BP Brauer et al. (1999b, 2000).

The chronologies are supported by AMS^{14}C and optically stimulated luminescence (OSL) dating and cross correlation using pollen analyses and tephra layers (Zolitschka et al. 1999).

The Eifel Late Glacial and early Holocene varvechronology is confirmed by the recently revised dendrochronology of southern Germany, the marine record from the Cariaco Basin, the corals from Southern Barbados and New Guinea and the record from Lake Suigetsu.

Besides detailed varve counting, determination of turbidites, homogenites and varve thickness enables the identification of four distinct periods, a pleniglacial pattern with a low varve thickness variability, Late Glacial – Holocene transition pattern with a high thickness variability and again low variability within the Holocene up to ~2800 cal. BP and a subsequent high variability and higher sedimentation rates due to human activity.

In time windows of low regular thickness variability within the clastic laminites and the diatomaceous gyttja (stable climate, diatom - dominated varves) time sequence analyses have been applied to varve thickness data, revealing quasi-periodicities of 11, 22, 88 and 208 years which are well known from other records like ^{14}C in tree rings and ice cores and possibly caused by solar forcing (Vos et al. 1997).

These small oscillations are not recognized in other parameters of the record because the temporal resolution of sampling is 10 – 50 years. Additionally, the varve chronology allows detailed dating of climate shifts as well as teleconnections with other independently dated high resolution sequences like the ice cores. YD started 12,680

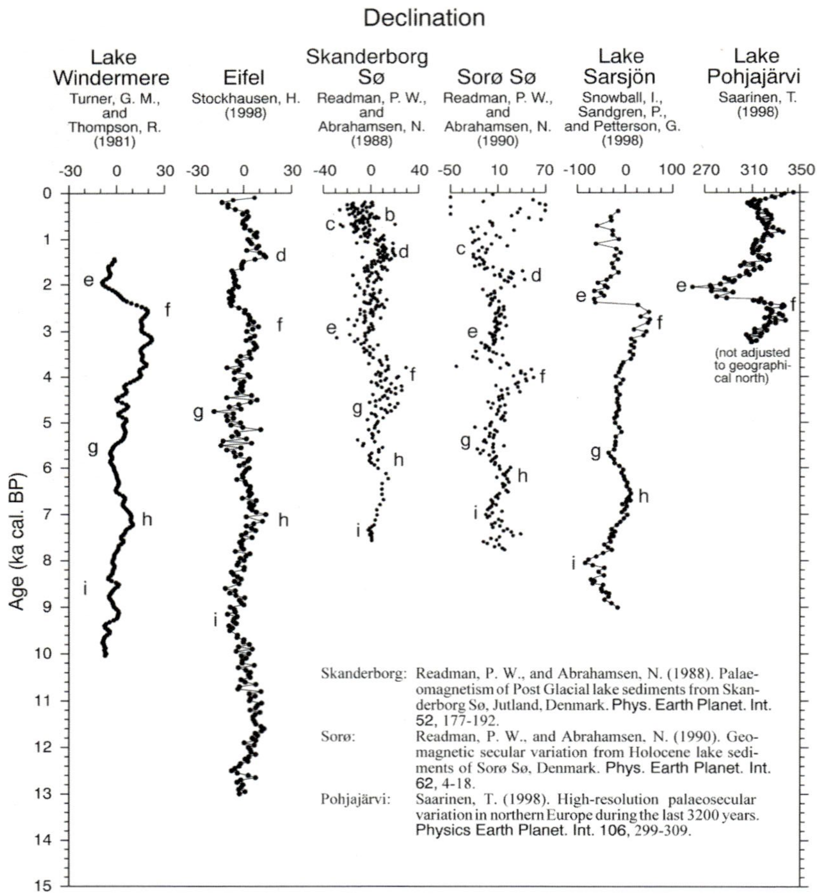

Fig. 3a

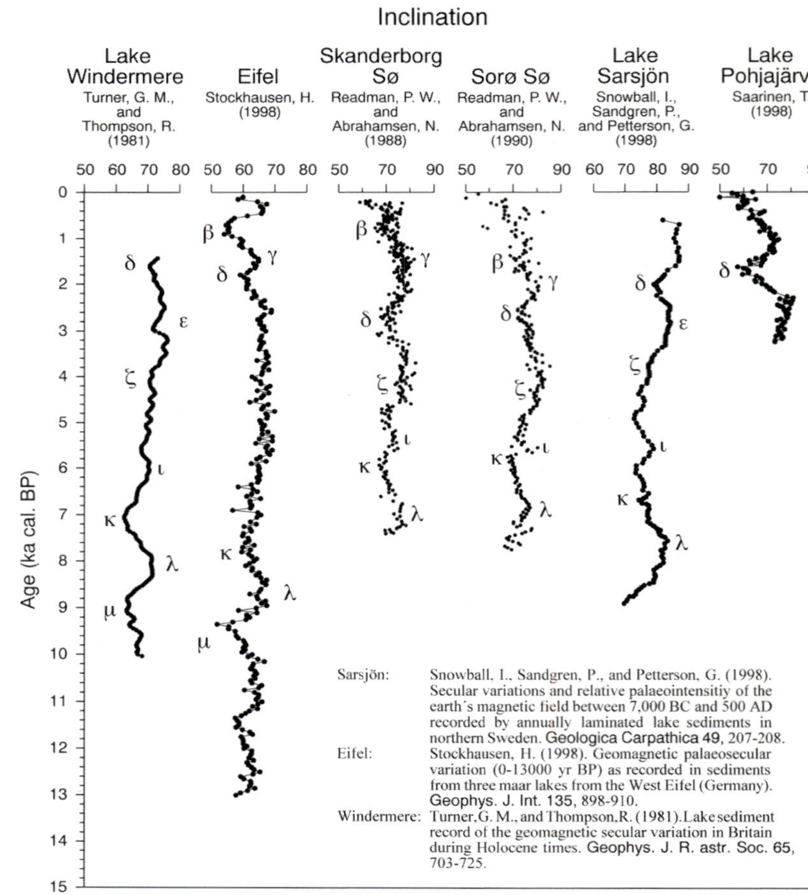

Fig. 3b

Fig. 3 (a, b). Comparison of the magnetic palaeosecular variation records (inclination, declination) of some locations in Europe with the standard curve of Lake Windermere (authors see Fig. 3).

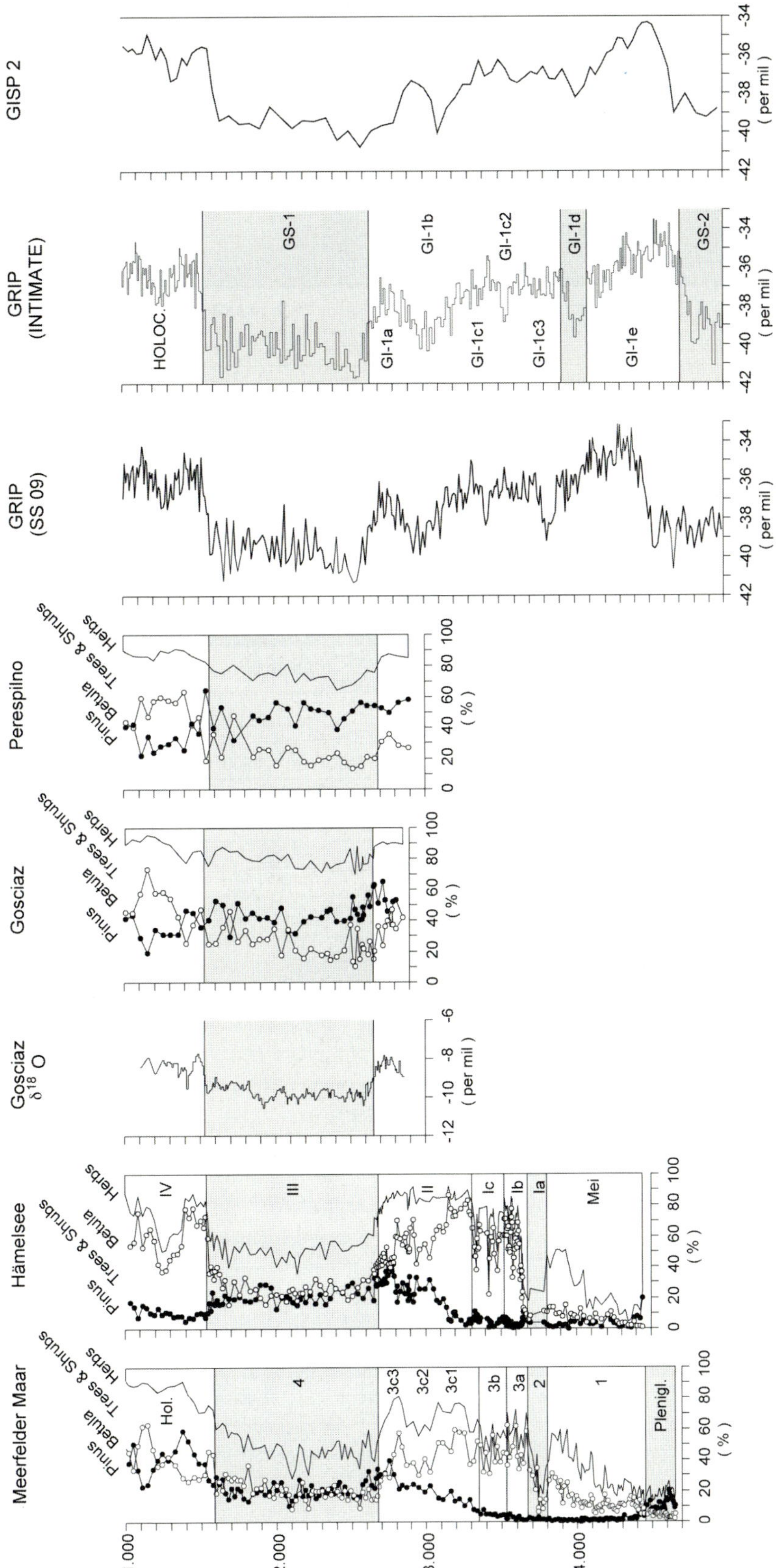

Fig. 4. (a-c) Pollen records of Late Glacial – Holocene

(a) Central Europe (Litt et al. 2000; Brauer et al. 1999b, 2000; Merkt and Müller 2000) plus ice core records

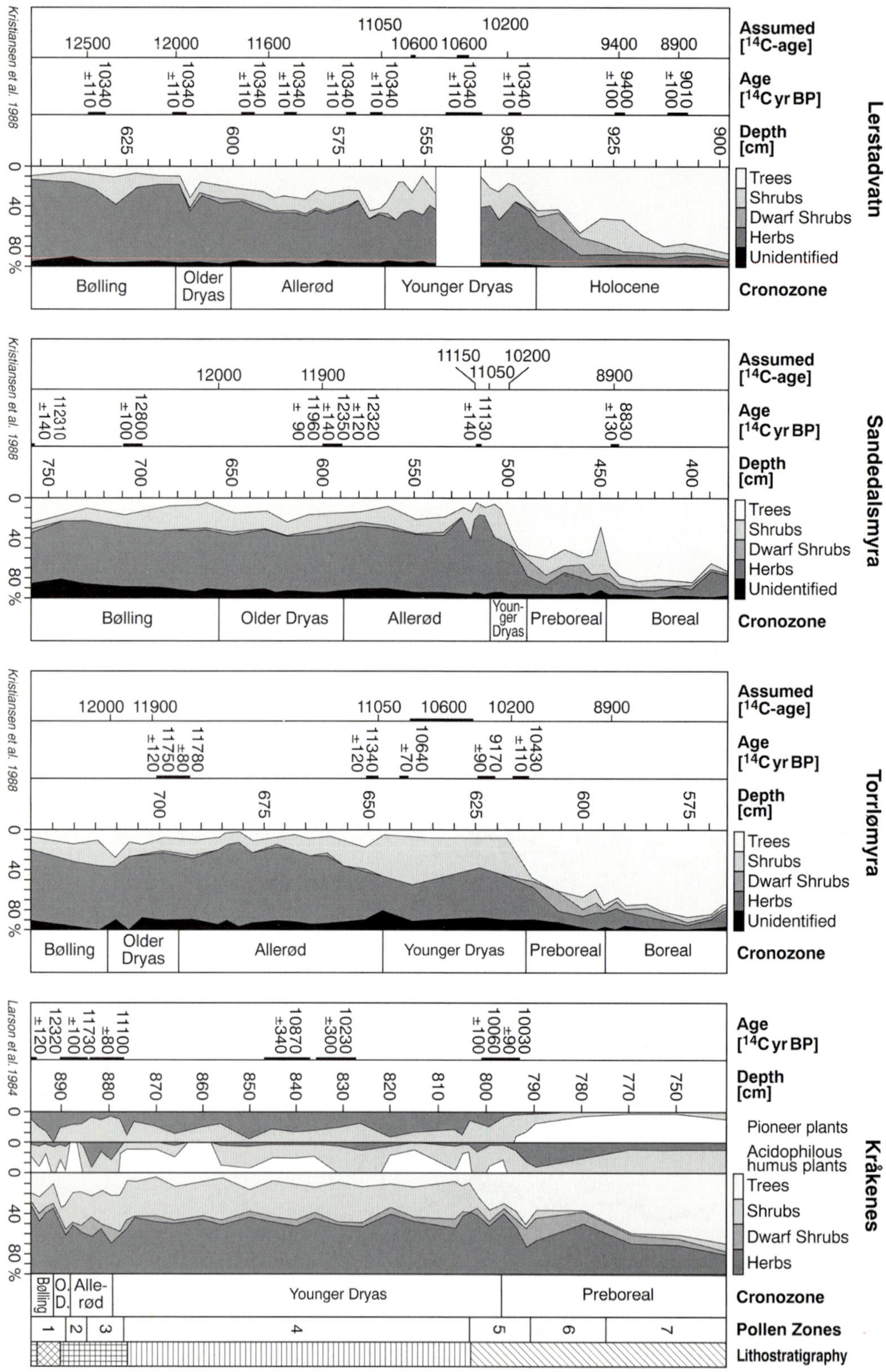

Fig. 4. (b) West Norway (Kristiansen et al. 1988; Larsen et al. 1984)

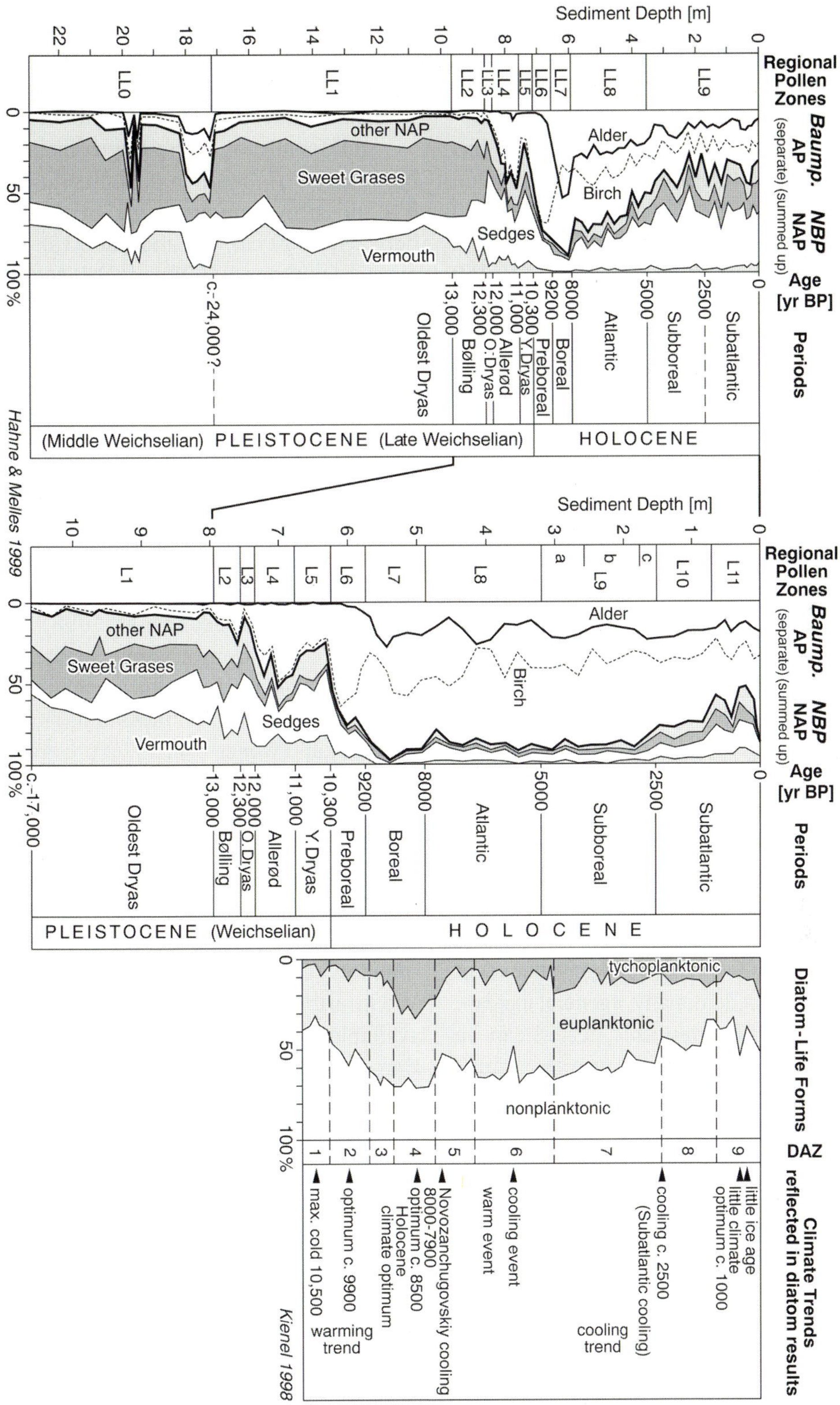

Fig. 4. (c) Siberia (Hahne and Melles 1999; Kienel 1998)

varve years BP in MFM and ended 11,590 varve years BP The climate deterioration took 25 years and the warming at the end of the 1090 year interval took ca. 50 years (Brauer et al. 1999b).

Detailed seasonal resolution of varves has revealed that the early Holocene warming trend was interrupted by two short periods (200 – 300 years) of climatic deterioration, the Preboreal (~11,360 – 11,100 varve years) and Boreal (~10,460 – 10,260 varve years BP) oscillations. Drastic changes in varve thickness and varve composition indicate cooler and wetter environmental conditions during the winter, probably connected to North Atlantic cooling events (Brathauer et al. 2000). Therefore, it is possible to discriminate a first and second Holocene warming from the final Holocene warming after 10,260 varve years BP.

However, detailed diatom analyses (Brüchmann 1998) in the time window of 10,1289,102 cal. BP demonstrate a lake ecology that shifted further to warmer climate conditions around 9680 cal. BP, which is also visible within the pollen record (*Quercus* increase, *Tilia* appearance).

Schettler et al. (1999) applied a multiproxy approach to analyse sediment cores from two Eifel maars (MFM and Schalkenmehrener Maar [SMM]). They reconstructed a higher discharge of groundwater from the lake's catchment (SMM) at around 10,200, 8500, 7150, 6000 and between 4050 and 3300 varve years BP and a significantly lower discharge around 11,000, 9500, 5000 and after about 3050 varve years BP Additionally, a general increase of discharge since the Neolithic period can be inferred from the MFM record due to anthropogenic forest clearance.

This multiproxy approach also demonstrates clear evidence for a change of the sedimentation rate around 2800 cal. BP, TOC, biogenic silica (BSi), dry density, palaeomagnetic parameters as well as in $\delta^{13}C$ values. This climate event is also described by van Geel et al. (1996, 1998): "A sudden and sharp rise in the ^{14}C content of the atmosphere, which occurred between ca. 850 and 760 calendar years BC (~2750 – 2450 BP on the radiocarbon time scale), was contemporaneous with an abrupt climate change. In northwest Europe (as indicated by palaeoecological and geological evidence) climate changed from relatively warm and continental (dry) to oceanic (wetter and cooler conditions). It is well known as the so-called 'Hallstadt plateau' in the ^{14}C calibration curve."

Additionally, an interesting coincidence exists between the *Tilia* pollen record and the record of $\delta^{13}C$ possibly demonstrating two warm phases between 9680 cal. BP and 2800 cal. BP, with a maximum around 6000 cal. BP. A minimum is recognized around 4700 – 5000 cal. BP (TOC, BSi, $\delta^{13}C$).

If we consider the general climatic evolution in Europe during the last 1000 years, then the medieval climate "optimum" as well as three cold spells – related to the Wolf, Spörer and Maunder minima in sunspots – are documented. It was the intention to use varve thickness data from this period for correlation with meteorological observations to reconstruct these fluctuations. Zolitschka and Negendank (1998) reported a tentative correlation of winter temperatures and winter precipitation with the varve thickness from HZM, valid for the last millennium. Three intervals of colder and wetter winters (expressed as thicker varves) were identified coinciding with possible climatic oscillations due to supposed Wolf, Spörer and Maunder sunspot minima. There is no evidence for the "medieval climatic optimum" in the Holzmaar record, possibly caused by a more continental climatic regime with exceptionally warm summers (Glaser 1995).

Unfortunately, no $\delta^{18}O$ values of diatoms are available for the Eifel maars. However, the Mid-European Decadal Isotope-Climate Record from 15,500 to 5000 yr. BP of southern Germany (Grafenstein et al. 1999) [Oxygen-isotope ratios of precipitation ($\delta^{18}O_p$) inferred from deep-lake ostracods from the Ammersee (southern Germany)] documents rapid climate shifts as seen in central Greenland ice cores during this interval.

Summarising the two described Middle European records reveals a tremendous set of data arranged by independent time scales enabling correlations using different parameters and showing a similar climatic history. However, the resolution of the different parameters determines the sensitivity for detecting environmental and climatic variations and not all information is available at each location on an annual or decadal basis to discuss the meaning of the variations of the different parameters and their mutual influences. Additionally, local signatures have to be disentangled from regional or global signals, e.g., the regional precipitation and evaporation reconstruction for the Eifel region (Schettler et al. 1999) during the Holocene.

In general, these two records are basic pillars for a new stratigraphic subdivision on a calendar time scale for Europe. However, a correlation of these two records with other, often fragmentary and not well dated, records (Figs. 1, 4b, c) is complicated. Therefore, the procedure of comparison with these two records is mainly based on pollen zones which do not document time slices.

First results using palaeosecular variation records (inclination, declination) for subdividing the Holocene seem to be promising (Fig. 3), but often other parameters are not available for these lake sediments. Thus, the documented records (Fig. 3) have been dated by different methods. However, more detailed studies on the variation of the geomagnetic palaeointensities will enable the establishment of a magnetostratigraphy on varved records being a synchronous global signal pattern. Additionally, geomagnetic palaeointensitiy minima and/or excursions/events coincide with ^{14}C and ^{10}Be maxima being coupled with solar activity changes (e.g. around 2800 cal. BP (van Geel et al. 1996, 1998)).

To describe the climate history of Europe on a regional scale some of the locations drawn in Fig. 1 will be cited from South to North without defined time slices, because the records document results over different time periods of the Holocene.

Central and Western Europe

In Central Europe the climatic "optimum" can be recognized during the Atlantic superimposed in younger times by human influences so that it is complicated to distinguish between climatic and human impacts for the youngest Earth history. First human impact can be traced around 7500 cal. BP in the Eifel region (Neolithic revolution), cereals appear around 5300 cal. BP, with increasing influence since the beginning of the Bronze age.

In general, the Atlantic in Central Europe is characterised by mixed *Quercus* forests with *Hedera*, *Ilex* and *Viscum* (Wiethold 1998). *Hedera* seems to indicate mild winters with average January temperatures of >1.7 – 2°C, whereas *Viscum* needs an average July temperature of >15.5°C. During the warm phase of the Atlantic the average July temperature has been 2°C higher than before and rich, mixed oak forests also con-taining *Ulmus* and *Tilia* dominated. Around ~5700 cal. BP the *Ulmus* decline at Belau Lake (N-Germany) started being a special event in Europe, possibly caused by a combination of disease and human activity as documented also in Diss Mere Lake, Norfolk. From ~3700 cal. BP on, a subboreal climatic decline towards more oceanic conditions with higher precipitation was recognized in the Bronze age.

Using indicator species of pollen records (*Hedera*, *Ilex*, *Viscum*, *Corylus*) Zagwijn (1994) reconstructed the Holocene climate change in Western and Central Europe. "Early in the Holocene, an influx of warm ocean water resulted in higher winter temperatures in the Gulf of Biscay, the Irish Sea, and areas east of Skagerrak-Kattegat. Temperature decline after the climate "optimum" was greatest in the north, i.e. at 60°N, where a depression in the order of 2°C in summer and 2-3°C in winter occurred. Temperature decline was less farther south, i.e. at 50°N, where a distinct west-east gradient in temperature change can be observed".

A comparison of the Holocene climatic warm phase with the present shows that a distinct north-south gradient is found for the temperature decline since the climate "optimum". Furthermore an east-west gradient can also be observed, which indicates increasing continentality during the last ca. 3000 years.

Scandinavia

Figure 4b documents the early Holocene pollen records of Norway showing a similar pattern like in Central Europe.

In general, climatic oscillations are better documented within extreme locations like the Holocene lake sediments of the subalpine Abisko area of

Northern Sweden (Berglund et al. 1996): "Since the deglaciation c. 9000 – 8500 ^{14}C yr. BP the vegetation of the Abisko area has been dominated by a subalpine birch woodland tundra with the exception of c. 5500 – 3500 BP when the boreal pine-birch forest zone reached the area. During the same period there seems to have been an altitudinal expansion of pine to an elevation of c. 100 – 150 m higher than today. Birch (probably mountain birch) has been the dominant tree species at the tree-limit since c. 8300 BP except for the pine-birch period (c. 5500 – 3500 BP). In this respect the area differs from the southern Scandes, where pine seems to have been more frequent during early Holocene.

The climate has been subarctic-oceanic since the regional deglaciation apart from the period c. 5500 – 3500 BP which seems to have been temperate-continental with warm and dry summers and cold winters. This is also associated with low lake levels which may indicate that the Holocene summer temperature optimum occurred during this period. No small-scale fluctuations of the climate have been recorded."

Korhola et al. (1999) reconstructed temperatures for a Holocene lake sequence in the mountainous region of northern Fennoscandia using pollen, chironomids, diatoms and cladocera and concluded that summer temperatures in this region varied continuously during the last 10,000 years with century- scale oscillations. During the Holocene "maximum" (6.5 – 6.2 kyr) the summers were warm with a clear mid-Holocene cooling after-wards and a temperature minimum between 4.5 – 2.8 kyr. In addition, distinct short cold spells have been recorded in these cores around 8.3, 7.2, 3.0, 2.8 and 0.4 kyr.

A comparison of the snowlines in Norway and the Swiss central Alps shows that whilst in Norway the snowline was about 400 m higher than at present in the period 8000 – 5000 BP, the snowline in the central Alps fluctuated around its present value throughout the Holocene, being on average slightly higher between 6000 and 4000 BP, but by less than 100 m. This contrasts to the view of Huntley and Prentice (1988) and Guiot et al. (1996) who – on the basis of pollen transfer functions – calculated temperature deviations of over 4°C for the Alpine region at 6000 BP, which are much higher than in the north.

Russia (Siberia)

In NW-Russia, records from Lake Onega dated the termination of the Younger Dryas (YD) event at 11,700 ± 100 cal. BP based on a varved record (Saarnisto and Saarinen 1999) which is in the range of ice cores and varved lake sediments from Central Europe.

The YD cold event has also been recognised in pollen records from Siberia (Lama and Levinson-Lessing Lakes, Fig. 4c), however, there are still problems related to the dating of these records and their correlation with records of northern Central Europe. On the basis of these records Hahne and Melles (1999) describe the vegetation and climate history of the unglaciated southern and northern Taymyr Peninsula since Late and Middle Weichselian times. The Pleistocene/Holocene transition at both sites is characterised by a climatic warming trend during the Bølling, Allerød, and Preboreal periods, with interruptions at the Older (post-Bølling) and YD events. The Holocene climatic "optimum" occurred during the late Preboreal and Boreal periods, larch forests dom-inated at Lama Lake, birch and dense alder boscages at Levinson-Lessing Lake. During the Atlantic period, a climatic deterioration took place favouring sedge and sweet-grass communities at Levinson-Lessing Lake, with no significant effect on the forest vegetation at Lama Lake. The cli-matic deterioration continued up to the middle Subboreal period, when present climate and vege-tation conditions established.

Temporal variations of arboreal pollen content (AP) at Lama Lake correlate with those at Levinson-Lessing Lake as well as with stable isotope ratios of an ice core from Severnaya Zemlya. This correlation reflecting summer and annual temperatures indicates that the continentality of the climate was regionally constant throughout the Holocene.

As result of the interpretation of the multi-proxies, it can be concluded that the general climate history with its trends and gradients obtained from the Central European sites (Gosciaz, Eifel region) appears to be reasonable even for northern Europe despite slight differences in Siberia. However, more detailed comprehensive studies are necessary to develop a reliable picture of the local and regional reactions of the environment and climate in different areas due to the influence of the North Atlantic Oscillation.

Conclusions

The climate history over the whole area of NW Europe during the Late Glacial and Holocene can be recognised in lake deposits. However, most records lack a reliable annual chronology. Correlations are mostly done by wiggle matching. These correlations, often using vegetation changes as temperature indicators, describe general climatic trends and gradients. Combined biological as well as geochemical (e.g. $\delta^{18}O$ records) parameters to derive climate variables (temperature, precipitation) by developing transfer functions are in 'status nascendi'. Both approaches – to establish a calen-dar year time scale and to derive climate variables from proxies – are basic requirements for investigating global teleconnections and for climate modelling.

In general, lake sediments are ideal terrestrial archives from which the environmental and climate history of the continents, the living place of mankind, can be recovered. Laminated records allow us to establish varve chronologies as independent time scales on a calendar basis as has been done for some lake records (e.g. Eifel maar lakes, Lake Gosciaz) and for the ice cores. This is the prerequisite for installing consistent and synthetic time scales for lake deposits and indeed for all palaeoclimate archives. Based on these high- resolution sequences (varved sediments and ice cores) it might be possible to use multiproxy studies to disentangle the millennial – century scale and shorter term (up to 100 years) climate oscillations, and to discover the pacemakers of climate beyond the Milankovitch cycles.

Therefore, future investigations have to focus primarily on laminated records using, in a joint effort, all available methods of dating, multiproxy analyses (biological, geochemical and physical) and correlation (magnetostratigraphy) to establish an absolute time scale for whole Europe, in correspondence with the ice cores to develop Holocene climate history and its variation on a regional to local scale in space and time.

References

Allen JRM, Brandt U, Brauer A, Hubberten H-W, Huntley B, Keller J, Kraml M, Mingram J, Negendank JFW, Nowaczyk NR, Oberhänsli H, Watts WA, Wulf S, Zolitschka B (1999) Rapid last glacial environmental fluctuations: new evidence from southern Europe. Nature 400:740–743

Berglund BE, Barnekow L, Hammarlund D, Sandgren P, Snowball IF (1996) Holocene forest dynamics and climate changes in the Abisko area, northern Sweden - the Sonesson model of vegetation history reconsidered and confirmed. Ecol Bull 45: 15-30

Björck S, Walker MJC, Cwynar LC, Johnsen S, Knudsen K-L, Lowe JJ, Wohlfarth B, INTIMATE members (1998) An event stratigraphy for the Last Termination in the North Atlantic region based on the Greenland ice-core record: a proposal by the INTIMATE group. J Quat Sci 13: 283-292

Brathauer U, Brauer A, Negendank JFW, Zolitschka B (2000) Rasche Klimaänderungen am Beginn der heutigen Warmzeit. Zweijahresbericht GeoForschugsZentrum Potsdam (1998 - 1999)

Brauer A, Günter C, Johnsen SJ, Negendank JFW (1999a) Land-ice teleconnections of cold climatic periods during the last Glacial/Interglacial transition. Clim Dyn 2/3:229–239

Brauer A, Endres C, Günter C, Litt T, Stebich M, Negendank JFW (1999b) High resolution sediment and vegetation responses to Younger Dryas climatic change in varved lake sediments from Meerfelder Maar, Germany. Quat Sci Rev 18:321-329

Brauer A, Endres C, Negendank JFW (2000) Lateglacial Calendar Year Chronology based on annually laminated sediments from Meerfelder Maar, Germany. Quat Int 61:17-25

Brüchmann C (1998) Diatomeen in frühholozänen Warven des Holzmaares (Eifel) und ihre Bedeutung für paläolimnologische Rekonstruktionen. Thesis, University of Potsdam.

Geel Bv, Buurman J, Waterbolk HT (1996) Archeological and paleoecological indications of an abrupt climate change in The Netherlands, and evidence for climatological teleconnections around 2650 BP. J Quat Sci 11:451-460

Geel Bv, Plicht Jvd, Kilian MR, Klaver ER, Kouwenberg JHM, Renssen H, Reynaud-Farrera I, Waterbolk HT

(1998) The sharp rise of ^{14}C ca. 800 cal BC. Possible causes, related climatic teleconnections and the impact on human environments. Radiocarbon 40:535-550

Glaser R (1995) Thermische Klimaentwicklung in Mitteleuropa seit dem Jahr 1000. Geowissenschaften 13:302-312

Grafenstein Uv, Erlenkeuser H, Brauer A, Jouzel J, Johnse SJ (1999) A mid-European decadal isotope-climate record from 15,500 to 5000 years B.P. Science 284:1654-1657

Guiot J, Cheddadi R, Prentice IC, Jolly D (1996) A method of biome and land surface mapping from pollen data: application to Europe 6000 years ago. Paleoclimates 1:311-324

Hahne J, Melles M (1999) Climate and vegetation history of the Taymyr Peninsula since middle Weichselian time - Palynological evidence from lake sediments. In: Kassens H, Bauch HA, Dmitrenko IA, Eicken H, Hubberten H-W, Melles M, Thiede J, Timokhov LA (eds) Land-Ocean Systems in the Siberian Arctic: Dynamics and History. Springer, Berlin Heidelberg New York, pp 407-423

Huntley B, Prentice IC (1988) July temperatures in Europe from pollen data, 6000 years before present. Science 241: 687-690

Kienel U (1998) Late Weichselian to Holocene diatom succession in a sediment core from Lama Lake, Siberia and presumed ecological implications. In: Kassens H, Bauch HA, Dmitrenko IA, Eicken H, Hubberten H-W, Melles M, Thiede J, Timokhov LA (eds) Land-Ocean systems in the Siberian Arctic. Dynamics and History. Springer, Berlin Heidelberg New York, pp 377-405

Korhola A, Weckström J, Seppä H, Birks HJ, Peglar SM, Toivonen H, Vasko K, Mannila H (1999) Quantitative Holocene climatic records from sedimentary remains of aquatic organisms and pollen in northern Fennoscandia. Terra Nostra 99/10:49-53

Kristiansen IK, Mangerud J, Lømo L (1988) Late Weichselian/Early Holocene pollen- and lithostratigraphy in lakes in the Álesund area, western Norway. Rev Palaeobot Palynol 53:185-231

Larsen E, Eide F, Longva O, Mangerud J (1984) Allerød-Younger Dryas climatic inferences from cirque glaciers and vegetational development in the Nordfjord area, western Norway. Arct Alp Res 2:137-160

Litt Th, Stebich M, Brauer A (1998) Bio- and chronostratigraphy of the Lateglacial in the Eifel region based on varred sediments. Terra Nostra 98/6:82-85

Litt T, Brauer A, Goslar T, Merkt J, Balaga K, Müller H, Ralska-Jasiewiczowa M, Stebich M, Negendank JFW (2000) Correlation and synchronisation of Lateglacial continental sequences in northern central Europe based on annually-laminated lacustrine sediments. Quat Sci Rev (in press)

Merkt J, Müller H (2000) Varve chronology and palynology of the Lateglacial in Northwest Germany from lacustrine sediments of Hämelsee in Lower Saxony. Quat Int 61:41-59

Ralska-Jasiewiczowa M, Goslar T, Madeyska T, Starkel L (eds) (1998) Lake Gosciaz, central Poland. A monographic study. W. Szafer Institute of Botany, Cracow, pp 1-340

Saarnisto M, Saarinen T (1999) Correlations between varves, paleomagnetic data and radiocarbon dates in the study of the deglaciation chronology of the Scandinavian Ice Sheet in NW Russia. Terra Nostra 99/10:78-81

Schettler G, Rein B, Negendank JFW (1999) Geochemical evidence for Holocene palaeodischarge variations in lacustrine records from Westeifel Volcanic Field, Germany: Schalkenmehrener Maar and Meerfelder Maar. The Holocene 9 4:381-400

Vos H, Sanchez A, Zolitschka B, Brauer A, Negendank JFW (1997) Solar activity variations recorded in varred sediments from the crater lake of Holzmaar - a maar lake in the Westeifel volcanic field, Germany. Surveys in Geophysics 18:163-182

Walker MJC, Lowe JJ, Cwynar LC, Johnsen, S, Knudsen K-L, Wohlfarth B, INTIMATE group (1999) Isotopic 'events' in the GRIP ice core: a stratotype for the Late Pleistocene. Quat Sci Rev 18:1143-1150

Wiethold J (1998) Studien zur jüngeren postglazialen Vegetations- und Siedlungsgeschichte im östlichen Schleswig-Holstein. Universitätsforschungen zur prähistorischen Archäologie 45

Zagwijn WH (1994) Reconstruction of climate change during the Holocene in western and central Europe based on pollen records of indicator species. Vegetation History and Archaeobotany 3:65-88

Zolitschka B, Negendank JFW (1998) A high resolution record of Holocene paleohydrological changes from Lake Holzmaar, Germany. In: Frenzel B (ed) Palaeohydrology as reflected in lake-level changes as climatic evidence for Holocene times. pp 37-52

Zolitschka B, Brauer A, Stockhausen H, Lang A, Negendank JFW (1999) Comparison of a multiple-dated lake site in Germany with Greenland ice cores for the last 23,000 calendar years BP. Terra Nostra 113-116

Climate Changes During the Holocene Recorded by Lakes from Europe

Sandy P. Harrison*[1,2], Ge Yu[2,3] and Juri Vassiljev[2,4]

[1]*Max-Planck Institute for Biogeochemistry, Postfach 100164, D-07701 Jena, Germany*
[2]*Dynamic Palaeoclimatology, Lund University, Postbox 117, S-221 00 Lund, Sweden*
[3]*Nanjing Institute of Geography and Limnology,*
Chinese Academy of Sciences, Nanjing, China
[4]*Institute of Geology at Tallinn Technical University, Estonia Pst. 7, 10143 Tallinn, Estonia*
**corresponding author (e-mail): sandy.harrison@bgc-jena.mpg.de*

Abstract: Lake records provide an important source of information about climate changes on sub-millenial to millenial timescales during the Holocene. Analyses of lake records from Europe show that there are regionally-coherent patterns of change in lake status (a qualitative index of changes in water depth, area or volume) through the Holocene, reflecting climatically-induced changes in regional water budgets. Rapid changes in lake status occur in many regions of Europe between 9500-9000, 7500-7000 and 4500-4000 yr BP This apparent synchroneity in the timing of changes, although the direction varies from region to region, indicates that the changes in lake status are driven by changes in regional water balance caused by major reorganisations of the atmospheric circulation. Modelling studies indicate that changes in precipitation are the most important cause of changes in lake behaviour across Europe, because the lake water balance in temperate humid regions is largely controlled by runoff from the catchment. Changes in winter precipitation produce a much larger change in runoff and hence in lake response, than changes during the summer growing season. Sensitivity to precipitation changes is enhanced where solar radiation receipts are high: a given change in precipitation will therefore produce a larger response in the lake water balance in southern and eastern Europe than in northern or western Europe. The sensitivity of the lake water balance to changes in precipitation is dependent on the ratio of lake area to catchment area: sensitivity increases as the ratio becomes larger. The lake area to catchment area ratio at which a lake become sensitive varies with the magnitude of the precipitation change. Lake studies have the potential to contribute significantly to documenting and explaining Holocene climate changes across Europe. Continued improvements to the existing lake data base (including improvements to the lake chronologies), coupled with the application of geostatistical and modelling techniques will be required in order to exploit these archives fully.

Introduction

All lakes, whether hydrologically closed or overflowing, fluctuate in depth and area in response to changes in the annual water balance (precipitation minus evaporation: P – E) over the lake and its catchment (Street-Perrott and Harrison 1985; Cheddadi et al. 1996). Sustained changes in lake depth and area affect the physical, chemical and biological characteristics of the lake system, including e.g. the patterns of sedimentation within the lake, water salinity, and biological productivity. Changes in such parameters, preserved in the geological record of the lake sediments, can be used to reconstruct past changes in lake depth and/or area (Harrison and Digerfeldt 1993). Changes in lake depth/area can be caused by non-climatic factors (Dearing and Foster 1986), operating either within an individual lake basin (e.g. changes to the inflows or outflows such as damming of the lake by landslides, human impacts on land use in the catchment) or at a regional scale (e.g. tectonism, sea-level changes which alter the regional hydrological base level). However, provided it is possible to rule out

From WEFER G, BERGER W, BEHRE K-E, JANSEN E (eds), 2002, *Climate Development and History of the North Atlantic Realm.* Springer-Verlag Berlin Heidelberg, pp 191-204

non-climatic factors as the cause of past changes in lake depth/area, the lake record can be used to infer past changes in the water balance at a regional scale.

Lakes have many advantages as a source of information about past climate changes in Europe. They are widely distributed, so it is possible to derive records of changes in lake depth/area that can be used to reconstruct climate changes across the whole of the continent. Furthermore, most European lakes have a continuous record of sedimentation covering the whole of the Holocene while lakes in central and southern Europe provide records that extend even further back in time. Lake sediments, with the possible exceptions of saline lakes in the semi-arid parts of southern Europe and lakes in predominantly limestone areas, are generally quite rich in organic matter and thus provide ample opportunities for radiocarbon dating. It is therefore possible to provide a good chronology for the re-constructed changes in lake depth/area. Finally, the climatic controls on lake fluctuations are relatively simple and well understood, and thus the interpretation of the records of reconstructed changes in lake depth/area in terms of changes in regional water budgets is straightforward.

The aims of this paper are to (a) review the theoretical basis for using lakes to reconstruct climatic changes, (b) review the lake record of regional changes in climate on sub-millenial to millenial time-scales through the Holocene, and (c) examine the factors that determine the sensitivity of individual lakes to climate changes. We conclude by summarising the remaining uncertainties in reconstructions of past climate changes in Europe from lake data and by suggesting avenues for future research in this area.

Lakes and the Water Balance

Lakes are sensitive to changes in the water balance, even in wet climates where there is an outflow (Fig. 1). The water balance of a lake (neglecting groundwater inputs and outputs for the time being) is given by:

$$R + A_L P = A_L E_p + D + \Delta V \qquad (1)$$

where R is runoff from the catchment, A_L is the area of the lake, P is precipitation (which is assumed to be constant across small basins), E_p is evaporation from the lake surface (for lakes up to a few km across this can be assumed equal to potential evaporation), D is discharge (outflow) from the lake, and ΔV is the net change in lake volume. Runoff is given by:

$$R = A_C (P - E) \qquad (2)$$

where A_C is catchment area and E is actual evapotranspiration over the catchment. Under most conditions, $E < E_p$. Combining equations (1) and (2) and dividing through by the total basin (catchment plus lake) area yields:

$$(1\text{-}\theta)(P - E) + \theta P = \theta E_p + (D + \Delta V)/A \qquad (3)$$

where $A = A_L + A_C$ and $\theta = A_L/A$. D increases with lake level, which generally increases with A_L. The relations between these variables are non-linear and governed by the topography of the basin and the outflow channel. There is generally a range from $\theta = 0$ up to a threshold where $D = 0$, after which D starts to increase steeply with θ.

The equilibrium value of θ (denoted θ^*) is obtained from equation (3) by setting ΔV to zero.

An implicit expression for θ^* in lakes with $D > 0$ is:

$$\theta^* = (P - E) + f\theta_1 / (E_p - E) + f \qquad (4)$$

where f is the slope $\delta D/\delta\theta$ (evaluated at θ^*) of the function relating discharge to θ, and θ_1 is the θ-intercept obtained by extrapolating this slope linearly back to $D = 0$ (Cheddadi et al. 1996).

In the case of closed lakes, where $D = 0$, equation (4) reduces to a simple explicit formula:

$$\theta^* = (P - E) / (E_p - E) \qquad (5)$$

These principles can be illustrated graphically (Fig. 1). Fig. 1 demonstrates why open (overflowing) lakes are always less sensitive than closed lakes to changes in the climatic water balance – while fluctuation in open lakes is nevertheless possible, through concurrent changes in A_L and D. Increasing precipitation leads to increased discharge (D), but increased D cannot occur unless there is also an increase in level.

This analysis neglects the role of groundwater inputs and/or outputs on the lake water budget. However, the groundwater contribution to the lake water balance in humid regions is generally small and usually responds to climate changes in a qualitatively similar way to changes in the surface water balance (P – E).

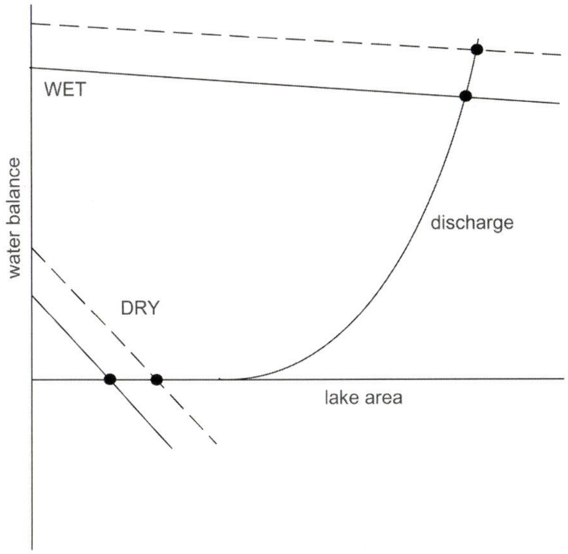

Fig. 1. Graphical representation of the dependence of equilibrium lake area on climate and topography. The heavy lines indicate the total water balance of the basin, A_C (P – E) + A_L (P – E_P), for a wet and a dry climate. The lines represent linear interpolation between A (P – E) and A (P– E_P). They slope downwards because E < E_P, especially in dry climates. The hatched lines indicate the effect of a small increase in P. The smallest lake area at which outflow occurs, and the shape of the curve relating discharge to water level and hence lake area, depend solely on the topography of the basin and outflow channel. The equilibrium area of the lake is the area at which the water balance is equal to discharge (shown by dots). In dry climates (where E_P > P) the water balance line can intersect the x-axis, allowing closed lakes to form. In wet climates (where E_P < P), the water balance line intersects the discharge curve. This implies the formation of an open (overflowing) lake. As the discharge curve is typically steep, the sensitivity of the equilibrium lake area to climate change (for e.g. an increase in precipitation as shown here) is less for open lakes than for closed lakes. Nevertheless, changes in lake level can occur even in open lakes, accompanied by increases in throughput and in the level of the outflow.

Regional Patterns of Lake-Level Changes in Europe During the Holocene

Stratigraphical investigations of lake sediments have been used to reconstruct past lake-level changes in Europe. The majority of such studies show only qualitative changes in lake level. However, in a series of studies starting in the early 1970's (see Digerfeldt 1986 and articles cited therein), Digerfeldt developed a method of reconstructing the absolute magnitude of past lake-level changes from transects of cores using multiple lines of evidence. Studies using this method indicate that European lakes have fluctuated by several meters during the Holocene (e.g. Hjelmroos-Ericsson 1981; Gaillard 1984, 1985; Digerfeldt 1988; Gaillard et al. 1991; Almquist-Jacobson 1995; Magny and Ruffaldi 1995; Saarse et al. 1995; Digerfeldt et al. 1997).

Curves showing lake-level changes through time have been reconstructed for a large number of lakes across Europe (e.g. Magny 1978; Hjelmroos-Ericsson 1981; Gaillard 1984, 1985; Digerfeldt 1988; Gaillard et al. 1991; Almquist-Jacobson 1995; Pazdur et al. 1995; Magny and Ruffaldi 1995; Saarse et al. 1995; Digerfeldt et al. 1997). Comparison of lake-level curves have revealed that there are distinct regional patterns in lake behaviour during the Holocene. Digerfeldt (1988), for ex-ample, compiled lake-level curves from ten lakes in southern Sweden and demonstrated that most lakes were low ca 9000 yr BP and rose between 9000 and 7000 yr BP All the lakes were high at 6500 yr BP and fell again soon after. Most lakes were at minimum levels between 4500 and 3500 yr BP The general rise in lake level after 3500 yr BP was interrupted by a brief interval of lowered lake levels centred on ca 1500 yr BP Similarly, compilations of data from 26 lakes in the Jura, France (Magny 1992; Magny and Ruffaldi 1995) show regionally con-sistent patterns. The lakes from the Jura were either high or increasing in level between 9000 and 8800 yr BP and between 7000 and 6000 yr BP, and there was a significant drop in lake levels between 8800 and 7000 yr BP and again between 4500 and 3500 yr BP.

There have been several attempts to establish whether there are regional differences in the temporal behaviour of lakes across Europe by comparing the lake-level patterns from different regions (e.g. Gaillard 1985; Magny 1992; Pazdur et al. 1995; Digerfeldt et al. 1997). The results of these comparisons have been inconclusive. For example, both Magny (1992), in a comparison of lake-level records from southern Sweden and the Jura, and Pazdur et al. (1995), in a comparison of records from these two regions and Lake Gosciaz in Poland, have suggested that the timing of lake transgressions and regressions during the Holocene is contemporaneous. Gaillard (1985), however, suggested that there were small differences between the timing of events in northern and central Europe. Furthermore, Digerfeldt et al. (1997) show that there is a distinct contrast in the timing of lake-level changes in southern Sweden and at Lac St Leger, in southern France.

Another way to analyse lake records is through mapping of patterns of the change in lake status (a qualitative index of changes in relative water depth, lake area or lake level) relative to present (e.g. Street-Perrott and Grove 1979; Street-Perrott and Harrison 1985). This approach has been applied to Europe (e.g. Harrison et al. 1991; Yu and Harrison 1995a; Harrison et al. 1996), drawing on data from various versions of the Global Lake Status Data Base (GLSDB: Kohfeld and Harrison 2000). The GLSDB contains continuous reconstructions of relative changes in lake status through time, based on geomorphic and biostratigraphic evidence, for 156 lakes from Europe (Yu and Harrison 1995b; Tarasov et al. 1994, 1996). These data have been screened to exclude lake records that are affected by non-climatic factors, including tectonism, basin-specific geomorphic events, hydroseral development or human modification of the catchment, or where the climate impact on the lake system operates indirectly through sea-level changes, glacier fluctuations or changes in the fluvial system. Nearly 200 individual lake records were examined and excluded from the data base for one or more of these reasons. The chronology for most of the lake records is based on radiocarbon-dating, although some lakes from the European part of Russia are dated by correlation of their pollen records with pollen records from nearby radiocarbon-dated sites.

According to these map analyses (see Harrison et al. 1996; Fig. 2), the lake records from Europe show regionally coherent patterns of changes during the Holocene. At ca. 9500 yr BP, conditions were wetter than today in the Mediterranean and drier than today in a narrow zone reaching from southern Finland through southern Sweden and westwards into southern Britain. Around 7000 yr BP, the circum-Baltic region was drier than today. However, the Atlantic coastal region, the far north, the inland part of Finland and the Russian Plain were wetter than present. A gradual transition to modern conditions occurred after ca. 5000 yr BP. However, this transition appears to have been briefly interrupted by a return to wetter conditions between 3000 and 2000 yr BP in the Mediterranean region. Harrison et al. (1996) have argued that the broad-scale changes in regional climate, and particularly in changes in precipitation patterns as a consequence of changes in atmospheric circulation patterns over Europe, can be explained as a consequence of known changes in insolation and glaciation. However, it is clear that the Holocene record of regional changes in lake status shows shorter-term fluctuations that are not obviously related to the smooth changes in insolation forcing.

Map analysis of lake data focuses on specific time periods and on regional differences in lake status at those specific times from present. Such an analysis is not designed to determine whether there are consistent regional differences in the temporal behaviour of lakes nor whether changes in lake behaviour, though of different sign, are synchronous between regions. However, the lake data in the GLSDB can also be analysed in terms of temporal patterns. We have extracted lake status data, subsampled at 250-yr intervals (i.e. for 41 time periods), for a subset of 65 European lakes with continuous Holocene records. The sampling in-terval (250 yr) was chosen to maximise the number of points of comparison between individual lakes; it is unrealistic to sample at a higher-resolution given the uncertainties inherent in the radiocarbon-dated chronologies. The status coding

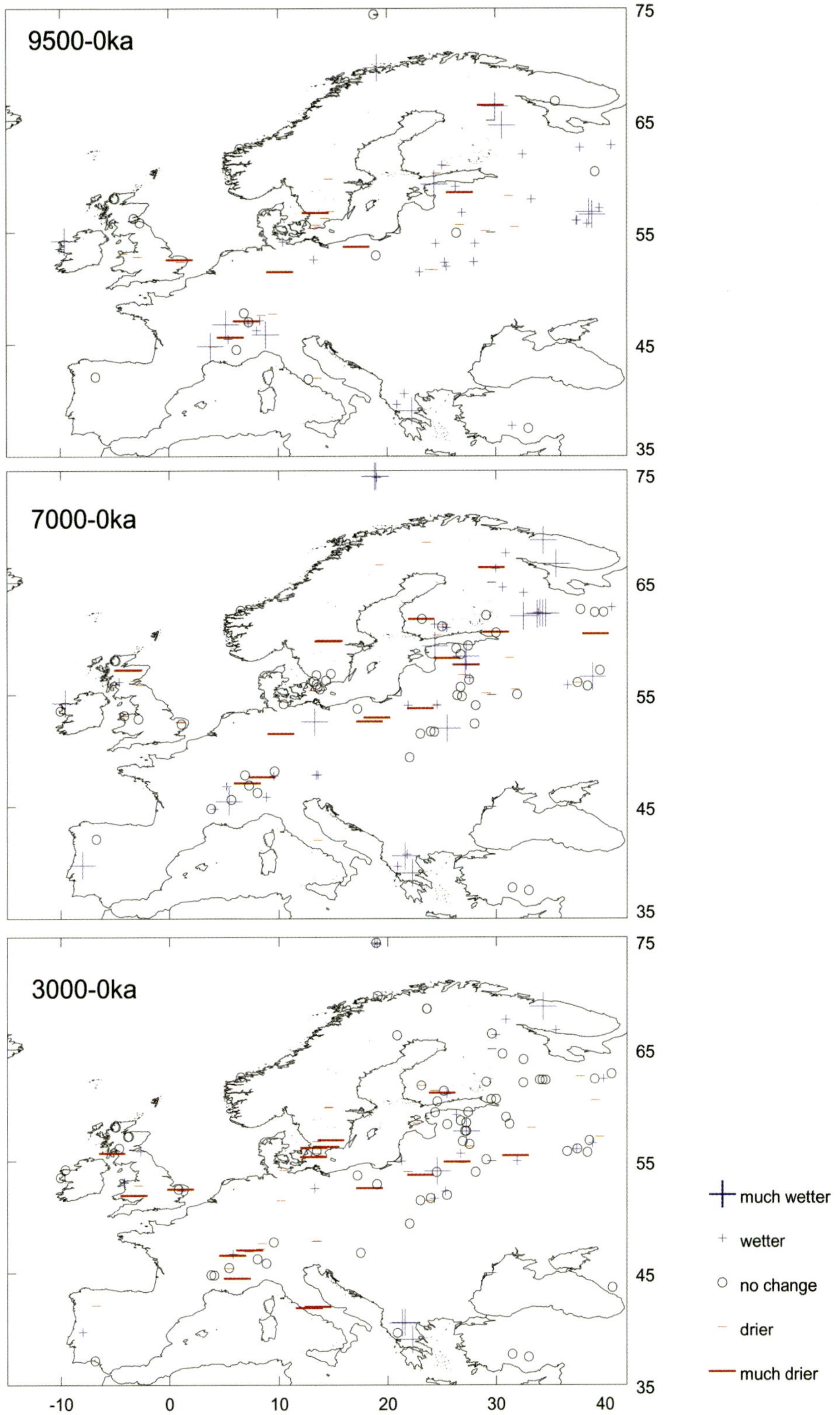

Fig. 2. Lake status, expressed as differences from present, at **(a)** 9500 yr BP, **(b)** 7000 yr BP and **(c)** 3000 yr BP. The data are derived from the European and Russian Lake Status Data Bases (Yu and Harrison 1995b; Tarasov et al. 1994, 1996).

scheme used in the data base varies from site to site, reflecting the fact that it is possible to obtain a finer discrimination of e.g. relative water depth at some sites than at others. We have therefore normalised the data as follows:

$$\%status = (status - status_{min})/(status_{max} - status_{min}) * 100$$

where status is the original lake status coding at each time period, $status_{min}$ is the minimum status class and $status_{max}$ is the maximum status class used in the coding for the site during the Holocene.

Site-by-site correlation of the normalised data shows that it is possible to identify seven groups of lakes which show the same temporal pattern of change through the Holocene (Fig. 3). The lakes in each group occupy a well-defined geographical location. Thus, lakes in group 1 are found in the high latitudes (including the Norwegian arctic islands and Finland). This group of lakes was low before 9000 yr BP, became deeper after 9000 yr BP, reached maximum levels between 7500 and 5000 yr BP, decreased in depth after 4000 yr BP and are relatively low at present. Similarly, lakes in group

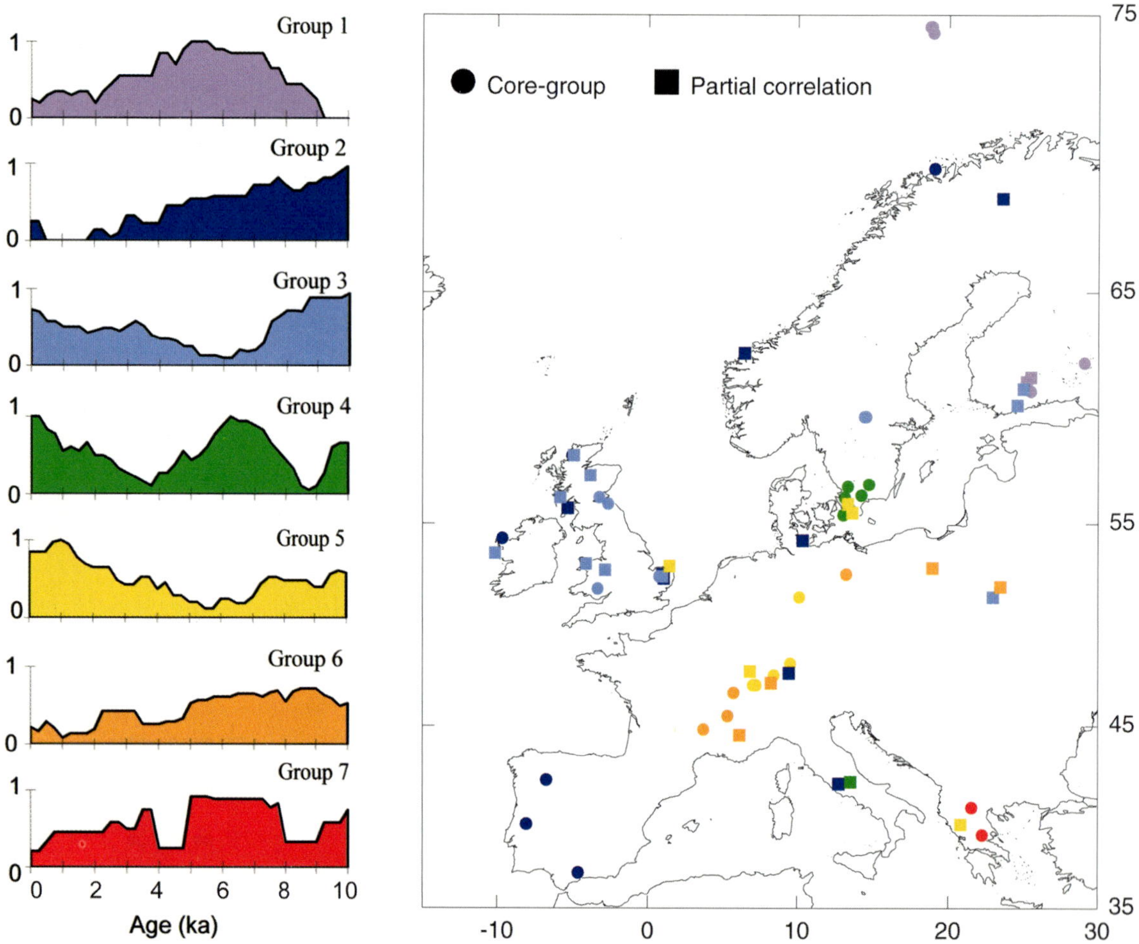

Fig. 3. Lake sites defined as correlated (closed circles) or partially correlated (open circles) in the site-by-site correlation analysis. The numbers indicate to which group each site belongs. Regional lake status histograms (where 1 equals high and 0 equals low) are shown for each of the core correlation groups.

2 are found along the Atlantic seaboard (Iberian peninsula, Ireland, northern Scotland, Norway). This group of lakes was generally high between 10,000 and 5000 yr BP, decreased in depth after 4000 yr BP to reach minimum levels between 1500 to 500 yr BP, and increased in depth towards the present day. Thus, it is possible to identify coherent geographical regions in which all lakes display a similar behaviour through time. Furthermore, the fact that 20 of the lakes included in this analysis show partial correlations with more than one group suggests that there are transitional zones between these coherent regions, where the lakes display patterns intermediate between those shown by lakes in the core regions. Group-by-group correlation analysis suggests that the changes registered by adjacent groups can occur at a similar time but with opposite sign (e.g. Groups 1 and 3). For example, when the high latitude lakes (Group 1) are low in the early Holocene, the lakes in central Finland, central Sweden and central Britain (Group 3) are high. The interval of maximum lake levels in the high latitude lakes between 7500 and 5000 yr BP corresponded to an interval of minimum lake levels in central Finland, Sweden and Britain. Such contrasting patterns between adjacent regions probably reflect shifts in atmospheric circulation patterns.

Major reorganisations of atmospheric circulation patterns imply that there are intervals in time when all of the lakes exhibit synchronous behaviour (regardless of the direction of change). We have looked at this using rate-of-change analysis (e.g. Jacobson et al. 1987), based on subsampling the lake data set over 500-yr intervals. The results (Fig. 4) suggest that there were intervals of rapid change in lake behaviour across most of Europe between 9500-9000, 7500-7000 and 4500-4000 yr BP. The intervening intervals (e.g. between 7000 and at least 5500 yr BP) appear to have been characterised by relative stability in lake (and hence climate) behaviour. It has been suggested that changes in ocean conditions could explain short-term events in continental records (see e.g. Duplessy et al. 1981; Street-Perrott and Perrott 1990). It is therefore interesting that there is an apparent similarity between the timing of intervals of rapid change shown by the lake data and records of sea-surface temperatures in the North Atlantic. Sarnthein et al. (1995), for example, suggested there were pronounced climate changes centred around 9000 yr BP and 7000 yr BP Similarly, records from individual marine cores from the Greenland-Iceland-Norwegian Seas show marked

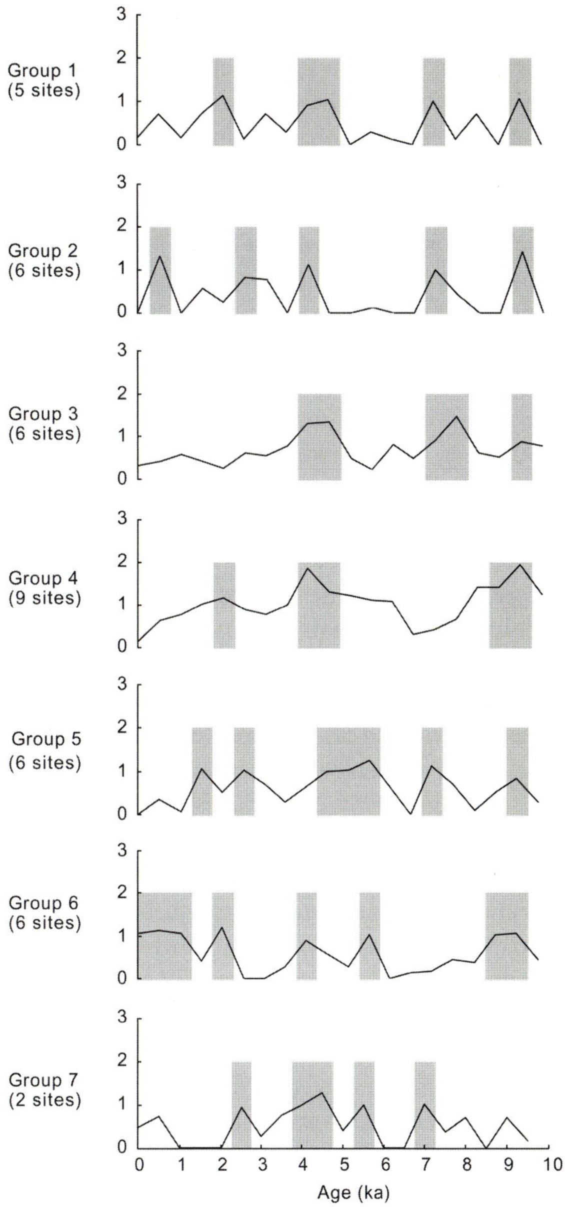

Fig. 4. Rates of change, measured using the chord distance, for the seven groups of lakes identified by correlation analysis and shown in Fig. 2. Intervals when the rates-of-change are significantly higher than average are marked by grey blocks.

decreases in sea-surface temperatures after 5000 yr BP and rapid changes after ca 4000 yr BP (e.g. Koç Karpuz and Schrader 1990; Koç et al. 1993).

Lakes as Sensors: Sensitivity to Changes in Individual Climatic Parameters

Regionally synchronous changes in lake levels have been interpreted in terms of climatically-induced changes in the regional water balance across Europe (e.g. Harrison and Digerfeldt 1993; Yu and Harrison 1995a; Harrison et al. 1996). However, changes in the regional water balance could be caused by changes in several climate variables, including insolation, temperature, cloudiness and precipitation. Harrison et al. (1993), on the basis of experiments with a simple water-balance model in which they systematically perturbed a number of climate variables in a way consistent with palaeo-observations, showed that known changes in insolation and inferred changes in temperature and cloudiness were unable to produce observed changes in lake level in Europe during the Holocene. The effects of changes in precipitation were not considered in these experiments because of the unreliability of existing estimates of palaeo-precipitation in most regions of Europe. Two further studies using simple water balance models have suggested that changes in the seasonality of precipitation could explain observed changes in lake level at individual sites in southern Europe. Evans and Trevisan (1995), used a simple soil water-balance bucket model to determine the causes of observed changes in water levels at Lagoni (northern Italy). They systematically perturbed modern climate on a seasonal basis with different combinations of insolation, temperature and precipitation. According to these simulations, changes of ±10% in the seasonal distribution of precipitation combined with changes in temperature of <1°C, caused changes in percolation and runoff of ±30% sufficient to explain all of the observed changes at this site. Prentice et al. (1992) used a simple water balance model forced on a seasonal basis with temperature, precipitation and cloudiness to show that a shift in precipitation to the winter season, without changing the total annual precipitation, was sufficient to maintain high lake levels at Lake Ioannina (northern Greece) at the last glacial maximum (LGM). Furthermore, Prentice et al. (1992) showed that such a shift in precipitation seasonality, combined with colder year-round temperatures, could plausibly explain the co-existence of high lakes and steppe vegetation (reflecting low growing-season moisture) in the eastern Mediterranean region at the LGM.

The most systematic analysis of the sensitivity of European lakes to changes in precipitation has been made using a physically-based coupled lake-catchment model (Vassiljev 1998). This model simulates evapotranspiration and runoff from the lake catchment, lake water temperature and evaporation from the lake surface, outflow from the lake and lake level (Vassiljev et al. 1994; Vassiljev et al. 1995). The model was used to simulate the impact of systematic changes in annual and seasonal precipitation on the lake level of three lakes from northern Europe (Lake Bysjön, southern Sweden, and Lakes Karujärv and Viljandi in Estonia). The experiments (Fig. 5) show that the level of lakes in temperate humid climates is more sensitive to decreases than increases in precipitation. In temperate humid climates, where actual evapotranspira-tion is close to potential evapotranspiration, increased precipitation results in increased runoff from the catchment. However, the increased runoff to the lake is largely compensated by increased outflow and the resulting change in lake level is small. Conversely, actual evapotranspiration is highly sensitive to a decrease in precipitation. Decreases in precipitation result in decreased evapo-transpiration but, because the catchment vegetation continues to take up moisture, the reductions in runoff are even larger. Reductions in winter (November-April) precipitation have a bigger impact on simulated lake level than changes in summer (May-October) precipitation, because runoff decreases more rapidly with a decrease in winter precipitation.

These various experiments indicate that the response of a lake and the response of its catchment to changes in precipitation are different. The thermal behaviour of a lake is largely controlled by factors affecting evaporation from the lake surface, including insolation, temperature, wind speed, atmospheric humidity and cloudiness. Simulated lake

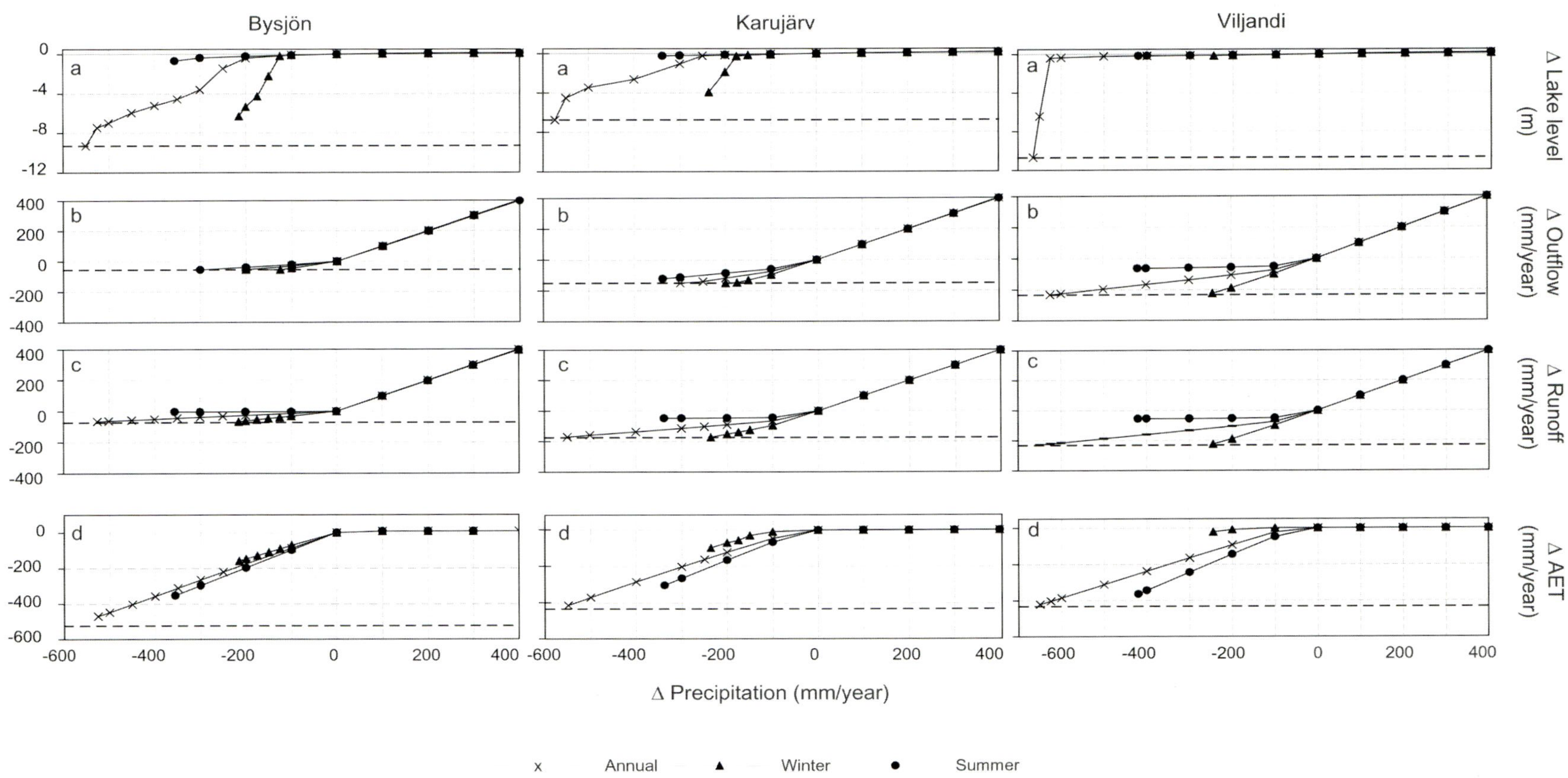

Fig. 5. Simulated changes in **(a)** lake level, **(b)** outflow, **(c)** runoff, and **(d)** actual evapotranspiration (AET) at Lake Bysjön, Sweden **(left)**, Lake Karujärv, Estonia **(centre)** and Lake Viljandi, Estonia **(right)**. The changes are expressed as anomalies from present, dashed lines in **(a)** mark the level at which the lake is dry, and in **(b)**, **(c)** and **(d)** the level at which outflow, runoff or evapotranspiration respectively is zero.

surface temperature and evaporation from the lake surface are relatively insensitive to changes in precipitation. The catchment is very sensitive to precipitation changes. The lake water balance in temperate humid regions is largely controlled by runoff from the catchment.

Lakes as Sensors: The Influence of Basin and Catchment Parameters on Sensitivity

Comparisons of lake records within a given region show that some lakes do not apparently respond to a climate change registered by other lakes from the region, although they do record climate-induced changes during other time intervals. Furthermore, even amongst the group of lakes that do record a regionally synchronous change, the magnitude of the response to a specific climate change varies from lake to lake (e.g. Digerfeldt 1988; Magny 1992). Thus, even when lakes show regionally synchronous behaviour, they do not show the same sensitivity to climate change. Since lakes vary in their morphology, sedimentation and water-balance type (Street-Perrott and Harrison 1985), it is plausible that the different response of lakes to the same change in climate could reflect basin parameters. A number of studies (see e.g. Dearing and Foster 1986; Magny 1992; Vassiljev 1998) have suggested that lake level is sensitive to the ratio of the lake area to the catchment area (A_L/A_C).

Modelling studies confirm that basin parameters are an important determinant of lake sensitivity to climate change. Vassiljev (1997) examined the behaviour of a set of 156 overflowing lakes under both modern and changed climatic conditions using a coupled lake-catchment model. This set of 156 lakes was created by systematically modifying lake bathymetry and basin morphometry, so as to change lake size, catchment size and lake volume independently. The simulations (Fig. 6) show that the response to a given change in precipitation is primarily determined by the A_L/A_C ratio. Small lakes in small catchments are the most sensitive recorders of a given precipitation change. The sensitivity of a moderate size lake to a given precipitation change will depend on the size of the catchment, such that as the magnitude of the change in precipitation increases it will be registered by lakes with progressively smaller A_L/A_C ratios. Thus, a decrease in mean annual precipitation of 200 mm produces a significant decrease in lake level when the A_L/A_C ratio is ≥1:10, whereas a decrease in precipitation of only 100 mm produces a significant decrease in lake level where the A_L/A_C ratio is greater than 1:2. Thus, the A_L/A_C ratio at which lakes become sensitive to climatic changes is not a fixed value (as suggested e.g. by Dearing and Foster 1986) but depends on the magnitude of the change in climate: as the decrease in precipitation (and therefore runoff) becomes larger, so lakes with smaller A_L/A_C ratios become sensitive to climate change. Lake bathymetry, which affects lake volume and area, does not appear to influence the response to precipitation changes except in extreme circumstances where the the A_L/A_C ratio is ≥1:1.

The results of these experiments help to explain earlier, empirical studies which classified the response of lakes to climatic change in terms of their hydrological characteristics (e.g. Street 1980; Street-Perrott and Harrison 1985; Mason et al. 1994), expressed as the relationship between runoff as a percentage of annual inputs and outflow as a percentage of annual outputs (Fig. 7). These empirical studies identified three kinds of lakes: atmosphere-controlled lakes (P – E), amplifier lakes (R – E) and reservoir lakes (R – O). The main source of inputs for atmosphere-controlled lakes is precipitation (P) and runoff (R) is the main source for amplifier lakes; the main way that water is lost in both types of lake is via evaporation (E). Runoff is the main input to reservoir lakes and water is lost primarily through outflow (O). Street-Perrott and Harrison (1985) suggested that reservoir lakes are of little use as palaeoclimatic indicators and amplifier lakes (in R – E and RP – E) are most sensitive to climatic changes. Vassiljev′s (1997) analysis shows that lake water-balance type and the A_L/A_C ratio are strongly related (Fig. 7). Lakes with relatively large catchments tend to be reservoir lakes (R – O) and lakes with relatively small catchments tend to be atmosphere-controlled lakes (P – E), because the total volume of runoff and hence outflow increases linearly as catchment area increases. Decreasing precipitation changes the lake water balance-type, shifting lakes to the left

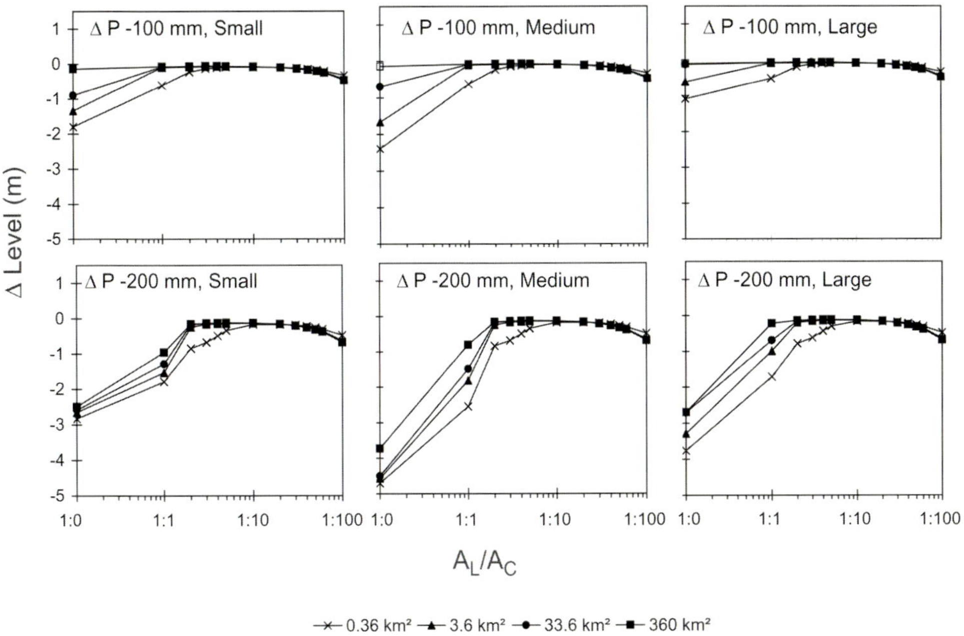

Fig. 6. The change in simulated lake level in response to a decrease in annual precipitation of 100 mm (top) and 200 mm (bottom) for lakes with different sizes, volumes and catchment areas. The A_L/A_C axis has a logarithmic scale.

of the diagram; the larger the decrease in precipitation the bigger the shift. However, the size of the shift is also related to the ratio of A_L/A_C, so that lakes with relatively large catchments shift less than lakes with relatively small catchments. The simulations show that lakes with large catchments (≤1:10) belong to the reservoir type of lakes (R – O) because the catchment acts as a collector of moisture: even a very small runoff per unit area can produce a significantly large total volume of runoff. Lakes with small catchments belong to the atmosphere-controlled lake group (P – E) because the total volume of runoff from the catchment is small and it is therefore less likely to offset the effects of increased evaporation from the lake. As precipitation decreases, there is also a tendency for lakes with small areas to be shifted further towards amplifier lakes than lakes with large areas. This helps to explain why the decrease in lake levels for a given change in precipitation is larger for small lakes.

The experiments carried out by Vassiljev (1997) are unrealistic because they consider changes in a single parameter (i.e. precipitation). However, precipitation changes are unlikely to occur without changes to other climatic parameters (such as cloudiness and hence radiation receipts), and changes in other climatic parameters (e.g. temperature) can also affect runoff and lake levels. When a given decrease in precipitation occurs in conjunction with e.g. an increase in temperature, one might expect the effect to be registered as a lake-level change by lakes with an A_L/A_C ratio smaller than those affected by precipitation alone. Given that there are systematic latitudinal variations in both total and seasonal radiation regime, the fact that radiation also influences runoff (via its effect on evapotranspiration) suggests that the sensitivity of lakes to a given change in precipitation should vary from northern to southern Europe. An increase in total radiation, or a decrease in the seasonal differences in radiation receipts, will both lead to an increase in lake sensitivity. This would suggest that the response to the same decrease in precipitation would be registered in lakes with much smaller A_L/A_C ratios in southern than northern Europe. The patterns of insolation receipts in the maritime climates of western Europe differ markedly from those experienced in the more continental climates of central and eastern Europe. The influence of the

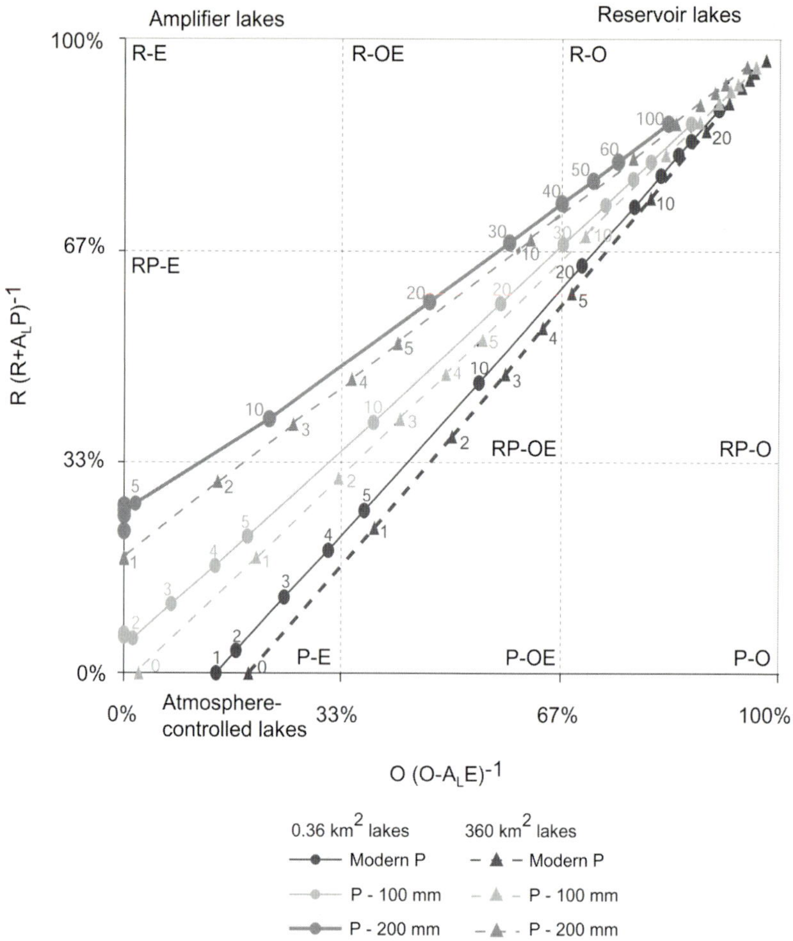

Fig. 7. Water-balance classification (according to the scheme given in Street-Perrott and Harrison 1985) of the set of theoretical lakes used by Vassiljev (1997) to examine the influence of lake size and catchment size on the response of lakes to changes in precipitation. Numbers in the figure represent A_C from the ratio of A_L/A_C; the numbers in italics belong to the lakes with an initial surface area of 0.36 km^2.

reduced total insolation (which tends to minimise the sensitivity to precipitation changes) seems to be more important than the reduction in radiation seasonality in maritime climates, suggesting that the response to the same decrease in precipitation would be registered in lakes with much smaller A_L/A_C ratios in central and eastern than western Europe.

Remaining Uncertainties and Future Possibilities

Lake status records from Europe show that there have been significant changes in regional water balance, driven primarily by changes in precipitation regimes, on the millenial timescale during the Holocene. Nevertheless much remains to be done to document these climatic changes and to understand the underlying mechanisms. We suggest that:

- the existing lake status data base for Europe needs to be expanded. New records from individual sites have been produced since the European component of the GLSDB was last updated in 1995. These records fill in important gaps in the spatial coverage of the data base, and need to be taken into consideration in future analyses
- the chronology of many of the European lake records included in the current version of the GLSDB is relatively poor. Advances in radiocarbon

dating (in particular the increased use of AMS dating for terrestrial records, and advances in radiocarbon calibration methods) make it possible to produce better chronologies for these lakes and will necessitate a re-examination of the timing of climatic changes across Europe

• advances in our knowledge of the role of lake and catchment parameters help to explain why not all lakes in a given region show a similar response to climatic changes. Differences in the response of individual lakes of differing A_L/A_C ratios could be used to quantify the magnitude of the underlying change in precipitation through inverse modelling

• comparisons of the lake status data with other terrestrial records and with marine records could help to diagnose the underlying causes of climatic changes in Europe during the Holocene. However, such comparisons need to be made on a rigorous statistical basis, rather than relying on subjective curve-matching. Advances in geostatistical methods, along with the existence of data bases for individual types of proxy data from Europe, make it possible to contemplate such analyses in the near future.

Acknowledgments

We thank Patrick Bartlein, Jean Grove, Colin Prentice and Bernd Zolitschka for helpful reviews of this manuscript, and Silvana Schott for cartographic assistance.

References

Almquist-Jacobson H (1995) Lake level fluctuations at Ljustjärnen, Central Sweden and their implications for the Holocene of Scandinavia. Palaeo Palaeo Palaeo 111:245-267

Cheddadi R, Yu G, Guiot J, Harrison SP, Prentice IC (1996) The climate of Europe 6000 years ago. Clim Dyn 13: 1-9

Dearing JA, Foster IDL (1986) Lake sediments and palaeohydrological studies. In: Berglund BE (ed) Handbook of Holocene Palaeoecology and Palaeohydrology. John Wiley & Sons, Chichester, pp 67-90

Digerfeldt G (1986) Studies on past lake-level fluctuations. In: Berglund, BE (ed) Handbook of Holocene Palaeoecology and Palaeohydrology. John Wiley & Sons, Chichester, pp127-143

Digerfeldt G (1988) Reconstruction and regional correlation of Holocene lake-level fluctuations in Lake Bysjön, South Sweden. Boreas 17:165-182

Digerfeldt G, de Beaulieu J-L, Guiot J, Mouthon J (1997) Reconstruction and paleoclimatic interpretation of Holocene lake-level changes in Lac de Saint Léger, Haute-Provence, southeast France. Palaeo, Palaeo, Palaeo 136: 231-258

Duplessy JC, Delibrias G, Turon JL, Pujol C, Duprat J (1981) Deglacial warming of the northeastern Atlantic Ocean: correlation with the paleoclimatic evolution of the European continent. Palaeo, Palaeo, Palaeo 35:121-144

Evans SP, Trevisan M (1995) A soil water-balance 'bucket' model for paleoclimatic purposes. 2. Model application to late Holocene climate. Ecol Model 82:131-138

Gaillard M-J (1984) Water-level changes, climate and human impact: A palaeohydrological study of Krageholmssjön (Scania, South Sweden). In: Mörner N-A, Karlén W (eds) Climatic Changes on a Yearly to Millenial Basis: Geological, Historical and Instrumental Records. Reidel, Dordrecht, pp 147-154

Gaillard M-J (1985) Postglacial paleoclimatic changes in Scandinavia and central Europe. A tentative correlation based on studies of lake level fluctuations. Ecologia Mediterranea 11:159-175

Gaillard M-J, Dearing JA, El-Daoushy F, Enell M, Håkansson H (1991) A late Holocene record of land use history, soil erosion, lake trophy and lake-level fluctuations at Bjäresjösjön (South Sweden). J Paleolimnol 6:51-81

Harrison SP, Digerfeldt G (1993) European lakes as palaeohydrological and palaeoclimatic indicators. Quat Sci Rev 12:233-248

Harrison SP, Saarse L, Digerfeldt G (1991) Holocene changes in lake levels as climate proxy data in Europe. Paläoklimaforschung 6:159-170

Harrison SP, Prentice IC, Guiot J (1993) Climatic controls on Holocene lake-level changes in Europe. Clim Dyn 8:189-200

Harrison SP, Yu G, Tarasov PE (1996) Late Quaternary lake-level records from Northern Eurasia. Quat Res 45:138-159

Hjelmroos-Ericsson M (1981) Holocene development of Lake Wielke Gacno area, northwestern Poland. Thesis, Lund University, Lund

Jacobson Jr GL, Webb III T, Grimm EC (1987) Patterns and rates of vegetation changes during the deglaciation of eastern North America. In: Ruddiman WF, Wright Jr H (eds) North America and adjacent Oceans during the Last Deglaciation. Geological

Society of America, Boulder. The Geology of North America Vol. 3:277-287

Koç Karpuz N, Schrader H (1990) Surface sediment diatom distribution and Holocene palaeotemperature variations in the Greenland, Iceland and Norwegian Sea. Palaeoceanography 5:557-580

Koç N, Jansen E, Haflidason H (1993) Paleoceanographic reconstructions of surface ocean conditions in the Greenland, Iceland and Norwegian Seas through the last 14ka based on diatoms. Quat Sci Rev 12:115-140

Kohfeld KE, Harrison SP (2000) How well can we simulate past climates ? Evaluating the models using global palaeoenvironmental datasets. Quat Sci Rev 19:321-346

Magny M (1978) La dynamique des dépôts lacustres et les stations littorales du Grand Lac de Clairvaux (Jura). C.R.A.S., C.R.A. Préhistoriques de France 77:17-25

Magny M (1992) Holocene lake-level fluctuations in Jura and the northern subalpine ranges, France: regional pattern and climatic implications. Boreas 21:319-334

Magny M, Ruffaldi P (1995) Younger Dryas and early Holocene lake-level fluctuations in the Jura mountains, France. Boreas 24:155-172

Mason IM, Guzkowska MAJ, Rapley CG, Street-Perrott FA (1994) The response of lake levels and area to climatic change. Climatic Change 27:161-197

Pazdur A, Fontugne MR, Goslar T, Pazdur MF (1995) Late Glacial and Holocene water-level changes of the Goscia Lake, central Poland, derived from carbon isotope studies of laminated sediments. Quat Sci Rev 14:125-135

Prentice IC, Guiot J, Harrison SP (1992) Mediterranean vegetation, lake levels and palaeoclimate at the Last Glacial Maximum. Nature 360:658-660

Saarse L, Veski S, Heinsalu A, Rajamäe R, Martma T (1995) Litho- and biostratigraphy of Lake Päidre, South Estonia. Proceedings of the Estonian Academy of Sciences. Geology 44:45-59

Sarnthein M, Jansen E, Weinelt M, Arnold M, Duplessy J-C, Erlenkeuser H, Flatoy A, Johannessen G, Johannessen T, Jung S, Koc N, Labeyrie L, Maslin M, Pflaumann U, Schulz H (1995) Variations in Atlantic surface ocean palaeoceanography, 50°-80°N: a time-slice record for the last 30,000 years. Paleoceanography 10:1063-1094

Street FA (1980) The relative importance of climate and hydrogeological factors in influencing lake-level fluctuations. Palaeoecol Afr 12:137-158

Street-Perrott FA, Grove AT (1979) Global maps of lake-level fluctuations since 30,000 yr B.P. Quat Res 12:83-118

Street-Perrott FA, Harrison SP (1985) Lake levels and climate reconstruction. In: Hecht AD (ed) Paleoclimate Analysis and Modeling. John Wiley & Sons, New York, pp 291-340

Street-Perrott FA, Perrott RA (1990) Abrupt climate fluctuations in the tropics: the influence of Atlantic Ocean circulation. Nature 343:607-612

Tarasov PE, Harrison SP, Saarse L, Pushenko MYa, Andreev AA, Aleshinskaya ZV, Davydova NN, Dorofeyuk NI, Efremov YuV, Khomutova VI, Sevastyanov DV, Tamosaitis J, Uspenskaya ON, Yakushko OF, Tarasova IV (1994) Lake status records from the former Soviet Union and Mongolia: Data base documentation. NOAA Paleoclimatology Publications Series Report 2: p 274

Tarasov PE, Pushenko MYa, Harrison SP, Saarse L, Andreev AA, Aleshinskaya ZV, Davydova NN, Dorofeyuk NI, Efremov YuV, Elina GA, Elovicheva YaK, Filimonova LV, Gunova VS, Khomutova VI, Kvavadze EV, Neustreuva IYu, Pisareva VV, Sevastyanov DV, Shelekhova TS, Subetto DA, Uspenskaya ON, Zernitskaya VP (1996) Lake status records from the former Soviet Union and Mongolia: Documentation of the Second Version of the Database. NOAA Paleoclimatology Publications Series Report 5, pp 1-224

Vassiljev J (1997) Simulating the palaeorecord of northern European lakes using a coupled lake-catchment model. Thesis, Lund University, Lund

Vassiljev J (1998) The simulated response of lakes to changes in annual and seasonal precipitation: implication for Holocene lake-level changes in northern Europe. Clim Dyn 14:791-801

Vassiljev J, Harrison SP, Hostetler S, Bartlein PJ (1994) Simulation of long-term thermal characteristics of three Estonian lakes. J Hydrol 163:107-123

Vassiljev J, Harrison SP, Haxeltine A (1995) Recent lake-level and outflow variations at Lake Viljandi, Estonia: validation of a coupled lake-catchment modelling scheme for climate change studies. J Hydrol 170:63-77

Yu G, Harrison SP (1995a) Holocene changes in atmospheric circulation patterns as shown by lake status changes in northern Europe. Boreas 24:260-268

Yu G, Harrison SP (1995b) Lake status records from Europe: Data base documentation. NOAA Paleoclimatology Publications Series Report 3, pp 1-451

The Post-Glacial Evolution of the Baltic Sea

Kay-Christian Emeis[1*], Rudolf Endler[1], Ulrich Struck[1] and Alexander Kohly[2]

[1]Baltic Sea Research Institute, Seestrasse 15, D-18119 Warnemünde, Germany
[2]Institute for Geosciences, University of Kiel,
Olshausenstrasse 40-60, D-24118 Kiel, Germany
[*]corresponding author (e-mail): kay.emeis@io-warnemuende.de

Abstract: Sediment records in the Baltic Sea image the development of late Glacial and Holocene climate in Northwestern Europe, the transgression of global sea level and the flooding of the shelf, and the effects of isostatic uplift of Scandinavia. Changes in salinity of surface and deep waters were pronounced to approximately 5,000 years before present and were mainly caused by sea level rise and sill subsidence. Density stratification between salty deep and fresher surface waters caused the development of anoxic conditions and deposition of laminated sediments on large parts of the sea floor in the Baltic proper. Blooms of cyanobacteria were more intense than today, and the Baltic Sea was a fertile environment that supported diverse and abundant coastal resources nourishing Mesolithic Man. At 5,000 years before Present, the climate deteriorated at the transition from the Atlantic to the Sub-Boreal climatic period. Early Neolithic agricultural communities spread from the South to the North and replaced foragers, initiating the clearing of forests in the Baltic Sea catchment. The salinity of the Baltic Sea decreased, oxygenation of deep seafloors was improved, and fertility of the sea surface was significantly reduced, as seen by low accumulation rates of organic carbon in bioturbated sediments. From approximately AD 1,000 on, salinity rises again and anoxic periods are more common. A clear impact of human activity in the catchment is seen in sediment records of the last 100 years, with increased burial rates of nutrient elements from fertilisers and trace elements. On the other hand, climate continues to govern changes in the salt budget and influences the balance between oxic and anoxic conditions at the seafloor that have a large impact on nutrient budgets in the water column.

Introduction

The Baltic Sea is a young shelf sea that is not quite part of the ocean and not quite a lake. Carved out by glaciers of the last glaciation, the basin has evolved into an almost land-locked big estuary that covers an area of 415,000 km^2. Because the Baltic Sea is shallow (on average, 54 m), the volume of water is only 21,000 km^3. The modern hydrography of the basin reflects the pronounced continental influence: Fed by river discharge that drain the large catchment (1,700,000 km^2), the surplus of pre-cipitation over evaporation is around 1,800 m^3 s^{-1}. In consequence, the portions farthest from the ocean are low in salinity and resemble freshwater environments. In the transition area near the ocean, located in the Skagerrak and Kattegat (Fig. 1a), high salinity environments dominate and support eco-systems of marine origin. Because the density of fresh water is lower than that of salt water, the sea surface slopes upward to the NE (approximately 60 cm) and the fresh water runs down that slope. To compensate the outflow, salt water flows in through the Kattegat and spreads out at depth, filling a series of basins that have been carved out by the glaciers (Fig. 1b). Today, the boundary between salty deep waters and fresh surface waters waters slopes shallows toward the entrance; it has depths of <20 m in the Kattegat, 25-30 m in Arkona Basin, 60 in Bornholm Basin, and around 80 m in the Eastern and NW Gotland Basin. Circulation of water is driven by the density difference of water masses and by wind. Because the individual basins are small and wind changes direction within a few days, the Baltic Sea has no stable large-scale hydro-

From WEFER G, BERGER W, BEHRE K-E, JANSEN E (eds), 2002, *Climate Development and History of the North Atlantic Realm.* Springer-Verlag Berlin Heidelberg, pp 205-221

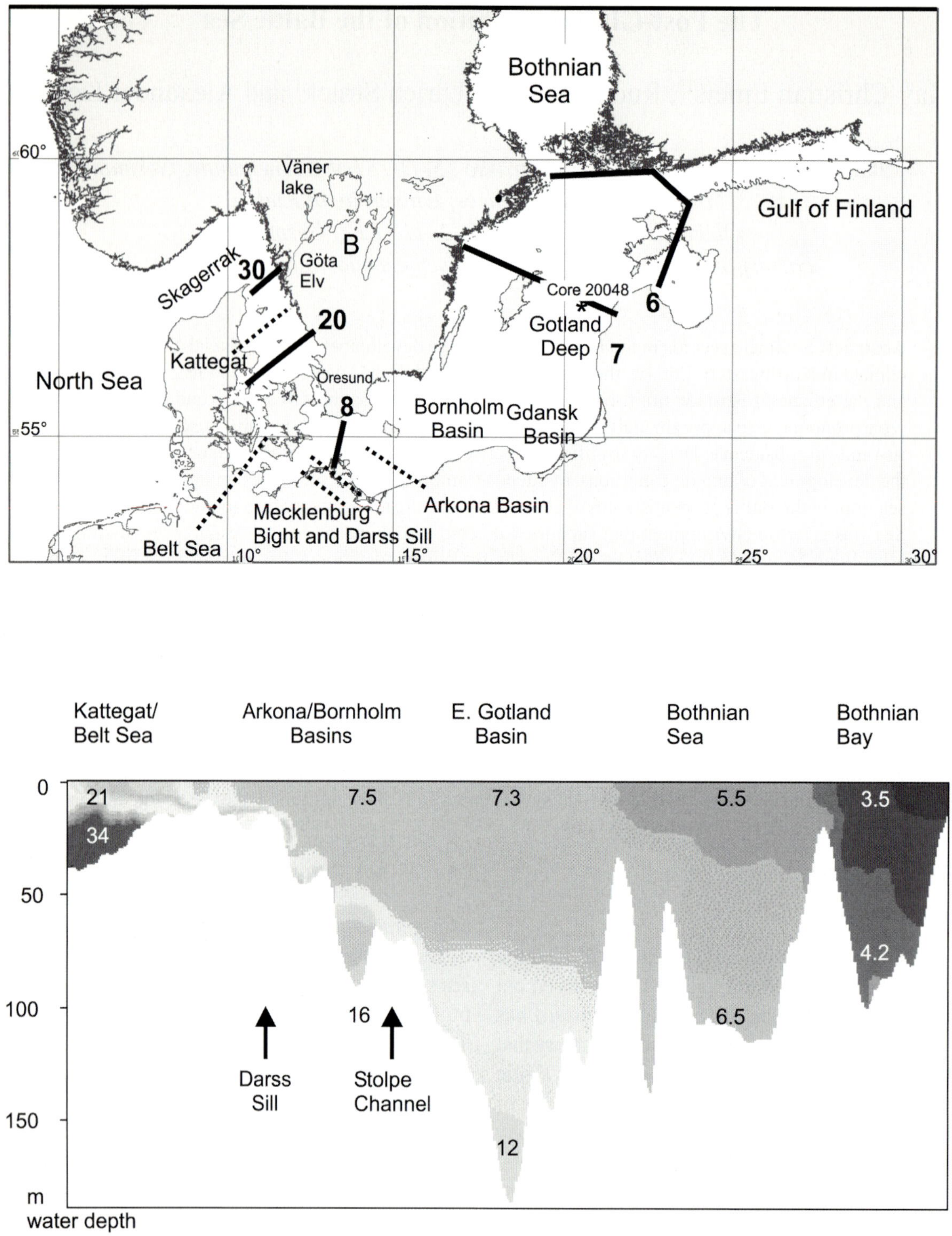

Fig. 1. (a) Map of the Baltic Sea area, giving modern shorelines, the modern salinity gradient in surface waters, and the location of Core 20048 in the Eastern Gotland Basin. The Billingen area is marked with a B. **(b)** The modern salinity gradient in the Baltic Sea along a transect from the Kattegat to the Bothnian Bay. Salinity values are given in psu.

dynamic circulation system comparable to the oceans. Its hydrographic features are transient and have a typical spatial extent of hundreds of kilometers and a temporal duration of days.

The vertical salinity gradient between deep and surface waters and the horizontal gradient between sub-basins is the single most important feature in determining biological assemblages, which have a minimum in species abundance in the salinity range from 5 to 10 psu. Their abundance rises sharply towards both the fresh water and the marine salinity range. The density contrast between inflowing salty waters and the low-salinity fresh water cap is also critical for the chemistry of the modern Gotland Basin (Fig. 1b). Anoxic conditions develop there over years to decades in deep water as a consequence of stable stratification, absence of bottom-water renewal, and oxygen demand by sinking organic matter. Any shift in the predominance of run-off versus deep-water inflow will shift the salinity gradient in the horizontal and will result in shifts of gradients in productivity, species composition, and sedimentary records of conditions in the water column.

Coasts of the Baltic Sea are either rocky (in the North, Northwest and Northeast) or sandy (in the South and Southeast). Freed of a thick glacial ice cap (up to 3,000 m thick) after the deglaciation, the continental lithosphere of Fennoscandia began to rise and continues to rise at rates of up to 9 cm per decade. The flow of mantle material into the rising region resulted in subsidence within the rim around that bulge (Fig. 2a). The rocky coasts to the North are on geologically very old crustal rocks that are rising, and new land is continuously being exposed. The southern rim of the modern Baltic Sea is subsiding and this creates space for accumulation of vast deposits of sand from cliff erosion and river loads.

Sediments on the modern floor mirror physical conditions in the water column. Wave energy is the main factor in determining sediment character, and the depositional environment is one where size sorting plays an overriding role. Shallow sea floors (to approximately 20 m water depth) do not accumulate fine-grained terrigenous materials delivered by rivers or biogenic material produced in the water column. The fine-grained material passes through the zone of shallow and agitated waters and accumulates in quiet deeper waters, usually below the halocline, where wave energy is dampened. The effect of water depth on sedimentation is a basic feature of any depositional environment and continuous records of Baltic Sea evolution thus can only be found in the clay and mud deposits of the basins. The sediment composition and facies show that the salinity levels changed considerably over the last 13,000 years. These changes are consequences of changes in global sea level (Fig. 2b), in sill depth and accessibility by the ocean, and of climate. Sediments under water and on land recorded all these processes, and have attracted generations of geologists studying the timing of ice recession and in-step formation of lakes in the glacial landscape, the isostatic rebound of the Fennoscandian shield freed from the load of the Scandinavian Ice Sheet, the formation of diverse coastal environments after the sea transgressed, and the timing of salinity changes in relation to global climatic events. A general sequence of stages in the development of the Baltic Sea is given in Fig. 3.

In recent years, interest has turned to the more subtle record of climatic variability encoded in the sediments. These influences became dominant after sea level rise leveled out and the isostatic adjust-ment in the transition area to the ocean subsided. Because the Baltic Sea watershed is strongly influenced by the atmospheric circulation of the North Atlantic Ocean (Rodwell et al. 1999), it acts as an amplifier for signals of changing hydrographic and climatic patterns there. The North Atlantic influence governs the water budget (Bergström and Carlsson 1994) and wind fields and thus determines the salt budget of the Baltic Sea (Lass and Matthäus 1996). The climatic influence may be dominant, but is only now being recognized as a major player in the Baltic Sea environment. In the 1980s anoxia greatly increased and so did the extent of anoxic sea floor, in reaction to a shoaling of the redox boundary. This was originally attributed to increased anthropogenic nutrient loading, increased biological production and an increased oxygen demand out-stripping the natural supply of oxygen by deep-water advection (Jonsson et al. 1990). Evidence is mounting, however, that anomalous climatic conditions

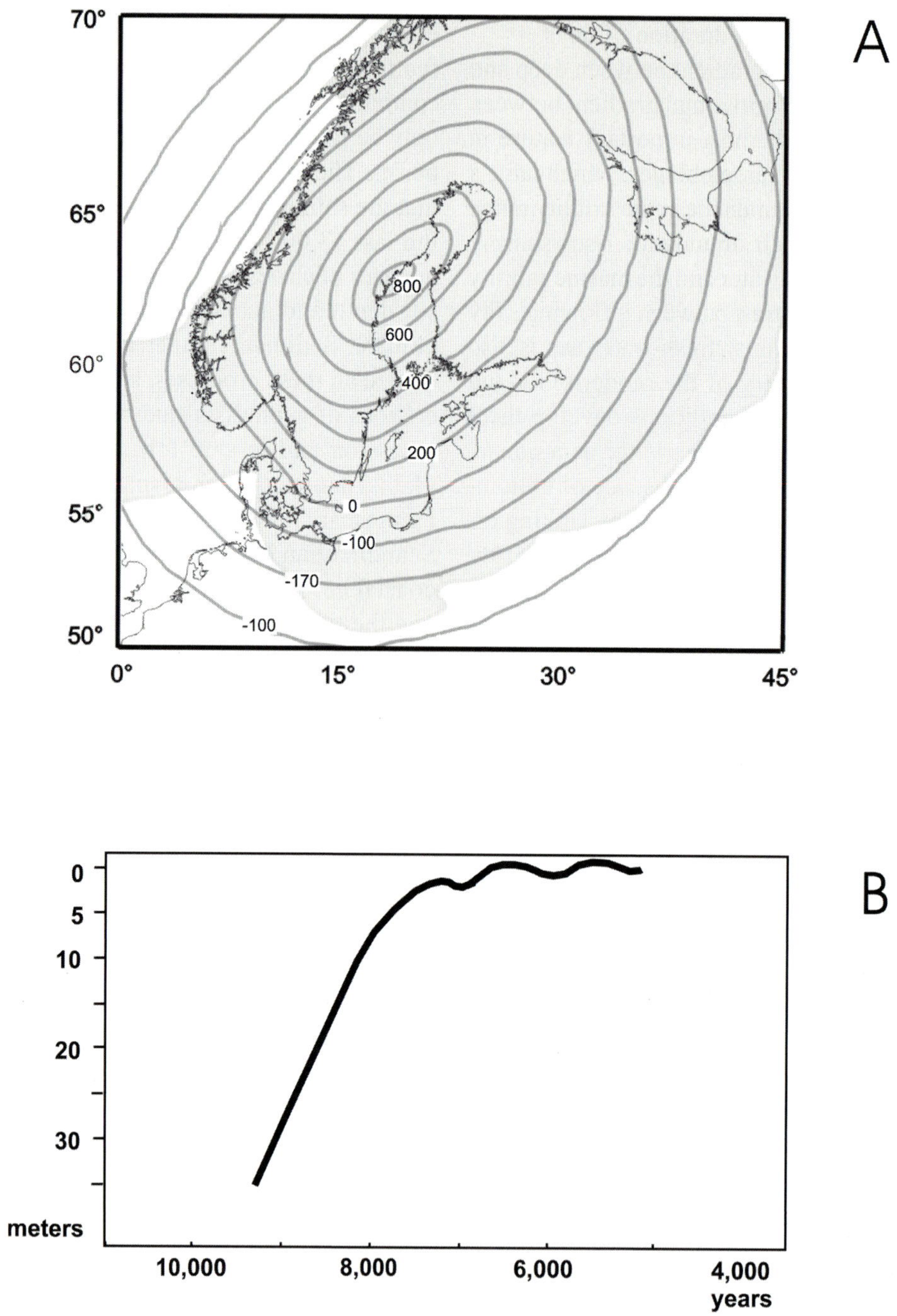

Fig. 2. (a) Approximate rates of crustal uplift (in mm per hundred years) and the approximate extent of ice during the last glaciation (grey) in NW Europe. **(b)** Sea level rise during the last ca. 10, 000 years.

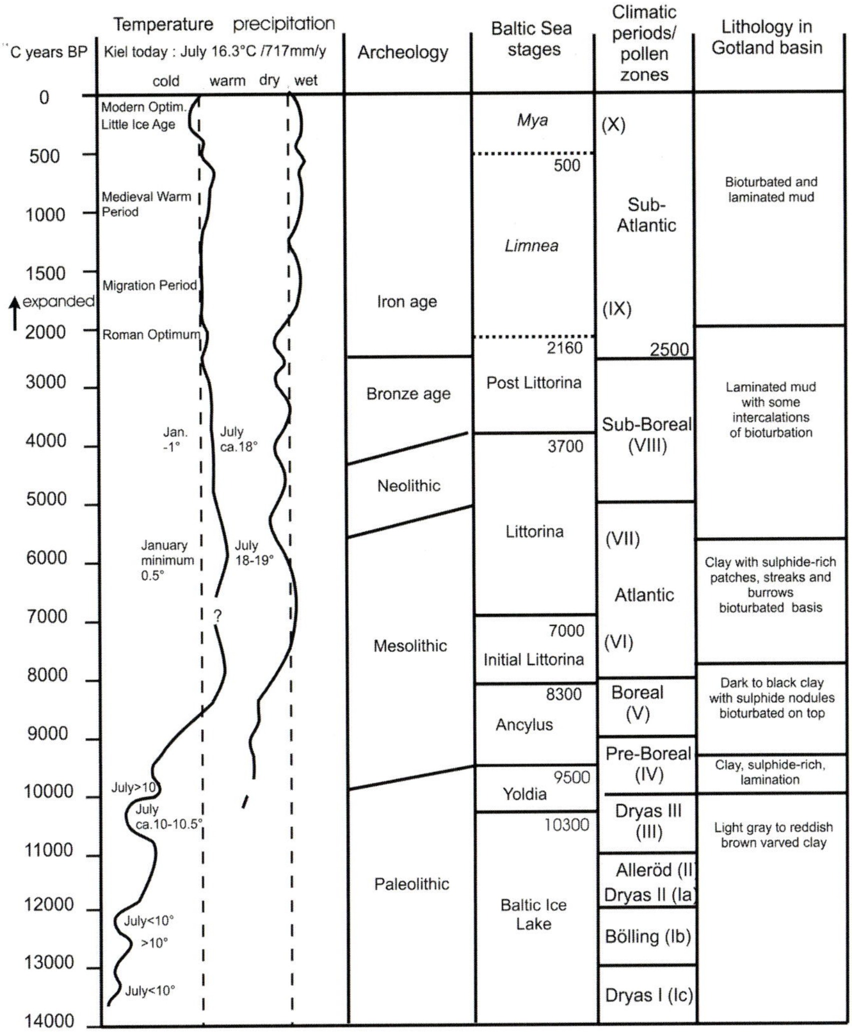

Fig. 3. Overview of chronology and environment in the evolution of the Baltic Sea. Modified after Overbeck (1975).

hindered inflow of salt water during the 1970s and 1980s and that natural causes therefore may in part have been responsible for environmental changes in the Baltic Sea. Similar changes (pre-dating human influence) are indicated by Holocene sediment intervals and suggest that considerable natural variations have always occurred in salinity and deep-water renewal of the Baltic Sea.

On the other hand, there is ample evidence that human influence is gaining in significance. Much of the history of the Baltic Sea was accompanied by human activity in the catchment regions and particularly near the coasts. Starting approximately with the cultural shift from the Mesolithic to the Neolithic and the onset of farming (approximately 5,000 years ago), the human impact on the Baltic Sea watershed has steadily increased. Since approximately the 100 years, increased industrialisation and eutrophication in the Baltic Sea area are unambiguously recorded in marine sediments. The challenge for scientists is to distinguish between natural and human impacts and to estimate the relative contributions to environmental variability.

In this contribution, we give a brief account on the development of the Baltic Sea during the late glacial and the Holocene and illustrate the changing environmental conditions with core images from the Gotland Deep, where a continuous and detailed diary of environmental conditions has been kept. Changing environmental conditions are accom-panied by changes in the biological environment, and we show how the biological environment reacted to salinity changes. All ages given in this text are uncorrected ^{14}C years, except the historical ages at the end of our account.

The Late Glacial Baltic Ice Lake (14,000 to 10,300 years)

The Scandinavian ice sheet began to retreat at around 14,000 (in the southern Baltic) to 13,000 yrs in the central Baltic, leaving a series of growing ice lakes in the glacially sculpted periglacial landscsape. The origin of the first of the Baltic Sea predecessors, the embryonic Baltic Ice Lake, was near the modern Arkona Basin between Mecklenburg and southern Sweden. By around 11,000 years, the ice margin had reached the vicinity of Stockholm (Björck et al. 1995). The growing ice lake was isolated from the Atlantic Ocean by a land bridge across the modern Danish Straits extending grow a barrier of rock and ice across Mount Billingen in south-central Sweden, and from the Arctic Ocean by ice over Finland. The ice lake surface was initially 7 meters below modern sea level and subsequently rose up to +28 m. This rise, caused by isostatic rebound, rapidly raised the northern shores of the Baltic until around 11,000 years ago the outlet of the lake — located in the Öresund area — was uplifted, damming the ice lake to increasingly higher levels.

The lake contained but few organisms and vegetation in the surrounding catchment area was sparse. The lake was cold and populated by few freshwater diatoms and haptophytes. Organic carbon content of the sediments from that time is very low. In the vicinity of ice, varved clays with dropstones and very little biogenic material were deposited (Fig. 4), in more distal areas the varves become thinner and less distinct, blending into homo-geneous clays. The intermittent lack of varves suggests that during the late ice lake stage, winter freezing was not a permanent feature in the Gotland basin, although ice persisted on land nearby.

Two drainage events of the Ice Lake occurred at Mt. Billingen in south-central Sweden and are attributed to the melting and later re-advance of glacial barriers there (Björck 1995). The first event occurred at 10,900 years, lowering the water level by 5 to as much as 15 m. The final drainage, triggered by the weakening ice giving way to 25 meters of water, occurred in distinct steps at around 10,300 years over a few years, and finally (but only shortly) established a connection between the Baltic Sea and the world ocean. In the marginal areas of the Baltic Ice Lake, the drainage events exposed the sediment surface to a highly dynamic shallow-water environment, indicated by distinct sandy layers in cores from the southwestern Arkona and Bornholm Basins. In the Gotland basin sediments and on raised lake bottoms on land, the event is marked by a colour change in the clays and by a change in sediment density, as is indicated by an increase in sound wave velocities (Fig. 5). At the time of drainage, the sills in the Danish Straits were still more than 20 m above modern sea level and the following short inflow of marine waters must have occurred across south-central Sweden. The final drainage of the Baltic Ice Lake (event Billingen 2 at 10,300 years) is coincident with the end of the Younger Dryas stadial and marks the transition to sediments of the Yoldia Sea/Ancylus Lake (Björck 1995). It also is the end of the glacial stages and coincides with rapid warming in the Baltic area and the beginning of a complex mosaic of transgressions and regressions of the ocean in the Baltic Sea area.

Yoldia Sea

The Yoldia Sea stage, which derives its name from marine macrofossils of the bivalve *Portlandia (Yoldia) arctica* found in land exposures in central Sweden, is the period after the final drainage of the Baltic Ice Lake and until a new barrier rose up to impede the flow of saline waters into the Baltic Sea. It brackets a time when the southern and eastern Baltic Sea — where eustatic sealevel rise exceeded isostatic lifting — experienced a transgression. At the same time the equilibrium line between uplift and sea-level rise moved northward (Björck 1995). The sedimentary deposits in the Gotland Basin do not indicate a drastic change after the relative sea level drop and sedimentation here and in many land exposures continued with varved sediments for at least 300 years until sediment

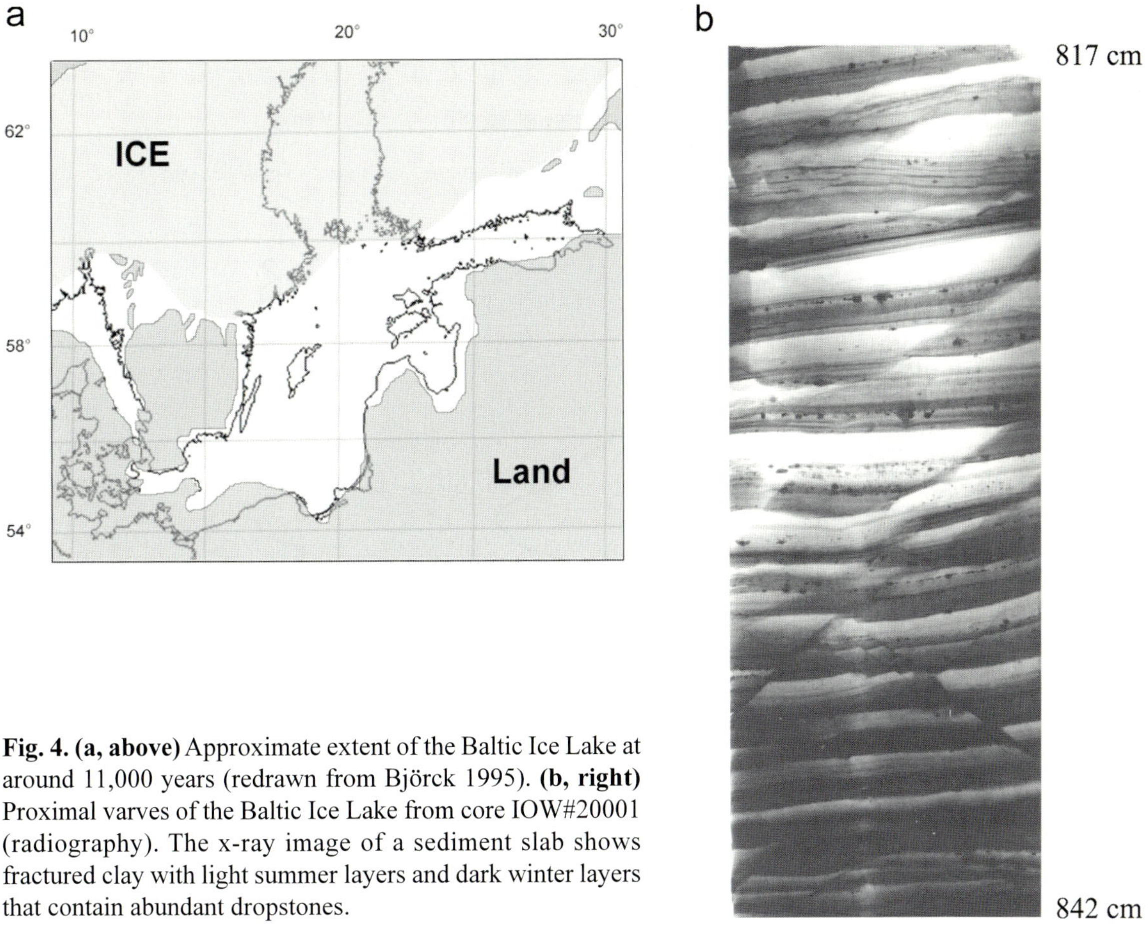

Fig. 4. (a, above) Approximate extent of the Baltic Ice Lake at around 11,000 years (redrawn from Björck 1995). **(b, right)** Proximal varves of the Baltic Ice Lake from core IOW#20001 (radiography). The x-ray image of a sediment slab shows fractured clay with light summer layers and dark winter layers that contain abundant dropstones.

facies, geochemistry, and fossil content mark a short spike of increased salinity (Fig. 5). In line with this sedimentary evidence, it has recently been proposed (Mörner 1995) that the classicial Yoldia stage is in fact divided into a basal lacustrine period which is characterised by indistinct varved sediments and lasting from 10,300 to 10,000 years, a brief marine phase, and a slow return to fresh conditions which lasted to around 9,500 years and blends into the following Ancylus Lake stage. Some decades after the final drainage of the Baltic Ice Lake (coincident with the colour change marked in Fig. 5), enough marine waters seeped over the shallow sill to fill the basin with salt water. Salt-water influence is reflected by iron sulphide mineralisation (giving rise to the high magnetic susceptibility), because marine waters contain abundant sulphate that is reduced by bacteria to sulphide in order to oxidise organic matter. Sulphide then reacts with iron and forms minerals such as the ferromagnetic mineral greigite, which grew within the sediment. Sediment properties also changed, resulting in increased sound velocity (Fig. 5). Organic carbon content is markedly increased in the marine interval (reached values up to 2%) and marks higher biological productivity; the diatom assemblages demonstrate that growth conditions for marine taxa were fairly good at elevated surface water salinity.

After the salinity spike, which has been correlated with the Pre-Boreal oscillation recognised in many terrestrial and marine climate records, diatom assemblages and sediment facies indicate a gradual freshening. The cooling associated with the Preboreal oscillation, which is well documented by the renewed onset of varving in the Gotland Basin

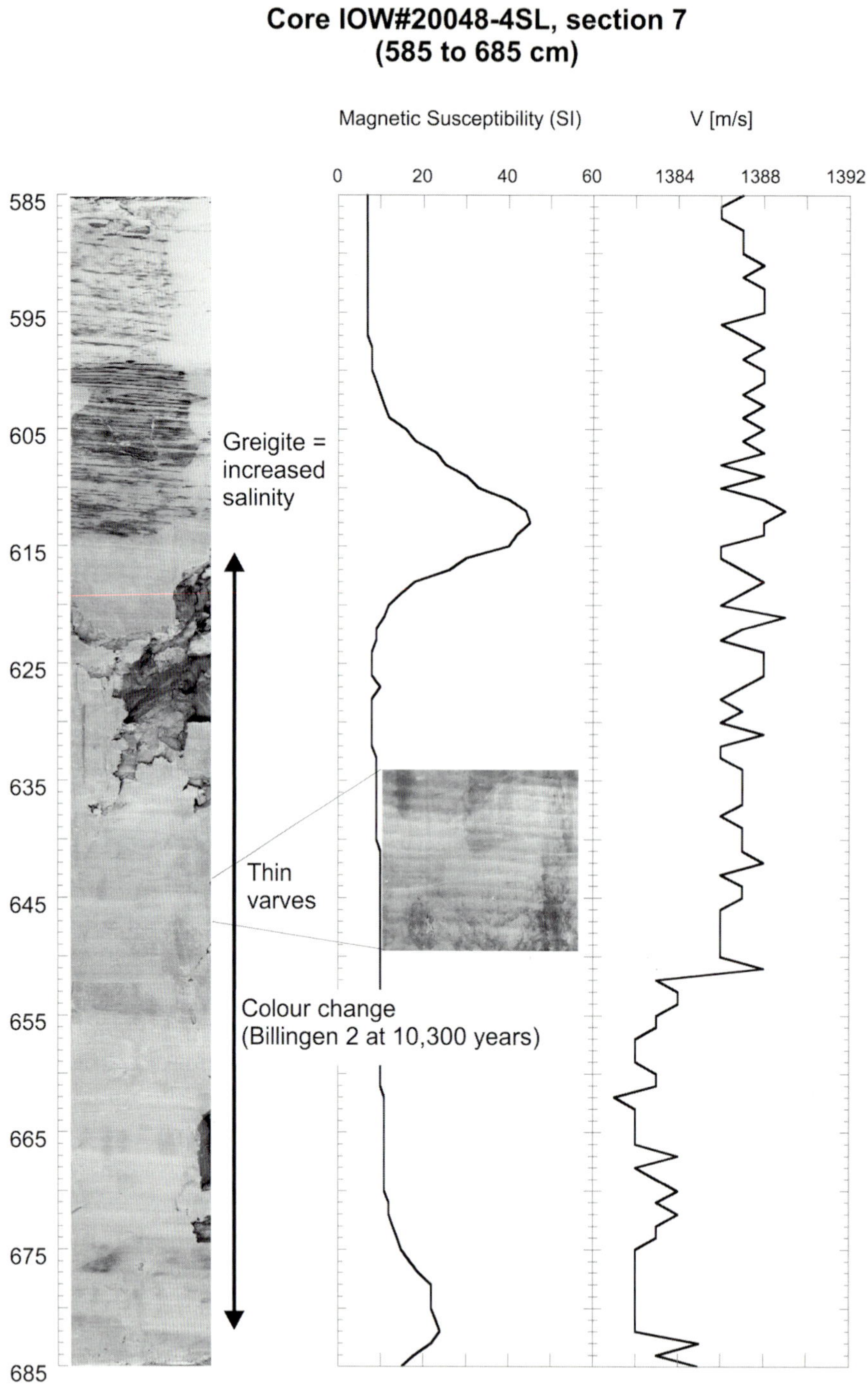

Fig. 5. Grey-scan of section 7 (585 to 685 cm core depth) of gravity core 20048-2 in the eastern Gotland Basin, which is the transition Baltic Ice Lake sediments to Yoldia sediments. Plotted are the magnetic susceptibility and p-wave velocities from multisensor core logs. The insert shows faintly varved sediments of the lacustrine (initial) Yoldia stage that suggest cooling (with complete winter freezing) of the central Baltic Sea coincident with the Pre-Boreal Oscillation cold event. The colour change marking the second outflow event at Mt. Billingen coincides with an increase in p-wave velocities. The marine spike is indicated by sulphide-stained sediments and increased magnetic susceptibility.

sediment record, was triggered by a large freshwater input to the Nordic Seas (Björck et al. 1996). The reduced surface water salinity there inhibited deep water formation in the North Atlantic, thus diminishing the release of heat and water vapor that fuel the west wind band. Mörner (Mörner 1980) describes a temperature deterioration of 0.5°C from lacustrine sediments from Gotland, while Norwegian glaciers re-advanced at approximately 3°C lower temperatures than modern (Nesje and Dahl 1993).

The Ancylus Lake Stage and the Transition to Brackish Conditions

The continued uplift of south-central Sweden again constricted the connection to the ocean and another phase of elevated basin level over sea level — the Ancylus Lake — occupied large areas of present-day land in Sweden, Finland, and the NE Baltic. In the southern Baltic, the lake level was at a maximum of –7 m (Kliewe and Jahnke 1991) between 9,300 to 8,700 years, and created a landscape of fjords that inundated the glacial relief previously occupied by smaller lakes. Its outflow was of order 15,000 to 20,000 m^3/s across the rising sills in Sweden and the lake level increased in pace with isostatic uplift (the water fall must have been quite a sight).

The global sea level rose at a rate of roughly 1.8m/100 years in the open ocean between 9,000 to 8,000 years (Fig. 2b). In the transition from the Skagerrak to the Baltic Sea, the increase in relative sea level was significantly higher, because the land subsided: The coasts were flooded at rates between 2.3 m per 100 years in the Storebaelt, 3.3 m in North Jutland, and up to 4 m in the Western Baltic Sea (Christensen et al. 1997). As pointed out by these authors, a rise of up to a meter in sea level over a human lifespan must have been an environmental change of enormous significance to the mesolithic inhabitants of the coastal area; rapid shoreline displacement and submergence of lowlands and islands would have required frequent relocation of habitations. Christensen et al. (1997) note that the transition from the Maglemose to the Kongemose culture in Denmark coincides with the peak in the transgression rate.

The sea increasingly inundated the Western Baltic transition area from around 9,000 years ago, but the global rise of sea level became much slower approximately 7,000 years ago because the large polar ice sheets had melted. The rapid climatic fluctuations that characterised the late glacial gave way to the more stable and benign climate of the Atlantic Period. In consequence, environmental change in the Baltic Sea area became more subdued. The link to the eastern basins was not yet firmly established. In consequence, the transition from lake sediments of the Ancylus stage and the brackish sediments of the early Littorina stage in the Gotland Basin was not abrupt. The marine influence is possibly seen as early as 8,900 years ago in the Bornholm Basin (Andrén et al. 2000), but most age datings of oldest sediments with marine influence cluster between 8,300 and 8,000 years ago in the western part of the Baltic Sea and between 8,350 to >7,860 years ago in the Central Baltic Sea. Based on radiometric dating, the transition from the Ancylus Lake stage to the brackish Littorina Stage lasted approximately 1,300 years.

Evidence for and Consequences of the Marine Transgression

During the transition time, the mid-water boundary between fresher and lighter surface water from rivers and saltier and denser bottom waters originating in the Kattegat slowly shoaled and an increasingly strong horizontal and vertical salinity gradient developed. The salinity inflow was forced by the increasingly stronger pressure head of the global sea level, and by continuing subsidence in the sill areas. By 7,000 years ago — judging from the geometry of oldest Littorina sediments in acoustic soundings of seafloor deposits — the halocline of the Gotland Basin had shoaled to its present-day water depth.

The transition from lacustrine to brackish conditions is best seen and is completely preserved in sediments of the Baltic proper (Fig. 6). During the transition, the bottom waters became increasingly salty. The influx of these waters brought nutrients into the basin which eventually rose to the surface and supported high biological productivity. The combination of high biological productivity and stable density stratification of the water column created anoxia in the bottom waters, because bacteria use the oxygen to decompose the sinking organic detritus. Laminated intervals depicted in Fig. 6 show that immediately after the transition from Ancylus to Littorina, oxygen-deficient bottom waters filled the deep Gotland Basin. The figure is a combination of a grey-scale scan (interval 388 to 488 cm in Core 20048 from the Central Gotland Basin), corresponding grey-scale values, and line plots of two variables (p-wave velocity and water content) that are physical properties of the sediments. Between 426 and 422 cm core depth (indicated by grey arrows and dated at approximately 7,600 years) and above black, sulphide-stained Ancylus clays, the lamination style changes from diffuse banding to sharp, millimetric laminae composed of alternating dark and light members.

The laminations are indicators for changes in the salt- and oxygen supply to the basin and they are pervasive in sediment cores from this locality and from neighbouring basins. In detail, the dense packages of laminations contain very thin layers of rhodochrosite, a manganese carbonate mineral. Its occurrence implies that the anoxic conditions must have been intermittently interrupted by oxic periods of several years´ duration, when new surges of oxygenated salt water reached the deep basins.

Geochemical, micropaleontological and isotopic measurements suggest that the marine influence had considerable consequences for the biological environment of the Baltic Sea (Fig. 7). The organic carbon accumulation rates in the sediment increased steadily from low values in the oligo-trophic (nutrient poor) Ancylus lake to moderately high values (up to 50 g m^{-2} a^{-1}) in the basal Littorina. Modern organic carbon accumulation rates at the same location are of order 10 g m^{-2} a^{-1}. The increase at the base of the brackish sediments was caused in part by higher biological production in surface waters and in part by better preservation of organic matter under anoxic conditions. The cause for enhanced biological production (which is also seen in microfossil assemblage composition) may have been linked to the development of anoxic conditions, as is suggested by the rapid change in the isotopic composition of the nutrient element nitrogen, denoted $\delta^{15}N$. The decrease of the $\delta^{15}N$ values from around 5‰ in sediments of the lake stage to 2‰ in the brackish sediments was caused by an addition of isotopically light nitrate, which is extracted from the atmosphere by nitrogen-fixing cyanobacteria. They thrive in environments that have excess phosphate in the surface waters, because only they are able to tap the unlimited atmospheric nitrogen pool. Upon sinking, the cyano-bacterial mats are oxidized by other bacteria and the nutrients released during this process fertilise the entire water column and bring the N:P ratios close to the optimal ratio of 16:1 that other algae need. This natural eutrophication by cyanobacterial nitrogen fixation is the likely reason for the increase in organic carbon concentrations during the transition, and may have been fuelled by phosphate leaking from the sediment under anoxic bottom water. The gradual decline of the $\delta^{15}N$ values mirrors the increasing area of seafloor becoming anoxic (providing increasing amounts of phosphate), and thus probably the rise of the halocline. Since brackish conditions were established in the Baltic Sea, the nitrogen isotope signature has remained light: cyanobacteria and, by inference, the presence of anoxic seafloors, have characterised the Baltic Sea ever since (Struck et al. 2000).

The change in salinity that led to the establishment of anoxic conditions is well imaged by increasing isotopic ratios of organic carbon ($\delta^{13}C_{TOC}$) and by changes in the diatom assemblages. Whereas the diatom assemblages found in sediments have been altered by opal dissolution during diagenesis and are not useful in determining absolute salinity levels, the $\delta^{13}C_{TOC}$ may be a more faithful proxy indicator of salinity. It reflects salinity effects on the amount of dissolved CO_2 and the different isotopic fractionation of dissolved CO_2 under freshwater and marine conditions. The organic carbon isotopic ratio $\delta^{13}C$ plotted in Fig. 7 increases from –29.5‰ in Ancylus Lake sediments to –25‰ in the Littorina. It co-varies with organic carbon concentrations, making it unlikely that much reworked terrestrial organic matter (which should be isotopically light) contributes to the increase in organic carbon. The increase begins approximately 15 cm into the transition stage (when $\delta^{15}N$ begins to decline) and coincides with a first increase in

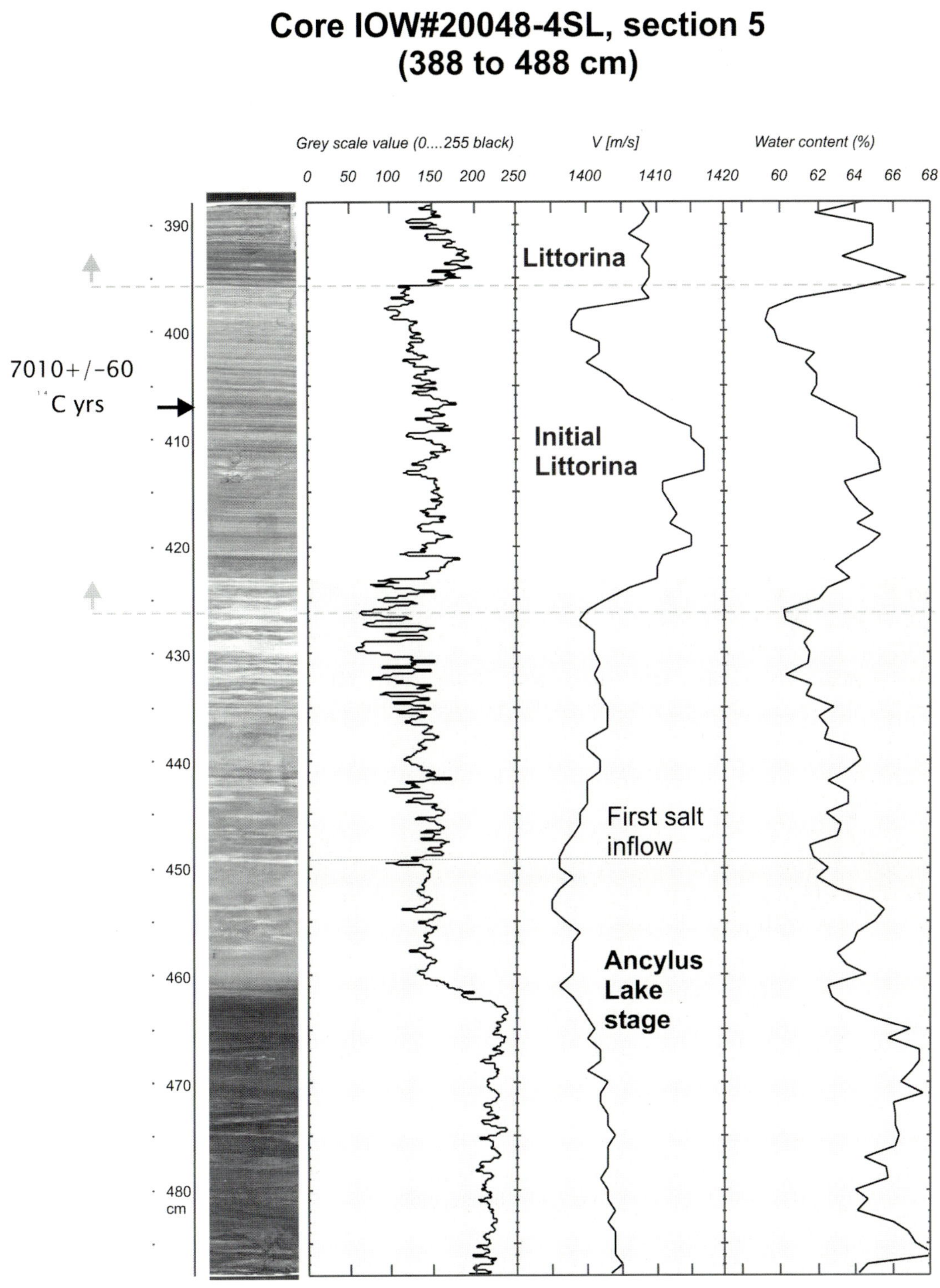

Fig. 6. Grey-scan of section 5 (388 to 488 cm core depth) of gravity core 20048-2 in the eastern Gotland Basin, which is the transition from sulphide-stained Ancylus Lake sediments to Littorina sediments. Plotted are the grey-scale values, p-wave velocities and water content from multisensor core logs. The age determination has been done on kastencore 20048-1 from the same locality; both cores are easily correlated by facies, water contents, and organic carbon concentrations.

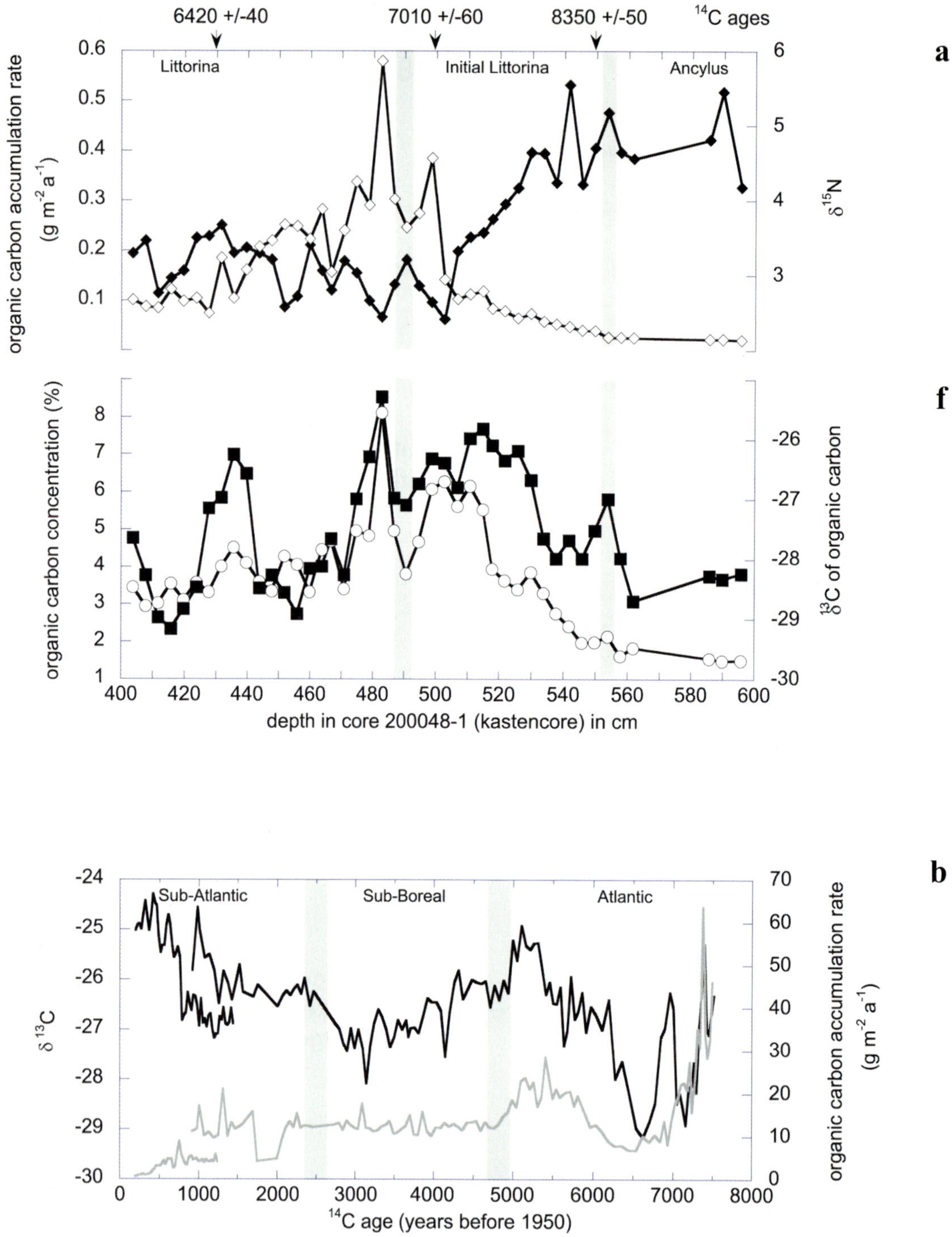

Fig. 7. (a) Plots of organic carbon accumulation rates (white rhombs) and δ^{15}N (black rhombs) in the upper panel, and δ^{13}C (black squares) and organic carbon concentrations (white circles) in the transition from lake sediments (Ancylus stage) to brackish sediments of the Littorina phase. The grey line at 490 cm marks the initial and massive occurrence of diatoms with saline affinities, which are rare to absent in the transition interval of the initial Littorina. **(b)** Values of δ^{13}C of organic carbon (a salinity proxy, black line) and organic carbon accumulation rate (a productivity proxy, grey line) plotted versus age in the Eastern Gotland Basin. Over the last 500 years, the central Baltic Sea was saltier, but less productive than at any time in the past.

Chaetoceros spores, which are a diatom species indicative of salty water. Highest values of $\delta^{13}C_{TOC}$ generally occur in the laminated intervals most of which have marine diatoms.

Littorina Sea Stage

The warmest postglacial times of the Atlantic period or Postglacial Climatic Optimum saw the emplacement of brackish conditions in much of the Baltic Sea; they characterise the environment until today. The transgression continued, at least in the western and southern Baltic Sea, until approximately 5,000 to 5,700 years ago (Christensen et al. 1997) with three main transgressive phases in the southern Baltic. During the sea level maximum, the so-called Littorina I Optimum, the area covered by water has been estimated as 26.5% larger than today, with a volume twice that of today around 6,500 years ago (Kliewe and Jahnke 1991). At the end of the transgression, the sea had inundated many of the glacial landforms of the southern rim of the Baltic Sea and, after a regressive phase around 5,000 to 4,500 years ago, the modern landforms with spits and lagoons developed, as did several generations of dunes. The main phase of Littorina in the southern Baltic Sea ended at approximately 5,700 years ago. On land, winter temperatures at 6,000 years ago have been estimated at 1-3°C warmer than present and summer warmth was greater, at reduced precipitation (50-250 mm less than present) (Overbeck 1975). Temperate deciduous forest reached 60°N latitude, occupying areas covered today by cool mixed and cool conifer forests to the North.

In the late stages of the transgression, when sea level was approximately 1-2 m below modern and salinity levels in the western Baltic Sea were high (Winn et al. 1998), the late Mesolithic Ertebölle culture replaced the foraging Kongemose culture. The Ertebölle sites on the coast are famous for thick layers of mollusc shells, and remains of fish, sea birds and sea mammals in so-called "kitchen middens". These artefacts attest to the fact that humans extensively exploited marine resources. Andersen (1995) points out that the role of the Sea was not purely that of a source for food, however but also for communication, transport, and social contacts. Oscillations of sea level in the area of modern Denmark continued and fluctuated around the modern level by 1 m (Christensen et al. 1997), but the environment had become stable enough to settle the coastal strip, with a convenient and practically inexhaustible source for food in the estuaries and silt-rich productive seafloors (Andersen 1995).

The Gotland Basin experienced quite substantial changes in salinity at the sea surface, in biological productivity, and in oxygen conditions in deep water during the Littorina stage. The sediment record indicates three periods of maximum salinity stratification in the basin — these are documented by bundles of laminated intervals each lasting a few hundred years. Three of them have been dated (Fig. 8), and correlate well with laminated intervals in other cores from the basin rim. Their diatom assemblages and $\delta^{13}C$ ratios are indicative of saline conditions at the sea surface, high diatom production and preservation, and all have high organic carbon concentrations. Around 5,000 years ago is the peak of lamination density, organic carbon accumulation rate, and $\delta^{13}C$ values.

In between the salinity maxima are intervals of reduced stratification (as indicated by non-laminated sediments in Fig. 8). They are low in organic carbon concentrations and $\delta^{13}C$ values, and the diatom assemblages are poor and dominated by frustules of typical freshwater species. Nested in the homogeneous layers are thinner bundles of laminae, which are clearly seen on the greyscale image; each of these minor bundles represents phases of several hundreds of years during which salinity stratification was stable and oxygen deficient conditions developed.

The end of the transgression and the transition from the Atlantic to the Sub-Boreal chronozone saw a phase of climatic deterioration. Lamb (1977) describes a period of cooling (The Piora Oscillation) at the end of the Atlantic period that lasted for approximately 500 years. It was the largest excursion to colder conditions since the Pre-Boreal Oscillation. The boundary between the Mesolithic and the Neolithic (transition from Ertebölle culture to Funnel Beaker culture, transition period lasting approximately 500 years) coincides with that cooling event, and the spreading of farming from the

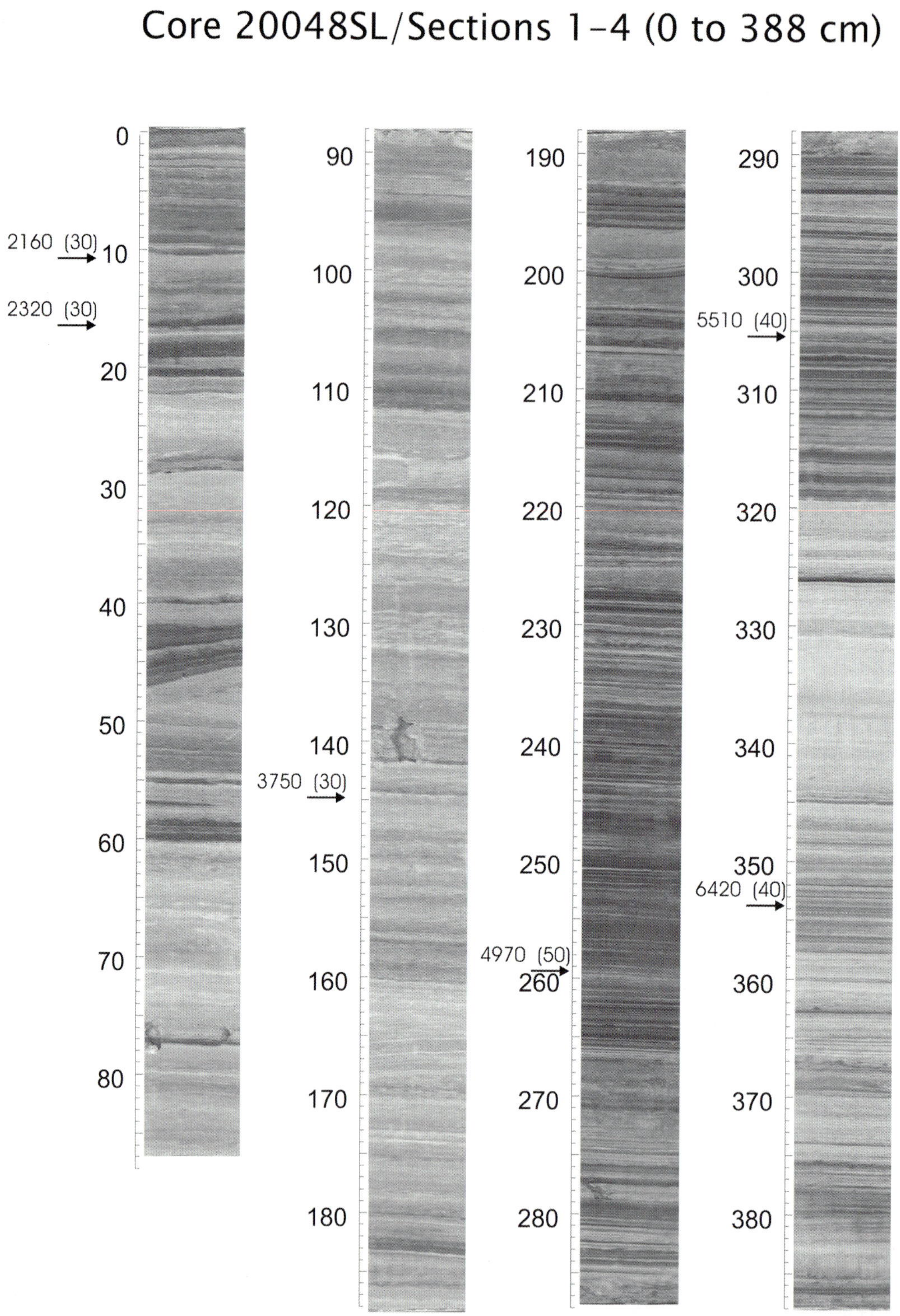

Fig. 8. Grey-scan of sections 1 to 4 of gravity core 20048-2 in the eastern Gotland Basin which covers the Littorina and part of the post-Littorina sequence. The age determination has been done on kastencore 20048-1 from the same locality.

Southeast to the area of the Baltic Sea (Zvelebil 1998). It has been argued that the transition may have been related to a decline in the marine ecosystem, possibly by a decrease in salinity that caused the mollusc population to decline (as seen by smaller shells towards the top of middens of the Ertebölle sites) (Christensen 1996). Aside from salinity, a reduction in the tidal range at the Atlantic/Preboreal transition has been postulated as the cause for declining mollusc populations (Christensen 1996). Whatever the reason, declining marine resources required the adoption of farming as a new method of food production The clearing of forests began, as documented by declining *Tilia* and increases in *Corylus* pollen and increasing in charcoal particles amounts of sediment records.

In the Gotland Basin, we see evidence for an abrupt decrease in the salinity proxy $\delta^{13}C$ at 5,000 years ago (Fig. 7b). The drop in $\delta^{13}C$ of organic carbon is accompanied by a significant decrease (by 50%) in the accumulation rate of organic carbon that indicates a drop in biological productivity. Frequency of lamination decreases in the cores from the Gotland Deep and is accompanied by an increasing abundance of species indicative of cool waters in the siliceous microfossil assemblage. Both observations suggest cooler conditions and decreasing tendency of stratification and anoxia. This trend lasts until approximately 3,700 years ago in the Gotland Basin, and to 2,800 years ago in the Bornholm Basin (Andrén et al. 2000).

Post-Littorina Environments

The transition from the Littorina to post-Littorina environments coincides roughly with the transition from the Neolithic to the Bronze Age in the Baltic Sea area and occurred in the upper part of the Sub-Boreal pollen zone. The Sub-Boreal was a period of warmth but fluctuating rainfall conditions. The boundary between the late Littorina stage and the beginning of the Post-Littorina stage (set at approximately 3,700 years ago) is diffuse in sediments of the Gotland Basin and is of different age in different basins of the Baltic Sea. The sediments are much lighter in colour, are bioturbated (as indicated by mottled core surface in Fig. 8) and lack distinct laminations. Siliceous microfossils indicative of marine conditions have disappeared from the record and $\delta^{13}C$ has a minimum at around 3,200 years ago (Fig. 7b). The freshening of the Baltic Sea is also recorded by rather low diatom amounts with high relative abundances of freshwater species. At that time the density contrast between deep and surface waters must have been reduced, permitting occasional or regular winter convection. Productivity in the water was stable over the entire period from 5,000 to 2,000 years ago, as indicated by uniform rates of organic carbon accumulation.

The following Sub-Atlantic period (from 2,500 years ago) was distinctly cooler (temperatures were about 2° lower than in the Sub-Boreal) and wetter, and the westerly air flow from the North Atlantic was increased (Lamb 1977). Glaciers advanced all over the world. Stuiver et al. (1995) postulate that reduced North Atlantic deepwater formation resulted in a reduction of surface-water transport by the Gulf Stream and caused a de-terioration of climate in western Europe and Greenland. The use of iron became widespread; it was mined locally in wetlands ("bog iron") and required large amounts of coal for smelting. Increasing humidity raised the level of ground water and favored growing crops on light soils that were easily tilled. During this period, the spread of agriculture advanced to the northern Baltic Sea area. Slash-and-burn cultivation reduced the primeval spruce forests, causing shifts in vegetation and pollen patterns, increases in land erosion, and in sedimentation rates of lakes. Land continued to rise out of the sea in the area around the Bothnian Sea, exposing fertile former sea floor areas that were settled and cultivated (Vuorela 1998).

The increasingly dark and laminated intervals in the core image above 60 cm (around 2,800 years ago) in the sediment record from the Gotland Deep (Fig. 8) are accompanied by a rise in $\delta^{13}C$. Also, diatom frustules of marine taxa again become common. This suggests that salinity has increased steadily since 3,000 years ago and reached a peak around 1,000 years ago. Differing from previous phases, when enhanced salinity caused an increase in biological production and organic carbon burial, the accumulation rate of organic carbon has decreased, albeit with some major fluctuations. From approximately 2,000 years ago to the present, productivity of the surface water apparently has been much lower than at any time in the older geological record, shedding some doubt on claims that the Baltic Sea is being fertilised by human activity.

Climatic fluctuations over the last 2,000 years were short-lived and subtle with reduced temperature fluctuations of less than 2°C (Briffa et al. 1990). The climatic improvement of the Roman Optimum ended the cold and wet conditions of the late Sub-Atlantic around 2,300 years ago and may have caused a decrease in precipitation and westerly winds in the Baltic Sea area. The long sediment record from the Gotland Basin is less helpful than historical climate records in detecting effects of the rapidly changing cold and warm climatic climatic phases on the Baltic Sea, in part because of dating uncertainties. We know from more highly resolved and dated records that sediments in the Gotland Basin are homogeneous, contain only few and weak rhodochrosite layers since approximately AD 1300 to around AD 1950 (Neumann et al. 1997) and host a diatom flora that indicates relatively cool conditions (Andrén et al. 2000). This is the period of the Little Ice Age, culminating in the period from 1675 to 1710 AD. Winter ice reconstructions mark the periods from 1540 to 1580, 1600 to 1640, and 1660 to 1700 AD as very cold, briefly interrupted by a phase of warm winters from 1700 to 1760 AD, and cold winters again from 1780 to 1860 AD (Koslowski and Glaser 1999).

The lack of laminations and the lack of rhodochrosite enrichment in sediments of that age are indicators of a relatively well oxygenated environment during this cold periods, and in our current understanding, the Little Ice Age (as a representative of colder climatic phases in the late Holocene) has been a period of enhanced salinity inflow into the Gotland Deep. This interpretation is consistent with the diatom flora, which marks cooler conditions and higher salinity (Andrén et al. 1999), and with the concept that low sea-level in the Baltic Sea is conducive for enhanced salt-water import to the Gotland Basin (Lass and Matthäus 1996). The same meteorological condition - a strong high-pressure zone over Scandinavia nearing Iceland during winter permitting cold arctic air to flow in from the East-enhances ice buildup in the western Baltic Sea and empties the Baltic Sea, pushing the salt gradient farther into the Baltic Sea. It blocks the westerly air flow from the North Atlantic and results in lower than average moisture transport and river runoff in the Baltic Sea catchment.

Man´s influence had been growing over the last millennia and is most clearly seen since approximately 100 years ago. Diatom records from near-shore environments show an increase in turbidity of the water column that results in a change from light-dependent benthic taxa to pelagic forms (Andrén et al. 1999). Approximately at the same time, the accumulation rate of heavy metals in the basin sediments increased significantly (Emeis et al. 1998). The three major depositional basins (the Gdansk, Bornholm and Eastern Gotland basins) of the Baltic Proper, which together account for >50% of the depositional areas in the entire Baltic Sea, have accumulated increased amounts of sediment, organic carbon, nitrogen and phosphorus during the last 50 years, partly buffering a four-fold increase in fertiliser consumption in the catchment (Emeis et al. 2000). Diatom assemblages have adjusted to increased nutrient abundance and turbidity (Andrén et al. 1999), and the increased use of fertilisers has shifted the $\delta^{15}N$ values (Struck et al. 2000).

Scientific and public interest in reconstructions of climatic variations on time scales of decades and their effects on the Baltic Sea ecosystem is mounting and has stimulated research into more highly resolved records than that presented here. It is likely that the impact of the human population in the Baltic Sea area will increase and will mask such natural variations, because even wind patterns and thus the physical motor of the Baltic Sea may be affected by global change. Future research will require combined observation and experiments with numerical models of atmospheric and hydrographic circulation and ecosystem responses. This will help to understand the physical mechanisms that determine the variability of the Baltic Sea over long-time scales. If successful, that information will then provide the possibility to assess and scale anthropogenic influence on the Baltic Sea, which may help in the sustainable development of this region.

References

Andersen SH (1995) Coastal adaptation and marine expoitation in Late Mesolithic Denmark - with special emphasis on the Limfjord region. In: Fisher A (ed) Man and The Sea in the Mesolithic. Oxbow Books, Oxford, pp 41-66

Andrén E, Shimmield G, Brand T (1999) Environmental changes of the last three centuries indicated by siliceous microfossil records from the southwestern Baltic Sea. The Holocene 9:25-38

Andrén E, Andrén T, Sohlenius G (2000) The Holocene history of the southwestern Baltic Sea as reflected in a sediment core from the Bornholm Basin. Boreas 29:233-250

Bergström S and Carlsson S (1994) River runoff to the Baltic Sea. Ambio 23:280-287

Björck S (1995) A review of the history of the Baltic Sea, 13.0 - 8.0 ka BP. Quaternary International 27:19-40

Björck S, Kromer B, Johnsen S, Bennike O, Hammarlund D, Lemdahl G, Possnert G, Rasmussen TL, Wohlfahrt B, Hammer CU (1996) Synchronized terrestrial-atmospheric deglacial records around the North Atlantic. Science 274:1155-1160

Björck S, Wohlfart B and Possnert G (1995) ^{14}C AMS measurements from the Late Weichselian part of the Swedish Time Scale. Quaternary International 27:11-18

Briffa KR, Bartholin TS, Eckstein D, Jones PD, Karlén W, Schweingruber FH and Zetterberg P (1990) A 1,400-year tree-ring record of summer temperatures in Fennoscandia. Nature 346:434-439

Christensen C (1996) The Littorina transgressions in Denmark. In: Fischer A (ed) Man and Sea in the Mesolithics, Coastal Settlements above and below Present Sea Level. Oxbow Books, Oxford, pp 15-22

Christensen C, Fischer A, Mathiassen DR (1997) The great sea rise in the Storebaelt. In: Pedersen L, Fischer A, Aaby B (eds) The Danish Storebaelt since the Ice Age - Man, Sea and Forest. A/S Storebaelt Fixed Link, Copenhagen. pp 45-54

Emeis K-C, Neumann T, Endler R, Struck U, Kunzendorf H, Christiansen C (1998) Geochemical records of sediments in the Gotland Basin—products of sediment dynamics in a not-so-stagnant anoxic basin? Applied Geochemistry 13:349-358

Emeis K-C, Struck U, Leipe T, Pollehne F, Kunzendorf H, Christiansen C (2000) Changes in the burial rates and C:N:P ratios in Baltic Sea sediments over the last 150 years. Marine Geology 167(1-2):43-59

Jonsson P, Carman R, Wulff F (1990) Laminated sediments in the Baltic - A tool for evaluating nutrient mass balances. Ambio 19:152-158

Kliewe H, Jahnke W (1991) Holozäner Küstenausgleich im südlichen Ostseegebiet unter besonderer Berücksichtigung der Boddenausgleichsküste Vorpommerns. Petermanns Geographische Mitt. 1/1991:1-14

Koslowski G, Glaser R (1999) Variations in reconstructed ice winter severity in the Western Baltic from 1501 to 1995, and their implications for the North Atlantic Oscillation. Climate Change 41:175-191

Lamb HH (1977) Climatic History and the Future. Princeton University Press, Princeton, pp 1-835

Lass HU, Matthäus W (1996) On temporal wind variations forcing salt water inflows into the Baltic Sea. Tellus 48A:663-671

Mörner NA (1980) A 10,700 years′ paleotemperature record from Gotland and Pleistocene/Holocene boundary events in Sweden. Boreas 9:283-287

Mörner N-A (1995) The Baltic Ice Lake - Yoldia Sea transition. Quaternary International 27:95-98

Nesje A, Dahl SO (1993) Lateglacial and Holocene glacier fluctuations and climate variations in western Norway: A review. Quaternary Science Reviews 12:255-261

Neumann T, Christiansen C, Clasen S, Emeis K-C, Kunzendorf H (1997) Geochemical records of saltwater inflows into the deep basins of the Baltic Sea. Cont Shelf Res 17:95-115

Overbeck F (1975) Botanisch-geologische Moorkunde. Wachholz, Neumünster, pp 1-719

Rodwell MJ, Rowell DP, Folland CK (1999) Oceanic forcing of the wintertime North Atlantic Oscillation and European climate. Nature 398:320-323

Struck U, Emeis KC, Voss M, Christiansen CC, Kunzendorf H (2000) Records of Baltic Sea eutrophication in δ^{13}C and δ^{15}N of sedimentary organic matter. Marine Geology 164:157-171

Stuiver M, Grootes PM and Braziunas TF (1995) The GISP 2 δ^{18}O climate record of the past 16,500 years and the role of the sun, ocean, and volcanoes. Quaternary Res 44:341-354

Vuorela I (1998) The transition to farming in Southern Finland. In: Zvelebil, M, Domanska, L and Dennell, R (eds) Harvesting the Sea, Farming the Forest. Sheffield Academic Publishers, Sheffield, pp 175-180

Winn K, Erlenkeuser H, Nordberg K, Gustafsson M (1998) Paleohydrography of the Great Belt, Denmark, during the Littorina Transgression: the isotope signal. Meyniana 50:237-251

Zvelebil M (1998) Agricultural frontiers, neolithic origins, and the transition to farming in the Baltic Basin. In: Zvelebil M, Domanska L, Dennell R (eds) Harvesting the Sea, Farming the Forest. Sheffield Academic Publishers, Sheffield, pp 9-27

Holocene Climatic History of Northern Europe as Recorded by Vegetation Changes: Possible Influences Upon Human Activity

Brian Huntley

Environmental Research Centre, University of Durham,
South Road, UK-Durham DH1 3LE, United Kingdom
e-mail: Brian.Huntley@durham.ac.uk

Abstract: The Holocene climate of northern Europe has exhibited changes and fluctuations at a variety of characteristic time scales. At the longest time scale of the Holocene as a whole a variety of 'slow' forcing factors has operated leading to long-term trends that in turn have elicited long-term and continental-scale responses in the structure and composition of the vegetation. At millennial time scales fluctuations in 'intermediate' forcing factors have resulted in climatic fluctuations that in turn have elicited vegetation responses on landscape to regional scales. At the shortest sub-centennial to annual time scales fluctuations in 'fast' forcing factors result in individual to plot-scale responses of vegetation. Brief accounts are given below of the principal forcing factors operating at each of these three time scales, of the nature of the resulting climate changes and of the vegetation responses that these changes elicited.

Introduction

Whereas today many human activities are carried on without regard to climatic influences, or even with strenuous efforts made to overcome unfavourable climate conditions, prehistoric humans had little option but to respond to climate and vegetation changes in much the same manner as all of the other components of the natural ecosystems of which they formed a part. Thus the major changes associated with the transition to the Holocene almost certainly played a major role by selecting the adaptations that resulted in the development of settled agriculture, although not at that time in northern Europe. The slow and intermediate changes during the Holocene almost certainly impacted upon human activities throughout almost all of that time; even some of the faster fluctuations are likely to have had noticeable impacts. The likely nature of some of these impacts, and how humans may have responded, will be discussed. Some general conclusions will then be presented as to the likely extent to which the Holocene history of human settlement and activity in northern Europe has been shaped, at least in part, by the continuing changes in the natural environment.

'Slow' Forcing and Continental-Scale Vegetation Responses

Over the time span of the Holocene as a whole the predominant forcing factor leading to long-term climatic trends has been the change in the seasonal pattern of insolation (Fig. 1). Whereas summer insolation in northern mid-latitudes peaked during the last deglaciation and has since that time progressively declined to its present level, winter insolation was at a minimum during deglaciation and has subsequently increased steadily to reach its present value (Berger 1978). Globally, although with its strongest influence perhaps being felt in the North Atlantic region, ice volume was a secondary but important long-term forcing factor during the early Holocene (Fig. 1). Although global ice volume decreased most rapidly during the deglaciation prior to the transition to the Holocene at 10,000 ^{14}C yr BP (ca. 9400 BC), substantial volumes of ice persisted into the Holocene covering extensive areas of Scandinavia and especially of north-eastern North America, the latter for several millennia. The third 'slow' forcing resulted from the combination of the eustatic sea-level rise (Fig. 1), itself a consequence of the melting of the ice sheets, and of the isostatic

From WEFER G, BERGER W, BEHRE K-E, JANSEN E (eds), 2002, *Climate Development and History of the North Atlantic Realm.* Springer-Verlag Berlin Heidelberg, pp 223-232

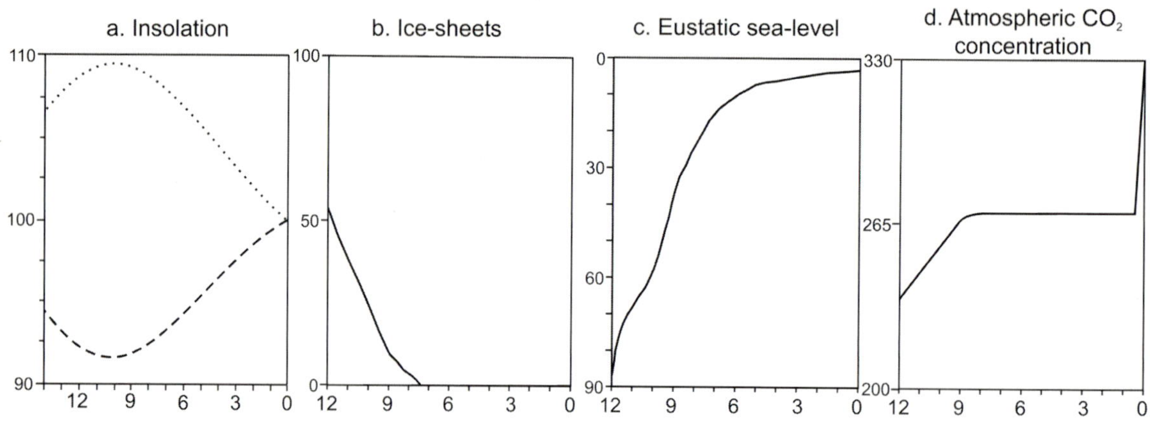

Fig. 1. 'Slow' forcing factors. **(a)** Insolation – Dotted line shows July insolation and dashed line January insolation (% of present value) (Berger 1978). **(b)** Ice-sheets – global ice-sheet extent relative to last glacial maximum (100%) and present (0%) (Kutzbach and Ruddiman 1993). **(c)** Eustatic sea-level – plotted in metres below modern sea-level (Fairbanks 1989). **(d)** Atmospheric CO_2 concentration – schematic representation (vertical axis ppmv) (Kutzbach and Ruddiman 1993). Note: horizontal axes all span an equivalent period of time, although that of **(a)** is in thousands of calendar years before present whereas those of **(b)**, **(c)** and **(d)** are in thousands of ^{14}C years before present.

rebound of land areas that had subsided during the preceding glacial stage as a consequence of ice loading onto continental areas. The interplay of these two resulted in changes in the arrangement of land versus ocean; such changes generally were of principally regional significance but were of particular relevance to northern Europe. The atmospheric concentration of carbon dioxide has also acted as a 'slow' forcing factor, although the changes in concentration were, like the changes in sea-level, intimately linked to other changes in the global system as part of a complex system of feedbacks. Good overall accounts of the Holocene history of these forcing factors can be found in various textbooks (see e.g. Lowe and Walker 1997; Roberts 1998)

Acting in concert, these four 'slow' forcing factors have resulted in a continuous trend of change in the climate of northern Europe during the Holocene. These climate changes have in turn resulted in continuous changes in the distribution of plant and animal species across the continent, and hence in changes in the overall vegetation patterns on the continent (Fig. 2). The following are amongst the most notable features of the sequence of maps of reconstructed biomes (Prentice et al. 1996) shown in Fig. 2:

- At 10,000 ^{14}C yr BP (ca. 9400 BC), forest cover was extensive only in mainland Europe, whereas most of the British Isles and Fennoscandia was occupied by treeless tundra-like vegetation.
- Much of the area occupied by forest biomes at 10,000 ^{14}C yr BP (ca. 9400 BC) had vegetation that is either absent or uncommon in Europe at the present day, notably 'cold mixed forest' and 'cold deciduous forest' (for descriptions of the biomes see Prentice et al. 1992; Prentice et al. 1996).
- Although 'temperate deciduous forest' was extensive by 8000 ^{14}C yr BP (ca. 6950 BC), the forest patterns were quite different from those of the present day; 'taiga' was principally found only around the Urals, whereas Fennoscandia was occupied mainly by 'cold mixed', 'cool conifer' and 'cold deciduous' forests.
- The extent of 'temperate deciduous forest' was greatest in the mid-Holocene, ca. 6000 ^{14}C yr BP (ca. 4850 BC); since that time it has been reduced in extent along its northern, eastern and southern limits. 'Taiga' has become extensive

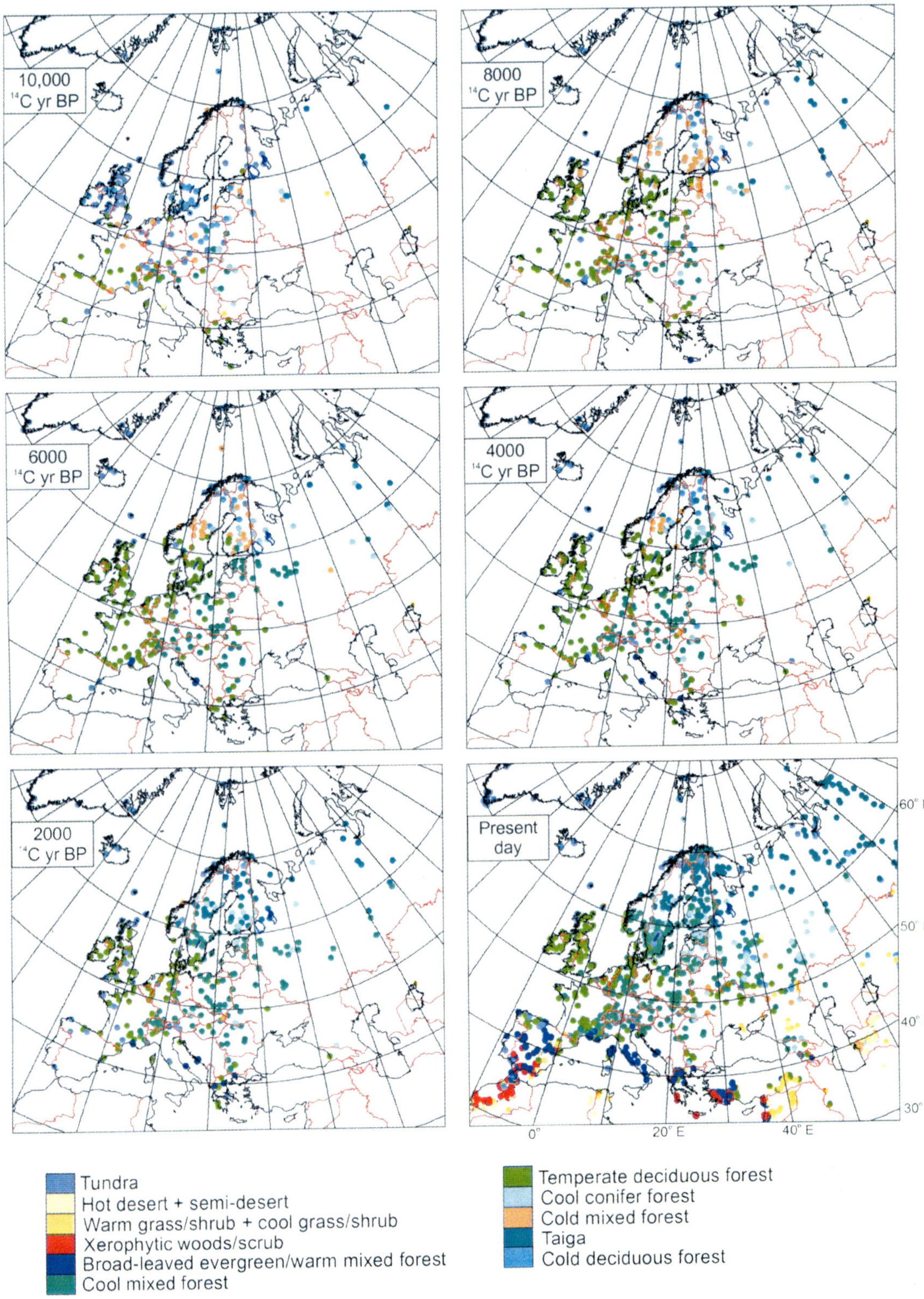

Fig. 2. Reconstructed biomes, based upon pollen data, for Europe. Maps are plotted at two thousand year intervals throughout the Holocene using data compiled by Huntley and Birks (1983) and Peterson (1983) and applying the 'biomisation' technique of Prentice et al. (1996). The dataset used for the map for the present day is that used by Prentice et al. (1996).

in Fennoscandia since the mid-Holocene, its expansion westwards being paralleled by the expansion of 'cool mixed forest' in eastern Europe and in the Alpine region.

- Although 'broad-leaved evergreen/warm mixed forest' has been present locally in southern Europe since the early Holocene, it has expanded considerably in the late Holocene, replacing 'temperate deciduous forest' in many parts of southern Europe.

- The expansion of treeless grassland and heathland vegetation in western Europe since the mid-Holocene is reflected by an increasing incidence of misclassification of pollen spectra from these regions as representing the 'tundra' biome.

These vegetation changes reflect independent changes in three principal aspects of the climate, namely:

1. annual temperature sum — the principal determinant of the latitudinal limits of forest growth and of the limits for the growth of important tree functional types such as temperate summergreen trees (e.g. oaks – *Quercus* spp.) that predominate in the temperate deciduous forest biome (for an account of the characteristics of the plant functional types comprising the various biomes see Prentice et al. 1992, 1996);
2. minimum temperature — the principal determinant of the eastward longitudinal or latitudinal limits for tree functional types such as the temperate summergreen or warm temperate evergreen trees (e.g. olive – *Olea europaea*) respectively; and
3. moisture availability — the principal longitudinal determinant of the limit of mid-latitude forests as the continental interior is approached and the limit upon the growth of mesic tree functional types such as temperate summergreen trees in regions of seasonal aridity in southern Europe that today are characterised by a Mediterranean climate type.

Thus at the outset of the Holocene northern Europe was characterised by low annual temperature sums and low winter temperature minima that together excluded trees from most of the British Isles and Fennoscandia. Thereafter, however, both the annual temperature sum and the winter temperature minima rose to values in excess of modern values throughout most of northern Europe, enabling forest to occupy most of Europe during the mid-Holocene and the temperate deciduous forest biome to extend both further eastwards and northwards than it does at the present day (Fig. 2). In terms of moisture availability, the indications are that northern Europe may have been somewhat drier than today whereas northern mainland Europe and the area around the southern North Sea was perhaps somewhat moister than at present (Huntley and Prentice 1993). These early to mid-Holocene changes cannot be accounted for by the principal insolation forcing alone, but reflect also the modifying influences of the ice-sheet and sea-level changes. Thus whereas summer insolation was declining toward the mid-Holocene, ice sheet extent was also declining to its minimum, leading to maximum summer warmth in the mid-Holocene rather than the early Holocene (Kutzbach et al. 1993). The relatively mild winter conditions of the early and mid-Holocene apparently conflict with the lower winter insolation of that time. However, they can be accounted for by the seasonal peak in ocean surface water temperatures lagging the summer peak in insolation, because of the high thermal mass of water, so that the oceans to the west and north of Europe act as a 'storage heater' that releases warmth during the winter months, so warming the adjacent continent. This effect was amplified during the early Holocene because of the higher input of warmth to the oceans during the summer as a consequence of the higher summer insolation at that time (Kutzbach and Gallimore 1988). In addition, it is likely that the greater northward extent of the forests provided a further amplifying feedback through the altered albedo and other land surface characteristics compared to areas today occupied by tundra (Foley et al. 1994).

Since the mid-Holocene, conditions have become generally more oceanic in north-western Europe, growing season warmth declining towards its present value and conditions becoming generally moister. This has been reflected in the progressive paludification of many north-west European landscapes with extensive development of blanket mires in upland areas. Anomalously, in

Fennoscandia especially, this trend has also been accompanied by cooler winter temperatures, and associated with this probably a more reliable and persistent snow cover. This results from the reduced summer heat storage in the oceans and consequent reduced winter release of warmth in the Norwegian and Barents Seas. The vegetation of northern Europe has responded with a progressive westward expansion of the distribution of *Picea abies* (Norway Spruce), in particular, and hence of the 'taiga' and 'cool mixed forest' biomes with which it is associated (Fig. 2).

Southern Europe provides an interesting contrast during the later Holocene, with the expansion of taxa sensitive to winter cold apparently associated with milder winter conditions in that region, and the parallel increase in drought-tolerant sclerophyllous taxa reflecting the intensification of the Mediterranean climate regime with its characteristic summer drought period. Although this latter change once again seems initially to be anomalous and contrary to the decline in summer insolation, it appears likely to reflect a reduced intensity of the local hydrological cycle in the Mediterranean region. This in turn is explicable as resulting from a reduction in the generation of local storm activity in summer that in turn would be a consequence of the reduced land–sea thermal contrast that resulted from the decreased summer insolation intensity.

'Intermediate' Forcing and Regional- to Landscape-Scale Vegetation Responses

Superimposed upon the overall long-term trends in climate and vegetation that have resulted from the operation of the various 'slow' forcing factors, there have been millennial-scale climate fluctuations throughout the Holocene that have also elicited vegetation responses at least at landscape scales and, in some cases, at regional scales. The principal well-understood candidate 'intermediate' forcing is that resulting from relatively low frequency periodic fluctuations in solar output. That such fluctuations occur is perhaps best evidenced by the fluctuations in the atmospheric abundance of ^{14}C (Fig. 3a) (Stuiver and Reimer 1993), a radioisotope of cosmogenic origin that is produced in the upper atmosphere at a rate that is dependent upon the flux of cosmic rays and hence, in turn, upon solar activity and the strength of the 'solar wind'. However, the magnitude of these fluctuations, at ca. 1% variation in insolation received at the top of the atmosphere, is such that they cannot alone account for the observed magnitude of climatic variations, estimated at 1-3°C shift in annual temperature in some regions (Lamb 1982); some form of amplifying mechanism is required, although what the nature of such an amplifier might be remains the subject of ongoing debate.

Notwithstanding the difficulty in demonstrating the causal factor(s) and associated mechanisms, there is abundant evidence of millennial-scale climate fluctuations during the Holocene. Evidence both from Greenland ice cores and from cores of ocean sediments from the North Atlantic and adjacent ocean basins (Bond et al. 1997) indicates a prevalent millennial-scale climatic variability in the North Atlantic region throughout the Holocene (Fig. 3b). The signals seen in the ice and ocean sediment cores appear to reflect principally temperature signals, although there is likely also to be a component related to changes in atmospheric circulation. Other proxies, for example changes in the predominant peat-forming *Sphagnum* spp. and other mire surface taxa record as macrofossils in the peat of ombrogenous mires (Barber et al. 1994), however, provide evidence principally of variations in moisture availability that may either reflect summer temperatures, moisture supply (and hence atmospheric circulation), or some combination of the two (Fig. 3c).

The two predominant terrestrial systems that record a signal related to millennial scale climate fluctuations are mountain glaciers and tree lines. Mountain glaciers in northern Europe and in the Alps exhibited their most recent millennial-scale readvance during the so-called 'Little Ice Age' of the last millennium, although a series of similar advances alternating with periods of retreat has occurred throughout the Holocene at approximately millennial intervals (Grove 1988). Although glacier extent is linked both to snowfall and to melting rates, both are principally temperature signals.

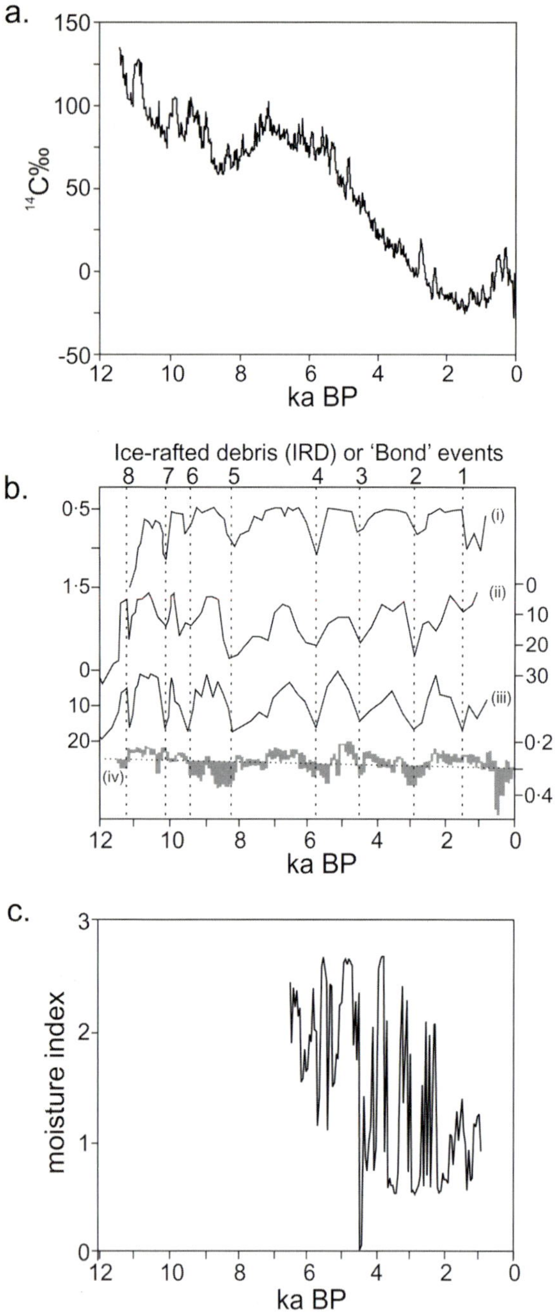

Fig. 3. (left) Millennial-scale fluctuations during the Holocene. **(a)** $\delta^{14}C$ curve showing marked fluctuations at millennial time scales superimposed upon the overall long-term trend (redrawn from Stuiver and Reimer 1993). **(b)** Ice (iv) and ocean sediment core (i – iii) proxies showing millennial variations during the Holocene: (i) VM 29–191 lithic grains (10^3 grains g^{-1}); (ii) VM 29–191 % Icelandic glass; (iii) VM 29–191 % haematite-stained grains (ash-free); (iv) GISP2 non-sea salt K flux (kg km^{-2} yr^{-1}) (redrawn from Bond et al. 1997). **(c)** Mid- and late-Holocene quasi-periodic fluctuations in an index of surface moisture, derived from quantitative plant macro-fossil analyses, on a mire in north-west England (high values indicating dry and low values moist conditions) (redrawn from Barber et al. 1994). Note: time scale in calendar years/calibrated ^{14}C years in all cases.

Times of glacier advance are thus generally characterised as times of relatively low mean annual temperatures.

Tree lines in the Alps of central Europe and in the Scandes mountains of Scandinavia have exhibited similar millennial scale fluctuations in their maximum elevation throughout the Holocene, extending as much as 200 m above their present elevation during early-Holocene maxima and reaching ca. 100 m above their present elevation less than a millennium ago during the 'Medieval Warm Period' that preceded the 'Little Ice Age'. Karlén and Kuylenstierna (1996) recently demonstrated that Holocene tree-line elevation fluctuations in the Scandes correlate inversely with $\delta^{14}C$ values that indicate the deviation of the atmospheric ^{14}C content from that of the present. Peak $\delta^{14}C$ values indicate peaks in cosmic ray fluxes (and hence in the generation of ^{14}C in the upper atmosphere); at least in the recent past these peaks in turn correlate with sunspot minima and thus with minima in solar radiation flux. Given that temperature, albeit expressed in various different ways, is generally regarded as the principal determinant of tree line elevation (Körner 1998), Holocene tree line elevation fluctuations can once again be taken as principally indicative of fluctuations in temperature, especially growing season temperature. Thus the $\delta^{14}C$ peaks that correlate with times of low tree-line elevation can be inferred to have corresponded to periods of cooler climate conditions; given the corresponding increase in extent of mountain glaciers at these times.

In addition to these fluctuations in tree lines and in glacier extent, more extensive regional vegetation responses have also occurred on millennial or sub-millennial time scales at some times and in some places during the Holocene. One of the most clearly documented of such vegetation responses was the rapid extension of the range of *Pinus sylvestris* (Scots Pine) across ombrogenous mires throughout far northern Scotland around four thousand years ago, followed a few centuries later by its demise and apparent retreat from the region (Gear and Huntley 1991; Huntley et al. 1997). Whereas mountain tree lines fluctuated in elevation by one to two hundred metres, this event involved a spatial shift of the species northern limit in Great Britain by 70 to 100 km and the transformation throughout an area of ca. 7000 km^2 from blanket mires and birch (*Betula pubescens*) woodlands to open stands of pine, followed by a revertence to mires and birch woodlands only 3 to 4 centuries later. This and similar such rapid events have been hypothetically linked to rapid changes in the $\delta^{14}C$ value and hence once again to variations in solar activity (Van Geel et al. 1996); Van Geel et al. (1996) also note that the rapid environmental change in the Netherlands at ca. 2650 yr BP was associated with the abandonment of some low-lying areas that had formerly been cultivated but that apparently became waterlogged and unsuitable for cultivation at that time.

'Fast' Forcing and Plot-Scale to Individual Vegetation Responses

Although rapid changes in solar output, such as those hypothesised to underlie the rapid, sub-millennial to centennial scale climate changes described above, might be included in the category of 'fast' forcing factors, it is principally the decadal to centennial scale variability in solar output that comes into this category. The climatic consequences of solar variability at these time scales are recorded by a variety of palaeoclimate proxies that have annual resolution; these include tree rings and annually-laminated lake sediments, both of which exhibit variations with the same characteristic periodicities as sun spot numbers and hence solar output. Such climatic variation, although important for example to the growth of individual trees and to the annual populations of plankton in lakes, is of relatively small magnitude and must be tolerated by any organism that lives more than one year. Nonetheless, at the extremes of its range such variation can result in the death of long-lived organisms that were living near their tolerance limits, thus generating disturbance in ecosystems that will result in recruitment of new individuals, often of different species to those that died. Variation will also lead to variation in which species of those in the local species pool are favoured with respect to recruitment in any one season, thus potentially resulting in changes in ecosystem structure and composition that may persist long after the event that triggered those changes.

Other important 'fast' forcing factors include such quasi-periodic phenomena as the North Atlantic Oscillation (NAO) and the El Niño — Southern Oscillation (ENSO) (Hulme and Barrow 1997), that each can result in extremes of inter-annual climate variability with the same consequences for ecosystem disturbance and recruitment as have been outlined above. Similar extremes can also result from the impacts upon the climate system of major volcanic eruptions; a number of such events during the historic and recent period is well documented (Lamb 1982) and tree ring sequences provide evidence of similar events sporadically throughout the Holocene.

Although the extreme seasons that can result from such 'fast' forcing factors can have climates outside the 'normal' range of inter-annual variability, these conditions are not persistent for more than a few years at most. In consequence, the ecological responses that they trigger are generally as described above and seen principally at the scale of individual plots or forest stands. Whereas such localised responses may then persist long after the event that triggered them, 'fast' forcing is unlikely ever to result in more spatially-extensive landscape- or regional-scale shifts in species ranges.

Influences Upon Hmans

Given this temporal and spatial hierarchy of climatic changes during the Holocene, it is instructive to ask how these changes might have impacted upon humans and what responses they might have elicited. At the time of deglaciation humans were probably absent or at least very sparsely present in northern Europe; those that were present in mainland Europe at that time were of the

Palaeolithic cultural grade. Contemporaneous with the environmental upheaval of deglaciation, however, the Mesolithic hunter-gatherers of the Middle East and south-west Asia first began to practice the cultivation of cereal type grasses, legumes and other crop plants, as well as to domesticate various animals. This development, that marks the transition to the Neolithic in that region, may itself have been a response to the environmental changes that were leading to a reduction in the availability of the wild grasses and other species upon which these people had come to depend during the preceding glacial stage.

With the transition to the Holocene the animals upon which the Palaeolithic people had preyed during the glacial, for example reindeer (*Rangifer tarandus*), shifted their range northwards, some of them soon becoming extinct at least in northern Europe, for example musk oxen (*Ovibos moschatus*) and woolly mammoth (*Mammuthus primigenius*). It seems likely that Palaeolithic humans showed a similar northward shift in their range, following their migrating prey and the retreating areas of treeless vegetation.

These northward shifts were associated with the climatic changes and the consequent northward expansion of forest cover that took place during the first millennia of the Holocene. Associated with the expansion of forest was expansion of forest animals, for example red deer (*Cervus elaphus*), and this in turn was associated with the expansion throughout most of Europe of Mesolithic hunter-gatherers. Such people remained closely linked to their environment and must have been subject to the influences of the millennial-scale climate fluctuations that would have affected the availability of food, whether originating from plants or from terrestrial or marine animals. It is likely that these people opportunistically expanded their ranges and increased their populations at times favourable for the foodstuffs upon which they depended, and suffered range and population declines during unfavourable periods. Such responses would have been facilitated by their nomadic and probably seasonally-migrant life style. It is arguable that the Sámi people, who practised such a nomadic and seasonally-migrant life style in marginal areas of northern Europe at least until the early twentieth century, represented the last vestige of a Mesolithic people in Europe; they certainly were subject to the influence of variability in the natural environment, especially as it impacted upon the principal natural resources upon which they relied for their survival.

When we turn to the spread of the Neolithic people and/or cultural grade across Europe, it is less easy to relate this directly to the influence of climate. Indeed, arguably, the spread of the practices of crop cultivation and of animal domestication mark the beginning of a trend whereby humans have progressively sought to isolate themselves from the influences of the natural environment, albeit with varying and often limited success. Thus the crops that the Neolithic farmers spread across Europe were grasses and other species most of which were native to the steppes of south-west Asia and thus adapted to a climate that was warm and dry in summer; the majority of them also were annuals and unable to compete either with perennial herbaceous species or tolerate shading by tall woody taxa. Neolithic farmers sought to cultivate these crops in northern Europe where the climate was cooler and moister and, as a result, forest vegetation was the naturally dominant land cover, with perennial herbaceous taxa predominant in naturally disturbed or open situations. Thus, in order to cultivate these crops, the Neolithic farmers had first to clear the forest and then were forced soon to abandon the cleared patches as their artificial disturbance was colonised by the perennial herbs of natural disturbed patches. This pattern of shifting agriculture is well recorded in the palynological record and was prevalent throughout northern Europe by 5000 ^{14}C yr BP (ca. 3750 BC).

Subsequently agriculture became more settled so that during the Bronze Age and the Iron Age there were persistent areas cleared of forest and growing crops. Nonetheless, the origins of the cereals especially must have rendered them vulnerable to crop failures and it seems likely that, as in more recent history (Parry 1978), settlements in marginal areas were particularly prone to abandonment as a consequence. This tendency is likely to have been exacerbated by the general trend toward cooler summers that was already underway at the time that Neolithic agriculture first reached northern Europe, and that accelerated during the Bronze

Age that coincided with a millennial fluctuation toward cooler wetter climate conditions that resulted in the initiation of peat growth in many upland areas of north-west Europe.

The Iron Age and the time of expansion of the Roman Empire, in contrast, coincided with a fluctuation toward warmer and drier conditions that would have been favourable for the more widespread cultivation of cereals upon which the Romans depended heavily. A further fluctuation followed with agricultural retreat and abandonment of settlements during the Dark Ages followed by the expansion of the Norse kingdoms during the subsequent Medieval Warm Period. The settlement of Iceland at that time, and even of Greenland, was favoured by the climate; cereals were cultivated even in Iceland. Thereafter the Little Ice Age saw the demise of the Greenland settlements, perhaps as a consequence of a failure to adapt to the changing climate as much as because of the climatic change in itself (Barlow et al. 1997), as well as a severe population reduction in Iceland as crops failed and even the fisheries suffered from marked cooling of the ocean surface in the northern North Atlantic (Lamb 1982). Elsewhere at that time various other impacts were felt by human settlements; glacier advance displaced farms in the Scandes mountains whilst some Alpine valleys were periodically devastated by floods resulting from glacier fluctuations (Grove 1988).

As recently as the early nineteenth century the vulnerability of European agriculture to natural climatic events was underlined by the crop failures associated with the 'year without a summer' that followed the eruption of the volcano Tamboro in the previous year (1815) (Lamb 1982). Even today, although agriculture now harnesses ever more complex technology in an effort to enable crops to be grown in areas that are fundamentally climatically unsuitable, food shortages follow unfavourable seasons.

Discussion

The Holocene has, like earlier geological periods, exhibited climatic changes at a wide range of temporal scales from interannual to multi-millennial. These changes have elicited responses by natural ecosystems that range from localised disturbances and changes in composition to continental-scale migrations and biome shifts. Both the climate changes themselves, and the ecosystem responses that they elicit, have undoubtedly impacted upon humans, principally effecting their capacity to obtain or, more recently, to cultivate sufficient food. These impacts are felt not only by human populations that depend principally upon 'natural' harvests, whether obtained from fisheries, hunting game, or collecting plant materials, but also by those that depend principally upon cultivated crops.

Prehistoric humans, especially Palaeolithic but also earlier Mesolithic peoples, were adapted to respond to environmental changes in much the same way as other ecosystem components and probably were limited principally by their caloric requirements, especially for over-winter survival. Although they had begun to wear clothing and to build or exploit natural shelters, thus modifying their caloric requirements and enabling them to expand the range of environmental conditions under which they were able to survive, they nonetheless remained essentially responsive to changes in the natural environment. Subsequently, however, with the development of agriculture and the translocation of crops and domesticated animals to areas beyond their natural geographic ranges, Neolithic humans began for the first time to render themselves at least partially independent of direct environmental constraints. Nonetheless, as events in the relatively recent historical past, such as the Little Ice Age population crash in Iceland (Lamb 1982), have shown, settlements dependent upon crops cultivated well beyond the natural limits of the crop species may be established during periods of favourable conditions, but will subsequently prove difficult or even impossible to sustain under less favourable conditions.

In the modern world the peoples of northern Europe are generally cushioned by technology from the direct effects of the environment in which they live and often are sustained by foodstuffs transported not just from elsewhere in Europe but in many cases from other continents. It is important when considering Holocene environmental history, however, to remember that as little as fifty years ago such marginal populations relied upon local produce to sustain themselves and heated their homes using locally gathered fuel. The relative independence of our modern culture from its environment is both a recent development and, perhaps, the greatest single threat to the long-term persistence of that culture.

References

Barber KE, Chambers FM, Maddy D, Stoneman, R, Brew JS (1994) A sensitive high-resolution record of late Holocene climatic change from a raised bog in northern England. The Holocene 4:198-205

Barlow LK, Sadler JP, Ogilvie AEJ, Buckland PC, Amorosi T, Ingimundarson JH, Skidmore P, Dugmore AJ, McGovern TH (1997) Interdisciplinary investigations of the end of the Norse Western Settlement in Greenland. The Holocene 7:489-499

Berger A (1978) Long term variations of caloric insolation resulting from the Earth's orbital elements. Quaternary Research 9:139-167

Bond G, Showers W, Cheseby M, Lotti R, Almasi P, deMenocal P, Priore P, Cullen H, Hajdas I, Bonani G (1997) A pervasive millennial-scale cycle in North Atlantic Holocene and glacial climates. Science 278:1257-1266

Fairbanks RG (1989) A 17,000-year glacio-eustatic sea level record: Influence of glacial melting rates on the Younger Dryas event and deep-ocean circulation. Nature 342:637-642

Foley JA, Kutzbach JE, Coe MT, Levis S (1994) Feedbacks between climate and boreal forests during the Holocene epoch. Nature 371:52-54

Gear AJ, Huntley B (1991) Rapid changes in the range limits of Scots Pine 4000 years ago. Science 251:544-547

Grove JM (1988) The Little Ice Age. Methuen, London

Hulme M, Barrow E (ed) (1997) Climates of the British Isles: Present, past and future. Routledge, London

Huntley B, Birks HJB (1983) An atlas of past and present pollen maps for Europe: 0-13000 B.P. Cambridge University Press, Cambridge

Huntley B, Prentice IC (1993) Holocene vegetation and climates of Europe. In: Wright Jr HE, Kutzbach JE, Webb III T, Ruddiman WF, Street-Perrott FA, Bartlein PJ (eds) Global Climates since the Last Glacial Maximum. University of Minnesota Press, Minneapolis, pp 136-168

Huntley B, Daniell JRG, Allen JRM (1997) Scottish vegetation history: The Highlands. Botanical Journal of Scotland 49:163-175

Karlén W, Kuylenstierna J (1996) On solar forcing of Holocene climate: evidence from Scandinavia. The Holocene 6:359-365

Körner C (1998) A re-assessment of high elevation treeline positions and their explanation. Oecologia 115:445-459

Kutzbach JE, Gallimore RG (1988) Sensitivity of a coupled atmosphere/mixed layer ocean model to changes in orbital forcing at 9000 years B.P. Journal of Geophysical Research 93:803-821

Kutzbach, JE, Ruddiman, WF (1993) Model description, external forcing, and surface boundary conditions. In: Wright Jr HE, Kutzbach JE, Webb III T, Ruddiman WF, Street-Perrott FA, Bartlein PJ (eds) Global. Climates since the Last Glacial Maximum, University of Minnesota Press, Minneapolis, pp 12-23

Kutzbach JE, Guetter PJ, Behling PJ, Selin R (1993) Simulated climatic changes: results of the COHMAP climate-model experiments. In: Wright Jr HE, Kutzbach JE, Webb III T, Ruddiman WF, Street-Perrott FA, Bartlein PJ (eds) Global Climates since the Last Glacial Maximum. University of Minnesota Press, Minneapolis, pp 24-93

Lamb HH (1982) Climate, history and the modern world. Methuen, London

Lowe JJ, Walker MJC (1997) Reconstructing Quaternary Environments (Second edition). Longman, Harlow, Essex

Parry ML (1978) Climatic change, agriculture and settlement. Dawson, Folkestone

Peterson GM (1983) Holocene vegetation and climate in the western USSR. PhD Thesis, University of Wisconsin - Madison

Prentice IC, Cramer W, Harrison SP, Leemans R, Monserud RA, Solomon AM (1992) A global biome model based on plant physiology and dominance, soil properties and climate. J Biogeogr 19:117-134

Prentice IC, Guiot J, Huntley B, Jolly D, Cheddadi R (1996) Reconstructing biomes from palaeoecological data: a general method and its application to European pollen data at 0 and 6 ka. Climate Dynamics 12:185-194

Roberts N (1998) The Holocene: An environmental history. Blackwell, Oxford

Stuiver M, Reimer PJ (1993) Extended ^{14}C database and revised CALIB radiocarbon calibration program. Radiocarbon 35:215-230

Van Geel B, Buurman J, Waterbolk HT (1996) Archaeological and palaeoecological indications of an abrupt climate change in The Netherlands, and evidence for climatological teleconnections around 2650 BP. Journal of Quaternary Science 11:451-460

Late Glacial and Holocene Glacier Fluctuations and Climatic Variations in Southern Norway

Atle Nesje

Department of Geology, University of Bergen, Allégt. 41, N-5007 Bergen, Norway
e-mail: atle.nesje@geol.uib.no

Abstract: Deglaciation chronologies for the Trondheimsfjord region, Nordfjord area, Bergen district, Sunnhordland/Ryfylke, and the Oslofjord/Romerike area are reviewed. Holocene glacier and climate variations in southern Norway have been reconstructed by using evidence from historical data, lichenometry, rock-surface weathering from Schmidt-hammer rebound values, palaeosols, palynology, and stratigraphic evidence from distal sites characterised by glaciofluvial and glaciolacustrine sediments. Due to extensive glaciers during the "Little Ice Age", almost all evidence from the period subsequent to the deglaciation was erased by the recent Neoglacial activity in the vicinity of the present glaciers. Records from continuous lacustrine sediment sequences downstream from glaciers are, however, considered to produce continuous, high-resolution evidence about Holocene glacier activity. The most significant climatic changes took place around the Younger Dryas/Holocene transition around 11,500 cal. yr BP (10,000 ^{14}C yr BP), at 10,300-9900 cal. yr BP (9100 ± 200 ^{14}C yr BP; the Erdalen event), around 8200 cal. yr BP (7600 ^{14}C yr BP; the Finse event), approximately 4200 cal. yr BP (3800 ^{14}C yr BP), and during the "Little Ice Age" (AD 1650-1930). The Holocene glacier and climate development in southern Norway is presented from sites in Sunnmøre, at Ålfotbreen, Jostedalsbreen area, Hardangerjøkulen area, Jotunheimen, and Dovre. Lithostratigraphic and palaeobotanical studies suggest that most of the south Norwegian glaciers disappeared periodically during the early and mid Holocene. The glacial activity increased significantly from about 4000 cal. yr BP and the glaciers in southern Norway reached their maximum Neoglacial extent during the "Little Ice Age", when the equilibrium-line altitude (ELA) was 100-150 m lower than at present.

Introduction

High-frequency climatic fluctuations during the last deglaciation (ca. 14,000-9000 ^{14}C yr BP) are well documented in terrestrial data (e.g. Birks et al. 1994 and references therein), marine records (e.g. Haflidason et al. 1995) and ice cores (e.g. Johnsen et al. 1992) from the North Atlantic region. These climatic changes occurred at a time of maximum solar radiation receipt in the Northern Hemisphere and cannot be explained by direct orbital forcing. The cause(s) of these climatic fluctuations must therefore be sought in the response of the ocean/ atmosphere/ climate system to such forcing.

Strong climatic gradients across southern Norway and its location relative to North Atlantic cyclone tracks and to arctic atmospheric and oce-anic polar fronts provide special information for detailed reconstruction of Late Glacial and Holocene glacier and climatic fluctuations in southern Norway. Data from this region provide important information relevant to the present debate about the magnitude, timing, and possible cause(s) of Late Glacial and Holocene glacier and climate variations.

Evidence of Late Glacial and Holocene glacier fluctuations and climatic change as deduced from marginal moraines, proglacial terrestrial and lacustrine sites, palynological data and tree-line fluctuations in southern Norway (Fig. 1) has been compiled. The data compilation indicates significant Late Glacial climate oscillations, glacier contraction and disappearance during the early and mid Holo-cene and subsequent Neoglacial events, including the "Little Ice Age".

From WEFER G, BERGER W, BEHRE K-E, JANSEN E (eds), 2002, *Climate Development and History of the North Atlantic Realm.* Springer-Verlag Berlin Heidelberg, pp 233-258

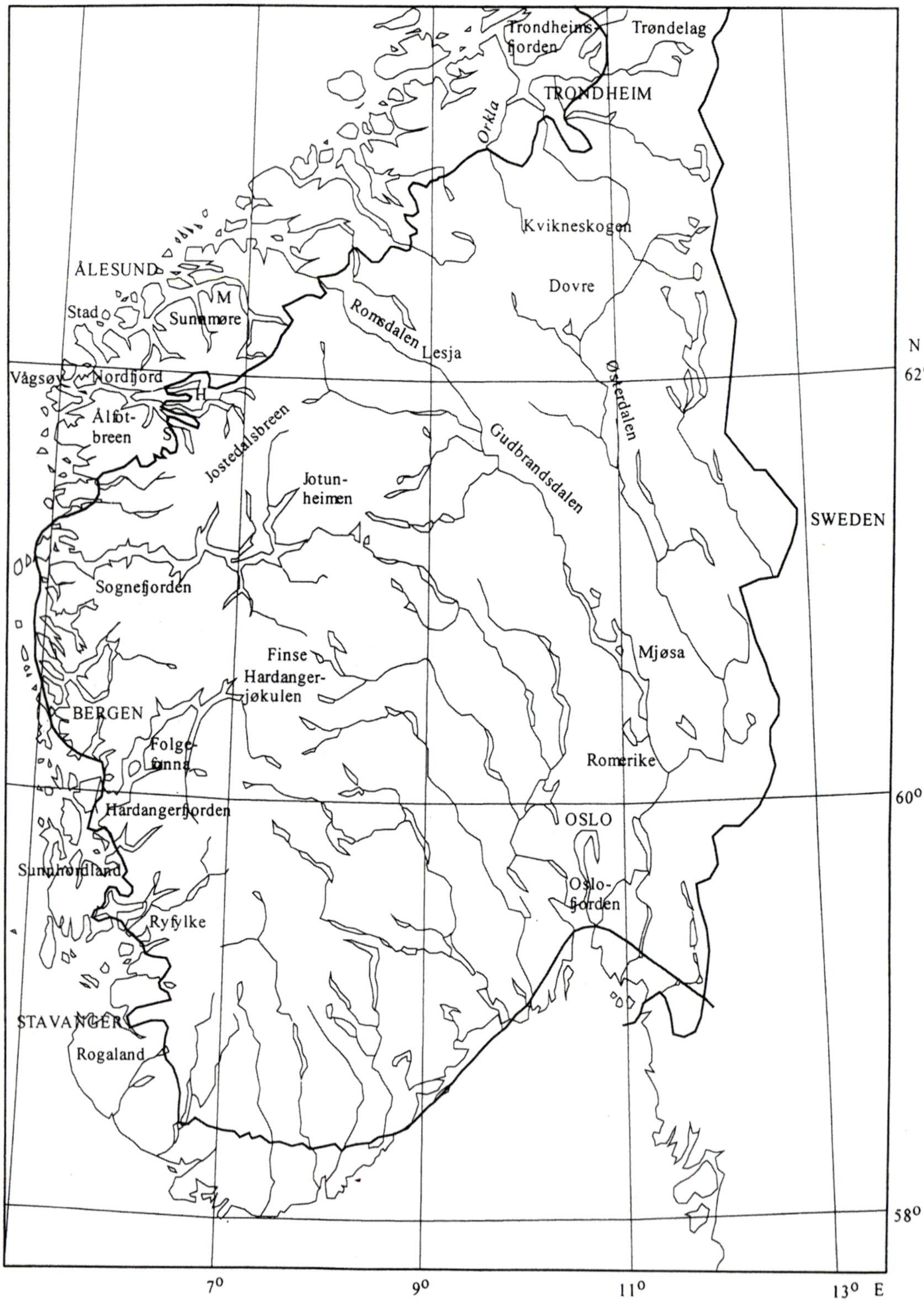

Fig. 1. Location map southern Norway. The bold line indicates the Younger Dryas ice marginal position around the coast. (H - Hornindalsvatnet, M - Midtivatnet, S - Sandane).

Methods

Holocene glacier and climate variations in southern Norway have been reconstructed by using evidence from historical data (e.g. Grove 1988), lichen-ometry (e.g. Bickerton and Matthews 1993), rock-surface weathering from Schmidt-hammer rebound values (e.g. Matthews and Shakesby 1984), palae-osols and palynology (e.g. Matthews and Caseldine 1987), and stratigraphic evidence from distal sites characterised by glaciofluvial and glaciolacustrine sediments (e.g. Dahl and Nesje 1994). Due to extensive glaciers during the "Little Ice Age", almost all evidence from the period subsequent to the deglaciation was erased by the recent Neoglacial activity in the vicinity of the present glaciers. Records from continuous lacustrine sediment sequences downstream from glaciers are therefore considered to produce one of the most reliable evidence about former glacier activity, due to variations in particle size (sedigraph), laminae/varves, sediment thickness, organic/minerogenic content (loss-on-ignition), and magnetic properties (magnetic susceptibility and intensity). However, lacus-trine sedimentary records from steep, alpine regions are commonly complicated by mass-flow avalanches and turbidity currents of non-glacial origin. Lacustrine sedimentary records therefore need to be studied by a multi-disciplinary approach and carefully evaluated, taking into account both climatic factors and local geomorphological conditions (e.g. Nesje et al. 1995).

Radiocarbon dates are reported as calibrated years BP (cal. yr BP) (BP=AD 1950) listed as the intercept age with 1 sigma in brackets, and as ^{14}C age BP (BP=AD 1950) ± 1 sigma. Calibration was carried out using the CALIB 3.0 ^{14}C age calibration program (Stuiver and Reimer 1993).

Late Glacial

Glacier Fluctuations

Trondheimsfjord Region

The coastal area of the Trondheimsfjord region was deglaciated during the Bølling Chronozone (Fig. 2), with a rapid glacier retreat during the Allerød, still with some local ice caps in the coastal mountains (e.g. Andersen et al. 1995). The Tautra ice marginal deposits (moraines and glaciofluvial deposits) were formed in the early Younger Dryas Chrono-zone. In the second part of the Younger Dryas, the ice sheet readvanced, forming the Hoklingen ice marginal deposits. Thereafter, the ice margin retreated further inland, however, at Vuku, the ice sheet made another halt/readvance during the early Preboreal (Fig. 2).

Nordfjord

The coastal district of Nordfjord was deglaciated at 12,300 ^{14}C yr BP (Mangerud et al. 1979; Larsen et al. 1984) (Fig. 2). A radiocarbon date of 11,360 ± 70 ^{14}C yr BP from the eastern part of Hornindalsvatnet suggests that at least the middle and inner fjord region of Nordfjord was deglaciated during the Allerød. In the middle and inner Nordfjord region, Fareth (1987) mapped prominent lateral moraines of Younger Dryas age. Beyond and above the Younger Dryas valley and fjord glaciers in Nordfjord and Sunnmøre, several cirque glaciers formed (e.g. Fareth 1987; Dahl and Nesje 1992). Fareth (1987) calculated the Younger Dryas ELA depression in middle and inner Nordfjord to 450 ± 50 m, while Larsen et al. (1984) calculated the Younger Dryas ELA depression of reconstructed cirque glaciers to about 700 m in the coastal district of Stad and Vågsøy. Dahl and Nesje (1992) estimated from a reconstructed Younger Dryas cirque glacier in inner Nordfjord that winter precipitation was reduced to approximately 60% compared to present values (100%), with a corresponding mean ablation-season temperature depression of 4-5°C from the present. In the inner Nordfjord region deglaciation during the mid Preboreal (9500 ± 200 ^{14}C yr BP) was characterised by vertical ice wastage (Rye et al. 1987), indicating a marked rise in ELA after termination of the Younger Dryas Chronozone (Rye et al. op. cit.; Nesje et al. 1991). In the inner Nordfjord region deglaciation during the mid Preboreal (9500 ± 200 ^{14}C yr BP) was characterised by vertical ice wastage, indicating a marked rise in ELA after termination of the Younger Dryas.

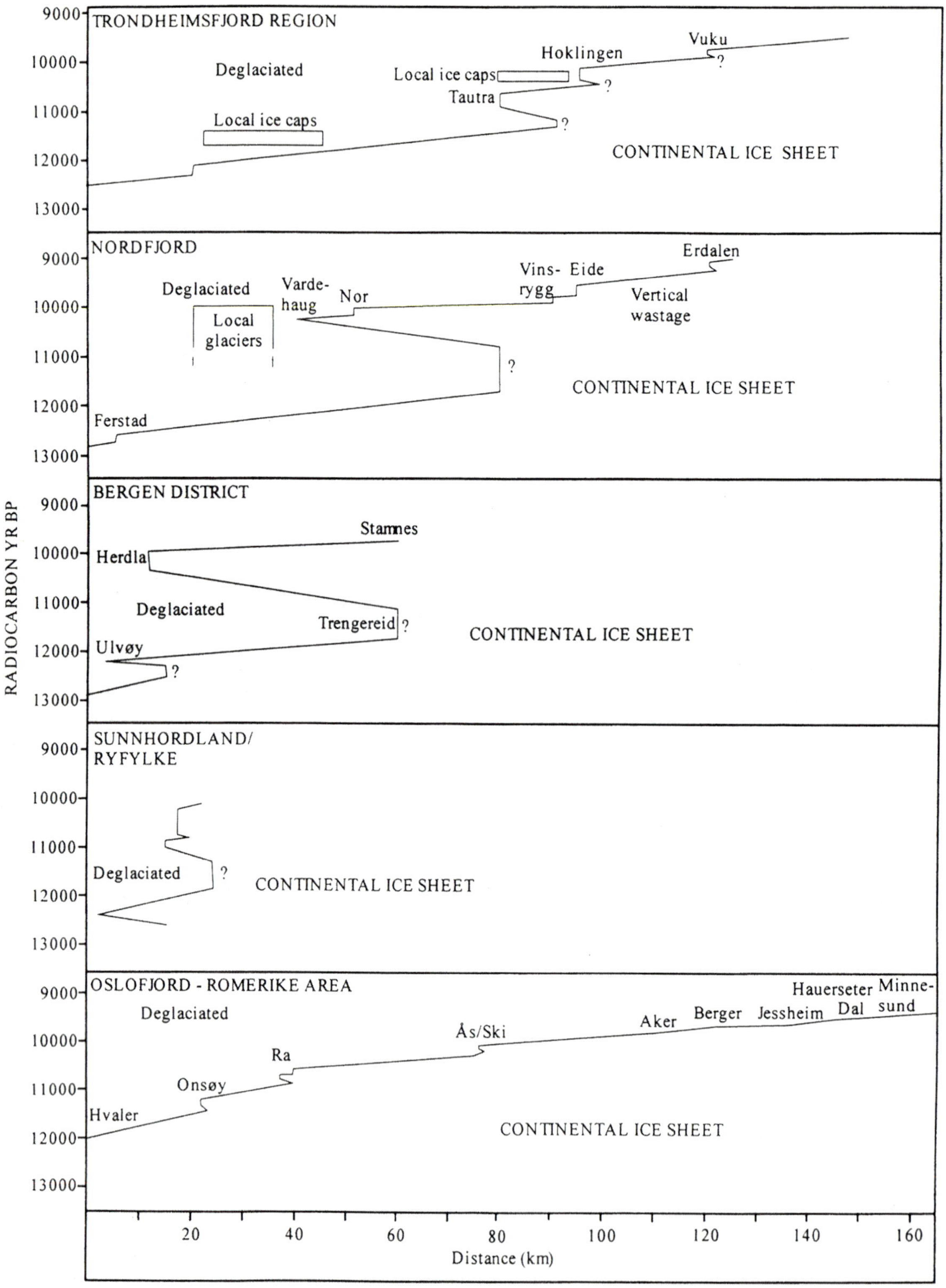

Fig. 2. Time/distance diagrams for the Late Glacial and early Holocene deglaciation from different parts of southern Norway. For location, see Fig. 1 and for references, see text.

Bergen District

The Late Glacial deglaciation of the coastal district of Bergen (Fig. 2) occurred at 12,600 ^{14}C yr BP with a subsequent readvance into the North Sea about 12,200 ^{14}C yr BP (Andersen et al. 1995 and references therein). During the Allerød Chronozone the continental ice sheet retreated at least 50 km inland. During the Younger Dryas the ice sheet readvanced to the coastal district; the Herdla ice-frontal marine delta marking the ice-marginal position (e.g. Andersen et al. 1995).

Sunnhordland/Ryfylke

In the Sunnhordland region, the deglaciation occurred during the Bølling or Older Dryas chronozones. The deglaciation chronology in SW Norway has been established by Anundsen, Follestad, Blystad, Andersen and collaborators (see Andersen et al. 1995) (Fig. 2).

Oslofjord/Romerike Area

The deglaciation chronology in the Oslofjord/Romerike area (e.g. Sørensen 1979) is shown in Fig. 2. The ice marginal zones are (^{14}C yr BP in brackets) (Fig. 3): 1. The Hvaler/Tjøme moraines (12,300-11,800), 2. The Onsøy/Slagen moraines (11,300), 3. The Ra moraines (11,000-10,700), 4. Ås/Ski (10,000), 5. Aker (9800), 6. Berger, 7. Jessheim, 8. Hauerseter (9500), 9. Dal, and 10. Minnesund (9400).

Palynological Evidence for Late Glacial Climate Change

Palynological evidence for Late Glacial vegetation and climatic change in western Norway has been reviewed by Nesje and Dahl (1993) and Birks et al. (1994). Four coastal sites in Rogaland, SW Norway (Fig. 1), (e.g. Paus 1992; Birks et al. 1994) provide a continuous record of the Late Glacial vegetation and climate development in western Norway from around 14,000 to 9500 ^{14}C yr BP. The pollen record shows a succession in six stages (1. From about 14,000 to 13,000 ^{14}C yr BP an *Artemesia*-dominated pioneer vegetation following deglaciation, 2. A *Salix* shrub consolidation phase between about 13,000 and 12,700 ^{14}C yr BP, 3. An open birch (*Betula*) vegetation between approximately 12,700 and 12,000 ^{14}C yr BP, 4. An open birch-forest phase from about 12,000 to 11,000 ^{14}C yr BP, 5. Re-establishment of an *Artemesia*-dominated pioneer vegetation during the first half of the Younger Dryas, and 6. Establishment of dense birch forests at the Younger Dryas/Holocene transition).

Final Deglaciation

Deglaciation after the termination of the Younger Dryas cold interval was characterised by rapid retreat of the fjord glaciers in western Norway primarily because of effective calving. Radiocarbon dates ranging in age between approximately 9800 to 9500 ^{14}C yr BP at the head of the fjords in western Norway suggest a rapid deglaciation of the fjord and valley glaciers in western Norway (see review by Nesje and Dahl (1993) and references therein).

Terminal moraines and ice-marginal deltas in the inner fjord region of western Norway, indicate halts and/or minor readvances of the glaciers. The ice-marginal deposits are located where the fjords become shallower, commonly on rock thresholds, and where the valleys/fjords are relatively narrow. The formation of the ice-marginal deposits occurred during a period of rapid warming and may be ex-plained in terms of glacial dynamics as a result of rapid calving of the fjord glaciers and unstable and dynamically active valley glaciers. During the succeeding stabilising process, when the glacier fronts became grounded, the glaciers either readvanced or at least halted their general retreat, forming frontal deposits (e.g. Kjenstad and Sollid 1982).

The deglaciation of eastern Norway (Gudbrandsdalen and Østerdalen) was characterised by vertical downwastage. During the final stage, the ice divide was located south of the main watershed. The ice thickness was largest in the valleys, while surrounding mountains and hills were ice free. Consequently, large ice-dammed lakes formed between the downmelting ice in the south and the pass-points along the main watershed in the north (e.g. Garnes and Bergersen 1980). In Gudbrandsdalen, *Store Dølasjø* drained NW across Lesja to Romsdalen. In Østerdalen, *Øvre Glomsjø*

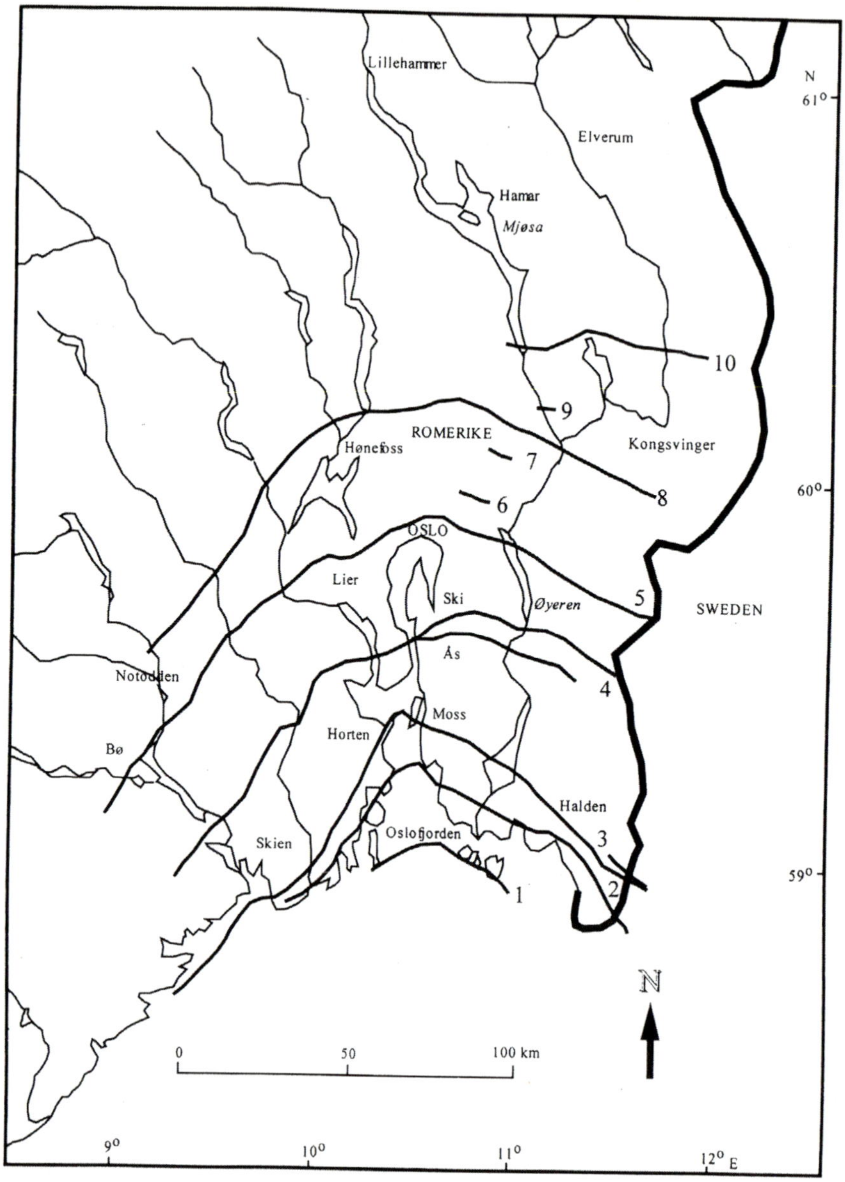

Fig. 3. Ice marginal lines from the deglaciation period in the Oslofjord - Romerike region. Adapted from Sørensen (1979) and later revisions. For numbering, see text.

drained N across Kvikneskogen toward the valley of Orkla (Fig. 1). The last, and lowest ice-dammed lake with northerly drainage was *Nedre Glomsjø*, the largest ice dammed lake in southern Norway (140 km long and ca. 120 km^3). The southward outburst of the lake (water discharge 350,000 m^3/sec) occurred around 10,200 cal. yr BP (9200 ^{14}C yr BP) (Longva and Thoresen 1991).

Investigations of cirque moraines in central-east Norwegian mountains indicate that the mean ablation-season temperature was 6-7°C lower than at present during their formation (Dahl et al. 1997). The authors propose that these moraines were formed in the Late Glacial period, most likely during the Younger Dryas. If these moraines were formed during the Younger Dryas, the ice-sheet surface was lower than 1100 m (relative to present sea level) in eastern central Norway.

Holocene Glacier and Climate Development

Early Holocene Climate Oscillations

The Erdalen Event

During the Erdalen event, terminal moraines were formed as far as 1 km beyond the "Little Ice Age" moraines that are present marginal to the present glaciers in western Norway (see review by Nesje and Dahl 1993). As inferred from the highest altitudes of lateral moraines formed during this readvance, and from calculations of the ELA depression based on an accumulation area ratio (AAR) of 0.6, the average ELA lowering is calculated to have been about 300 m (adjusted for land uplift). Dahl and Nesje (1992) calculated from a reconstructed Erdalen event cirque glacier in inner Nordfjord that winter precipitation was reduced to approximately 70% compared to present values (100%), with a corresponding mean ablation-season temperature depression of ca. 3.0°C from the present. The Erdalen event has been identified stratigraphically in front of Nigardsbreen, Jostedalsbreen region, and dated at 10,050-9700 cal. BP. In addition, the Erdalen event has been identified as gyttja silt in lacustrine sediments deposited from a local cirque glacier in Sunnmøre, yielding maximum and minimum dates of 10,450 (10,900-10,215) cal. yr BP (9460 ± 185 ^{14}C yr BP) and 10,005 (10,285-9895) cal. yr BP (9060 ± 190 ^{14}C yr BP), respectively. In lake Bøvertunsvatnet, Jotunheimen, the Erdalen event, recorded as high minerogenic content (high magnetic susceptibility), was terminated at 9860 (9910-9650) cal. yr BP (8805 ± 80 ^{14}C yr BP). At Finse, the Erdalen event terminated at 9530 (9805-9485) cal. yr BP (8630 ± 95 ^{14}C yr BP) (Dahl and Nesje 1994).

The Finse Event

In glacier records from southern Norway, an episode between 8300 and 8000 cal. yr BP is clearly the same widespread event as recorded in the GRIP and GISP2 Greenland ice cores, in lacustrine and proglacial sites, and marine sediments. In both the GRIP and GISP2 Greenland ice cores, as well as in terrestrial peat sections, in lacustrine sequences and marine records, this widespread event took place between 8400 and 8000 cal. yr BP [centred around 8200 cal. yr BP (7600 ^{14}C yr BP)] (see Alley et al. 1997). The 8400-8000 cal. yr BP oscillation has in central south Norway (at Hardangerjøkulen) has been termed the Finse event by Dahl and Nesje (1994, 1996).

Regional Studies

Sunnmøre

A core through lacustrine sediments from Midtivatnet showed four silt/clay bands (Fig. 4) indicative of glacier activity in the upper catchment (Karlén and Matthews 1992; Matthews and Karlén 1992). Radiocarbon dates indicate that the glacier expansion periods occurred prior to 7390 (7520-7290) cal. yr BP (6540 ± 100 ^{14}C yr BP), some time between 4530 (4805-4435) cal. yr BP (4070 ± 70 ^{14}C yr BP) and 2870 (2960-2790) cal. yr BP (2800 ± 70 ^{14}C yr BP), and subsequent to 910 (930-780) cal. yr BP (950 ± 60 ^{14}C yr BP). As evident from Fig. 4, however, the silt bands are not accurately dated.

A core described by Karlén and Matthews (1992) from Storevatnet, 84 m lower and 6 km to the NW of Midtivatnet, shows more restricted glacial activity than in Midtivatnet. A basal radiocarbon date of 9400 (9490-9265) cal. yr BP (8420 ± 110 ^{14}C yr BP) gives a minimum deglaciation age. Except for a silt-clay band 10-20 cm below surface deposited from glacial meltwater entering the lake during the "Little Ice Age", no sediment properties indicate glacial activity in the catchment between the early Holocene deglaciation and the "Little Ice Age".

Ålfotbreen

Proglacial lake sediments from Grøndalsvatn (Fig. 5) south of Ålfotbreen, the westernmost glacier in southern Norway, show evidence of avalanche activity into the lake between at least 3020 (3120-2950) cal. yr BP (2910 ± 60 ^{14}C yr BP) and 2740

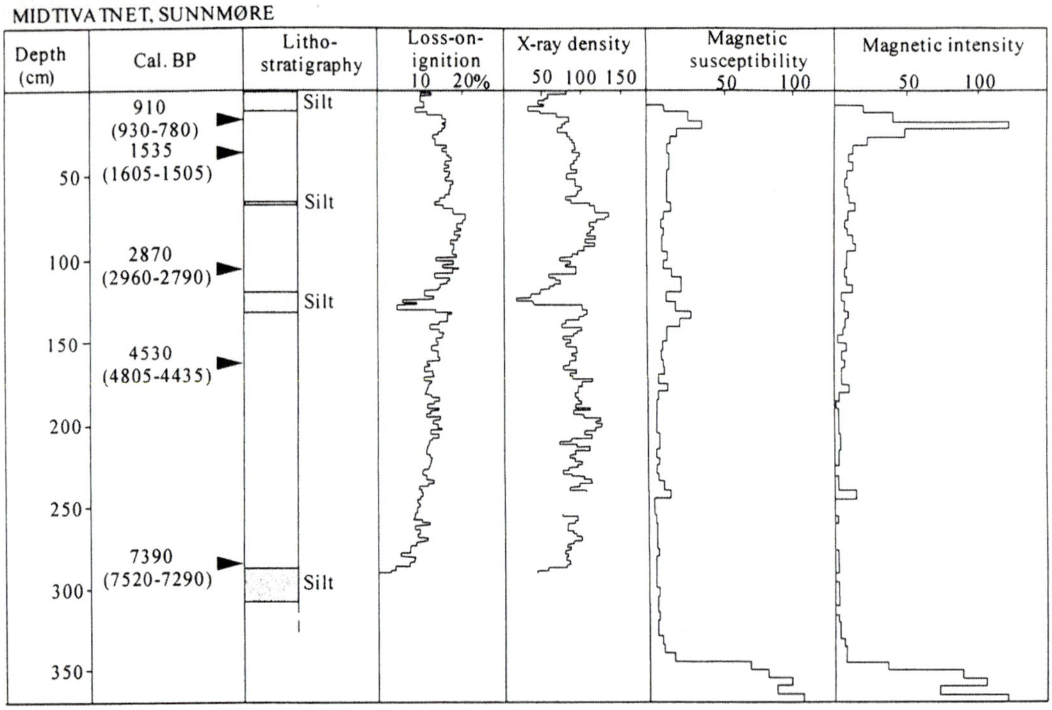

Fig. 4. Stratigraphy, radiocarbon dates [in calendar years BP (1 sigma in brackets)] and sediment properties for the Midtivatnet core (adapted from Karlén and Matthews 1992).

(2760-2720) cal. yr BP (2590 ± 60 ^{14}C yr BP) (Nesje et al. 1995). A glacier episode occurred between 2740 (2760-2720) cal. yr BP (2590 ± 60 ^{14}C yr BP) and 2120 (2280-2010) cal. yr BP (2140 ± 60 ^{14}C yr BP). Renewed glacial activity in the lake's catchment occurred from about 1540 (1580-1500) cal. yr BP (1650 ± 60 ^{14}C yr BP). Increased Neoglacial activity which began at 790 (920-730) cal. yr BP (910 ± 70 ^{14}C yr BP).

Jostedalsbreen Area

Palynological Evidence

Palynological studies in Sygneskardet (680 m), at the NW margin of Jostedalsbreen (Fig. 6), show that the basal gyttja layers [deglaciation date of 10,250 (10,380-10,040) cal. yr BP (9280 ± 130 ^{14}C yr BP)] are dominated by *Betula, Juniperus, Salix,* and *Empetrum* and a rapid increase of pine (*Pinus* sp.) after 10,000 cal. yr BP (9000 ^{14}C yr BP) (Nesje et al. 1991). The pine maximum occurred about 9400 cal. yr BP (8400 ^{14}C yr BP), suggesting that modern mean summer temperatures were reached near the Preboreal/Boreal Chronozone transition. The pollen record also shows that between about 7400 cal. yr BP (6500 ^{14}C yr BP) and 5800 cal. yr BP (5000 ^{14}C yr BP), stands of elm (*Ulmus* sp.) grew close to the modern birch forest limit. Based on modern summer temperature requirements of elm and birch (11°C and 7°C, respectively), the mean summer temperature was nearly 4°C warmer than at present. In the absence of the cooling effect imposed by the present Jostedalsbre ice cap, that is evident at the birch forest limit at Sygneskardet, the difference between the Atlantic Chronozone mean summer temperature and the modern mean summer temperature is calculated to have been 1.5-2.0°C (corrected for glacio-isostatic recovery).

Palynological studies of peat at three sites at altitudes between 565 and 700 m along a transect in Sprongdalen, a north-easterly tributary valley to Jostedalen, revealed a significant alder (*Alnus* sp.)

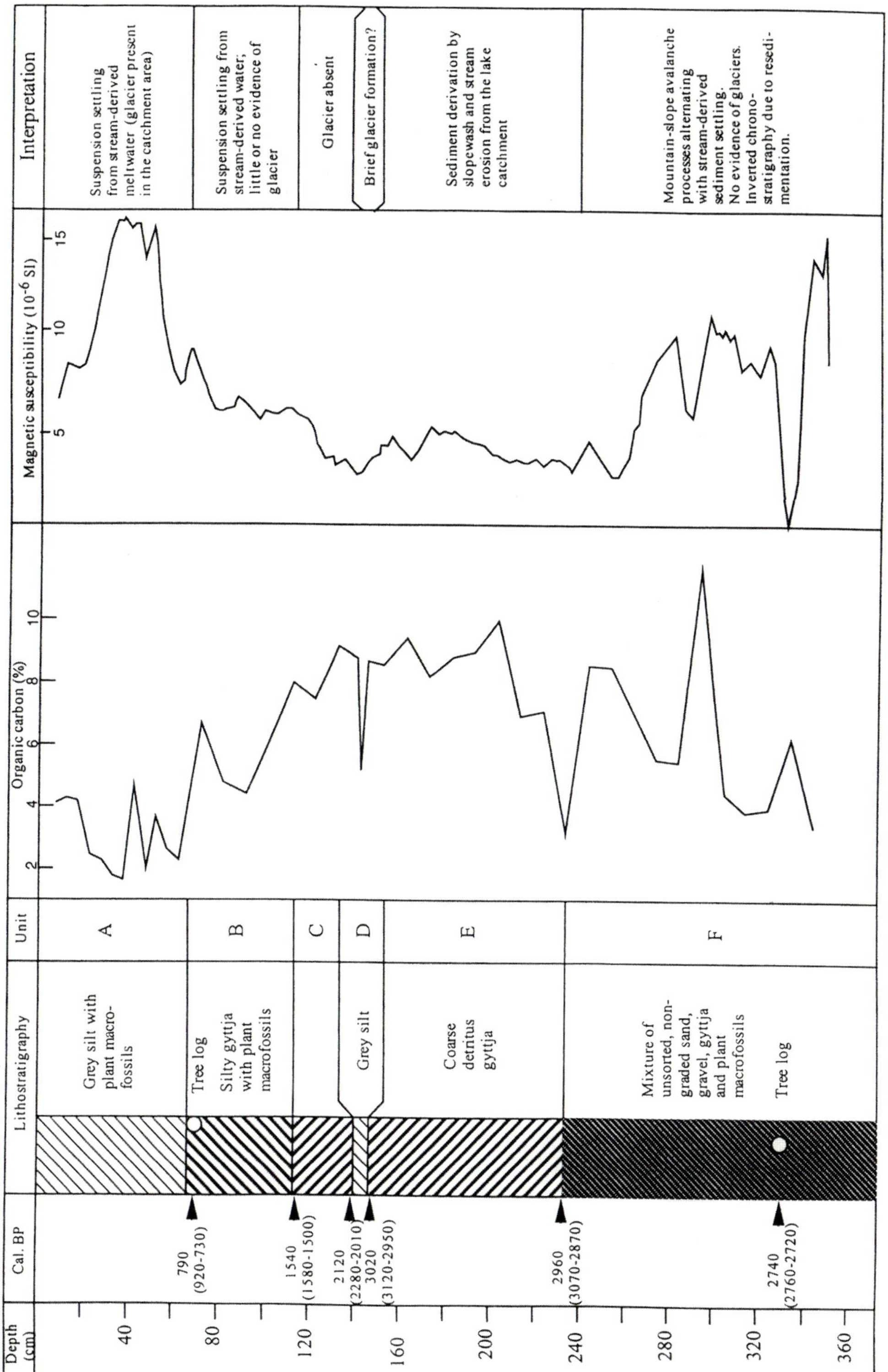

Fig. 5. Lithostratigraphy, radiocarbon dates [in calendar years BP (1 sigma in brackets)], organic carbon content and magnetic susceptibility in the Grøndalsvatn core (adapted from Nesje et al. 1995).

decline approximately 7300 cal. yr BP (6300 ^{14}C yr BP), reflecting a climatic deterioration following the Hypsithermal interval (Nesje et al. 1991). After ca. 6200 cal. yr BP (5300 ^{14}C yr BP), birch (*Betula* sp.) expanded at the expense of other tree species, mainly pine and alder. This expansion is interpreted to have occurred in response to cooling caused by glacier formation on the Jostedalsbre Plateau.

At Haugabreen (Fig. 6), Caseldine and Matthews (1987) suggested that a shift to cooler and possibly wetter conditions began around 6015 (6175-5935) cal. yr BP (5265 ± 70 ^{14}C yr BP). Between 4000 cal. yr BP (3600 ^{14}C yr BP) and 3600 cal. yr BP (3300 ^{14}C yr BP) the tree line dropped below 660 m. Between 3600 cal. yr BP and 750 cal/^{14}C BP the tree line remained below the site, except for two intervals at 2700 cal. yr BP (2600 ^{14}C yr BP) and 2200-2000 cal. yr BP/^{14}C yr BP. After the post-Medieval climatic deterioration around AD 1270, the tree line fell below 660 m.

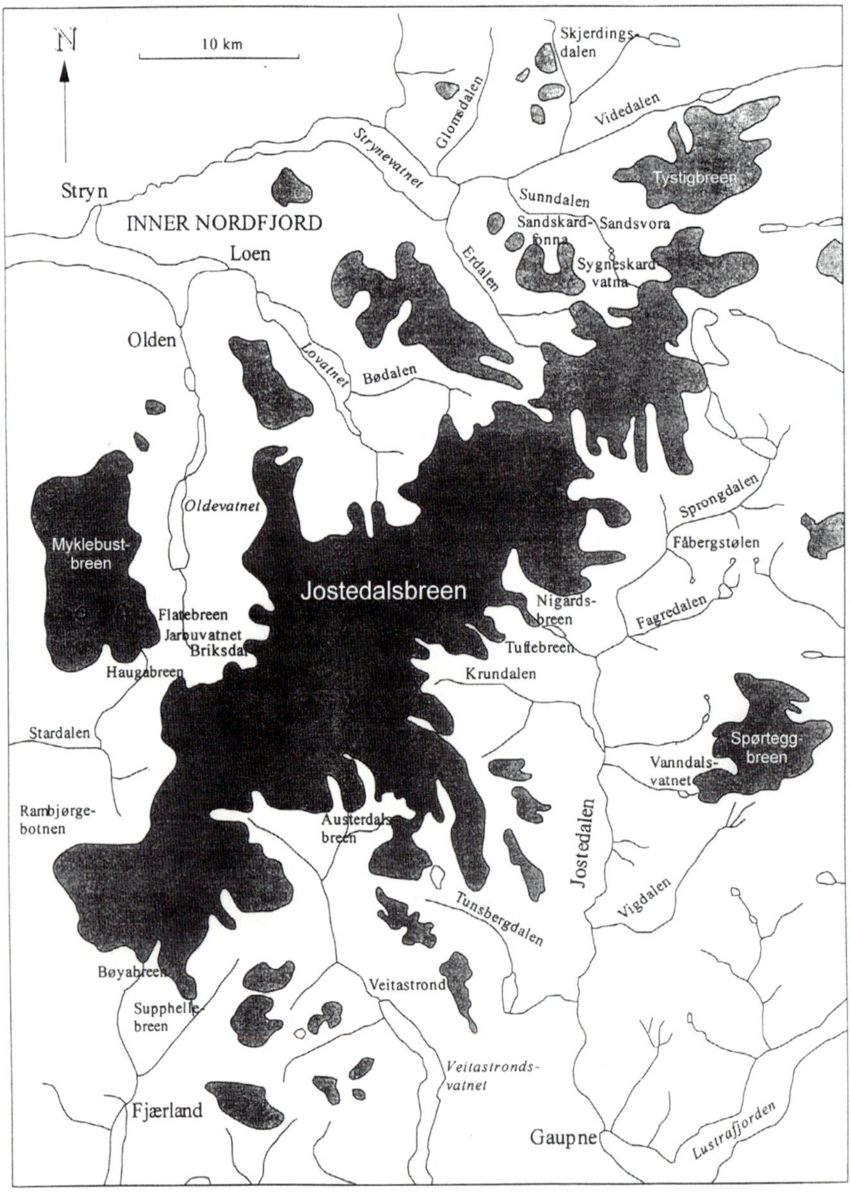

Fig. 6. The Jostedalsbre region.

A pollen site in Rambjørgebotnen at the SW part of Jostedalsbreen (Fig. 6), showed that *Pinus* reached its maximum altitude around 9120 cal. yr BP (8200 ^{14}C yr BP) (Torske 1996) and that pine was an element of the subalpine forest during the early and middle Atlantic Chronozone. In the first part of the Atlantic, a tetratherm (June-September) of 1.4°C is inferred. Torske (op. cit.) concluded that the major SW part of Jostedalsbreen disappeared during the late Boreal Chronozone. A gradual shift to cooler climate probably occurred in the late Atlantic and Subboreal chronozones, with re-establishment of local glaciers around 5780 cal. yr BP (5100 ^{14}C yr BP). At ca. 5700 cal. yr BP (5000 ^{14}C yr BP) *Betula* became the dominant tree species in the subalpine forest. Finally, the pollen data indicates that the southern part of Jostedalsbreen became re-established ca. 3800 cal. yr BP (3500 ^{14}C yr BP).

Glacier Fluctuations

Lacustrine sediments from Vanndalsvatnet, draining the SW part of Spørteggbreen (Fig. 6), a plateau glacier east of Jostedalsbreen, covers the last 8900 cal. yr BP (8000 ^{14}C years). The sediments are glacially laminated in the upper ca. 25 cm of the core (Fig. 7). The transition to the glacially laminated sediments in the top of the core has been dated at 1415 (1530-1355) cal. yr BP (1570 ± 70 ^{14}C yr BP). Numerous thin bands or layers of massive to graded, non-laminated minerogenic sediments mainly in the upper part of the core are interpreted as turbidity currents which most likely entered the lake during single climatic events, mainly subsequent to 3615 (3810-3460) cal. yr BP (3360 ± 130 ^{14}C yr BP). Horizons of isolated clasts, mostly of gravelly size, mainly in the upper part of core, are interpreted as ice-rafted particles from the lake shore (Nesje et al. 1991).

Two lacustrine sediment cores from Sygneskardvatnet, Sunndalen, at the NW side of Jostedalsbreen (Figs. 6 and 8), for the first time demonstrate that most of, or possibly the whole ice cap was totally melted away during the Atlantic Chronozone (Nesje et al. 2000). Two periods of increased glacier activity are recorded at ca. 8300 and 7800 cal. yr BP (ca. 7600 and 7000 ^{14}C yr BP). The former of the two might be the same widespread event as recorded in Greenland ice cores, in lacustrine and proglacial sites, and marine sediments. Radiocarbon dates show that the glacier in the lake catchment disappeared completely be-

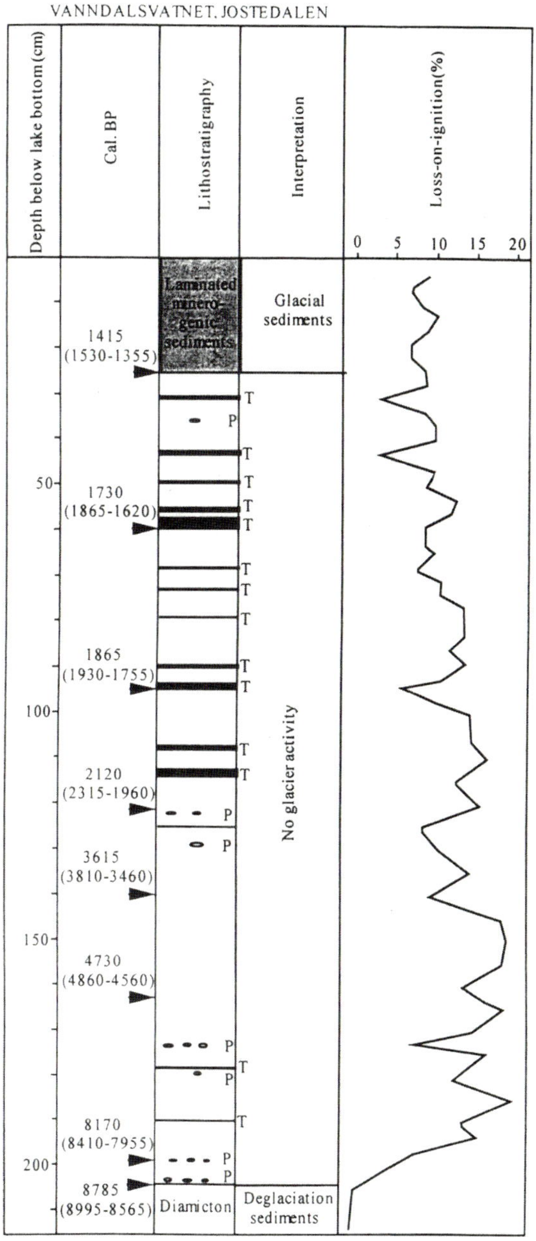

Fig. 7. Lithostratigraphy, loss-on-ignition and radiocarbon dates [in calendar years BP (1 sigma in brackets)] from one of three cores retrieved from Vanndalsvatnet (adapted from Nesje et al. 1991). P - Single minerogenic particles, mostly gravel, T - Massive to graded, none-laminated minerogenic-rich sediments.

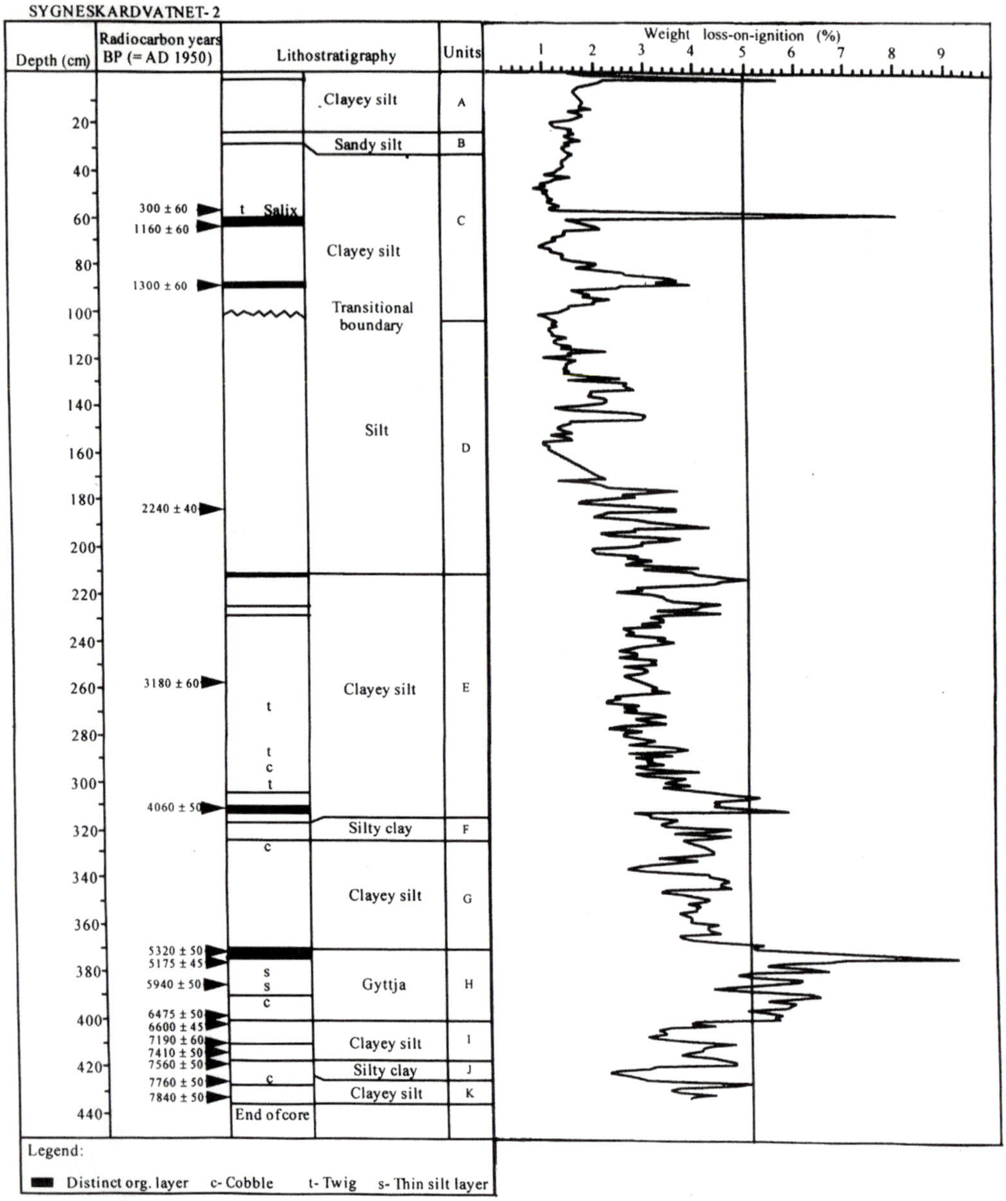

Fig. 8. Lithostratigraphy, radiocarbon dates, and weight loss-on-ignition in the Sygneskardvatnet-2 core. The vertical line at 5% on the weight loss-on ignition panel marks the value above which the sediments were brown-coloured.

tween 7500 and 6100 cal. yr BP (6600 and 5300 ^{14}C yr BP), interrupted by two minor glacier oscillations around 7000 and 6600 cal. yr BP (6200 and 5700 ^{14}C yr BP). The sedimentary record shows that the glacier has existed continuously since 6100 cal. yr BP (5300 ^{14}C yr BP). Loss-on-ignition analyses at 1-cm intervals and thin gyttja layers indicate that periods of glacier contraction occurred approximately 4500, 2700, and 2000 cal. yr BP (4000, 2600, and 2000 ^{14}C yr BP, respectively) and at 3 cm in the core (probably during the AD 1930s and 1940s with high summer temperatures). Magnetic susceptibility peaks in the upper half of the core may reflect increased glacier activity around 2000, 1500, 1200, 1100, and 700 cal. yr BP/^{14}C yr BP (Nesje et al. 2000).

In Jarbuvatnet [named Flatebrevatnet II by Karlén and Matthews (1992)], located close to Briksdal at the western side of Jostedalsbreen (Fig. 6), Karlén and Matthews (op.cit.) obtained a 1.7 m long core comprising stiff, homogeneous grey silt/clay and eight thin, but distinct moss layers between 130 and 170 cm depth in core. Radiocarbon dates of two 2-3 cm thick moss layers at 152-154 and 167-170 cm indicated the presence of Flatebreen, a minor glacier (1.55 km^2, altitudinal range from 1680 to 1090 m) at the SE margin of Myklebustbreen, at least since 5610 (5715-5585) cal. yr BP (4890 ± 80 ^{14}C yr BP), with intensified glacial activity after 5305 (5445-5085) cal. yr BP (4600 ±70 ^{14}C yr BP) and in particular subsequent to about 4800 cal. yr BP (4200 ^{14}C yr BP).

The history of Holocene glacier variations of Flatebreen at the SW part of the Jostedalsbreen ice cap has recently been reconstructed from new cores retrieved from the lacustrine sediments in the proglacial lake Jarbuvatnet (Nesje et al. 2001). The sedimentary succession shows evidence of three main episodes of Holocene glacier expansion (Fig. 9). The first is recorded in the basal part of the core

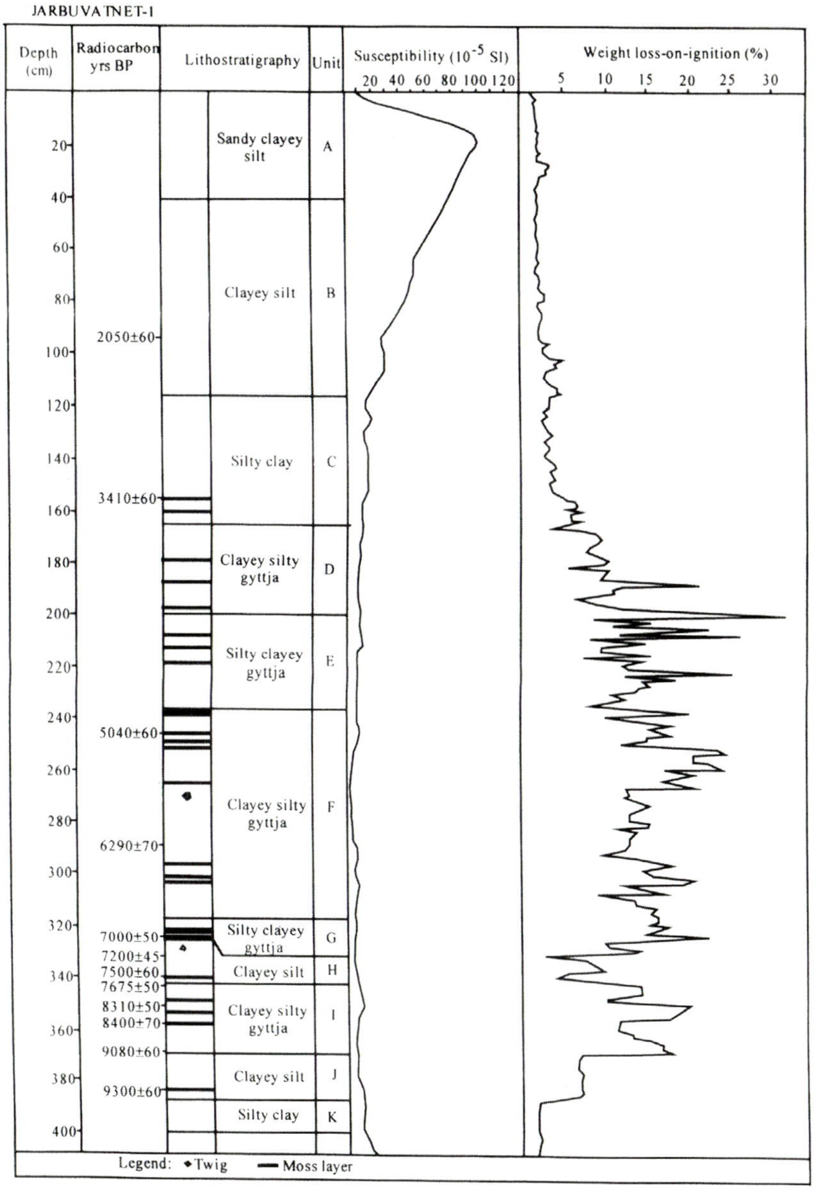

Fig. 9. Lithostratigraphy, radiocarbon dates, magnetic susceptibility and weight loss-on-ignition in core Jarbuvatnet-1.

up to 370 cm. According to the age/depth relationship in the sediment core (based on twelve AMS radiocarbon dates), this glacier expansion episode terminated about 10,000 cal. yr BP (9000 ^{14}C yr BP). The second major glacier phase occurred between 340 and 330 cm, and lasted from 8300 to 8000 cal. yr BP (7600 to 7200 ^{14}C yr BP), while the third is represented in the upper 170 cm in the core and was initiated at 4100 cal. yr BP (3700 ^{14}C yr BP). In the intervening periods, when the glacier was small and periodically melted from the lake catchment, the ELA may have been at least 260 m higher than at present. At 43 cm in the core, the medium silt content increases significantly, accompanied with a minor increase in the sand content. This textural change is interpreted as the first time that the terminus of Flatebreen reached an upstream lake at 1083 m a.s.l. The age model suggests that this event took place around 800 cal. yr BP (~AD 1150), as a response to the initial "Little Ice Age" glacier expansion after the 'Medieval Warm Period'. By using a Holocene summer temperature curve from central southern Norway in an exponential relationship between accumulation season precipitation and ablation-season temperature at the ELA, periods of more winter precipitation than the 1961-90 normal in the Jostedalsbreen region occurred 9700-9400, 9200-8300, 8200-6500, 5700-5100, 4700-4600, 4500-4300, 3800-3000, 2100-1800, 1600-1300, 1200-1000 cal. yr BP, and 900 cal. yr BP to the present. Periods of less winter precipitation than the 1961-90 normal occurred >9700, 9400-9200, 8300-8200, 6500-5700, 5100-4700, 4600-4500, 4300-3800, 3000-2100, 1800-1700, 1300-1200, and 1000-900 cal. yr BP. Periods of low winter precipitation during the Holocene in the Jostedalsbreen region correlate with periods of enhanced ice-rafting in the North Atlantic.

In a section in Glomsdalen (Fig. 10) (altitude 730 m, above the modern birch tree line), a northern tributary to the Stryn valley, a thin sand layer in the basal part may indicate a glacier episode between 6890 (7010-6795) cal. yr BP (6060 ± 80 ^{14}C yr BP) and 6505 (6665-6415) cal. yr BP (5740 ± 90 ^{14}C yr BP). Birch twigs are present in the basal part of the section, but are absent above a level dated at 5740 (5900-5650) cal. yr BP (5020 ± 90 ^{14}C yr BP). The disappearance of birch was apparently caused by a regional lowering of tree limit below the site (Nesje et al. 1991). Weakly laminated sand and silt, deposited between 4030 (4130-3935) cal. yr BP (3710 ± 50 ^{14}C yr BP) and 3280 (3385-3175) cal. yr BP (3100 ± 90 ^{14}C yr BP) reflect seasonal/annual rhythmic glaciofluvial sediment transport from a local cirque glacier northwest of Jostedalsbreen.

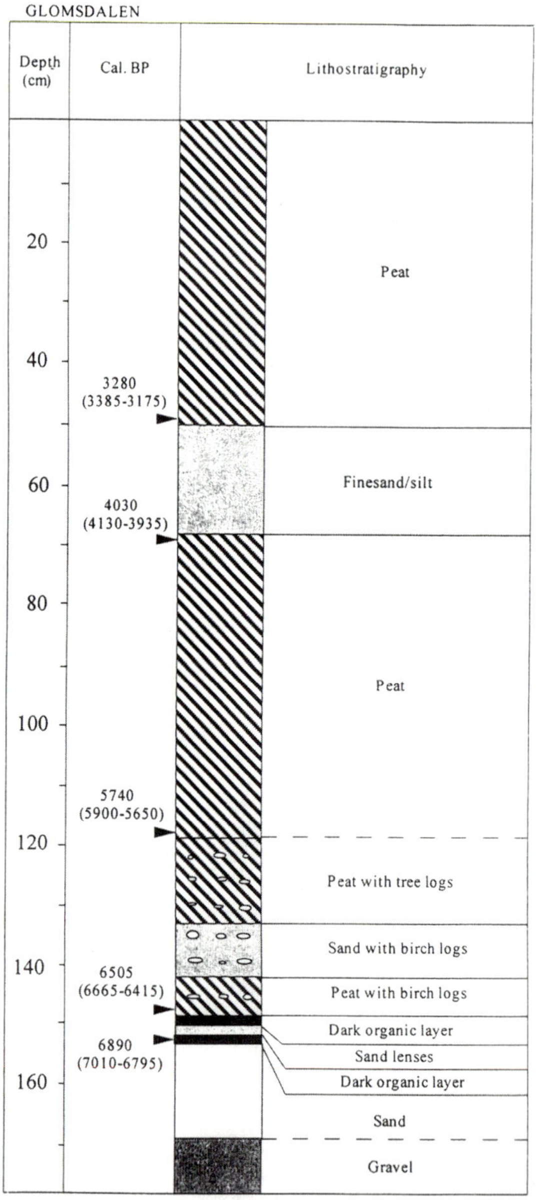

Fig. 10. Lithostratigraphy and radiocarbon dates [in calendar years BP (1 sigma in brackets)] from the Glomsdalen site (adapted from Nesje et al. 1991).

In Tunsbergdalen (at the SE margin of Jostedalsbreen, Fig. 6), radiocarbon dates from the base and top of a peat exposed by glacier retreat between AD 1962 and 1966 show that the glacier front was located upvalley from the site between 8985 (9190-8730) cal. yr BP (8083 ± 100 ^{14}C yr BP) and 4250 (4405-4150) cal. yr BP (3855 ± 55 ^{14}C yr BP).

In Rambjørgebotnen at the southwestern margin of Jostedalsbreen (Fig. 6), two sections (section I 265 cm deep and section II 130 cm deep) through Holocene clastic sediments and peat are exposed along a meltwater stream that drains two minor cirque glaciers SW of the sites. In section I, which mainly consists of peat (see pollen evidence above), Torske (1996), described four greyish-blue sand/silt horizons at 63, 158, 173, and 182 cm depth, which by linear interpolation between four radiocarbon-dated levels, were deposited at ca. 5750, 9000, 9250, and 9400 cal. yr BP (ca. 5000, 8200, 8300, and 8400 ^{14}C yr BP). Four clastic sedimentary units in section II consist of bluish-grey sand and silt. The blue-grey silt/sand layers in both sections, similar in texture to glacially eroded rock flour, are interpreted to have been deposited during periods of glacier activity. The intervening peat units are interpreted to have been deposited during periods when the glaciers were very small and dynamically inactive or absent in the catchment (Nesje and Dahl 1991). Radiocarbon ages from section II indicate that glaciers were present between 2745 (2770-2550) cal. yr BP (2595 ± 85 ^{14}C yr BP) and 2350 (2460-2325) cal. yr BP (2360 ± 80 ^{14}C yr BP), between 2225 (2340-2145) cal. yr BP (2250 ± 65 ^{14}C yr BP) and 2130 (2305-2000) cal. yr BP (2150 ± 80 ^{14}C yr BP), between 1670 (1720-1540) cal. yr BP (1740 ± 75 ^{14}C yr BP) and 1675 (1715-1535) cal. yr BP (1730 ± 75 ^{14}C yr BP), and subsequent to 1310 (1345-1290) cal. yr BP (1430 ± 45 ^{14}C yr BP).

A section at Sandsvora at the northwestern margin of Jostedalsbreen (Fig. 6) revealed five palaeosols containing tree stumps (Fig. 11), interlayered with distal glaciofluvial fan material deposited by glacial meltwater from a local cirque glacier, Sandskardfonna (Nesje et al. 1991). Radiocarbon ages of tree stumps from the palaeosols are 1220 (1275-1090) cal. yr BP (1270 ± 60 ^{14}C yr BP), 1065 (1170-985) cal. yr BP (1180 ± 70 ^{14}C yr BP), 985 (1065-945) cal. yr BP (1110 ± 60 ^{14}C yr BP), 780 (910-725) cal. yr BP (890 ± 60 ^{14}C yr BP), and 605 (655-540) cal. yr BP (620 ± 70 ^{14}C yr BP). The youngest palaeosol predate the initial "Little Ice Age" glacier expansion of the cirque glacier.

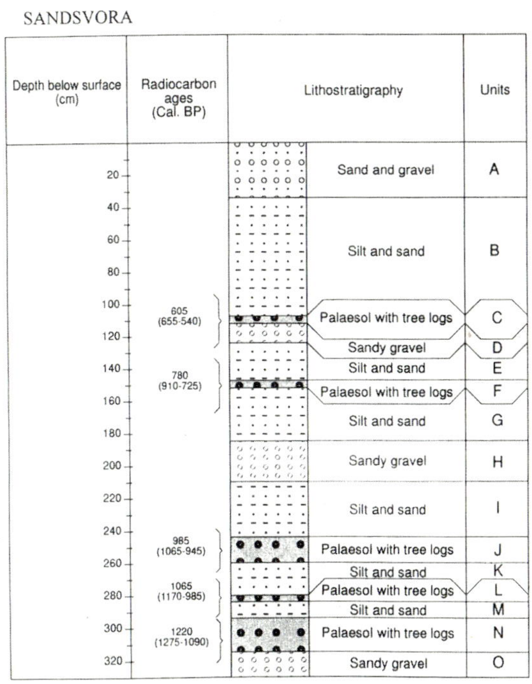

Fig. 11. Lithostratigraphy and radiocarbon dates [in calendar years BP (1 sigma in brackets)] from the Sandsvora site (adapted from Nesje et al. 1991).

Hardangerjøkulen

Northern Sector

Holocene glacier fluctuations along the northern margin of Hardangerjøkulen have been reconstructed from four lake and terrestrial sites (Dahl and Nesje 1994, 1996). Between about 10,100 cal. yr BP (9100 ^{14}C yr BP) and 9500 cal. yr BP (8600 ^{14}C yr BP) the glacier readvanced and deposited terminal moraines beyond the "Little Ice Age"

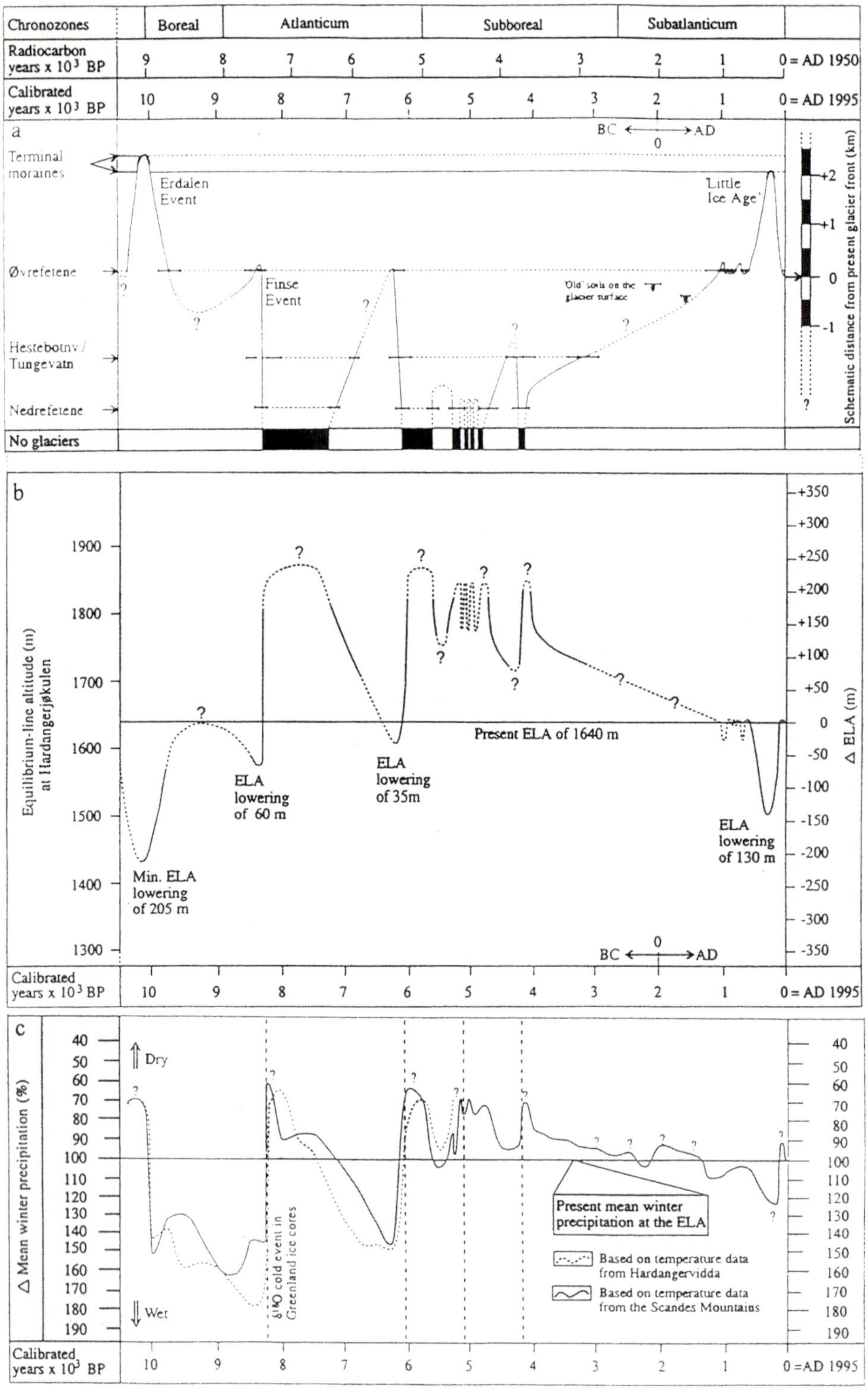

Fig. 12. **(a)** Holocene glacier fluctuations at the northern sector of Hardangerjøkulen inferred from radiocarbon dated lacustrine and terrestrial sites in the Finse area (adapted from Dahl and Nesje 1994). **(b)** Holocene ELA fluctuations at the northern sector of Hardangerjøkulen (from Dahl and Nesje 1996). **(c)** Holocene variations in winter precipitation at the present ELA at Hardangerjøkulen (adapted from Dahl and Nesje 1996).

Synthesis of Holocene Glacial Activity in Southern Norway

Evidence of Holocene glacial activity from sites described in the text is shown in Fig. 15. The first half of the Holocene was characterised by few, and mostly asynchronous glacial episodes, however, with some increased glacial activity around 10,500-10,000, 8800 (Fig. 16). Subsequent to 4200 cal. yr BP, the glacial activity increased significantly in southern Norway, peaking around 3500, 2600, and 2200 cal. yr BP and in particular during the last centuries (the "Little Ice Age").

Comparison with the Northern Scandinavian Record

Because Scandinavia is a climatologically heterogeneous region and the records of Holocene glacier variations are based on different types of evidence, there has been some discussion concerning the age, magnitude and degree of synchronism of Holocene glacier fluctuations. Mass balance studies in Scandinavia show, for example, that the net balance on present maritime glaciers is controlled mainly by winter precipitation, whereas the net balance on continental glaciers are controlled mainly by summer temperature (Fig. 17). Karlén and Kuylenstierna (1996) and Karlén et al. (1995) suggested that a number of Neoglacial advances occurred in Scandinavia during the Holocene, the most significant glacier advances occurring 8500-7900, 7400-7200, 6300-6100, 5900-5800, 5600-5300, 5100-4900, 4600-4100, 3400-3200, 3000-2800, 2700-2000, 1900-1600, 1200-1000, and 700-200 cal. yr BP (7700-7200, 6700-6400, 5600-5300, 5100-5000, 4800-4600, 4450-4300, 4050-3800, 3300-3050, 2900-2700, 2600-2050, 1950-1600, 1300-1050, and 800-200 ^{14}C yr BP) (Fig. 16). Matthews (1991) demonstrated, however, that there so far is no moraine stratigraphic evidence for Neoglacial advances of similar magnitude as the "Little Ice Age" in southern Norway.

The data presented here also clearly demonstrates that the most extensive glacial activity occurred during the second half of the Holocene. The evidence also suggests that the early Holocene glacial episodes in southern Norway were partly asynchronous with those reported from elsewhere in Scandinavia by Karlén and Kuylenstierna (1996) (Fig. 16). Their early Holocene glacier expansion phases seem, however, to be more-or-less synchronous with episodes of colluvial activity re-corded from Møre (Blikra and Nesje 1997) and Jotunheimen (Matthews et al. 1997) (Fig. 16). Periods of colluvial deposition are considered to reflect increase in the magnitude and frequency of debris-flow activity due to increased rainfall and/or snowmelt intensity. Karlén's evidence for early Holocene glacier advances in northern Scandinavia has been challenged by multi-disciplinary studies by Snowball and Sandgren (1996). They suggested that the Kårsa glacier disappeared during the early and mid-Holocene and was reformed at ca. 3300 cal. yr BP (ca. 3000 ^{14}C yr BP).

Future Research Strategies

Differences in magnitude and timing of Late Glacial and Holocene ice advances have the potential to establish climatic gradients and circulation patterns and provide information about underlying climatic factors. Glacial and climate reconstructions in southern Norway are particularly important in this respect because it is possible to record changes in oceanic and atmospheric circulation patterns in the North Atlantic region. Controls on Late Glacial deglaciation and the following Holocene period merit further study. So far, there has been a tendency to produce regional correlations of glacier and climate variations. The data compilation presented here, however, demonstrates significant intraregional differences in climate (temperature/precipitation) and glacier response. Future research should therefore focus on establishing representative, multi-disciplinary, well-dated records from individual glaciers to study similarities and dissimilarities in glacier response to topographical and climatic parameters (e.g. summer temperature, winter precipitation, wind direction, glacier aspect, radiation).

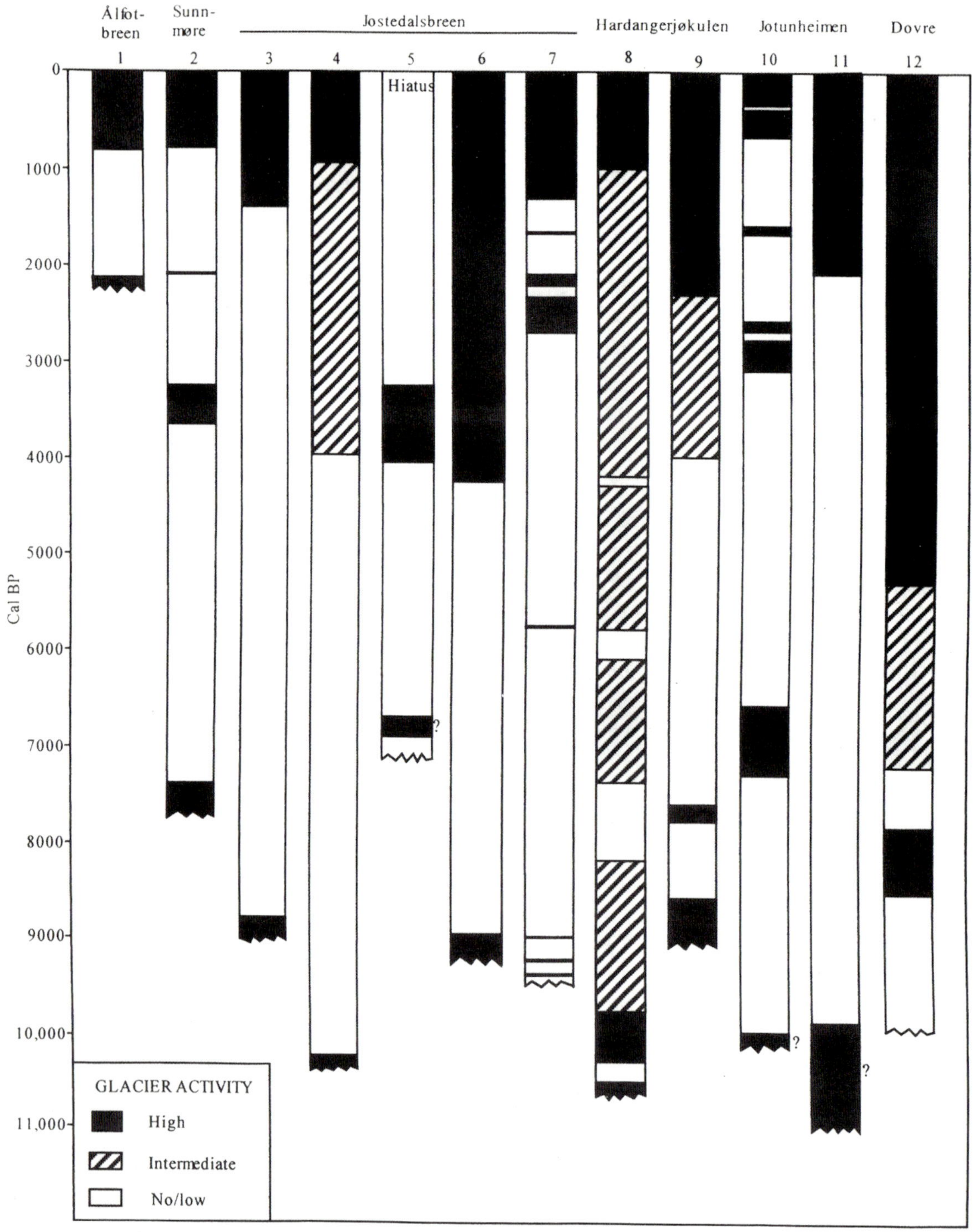

Fig. 15. Holocene glacial activity at sites described in the text; 1. Grøndalsvatn, 2. Midtivatnet, 3. Vanndalsvatnet, 4. Jarbuvatnet, 5. Glomsdalen, 6. Tunsbergdalen, 7. Rambjørgebotnen, 8. Finse, 9. Isdalsvatnet, 10. Gjuvvatnet, 11. Bøvertunsvatnet, 12. Snøheim. For references, see text.

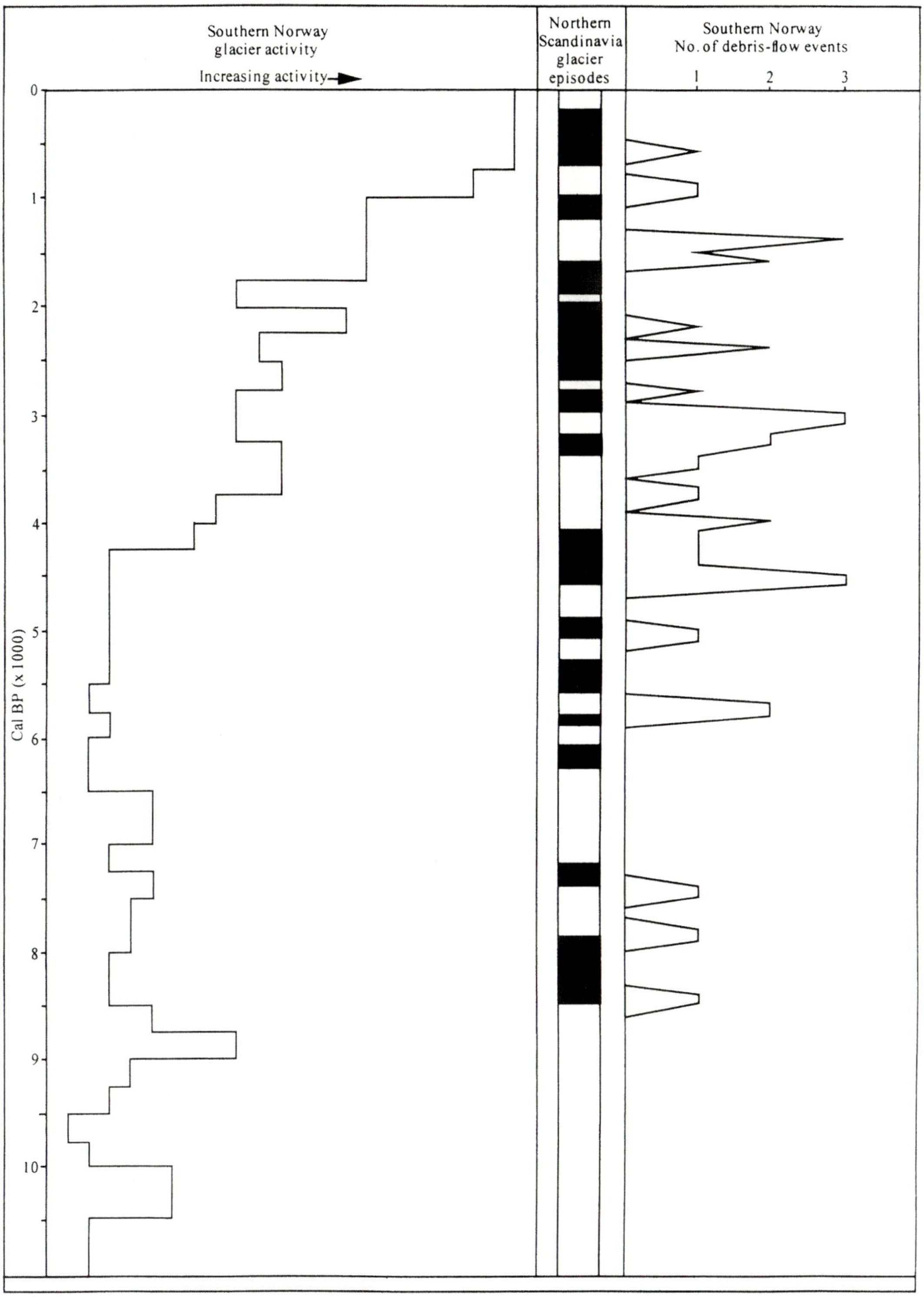

Fig. 16. Holocene glacier activity in southern Norway based on the evidence presented in Fig. 13, Holocene glacier episodes in northern Scandinavia (Karlén and Kuylenstierna 1996), and Holocene colluvial (mainly debris flows) activity based on data from Møre and Jotunheimen (adapted from Matthews et al. 1997). Note timescale in calendar yr BP.

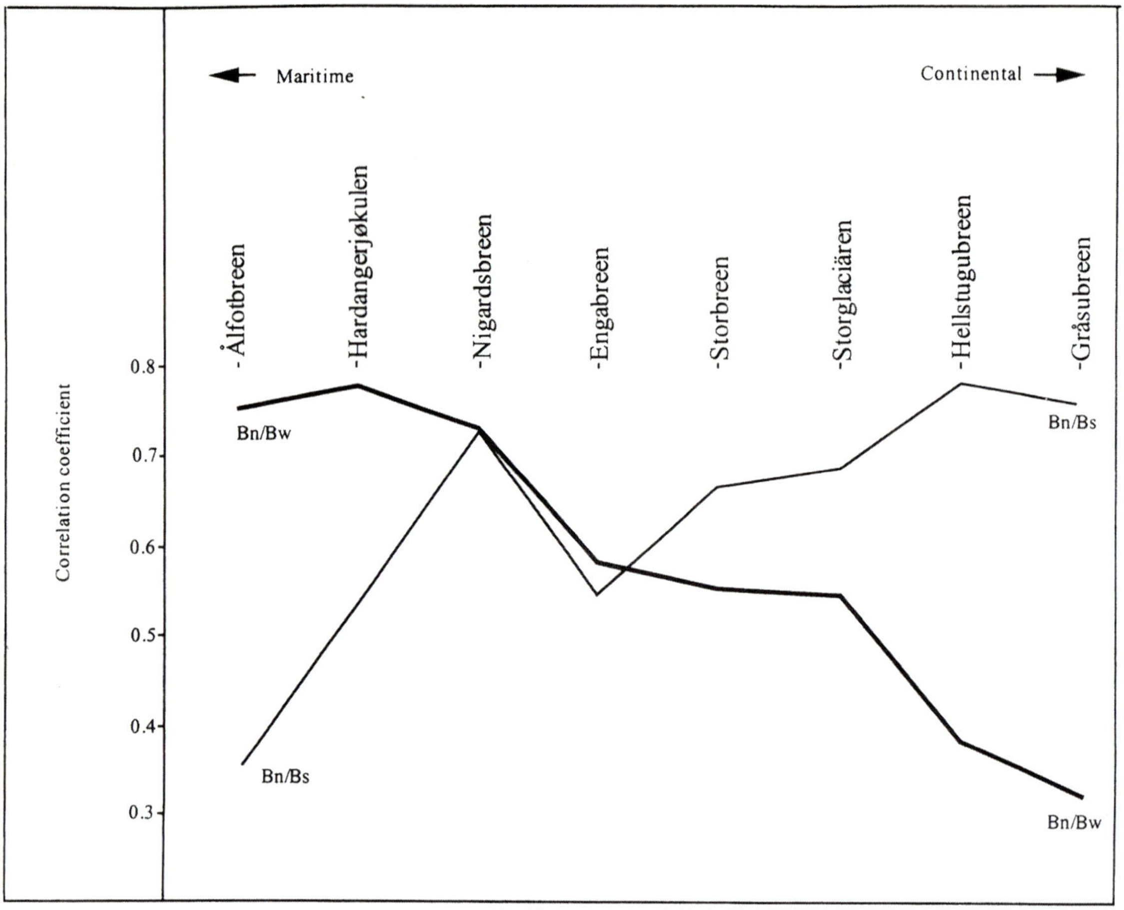

Fig. 17. Correlation coefficient between net balance (Bn) and winter (Bw)/summer (Bs) balance at Scandinavian glaciers. The calculations on the Norwegian glaciers are based on data provided by Glaciology Section, Norwegian Water Resources and Energy Administration (NVE), while the calculations on Storglaciären is based on Holmlund et al. (1996).

References

Alley RB, Mayewski PA, Sowers T, Stuiver M, Taylor KC, Clark PU (1997) Holocene climate instability: A prominent, widespread event 8200 yr ago. Geology 25:483-486

Andersen BG, Mangerud J, Sørensen R, Reite AJ, Sveian H, Thoresen, M, Bergstrøm B (1995) Younger Dryas ice-marginal deposits in Norway. Quat Int 28:147-169

Andersen JL, Sollid JL (1971) Glacial chronology and glacial morphology in the marginal zones of the glaciers Midtdalsbreen and Nigardsbreen, South Norway. Nor geogr Tidsskr 25:1-38

Bakke J (1999) Rekonstruksjon av bre- og klimavariasjonar på Nordre Folgefonna med kringliggjande botnar. Thesis Department of Geography, University of Bergen, pp 1-172

Bickerton RW, Matthews JA (1993) «Little Ice Age» variations of outlet glaciers from the Jostedalsbreen ice cap, southern Norway: a regional lichenometric-dating study of ice-marginal moraine sequences and their climatic significance. Journ Quat Sci 8:45-66

Birks HH, Paus Aa, Svendsen JI, Alm T, Mangerud J, Landvik J (1994) Late Weichselian environmental change in Norway, including Svalbard. Journ Quat Sci 9:133-145

Blikra LH, Nesje A (1997) Holocene avalanche activity in western Norway: chronostratigraphy and palaeoclimatic implications. In: Matthews JA, Brunsden D, Frenzel B, Gläser B, Weiss MM (eds)

Rapid mass movement as a source of climatic evidence for the Holocene. Palaeoclim Res 19:299-312

Caseldine CJ, Matthews JA (1987) Podzol development, vegetation change and glacier variations at Haugabreen, south Norway. Boreas 16:215-230

Dahl SO, Nesje A (1992) Palaeoclimatic implications based on equilibrium-line altitude depressions of reconstructed Younger Dryas and Holocene cirque glaciers in inner Nordfjord, western Norway. Palaeogeog, Palaeoclimat, Palaeoecol 94:87-97

Dahl SO, Nesje A (1994) Holocene glacier fluctuations at Hardangerjøkulen, central-southern Norway: a high resolution composite chronology from lacustrine and terrestrial deposits. The Holocene 4:269-277

Dahl SO, Nesje A (1996) A new approach to calculating Holocene winter precipitation by combining glacier equilibrium-line altitudes and pine-tree limits: a case study from Hardangerjøkulen, central southern Norway. The Holocene 6:381-398

Dahl SO, Nesje A, Øvstedal J (1997) Cirque glaciers as morphological evidence for a thin Younger Dryas ice sheet in east-central southern Norway. Boreas 26:161-180

Erikstad L, Sollid JL (1986) Neoglaciation in South Norway using lichenometric methods. Nor geog Tidsskr 40:85-105

Fareth OW (1987) Glacial geology of the middle and inner Nordfjord. Nor geol unders Bull 408:1-55

Garnes K, Bergersen OF (1980) Wastage features of the inland ice sheet in central South Norway. Boreas 9:251-269

Grove JM (1988) The Little Ice Age. Methuen, London, pp 498

Gunnarsdottir H (1996a) Holocene vegetation history and forest-limit fluctuations in Smådalen, eastern Jotunheimen, South Norway. Palaeoclimate Research 20:233-255

Gunnarsdottir H (1996b) Holocene vegetation history in the northern parts of the Gudbrandsdalen valley, south central Norway. Dr. Thesis, University of Oslo (separate papers)

Haflidason H, Sejrup HP, Klitgaard Kristensen D, Johnsen S (1995) Coupled response of the late glacial climatic shifts of northwest Europe reflected in Greenland ice cores: Evidence from the northern North Sea. Geology 23:1059-1062

Holmlund P, Karlén W, Grudd H (1996) Fifty years of mass balance and glacier front observations at the Tarfala Research Station. Geog Ann 78A, 105-114

Johnsen SJ, Clausen HB, Dansgaard W, Fuhrer K, Gundestrup N, Hammer, CU, Iversen P, Jouzel J, Stauffer B, Steffensen JP (1992) Irregular glacial interstadials recorded in a new Greenland ice core. Nature 359:311-313

Karlén W, Kuylenstierna J (1996) On solar forcing of Holocene climate: evidence from Scandinavia. The Holocene 6:359-365

Karlén, W, Matthews JA (1992) Reconstructing Holocene glacier variations from glacial lake sediments: studies from Nordvestlandet and Jostedalsbreen-Jotunheimen, southern Norway. Geog Ann 74:327-348

Karlén W, Bodin A, Kuylenstierna J, Näslund J-O (1995) Climate of northern Sweden during the Holocene. Journ Coast Res Special Issue 17: Holocene Cycles: Climate, Sea Levels, and Sedimentation: 49-54

Kjenstad K, Sollid JL (1982) Isavsmeltingsforløpet i Trondheimsfjorden. Glasialdynamiske prinsipper. Nor Geol Tidsskr 36:153-162

Kvamme M (1993) Holocene forest limit fluctuations and glacier development in the mountains of Southern Norway, and their relevance to climate history. Pal Res 9:99-113

Larsen E, Eide F, Longva O, Mangerud J (1984) Allerød - Younger Dryas climate inferences from cirque glaciers and vegetational development in the Nordfjord area, western Norway. Arc Alp Res 16:137-160

Longva O, Thoresen M (1991) Iceberg scours, iceberg gravity craters and current erosion marks from a gigantic Preboreal flood in southeastern Norway. Boreas 20:47-62

Mangerud J, Larsen E, Longva O, Sønstegaard E (1979) Glacial history of western Norway 15,000-10,000 B.P. Boreas 8:179-187

Matthews JA (1977) A lichenometric test of the 1750 end-moraine hypothesis: Storbreen gletschervorfeld, southern Norway. Nor geogr Tidsskr 31:129-136

Matthews JA (1991) The late Neoglacial («Little Ice Age») glacier maximum in southern Norway: new ^{14}C-dating evidence and climatic implications. The Holocene 1:219-233

Matthews JA, Caseldine CJ (1987) Arctic-alpine brown soils as a source of palaeoenvironmental information: further ^{14}C dating and palynological evidence from Vestre Memurubreen, Jotunheimen, Norway. Journ Quat Sci 2:59-71

Matthews JA, Karlén W (1992) Asynchronous Neoglaciation and Holocene climatic change reconstructed from Norwegian glaciolacustrine sedimentary sequences. Geology 20:991-994

Matthews JA, Shakesby RA (1984) The status of the «Little Ice Age» in southern Norway: relative age-dating of Neoglacial moraines with Schmidt hammer and lichenometry. Boreas 13:333-346

Matthews JA, Dahl SO, Berrisford MS, Nesje A, Dresser P Q, Dumayne-Peaty L (1997) A preliminary history of Holocene colluvial (debris-flow) activity, Leirdalen, Jotunheimen, Norway. Journ Quat Sci 12:117-129

Matthews JA, Dahl SO, Nesje A, Berrisford MS, Andersson C (2000) Holocene glacier variations in central Jotunheimen, southern Norway based on distal glaciolacustrine sediment cores. Quat Sci Rev 19:1625-1647

Nesje A, Dahl SO (1991) Late Holocene glacier variations in Bevringsdalen, Jostedalsbreen region, western Norway (ca 3200-1400 BP). The Holocene 1:1-7

Nesje A, Dahl SO (1993) Lateglacial and Holocene glacier fluctuations and climate variations in western Norway: A review. Quat Sci Rev 12:255-261

Nesje A, Kvamme M, Rye N, Løvlie R (1991) Holocene glacial and climate history of the Jostedalsbreen region, western Norway; evidence from lake sediments and terrestrial deposits. Quat Sci Rev 10:87-114

Nesje A, Dahl SO, Løvlie R, Sulebak J (1994) Holocene glacier activity at the soutwestern part of Hardangerjøkulen, central-southern Norway: evidence from lacustrine sediments. The Holocene 4:377-382

Nesje A. Dahl SO, Løvlie R (1995) Late Holocene glacier and avalanche activity in the Ålfoten area, western Norway: evidence from a lacustrine sedimentary record. Nor Geol Tidsskr 75:120-126

Nesje A, Dahl SO, Andersson C, Matthews JA (2000) The lacustrine sedimentary sequence in Sygneskardvatnet, western Norway: a continuous, high-resolution record of the Jostedalsbreen ice cap during the Holocene. Quat Sci Rev 19:1047-1065

Nesje A, Matthews JA, Dahl SO, Berrisford MS, Andersson C (2001) Holocene glacier fluctuations of Flatebreen and winter precipitation changes in the Jostedalsbreen regions western Norway, based on glaciolacustrine records. The Holocene 11:267-280

Paus Aa (1992) Late Weichselian vegetation, climate, and floral migration in Rogaland, soutwestern Norway; pollen analytical evidence from four lateglacial basins, Dr. Thesis University of Bergen, pp 1-87

Rye N, Nesje A, Lien R, Anda E (1987) The Late Weichselian ice sheet in the Nordfjord-Sunnmøre area and deglaciation chronology for Nordfjord, western Norway. Nor geog Tidsskr 41:23-43

Selsing L (1996) The climatic interpretation of Holocene megafossils of pine (Pinus sylvestris L.) from the mountain area of southern Norway; the importance of the precession in controlling Holocene climate. Palaeoclim Res 20:147-156

Snowball I, Sandgren P (1996) Lake sediment studies of Holocene glacial activity in the Kårsa valley, northern Sweden: contrast in interpretation. The Holocene 6:367-372

Stuiver M, Reimer PJ (1993) Extended ^{14}C data base and revised CALIB 3.0 ^{14}C age calibration program. Radiocarbon 35:215-230

Sørensen R (1979) Late Weichselian deglaciation in the Oslofjord area, south Norway. Boreas 8:241-246

Torske N (1996) Holocene vegetation, climate and glacier histories in the Jostedalsbreen region, western Norway - palaeoecological interpretations from an alpine peat deposit. Palaeoclim Res 20:215-232

Torsnes I, Rye N, Nesje A 1993 Modern and Little Ice Age equilibrium-line altitudes on outlet valley glaciers from Jostedalsbreen, western Norway: an evaluation of different approaches to their calculation. Arct Alp Res 2:106-116

Tvede AM (1973) Folgefonni, en glasiologisk avviker. Naturen 97:11-15

Tvede AM, Liestøl O (1977) Blomsterskardbreen, Folgefonni, mass balance and recent fluctuations. Nor Polarinst Årb 1976:225-234

Holocene Palaeoenvironmental Changes in North-West Europe: Climatic Implications and the Human Dimension

Brian Huntley*[1], Mike Baillie[2], Jean M. Grove[3], Claus U. Hammer[4], Sandy P. Harrison[5], Stefanie Jacomet[6], Eystein Jansen[7], Wibjörn Karlén[8], Nalân Koç[9], Jürg Luterbacher[10], Jörg Negendank[11], Jörg Schibler[12]

[1] *Environmental Research Centre, University of Durham, South Road, UK-Durham DH1 3LE, United Kingdom*
[2] *The Queen's University of Belfast, School of Archaeology and Palaeoecology, University Road, UK-Belfast BT7 1NN, Northern Ireland*
[3] *Girton College, UK-Cambridge CB3 OJG, United Kingdom*
[4] *University of Copenhagen, Niels Bohr Institute of Astronomy, Physics and Geophysics, Dept. of Geophysics, Juliane Maries Vej 30, DK-2100 Copenhagen O, Denmark*
[5] *Max-Planck Institute for Biogeochemistry, Postfach 100164, D-07701 Jena, Germany*
[6] *Universität Basel, Archäobiologische Abteilung, Seminar für Ur- und Frühgeschichte, Petersgraben 9-11, CH-4051 Basel, Switzerland*
[7] *Department of Geology, University of Bergen, Bergen, Norway*
[8] *University of Stockholm, Department of Physical Geography and Quaternary Geology, S-10691 Stockholm, Sweden*
[9] *Norwegian Polar Institute, Polar Environmental Centre, N-9296 Tromsø, Norway*
[10] *Institute of Geography, Climatology and Meteorology, University of Bern, Hallerstraße 12, CH-3012 Bern, Switzerland*
[11] *GeoForschungsZentrum Potsdam, Telegrafenberg, D-14473 Potsdam, Germany*
[12] *Universität Basel, Seminar für Ur- und Frühgeschichte, Abt. Archäologie, Petersgraben 9-11, CH-4051 Basel, Switzerland*
**corresponding author (e-mail): Brian.Huntley@durham.ac.uk*

Abstract: Holocene climate changes and variability in Europe are outlined on three time scales: long-term changes throughout the period as a whole; shorter-term fluctuations at centennial to millennial scales; and events with an annual to multi-decadal duration. Human population history in Europe during the Holocene is considered in relation to this history of climatic change and variability; in particular, evidence is reviewed that may indicate that human populations responded to, and perhaps also contributed to, climatic changes. The limitations of the data sources and chronological techniques available are briefly outlined.

The prevalent long-term pattern is one of relatively warmer conditions during the early to mid-Holocene, temperature subsequently declining towards its present values. Superimposed upon this pattern, however, are marked spatial patterns in the strength of the overall changes, as well as evidence that temperature seasonality varied systematically during the Holocene. No clear patterns in precipitation change emerge at the scale of Europe as a whole; long-term precipitation changes varied both between regions and between seasons. The extent to which human populations responded to these long-term changes, in particular through changes in agricultural practices or settlement patterns, is a matter of ongoing debate.

From WEFER G, BERGER W, BEHRE K-E, JANSEN E (eds), 2002, *Climate Development and History of the North Atlantic Realm.* Springer-Verlag Berlin Heidelberg, pp 259-298

At centennial to millennial scales, numerous independent palaeoclimate sensors provide evidence of variability, with some fluctuations being recorded by many different sensors. A marked cooling event at ca. 8.2 ka BP is widely recorded. Although other fluctuations correlate with less certainty between sensors, a common pattern emerges from many of these records. There is some evidence that these fluctuations may be periodic in character. Although chronological limitations generally preclude making definitive links between records of human populations and those of climatic fluctuations, there are notable coincidences between dates for archaeological events and for climatic fluctuations. In the more recent period, however, especially the last millennium, the link is quite secure between the climatic fluctuations of the 'Mediæval Warm Period' and 'Little Ice Age', and impacts upon human populations.

At the shortest time scales, annual to multi-decadal events are recorded in tree rings and other records that offer annual resolution. The most striking of these events are dramatic, albeit generally short-lived, cooling events; furthermore, at least some of these events were apparently global in extent. Although the cause of these events remains a matter of debate, it seems most probable that they reflect catastrophic events of some nature that resulted in substantial quantities of material being injected into the upper atmosphere, thus reducing penetration of solar radiation for a period of years to a decade or more. Some of these events appear to be recorded in human mythology, and it is apparent that they have often coincided with major events in human history.

Although much remains to be discovered, and much further research is required, it is apparent that human populations have been affected by climatic change, variability and events during the Holocene. It is also apparent that, when considered in the context of the Holocene, the twentieth century period of increase in atmospheric concentrations of carbon dioxide and other greenhouse gases, and of the onset of consequent global climatic warming, is anomalous. The predicted rate and magnitude of twenty first century global mean temperature increase will probably exceed anything experienced during the Holocene.

Introduction

The beginning of the Holocene period, some 11,400 years ago (10,000 ^{14}C yr BP; Radiocarbon years before present where 'present' is conventionally taken to be AD 1950), was marked by a rapid global climate change of large magnitude. Subsequently, the climate has exhibited variability upon time scales ranging from years to millennia, as well as long-term trends spanning the period as a whole. North-west Europe, the adjacent ocean basins and Greenland together offer an unrivalled density of records of Holocene palaeoenvironmental conditions. In ad-dition, the history of humans in Europe has been intensively studied by archaeologists, providing a unique opportunity to examine the extent to which humans may have responded to, and perhaps also affected, the changing climate of the region. The overall aims of the present paper are to review what is known of Holocene climate changes in north-west Europe and to consider human history in the region in relation to those changes. The review is organised into three sections that deal in turn with the long-term changes across the period as a whole, with shorter-term fluctuations that span centuries to millennia, and with events that have a duration of between a year and several decades. A final discussion then considers the extent to which the late twentieth century is climatically anomalous in the context of the record of Holocene climate variability.

Before commencing the review it is necessary to deal with three general issues: firstly, the terminology to be used for chronological periods; secondly, the chronological framework to be used; and thirdly, the nature and inherent limitations both of the data sources that provide palaeoenvironmental evidence and of the reconstruction techniques used to infer past climate conditions from these data. It is also appropriate to discuss the key questions that we attempt to address in our review.

Terminology

Two series of terms continue to be widely (mis-)used in the literature as if they refer to fixed chronological sub-divisions of the Holocene: firstly, the scheme of sub-divisions of the Holocene originally introduced by Blytt and Sernander (Blytt 1876, 1881; Sernander 1908, 1910), and subsequently widely applied by palynologists (Pre-Boreal, Boreal, Atlantic, Sub-Boreal, Sub-Atlantic); and secondly, the sequence of terms used by archaeologists to refer to the 'cultural grade' of human groups (Palaeolithic, Mesolithic, Neolithic, Bronze Age, Iron Age, Roman, Post-Roman, Early Mediæval, Mediæval, etc.) (Fig. 1). Whereas the first series of terms were appropriate when introduced, and in the geographical area of southern Scandinavia where they were first used, and the second series are entirely appropriate when used to refer strictly to different cultural grades, we avoid using any of these terms to refer to chronological periods for two reasons. Firstly, both sets of terms refer to, or are defined by, phenomena that are diachronous between regions within north-west Europe. The transition to the Bronze Age, for example, defined by the first evidence of bronze artefacts, occurs several centuries earlier in the south of the region than in the north (Fig. 1 and see Hochuli et al. 1998). The 'Boreal-Atlantic Transition', as defined by palynological criteria, is characterised by a rapid increase in the abundance of *Alnus* pollen, an event that is markedly diachronous even within northern Europe (Huntley and Birks 1983). Secondly, there are regional differences in climate and/or human history within north-west Europe: for example, there are no Roman archaeological sites in northern Germany or Fennoscandia; climate history, as reflected by the palynological record, differs markedly between the Alpine region and Fennoscandia (Huntley and Prentice 1993). For these reasons we would urge that others should also in future avoid the misuse of these terms as if they refer to fixed chronological periods.

We also avoid using the term 'neoglacial', principally because its use could be misleading as a result of it being inadequately defined. We do use the terms 'Little Ice Age' (Grove 1988) and 'Mediæval Warm Period' because they are convenient and useful labels for two intervals in the historic period characterised by particular patterns of deviation of their mean climatic conditions from the mean conditions of the mid twentieth century. In order that our usage is clear, however, we first define how we are using these terms.

The Little Ice Age (LIA) we define as: "the globally relatively cool period during the second millennium AD, characterised by the advance of non-surging mountain glaciers in many parts of the world". It should be noted that the period was not one of uniform cool conditions, but rather was characterised by shorter-term fluctuations, as in any other period, to which the glaciers responded with fluctuations, albeit about a relatively 'forward' position. Data are available showing that such LIA glacier fluctuations occurred in Europe (Scandinavia and the Alps), Asia (The Urals, Altai and Kamchatka), North America (Alaskan mountains, Rockies, Coast Ranges of British Columbia), South America (both tropical and temperate mountain ranges), The Southern Alps of New Zealand, the Sub-Antarctic Islands (especially South Shetlands) and, less certainly, Antarctica. Evidence of initiation of LIA advances in the 13th and 14th centuries is widespread (Grove, in press)

The Medieval Warm Period (MWP) (Hughes and Diaz 1994; Pfister et al. 1998) we define as: "the relatively warm period, centred around AD 1000, that preceded the Little Ice Age and followed a previous cooler period during the 8th and 9th centuries AD".

We deliberately avoid assigning any particular dates to the beginning and end of these periods, or to the transition between the MWP and the LIA, in large part because there are clear regional differences in the expression and timing of the sequence of cooling and warming events during the last millennium even within north-west Europe. Understanding the dynamics of the climate system that these complex changes reflect requires careful assessments of their relative timing and duration and of their spatial patterns. Rigid and artificial definitions of their chronology contribute nothing to our understanding of how these fluctuations came about.

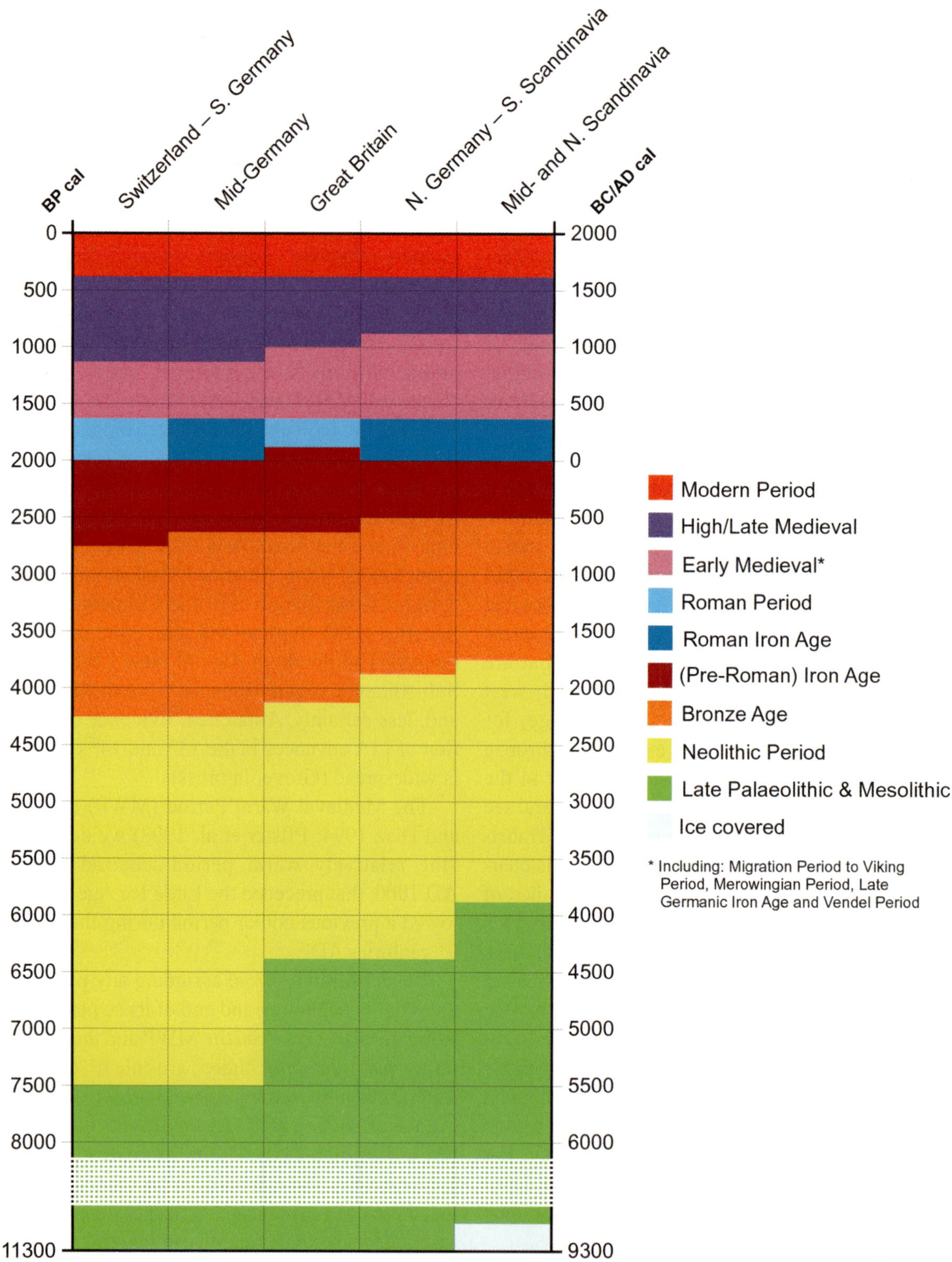

Fig. 1. Chronology of pre-historic and historic periods in various regions of Europe. (Based upon: Boaz (1998); Darvill (1987); Dyer (1990); Gibson (1986); Hoika and Meurers-Balke (1994); Hoops (1981); Hvass and Storgaard (1993); Lüning (1996); Müller et al. (1999); NNA (1995); Prescott (1995); Randsborg (1996); Regnell (1998); Rychner (1998); Stöckli (1995); *plus* personal communications from F. Siegmund and A. Zimmermann)

It is more difficult to identify an appropriate term by which to refer to the cool period that preceded the MWP. Amongst the terms used in the literature, the 'Dark Ages' and the 'Early Medieval' are defined in terms of human history and not in terms of the record of climatic variability, although it seems increasingly likely that the Dark Ages were triggered by a global environmental event in the mid sixth century (Baillie 1994). Other terms sometimes used have regional rather than general relevance, often being defined in terms of one particular climatic record. We prefer not to apply any of these terms in our discussion; nor do we propose any new term, although we recognise the value and convenience of developing a set of terms that could be widely and unambiguously applied to the millennial-scale periods of warmer and cooler climatic conditions that appear to characterise the Holocene. For the present we refer to all periods prior to the MWP in terms only of their chronology.

Chronology

Several distinct methods are available for determining the timing and duration of Holocene palaeoenvironmental changes; broadly, however, these fall into two classes. The first is based upon counting annual layers in appropriate materials, including, for example, tree rings in wood and accumulation layers in ice, lake sediments and calcite deposits. Such methods approximate more or less closely to a calendar year time scale, although each is characterised by some degree of uncertainty. The second class of methods are those based upon measuring some physical, or sometimes chemical, property of a material as the basis for estimating its age. Of these, radiocarbon (^{14}C) dating is the most widely used for dating organic materials of Holocene age. The basis of this technique requires an assumption that the ratio of ^{14}C to ^{12}C in the atmosphere has been constant; this, however, has been demonstrated not to have been the case. However, ^{14}C age estimates may be transformed into calendar year age estimates using calibrations obtained from numerous ^{14}C measurements upon wood of known dendrochronological age (Stuiver and Reimer 1993; Stuiver et al. 1998).

Throughout the remainder of this review the time scale used is in calendar years unless otherwise indicated. The age estimates used to obtain this time scale include calibrated ^{14}C age estimates as well as ages determined from materials with annual layers. Dates are expressed informally as, for example, '3000 years ago', or formally as, for example, '3000 BP' or '3 ka BP' (i.e. 'before present', where present is conventionally defined as AD 1950). In general dates are not cited as 'BC', although 'AD' dates are cited as such (e.g. AD 1000) where this facilitates comprehension. Where the only chronology available in the literature was based upon uncalibrated ^{14}C age estimates, those ages have been calibrated and are cited as, for example, '3000 cal. yr BP'.

Limitations of Data Sources and Climate Reconstruction Approaches

All of the different techniques and data sources used for the reconstruction of palaeoclimate have potential limitations with respect to their inherent temporal and spatial resolution, and to the accuracy and precision with which palaeoclimate values may be reconstructed (Jones et al. 1998; Mann et al. 1998, 1999). In addition, many approaches are beset with potential pitfalls relating, for example, to the need to correctly identify the climatic variable(s) to which the palaeosensor in question responds, as well as to recognise that some palaeosensors may exhibit systematic patterns of changing relative sensitivity to different climate variables as the general climatic regime changes. This is not the appropriate place for a systematic review of the limitations of all of the palaeosensors currently in use; instead we limit ourselves to some comments relevant to those particular palaeosensors upon which we have drawn for our review.

Limitations upon temporal resolution may either be practical limitations imposed, for example, by sedimentation rates and/or the time taken to analyse samples, or theoretical limitations imposed, for example, by mixing depths of sediments or response time of the palaeosensor. Nonetheless, a variety of records can at least theoretically approach annual resolution. Tephrachronology, where it allows the injection of several fixed points into pollen records, is increasingly allowing direct comparisons between sections of the pollen record and data from other sources such as dendrochronology, varves, ice-cores and even historical/archaeological records. Although for many techniques of microfossil analysis (e.g. palynology) this would only ever be practically possible for short intervals within the Holocene, for others, such as tree rings, annual resolution is routinely available throughout most of the Holocene, enabling them to reveal inter-annual to decadal fluctuations (Briffa et al. 1992) with a chronological precision rivalled only by other laminated records (e.g. varved lake sediments, ice cores).

Notwithstanding the theoretical possibility of an annually-resolved palynological record, the value of such a record would be limited. Although contemporary aeropalynological and pollen trapping studies demonstrate inter-annual variability in the relative abundance of pollen of different taxa (see e.g. Hicks 1985), such variability does not reflect shifts in vegetation composition in response to climate but rather inter-annual variability in plant performance related to seasonal weather patterns. Vegetation composition generally responds qualitatively only to climatic fluctuations at decadal or longer time scales, the two principal exceptions being extreme climatic events of sufficient magnitude to cause death of some plant taxa, and hence vegetation disturbance, and communities dominated by plant species with an annual life cycle. The palaeovegetation record for Europe during the Holocene has two particular characteristics that limit its ability to record responses to sub-decadal climatic fluctuations. Firstly, the record is dominated by trees over much of Europe during the Holocene, reflecting the dominant role of forest ecosystems that have considerable ecological 'inertia'. Relatively rapid changes in canopy composition only occur following some form of disturbance of the forest; such disturbances can result from extreme climatic events that kill trees. Without such disturbances many forests can persist for centuries in the face of climatic conditions that would favour a forest of different composition were it able to develop. Secondly, for much of the Holocene there are, as a result of the general predominance of forests, no ecotones within most of Europe that are characterised by marked life-form changes (e.g. the forest – tundra ecotone). The importance of ecotones characterised by marked life-form changes is that such ecotones are generally more sensitive to short-term climatic fluctuations; this is illustrated by the extremely sensitive and rapidly fluctuating palaeovegetation record of the last glacial stage from Lago Grande di Monticchio (Allen et al. 1999) where the fluctuations reflect shifts from steppe communities, dominated by grasses and herbaceous taxa, to areas of woodland dominated by a variety of trees. Trees are able rapidly to invade steppe if moisture availability increases and/or thermal conditions become more favourable, but are killed when drought conditions and/or extreme temperatures return.

The possibility cannot be ignored that the responses of biological palaeosensors may lag sufficiently far behind the most rapid and/or large magnitude of climatic changes that these palaeosensors might fail to record some millennial scale features of the record. Such a lag could arise if the migratory capabilities of species, especially longer-lived plants such as forest trees, were insufficient to enable them to maintain their ranges in equilibrium with the changing climate. Although this may be a problem immediately following the transition from the Younger Dryas, any such lag, if it did occur, is unlikely to have extended much beyond a millennium or so, and certainly did not influence the composition of mid-Holocene forests. Furthermore, studies of the contemporary rates of range extension of a wide range of organisms, including relatively strongly flying butterflies (Hill et al. 1999; Parmesan et al. 1999; Thomas 1991), indicate remarkable consistency of the rates achieved both amongst disparate groups of organisms and between contemporary range extensions and those of the early Holocene. Range extension rates are typically between a few hundred metres and ca. 2 km per year in the long-term average (Huntley 1989, 1991; Huntley and Webb 1989; Preece 1997). It should be remembered that such long-term averages obscure the underlying mechanism of range expansion that seems always to require relatively rare long-distance dispersal events over distances that often exceed 10 km (Collingham et al. 1996); such events are likely also to depend upon dispersal mechanisms for which the propagule is not adapted (Giannakos 1997), propagule adaptations apparently being relevant principally to local dispersal over distances of often much less than a kilometre (Wilkinson 1997).

Lake-level records also suffer from limitations upon their temporal resolution, being inherently temporally smoothed because the changes in precipitation to which north-west European lakes principally respond (Harrison et al. 1993) must persist sufficiently long to result in a water-level change that is recorded in the sedimentary record. Furthermore, since the chronological frameworks for most lake-level records depend upon ^{14}C age measurements, further temporal smoothing results from the inherent uncertainties in the ^{14}C ages. The majority of lake-level records, therefore, record climate changes with at best a centennial to millennial resolution (Harrison and Digerfeldt 1993), the two principal exceptions being records from lakes with laminated sediments and records derived from lake dwelling sites, timbers from which have been dendrochronologically dated (see e.g. Gross-Klee and Maise 1997); in these cases climate changes can be recorded with decadal resolution. Furthermore, changes in level registered by a particular lake are dependent not only upon the magnitude of the precipitation change, but also upon the relationship between the area of the lake surface and of its catchment, as well as upon the initial level. In addition, as Baillie (1999a) has suggested, lake levels may suffer catastrophic changes as a consequence of tectonic events.

Records of the history of the altitudinal tree limit are subject to limitations of the available data that arise from the vagaries of fossilisation and of locating sub-fossil wood above the present tree limit. Being also dependent primarily upon ^{14}C dating of the sub-fossil wood, these records are probably best viewed as providing evidence of centennial mean conditions. Furthermore, although dating fossil wood from sites above the present tree limit allows the minimum upward extension of the tree limit to be estimated for periods when the tree limit was higher than at present, it is generally impractical to apply this approach to estimate past depression of the tree limit because this would require ^{14}C dating of very large numbers of fossil wood samples from below the present tree limit. Thus tree limit changes have only been used to infer climatic conditions during intervals that were warmer than the present. It must also be remembered that, as a result of the vagaries of fossilisation, the difficulties in locating fossil wood above the present tree limit and the limitations upon the number of samples that it has been possible to ^{14}C date, the magnitude of tree limit elevation during individual warm intervals may be under-estimated, as may the duration of the warm intervals.

Tree limit data have generally been used to infer elevations of the summer temperature. The magnitude of the temperature elevation has been estimated from the altitudinal increase in tree limit compared to present using an arbitrary value for the mean adiabatic free-air temperature lapse rate of 6 (5-7) °C km^{-1}. However, many lines of evidence (Körner 1998) point to accumulated temperature during the growing season, rather than mean annual temperature or temperature at any particular season, as the factor that determines tree limit. Seasonal or annual temperature reconstructions from tree limit data thus must be viewed with caution.

In contrast to the tree limit data, the record of advances by non-surging mountain glaciers can be used to infer the magnitude and duration of climatic changes principally during intervals that were 'cooler' than present. Caution is required in deducing the nature of the climate changes, however, because glacier volume has a dual dependency upon accumulation and ablation rates, and hence upon precipitation that falls as snow as well as summer temperature. Despite this, there is a general tendency for periods of glacier advance to alternate with periods of elevated tree limit in any given mountainous region, at least in areas with a relatively continental climate. The record of glacier advances obtained from moraines has limited chronological resolution; it also will potentially fail to record evidence of earlier Holocene advances of lesser magnitude than later Holocene advances because the latter have destroyed the evidence of the former. A more complete record of glacier activity is obtainable from the sediments of lakes downstream from the glacier terminus (Dahl and Nesje 1994; Karlén 1981; Karlén and Matthews

1992), although these depend upon ^{14}C chronology and thus inherit the limitations of this dating method. Seasonal, annual and decadal variations in glacier activity influence sediment production and deposition in proglacial lakes. If unaffected by bio-turbation, the seasonal depositional variation in such lakes may result in the annual production of laminae/varves. However, in many cases records of changes in glacial input must be ^{14}C dated and thus can provide at best a centennial-scale record.

Records of the occurrence of particular plant or animal species in samples from archaeological sites have enormous, but as yet largely unrealised, potential as a source of information about biological palaeosensors. Regional to continental scale databases urgently need to be developed in order to capitalise upon this as yet largely untapped resource (see e.g. Benecke 1999). Such records, however, must be used with caution for several reasons; as a primary constraint, only presence of remains of a species should be considered to contribute evidence. Not only do preservation conditions influence what remains have persisted (Jacomet and Kreuz 1999), a species may also be absent from a particular sample or site for cultural reasons (e.g. a range of grassland plant species present at one site but absent at another may reflect the presence of material derived from hay cutting at the first but not at the second), or it may go unrecorded for methodological reasons (e.g. most archaeozoologists process samples using relatively coarse sieves, so obtaining no record of smaller mammals such as *Apodemus* spp. or *Microtus* spp.).

Speleothems, underground deposits of calcium carbonate that accumulate slowly through time, provide isotopic evidence of past climate and can now be dated with a precision less than 1% of the age using TIMS uranium series measurements. They can thus provide more accurate timing of climatic events than records based upon ^{14}C chronologies. Inferring climate from oxygen isotope studies of speleothems is complicated, however, by the fact that both cave temperature and isotopic composition of the regional precipitation combine to produce the recorded signal.

Marine climate records, primarily of sea surface temperature (SST), can be inferred from plankton community changes, using transfer functions, and from stable isotope studies of carbonate shells in sediment cores. As a result of ocean mixing, these records are of regional rather than local significance. With the rare exceptions of annually layered (varved) marine sediments, that also seldom contain remains of plankton communities that can be used for SST estimates, the temporal resolution of such records is hindered by their reliance upon ^{14}C chronologies. Mixing of sediments through bioturbation also reduces resolution, although high sediment accumulation rates can in part compensate for this, allowing decadal events to be recognised in some records.

Ongoing projects that are compiling, for example, sea surface temperature, palynological and lake-level data for Europe and elsewhere will enable spatial patterns in these palaeosensors to be examined without any further loss in temporal resolution compared to the attributes of individual site records for the particular palaeosensor. Those large-scale data compilations currently available for north-west Europe, however, typically have been compiled in terms of time slices at intervals of 250 years at best, more often of 500 or 1000 years. In addition, these data sets are constructed from irregularly sampled records that are generally dependent upon ^{14}C for their chronology. In a worst-case scenario, the chronological precision may be reduced to ±500 yr, resulting in the potential for data compiled for a given time slice to be derived from samples that, at different localities, represent contrasting phases of a millennial fluctuation. This issue must be borne in mind when mapped syn-theses are examined, although is it to some extent overcome by the spatial density of sites from which data are compiled. The likelihood is extremely small of all sites in one region being accidentally sampled in one phase of a millennial fluctuation and all those in another being sampled in the opposite phase. Furthermore, the use of indices of chronological reliability for the data compiled allows 'rogue' points to be examined and their reliability assessed in comparison with those of surrounding points.

Principal Questions Addressed

In the following review of Holocene climatic history in north-west Europe, and of the relationships between this and human history in the region, we attempt to address the following principal questions, insofar as each is relevant, in relation to each of the three time scales of climate change around which we organise our discussion:

- What was the range of climatic change?
- What were the spatial patterns of the climatic anomalies at different times?
- What were the principal climatic events?
- Does the climatic record exhibit recurrent temporal patterns that may relate to particular forcing factors?
- Did the climatic changes or events have effects upon human populations?

These questions inevitably give rise to a variety of subsidiary questions. Some of these are relatively easy to address: for example, when we compare climatic conditions with those of 'the present' what do we mean by 'the present'? In answer, we define 'the present' to be the mid twentieth century and make quantitative comparisons with climatic data for the climatic normal period 1931-60. Other questions that arise are more difficult to answer: for example, what is meant by a climatic event? We define a climatic event to be "a short-lived large-magnitude departure of climatic conditions from the previous long-term mean". Appropriately, this definition is dependent upon the time scale being considered; thus when we examine the entire Holocene a large magnitude fluctuation that endures for a few centuries, such as that centred upon 8·25 ka BP (Alley et al. 1997), is a climatic event within that frame of reference – in the time frame of the last 2000 years, however, the LIA, that also endures for a few centuries, does not qualify to be termed a climatic event, whereas the period of a decade or more of cooler conditions centred upon AD 540 is considered a climatic event.

When we turn to the last of our principal questions, and ask what, if any, effects the climatic changes or events had upon human populations, then we also must ask two subsidiary questions. Firstly, what is known about the overall spatial and temporal patterns of human settlement and land use activities, including agriculture, throughout north-west Europe during the Holocene? Secondly, what do we consider to be an effect upon human populations?

A prerequisite for the reconstruction of changes in settlement patterns in time and space are regions that have been systematically prospected for archaeological sites. Despite the long tradition of field archaeology in some countries, only very few regions of Europe have been systematically searched in this way, and even fewer have a sufficient number of excavated sites within given periods. Nonetheless, a few examples can be identified, including: Early Neolithic 'Linear Band Keramik' (LBK) settlements of the Aldenhovener Platte in Western Germany (Lüning and Stehli 1994); Neolithic and Bronze Age settlement history in the middle Rhine valley (Frank 1998); Neolithic and Bronze Age sites around Alpine Foreland lakes, for example, the Swiss side of Lake Constance (Winiger and Hasenfratz 1985) or Lake Biel (Winiger 1989); Iron Age sites in southern (Maise 1998) and in northern (see e.g. Behre and Kucan 1994; Behre and Schmid 1998; Kossack et al. 1984) Germany; and Late Bronze Age and Early Iron Age settlements in the Netherlands (Van Geel et al. 1996, 1998). Multi-period compilations are extremely rare, although Pratt (1997) made a GIS-based compilation of the available evidence for archaeological sites of all periods in the north east of England. Evidence of human agricultural activities can also in some cases be obtained from palynological data. Although various workers have attempted to use the palynological record to assess the extent of the landscape impacts of human activities (see e.g. Berglund 1991; Brostrom et al. 1998; Gaillard et al. 1994; Pratt 1997; Wiethold 1998), and others have attempted to infer these from data recording plant and animal remains found when archaeological sites are excavated (see e.g. Hüster-Plogmann et al. 1999a; Schibler and Jacomet 1999), as yet no systematic method is available for the quantitative reconstruction of different land-cover types in a landscape, and hence for evaluating the extent, for example, of forest destruction at different times (Brostrom et al. 1998). Whereas records of plant macrofossils and animal bones from individual archaeological sites ("on-site" analyses) provide evidence principally of the relatively short-term and localised nature of the impacts of humans on their environment, regional or more extensive compilations of such results, as well as conventional palynological studies ("off-site" analyses), are required in order to reveal longer-term regional impacts, as well as to identify any trends in these impacts.

Animal husbandry and large scale plant cultivation began in central Europe between 8000 and 7500 BP (see e.g. Kreuz 1990; Uerpmann and Uerpmann 1997), spreading northwards and westwards to most of the remainder of north-west Europe during the following two millennia and marking the transition to the Neolithic cultural grade (Fig. 1). Neolithic economies depended upon subsistence agriculture, an important component of which was the cultivation of oil-seed and fibre crops, as well as of cereals; the latter were by far the most important, apparently being the principal source from which the people met their caloric needs (Ebersbach 1999; Gross et al. 1990). Animal husbandry involved the keeping and use of domesticated cattle (*Bos taurus*), pigs (*Sus scrofa f. dom.*), goats (*Capra hircus*) and sheep (*Ovis aries*). After the LBK, Neolithic cultures display a strong tendency toward regional patterns (Strahm 1994), with contrasting influences from south-west and Central Europe meeting in the northern Alpine Foreland. Later Neolithic cultures were more homogeneous, however, widely characterised throughout most of central Europe by their use of pottery of the type referred to as "corded ware". Neolithic agriculture was not very intensive and had limited impact upon the landscape; the extent of forest clearance was quite small. There are, however, clear signs of intensification in the course of the period of prevalence of Neolithic cultures (see e.g. Schibler and Jacomet 1999); cleared areas were probably being used for longer periods and cultivation was becoming more settled rather than shifting (Brombacher and Jacomet 1997).

With the spread of metal working, and the subsequent development and use of bronze that marked the transition to Bronze Age cultures, a substantial intensification of land use is recorded in several parts of Europe (see e.g. pollen diagrams from the Lake Constance area in Rösch 1985, 1990, 1992, 1993). Settlements extended to higher elevations in mountain areas (Hochuli et al. 1998) where Neolithic peoples had probably practised only seasonal summer farming (Höpfel et al. 1992; Südtiroler Archäologiemuseum 1999). Lower elevation coastal areas, for example, the North Sea coast, were also settled, apparently for the first time, and animals began to be kept in byres/stables (Andraschko 1995). The subsequent development and spread of Iron Age cultures is associated with evidence of further intensification of animal husbandry, especially cattle breeding. Associated with the keeping of cattle under cover, and the overall intensification, there is also the first evidence of the regular management of grasslands as meadows to provide crops of hay (Jacomet et al. 1999). Forest clearance was in many regions much more extensive then in any previous period of the Hol-ocene and remaining forests were also often extensively used (Bunnik et al. 1995). During the period of widespread expansion of Roman influence (ca. 15 BC-AD 400), land use within the frontiers of the Roman Empire was intensive in some regions, often associated with intensification of cereal cultivation (e.g. Bunnik et al. 1995); in other regions there is evidence of more extensive deforestation at this time than previously, although the precise timing of clearance episodes is often unclear and in some cases may have preceded Roman occupation and been associated with later Iron-Age agricultural intensification (Huntley and Stallibrass 1995; Manning 1997). Forests were also extensively used as a source of wood for building and for many other purposes, for example, the frequent use of wood of fir (*Abies alba*) to construct barrels in which wine and other fluids were transported from their areas of production to areas of consumption elsewhere within the Roman Empire (see e.g. Huntley 1999, 2000; Küster 1994). During the subsequent "Migration Period" the human population was reduced and anthropogenic activities, as well as their impacts upon the landscape, decreased throughout Europe (Frenzel et al. 1992, 1994). After ca. AD 900 the onset of the MWP is associated with a renewed intensification of land use in many regions of Europe.

Turning to the issue of "effects upon human populations", it is clear that short-term climatic cooling in particular can have major impacts upon agricultural production in pre-industrial societies, leading to crop failure and the potential for famines and starvation. Such events almost certainly also impacted heavily upon prehistoric populations, whether these were culturally of the Mesolithic hunter-gatherer grade or of Neolithic, Bronze Age or Iron Age subsistence farming grades (Schibler et al. 1997a, 1997b). Substantial fluctuations have been detected in the subsistence economy of some Neolithic lake-shore settlements of the Alpine Foreland (see e.g. Schibler and Chaix 1995). These fluctuations relate especially to the varying intensity of exploitation of wild animals (hunting) and collection of wild plants (gathering) (Schibler et al. 1997a, 1997b) and may indicate reactions of the people to the impacts of short-term climatic fluctuations upon their subsistence agriculture. More recent cultural grades, however, had an improved subsistence agriculture with a broader crop base that probably rendered it less susceptible to massive crop failures during periods of unfavourable climate. Bronze Age agriculture, for example, utilised spelt wheat (*Triticum spelta*), millet (*Panicum miliaceum*) and Celtic bean (*Vicia faba*), all of which were ground to produce flour. Fluctuations in the intensity of exploitation of wild animal and plant resources recorded in studies of Bronze Age lake-shore settlements are of smaller magnitude than those recorded from the Neolithic (Schibler and Studer 1998) and during later periods wild animal resources are no longer exploited intensively (Schibler et al. 1999).

Multi-Millennial Changes and Long-Term Trends

Individual time series for the entire Holocene are available from various parts of north-west Europe, Greenland and the surrounding ocean basins, and for a variety of palaeosensors that record variously mean annual temperature, seasonal temperatures, both over land and for the sea surface, temperature sum, moisture availability and precipitation. A selection of these is illustrated in Fig. 2. For three of the palaeosensors, namely the palynological record, lake-levels and sea surface temperatures (SST), spatial compilations are available. These enable assessments either of the continental-scale spatial patterns in the climatic anomalies for particular time slices, or of the change through the Holocene in the anomaly values for a given climatic variable, for example, along a latitudinal transect.

Ice bore-hole temperature logs from the Greenland ice cores (Fig. 2a) provide probably the most reliable estimates of ice-sheet surface temperatures during the Holocene. The estimates obtained are of mean annual temperature integrated over periods of years to decades. They indicate that mean annual temperatures in Greenland peaked around 5 ka BP at a value ca. 2·5°C higher than the present day value. A subsequent decrease commenced ca. 4·1 ka BP and reached its minimum at around 2·2 ka BP. Thereafter the record indicates smaller magnitude fluctuations in temperature with the LIA being no more than ca. 1-1·5°C cooler than the present day at its extreme during the Late Maunder Minimum (Luterbacher et al. 2000), whereas the MWP was about as warm as the present day (Pfister et al. 1998). Although this relatively smoothed record arguably obscures extremes that may deviate from the longer-term mean by values as much as twice as large as do the recorded values, the overall trends that it records in the last four millennia are supported by evidence of the history of the Hans Tausen Ice Cap. This ice cap, located in the far north on Peary Land, began to develop at about 3·5 ka BP when the ice bore hole temperature record from central Greenland indicates temperatures about the same as the MWP and also close to present. The ice cap subsequently survived the MWP and is currently still growing at ca. 4 cm yr^{-1}, not yet having reached equilibrium. Despite this, however, in the event of significant further warming over the next century it is a likely candidate ice sheet to disappear; monitoring its future development would provide valuable evidence of the impact, if any, of the present warming trend seen in the instrumental record when averaged over the entire northern hemisphere (Mann et al. 1998).

In contrast to the very clear long-term temperature signal obtained from the ice bore-hole temperature logs, the Holocene $\delta^{18}O$ record from the Greenland ice cores is characterised by its overall blandness and shows no evidence of an early or mid-Holocene thermal maximum. The lack of a clear long-term signal is most probably accounted for by shifts in the seasonal distribution of precipitation. A 20% increase in the proportion of winter snowfall (that is 15‰ lighter than summer precipitation) could overcome any effect of the warmer temperatures at the Holocene thermal maximum. Such a change in seasonality of precipitation would be consistent with the seasonality of the orbital forcing to which the long-term Holocene climatic trends are a response, and would also be consistent with evidence from other palaeosensors. The hypothesis of changed seasonal distribution of precipitation could in principle be tested if the seasonal layers within the annual ice layers could be distinguished and separate measurements made of their thickness. Furthermore, if these layers could be distinguished, then a seasonal temperature record could also potentially be extracted by measuring the $\delta^{18}O$ values for the ice from the summer and winter seasons. Although in principle the seasonal components of the annual ice layers might be distinguished using markers such as the nitrate or dust indices, this is likely only to be possible for the last millennium or so.

The hypothesised impact of changed precipitation seasonality upon temperature estimates derived from ice $\delta^{18}O$ values brings into question the magnitude of reconstructed pre-Holocene temperature changes, especially the Younger Dryas to Holocene temperature rise that is estimated at ca. 10°C on the basis of the $\delta^{18}O$ values. It is suggested that this may be an underestimate of the actual temperature rise that may have been closer in magnitude to 25°C. Although this value may in part reflect the breakdown of an inversion that is believed to have persisted over the ice cap during the glacial, such a larger value (i.e. >10°C) would better accord with estimates made from biological palaeosensors in, for example, the British Isles (Atkinson et al. 1987; Huntley 1994).

Changes in the altitudinal tree limit in the Scandinavian mountains provide one time series of Holocene temperatures from Scandinavia (Fig 2d).

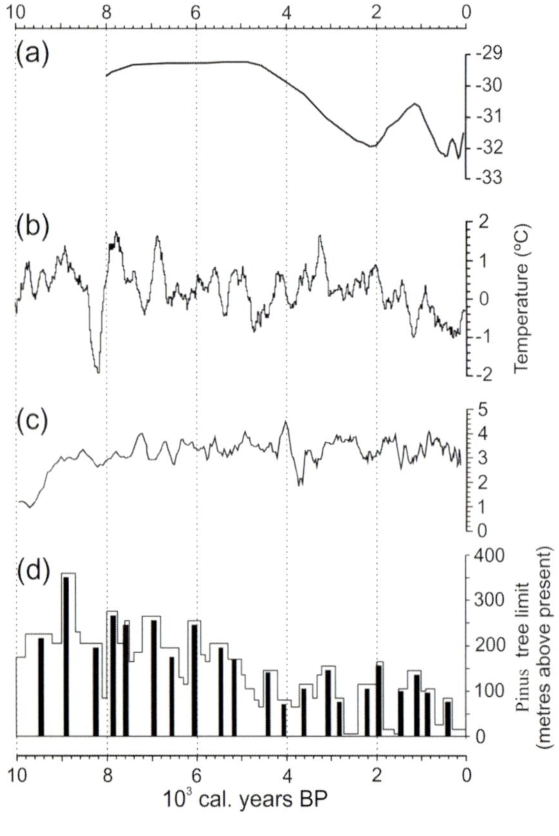

Fig. 2. (left) Holocene thermal trends in a series of palaeosensors. **(a)** Ice-sheet surface palaeotemperature record deduced from temperature measurements made down the GRIP ice core borehole (redrawn from Dahl-Jensen et al. 1998). **(b)** Reconstructed annual mean surface temperature anomaly for Greenland Summit. Values are derived from the GISP2 ice core $\delta^{18}O$ record using a transformation of 0·6‰ per °C (redrawn from Lauritzen and Lundberg 1999 and based upon data obtained from WDC-A, Boulder). **(c)** Reconstructed palaeotemperature record derived from stable oxygen isotope measurements made upon calcite from a speleothem (SG93) obtained from Søylegrotta, near Mo I Rana, Norway (redrawn from Lauritzen and Lundberg 1999). Values are local annual mean temperatures. **(d)** Relative elevation of the *Pinus sylvestris* tree limit in the Scandes. Line indicates inferred tree limit altitude relative to present tree limit for *P. sylvestris* and bars denote the most likely periods of establishment of *P. sylvestris* above the present tree line (W. Karlén, unpublished data).

The maximum tree limit of the early to mid-Holocene lies about 300 m above the present tree limit (with the exception of some questionable early Holocene records from as much as 600 m above the present tree limit, see Kullman 1999). Although inferring the associated temperature difference, based upon the present typical lapse rate for the region of 6°C km-[1], is complicated by land uplift since the early Holocene, it appears that summer temperature was generally warmer during the early to mid-Holocene and was optimal between 6 and 5 ka BP, at which time it was ca. 1·5°C (±0·5°C) higher than present values. Subsequently there was a quite marked decrease in the elevation reached during tree-limit maxima; although a series of tree-limit maxima occur after 5 ka BP, none reaches to an altitude comparable to those of the tree-limit maxima before that date (Karlén and Kuylenstierna 1996).

When the evidence of glacier limits in the Scandinavian mountains is examined then, in contrast to the tree-limit data, these data reveal principally evidence relating to temperature minima that alternate with the maxima indicated by the tree-limit data. Although similar data exist for the glaciers in the more oceanic mountains of western Norway, interpretation of these records is rendered more complicated by the greater extent of their dependency upon precipitation rather than summer temperature. Nonetheless, by combining temperature data with glacier equilibrium line altitude (ELA) fluctuations it is possible to extract the winter precipitation component of the record (Dahl and Nesje 1994). An exponential relationship exists between mean ablation-season temperature (1st May-30th September) and winter accumulation (1st October-30th April) at the ELA (Sissons 1979; Sutherland 1984). These two variables are positively correlated amongst different glaciers because greater mass turnover at the ELA requires greater ablation, and hence higher summer temperatures, to balance the annual mass budget. Mean winter precipitation can be quantified if mean ablation-season temperature is known. For southern Norway this approach indicates early Holocene glacier activity was governed by much greater quantities of winter precipitation than at the present day (see Dahl and Nesje 1996 for further details).

Further evidence of warmer early- to mid-Holocene conditions comes from $\delta^{18}O$ values measured in the U/Th dated calcite of a speleothem from Nordland in Northern Norway (Lauritzen and Lundberg 1999). These values are considered to reflect mean annual temperatures and indicate mean values almost 1°C warmer, with peak temperatures ca. 1·5°C warmer, during the Holocene maximum (ca. 8-4 ka BP in this record) than at present (Fig. 2c). A subsequent marked drop in temperature to a minimum between 4·0 and 3·5 ka was succeeded by a period of relative stability of the mean that persists until the present.

Knowledge of tree line history in the Alps is currently incomplete. Glaciologists meeting in 1991 recommended that studies of the extent and timing of past tree line fluctuations, together with pollen analyses of sites associated with moraines of known date, should be made in order to increase understanding of Holocene climatic history (Grove 1997).

Although a number of dates on fossil wood from glacier fore-fields, believed to indicate times when tree lines in the western Alps were higher than in the twentieth century, have been published (see e.g. various reports in *Radiocarbon* vols. 15 (1973) and 17 (1975)), no details are given of differences in height between the samples dated and the local tree lines. Identification of tree line fluctuations in the Alps is probably complicated by the complexity of the topography, with differences in factors such as aspect, slope, duration of shade during the day and extent of snow drifting causing variations in microclimate, both above and below the ground surface, within short distances. Tinner et al. (1996) found that the highest Holocene tree line in the Val d' Heremence in the central Alps of the Swiss Valais was 50-100 m higher between 10,200 and 5400 cal. yr BP than the uppermost single specimen of *Pinus cembra* today. (This falls within the range of ±100 m accepted by Körner (1998) as the full range of post-glacial tree line oscillations at local scales.) They also found evidence of a strong climatic deterioriation around 7700 cal. yr BP. Tinner et al. (1996) recommended that the presence of ancient forest soils outside the modern forested area, as well as occurrences of plant macrofossils, may be used, in addition to pollen studies, as indicators of higher tree limits in the past. They noted that minima of *P. cembra* macro-fossils coincide with cold phases recorded in the Alps by glacier advances, as well as to minima in the ratio of arboreal to non-arboreal pollen and maxima in solifluction activity, their results being in accord with those of a number of earlier inves-tigators including Zoller (1977) and Gamper and Suter (1982).

This work was extended by examining pollen and plant-macrofossil content of sediments in two lakes in the Italian and Swiss central Alps (Wick and Tinner 1997). Their results indicate that the timber line was at its highest between 9650 and 5700 cal. yr BP, after which it retreated towards its present-day level, being below its potential elevation since about 3400 cal. yr BP as a consequence of human influence. They assessed the magnitude of Holocene fluctuations in tree line altitude caused by climate as probably not extending at their maximum more than 100-150 m above the present. Using the same approach, but taking into account data from Alpine lake sediments and reconstructed lake level fluctuations in two lakes on the Swiss Plateau, this estimate was later amended to not more than 100-130 m above the present potential timberline (Haas et al. 1998). Eight synchronous cold phases were found to have occurred at 10,900-10,350, 9550-9150, 8350-7750, 7500-7100, 6100-5600, 5350-4950, 3750-3400 and 2750-2350 cal. yr BP. These overall results were based only upon identification of events that occurred synchronously on the Swiss Plateau and at the Alpine timberline. Cooling events identified at only one location were not taken into account. It appears that, before a more complete picture of the Holocene history of Alpine tree lines and their subregional variations can be obtained, many further detailed investigations and reconstruction of fluctuations in the Alpine timber line are required from the eastern as well as western Alps.

The magnitude of the temperature fluctuations inferred from such tree limit studies in the Alps appears to conflict with the reconstruction by Huntley and Prentice (1988) from palynological data of mid-Holocene summer temperatures ca. 4°C above present values, and the associated inference of reduced lapse rates, although their results are independently confirmed by Guiot (1987), who reconstructed mid-Holocene temperatures at high elevations in the French Alps as much as ca. 5°C above present values. The inferred reduced lapse rate, however, may be accounted for by the combined effect of increased humidity and the reduction in the extent of ice- and snow-covered areas at high elevation affecting the albedo. Independent evidence indicates that snow lines in the Swiss Alps were generally higher than at present between 8·0 and 6·0 ka BP. In addition, it seems likely, as discussed above, that tree line elevations cannot be interpreted directly in terms of temperatures because they are determined principally by temperature sums.

As the foregoing examples show, an early to mid-Holocene 'thermal maximum' is seen in Greenland, Scandinavia and central Europe and appears to be a quite general feature of the Holocene climatic history of much of Europe. Although some palaeosensors provide estimates only of the annual mean temperature, the biological palaeosensors enable the winter and summer thermal conditions to be distinguished. Winter minimum temperatures during the early Holocene as much as 4 to 5°C above present day values in some regions are indicated by macrofossil remains of *Corylus avellana* (hazel) far to the north of its present limits in Sweden and Norway (Anderson 1902), these limits being determined by winter cold (Dahl 1998). Evidence for a maximum in both growing season warmth and summer temperatures in northern Europe during the mid-Holocene comes from remains of other plant taxa (e.g. *Viscum album* – mistletoe, *Trapa natans* – water chestnut) north of present limits that are determined by their requirements for summer warmth. Similar evidence of warmer summer conditions comes from mid-Holocene remains of *Emys orbicularis* (European Pond Tortoise) found north of its present range limits; this species northern limit is determined by the thermal requirements of its reproductive cycle, successful incubation of its eggs occurring only above a threshold value for summer warmth. Thus it is evident that both winter and summer conditions were warmer than the present during the first half of the Holocene. In northern Europe at least, however, the maximum in winter temperatures occurred during the early Holocene, whereas the maximum in summer warmth occurred somewhat later, during the mid-Holocene.

Turning now to the oceans surrounding north-west Europe, recent records from the eastern Norwegian Sea offer high resolution records of Holocene summer (August) sea surface temperature (SST_{AUG}). These values are reconstructed from diatom assemblages using transfer functions; the relatively warmest SST_{AUG} values relate to assemblages that include diatom species not found today in the Norwegian Sea but having their northern limits in the northern North Atlantic. The reconstruction indicates a sharp temperature increase of ca. 9°C between the Younger Dryas and the Holocene (Koç Karpuz and Jansen 1992; Koç and Jansen 1994; Koç et al. 1996). Thereafter, SST_{AUG} peaked in the mid Holocene at values of 15-16°C, compared to recent values of 10-11°C. Temperatures fell progressively between ca. 6 ka BP and 3·5 ka BP and after that time fluctuated about a relatively stable mean. Interestingly, a parallel temperature reconstruction from al-kenones in the same core (E. Calvo, pers. comm.) gives a Holocene temperature maximum only about half the relative magnitude of that reconstructed from the diatoms. However, rather than being seen as conflicting evidence, that might result from differences in the calibration range of the methods of SST estimation, this apparent discrepancy may be revealing important aspects of the seasonality of SST at that time. Whereas the diatom assemblage reflects a summer planktonic community, the al-kenones are derived from spring-blooming communities of planktonic coccoliths. The different SST signals thus may indicate that spring SSTs were increased less than summer SSTs, a result that might be expected given the seasonal pattern of the insolation anomalies during the early Holocene.

Although further apparently conflicting evidence comes from $\delta^{18}O$ values measured in foraminifera in the same core that appear to indicate a marked warming ca. 3 ka BP, this is interpreted as re-flecting a change in source water rather than a conflicting temperature signal, and may indicate a changed precipitation regime over the Norwegian Sea and the adjacent land area after ca. 3 ka BP. Such a change would be consistent with a change in atmospheric circulation at that time that is also hinted at by other palaeosensors. Furthermore, the SST_{AUG} reconstructions for the early to mid-Holocene are also supported by evidence of northward range extensions of a number of molluscs, notably *Mytilus edulis* (common mussel), in the Norwegian–Greenland Seas at that time. A mid-Holocene summer SST maximum is also recorded close to Greenland in the western Norwegian Sea, as well as in the northern North Atlantic (Koç and Jansen 1994; Koç et al. 1993; Koç et al. 1996). The duration of the thermal maximum is, however, shorter in the western Norwegian Sea close to Greenland, and also in the north close to Svalbard, than in the east close to Norway.

In addition to these records of palaeotemperature, a number of palaeosensors can provide evidence of past precipitation and/or moisture availability. Reconstructions of one or other of these variables can be made from palaeovegetation data, from lake-level data, and from data relating to the varying thickness of the annual layers in Greenland ice cores. Such moisture balance reconstructions show much less spatially consistent anomaly patterns across north-west Europe as a whole than do the temperature reconstructions. Although this may in part reflect the inherently finer scale of spatial and temporal variability in precipitation patterns, it is likely to arise principally because changes in atmospheric circulation patterns will often result in precipitation changes of opposite sign in adjacent regions. In addition, vegetation responds to moisture availability, and lake status to the balance between precipitation and evaporation; both are thus influenced by temperature as well as by precipitation. It should also be noted that the signal output by a particular lake, in terms of the change in level, is dependent upon not only the magnitude of the precipitation change, but also upon the relationship between the area of the lake surface and that of its catchment, as well as upon the initial level of the lake. Although some of the recorded changes in lake level are substantial, water level in some south Swedish lakes being as much as 9 m below present levels between 10,850 and 10,350 cal. yr BP, such relative lowering of lake level may result from quite modest changes in precipitation. Vassiljev et al. (1998) have shown that during the early Holocene, when winter temperatures were warmer than at present, an observed lowering of relative water level by 9 m could have resulted from a reduction in annual precipitation of <75 mm. A subsequent increase in precipitation resulted in these lakes being near to their present levels 8900-7850 cal. yr BP. Although they show further water-level minima at ca. 6450 and 5350 cal. yr BP, these may reflect relatively small changes in precipitation. Unfortunately, because under present conditions they are close to their maximum level and overflowing, these and similar lakes are unable to provide any indication of times when precipitation minus evaporation (P-E) was higher than present. Nonetheless, there seems to be general consistency between several regions of Europe, and amongst the various proxies for moisture availability, indicating that the magnitude of changes in water budget during the Holocene was relatively small. The ice-layer thickness data from Greenland similarly indicate that annual precipitation did not vary by more than ca. 25% in the course of the Holocene, although, as discussed above, the seasonal distribution of precipitation may have varied by a similar magnitude.

Turning now to those palaeosensors for which sufficient data are available to make it possible to explore the spatial patterns in the Holocene palaeoclimate of north-west Europe, we should note the necessity for this purpose of databases containing extensive compilations of comparable data from many individual localities. At present only three data types offer information from sufficient localities: pollen analytical data, lake-level data and data recording sea surface conditions. The only published large-scale compilation of pollen analytical data for north-west Europe (Huntley and Birks 1983; Huntley and Webb 1988; Huntley 1990a, b) has limited chronological resolution, data having been compiled only for discrete time slices at intervals of 500 ^{14}C years; the currently ongoing efforts to compile all of the available data from numerous localities throughout Europe into a European Pollen Database will, however, overcome this problem. Lake-level records are inherently temporally smoothed, because a persistent change in precipitation is required to bring about a persistent change in mean lake level that will be recorded in the sedimentary record. As a result, the available compilation of lake-level data provides a record of relatively long-term (centennial to millennial) changes in precipitation patterns. In addition to the specific limitations of these data sets, these and the available compilation of data relating to sea surface conditions all depend principally or even solely upon a ^{14}C chronology to establish the time scale for individual time series. As a result, all three data sets have an inherent chronological precision of typically no better than ca. ±250 yr at 6000 ^{14}C yr BP (6800 cal. yr BP). In extreme cases this may result in time slice compilations that are mixtures of records from contrasting phases of millennial fluctuations; this issue must be borne in mind when mapped syntheses are examined because individual anomalous data points may reflect this problem.

These limitations of the data compilations are probably minimal at around 6800 cal. yr BP because this is a period of relative stability of lake levels and is within the general Holocene thermal maximum, during which the palaeovegetation was relatively stable in many regions. The mapped patterns for this interval are very consistent in reconstructions made from pollen data using different techniques (constrained analogues and constrained response surfaces) and with and without constraints derived from lake-level data. Summer warmth (degree days above a threshold of 5°C) is greater than present throughout Europe north of ca. 50°N, with a maximum anomaly in the more maritime regions of north-west Europe close to the influence of the oceanic warm currents. Winter cold (coldest month mean temperature) is lower than present in the Mediterranean region but generally somewhat warmer than today north of the Alps, regionally by as much as 4-5°C above present day values and with strong anomalies in north-eastern Europe. Moisture availability was greater in southern Europe, but reduced in other areas, including northern Fennoscandia (Cheddadi et al. 1997). A latitudinal compilation of data relating to sea surface conditions during the Holocene thermal maximum shows evidence of an SST anomaly west of Europe that increased northward and westward in the Nordic Seas (Koç et al. 1996). The winter SST anomaly for the same period exhibits a similar latitudinal pattern.

A reconstruction of European palaeoclimate at 3000 ^{14}C yr BP (3175 cal. yr BP) made from pollen-analytical data (B. Huntley, unpublished data) exhibits broadly similar seasonal thermal patterns and moisture availability patterns to those seen for 6000 ^{14}C yr BP (6800 cal. yr BP), although the magnitude of the anomalies generally is less than at that time. The lake status patterns for 3000 ^{14}C yr BP are also rather similar to those at 6000 ^{14}C yr BP (Harrison et al., this volume), although the lake status records reveal that different patterns prevailed during the intervening period.

Comparing the spatial patterns of the mid-Holocene anomalies in thermal conditions and moisture availability with modern patterns of annual anomalies from the long-term climatic normal values is difficult because the available mapped reconstructions relate to long-term mean conditions that appear to have been strongly influenced by changes in the adjacent oceans. The limited inter-annual variability of the oceans renders it unlikely that such long-term mean anomalies will match any single annual anomaly pattern in the recent record.

We can now ask how these long-term trends in Holocene climatic conditions, and the spatial patterns of the climatic anomalies during the early to mid-Holocene, may have impacted upon or influenced human populations? There is evidence of localised cereal cultivation, probably by Mesolithic people, from about 8·2 ka BP onwards (see e.g. Erny-Rodmann et al. 1997). Evidence from the Alpine Foreland lakes shows the onset of Neolithic lake shore settlements in that region to have been at ca. 6·3 ka BP, and hence following the thermal maximum. This may indicate a response to either reduced or more stable lake levels. There is also evidence from the northern Alpine foreland of the extensive cultivation of relatively warmth demanding crops, notably *Papaver somniferum* (Opium Poppy), prior to ca. 5 ka BP. This species, grown as an oil seed crop as well as for 'medicinal' purposes, appears never subsequently to have been cultivated in such large amounts in this region; it is considered to require warmer and drier summers than prevail in the region at present (Jacomet et al. 1989). The likelihood that climate favoured its cultivation, rather than that it was displaced by an alternate crop introduced at a later date, is supported by the occurrence of *Linum usitatissimum* (Flax, Linseed) prior to, along with and subsequent to the period of poppy cultivation. A tetraploid naked wheat (*Triticum durum* or *T. turgidum*) was also cultivated extensively in the northern Alpine foreland between 6·3 and 5 ka BP, after which date it was not subsequently cultivated on a large scale. Although this wheat cannot be identified to varietal level with sufficient precision to allow detailed climatic inferences to be made, it is notable that neither of the tetraploid naked wheats is very cold tolerant (Jacomet et al. 1989). With the onset of the Bronze Age in Central Europe at ca. 4 ka BP, spelt (*T. spelta*), a hexaploid glume wheat that is relatively more tolerant of cold and wet conditions than the tetraploid species, became very important as a new cereal crop. Whether this change reflects a response to long-term climatic changes or cultural changes that were independent of environmental changes remains a matter of debate.

Centennial to Millennial Fluctuations

Numerous independent palaeosensor time series show evidence of centennial to millennial climatic fluctuations superimposed upon the overall long-term trends of the Holocene as a whole. Furthermore, at least regionally some of these fluctuations were comparable in magnitude to, or even exceed in magnitude, the overall Holocene climatic changes.

A marked cooling event is recorded at ca. 8·2 ka BP by several palaeosensors. Taking into account the extent of land uplift since that time, evidence from moraines in the Scandinavian mountains indicates that this event was the most extreme cooling event during the Holocene in that region, with annual mean temperatures >1°C cooler than the recent past. The subsequent Holocene thermal maximum appears also to have been a period of more subdued millennial fluctuations. Thereafter, a strong increase in avalanche frequency in western Norway occurred between 5·0 and 3·0 ka BP, following a minimum in their frequency during the period of the thermal maximum (Blikra and Fjeldstad Selvik 1998). Glaciological evidence from southern Norway then indicates a series of maxima of glacial activity at ca. 3·5, 2·6 and 2·2 ka BP

(Nesje and Dahl 1994). A protracted cooling event between 2·5 and 2·0 ka BP is also expressed as a tree limit minimum (Karlén and Kuylenstierna 1996). It is notable that some Scandinavian glaciers reached their Holocene maximum extent at this time, indicating that in some regions this event may have been cooler and/or more humid than the LIA (Karlén 1973). Tree-limit and glaciological evidence from Switzerland also indicates cool periods at or just following 3·5, 2·8 and 2·4 ka BP (see e.g. Maise 1999). A further extension of north Swedish glaciers took place later, between AD 550 and AD 850; tree-ring evidence from the same region suggests two periods of cooling within this interval (Briffa et al. 1992).

Glacier fluctuations during the Holocene have been traced with greater precision in the Alps than in any other part of the world (see e.g. Burga and Perret 1998). In the Alps, glacier tongues have frequently extended to positions below contemporary tree-lines so that, if careful searches are made, fossil wood within or closely associated with moraines left by expanded glaciers may be found. Patzelt (1974), identified nine intervals of glacier expansion during the Holocene in the eastern Alps. In the western Alps, Röthlisberger et al. (1980) traced a similar number of phases of Holocene expansion, some of them compound, using the known positions of glacier fronts in 1850, 1890, 1920 and 1876 to indicate the scale of earlier oscillations. This revealed that many periods of advance earlier in the Holocene were on scales roughly, if not exactly, similar to the LIA. Gamper and Suter (1982) showed that the rapid switches between warm and cold conditions demonstrated by dated moraines are supported by alternating phases of solifluction and soil formation.

The consensus view amongst glaciologists in 1991 (Grove 1997) was that an advance period may have occurred between 10,650 and 10,200 cal. yr BP. A complex period of recession between 10,200 and 6800 cal. yr BP was possibly interrupted by advances between 9400 and 9100 cal. yr BP, and between 8100 and 7500 cal. yr BP, as well as more certainly between 7500 and 6800 cal. yr BP when the Tour Glacier in the western Alps and the Frosnitz in the eastern Alps both advanced. Tree lines in the Alps did not rise more than 150 m above their present height of ca. 2300 m during this interval, although several palynological studies indicate at least two phases of tree line depression within this period (see e.g. Haas et al. 1998). Knowledge of conditions between 10,200 and 6800 cal. yr BP is less securely based than for the later part of the Holocene.

During the period from 6800-5350 cal. yr BP several advances on scales similar to those of the LIA took place, affecting, for example, the Oberaar, Rotmoosferner and Rutor glaciers. Tree lines in the Alps were high during warm conditions in the period between 5350 and 3900 cal. yr BP (Burga and Perret 1998); in the absence of direct evidence of glacier snout positions this evidence has been used to infer a period of major recession. Major glacier advances, for instance of the Allalin, Mont Miné, Aletsch, Reid and Stein glaciers, followed during the Löbben oscillation of 3900 to ca. 3200 cal. yr BP (1986; Röthlisberger et al. 1980); glaciers in both the eastern and western Alps were affected. In their extent and duration, and in the occurrence of several maxima, glacial phenomena were comparable to those that characterise the LIA, although conditions were probably more extreme. In the Venedigergruppe and the Stubaier Alps, glaciers advanced to positions 100-150 m outside LIA moraines (Mayr 1968; Patzelt 1973).

The period 3200-1000 cal. yr BP was generally warm; Alpine glaciers experienced long periods during which they were more contracted than in the late twentieth century, although advances ca. 2750-2050, ca. 1620-1280 and ca. 1030 cal. yr BP caused some termini to reach positions probably comparable with those of the early twentieth century (Holzhauser and Zumbuhl 1988). The latter period of advance, ca. AD 800-900, affected many glaciers in the Alps and elsewhere (Grove and Switsur 1994); the extent of the Grosser Aletsch at that time was comparable to that of 1880 (Holzhauser and Zumbuhl 1988). During the MWP (ca. AD 900 to 1250-1300) glaciers withdrew to positions comparable to those of the late twentieth century, although the period was interrupted by advances of some glaciers, including the Grosser Aletsch and Unterer Grindenwald ca. AD 1050-1150, albeit not to the same extent as during either the preceding period or the LIA.

The wealth of fossil wood in the western Alps makes it possible to trace the behaviour of the glaciers in greater detail than would otherwise be possible, especially where fossil stumps killed by advancing ice can be absolutely dated by reference to long tree-ring series (Holzhauser 1984a, 1997). Much more precise histories of individual glacier tongues in the western Alps during the later Holocene are being reconstructed by Holzhauser (see e.g. Holzhauser 1984b, 1997), the majority of the available ^{14}C age determinations and historical evidence coming from the last 3200 yr. During this time the Grosser Aletsch has had ten major advances, at least three of which were comparable in scale with its extension in 1850. Major Grosser Aletsch extensions culminated ca. 3400-3300, 2900-2800, 2400-2350, 1730-1620 and 1310-1230 cal. yr BP, the glacier being substantially smaller than today during the interval between the earliest two of these phases of extension (Holzhauser 1997). These advances correlate with advance phases identified by Patzelt (1974), suggesting that the history of the Grosser Aletsch may be considered representative of the history of Alpine glaciers generally (notably, since systematic monitoring of glaciers began in Switzerland in 1880, tongues of similar size have fluctuated in concert with each other).

The MWP was interrupted by small, short-lived advances of the Gorner and Grosser Aletsch Glaciers in the twelfth century, during a period with winter temperatures comparable to those during the LIA (Pfister et al. 1998), and by less precisely dated advances of some glaciers in the French and Italian Alps. During the MWP, settlements expanded, population increased, and mortality due to harvest failure was less than during the LIA that followed. During the transition into the LIA (first half of the 14th century) cultivated areas retracted, incidence of crop failures and famine increased, and mortality rose (Campbell 1993). The climatic conditions responsible for LIA glacial expansion caused shortening of the growing season, more frequent incidence of extremes and increased risk of harvest failure. The upper limit for viable cultivation decreased in upland areas of Great Britain (Parry 1975; Parry and Carter 1985), as well as in continental Europe (Pfister 1996; Pfister et al. 1999). Lowland regions, from Finland to southern England, suffered decreased productivity, and consequent social and financial adjustments (Campbell 1993; Grove 1996; Jutikkala 1955). LIA extensions of the Alpine glaciers caused destruction of high settlements near advancing tongues and reduction in dues paid on the use of mountain pastures. High passes were blocked by snow and ice (Röthlisberger 1974). Greater frequency of landslides, rock falls and floods caused substantial damage to farmland (Grove and Battagel 1983). More widespread destruction, extending down valleys, was caused by the emptying of lakes dammed by advancing ice (Grove 1988; Patzelt 1994). Glacier advance in the Alps clearly shows three culminations of similar magnitude during the LIA; these occurred during the fourteenth, seventeenth and nineteenth centuries. The earliest of these, during the fourteenth century, affected the Aletsch, Rhone and Gorner glaciers, as well as the Gurgler Ferner in the eastern Alps (Patzelt 1974).

Turning to evidence from other sources, higher frequency fluctuations, with a magnitude of ca. 1-1·5°C, are superimposed upon the overall Holocene trend in the Norwegian Sea SST_{AUG} record, as well as upon those in oxygen isotope records from fjord basins (Mikalsen and Sejrup, submitted, N. Koç, C. Andersen, C. Birks unpublished data). The $\delta^{18}O$ record obtained from the U/Th dated calcite of a speleothem from Nordland in Northern Norway (Lauritzen and Lundberg 1999) similarly provides evidence of centennial to millennial temperature fluctuations of ca. 1-1·5°C about the general trend throughout the Holocene, as well as of a short-lived but extreme cooling event between 4·0 and 3·5 ka BP (Fig. 2c).

The data available for the last three millennia generally indicate little or no overall trend in thermal conditions, but a remarkably consistent series of fluctuations between relatively warmer and relatively cooler conditions recorded by a series of palaeosensors. This can be illustrated using data from a variety of sources, including tree rings (Fig. 3a), speleothem calcite (Fig. 3b) and ocean cores (Fig. 3c). The Norwegian Sea SST_{AUG} record shows marked temperature minima during the periods 500 BC to AD 1 and AD 700-850 (Fig. 3c), as well as a shorter-lived event at around AD 550. In addition, the MWP and LIA are clearly seen in this record. A foraminiferal δ⁸O record from a core taken in a fjord on the west Norwegian coast shows indications of peak-to-peak matches with the tree-ring $\delta^{14}C$ curve over the last three millennia (Mikalsen and Sejrup, submitted). The tree-ring record from the Torneträsk area of northern Sweden (Fig. 3a) shows a very similar temporal pattern. Notably in this record, as in glaciological records from both Scandinavia and the Alps, the LIA is not a single event. The temperature minima indicated by the tree-ring record more or less equate to a series of glacier maxima during the LIA that were determined by interacting fluctuations in temperature and precipitation. At many Alpine sites there are three culminations of the LIA glacial advance. In Sweden, lacustrine sediment records of lakes with glaciers in their catchments indicate a series of peaks in glacial activity at around AD 1140, during the early 16th century, during the late 16th century and during the 17th century. Furthermore, 14th century moraines were later over-run, indicating that the later events were of greater magnitude. In the Swiss Alps glacier expansions are recorded with peaks at AD 1360, 1660 and 1840 (Wanner et al. 1999). These quasi-periodic glacier advances have been referred to as 'Little Ice Age Type Events' and are typical of Alpine glacier dynamics during the last 3 millennia (Wanner et al. 2000).

A number of studies have sought to identify characteristic periodicities in records of Holocene climatic variability on centennial to millennial time scales. Various workers have identified millennial-scale periodicities in North Atlantic ocean records of SST, ice berg flux and deep water current strength with a cycle length of ca. 1500 yr ('Bond cycles', see Bianchi and McCave 1999; Bond et al. 1997). Others, studying the $\delta^{18}O$ record from the GISP2 ice core, identify periodic variability with a cycle length of ca. 900 yr (Grootes and Stuiver 1997; Schulz and Paul, in press), although there may be methodological reasons why the analyses of the marine and ice-core records give different cycle lengths (Schulz and Paul, this volume). On the continents there are hints of multi-centennial to millennial cycles in some palaeoecological records, for example, a periodicity of ca. 1500 yr in reconstructed precipitation in central Italy (Kelly and Huntley 1991) and 'quasi-periodic' fluctuations in moisture status of raised mires in northern England with a cycle length of ca. 800 yr (Barber et al. 1994). The $\delta^{14}C$ tree-ring curve also has a re-ported periodicity of ca. 830 (787-877) years (J.F.W. Negendank, pers. comm.), whilst Haas et al. (1998) found indications of a millennial cycle in their studies of the Alpine region.

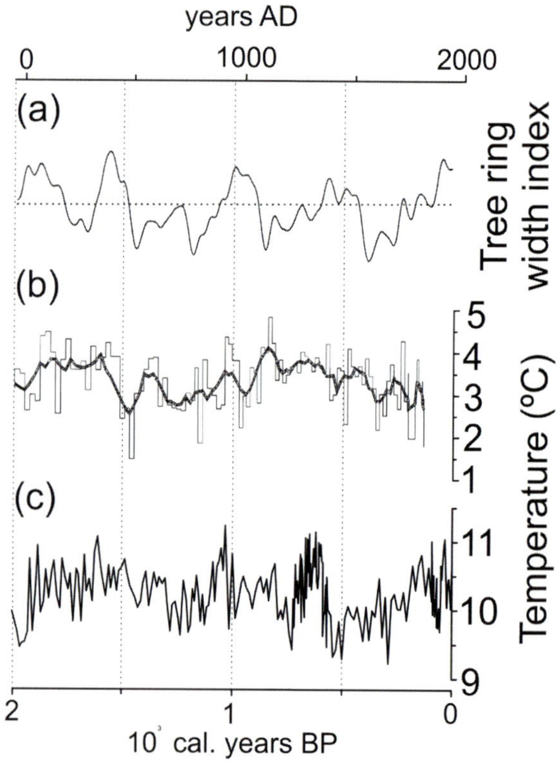

Fig. 3. Climatic records for the last two millennia from a series of palaeosensors. **(a)** Smoothed curve of tree ring width index for Torneträsk, Northern Sweden (redrawn after Grudd, unpublished data – Karlén, pers. comm.). **(b)** Reconstructed palaeotemperature record derived from stable oxygen isotope measurements made upon calcite from a speleothem (SG93) obtained from Søylegrotta, near Mo I Rana, Norway (redrawn from Lauritzen and Lundberg 1999). Values are local annual mean temperatures. **(c)** Reconstructed sea surface temperatures for core MD952011 from the Norwegian Sea (N. Koç, unpublished data).

Spectral analyses of the thickness of annual laminations in lake sediments of Holzmaar indicate a series of periodicities at frequencies that include some known to be characteristic of solar periodicities (Vos et al. 1997). So as to avoid interference that results from human activities around the lake in the later Holocene and from instabilities during the late-glacial interstadial and Younger Dryas, the spectral analyses have been confined to the mid-Holocene interval of dominance of 'mixed oak forest' in the pollen record from the site and to the full glacial period. The cycles detected have periodicities of 11, 22, 88, 208-210 and 490-500 yr; of these, those at 11 and 208-210 yr can reasonably confidently be assigned to solar forcing. The strongest cycle during the Holocene is that at 11 yr; in the clastic varves of the glacial, however, the 88 yr cycle is stronger. The 11 yr cycle is the most robust, however, being seen at a significant level even during times of rapid change and human disturbance; the other cycles, although apparent, do not reach significance at such times. Analyses of the phase relationships of these various cycles reveals a phase shift at around 9650 BP that is associated with changes in the algal flora within the lake and with a major change in the surrounding vegetation. Spectral analyses of ice-core data from GISP2 provides evidence of a similar phase shift at about this time, as well as revealing cyclic variations with periodicities of 11, 22, 33 and 208 yr (J.F.W. Negendank, pers. comm.). Oxygen isotope records from a Norwegian fjord core also reveal an apparent periodicity of ca. 210 yr (Mikalsen and Sejrup, submitted). Spectral analysis of several stacked tree-ring series reveals a particularly strong 37 yr periodicity during several centuries of the later third millennium BC (M. Baillie, pers. comm.); although the periodicity is present before and after this interval, it becomes amplified and then decays once again during this time.

Proxy and early instrumental data able to reveal the spatial patterns of climatic anomalies during these fluctuations are limited principally to those for the historic period, and hence for the last part of the LIA (i.e. the Late Maunder Minimum, AD 1675, onwards). Instrumental measurements of pressure, temperature and precipitation, as well as proxy data, have been used to statistically reconstruct monthly time series of the North Atlantic Oscillation (NAO) and the Eurasian (EU) circulation indices extending back to AD 1675 (Luterbacher et al. 1999; see also report of working group 1 in this volume). The NAO is a meridional oscillation in atmospheric mass between the area of high surface pressure near the Azores and the sub-polar low surface pressure near Iceland; this in turn is an expression of the strength of the westerly circulation over the eastern North Atlantic and western Europe. The EU index measures the pressure difference across central Europe (i.e. between Great Britain and the Black Sea); it reflects variability of western Eurasian climate, especially during the winter season. These reconstructions exhibit strong decadal to multi-decadal variations from the Late Maunder Minimum to the 1990s (Luterbacher et al. 1999), with some periods of the LIA (e.g. parts of the Late Maunder Minimum: AD 1675-1715) being characterised by stronger northerly to easterly air flow over central and eastern Europe linked to advection of cold dry air (strong presence of the Scandinavian high pressure cell), especially during winter and spring (Luterbacher et al. 2000). At the same time the westerlies over the eastern North Atlantic were weaker (negative NAO). The data also provide hints of a west-east differentiation of the climatic signals within the Alpine region. The Late Maunder Minimum was one of the few very cold periods (apart from the period 1880 to 1910) during recent centuries that persisted for several decades; it was characterized both by the eruption of several large volcanoes (Briffa et al. 1998) and by reduced solar activity (Eddy 1976).

Stronger westerlies over the eastern North Atlantic were prevalent in winter from 1830 to 1870 and in the early 20th century (Luterbacher et al. 1999), whereas periods with reduced northerlies over central Europe were observed from 1760 to 1850. For the winter and autumn season a running 30-year correlation between the EU and the NAO indices was computed to investigate low-frequency co-variability over the last 315 years (Fig. 4). In both seasons the correlations were mostly positive, suggesting that strong (weak) westerlies over the North Atlantic tend to be associated with strong (weak) northerly flow over central Europe. The correlations varied considerably, however, throughout the entire 315 year period (Luterbacher et al. 1999). Notably, mean winter correlation for the period 1840-60, sometimes considered to mark the end of the LIA in the North Atlantic region, becomes slightly negative. This break down of the previously generally positive pattern of correlation between winter NAO and EU indices in the middle of the nineteenth century suggests that circulation over the eastern North Atlantic at sea level is temporally decoupled from the continental flow, hence the NAO is not always the dominant mode for the European climate (Luterbacher et al. 1999). This may reflect a systematic shift in the dominant planetary wave number and/or in the location of

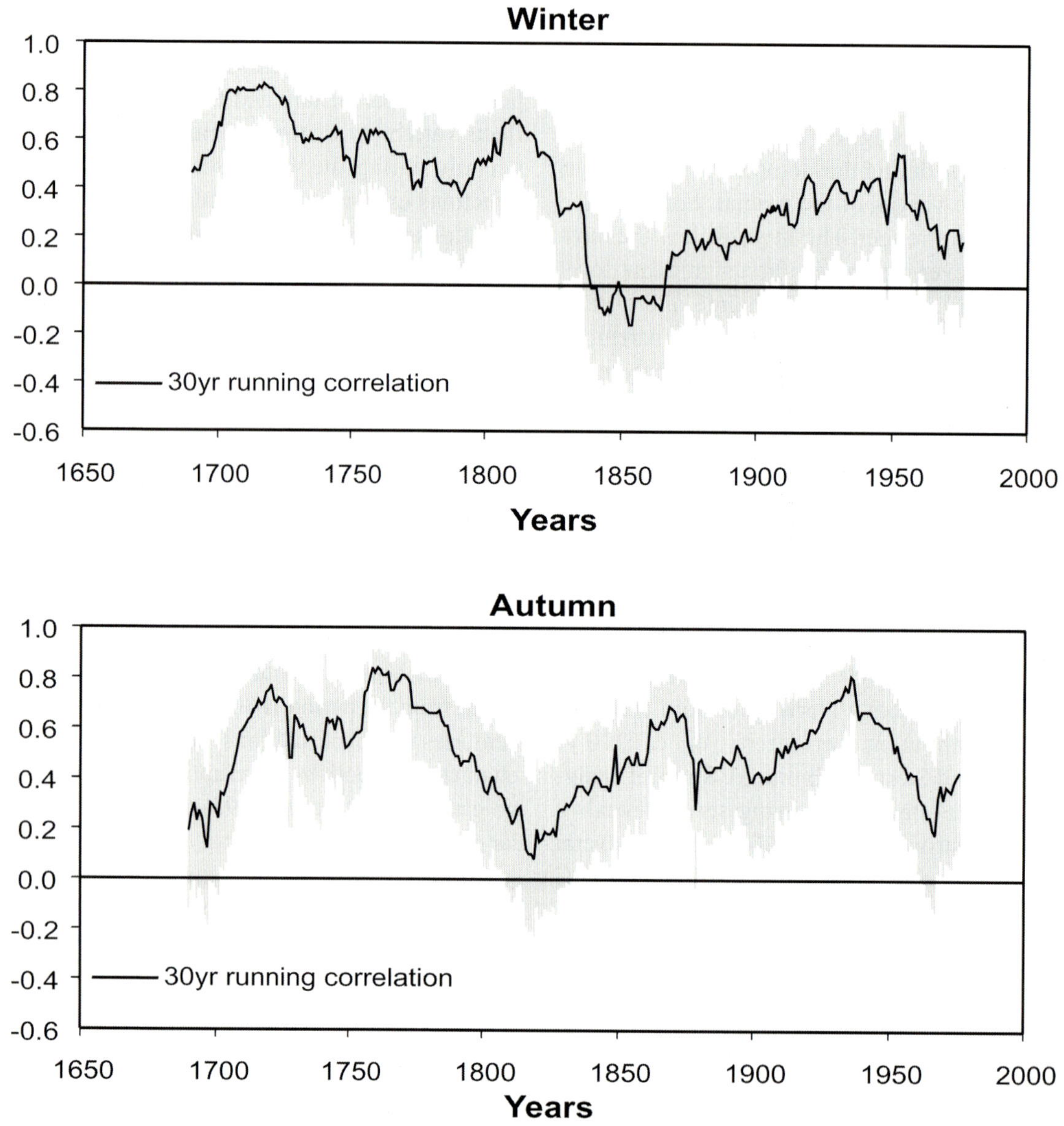

Fig. 4. 30-year running correlation between reconstructed winter and autumn NAO and EU indices. Reconstructed winter (top) and autumn (bottom) NAO and EU indices are shown; bars indicate 90% confidence intervals for the correlation coefficients (from Luterbacher et al. 1999).

planetary waves. Further analyses are required to investigate this observation (Luterbacher et al. 1999).

The potential impacts of these centennial to millennial scale climatic fluctuations upon human populations are difficult to resolve during most of the Holocene because the reactions of human societies to such fluctuations are difficult to distinguish from cultural changes arising from other causes, as well as because of the chronological limitations of the archaeological record. Nonetheless, as Table 1 indicates, there are tantalising coincidences between the dates for some of the archaeological events and the dates of some of the climatic fluctuations. For the historic period, however, the impacts of the MWP and LIA upon human pop-ulations are documented (Ogilvie 1991); in some cases even the proximal mechanism by which the climatic fluctuation had its regional impact upon a human population is known. Ogilvie (1984) has, for example, compiled an historical record of sea-ice extent for the area around Iceland. The prevalence of extensive sea-ice cover during the LIA was associated with a decline in fisheries around Iceland that, together with the impact of LIA climatic conditions upon crop cultivation, largely accounted for a substantial decline in the population of Iceland during the LIA. The extinction of the Norse settlements in Greenland during the LIA is also well-documented and can be related to the changing climatic conditions that rendered the area no longer suitable even for the pastoralism practised there during the MWP (Buckland et al. 1996). Malaria, which was endemic around the coasts of southern England during the LIA, declined in prevalence and eventually disappeared during the period of general warming that spanned the nineteenth and twentieth centuries. This seemingly anomalous phenomenon may, at least in part, be attributed to the regional prevalence of relatively cooler summers in England during the nineteenth century that contrasts with the general warming trend in annual temperatures (Lindsay and Joyce 2000).

Multi-Annual to Multi-Decadal Events

Evidence for events with a duration of less than a century must be sought in records that have annual temporal resolution and their own inherent time scale; such records include tree rings, ice cores and other laminated deposits such as varved lake sediments. When such records are examined it is striking that the most prominent sub-centennial climatic events during the Holocene are in many cases associated with evidence of often dramatic, albeit short-lived, cooling. The clearest records of at least the more extreme of these short-lived cold events are those obtained from tree rings and ice-cores. Furthermore, at least some of the extreme events recorded are apparently global in extent, being seen in several records. In the case of, for example, the event at AD 540 (1410 BP), effects are seen in tree-ring sequences from many different parts of the world (e.g. Ireland, Germany, Scandinavia, Siberia, North America and South America): AD 540 is the joint second worst growth year in 1500 yr in the European oak (*Quercus*) tree-ring sequences; the Fennoscandian pine (*Pinus sylvestris)* tree-ring sequence records AD 536 as the second coldest year in 1500 yr; AD 540 has the worst growth in 1500 yr in pines (*P. sylvestris*) from northern Russia; AD 535, 536 and 541 are recorded as the second, third and fourth coldest years in 2000 yr by the foxtail pine (*P. balfourana*) tree-ring sequence from the Sierra Nevada of western North America; the period around AD 540 includes the coldest year in 1500 yr of the Chilean *Fitzroya cupressoides* tree-ring sequence; the *F. cupressoides* tree-ring sequence from Argentina records the fourth worst growth conditions in 1500 yr in the years just after AD 540.

The most marked of these events, recorded as 'narrowest ring events' in the Irish tree-ring sequences, seem to approach a decade in duration; they are as follows:

2345 BC - i.e. 4295 BP
1628 BC - i.e. 3578 BP
1159 BC - i.e. 3109 BP
207 BC - i.e. 2157 BP
AD 540 - i.e. 1410 BP

The cause of these abrupt events is still a matter of debate. It is well understood that large explosive volcanic eruptions that deliver material to the stratosphere can cause climatic changes, although

Age (BC/AD)	Age (BP – i.e. before AD 1950)	Climatic event or change; Human population response; Palaeosensor	Duration	Spatial extent	Nature and magnitude of climatic change
	9650	Phase shift in Holzmaar varve periodicities and in ice core periodicities		N. Germany and Greenland	
	9600–8000	Tree limit in northern Sweden fluctuating but generally indicating warm conditions		Sweden	
	8500–7900	Tree limit depressed in northern Sweden; cold		Sweden	
	8200	Cooling event peaking at 8250 but extending between 8400 and 8000 in the ice core records. Similar duration in ocean core records. Duration of *ca*. 200 yr in Ammersee record. Most marked cooling event in Scandinavia during the Holocene with substantial recession of the tree limit and glaciers advanced to a limit similar to that of the LIA at a time when isostatic depression of the land was still substantial.	200 – 400 yr	Entire region?	Marked cooling 6 ± 2°C in GRIP $\delta^{18}O$ Norwegian Sea SST *ca*. 2°C Ammersee $\delta^{18}O$ *ca*. 3°C
	7900–7400	Tree limit high in northern Sweden; warm		Sweden	
5500 BC	7450	Beginning of agriculture (Central Europe)		Central Europe	
	7400–7200	Tree limit depressed in northern Sweden; cold		Sweden	
	7200–6700	Tree limit high in northern Sweden; warm		Sweden	
5000 BC	6950	End of the linear band ceramic pottery settlement period (Central Europe)		Central Europe	
	6700–6000	Tree limit in northern Sweden relatively low; relatively cold		Sweden	
4300 BC	6250	Earliest Neolithic lake-shore settlements		Central Europe	
	6000–5100	Tree limit relatively high in northern Sweden but fluctuating; warm but with cooler intervals		Sweden	

Table 1. Principal Holocene climatic changes and events, and major changes in human settlement patterns and activities.

Age (BC/AD)	Age (BP – i.e. before AD 1950)	Climatic event or change; Human population response; Palaeosensor	Duration	Spatial extent	Nature and magnitude of climatic change
	6000–4000	Thermal maximum in many proxies		North-west Europe	
ca. 4000 BC	5950	break in the economy of lake-shore settlements (Central Europe)		Central Europe	
	5800	Tree limit decrease in northern Sweden; cooling		Sweden	
	5600	Tree limit decrease in northern Sweden; cooling		Sweden	
ca. 3600 BC	5550	break in the history of lake-shore settlements and an associated impact upon the subsistence economy (Central Europe)		Central Europe	
3195 BC	5145	Narrowest ring event; tree limit decrease in northern Sweden; cooling		North Atlantic, probably global	
	5000	Marked end of relatively warm conditions in northern Sweden		Northern Sweden	
ca. 2900 BC	4850	cultural changes (Central Europe); tree limit minimum and glacier maximum in Scandinavian mountains; cold		Central Europe and Sweden	
2345 BC	4295	Narrowest ring event; short cold event indicated by tree-limit and glacier extent data from Scandinavian mountains		probably global	
	4000 – 3000	Temperature minimum in the Nordland speleothem carbonate $\delta^{18}O$ record	*ca.* 400 – 500	Norway	
	3900–3600	Glacier extent maximum in Scandinavian mountains; cold		Scandinavia	
	3625	Extreme cold event in Nordland speleothem record		Nordland	
1628 BC	3578	Narrowest ring event (Santorini)		probably global	

Table 1. (cont.)

Age (BC/AD)	Age (BP – i.e. before AD 1950)	Climatic event or change; Human population response; Palaeosensor	Duration	Spatial extent	Nature and magnitude of climatic change
ca. 1650 – 1500 BC	3600 – 3450	early Bronze Age lake-shore settlements (Central Europe)		Central Europe	
	3600–3500	Tree limit increasing in northern Sweden; warming		Sweden	
	3500	Onset of cool phase in Switzerland; peak in glacial activity in southern Norway		Central Europe and southern Norway	
ca. 1500 – 1400 BC	3450 – 3350	Middle Bronze Age begins (Central Europe)		Central Europe	
	3400	Glacier advance in Scandinavian mountains; short cold interval		Scandinavia	
1159 BC	3109	Narrowest ring event; cold in Scandinavia		probably global	
	3000	End of downward temperature trend in Norwegian Sea and Nordland		Norwegian Sea and Nordland	
ca. 1050 – 850 BC	3000 – 2800 BP	late Bronze Age lake-shore settlements (Central Europe); short-lived warming indicated by modest increase in tree limit in northern Sweden		Central Europe and Sweden	
ca. 850 BC	2800	Abandonment of settlements and cultivated areas in the Netherlands that subsequently were overgrown by peat; onset of cool phase in Switzerland; evidence of cooling in northern Scandinavia		Netherlands, Central Europe and northern Scandinavia	
ca. 800 BC	2750	transition from the Bronze Age to the Iron Age; end of the period of lake-shore settlements (Central Europe)		Central Europe	
	2600	Peak in glacial activity in southern Norway and the Scandinavian mountains; cold		Southern Norway and northern Scandinavia	

Table 1. (cont.)

Age (BC/AD)	Age (BP – i.e. before AD 1950)	Climatic event or change; Human population response; Palaeosensor	Duration	Spatial extent	Nature and magnitude of climatic change
500 BC – AD 1	2450 – 1950	Summer SST minimum in eastern Norwegian Sea; glacier advance in Scandinavian mountains; cold		Norwegian Sea and Scandinavia	
	2500 – 2000	Cooling event in Scandinavia marked by tree limit minimum and glacier maximum; temperature minimum in Greenland	*ca.* 500	Scandinavia and Greenland	
	2400	Onset of cool phase in Switzerland		Central Europe	
ca. 400 BC	2350	transition from the Early to the Late Iron Age (Central Europe)		Central Europe	
	2200	Peak in glacial activity in southern Norway		Southern Norway	
207 BC	2157	Narrowest ring event		global	
	1500	Consistent temperature decrease indicated by Greenland ice core $\delta^{18}O$ records; glacier advance in Scandinavian mountains; cooling		Greenland and Scandinavia	
AD 540	1410	Narrowest ring event (Justinian Plague)		global	
AD 550	1400	Summer SST minimum in eastern Norwegian Sea		Norwegian Sea	
AD 550 – 850	1400 – 1100	Glacier advances in northern Sweden; tree-ring evidence suggests that there were two cold periods		Northern Sweden	
AD 700 – 850	1250 – 1100	Summer SST minimum in eastern Norwegian Sea		Norwegian Sea	
AD 1000	950	Peak of MWP: tree limit increase in northern Sweden; warm		throughout north-west Europe	
AD 1700	250	Extreme of LIA; glacier advance in Scandinavian mountains; cold		throughout north-west Europe	

Table 1. (cont.)

these persist normally over a period of not more than a few years. Such eruptions usually leave traces of acid, and sometimes tephra, in ice-cores. Recently, attention has been drawn to the possibility that similar abrupt climatic effects may be induced by cometary debris and that some Holocene events may be related to this mechanism (Baillie 1998, 1999a). Although the evidence remains circumstantial, the very marked environmental downturn centred upon AD 540 (Baillie 1994) corresponds to a period (AD 400-600) identified as one during which the Earth was at increased risk of bombardment (Bailey et al. 1990). The principal hazard identified by Bailey et al. (1990) is that arising from the fragments of dead comets found orbiting the sun as annuli that give rise to the well-known meteor showers each year; these latter phenomena result from collisions with tiny fragments and dust particles. At times when the Earth's orbit intersects the centre of such a debris stream the risk of impacts by larger objects is increased. The model for such larger objects is the Tunguska object of 30th June 1908 that exploded in the atmosphere at an estimated altitude of ca. 6 km and with an estimated magnitude equivalent to 12-30 megatons (of TNT). Bailey et al. (1990) hypothesise that the major hazard is collision with a large number of such objects within a relatively short period (minutes to years); they suggest the term 'cosmic swarm' for such a hazard. Baillie (1999a) has subsequently hypothesised that a very close approach by a cometary nucleus might have very similar consequences; during such an approach cometary fragments travelling with the main body would be likely to collide with the Earth giving rise to detonations in the atmosphere similar to the Tunguska event.

These hypotheses raise interesting questions relating to the 'apparency' of such impacts in the records from various palaeosensors. Detonations in the atmosphere leave no obvious impact craters, unlike collisions with objects of sufficient size to penetrate the atmosphere and impact upon the Earth's surface. Even collisions with such very large objects leave no crater if the object impacts upon an ocean basin, although traces of the impact may be recorded by tsunami deposits on surrounding land masses. Even if atmospheric detonations destroy vegetation, as in the case of the Tunguska object that flattened some 2000 km^2 of forest, such effects are relatively short-lived and likely to be difficult to detect in the Holocene palaeovegetation record. Comets being composed principally of ice, they will also leave few traces in deposits such as ice layers or lacustrine varves. Thus the hypothesis of cometary impacts as the cause for at least some sub-centennial climatic events during the Holocene is extremely difficult to test by means of conventional palaeoenvironmental investigations. Nonetheless, the number of impacts on the upper atmosphere recorded by Earth-orbiting satellites in the few decades since such observations have been possible renders it unwise to dismiss the hypothesis. Furthermore, a study of mythology reveals a recurrent association between comets (bright "sky gods") and environmental catastrophes, including events dated to the times represented by several of the 'narrowest ring events'; such evidence, although unlikely to be scientifically verifiable, is difficult to dismiss as simply coincidence (Baillie 1999a).

If such impacts with cometary fragments have occurred during the Holocene, then the principal mechanism that would lead to consequent environmental effects is loading of the stratosphere, and possibly also the troposphere, with cometary dust derived from the impactors themselves, as well as potentially from the parent comet and/or the debris ring. Such material injected into the stratosphere from the top of the atmosphere will have a longer residence time than do volcanic dust or gases that are injected from below. The associated climatic effects would thus be expected to persist longer then do the climatic effects of explosive volcanic eruptions. Most agricultural societies would be highly vulnerable to any persistent cooling that lasted longer than two years; three successive failed harvests will normally result in the collapse of any agricultural society. Such collapses are recorded for prehistoric societies by, for example, Schibler et al. (1997a) and Maise (1998).

Returning to the series of 'narrowest ring events' recorded in the Irish tree-ring chronology, it is striking firstly that several of these persist for periods approaching a decade, apparently ruling out a single explosive volcanic eruption as their cause. Nonetheless, there is good evidence for the involvement of a major volcano (Santorini) in the 1628 BC event. This event, seen in North American bristle-cone pines (*Pinus longaeva*) (LaMarche and Hirschboeck 1984) as well as in Irish, English and German oaks, has also been postulated to have caused anomalous growth in Anatolian trees (Kuniholm et al. 1996). Although there is continued debate about the possible identification of Santorini tephra in the seventeenth century BC acid layers of the GISP2 ice core (Manning 1998; Zielinski and Germani 1998), it seems increasingly probable that both Santorini acid fallout and tephra are present in the GRIP ice core (C.U. Hammer, pers. comm.). Although volcanic triggers cannot be ruled out for any of the events prior to AD 1, nor can impacts with cometary debris.

Whatever their cause, these events appear to coincide with major events in human history, often presaging the fall of dynasties in the civilised world and being associated with mythical happenings in many regions. The most recent of the events, at AD 540, falls within the historical period and is associated with a variety of documented impacts upon humans including: the Justinian Plague, that is estimated to have killed one-third of the population of Europe; severe famines in China and the Mediterranean area; and, impacts upon other peoples including the Maya in Central America and native populations of southern North America. It is worth noting that there are hints of changes in the solar disc associated with this event; contemporary writers refer to the sun 'appearing as in eclipse', a statement implying a darkening of a significant proportion of the solar disc.

Interestingly there are hints that some of the events may also be recorded by other palaeosensors that record a series of major short-term climatic events during the Holocene, often with marked impacts upon humans. $\delta^{18}O$ values measured on U/Th-dated calcite of a speleothem from Nordland, northern Norway, for example, record an extremely cold event at ca. 3625 BP (on a time scale that is estimated to be accurate to within 1%, i.e. ± 36 yr for this age), a date that lies tantalisingly close to the 1628 BC 'Santorini' event (Lauritzen and Lundberg 1999). Furthermore, all of the listed 'narrowest ring events' correspond to periods of cold conditions in Scandinavia. In several regions changes in settlement patterns and economies are reported that apparently are the consequences of such climatic events. Van Geel et al. (1998), for example, report an event at 2800 BP in the Netherlands that is associated with the establishment of markedly moister conditions (that may reflect cooling or increased precipitation, or a combination of the two) and that led to the abandonment of settlements and cultivated areas that subsequently became overgrown by peat. In the Alpine foreland Hochuli et al. (1998) report the abandonment of lake dwelling settlements concident with this event. Schibler et al. (1997a; 1997b) report a similar coincidence between a climatic shift towards cooler and/or moister conditions and an economic decline from Switzerland during the century prior to 3600 BC (i.e. 5550 BP); this event is succeeded by a period of cooler conditions that had effects upon cereal cultivation and hence upon the whole economy.

The Central European archaeological record prior to the Roman occupation is characterised by a series of major events, as follows:

- ca. 5500 BC (7450 BP) beginning of agriculture
- ca. 5000 BC (6950 BP) end of the Linearbandkeramik settlement period
- ca. 4300 BC (6250 BP) earliest Neolithic lake-shore settlements in the Alpine Foreland
- ca. 4000 BC (5950 BP) changes in economy of lake-shore settlements
- ca. 3650 BC (5600 BP) strong impact upon the subsistence economy in lake dwelling sites of the Alpine foreland
- ca. 3500-3300 BC (5450-5250 BP) break in the history of lake shore settlements with only few such settlements recorded during this interval
- ca. 2900 BC (4850 BP) agricultural changes
- ca. 2400-1700 BC (4350-3650 BP) few settlements known; almost no lake shore settlements
- ca. 1650-1500 BC (3600-3450 BP) early Bronze Age lake-shore settlements
- ca. 1500-1400 BC (3450-3350 BP) Middle Bronze Age begins; no lake-shore settlements
- ca. 1050-850 BC (3000-2800 BP) late Bronze Age lake-shore settlements

• ca. 800 BC (2750 BP) transition from the Bronze Age to the Iron Age; end of the period of lake-shore settlements
• ca. 400 BC (2350 BP) transition from the Early to the Late Iron Age
• ca. 15 BC (1965 BP) Roman occupation of regions north of the Alps

Although the extent to which these events relate to climate remains open to debate, many of them coincide with events in the $\delta^{14}C$ curve. In the records from dendrochronologically dated lake-shore settlements from Switzerland a striking coincidence is seen between periods of intensive exploitation of wild animal resources, especially red deer *(Cervus elaphus)*, and times with relatively high $\delta^{14}C$ values (Fig. 5). Notably, periods of lake-shore settlement generally are associated with periods of 'warmer and drier' climate, whereas periods of abandonment and/or lack of such settlements correspond to periods of 'cooler and wetter' climate (Gross-Klee and Maise 1997). The potential role of climate in bringing about these major archaeological events is further highlighted when we note that many of the events can be augmented by parallel evidence from palaeoecological studies. For example:

• ca. 5500 BC (7450 BP) initiation of tree-ring chronologies: Ireland, Britain, N Sweden, Finland
• ca. 4000 BC (5950 BP) lake level rise at Ballynagilly, Ireland
• ca. 3900 BC (5850 BP) postulated wet event; problems in oak chronology construction; W. English bog oak hiatus, Sweet Track 3807 BC, river gravel oaks
• 2911 BC (4861 BP) tree-ring event, Irish pines (Baillie 1999b)
• ca. 2350 BC (4300 BP) narrowest-ring event; widespread destruction in Near East (Courty 1998; Peiser 1997)
• ca. 2310 BC (4260 BP) widespread occurrence of Hekla 4 tephra as marker horizon in north western Europe (Blackford et al. 1992; Daniell 1997; Hall et al. 1994; Huntley et al. 1997; Pilcher and Hall 1992).
• ca. 430 BC (2380 BP) widespread tree-ring downturn (Baillie 1995)

A principal dilemma in the identification and/or accumulation of evidence for widespread environmental events is the necessary dependence upon ^{14}C dating in many disciplines. ^{14}C age deter-minations 'smear' dates through time so that it is possible that evidence for some of the abrupt events, noted in sequences dated by dendrochronology, is rendered invisible in data dated only by calibrated ^{14}C measurements.

Discussion – Are our Times Anomalous?

This question requires some further definition. In particular, 'our times' is taken to refer to the last few decades of the twentieth century AD, and the extent of their possible anomaly is considered in relation only to the Holocene and not, for example, to earlier interglacials. The question is also understood to refer to climate, globally as well as in north-west Europe.

Whereas the instrumental part of the Mann et al. (1998) temperature record may provide a reliable record of northern hemisphere mean temperature, it is not representative of the situation in most of north-west Europe. Parts of northern Fennoscandia, for example, experienced no net change in mean annual surface temperature during the period 1961-90 (Maxwell 1997). In contrast, whereas a steep temperature increase is recorded in the Greenland ice core records during the first 50 yr of the 20th century, where it is clearly unique within the last 2000 yr, a cooling trend is recorded for the subsequent 40 yr (Dahl-Jensen et al. 1998). Furthermore, other parts of Europe, and even different parts of the British Isles (Jones and Hulme 1997), have experienced different patterns of twentieth century temperature change. A regional synthesis of temperature data for north-west Europe, although beyond the scope of the present contribution, is clearly needed. Any such synthesis must address the need to ensure representative sampling of the climate record and is likely to need to subdivide north-west Europe if it is to produce useful and meaningful patterns.

The Mann et al. (1998) curve for the 'pre-instrumental' period is based upon only a small number of unevenly distributed climate palaeosensors. The curve is not representative of the situ-

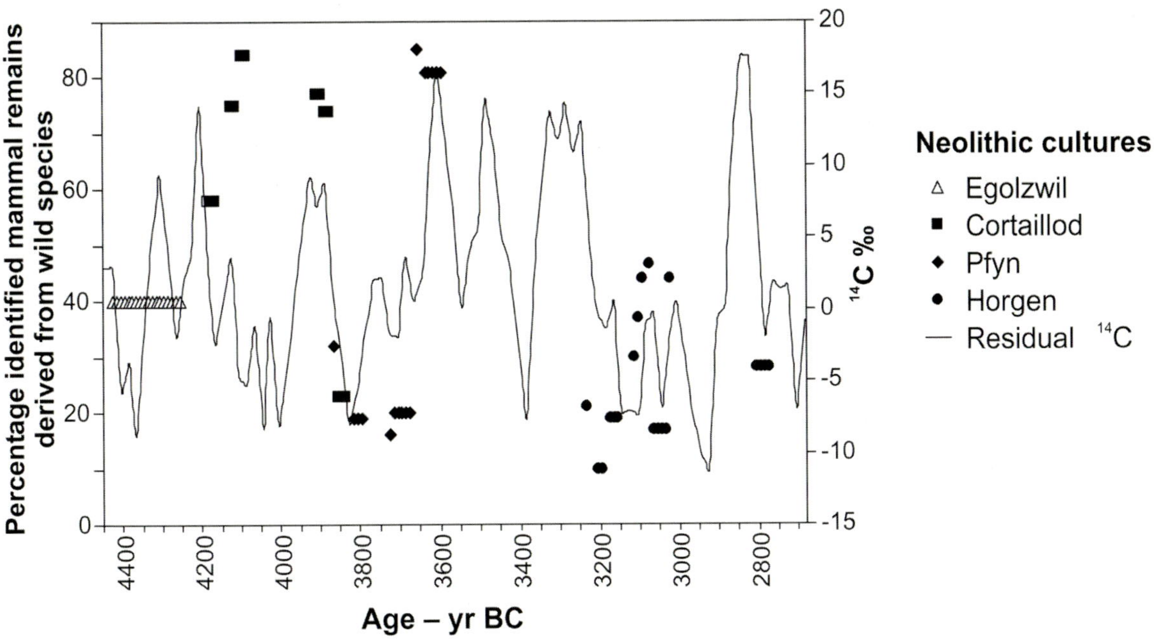

Fig. 5. Use of wild animal resources by Neolithic cultures in relation to atmospheric $\delta^{14}C$. Atmospheric $\delta^{14}C$ ‰ between 4450 and 2700 BC and the percentage of identified mammalian remains derived from wild species at selected precisely dated sites on lower Lake Zurich (after Hüster-Plogmann et al. 1999b). Filled symbols indicate dendrochronologically dated assemblages; open symbols indicate ^{14}C dated assemblages.

ation in most of north-west Europe, and is probably not an adequate representation of the northern hemisphere as a whole. One particularly noticeable discrepancy is the lack of evidence of the temperature fluctuations of the LIA that are clearly recorded in most north-west European palaeosensors (Fig. 3).

A more appropriate synthesis for the present purpose would be obtained by compiling records from north-west Europe spanning the last 2 millennia for a selection of the climate palaeosensors available. Fig. 3 illustrates three examples of such records. Although the records shown all relate to Scandinavia and the adjacent Norwegian Sea, and exhibit a considerable degree of homogeneity with respect to their principal features, we have not attempted to synthesise an overall mean from these very disparate palaeotemperature sensors. It is clear, however, that the magnitude and rate of climatic variability during the past two millennia have been considerable. It is notable that, for various reasons, none of these records provides evidence of the twentieth century warming.

Until appropriate regional syntheses of instrumental and palaeosensor data have been made, it is impossible to determine definitively whether or not 'our times' are anomalous in relation to Holocene climatic variability of north-west Europe. Nonetheless, it is clear that atmospheric concentrations of CO_2 and other 'greenhouse gases' are anomalous with respect to the Holocene. It is also clear that the recent trend in the instrumental records of global temperature provides strong evidence of a warming to mean global temperatures that are anomalously warm. Furthermore, this warming is taking place at a rate that is at least as fast as the most rapid Holocene temperature increases, and is probably more rapid than any temperature increase of comparable magnitude during the Holocene.

Recommendations for Future Work

In attempting to address our overall question, as well as our other principal questions, we have identified a number of gaps in present knowledge and needs for future research. These are summarised below:

- There is a general need for improved chronologies. Although major improvements in this respect have been made during the past few years, there are still major problems in assessing the synchroneity or phase relationships between records. These problems relate especially to the limitations of precision and accuracy of ^{14}C measurements, as well as to the inherent biases in the ^{14}C chronology that arise from past variations in the abundance of ^{14}C in the atmosphere. A key task is to reconstruct the spatial structure of climatic events or of specific periods of climate history. Prior to the last 500 years, however, and using currently available data, this can only be achieved for the most pronounced millennial to centennial scale events.
- In order to improve the reliability of the reconstructions of climatic variables for the Holocene, there is an urgent need to improve the quality of the palaeo-observation methods. This is particularly important for this period because the climatic signal is rather weak when compared, for example, with overall glacial–interglacial variability. In addition to improved statistical methods, strategies are required for increasing the signal to noise ratio and for more directly relating the reconstructions (or transfer functions) to the processes that lead to changes in the variable that is recorded (e.g. tree ring characteristics, pollen assemblages, lake levels, stable isotope ratios, planktonic assemblages).
- In order to understand the dynamics of climatic changes, a wider range of palaeo-observations needs to be synthesised to enable reconstructions of the spatial patterns in the climatic anomalies. These anomaly patterns can then be compared with modern modes of variability, and will be an important asset both for determining the dynamics behind the climatic changes and for the identification of the range of modes in which the climate system operates.
- In order to evaluate the nature and extent of the impacts of climate change upon human populations there is a parallel need for a spatially and temporally extensive compilation of data derived from archaeological studies throughout north-west Europe. Such a compilation would enable the changing spatial patterns of settlement and of various human activities (e.g. cultivation of particular crops, use of particular natural resources) to be determined and thus pave the way for comparisons with the spatial patterns in the climatic anomalies.
- With respect especially to short-term climatic events, there is a need to determine the relative importance of changes in the return periods between extreme events as opposed to changes in the long-term mean. Climate model simulations show that modes of climate variability, and the return periods between extreme events, are not linked in a simple manner to changes in mean climate. It is also clear that changes in the frequency of extreme events are likely to have greater impacts upon both natural ecosystems and human populations than do modest changes in mean climate.
- There is a need for continued emphasis upon research that can unravel the history of climatic forcing factors (e.g. volcanism, extra-terrestrial impacts, aerosols, solar variability), and can evaluate the possible climatic effects that these may have had.

This list is by no means comprehensive, but attempts only to highlight those areas in which research would enable the questions that we have addressed to be answered more completely and with greater confidence.

Acknowledgments

The authors are grateful to the organisers of the Hanse conference on "Past climate and its significance for human history in NW Europe, the last 10,000 years", held at the Hanse Institute for Advanced Study in Delmenhorst, October 9-14th 1999, for inviting them to participate in this conference and for the financial support that made this possible. Mats Regnell (Lund), Frank Siegmund (Basel) and Andreas Zimmerman (Köln) kindly assisted with information about archaeological data. Lucia Wick (Bern) provided valuable additional information about Alpine tree line history.

References

Allen JRM, Brandt U, Brauer A, Hubberten H, Huntley B, Keller J, Kraml M, Mackensen A, Mingram J, Negendank JFW, Nowaczyk NR, Oberhänsli H, Watts WA, Wulf S, Zolitschka B (1999) Evidence of rapid last glacial environmental fluctuations from southern Europe. Nature 400:740-743

Alley RB, Mayewski PA, Sowers T, Stuiver M, Taylor KC, Clark PU (1997) Holocene climatic instability: A prominent, widespread event 8200 yr ago. Geology 25:483-486

Anderson G (1902) Hasseln i Sverige fordom och nu. Sveriges Geologiska Undersökning Ser. Ca, 3, 168

Andraschko FM (1995) Studien zur funktionalen Deutung archäologischer Siedlungsbefunde in Rekonstruktion und Experiment: Vol. 1, Duderstadt

Atkinson TC, Briffa KR, Coope GR (1987) Seasonal temperatures in Britain during the past 22,000 years reconstructed using beetle remains. Nature 325:587-593

Bailey ME, Clube SVM, Napier WM (1990) The Origin of Comets. Pergamon Press, London

Baillie MGL (1994) Dendrochronology raises questions about the nature of the AD 536 dust-veil event. The Holocene 4:212-217

Baillie MGL (1995) Dendrochronology and the chronology of the Irish Bronze Age. In: Waddell J, Shee-Twohig E (eds) Ireland in the Bronze Age. Stationery Office, Dublin, pp 30-37

Baillie MGL (1998) Hints that cometary debris played some role in several tree-ring dated environmental downturns in the Bronze Age. BAR International Series 728:109-116

Baillie MGL (1999a) Exodus to Arthur: catastrophic encounters with comets. Batsford, London

Baillie MGL (1999b) A View From Outside: recognising the big picture. Quaternary Proceedings 7:625-635

Barber KE, Chambers FM, Maddy D, Stoneman R, Brew JS (1994) A sensitive high-resolution record of late Holocene climatic change from a raised bog in northern England. The Holocene 4:198-205

Behre K-E, Kucan D (1994) Die Geschichte der Kulturlandschaft und des Ackerbaus in der Siedlungskammer Flögeln, Niedersachsen, seit der Jungsteinzeit. Vol. 21. Verlag Isensee, Oldenburg

Behre K-E, Schmid P (1998) Das Niedersächsische Institut für Historische Küstenforschung, 60 Jahre Forschungstätigkeit im Küstengebiet. Brune, Wilhelmshaven

Benecke N (ed) (1999) The Holocene History of the European Vertebrate Fauna – Modern Aspects of Research, Vol. 6. Verlag Marie Leidorf, Rahden

Berglund BE (ed) (1991) The cultural landscape during 6000 years in southern Sweden – the Ystad-Project, Vol. 41, Munksgaard, Kopenhagen

Bianchi GG, McCave IN (1999) Holocene periodicity in North Atlantic climate and deep-ocean flow south of Iceland. Nature 397:515-517

Blackford JJ, Edwards KJ, Dugmore AJ, Cook GT, Buckland PC (1992) Icelandic volcanic ash and the mid-Holocene Scots pine *(Pinus sylvestris)* pollen decline in northern Scotland. The Holocene 2:260-265

Blikra LH, Fjeldstad Selvik S (1998) Climatic signals recorded in snow avalanche-dominated colluvium in western Norway: depositional facies successions and pollen records. The Holocene 8:631-658

Blytt A (1876) Essay on the immigration of the Norwegian flora during alternating rainy and dry periods. Cammermeyer, Kristiana

Blytt A (1881) Die Theorie der wechselnden kontinentalen und insularen Klimate. Englers Bot. Jahrb., 2, 1-50, 177-184

Boaz J (1998). Hunter-Gatherer Site Variability. Changing patterns of site utilization in the interior of eastern Norway, between 8000 and 2500 B.P. Universitetets Oldsaksamlings Skrifter, Ny rekke, 20. Oslo

Bond G, Showers W, Cheseby M, Lotti R. Almasi P, deMenocal P, Priore P, Cullen H, Hajdas I, Bonani G. (1997) A pervasive millennial-scale cycle in North Atlantic Holocene and glacial climates. Science 278:1257-1266

Briffa KR, Jones, PD, Bartholin TS, Eckstein D, Schweingruber FH, Karlén W, Zetterberg P, Eronen M (1992) Fennoscandian Summers from Ad-500 - Temperature-Changes on Short and Long Timescales. Climate Dynamics 7:111-119

Briffa KR, Jones PD, Schweingruber FH, Osborn TJ (1998) Influence of volcanic eruptions on Northern Hemisphere summer temperature over the past 600 years. Nature 393:450-455

Brombacher C, Jacomet S (1997) Ackerbau Sammelwirtschaft und Umwelt: Ergebnisse archäobotanischer Untersuchungen. In: Schibler J, Hüster-Plogmann H, Jacomet S, Brombacher C, Gross-Klee E, Rast-Eicher A (eds) Ökonomie und Ökologie neolithischer und bronzezeitlicher Ufersiedlungen am Zürichsee. Verlag Fotorotar, Zürich und Egg, Vol. 20, pp 220-299

Brostrom A, Gaillard MJ, Ihse M, Odgaard B (1998) Pollen-landscape relationships in modern analogues of ancient cultural landscapes in southern Sweden - a first step towards quantification of vegetation openness in the past. Vegetation History and Archaeobotany 7:189-201

Buckland PC, Amorosi T, Barlow LK, Dugmore AJ, Mayewski PA, McGovern TH, Ogilvie AEJ, Sadler JP, Skidmore P (1996) Bioarchaeological and climatological evidence for the fate of Norse farmers in medieval Greenland. Antiquity 70:88-96

Bunnik FPM, Kalis AJ, Meurers-Balke J, Stobbe A (1995) Archäopalynologische Betrachtungen zum Kulturwandel in den Jahrhunderten um Christi Geburt. Archäologische Informationen 18:169-185

Burga C, Perret R (1998) Vegetation und Klima der Schweiz seit dem jüngeren Eiszeitalter, Thun

Campbell BMS (ed) (1993) Before the Black Death: Studies in the "crisis" of the Early Fourteenth Century. Manchester University Press, Manchester

Cheddadi R, Yu G, Guiot J, Harrison SP, Prentice IC (1997) The climate of Europe 6000 years ago. Climate Dynamics 13:1-9

Collingham YC, Hill MO, Huntley B (1996) The migration of sessile organisms: a simulation model with measurable parameters. Journal of Vegetation Science 7:831-846

Courty M-A (1998) The Soil Record of an Exceptional Event at 4000 B.P. in the Middle East. In: Peiser BJ, Palmer J, Bailey ME (Guest eds) Natural Catastrophes During bronze Age Civilizations. BAR International Series 728:93-108

Dahl E (1998) The phytogeography of northern Europe. Cambridge University Press, Cambridge

Dahl SO, Nesje A (1994) Holocene glacier fluctuations at Hardangerjøkulen, central southern Norway; a high-resolution composite chronology from lacustrine and terrestrial deposits. The Holocene 4:269-277

Dahl SO, Nesje A (1996) A new approach to calculating Holocene winter precipitation by combining glacier equilibrium-line altitudes and pine-tree limits: a case study from Hardangerjøkulen, central southern Norway. The Holocene 6:381-398

Dahl-Jensen D, Mosegaard K, Gundestrup N, Clow GD, Johnsen SJ, Hansen AW, Balling N (1998) Past Temperatures Directly from the Greenland Ice Sheet. Science 282:268-271

Daniell JRG (1997) The late-Holocene palaeoecology of Scots Pine (*Pinus sylvestris* L.) in north-west Scotland. Ph.D. thesis, University of Durham

Darvill T (1987) Prehistoric Britain. B.T. Batsford Limited, London

Dyer J (1990) Acient Britain. B.T. Batsford Limited, London

Ebersbach R (1999) Viehwirtschaftssysteme neolithischer Seeufersiedlungen der Schweiz – Modelle unter Einbezug ethnographischer und historischer Daten. PhD Thesis, Basel University

Eddy JA (1976) The Maunder Minimum. Science 192:1189-1202

Erny-Rodmann C, Gross-Klee E, Haas JN, Jacomet S, Zoller H (1997) Früher human impact und Ackerbau im Übergangsbereich Spätmesolithikum-Frühneolithikum in schweizerischen Mittelland. Jahr-buch der Schweizerischen Gesellschaft für Ur- und Frühgeschichte, Annuaire de la Société de Préhistoire et d'Archéologie 80:27-56

Frank T (1998) Die neolithische Besiedlung zwischen Köln-Bonner Reinebene und den Bergischen Hochflächen, Vol. 10. Verlag Deutsche Gesellschaft für Ur- und Frühgeschichte in Kommission bei Habelt Verlag, Bonn

Frenzel B, Reisch L, Gläser B (eds) (1992) Evaluation of land surfaces cleared from forests by prehistoric man in Early Neolithic times and the time of migrating Germanic tribes, Vol. 8. Gustav Fischer Verlag, Stuttgart/Jena/New York

Frenzel B, Andersen ST, Berglund BE, Gläser B (eds) (1994) Evaluation of land surfaces cleared from forests in the Roman Iron Age and the time of migrating Germanic tribes based on regional pollen diagrams, Vol. 12. Gustav Fischer Verlag, Stuttgart/Jena/New York

Gaillard MJ, Birks HJB, Emanuelsson U, Karlsson S, Lageras P, Olausson D (1994) Application of Modern Pollen Land-Use Relationships to the Interpretation of Pollen Diagrams - Reconstructions of Land-Use History in South Sweden, 3000-0 Bp. Review of Palaeobotany and Palynology 82:47-73

Gamper M, Suter J (1982) Postglaziale Klimageschichter der Schweizer Alpen. Geographica Helvetica 37:105-114

Giannakos P (1997) Frugivory and seed dispersal by carnivores in the Rhodopi mountains of northern Greece. PhD thesis Durham

Gibson A (1986) Neolithic and Early Bronze Age Pottery. Shire Publication Ltd, Bucks

Grootes PM, Stuiver M (1997) Oxygen 18/16 variability in Greenland snow and ice with 10^3 to 10^5-year time resolution. Journal of Geophysical Research 102:26455-26470

Gross E, Jacomet S, Schibler J (1990) Stand und Ziele der wirtschaftsarchäologischen Forschung an neolithischen Ufer- und Inselsiedlungen im unteren Zürichseeraum (Kt. Zürich, Schweiz) In: Schibler J, Sedlmeier J, Spycher HP (eds) Festschrift für HR Stampfli, Helbing&Lichtenhahn, Basel, pp 77-100

Gross-Klee E, Maise C (1997) Sonne, Vulkane und Seeufersiedlungen. Jahrbuch der Schweizerischen Gesellschaft für Ur- und Frühgeschichte/Annuaire de la Société de Préhistoire et d'Archéologie 80:85-94

Grove JM (1988) The Little Ice Age. Methuen, London, pp 1-498

Grove JM (1996) The century Time-Scale. In: Driver SD, Chapman GP (eds) Time-Scales and Environmental Change. Routledge, London & New York, pp 39-87

Grove JM (1997) The spatial and temporal variations of glaciers during the Holocene in the Alps, Pyrenees, Tatra and Caucasus. Palaoklimaforschung Palaeoclimate Research 24:95-103

Grove JM (in press) The Initiation of the Little Ice Age: a global scale review. In: Jones PD, Davies TD, Ogilvie AEJ, Briffa KR (eds) Climate and Climatic Impacts through the Last 1000 years, Climatic Research Unit. University of East Anglia, Norwich

Grove JM, Battagel A (1983) Tax Records from Western-Norway, as an Index of Little Ice-Age Environmental and Economic Deterioration. Climatic Change 5:265-282

Grove JM, Switsur R (1994) Glacial geological evidence for the Medieval Warm Period. Climatic Change 26:143-169

Guiot J (1987) Late-Quaternary climatic change in France estimated from multivariate pollen time series. Quaternary Research 28:100-118

Haas JN, Richoz I, Tinner W, Wick L (1998) Synchronous Holocene climatic oscillations recorded on the Swiss Plateau and at timberline in the Alps. The Holocene 8:301-309

Hall VA, Pilcher JR, McCormac FG (1994) Icelandic volcanic ash and the mid-Holocene Scots pine *(Pinus sylvestris)* decline in the north of Ireland: no correlation. The Holocene 4:79-83

Harrison SP, Digerfeldt G (1993) European lakes as palaeohydrological and palaeoclimatic indicators. Quaternary Science Reviews 12:233-248

Harrison SP, Prentice IC, Guiot J (1993) Climatic controls on Holocene lake-level changes in Europe. Climate Dynamics 8:189-200

Hicks S (1985) Modern pollen deposition records from Kuusamo, Finland. 1. Seasonal and annual variation. Grana 24:167-184

Hill JK, Thomas CD, Huntley B (1999) Climate and habitat availability determine 20th century changes in a butterfly's range margin. Proceedings of the Royal Society of London 266:1197-1206

Hochuli S, Niffeler U, Rychner V (eds) (1998) Bronzezeit/Age du Bronze/Età del Bronzo. Die Schweiz vom Paläolithikum bis zum Frühen Mittelalter/ La Suisse du Paléolithique à l'aube du Moyen-Age: Vol. 3. Verlag Schweizerische Gesell-schaft für Ur- und Frühgeschichte, Basel

Hoika J, Meurers-Balke J (eds) (1994) Beiträge zur frühneolithischen Trichterbecherkultur im westlichen Ostseegebiet. 1. Internationales Trichterbechersymposium in Schleswig vom 4. bis 7. März 1985. Wachholz Verlag, Neumünster

Holzhauser H (1984a) Rekonstruktion von Gletscherschwankungen mit Hilfe fossiler Holzer. Geographica Helvetica 39:3-15

Holzhauser H (1984b) Zur Geschichte des Aletsch-und des Fieschergletschers. Physische Geographie 13:1-448

Holzhauser H (1997) Fluctuations of the Grosser Aletsch Glacier and Gorner Glacier during the last 3200 years: new results. Glacier Fluctuations during the Holocene. Palaoklimaforschung Palaeoclimate Research 24:35-58

Holzhauser H, Zumbuhl HJ (1988) Gletsch-Klima-Kunst. Die Alpen 64:292-322

Hoops J (1981) Reallexikon der germanischen Altertums-kunde. Vol. 4. Walter de Gruyter, Berlin

Höpfel F, Platzer W, Spindler K (eds) (1992) Der Mann im Eis. Volume 1: Vol. 187. Verlag Universität, Innsbruck

Hughes MK, Diaz HF (1994) Was there a 'Medieval Warm Period', and if so, where and when? Climatic Change 26:109-142

Huntley B (1989) European post-glacial vegetation history: a new perspective In: Ouellet H (ed) Proceedings of the XIX International Ornithological Congress: Vol. 1, University of Ottawa Press, Ottawa, pp 1060-1077

Huntley B (1990a) European post-glacial forests: compositional changes in response to climatic change. Journal of Vegetation Science 1:507-518

Huntley B (1990b) European vegetation history: Palaeovegetation maps from pollen data - 13000 yr BP to present. Journal of Quaternary Science 5:103-122

Huntley B (1991) How plants respond to climate change: migration rates, individualism and the consequences for plant communities. Annals of Botany 67:15-22

Huntley B (1994) Late Devensian and Holocene palaeoecology and palaeoenvironments of the Morrone Birkwoods, Aberdeenshire, Scotland. Journal of Quaternary Science 9:311-336

Huntley JP (1999) Environmental evidence from Hadrian's Wall. In: Bidwell P (ed) Hadrian's Wall 1989-1999. Cumberland and Westmorland Antiquarian and Archaeological Society & The Society of Antiquaries of Newcastle upon Tyne, pp 224

Huntley JP (2000) Non-structural wood. In: Buxton K, Howard-Davis C (eds) Bremetenacum. Excavations at Roman Ribchetser 1980, 1989-1990. Vol. 9. Lancaster University Archaeology Unit, Lancaster, pp 359-366

Huntley B, Birks HJB (1983) An atlas of past and present pollen maps for Europe: 0-13000 B.P. Cambridge University Press, Cambridge

Huntley B, Prentice IC (1988) July temperature in Europe from pollen data 6000 years before present. Science 241:687-689

Huntley B, Prentice IC (1993) Holocene vegetation and climates of Europe. In: Wright HE, Jr, Kutzbach JE, Webb T, III, Ruddiman WF, Street-Perrott FA, Bartlein PJ (eds) Global Climates since the Last Glacial Maximum. University of Minnesota Press, Minneapolis, pp 136-168

Huntley JP, Stallibrass S (1995) Plant and vertebrate remains from archaeological sites in northern England. Research Report No. 4, Architectural and Archaeological Society of Durham and Northumberland, Durham

Huntley B, Webb T, III (eds) (1988) Vegetation History: Vol. 7. Kluwer Academic Publishers, Dordrecht

Huntley B, Webb T, III (1989) Migration: species' response to climatic variations caused by changes in the earth's orbit. Journal of Biogeography 16:5-19

Huntley B, Daniell JRG, Allen JRM (1997) Scottish vegetation history: The Highlands. Botanical Journal of Scotland 49:163-175

Hüster-Plogmann H, Schibler J, Jacomet S (1999a) The Significance of Aurochs as hunted Animal in the Swiss Neolithic. In: Weniger G-C (ed) Archäologie und Biologie des Aurochsen: Vol. 1, Verlag Neanderthal Museum, Mettmann, pp 151-160

Hüster-Plogmann H, Schibler J, Steppan K (1999b) The relationship between wild animal exploitation, climatic fluctuations, and economic adaptations. A transdisciplinary study on Neolithic sites from lake Zurich, Southwest Germany and Bavaria. In: Becker C, Manhart H, Peters J, Schibler J (eds) Historia animalium ex ossibus. Festschrift für Angela von den Driesch. Verlag Marie Leidorf, Rahden (Westf.) , pp 189-200

Hvass S, Storgaard B (eds) (1993) Da klinger i muld...25 års arkaeologi i Denmark (It rings in the soil...25 years of archaeology in Denmark). Aarhus University Press, Aarhus

Jacomet S, Kreuz A (1999) Archäobotanik. Aufgaben, Methoden und Ergebnisse vegetations- und agrargeschichtlicher Forschung UTB. Ulmer Verlag, Stuttgart

Jacomet S, Brombacher C, Dick M (1989) Archäobotanik am Zürichsee. Ackerbau, Sammelwirtschaft und Umwelt von neolirhischen und bronzezeitlichen Seeufersiedlungen im Raum Zürich. Vol. Monographien 7. Orell Füssli Verlag, Zürich

Jacomet S, Jacquat C, Winter M, Wick L (1999) Umwelt, Ackerbau und Sammelwirtschaft. In: Müller F, Kaenel G, Lüscher G (eds) Die Schweiz vom Paläolithikum bis zum frühen Mittelalter, SPM IV Eisenzeit, Verlag Schweizerische Gesellschaft für Ur- und Frühgeschichte, pp 98-115

Jones P, Hulme M (1997) The changing temperature of 'Central England'. In: Hulme M, Barrow E (eds) Climates of the British Isles: Present, past and future. Routledge, London, pp 173-196

Jones PD, Briffa KR, Barnett TP, Tett SFB (1998) High-resolution palaeoclimatic records for the last millenium: interpretation, integration and comparison with General Circulation Model control-run temperatures. The Holocene 8:455-471

Jutikkala E (1955) The great Finnish famine of 1696–97. Scandinavian Economic History Review 3:48-63

Karlén W (1973) Holocene glacier and climatic variations, Kebnekaise mountains, Swedish Lappland. Geografiska Annaler 55A:29-63

Karlén W (1981) Lacustrine sediment studies. Geografiska Annaler 63A:273-281

Karlén W, Kuylenstierna J (1996) On solar forcing of Holocene climate: evidence from Scandinavia. The Holocene 6:359-365

Karlén W, Matthews JA (1992) Reconstructing Holocene glacier variations from glacial lake sediments: studies from Nordvestlandet and Jostedalsbreen-Jotunheimen, southern Norway. Geografiska Annaler 74A:327-348

Kelly MG, Huntley B (1991) An 11,000-year record of vegetation and environment from Lago di Martignano, Latium, Italy. Journal of Quaternary Science 6:209-224

Koç Karpuz N, Jansen E (1992) A high resolution diatom record of the last deglaciation from the SE Norwegian Sea: Documentation of rapid climatic changes. Paleoceanography 7:499-520

Koç N, Jansen E (1994) Response of the high-latitude Northern Hemisphere to orbital climate forcing: evidence from the Nordic-seas. Geology 22:523-526

Koç N, Jansen E, Haflidason H (1993) Paleoceanographic reconstructions of surface ocean condi-tions in the Greenland, Iceland and Norwegian seas through the last 14 ka based on diatoms. Quaternary Science Reviews 12:115-140

Koç N, Jansen E, Hald M, Labeyrie L (1996) Late glacial–Holocene sea surface temperatures and gradients between the North Atlantic and the Norwegian Sea: implications for the nordic heat pump. In: Andrews JT, Austin WEN, Bergsten H, Jennings AE (eds) Late Quaternary Palaeoceanography of the North Atlantic Margins. Geological Society Special Publication No. 111. The Geological Society, London, pp 177-185

Körner C (1998) A re-assessment of high elevation treeline positions and their explanation. Oecologia 115:445-459

Kossack G, Behre K-E, Schmid P (1984) Archäologische und naturwissenschaftliche Untersuchungen an Siedlungen im deutschen Küstengebiet vom 5 Jahrhundert v.Chr. bis zum 11 Jahrhundert n. Chr. 1: Ländliche Siedlungen, Weinheim

Kreuz A (1990) Die ersten Bauern Mitteleuropas. Eine archäobotanische Untersuchung zu Umwelt und Landwirtschaft der ältesten Bandkeramik. Vol. 23, Leiden

Kullman L (1999) Early holocene tree growth at a high elevation site in the northernmost Scandes of Sweden (Lapland): A palaeobiogeographical case study based on megafossil evidence. Geografiska Annaler Series a-Physical Geography 81A:63-74

Kuniholm PI, Kromer B, Manning SW, Newton M, Latini CE, Bruce MJ (1996) Anatolian tree rings and the absolute chronology of the eastern Mediterranean 2220-718 BC. Nature 381:780-783

Küster H (1994) The economic use of *Abies* wood as timber in central Europe during Roman times. Vegetation History and Archaeobotany 3:25-32

LaMarche VC, Jr, Hirschboeck KK (1984) Frost rings in trees as records of major volcanic eruptions. Nature 307:121-126

Lauritzen S-E, Lundberg J (1999) Calibration of the speleothem delta function: an absolute temperature record for the Holocene in northern Norway. The Holocene 9:659-669

Lindsay S, Joyce A (2000) Climate change and the disappearance of malaria from England. Global Change and Human Health 1:184-187

Lüning J (1996) Erneute Gedanken zur Benennung der neolithischen Perioden. Vol. 74/1, Mainz

Lüning J, Stehli P (eds) (1994) Die Bandkeramik im Merz-bachtal auf der Aldenhovener Platte. Beiträge zur neolithischen Besiedlung der Aldenhovener Platte. Vol. 5: Vol. 36. Habelt Verlag, Bonn

Luterbacher J, Schmutz C, Gyalistras D, Xoplaki E, Wanner H (1999) Reconstruction of monthly NAO and EU indices back to AD 1675. Geophysical Research Letters 26:2745-2748

Luterbacher J and 30 co-authors (2000) Monthly mean pressure reconstruction for the Late Maunder Minimum Period (AD 1675-1715). International Journal of Climatology 20:1049-1066

Maise C (1998) Archäoklimatologie – Vom Einfluss nacheiszeitlicher Klimavariabilität in der Ur- und Frühgeschichte. Jahrbuch der Schweizerischen Gesellschaft für Ur- und Frühgeschichte/Annuaire del Société de Préhistoire et d'Archéologie 81:197-235

Maise C (1999) Die Klimageschichte. In: Müller F, Kaenel G, Lüscher G (eds) Die Schweiz vom Paläolithikum bis zum frühen Mittelalter, SPM IV Eisenzeit, Verlag Schweizerische Gesellschaft für Ur- und Früh-geschichte, pp 43-47

Mann ME, Bradley RS, Hughes MK (1998) Global-scale temperature patterns and climate forcing over the past six centuries. Nature 392:779-787

Mann ME, Bradley RS, Hughes MK (1999) Northern Hemisphere Temperatures during the past Millenium: Inferences, Uncertainties, and Limitations. Geophysical Research Letters 26:759-762

Manning A, Birley R, Tipping R (1997) Roman impact on the environment at Hadrian's Wall: precisely dated pollen analysis from Vindolanda, northern England. The Holocene **7**:175-186

Manning SW (1998) Correction: New GISP2 ice-core evidence supports 17th century BC date for the Santorini (Minoan) Eruption: Response to Zielinski, Germani (1998). Journal of Archaeological Science 25:1039-1042

Maxwell B (1997) Recent climate patterns in the Arctic. In: Oechel WC, Callaghan T, Gilmanov T, Holten J, Maxwell B, Molau U, Sveinbjörnsson B (eds) Global change and Arctic terrestrial ecosystems: Vol. 124. Springer, New York, pp 21-46

Mayr F (1968) Postglacial glacier fluctuations and correlative phenomena in the Stubai Mountains, eastern Alps, Tyrol. In: Richmond GM (Guest ed) INQUA 1965, Glaciation of the Alps. University of Colorado Studies, series in Earth Science **7**:143-165

Mikalsen G, Sejrup HP (submitted) A high resolution 25,000 year temperature record from Sognesjøen, western Norway; possible correlations to atmospheric $\delta^{14}C$ cycles. Geology

Müller F, Lüscher G, Niffeler U (eds) (1999) Eisenzeit/Age du Fer/Età del Ferro. Vol. 4. Verlag Schweizerische Gesellschaft für Ur- und Frühgeschichte, Basel

Nesje A, Dahl SO (1994) Lateglacial and Holocene glacier fluctuations and climate variations in western Norway, A review. Quaternary Science Reviews

12:255-261

NNA (1995) Nordic Archaeological Abstract 1995. Hojbjerg

Ogilvie A (1991) Climatic changes in Iceland A.D. 865 to 1598. In: Bigelow G. (Guest ed) The Norse of the North Atlantic. Munksgaard, Köbenhavn, Acta Archaeologica 61:233-251

Ogilvie AEJ (1984) The past climate and sea-ice record from Iceland. Climatic Change 6:131-152

Parmesan C, Ryholm N, Stefanescu C, Hill JK, Thomas CD, Descimon H, Huntley B, Kaila L, Kullberg J, Tammaru T, Tennent J, Thomas JA, Warren M (1999) Poleward shifts in geographical ranges of butterfly species associated with regional warming. Nature 399:579-583

Parry ML (1975) Secular climatic change and marginal land. Transactions of the Institute of British Geographers 64:1-13

Parry ML, Carter TR (1985) The Effect of Climatic Variations on Agricultural Risk. Climatic Change 7:95-110

Patzelt G (1973) Die neuzeitlichen Gletscherschwankungen in der Venedigergruppe (Hohe Tauern, Ostalpen). Zeitschrift für Gletscherkunde und Glazialgeologie 9:5-57

Patzelt G (1974) Holocene variations in the Alps) Les méthods quantitative d'étude des variations du climat au cours du Pléistocene. Colloques Inter-nationaux du Centre de la Récherche Scientifique 248-259

Patzelt G (1994) Former lakes dammed by glaciers and resulting floods in the Ötztal, Tyrol. Mountain Research and Development 14:298-301

Peiser BJ (1997) Comets and disaster in the Bronze Age. British Archaeology 30:6-7

Pfister C (1996) European weather conditions during the Little Ice Age. In: Pedersen ES (ed) North European climate data in the latter part of the Maunder Minimum period A.D. 1675-1715. Stavanger, pp 51-53

Pfister C, Luterbacher J, Schwarz-Zanetti G, Wegmann M (1998) Winter air temperature variations in western Europe during the Early and High Middle Ages (AD 750-1300). The Holocene 8:535-552

Pfister C, Brázdil R, Glaser R (eds) (1999) Climatic Variability in Sixteenth Century Europe and its Social Dimension. Climatic Change Vol. 43/1

Pilcher JR, Hall VA (1992) Towards a tephrochronology for the Holocene of the north of Ireland. The Holocene 2:255-259

Pratt KE (1997) Development of methods for investigating settlement and land-use using pollen data : a case-study from north-east England, circa 8000 cal. BC–cal. AD 500. PhD thesis, Durham

Preece RC (1997) The spatial response of non-marine Mollusca to past climate changes. In: Huntley B, Cramer W, Morgan AV, Prentice HC, Allen JRM (eds) Past and future rapid environmental changes: The spatial and evolutionary responses of terrestrial biota: Vol. 47. Springer-Verlag, Berlin, pp 163-177

Prescott C (1995) From Stone Age to Iron Age. A study from Sogn, western Norway. BAR International Series, 603. Oxford

Randsborg K (ed) (1996) Absolute Chronology. Archaeological Europe 2500-500 BC. Acta Archaeologica (Acta Archaeologica Supplementa Vol. I), 67. Munksgaard, Kopenhagen

Regnell M (1998) Archaeobotanical finds from the Stone Age of the Nordic countries. A catalogue of plant remains from archaeological contexts. Lundqua Report 36, Lund University, Dept. Of Quaternary Geology, Lund

Rösch M (1985) Ein Pollenprofil aus dem Feuenried bei Ueberlingen am Ried. Stratigraphische und landschaftsgeschichtliche Bedeutung für das Holozän im Bodenseegebiet. Materialhefte zur Vor- und Frühgeschichte in Baden-Württemberg 7:43-79

Rösch M (1990) Vegetationsgeschichtliche Untersuchungen im Durchenbergried. Siedlungsarchäologie im Alpenvorland II. Forschungen und Berichte zur Vor- und Frühgeschichte in Baden-Württemberg 37:9-64

Rösch M (1992) Human impact as registered in pollen record: some results from the western Lake Constance region, Southern Germany. Vegetation History and Archaeobotany 1:101-109

Rösch M (1993) Prehistoric land use as recorded in a lake-shore core at Lake Constance. Vegetation History and Archaeobotany 2/4:213-232

Röthlisberger F (1974) Études des variations climatiques d'apres l'histoire des cols glaciaires. Le Col d'Herens,(Valais, Suisse). Bolllettino del Comitato Glaciologico Italiano (ser.2) 22:9-34

Röthlisberger F (1986) 10'000 Jahre Gletschergeschichte der Erde. Sauerländer, Aarau

Röthlisberger F, Haas P, Holzerhauser H, Keller W, Bircher W, Renner F (1980) Holocene climatic fluctuations-radiocarbon dating of fossil soils (fAh) and wood from moraines and glaciers in the Alps. Geographica Helvetica 35:21-52

Rychner V (1998) Chronologie: Einleitung. In: Hochuli S, Niffeler U, Rychner V (eds) pp 13-19

Schibler J, Chaix L (1995) Wirtschaftliche Entwicklung aufgrund archäozoologischer Daten/ L'évolution économique sur la base de données archéozoologiques. In: Die Schweiz vom Paläolithikum bis zum Frühen Mittelalter/ La Suisse du Paléolithique à l'aube du Moyen-Age. Volume 2. Verlag Schweizerische Gesellschaft für Ur- und Frühgeschichte, Basel, pp 97-120

Schibler J, Jacomet S (1999) Archaeozoological and archaeobotanical evidence of human Impact on Neolithic environments in Switzerland. In: Benecke N (ed) The Holocene History of the European Vertebrate Fauna – Modern Aspects of Research. Vol. 6. Verlag Marie Leidorf, Rahden, pp 339-354

Schibler J, Hüster-Plogmann H, Jacomet S, Brombacher C, Gross-Klee E, Rast-Eicher A (1997a) Ökonomie und Ökologie neolithischer und bronzezeitlicher Ufersiedlungen am Zürichsee. Vol. 20. Verlag Fotorotar, Zürich und Egg

Schibler J, Jacomet S, Hüster-Plogmann H, Brombacher C (1997b) Economic crash in the 37th and 36th century BCcal in neolithic lake shore sites in Switzerland. Anthropozoologica Vol. 25/26. Paris, pp 553-570

Schibler J, Stopp B, Studer J (1999) Haustierhaltung und Jagd - Élevage et chasse. In: Müller F, Kaenel G, Lüscher G (eds) Eisenzeit/Age du Fer/Età del Ferro. Vol. 4. Verlag Schweizerische Gesellschaft für Ur- und Frühgeschichte, Basel, pp 66-86

Schibler J, Studer J (1998) Haustierhaltung und Jagd während der Bronzezeit der Schweiz / Elevage et chasse à l'âge du Bronze en Suisse. In: Hochuli S, Niffeler U, Rychner V (eds) Die Schweiz vom Paläolithikum bis zum frühen Mittelalter: Bronzezeit (SPM 3). Verlag Schweizerische Gesellschaft für Ur- und Frühgeschichte, Basel, pp 171-191

Sernander R (1908) On the evidence of post-glacial changes of climate furnished by the peat mosses of northern Europe. Geologiska Föreningens i Stockholm Förhandlingar 30:365-478

Sernander R (1910) Die schwedischen Torfmoore als Zeugen postglazialer Klimaschwankungen, Stockholm

Sissons JB (1979) Palaeoclimatic inferences from former glaciers in Scotland and the Lake District. Nature 278:518-521

Stöckli WE (1995) L'évolution du Néolithique Suisse. In: Stöckli WE, Niffeler U, Gross-Klee E (eds) SPM II, Neolithikum. Die Schweiz vom Paläolithikum bis zum frühen Mittelalter. Schweizerische Gesellschaft für Ur- und Frühgeschichte, Basel, pp 19-52

Strahm C (1994) Die Anfänge der Metallurgie in Mitteleuropa. Helvetia Archaeologica 97:2-39

Stuiver M, Reimer PJ (1993) Extended ^{14}C database and revised CALIB radiocarbon calibration program. Radiocarbon 35:215-230

Stuiver M, Reimer PJ, Bard E, Beck JW, Burr GS, Hughen KA, Kromer B, McCormac G, van der Plicht J, Spurk M (1998) INTCAL98 Radiocarbon Age Calibration, 24,000-0 cal BP. Radiocarbon 40:1041-1083

Südtiroler Archäologiemuseum (1999) Die Gletschermumie aus der Kupferzeit / La mummia dell'età del rame. Folio-Verlag, Wien/Bozen

Sutherland DG (1984) Modern glac ier characteristics as a basis for inferring former climates with particular reference to the Loch Lomond Stadial. Quaternary Science Reviews 3:291-309

Thomas JA (1991) Rare species conservation: case studies of European butterflies. In: Spellerberg IF, Goldsmith FB, Morris MG (eds) The Scientific Management of Temperate Communities for Conservation. Blackwell Scientific Publications, Oxford, pp 149-197

Tinner W, Ammann B, Germann P (1996) Treeline fluctuations recorded for 12,500 years by soil profiles, pollen and plant macrofossils in the central Swiss Alps. Arctic and Alpine Research 28:131-147

Uerpmann M, Uerpmann H-P (1997) Remarks on the faunal remains of some early farming communities in central Europe. L'homme et l'Animal. Anthropozoologica, Paris

Van Geel B, Buurman J, Waterbolk HT (1996) Archaeological and palaeoecological indications of an abrupt climate change in The Netherlands, and evidence for climatological teleconnections around 2650 BP. Journal of Quaternary Science 11:451-460

Van Geel B, Van der Plicht J, Kilian MR, Klaver ER, Kouwenberg JHM, Renssen H, Reynaud-Farrera I, Waterbolk HT (1998) The sharp rise of delta-14C ca. 800 cal BC: possible causes, related climatic teleconnections and the impact on human environments. Radiocarbon 40:535-550

Vassiljev J, Harrison SP, Guiot J (1998) Simulating the Holocene lake-level record of Lake Bysjön, southern Sweden. Quaternary Research 49:62-71

Vos H, Sanchez A, Zolitschka B, Brauer A, Negendank JFW (1997) Solar activity variations recorded in varved sediments from the crater Lake of Holzmaar - A Maar Lake in the Westeifel volcanic field, Germany. Surveys in Geophysics 18:163-182

Wanner H, Holzhauser H, Pfister C, Zumbühl H (2000) Interannual to centennial scale climate variability in the European Alps? Erdkunde 54:62-69

Wick L, Tinner W (1997) Vegetation changes and Timberline fluctuations in the Central Alps as indicators of Holocene climatic oscillations. Arctic and Alpine Research 29:445-458

Wiethold J (1998) Studien zur jüngeren postglazialen Vegetations- und Siedlungsgeschichte im östlichen

Schleswig-Holstein. Vol. 45, Bonn

Wilkinson DM (1997) Plant colonization: are wind dispersed seeds really dispersed by birds at larger spatial and temporal scales? Journal of Biogeography 24:61-65

Winiger J (1989) Bestandesaufnahme der Bielerseestationen als Grundlage demographischer Theoriebildung. Vol. 1. Staatlicher Lehrmittelverlag Bern

Winiger J, Hasenfratz A. (1985) Ufersiedlungen am Bodensee – Archäologische Untersuchungen im Kanton Thurgau 1981-1983. Vol. 10. Verlag Schweizerische Gesellschaft für Ur- und Früh-geschichte Basel

Zielinski GA, Germani MS (1998) New Ice-Core Evidence Challenges the 1620 BC Age for the Santorini 'Minoan' Eruption. Journal of Archaeological Science 25:279-289

Zoller H (1977) Alter und Ausmass postglazialer Klimaschwankungen in dem Schweizer Alpen. In: Frenzel B (ed) Dendrochronologie und postglazialer Klimaschwankungen in Europa. Erdwissenschaftliche Forschung 13:271-281

Landscape Development and Occupation History Along the Southern North Sea Coast

Karl-Ernst Behre

Niedersächsisches Institut für historische Küstenforschung, Viktoriastraße 26/28, D26382 Wilhelmshaven, Germany
e-mail: karl-ernst.behre@nihk.terramare.de

Abstract: Due to sea level fluctuations and the consequences for human activity, there is a strong dependency between landscape development and the history of occupation in the coastal area. In the early Holocene, large parts of the southern North Sea were dry and Britain was connected with the European mainland. Forests covered this region as far north as the Dogger Bank and beyond, and there is evidence for a Mesolithic presence in this area. The rise in sea-level was very fast at that time so that by the Neolithic period, i.e. ca. 5000 BC, the North Sea transgression resulted in a coastline that corresponded approximately to that of today. Subsequently, the rate of sea-level rise slowed down and several static or regression phases are known. These regression phases provided opportunities for settlement in the Clay District "Marsch". The oldest settlements known so far in the German Clay District are dated to late Bronze and early Pre-Roman Iron Age. These early settlements were destroyed by another transgression phase which lasted until around the Birth of Christ when another, even stronger, regression occurred. At this time, large scale settlement of the Clay District took place. Already, early in the first millennium, the next rise in sea-level forced the people either to abandon their settlements or to adapt to the new situation. Adaptation took the form of dwelling mounds, i.e. "Wurten", which were continuously raised with increasing storm-flood levels. These "Wurten" of the Roman Iron Age were abandoned during the Migration period. After a 200 years break in habitation, a new occupation phase started in the 7th century, when Frisians from the west immigrated into the German Clay District. Again they started their settlements on the level ground and were later forced to build dwelling mounds. Starting in the 11th century, the construction of dikes began and, around AD 1300, a continuous dike protected the hinterland. However, diking led to a further increase of the stormflood level and breaches occurred resulting in the formation of several large bays such as the Jade bay in the Late Middle Ages. From about AD 1500, the technique of diking was improved and reclamation of new land prevailed.

Introduction

There are very few regions in Europe where the history of habitation is as closely linked to the changes in the natural landscape as along the southern North Sea coast. This is due to the rise and the subsequent oscillations of the sea-level which have delimited the space for occupation and frequently transformed the environment from fresh-water to brackish or saline conditions. The resulting prehistoric occupation history is thus the visible expression of the climatic changes responsible for the evolution of the coastal landscape. In this contribution, the history of the coastal landscape and its occupation is elucidated on the basis of archaeological, geological and geobotanical evidence. Similar investigations have been carried out in the adjoining Netherlands in the west as well as in Schleswig-Holstein to the north-east of Lower Saxony (Niedersachsen).

While on one hand the North Sea determined the occupation area before the start of diking, the posi-

From WEFER G, BERGER W, BEHRE K-E, JANSEN E (eds), 2002, *Climate Development and History of the North Atlantic Realm.* Springer-Verlag Berlin Heidelberg, pp 299-312

tion of dated prehistoric settlements on the other hand often provides important indications as to the location of former coastlines.

For older periods, in particular pre-Neolithic periods, other sources of evidence are required. Here geological methods are used and, in particular, submerged peats of known depth and age from the coastal area and the North Sea itself provide the dates for the early North Sea transgression. The resulting changes in the environment can be traced by the investigation of plant macrofossil remains which are preserved in the peat layers as well as in the prehistoric or medieval dwelling sites which yield remains of plants that have been collected in the surrounding area.

Early and Middle Holocene: The Southward Advance of the North Sea

At the maximum of the Weichselian glaciation the sea level had dropped by more than 120 m below its present level. Of these early stages no dates are available from the North Sea region itself. Until Preboreal times (calibrated 10 000-8000 years BC) most parts of the southern North Sea basin were dry land and Britain was still connected to the European mainland (Fig. 1). The oldest known transgressional contact was found at a depth of 46 m in a large depression south of the Dogger Bank, dated as Preboreal (Behre and Menke 1969). The sea level curve initially shows a rapid rise of the mean high

Fig. 1. During the last glaciation large parts of the North Sea area were dry. The map shows the coastline, based on the best evidence, at the onset of the Holocene (1: 65 m depth contour, around 10000 BC) and in the late Boreal (2: 40 m depth contour, 7000 BC; both time-slices lie within the Mesolithic period).

water level, reaching a depth of 15 m below NN (= German ordnance datum) at about 5000 BC (e.g. Behre 1987) (Fig. 2).

Until then the area of the southern North Sea was covered with forests similar to those found in northern Germany at that time. This was shown by pollen diagrams from submarine peat layers (e.g. Behre and Menke 1969, Behre, Dörjes and Irion 1984). Mesolithic occupation of this region is documented by the artifacts left behind by particular prehistoric cultures, e.g. the core-and flake-axe culture during Boreal times (8000-6800 BC) which was present in England and Denmark, but in Germany only north of the Elbe river (Schwabedissen 1951).

The rising sea hampered the terrestrial drainage of the coast, as a result of which extensive mires formed. In the course of the transgression this belt of mires expanded southwards, while the parts near the sea were flooded as sea level rose. The latter were covered by shallow-marine deposits to form the so-called basal peats which represent the beginning of the Holocene sequence. These transgressional contacts, which can be dated by ^{14}C and pollen analysis, serve as fixed points in the sea-level curve.

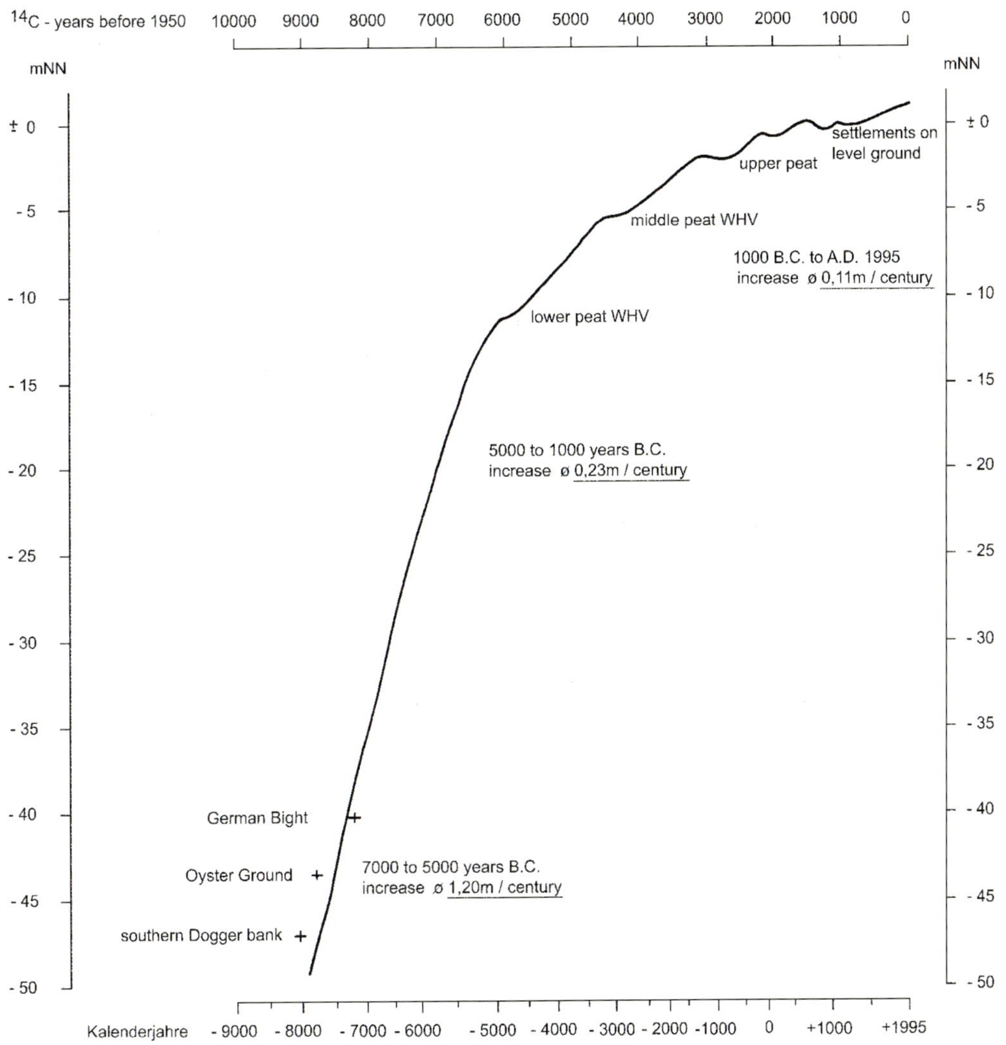

Fig. 2. Sea-level curve for the southern North Sea area (after Behre 1987).

The Holocene transgression of the North Sea (the Flandrian transgression) did not advance continuously but was interrupted by temporary still stands and, in the course of the later Holocene, also by regressional phases. These regressions can be recognized by the occurrence of intercalated peat layers within the stratigraphic column (Fig. 3).

From the Neolithic to the Birth of Christ: Occupational Reaction to Sea Level Oscillations

Whereas the clastic sediments (clay, silt and sand) document marine conditions, the intercalated peat layers indicate freshwater environments. During the regressional phases the peat formation advanced seawards. As a result, some coastal areas dried up and thus provided opportunities for occupation. The oldest settlements in the former marine area belong to the Vlaardingen culture of middle Neolithic age (3000-2500 BC). They are known mainly from beach barriers and older dunes in the western Netherlands. No settlement records of this period have thus far been found along the German coast, although they may occur on the fossil beach barriers of Schleswig-Holstein several meters below the modern land surface.

The most important intercalated peat layer is the so-called Upper Peat which was formed between 1500 and 600 BC. This peat layer normally commences as reed peat. It is preceded by a brackish transition phase and followed by sedge and brushwood peat. In some areas the sequence is topped by *Sphagnum* peat which indicates a considerable lowering of the groundwater table and consequently also of the mean sea level. The first recorded habitation with settlements in the German Clay District (Marschen) took place during this regressional phase: good examples are found at Rodenkirchen/Weser and along the lower Ems from the late Bronze Age until the early Pre-Roman Iron Age.

These early settlements were located on the muddy floodplains of rivers. Covering extensive areas beyond the river embankments, these floodplains to a large extent determined the vegetation as well as settlement patterns and land use. Natural deposition on the river banks created elevated levees which not only lined the large rivers but also the smaller creeks. Such levees consist of clay or sand

Fig. 3. Schematic cross-section through Holocene deposits in the area of Wilhelmshaven.

and were deposited during high storm surges when the rivers left their beds and flooded the river banks. Due to the rapid decrease in flow velocity with increasing distance from the river, the transport capacity of the water decreased and the mineral particles settled out. In the freshwater to slightly brackish river clay areas, this process was also favoured by the existence of wide reed beds along the rivers. These further slowed down the water flow and acted as filters. As a consequence, deposition of sediment was restricted to the levees which were continuously raised with the rising water level.

The elevated levees have an average width of 1,5 km along the lower Ems river (Fig. 4) and are even wider along the lower Weser and Elbe rivers. Along tributaries and creeks they are much less extensive. Behind the elevated levees there was a permanent deficit of sediment, although the water penetrated far beyond them. As a consequence, vast mires developed and characterized the so-called 'backswamp area'. Clay was deposited in these areas only along some old creeks.

Apart from temporary phases of brushwood communities the huge backswamp areas were tree-

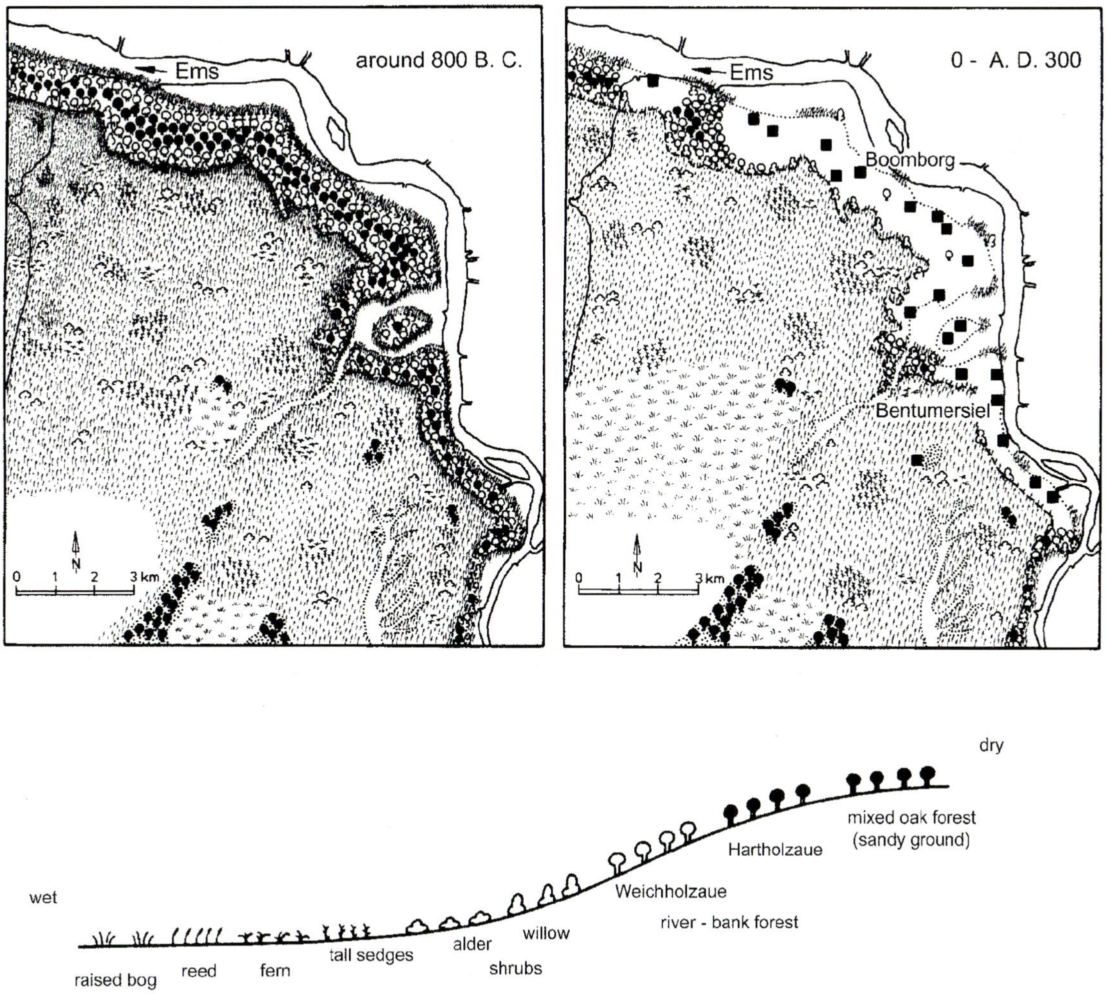

Fig. 4. Map of the lower Ems river as an example of a river system from the Clay District ("Marsch"). Left: natural river bank forests on the elevated levees and the adjoining back-swamp area (around 800 BC). Right: The same area during the Roman Iron Age after forest clearance (squares represent settlements).

less for millennia. This stands in sharp contrast to the elevated levees which were covered with woodland from the very beginning. The first reconstructions of these forests were made for the lower Ems area (Behre 1985). The archaeological excavations of the riverside settlements along the Ems from the Early Iron Age yielded thousands of wood samples from buildings etc. but also from a considerable number of tree stumps, remnants of the clearings in the occupation phase. The composition of the wood samples allowed a reliable reconstruction of the river-bank forests, starting with the willow belt (divided into willow shrubs and willow forest) and ending with the upper floodplain forest, the Fraxino-Ulmetum (obere Hartholzaue). This elm-oak-ash forest occupied the highest parts of the elevated levees and was therefore cleared first by the colonizing settlers who wanted safe houses and dry fields. Already in the Roman period, the Fraxino-Ulmetum of this area was almost completely destroyed, becoming extinct in the Early Middle Ages.

On the elevated levee of the lower Weser river near Rodenkirchen the occupation started as early as the late Bronze Age around 900 BC (Strahl 1998). The excavations here are still in progress, large farm houses being uncovered and evidence for pastoral as well as arable farming in this moist environment coming to light. This habitation period ended around 300 BC at the latest. A new transgression of the North Sea destroyed the settlements and covered them with sand and clay. As a result the Clay District was abandoned. Also the Upper Peat was flooded and covered by clastic deposits, now concealed beneath a 1-2 m thick clay layer. Only in the backswamp areas, in the rear of the elevated levees of the main rivers, the bogs were not affected and were able to continue their growth.

This devastating transgression lasted about 200 years and ended around 100 BC It was followed by an extensive regression registered in all records of sea level change (mean sea level, mean high water levels and storm surge levels; cf. Behre 1986). A characteristic feature of this regression is the development of soils over large areas of the Clay District, as well as changes in the organic matter accumulation in the peaty hinterland of the peri-marine areas. In many present day exposures in the Clay District, particularly in newly cleared ditches, a thin dark line can be observed which, in most cases, indicates the regressional phase between 100 BC and AD 100 (Fig. 5). During this regression phase, the first extensive colonization of the German Clay District took place. The settlements were founded on flat ground and hence called "Flachsiedlungen". Since they were evidently not threatened by the sea, they serve to establish an upper limit for storm surge levels. However, this regressional phase was only of short duration and, in most places, too brief and even too dry for the formation of peat which is encountered only in a few areas.

The available datings show that the regressional phase was synchronous from the northern Netherlands across Niedersachsen to Schleswig-Holstein and took place between 100 BC and AD 100. Colonization of the clay flats, however, was not synchronous. Due to population movements there was a considerable delay from west to east.

The First Millenium AD: Active Response to the Challenge of the Sea

A renewed rise in storm flood levels can be observed in the first century AD. Some of the "Flachsiedlungen" were abandoned, others remained occupied. The inhabitants now protected themselves against the storm floods by constructing their dwelling places on artificially raised mounds, a method, which had been employed in the western Netherlands already in the 5th century BC. These mounds had to be further raised in succeeding years to cope with the continuously rising storm flood levels. In this way the "Wurten" or dwelling mounds were created, some of which reached an altitude of more than +5 m NN (even +7 m NN have been recorded) and have an average diameter of more than 200 m (Fig 6a). These village mounds of the first "Wurten" phase in Germany are widely distributed over the Clay District. They contain several habitation layers. The well-known ex-cavation of the Feddersen Wierde (Haarnagel 1979; Schmid 1984), for example, produced seven complete villages in vertical succession on top of the initial "Flachsiedlungen" (Fig. 6b).

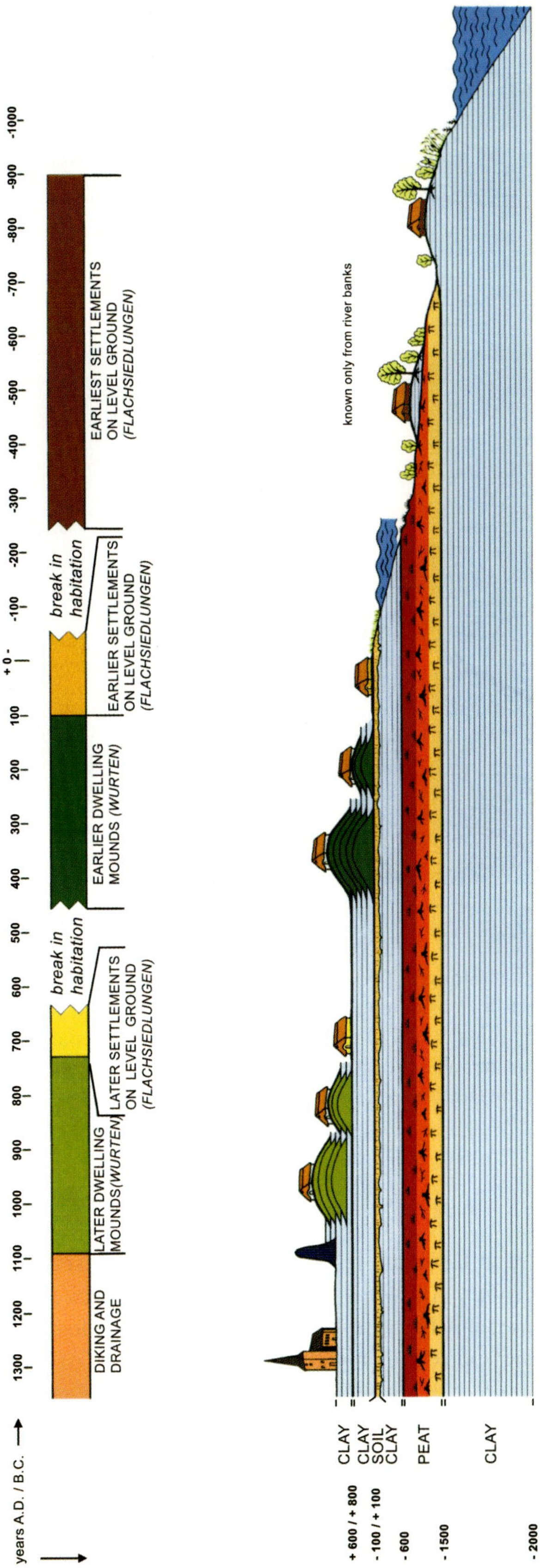

Fig. 5. Schematic illustration of deposition and occupation in the German Clay District ("Marsch").

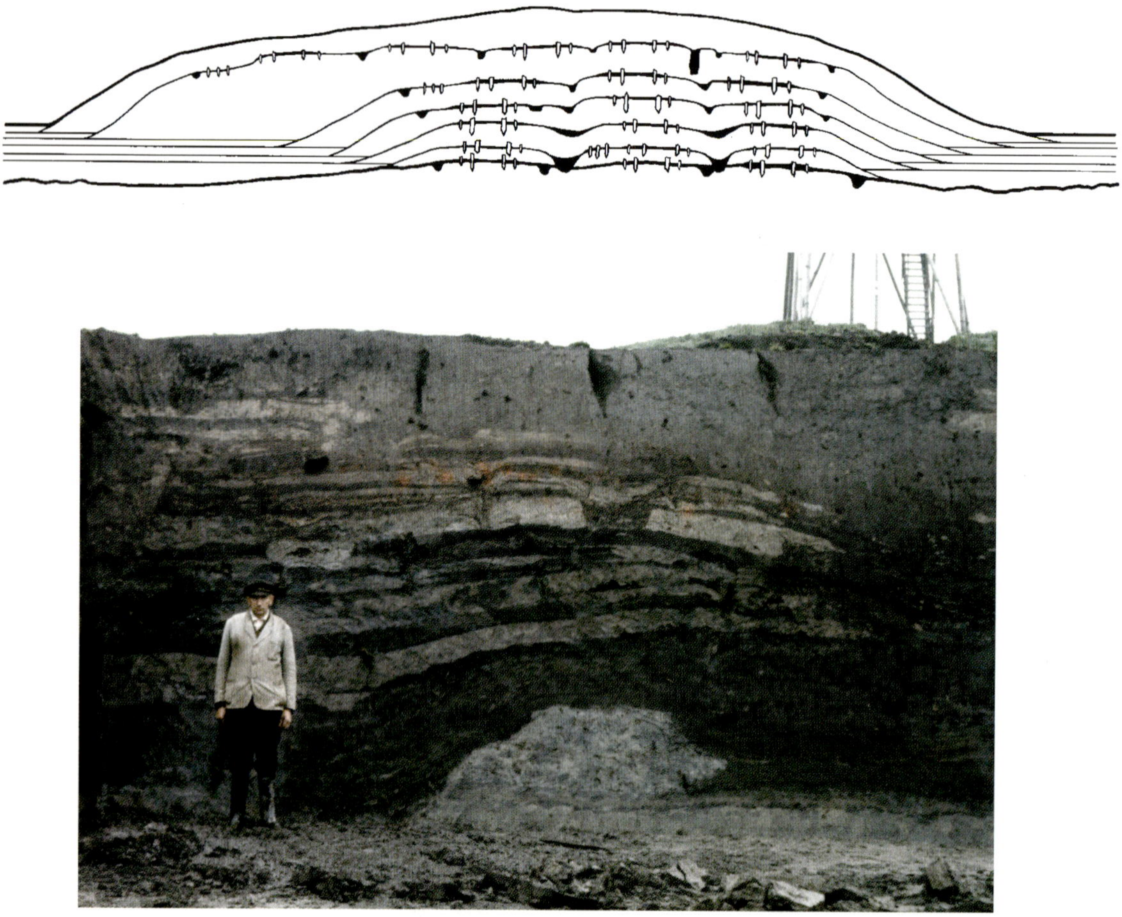

Fig. 6. (a) Schematic cross section through a "Wurt", i.e. a dwelling mound, of the Roman Iron Age. Shown at the bottom, is a natural levee with a "Flachsiedlung" (settlement on non-elevated ground) above which are several habitation layers containing the foundations of buildings. Each horizon is separated by clay layers which were brought onto the site so as to further raise the dwelling mound above the prevailing storm flood-level. **(b)** Cross section from the excavation of the Roman Iron Age "Wurt" Feddersen Wierde.

This first "Wurten" phase in Germany lasted from the first to the fifth century AD when the dwellings were abandoned, probably because their inhabitants emigrated to England together with their neighbours from the more elevated sandy areas of Pleistocene origin. A remarkable feature of these "Wurten" is the excellent preservation of the as-sociated organic remains. Archaeologists found intact posts, wattle and daub structures, as well as other wooden parts of houses, whereas in the prehistoric settlements of the adjoining Pleistocene the buildings can barely be traced.

For palaeoecological investigations relating to the environment of these settlements, the well-preserved plant remains from the habitation layers of these waterlogged sites represent archives for the history of landscape development and the evolution of a the prehistoric farming economy. Such layers provide material that is more or less continuous over time and which can be accurately dated by archaeological artifacts and also by dendrochronology.

Particularly in the "Wurten" of the Roman period and of the Early Middle Ages, the habitation layers mostly consist of large quantities of so-called manure. This term includes a wide variety of plant debris such as litter from the stables, remains of thatched roofs, fossil hay, remains of fodder and threshing and, of course, digested material. Even small samples may contain many thousands of fruits and seeds, but also stems, leaves, flowers etc., all in excellent state of preservation. A special feature of these waterlogged sites is the preservation of non-carbonized plant remains, this being an ad-ditional advantage compared with sites on dry ground where only charred remains survive. Thus, almost all plant species which grew around the settlements show up in seed analyses. In some cases homogenous plant layers, for instance pure samples of fossil hay or bean straw, allow the reconstruction of individual fossil plant communities such as the Juncetum gerardii which has been described in detail, and even been subdivided into subassociations, from the Feddersen Wierde in Niedersachsen (Körber-Grohne 1967) and from Elisenhof in Schleswig-Holstein (Behre 1976).

These palaeoecological investigations have facilitated the reconstruction of the vegetation around the settlements and also documented the changes which occurred in the course of the habitation period. Special attention is given to salinity levels in the area, the relative heights of the respective layers with regard to mean high water levels and, of course, storm flood levels, and the agricultural potential of the soil. Very thorough botanical investigations of a Roman period "Wurt" were carried out by Körber-Grohne (1967) at the Feddersen Wierde in the Land Wursten, north of Bremerhaven (Fig. 7). The dwelling was situated in a fully developed salt marsh, though in a somewhat sheltered position which occasionally experienced brackish conditions. Flourishing arable farming was recorded and crops like *Vicia faba* (broad or celtic bean), *Hordeum vulgare* (barley), *Avena sativa* (oats), *Linum usitatissimum* (flax) and *Camelina sativa* (gold-of-pleasure, an old oil plant) were grown. However, the main basis of the economy was cattle breeding and large amounts of bones from domesticated animals such as cattle, goat/sheep, pig, horses and also dogs have been recovered (Reichstein 1991). Apart from the large-scale excavation of the Feddersen Wierde, several other "Wurten" of the Roman Iron Age have been investigated and show similar evidence with respect to structure as well as to their economy.

In the course of the Migration period in the 4th/5th century AD a substantial part of the Saxon as well as of the Anglian population left the coastal area to settle in England and it was particulary the Clay District of northern Germany that was abandoned. As far as is known not a single "Wurt" continued to exist and a 200 year gap in the habitation of this area is recorded. Thereafter, renewed occupation took place by reactivating existing older "Wurten", but numerous new dwelling mounds were also built (Schmid 1988). New settlements were founded on the natural levees that were sufficiently high to avoid artificial elevation of the ground. "Flachsiedlungen" of this time have been found in several places. This indicates the onset of another regression or at least a lowering of storm flood levels. During this period the Frisian people migrated eastwards to once again colonize the Clay District. According to dendrochronology the oldest Frisian village is Oldorf, north of Wilhelmshaven, where recent excavations took place and which dates back to AD 630 (Schmid 1994). The favourable conditions, however, were of short duration. Storm flood levels increased again, forcing the inhabitants of the Clay District to respond by constructing dwelling mounds once more, known as the second "Wurt" phase along the German coast.

Apart from those already mentioned, other large scale excavations have been carried out in the Early Medieval "Wurt" of Niens/Butjadingen (Brandt 1991), accompanied by botanical investigations. These too show marine to brackish conditions and limited arable farming: mainly *Vicia faba, Hordeum vulgare, Linum usitatissimum* and some *Triticum dicoccon* were grown (Behre 1991), the main economic basis also being cattle breeding. Another important investigation of a contemporary "Wurt" was made at the "Wurt" Hessens in the town of Wilhelmshaven. This settlement was situated between the salt marshes and the landward bogs (Haarnagel 1941).

Fig. 7. Photograph from the excavation of the Roman Iron Age "Wurt" at Feddersen Wierde. It shows the wooden remains and walls of three-aisled farmhouses arranged radially on the "Wurt".

The Frisian occupation extended as far north-east as Eiderstedt and northern Friesland in Schleswig-Holstein, where Bantelmann (1975) excavated the dwelling mound of Elisenhof near Tönning. In other regions, e.g. in Dithmarschen, native Saxons also resettled the Clay District in this period (Meier 1996).

The Second Millenium AD: Coastal Protection by Land Reclamation and Diking

In the first millenium AD the inhabitants of the coastal fringe responded to increasing flood levels by elevating their settlements to dwelling mounds. In the second millenium, by contrast, they began to protect their livelyhood by the construction of dikes. This new phase began at the end of the 11th century when the inhabitants of the "Wurten" started to build ring dikes around their fields and some of their meadows, intending to improve the conditions for farming. These first dikes were low summer dikes, meant to protect the area during the vegetation period in the summer. They did not, however, provide shelter from storm floods in winter time.

In the course of the 13th century these ring dikes were inter-connected and raised, thereby creating the first continuous line of dikes, which now protected the land behind them also against the winter storms. This new form of protection spawned new settlements in the clay district, now mainly as farms outside the villages. The construction of dwelling mounds, which required a great deal of effort was not longer necessary, new houses now being erected on slightly elevated ground.

Until the end of the first millennium AD the coastal zone was characterised by a great variety of vegetation units, ranging from salt marshes to freshwater communities with many intermediate

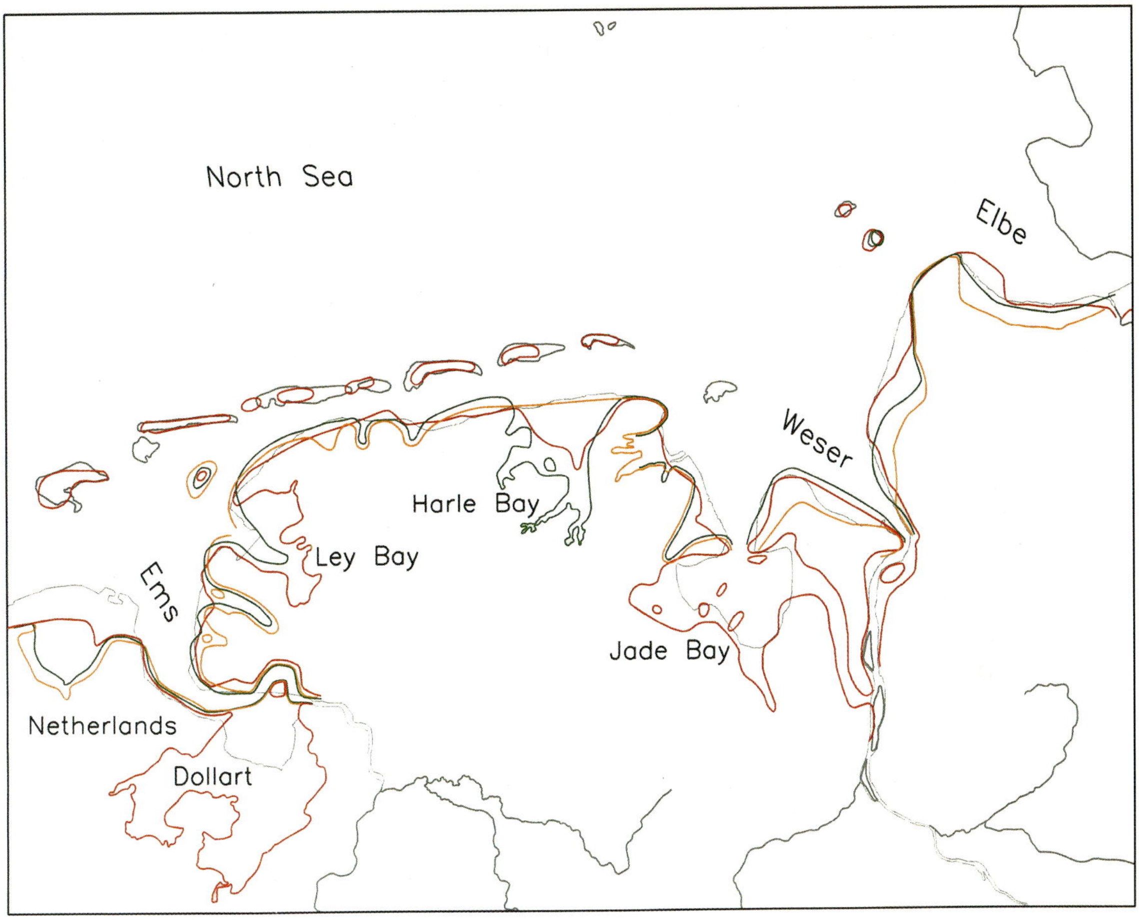

Fig. 8. Former coastlines in Niedersachsen/Lower Saxony. Yellow: around the Birth of Christ, green: around AD 800 (before diking), red: around AD 1500 when the bays had reached their maximum extent.

forms. Now, with the onset of diking, a dramatic change took place. The dikes sharply separated the seaward marine and brackish area from the landward freshwater area.

The aim of diking was to secure the coastline. Although this aim was largely achieved, it also created new problems. One consequence of diking was the reduction of the flooding space for storm waters and this led to an elevation of the storm flood levels. Particularly in the estuaries of the Ems, Weser and Elbe rivers this had severe consequences due to the funneling effect which greatly magnified the flood levels. Especially northwesterly storms caused the water to pile up against the dikes, thereby forcing more water into the estuaries. Breaches of the dikes were common and, in response to this, the dikes were raised and new areas were reclaimed. This further enhanced the storm flood levels. Several catastrophic floods damaged the coastline, creating new bays and causing terrible losses of people and cattle (Fig. 8). Noteworthy are the following extreme floods: the Lucia flood of 1287, the Marcellus flood of 1362, the Antonius flood of 1511, the 2nd Mannestränke of 1634, and the Christmas flood of 1717 AD.

In Lower Saxony the Jade Bay first formed in the 13th Century. About 1500 AD it reached far beyond its present limits. At about the same time the Dollart came into existence, also reaching its maximum size around 1500 AD and extending far into the Netherlands. The Ley Bay too was considerably larger than today. On the other hand, former bays like the Harle Bay in East Frisia and the Maade Bay in Wilhelmshaven, which formed in early Medieval times before diking, were reclaimed and have completely disappeared since then (Behre 1999).

Schleswig-Holstein also suffered great losses of land during the late Middle Ages and in early Modern time. During the storm floods of 1362 and 1634 AD, northern Friesland between Eiderstedt and Sylt, an area of about 1300 km², was completely lost except for a few islands. After 1362 the shoreline was situated along the edge of the Pleistocene from where subsequent reclamation started (Prange 1986). On the other hand, in the more southern Dithmarschen and in parts of Eiderstedt, the marshes have remained nearly unchanged since the Birth of Christ.

Another important anthropogenic factor which contributed to the huge land losses in the Middle Ages were the drainage measures adopted in the course of diking. Particularly in the backswamp areas of the hinterland, drainage led to a compaction of the bogs, often lowering their surface below the Mean High Water (MHW) level. When the dikes were breached the flood waters passed the wide and firm levees along the coast and the lower rivers to reach these artificially lowered areas. It was subsequently difficult or impossible to get rid of the flood waters. This is illustrated by the reconstruction of the coastline at 1500 AD, around the Jade Bay. It clearly demonstrates that major land losses occurred in the low-lying hinterland, the elevated coastal zone being far less affected.

Especially in northern Friesland, but also in other regions bordering the southern North Sea the ex-cavation of clay-covered peat for salt extraction contributed to the lowering of the land surface. Having penetrated into the layers of the upper peat, the salt was subsequently covered by the deposition of clay (see chapter 3). This peat , occurring both landwards and seawards of the dikes, was excavated and burnt to extract the soluble salt from the ash by evaporation. The so-called Frisian salt was widely traded during Medieval and early Modern times. Even today such former salt extraction areas are visible in the North Frisian Wadden Sea region, clearly demonstrating that this industry strongly contributed to the loss of land.

From the 16th century onwards, the technique of diking improved and active reclamation took place although there were still considerable setbacks by severe storm damage, e.g. during the surges of 1717, 1825 and 1905. The highest storm flood level ever reached in northern Germany was that of 1962 (in Wilhelmshaven +5,22 m NN, i.e. 3,55 m above the mean high water level).

The current rate of the mean sea level rise measures less than 20 cm per century, although the increase of the mean high water level in the German Bight amounts to nearly 30 cm per century due to an increase of the tidal range.

Conclusions

The Holocene rise of the global sea level and its oscillations are mainly responsible for the formation of the landscape in and along the southern section of the North Sea. Crustal movements are of no importance in this region. The main factor controlling the sea level is the changing of global climate which can be traced by sea level investigations.

Submerged basal peats in the North Sea area and peats intercalated in the Holocene sequences of the Clay District can be dated by the radiocarbon method as well as by pollen analysis, thus providing data to construct the older part of the sea level curve. For younger periods the record of human occupation, which is closely linked to sea level changes, contributed additional dates.

Prehistoric settlements of known depth and age are often regarded as prehistoric tide gauges (e.g. Brandt 1980). However, one has to strictly distinguish between the mean high water (MHW) level, which is expressed in the sea-level curve and marks the border between salt marshes and tidal flats, and storm flood levels which are much higher. The existence of permanent settlements indicates that their altitude was above the maximum storm flood level of the respective period, the settlement layers inside the dwelling mounds marking increases in storm flood levels, not of the MHW. On the other hand, a temporary lowering of the storm flood level is not reflected in the dwelling mounds. However, a reconstruction of MHW levels may be achieved by evaluating the evidence from "Wurten". Thus landing ramps or wooden bridges across a gully along the side of a "Wurt" may indicate the MHW level. Another source are botanical remains which are often well preserved. They allow the reconstruction of the former salt marsh plant communities which are reliable indicators for a precise position above or around the MHW level.

The history of diking and the subsequent positive and negative changes to the coastline also have climatic backgrounds, although in different ways. The climatic optimum in the Early and High Middle Ages (900-1100 AD) is well documented. It is followed by the so-called "Little Ice Age" from the 14th to the 18th century AD. This global deterioration in climate resulted in a slight decrease of the mean sea level, as shown by Streif (1990). One would assume that such a lowering of mean sea level would benefit human settlement along the coast. However, a climatic deterioration is commonly associated with an increase in storm frequency and this endangered the settlements even more. During the Little Ice Age there was thus a decrease of MHW coupled with an increase of storm flood levels. To make matters worse, the construction of dikes further amplified the storm flood levels and hence posed an increasing threat to human occupation which was only alluviated in the course of the 20th Century by the application of modern dike construction techniques.

References

Bantelmann A (1975) Die frühgeschichtliche Marschensiedlung beim Elisenhof in Eiderstedt. Landschaftsgeschichte und Baubefunde. Lang, Bern Frankfurt/M. Studien z. Küstenarchäol. Schlesw.-Holsteins, Ser. A, vol 1, pp 1-172

Behre KE (1976) Die Pflanzenreste aus der frühgeschicht-lichen Wurt Elisenhof. Lang, Bern Frankfurt/M. Studien z. Küstenarchäol. Schlesw.-Holsteins, Ser. A, vol 2, pp 1-144

Behre KE (1985) Die ursprüngliche Vegetation in den deutschen Marschgebieten und deren Veränderung durch prähistorische Besiedlung und Meeresspiegelbewegungen. Verhandlungen der Gesellschaft für Ökologie 13:85-96

Behre KE (1986) Meeresspiegelverhalten und Besiedlung während der Zeit um Christi Geburt in den Nordseemarschen. Offa 43:45-53

Behre KE (1987) Meeresspiegelbewegungen und Siedlungsgeschichte in den Nordseemarschen. Holzberg, Oldenburg. Vorträge der Oldenburgischen Landschaft, vol 17, pp 1-47

Behre KE (1991) Umwelt und Ernährung der frühmittelalterlichen Wurt Niens in Butjadingen nach den Ergebnissen der botanischen Untersuchungen. Probleme der Küstenforschung im südlichen Nordseegebiet 18:141-179

Behre KE (1999) Die Veränderungen der niedersächsischen Küstenlinien in den letzten 3000 Jahren und ihre Ursachen. Probleme der Küstenforschung im südlichen Nordseegebiet 26:9-33

Behre KE, Menke B (1969) Pollenanalytische Untersuchungen an einem Bohrkern der südlichen Doggerbank. Beiträge zur Meereskunde 24/25:122-129

Behre KE, Dörjes J, Irion G (1984) Ein datierter Sedimentkern aus dem Holozän der südlichen Nordsee. Probleme der Küstenforschung im südlichen Nordseegebiet 15:135-148

Brandt K (1980) Die Höhenlage ur- und frühgeschichtlicher Wohnniveaus in nordwestdeutschen Marschengebieten als Höhenmarken ehemaliger Wasserstände. Eiszeitalter und Gegenwart 30:161-170

Brandt K (1991) Die Ergebnisse in den Grabungen der Wurten Niens und Sievertsborch (Kreis Wesermarsch). Probleme der Küstenforschung im südlichen Nordseegebiet 18:89-140

Haarnagel W (1941) Die Grabung auf der Wurt Hessens und ihr vorläufiges Ergebnis. Probleme der Küstenforschung im südlichen Nordseegebiet 2:117-156

Haarnagel W (1979) Die Grabung Feddersen Wierde: Methode, Hausbau, Siedlungs- und Wirtschaftsformen sowie Sozialstruktur. Steiner, Wiesbaden . Feddersen Wierde, vol 2, pp 1-364

Körber-Grohne U (1967) Geobotanische Untersuchungen auf der Feddersen Wierde. Steiner, Wiesbade Feddersen Wierde, vol 1, pp 1-357

Meier D (1996) Landschaftsgeschichte und Siedlungsmuster von der römischen Kaiserzeit bis in das Mittelalter in den Küstengebieten Eiderstedts und Dithmarschens. Siedlungsforschung 14:245-276

Prange W (1986) Die Bedeichungsgeschichte der Marschen in Schleswig-Holstein. Probleme der Küstenforschung 16:1-53

Reichstein H (1991) Die Fauna des germanischen Dorfes Feddersen Wierde. Steiner, Wiesbaden. Feddersen Wierde, vol 4, pp 1-346

Schmid P (1984) Siedlungsstrukturen. In: Kossack G, Behre KE, Schmid P (eds) Archäologische und naturwissenschaftliche Untersuchungen an ländlichen und frühstädtischen Siedlungen im deutschen Küstengebiet vom 5. Jahrhundert v. Chr. bis zum 11. Jahrhundert n. Chr., vol 1: Ländliche Siedlungen. Acta humaniora der Verlag Chemie GmbH,

Weinheim, pp 193-244
Schmid P (1988) Die mittelalterliche Neubesiedlung der niedersächsischen Marsch. In: Bierma M, Harsema OH, Zeist, W van (red) Archeologie en landschap (Festschrift HT Waterbolk). Biologisch-Archaeologisch Instituut, Groningen, pp 133-164
Schmid P (1994) Oldorf – eine frühmittelalterliche friesische Wurtsiedlung. Germania 72: 231-267
Schwabedissen H (1951) Zur Besiedlung des Nordseeraumes in der älteren und mittleren Steinzeit. In: Kersten K (ed) Festschrift für Gustav Schwantes. Wachholtz, Neumünster, pp 59-77
Strahl E (1998) Die älteste Siedlung an der deutschen Nordseeküste. Archäologie in Niedersachsen 1:54-56
Streif H (1990) Das ostfriesische Küstengebiet, 2nd edn.. Borntraeger, Berlin Stuttgart. Sammlung Geologischer Führer, vol 57, pp 1-376

Climatic Change in Northern Europe Over the Last Two Thousand Years and its Possible Influence on Human Activity

Jean M. Grove

Girton College, UK-Cambridge CB3 OJG, United Kingdom
e-mail: dgrove@talk21.com

Abstract: A brief overview of European climate during the last two millennia, and of modern improvements in climatic and glaciological methodology, is followed by discussion of some of the many instances when adverse weather conditions have affected human behaviour and prosperity. Such conditions were particulary frequent during the principal episodes of glacial advance in the course of the Little Ice Age, but have also occurred during intervening warmer periods.

Introduction

The climate of a place is more than its average weather, it is the sum total of the weather including its extremes. A change in climate involves not simply a change in mean conditions but also a change in the frequency of occurrence of extreme weather. Extreme weather events may last for a day, a season, or a year, or they may occur in clusters of years with several departures from the mean being at short intervals, thereby changing the mean values of temperature or precipitation for periods perhaps lasting a decade or more.

A change in climate, and especially in the scale and frequency of extremes, has important effects on plant growth, on disease risks, on fisheries, on flooding, and on agriculture. When most of the population of northern Europe was directly dependent on agriculture, pastoralism and fishing for its livelihood, as in poor countries today, the influence of such changes on human activity was inevitably considerable.

Glaciers respond to the balance between the accumulation of snow, mostly in winter, and ablation (loss of volume), mainly in summer. Dated evidence of past changes in the behaviour of glaciers provides a useful index of climate change over Europe. Climatic changes on the century timescale within the Little Ice Age, have been deduced in both the Alps and Scandinavia by tracing the oscillations of large valley glaciers.

Improvements in recent years in the methodologies of both glaciology and historical climatology have made it possible to identify numerous occasions in the last two millennia when climatic changes had important influences on human activity. All of northern Europe has been affected but not precisely in the same manner everywhere owing to differences in altitude, latitude, continentality and aspect. Some very simple hypothetical examples of the effects of climate change in different geographical situations illustrate the reasons:

(i) A decrease in annual precipitation by 200 mm in eastern England, where the mean is about 500 mm, will have a far greater impact on river flow, water-tables and agriculture than a similar decrease in the west where the mean annual precipitation is about 1500 mm.

(ii) Whereas in Scotland the upper limit for cultivation may be 300 m above sea-level, in the Alps it is likely to be nearer 1000 m. A change in annual mean temperature in Scotland of 1.5°C might then reduce the cultivation limit to near sea-level, while the same change in the Alps would have far less drastic consequences.

(iii) A reduction of mean annual temperature by 1°C in a maritime situation may reduce the growing season from, say, 200 days to 60 days. The same change in a continental situation may reduce the growing season by only a few days, though the ef-

From WEFER G, BERGER W, BEHRE K-E, JANSEN E (eds), 2002, *Climate Development and History of the North Atlantic Realm.* Springer-Verlag Berlin Heidelberg, pp 313-326

fect may be more drastic near the northern limit for agriculture.

(iv) Within a period of twenty years, a reduction of temperature at a place by 5 degrees in each of four spring seasons will have very serious consequences for agriculture, especially if the years concerned are close together, yet the reduction in mean annual temperature over the 20 year period is only 0.25°C.

Before the industrial revolution and the expansion of world trade that took place in the course of the nineteenth century, local communities and even entire countries were liable to suffer severe food shortages, famines and associated diseases on account of climatic events - droughts, severe winters, cold summers, excessive rains and floods. Poor countries, such as those of the Sahel zone, still do. No part of northern Europe has been unaffected by such events and by climatic changes during the last two millennia, yet historians have frequently ignored or discounted the possibility that they may have played an important part in human affairs.

Le Roy Ladurie (1971) and other historians and economists rejected the suggestions made by certain Scandinavian historians, particularly Utterstrom, that climatic deterioration during the fourteenth, sixteenth and seventeenth centuries had disastrous economic consequences for Europe. Ladurie considered that such phenomena as the switch from grain-growing to fishing in Iceland between 1300 and 1350, and the decline of English vineyards in the fourteenth century, could be adequately explained by economic circumstances. He pointed to the need for careful examination of harvest yields and the scientific study of climate; "both... concern the historian, with his traditional rules governing the rigorous use of documentation, and his new norms of quantitative elaboration". He found evidence himself, in the dates of grape harvests for climatic fluctuations. Nevertheless he considered that social and economic processes overwhelmed the influence of climate.

When Ladurie was writing in 1971 the accuracy of glacial and hence climatic chronologies was deficient. The chronologies of early climatic historians, including Britton and Brooks, are now known to have been inaccurate because they were insufficiently rigorous in their use of sources. The mistaken coincidence of the early medieval period with "one of Europe's cold cycles", as identified by Brooks, led Postan to conclude that the bearing of climate on the desertion of English villages in later medieval times was minimal. Now it is known from the Swiss Alps, where the dating is most precise, that the Little Ice Age had started by the thirteenth century, with glacier advance episodes culminating in the fourteenth, seventeenth and nineteenth centuries (Holzhauser 1997).

Analyses of ice cores from the summit of the Greenland Icecap have helped to revolutionise knowledge of the range and timing of climatic fluctuations; they have demonstrated the natural instability of climate and the rapidity with which changes can occur. Detailed investigations of the former extent of glaciers in Scandinavia and especially the Alps have provided far more reliable chronologies of glacial advance and retreat than were available in the 1970s. Recession of the ice during the twentieth century has revealed much previously concealed evidence. Radiocarbon dating of organic samples has been of outstanding importance, and improved conversion of radiocarbon to calendar dates now allows more reliable comparisons to be made of events in widely separated regions. In the Alps, identification of the calendar dates at which trees, found in their growth positions, were killed by advancing ice, and their ages at death, has been achieved by matching their ring sequences with long dendrochronological series (Holzhauser 1997). Careful analysis of written records, paintings and drawings has revealed the details of more recent fluctuations. In Scandinavia, intensive field work, identification and dating of plant remains, and studies of sedimentary sequences within moraines and in lakes, have greatly increased knowledge of the course of events.

The longest and most pronounced long term climatic fluctuations during the last two thousand years were the Medieval Warm Period and the subsequent Little Ice Age, together lasting a thousand years (Grove 1996). The Medieval Warm Period is defined here as the time interval of general glacial withdrawal, between about 900 and 1300 AD. In the succeeding Little Ice Age when glaciers remained enlarged (although their fronts are known to have oscillated about their forward positions), the climate was such as to cause snow and ice to accumulate, pointing to generally cooler summers and snowier winters. Short term fluctuations of temperature and precipitation, similar in scale to those with which we ourselves are familiar, and which have been measured ever since European meteorological stations began instrumental records some 200 years ago (Pfister et al. 1996, 1998), were superimposed on both major periods.

In the past, publication of inaccurate or erroneous climatic reconstructions based on misdated documents, or documents written long after the event, or in places far away, have caused confusion. In recent years, detailed information about the nature of short term climatic oscillations, or anomalies, superimposed on the major warming and cooling periods has been obtained by cross-dating different kinds of documentary proxy data with descriptive evidence. By using verified, contemporary documents, our colleagues are obtaining quantitative estimates of temperature and precipitation and are producing increasingly detailed accounts of past weather (Pfister et al. 1996).

These methodological advances have allowed many instances of climate change to be identified which are significant for human social and economic history. Although many of the fluctuations in agricultural productivity and demographic conditions can be attributed at least in part to social, political and economic factors associated, for example, with the growth of towns, the establishment or interruption of trade routes, wars, migrations and colonisation, the existence of such circumstances does not necessarily explain the change in fortune of particular areas. Climatic factors are also important but their influences have to be disentangled from the social and economic factors operating contemporaneously.

Major Climatic and Glaciological Characteristics of the Last Two Millennia

Severe winters were less frequent during the Medieval Warm Period and more frequent in the Little Ice Age than over the millennium as a whole. Nevertheless, warm individual years or clusters of years occurred during major periods of glacial expansion and cold years and clusters of years during periods when the glaciers were withdrawn high up their valleys (Pfister et al. 1998). Cold years have followed volcanic eruptions in Iceland, notably those of Katla in 1660 and Hekla in 1693, and eruptions further afield in Indonesia, notably Tambora in 1815 and Krakatoa in 1883 (Grove 1988, p. 370). Cold episodes in northern Europe have been associated with the North Atlantic Oscillation and even with ENSO events.

The incidence of severe anomalies is of great importance to plant growth and therefore to agriculture. Persistent droughts or successive wet years or seasons greatly affect agricultural and pastoral activity. The impact of temperature departures from the mean is greatest in maritime situations, especially near limiting altitudes and latitudes for cultivation and tree growth, because in such critical zones their effect on the length of the growing season is at a maximum.

Cold Phases in the First Millennium AD and Their Impact

Evidence of the occurrence of several cold phases in the first millennium AD is provided by records of glacial expansion in both the Alps and Scandinavia (e.g. Holzhauser 1997; Nesje et al. 1991; Nesje and Rye 1993). An excellent example is shown in Fig. 1.

In the Alps the Grosser Aletsch advanced in the third to fourth century (around cal 240-380 AD) and again in the late seventh century (around cal 650-690 AD) reaching well separated culminations. Data from the Gorner and Reid Glaciers are in agreement. The Gurgl in Austria is also known to have advanced in the fourth century AD.

In Scandinavia, although dating is less precise than in the Alps, and no history of an individual glacier is as complete as that of the Grosser Aletsch, dating of moraines and sediment sequences in lake basins shows that several glacial advances also occurred there during the first millenium AD. Radiocarbon dates on trees sheered off or damaged by advancing ice indicate that Engabreen pushed forward several times between cal 600 and 1200 AD. Tree ring records document cold summers around 550 AD and between 780 and 970 AD - and also 20-30 years of summer warmth around 930

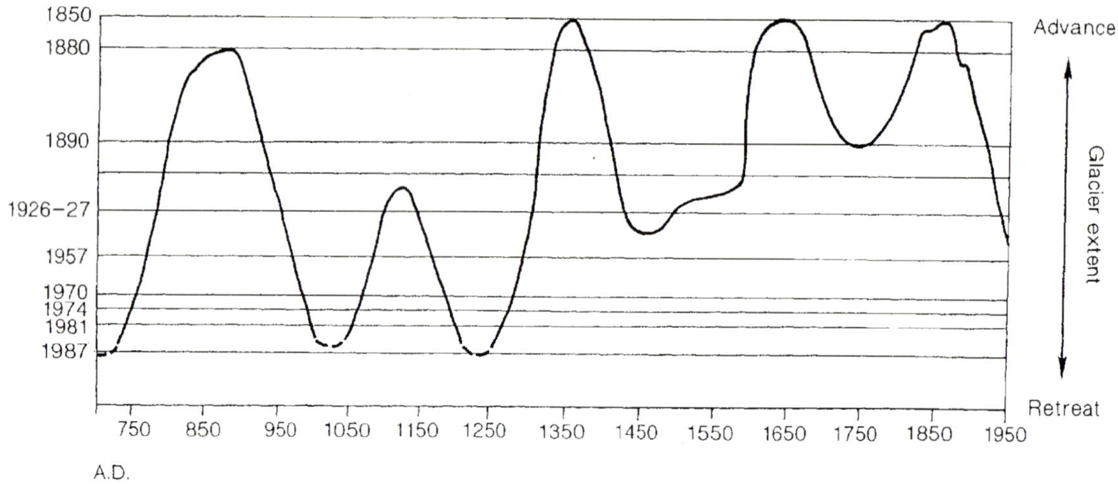

Fig. 1. The Oscillations of the Grosser Aletsch Glacier (After Holzhauser 1984, from Grove 1996).

AD. (Karlén 1990; Boulton et al. 1997). Evidence from Sandskardfonna in southern Scandinavia reveals several minor glacial advances between cal 657 and 860 AD, well before the first advances of the Little Ice Age between cal 1290 and 1410 AD. Dated sediment layers in lake Gjuvatnet show that its adjacent cirque glacier expanded about cal 145-335 and cal 770-818 AD as well as about cal 1225-1300 AD. Such cirque glaciers are sensitive indicators, reacting faster to minor climate fluctuations than valley glaciers and providing more detailed information. The times of glacial expansion of the outlets of ice caps near the western coast, it should be noted, differed somewhat from those further inland because of differences in aspect and weather systems bringing precipitation (Dahl and Nesje 1994).

Documentary evidence of cold weather anomalies in central and western Europe before 750 AD is sparse; warm anomalies are not detectable from written evidence before the twelfth century (Pfister et al. 1998). The winter of 763/4 (when the Grosser Aletsch was advancing rapidly) seems to have been one of the coldest in the last 2000 years. In the ninth century there were at least four extremely severe winters: in 821/2 when snow cover in the Rhineland lasted from November to April, and the Rhine, Danube and Seine froze for over 30 days; in 845/6, northeasterly winds in Champagne damaged cereal crops; the winter of 859/60 was amongst the longest of the last 1300 years with snow and frost in the Rhineland lasting from November till early April; in the 873/4 winter, snow in parts of France lasted 130 days and grain seeds were killed; in 874/5 the Baltic froze. Such extreme conditions no doubt brought trials and tribulation, but contemporary records are insufficient to assess their impacts on human activity.

The Impact of Medieval Warm Period Conditions and the Shock of the Early Little Ice Age

During the Medieval Warm Period, between about AD 900 and 1300 environmental conditions were predominantly, but not unbrokenly, benign. In western and central Europe extremely cold winters were less frequent than in the Little Ice Age and the climate was at least as warm, if not rather warmer, than in the twentieth century. A few tenth century famines recorded in Irish annals, as in 963, and ascribed to "cold and scarcity of corn", seem to relate to single episodes of crop failure (Lyons 1989). In contrast, during the Little Ice Age conditions were predominantly, but not unbrokenly, harsher, and in some years and places, very severe. In western and central Europe winters in the period 1300-1329 were 1°C colder than in the thirteenth or twentieth century (Pfister et al. 1998); a drop of temperature of 1°C is equivalent, it might be noted, to an altitudinal difference of 200 m.

Prolonged periods of benign weather with few crop failures might be expected to lead to population increases amongst the peasant communities spread across Medieval Europe, and to the successful occupation of areas marginal for agriculture, yet such a possible influence is frequently ignored by historians. Were the Benedictine agricultural improvers aided by the warmer conditions? Did rising populations during the warm centuries play a part in the arrival of Scandinavians in other parts of Europe, including England and Normandy?

The change of climate from the Medieval Warm Period to the early Little Ice Age constituted a deterioration of the environment so far as the inhabitants of most of northern Europe were concerned.

Greenland

The Vikings could not have colonised western Greenland in the tenth century, and survived there for several centuries, living by herding and hunting, and maintaining their connections with Scandinavia and from one coastal settlement to another along the west coast, except for the benign conditions of the Medieval Warm Period. Depending on reliable grass crops for their beasts and regular arrival of migratory harp seals along the coast, they were living very near the margin. They were very vulnerable to environmental change, particularly as it is believed that grazing pressure was gradually reducing the fertility of their grasslands, which were very restricted in area. Lower temperatures in the early and middle fourteenth century, identified in the GISP2 ice core, with an accompanying extension of sea ice and greater storminess, undermined the Greenlanders' economy. (Barlow et al. 1997). Clusters of cold years in 1308-18, 1324-29, 1343-62 and 1380-84, impoverished a society living so precariously. The close spacing of abnormally cold summers, especially in 1351 and 1352, and of exceptionally cold winters in 1351 and 1355, are seen to have triggered the abandonment of the Western Settlement.

The impact of a harsher climate, it has been argued, would not have eliminated the Viking colonists if they had been able and willing to adapt their lives and equipment to changing circumstances. They did not use skins for clothing or boat covers, or employ harpoons for hunting, usages which enabled the Inuit to extend their occupation of Greenland just at the time when the Viking settlements were failing.

British Isles

The taxable male population of Taunton district, within the Winchester estates, rose from 612 in 1290 to 1448 in 1311. Similar increases in population have been recorded elsewhere in England, and in other parts of Europe during the Medieval Warm Period. Population increase is often attributed by historical demographers to changes in marriage age, extension of cultivation and improvements in agricultural techniques. Such social and economic factors were undoubtedly relevant, but many historians have overlooked the importance of the relative absence of difficult weather and harvest failures which made them possible.

A series of agricultural catastrophes brought distress to England in the early fourteenth century, at the very time when Swiss glaciers, including the Grosser Aletsch, were advancing swiftly and the Greenlanders were beginning to suffer from food shortages. Conditions between 1314 and 1322 were especially difficult. Because of excessive rain, the meagre crops of 1314 were harvested with great difficulty. In 1315 torrential rain and floods ruined hay and cereal crops. Famine followed immediately and was accompanied by a virulent epidemic which greatly increased mortality. England may already have been suffering from inflation, which has been attributed to over-much silver gained by wool exports, but the rise in grain prices between 1315 and 1316, from 8s per quarter to 26s was a direct result of the weather (Harvey 1993). Widespread sheep murrain caused serious losses of lambs and yearlings, especially on clay soils and flooding grasslands, and then in 1319 an epidemic of cattle disease broke out, causing widespread losses. The incidence and spread of cattle and sheep diseases was probably due to the temperature and humidity conditions favouring the pathogens concerned. A mediocre harvest in the wet summer of 1320, was followed by a disastrous harvest due to drought in 1321 and there was also serious coastal flooding.

These weather-related disasters during the early Little Ice Age certainly had major economic and social effects, the impact varying from place to place, and class to class. Low-lying ground suffered more than well-drained land at moderate altitude. Livestock disease reduced food supplies directly and caused serious shortages of manure vital for arable production. Lack of draft animals prevented land being tilled. It took twenty years for stock levels to recover.

The poor suffered most, but the rich were not immune. Wool exports declined by nearly a third between 1215 and 1345, affecting better-off farmers and merchants and the national economy. Demesne production was reduced on many estates. With labour short, some arable reverted to pasture, some pasture to waste. The climatic disasters accelerated the disintegration of the feudal tenurial structure. For a few people the crisis brought profits, but many small holders had to sell out, allowing richer farmers to increase their holdings. In the worst hit regions, as land went out of cultivation, villages were deserted, their abandonment being well under way before the arrival of the Black Death.

The historian Titow, after analysing data from the estates of the Bishop of Winchester for 1209-1349, considered it "impossible to isolate and even more quantify the effects of changing weather conditions on the level of productivity". He thought the references to bad weather in the account rolls did not provide an adequate guide to climate because they were only incidentally included. The basic cause of deteriorating yields, he believed, was soil exhaustion caused by the preceding increases in the population. But this view conflicts with the detailed evidence . The Winchester documents record sharp fluctuations in yields from year to year, not the progressive decline which might be expected to result from soil exhaustion (Fig. 2).

The uplands of Britain suffered as much as the poorly drained lowlands. On Dartmoor, in south-west England, many medieval farmsteads were deserted in the late thirteenth and fourteenth centuries, and by the midddle of the fourteenth century all settlements above 300 m had been abandoned. Even in the lowlands abandonment had begun well before the Black Death (Lamb 1977) In uplands, such as the Lammermuir Hills of southern Scotland, a greater frequency of very severe years increased the incidence and risks of crop failure, lowering the limits of viable cultivation, adding to the impact of the Black Death and causing high altitude settlements to be abandoned (Parry 1978).

In Ireland, three exceptionally wet years beginning in 1315, caused steep rises in grain prices. A good harvest in 1318 was followed by cattle murrain in the 1320s and by more crop failures which continued into the 1330s. The population ceased

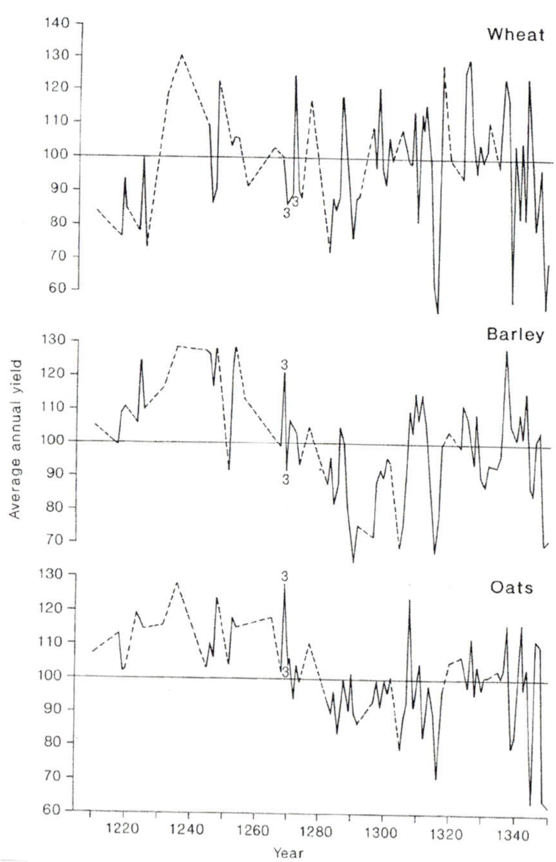

Fig. 2. Annual yields of grain crops from the Winchester estates expressed as percentages of the 1209-1350 mean yields. 3 indicates individual years when it happens that figures for the yields of only 3 manors were available. From Grove (1996).

to grow and well before the Black Death decline began. Climatic deterioration here in Ireland coincided with war and political unrest and making it impossible to disentangle the effects of one from the others (Lyons 1989).

The course of events elsewhere in Europe was similar to that in the British Isles. Famines in 1315, 1316 and 1317 are recorded in a great number of documents from northern Europe as far east as Poland and Silesia. (Lucas 1962). At the height of the famines, diseases were widespread and cattle murrain disastrous. The epidemics amongst people and animals were probably associated with inadequate diets, the use of unclean water and environmental conditions, especially climate, favourable to pathogens.

Information about this period in Scandinavia is sparse. To what extent can the widespread desertion of settlements in the early Little Ice Age in Norway and other parts of Scandinavia be attributed to declining productivity associated with climatic stress before the onset of the Black Death?

Climatic Impacts later in the Little Ice Age

The Sixteenth, Seventeenth and Eighteenth Centuries.

As glaciers in the Alps swelled towards their second Little Ice Age culmination widespread harvest failures, food shortages and famines occurred across Europe, triggering social unrest, and increasing numbers of vagrants. Especially severe conditions in the 1590s and 1690s brought crop failures and social disruption to many parts of Europe. In 1740 the mean annual temperature for central England was the lowest in Manley's (1974) series. With harvest failures, grain prices rose in western and central Europe in 1769 and 1771 and in northern Europe in 1771 and 1772.

The Alps

The difficulties faced by rural populations are well illustrated by events in the Chamonix area, west of Mont Blanc, where glaciers expanded rapidly between 1580 and 1600, covering farmland and forests. High settlements were abandoned, or destroyed; meltwater streams carried away houses and barns and ruined lower land. By 1600 the glaciers had spoilt 195 'journaux' of land in the parish of Chamonix. The local economy was severely affected; yields of crops in the valleys were meagre (Ladurie 1971). The glaciers continued to swell intermittently, causing further direct damage. A document of 1640 referred to the 'third of good and cultivable land lost in the last ten years through avalanches, falls of snow and glaciers' (Ladurie 1971, p.340). A series of applications for tax relief were found to be justified. Processions to exorcize the glaciers were made between the 1640s and 1690. The Bishop of Geneva visited the Chamonix area to exorcise the swelling ice fronts in 1664 and again in 1690 (Grove 1988, p.116-8).

Similar events occurred elsewhere in the Alps, glacier advances being particularly damaging when they blocked tributary valleys, forming lakes. These eventually broke out, sending flood waters far down valleys. The blocking of Saastal by the Allalin in 1589, forced half the inhabitants to migrate when the ice dam broke in 1633. The Gietroz dammed the Val de Bagnes in 1595, and flooded a newly opened silver mine;140 people were drowned in Bagnes itself and over 500 buildings destroyed. Grape yields in Switzerland were below the long term trend in every year except one between 1585 and 1597.

The Swiss Pre-Alps experienced a severe famine in 1770-71, when the glaciers were again advancing.Wet autumns had affected lowland wheat yields, long winters damaged upland crops and depleted hay stocks.Wet summers reduced yields of hay, potatoes and grain, and snowy summers diminished cheese production. Dairy production was low on account of pasture being short, hay of poor quality and herds reduced in size. Seed potatoes were scarce and expensive, government grain stores empty, and imports hindered by embargoes placed elsewhere to counter harvest failures and high prices (Pfister 1978). In years of heavy snowfall, such as 1755, 1770 and 1785, with snowcover lasting into late spring on the Swiss alpine pastures, winter crops were seriously damaged by a snow mould, *Fusarium nivale*. The

Gurglerferner, in Austria, which had been advancing for some years, blocked Langtal in 1717, forming a lake, which emptied and filled intermittently during the eighteenth century. In Ötztal barley and oats ripened completely only in warmer years, and grass could be cut only every other year in the rich pastures near the head of the valley. Despite the eventual disappearence of the dam without major floods, the cost of dismantling twenty two bridges in the early 1770s and rebuilding them in 1779 cost 1920 florins.

France

In northern France crops were deficient in 1738, and mediocre in 1739. In 1740 winter grains, supplying most of the bread cereals, were destroyed by deep frosts, especially in areas without protective snow cover. The region round Paris was the worst affected; harvests were not completed till late September rather than mid-August. Late summer rains not only delayed the cereal harvest of 1740 but set back sowing of the subsequent crops. Harvests were mediocre in 1741, scarcity persisting until 1742, when there were good yields. Mortality in France, as in Scotland and Ireland, but not in England, paralleled the rise and fall of cereal prices. In the Arve valley the tongue of the Mer de Glace and its neighbours expanded again in the eighteenth century, reaching nearly to the Arve in 1780.

Germany

In southwest Germany rain hindered reaping and harvests were delayed for over a month in 1740; some crops were still uncut in mid-October when severe frosts were followed by snow which covered more than two thirds of the oats grown on high ground so that it rotted. Many cattle and sheep died, though arable production was apparently little affected.

Scotland

In southern Scotland, more settlements were abandoned as climatic limits to cultivation fell further (Fig.3). Parry (1978) reckoned that 2990 hectares inn the Lammermuirs were abandoned to moorland between 1600 and 1800. More than 80% of this area lies between the isopleths he calculated for the climatic limit for cereal cultivation for 1530 and 1600; apparently it became submarginal for cultivation in the sixteenth century. The retreat of high-level cultivation in southern Scotland may have been accelerated by such factors as English invasions and civil war, but the effects of sequences of cold summers in the 1590s and 1690s were of fundamental importance.

Scandinavia

Seventeenth and eighteenth centuries taxation records in western Norway reflect the impact of climatic events on the local economy with unusual precision (Grove and Battagel 1983). Land rents were separately assessed for each farm, and recorded in contemporary documents. In 1670 a uniform system was adopted, taxes being assessed on the number of cattle wintered and the quantity of grain sown and harvested. Alterations to the assessments required public on-site evaluation, involving local lay assessors, the Sheriff and the farmers involved. No tax relief was granted without specific explanations.

Farms in Sunnfjord administrative district (about 61 1/2°N), none at altitudes over 400 m, and very few over 200m, depended primarily on cattle and the grass crop. Detailed valuations were made in 1667 and 1723. Voluminous records of appeals for tax reduction show that environmental deterioration set in within three years of the 1667 assessment. While glaciers were rapidly swelling, the higher farms were damaged by landslides and avalanches, and lower ones by flooding. Not only was the yield of pastures and hayfields immediately reduced by the inclement weather, but streams eroding and spreading sediment over floodplains caused long-lasting reductions in fertility. The number of cattle that could be kept dropped substantially; out of six parishes in the Nordfjord area five had lost more than 20% of their cattle by 1723 (Fig. 4). Some coastal farms suffered from failure of fisheries or sealing, associated with lower water temperatures.

Many further tax reductions were made on account of physical damage to farmland later in the eighteenth century, especially between 1740 and

1750, when glaciers were reaching their maximum Little Ice Age extent. Grain harvests failed over large parts of the country in 1741-2. More than 12% of the population of Swedish Varmland died in 1742. In Finland actual famine was caused. In 1743, 174 farms scattered over Vestlandet in south-west Norway had their taxes reduced because of serious damage caused by heavy rains in December. The harvest failed again in 1748 causing peaks in the death rate. After 1750 the glaciers began to retreat but grain harvests failed again over large parts of the country in 1773.

Iceland

In Iceland livelihoods depended firstly on fishing and secondly the hay harvest on which stock were fed through the winter. Lower temperatures in the Little Ice Age had disastrous effects on both activities. Ogilvie's (1981) survey of primary sources, has shown that fishing declined, especially in 1685-1704 and 1744-1759, because of increased ice round the coast, stormy weather and withdrawal of the cod southwards with the arrival of colder water. Contemporary documents show that livestock died

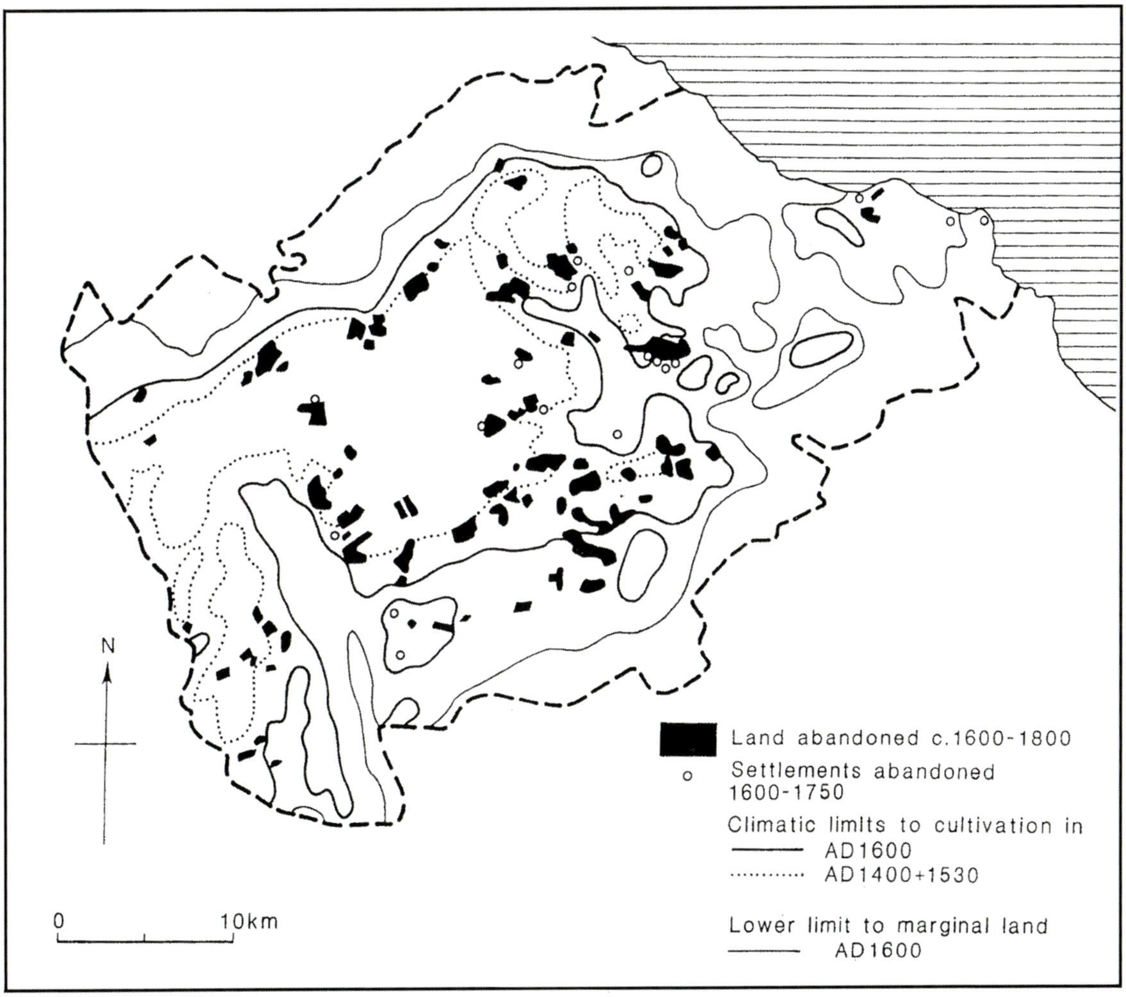

Fig. 3. Abandoned farmland and lowered climatic limits to cultivation in the Lammermuir Hills, southeastern Scotland between AD 1600 and 1800. From Parry (1978).

of hunger and cold in 24 out of 36 years between 1730 and 1766. Epidemics, including a smallpox epidemic in 1707, accentuated suffering, as did also volcanic eruptions, and associated ash and lava, notably from the enormous Eldeyjar eruption of 1783, which laid waste large areas. The outstandingly cold 1780s, probably attributable to the volcanic activity, saw the greatest incidence of sea-ice between the sixteenth and nineteenth centuries. The rigours of the Danish monopoly of trade remained unrelaxed and must share the blame, but the primary cause of Iceland's suffering and impoverishent at this time was climatic deterioration.

Nineteenth Century Harvest Failures in Europe

In the second decade of the nineteenth century, as the Grosser Aletsch swelled towards its third culmination, a run of poor harvests afflicted northern Europe. Especially severe was 1816, the "year without a summer", which brought hunger, want and famine to much of Europe, especially Switzerland. According to J.D. Post, it was but one component of a hemispherical crisis, attributable to the eruption of Tambora in 1815.

The Early Nineteenth Century

In 1816 and 1817 harvests were late and poor in the British Isles, the Netherlands, France, Germany, Switzerland, and Austria. Widespread drought added to the difficulties caused by cold. In the Arve valley near Chamonix, mid-nineteenth century glacier advances panicked the local population in 1850-55. By 1850 the tongue of the Mer de Glace had destroyed a larch wood in the valley bottom and approached close to the village of Les Bois. Farmland was covered by the ice, and some hamlets were buried. But the crisis was not longlastiing, retreat soon followed. In southwest Germany rain hindered reaping and harvests were delayed for over a month; some crops were still uncut in mid-October, when severe frosts were followed by snow which covered more than two thirds of the oats on high ground so that it rotted.

In the 1840s adverse climatic conditions for plant growth recurred over much of northern and western Europe. In central England mean air temperatures for January-February were more than 6°C below the decadal average (Manley 1974).

The failure of the potato crop in Ireland in the late 1840s was the greatest agricultural disaster of nineteenth century Europe. Potatoes, well suited to the moist climate, had been long established as the staple crop, allowing the population to rise during the period before the crop failed. Potato blight, *Phytophthora infestans*, was brought to Europe with new varieties introduced to Flanders from the Americas. Ideal conditions for the blight were believed to be long spells of weather with saturated or near saturated air and temperatures above 10°C, and preferably nearer 15°C, with free moisture on the leaves (Bourke 1984). A sequence of cyclonic depressions brought such weather encouraging the blight to spread. Yields of potatoes in 1845 were less than two thirds the average. In 1846 the crop was devastated everywhere, yields being only about a quarter of those normally expected. In 1847 the crop was again deficient because so little seed had been available for sowing. In 1848 blight struck again, reducing yields by about a third. Blight continued to reappear at irregular intervals until around 1900 when effective treatment began. The Irish famine caused very high mortality, emigration reduced the population further, and the resource base was damaged for many years. The social and economic consequences of the potato blight were certainly magnified by the original high population density, reliance on monoculture, the actions of landowners and by political attitudes, particularly the operation of the corn laws by the British government, but the underlying climatic factors cannot be completely ignored.

Harvest Failures in the Late Nineteenth Century

Lowered prices due to massive imports from overseas were the main reason for the great depression affecting British agriculture at the end of the nineteenth century, but adverse weather conditions and accompanying disease were an important secondary factor stressed by contemporary investigators. Runs of poor harvests in England in the 1870s and 1880s coincided with fleeting advances of the Mont Blanc tongues and the approach of some of the Icelandic glaciers to their greatest extensions in historic times. Although the Grosser Aletsch was then retreating, the Unterer Grindelwald advanced in 1855-6 (Messerli et al. 1978). Yields fell on upland farms in Britain, land was left uncultivated, especially at the edge of moorland, and the upper limit of cultivated land fell (Parry 1978).

Warming Since the End of the Little Ice Age

Since the end of the Little Ice Age temperatures have risen so that conditions resemble those of the Medieval Warm Period, and are very likely to surpass them, during the next millennium (Houghton et al. 1996). But short term fluctuations have not ceased. Athough values of carbon dioxide and other greenhouse gases continued to increase, global, as well as northern hemisphere temperatures dipped unexpectedly between 1945 and 1976 (Jones 1994), as did sea surface temperatures (Houghton et al. 1996).

Glaciers have withdrawn, and sea ice presence near Greenland and Iceland has diminished. The distribution of some species of fauna and flora has responded. Thus northward extension of bird species led to the arrival of swallows in the Faroe Islands and Iceland in the 1930s, and fieldfares in Greenland and Jan Mayen after 1936. Since 1850 more than fifty species of butterflies and moths

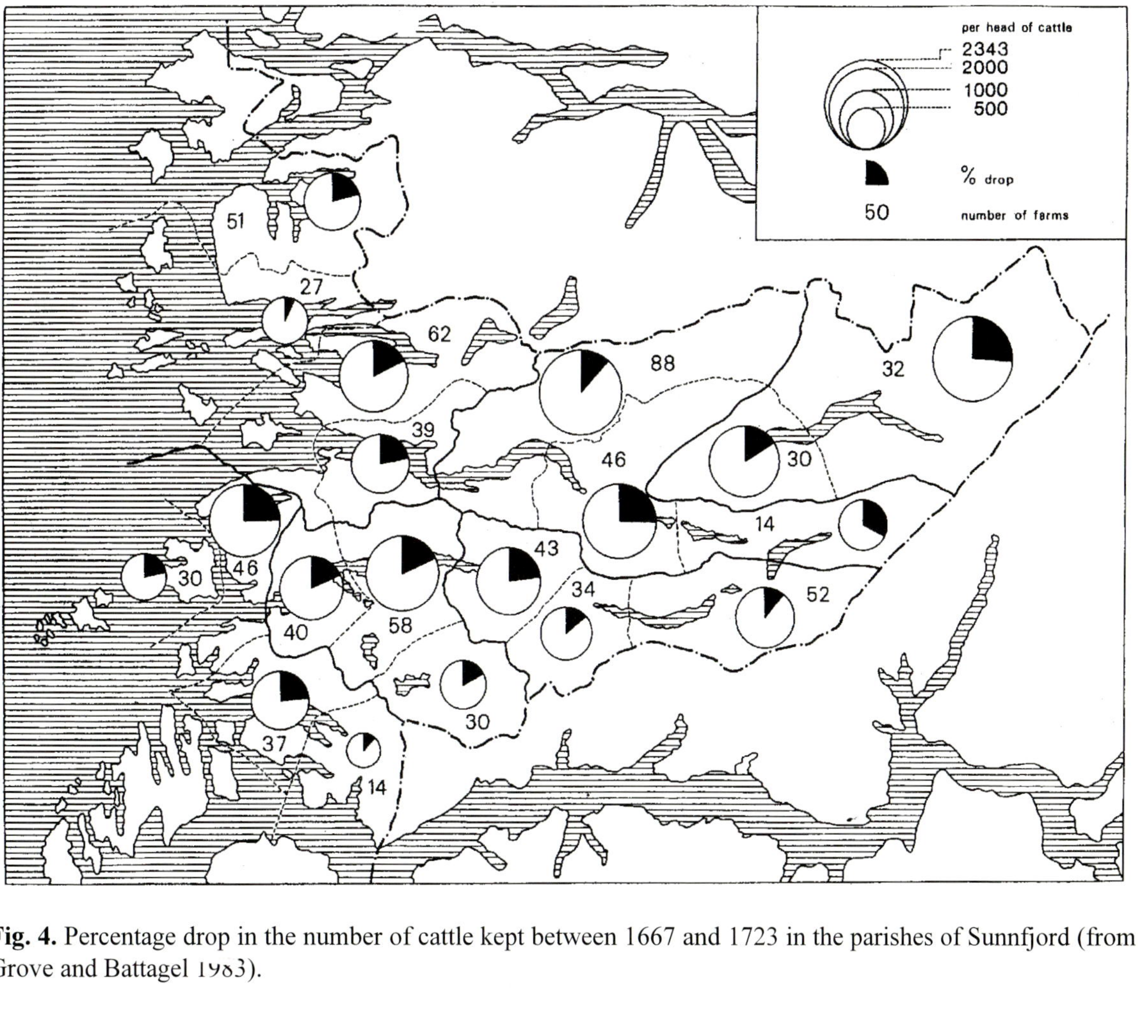

Fig. 4. Percentage drop in the number of cattle kept between 1667 and 1723 in the parishes of Sunnfjord (from Grove and Battagel 1983).

have appeared in the British Isles from further south on the continent. The ranges of animal species increased more than those of plants because of their greater mobility. But movement has not been unbrokenly northward. Cooling, after 1960 was sufficiently marked to cause some temporary movements of terrestial species southwards, while readjustments of ocean currents, such as the incursion of the cold East Greenland current round the south east coast of Greenland after 1950. affected fish catches (Grove 1988, p. 391-4).

The current warm period, has been interrupted, as was the Medieval Warm Period. Short term episodes of more severe weather have been sufficient to cause measurable agricultural and economic consequences in some parts of Europe.

The Impact of a Cold Spell in Iceland During the Twentieth Century Warming

In the 1960s Iceland temperatures dropped and; although no apparent trend in precipitation took place, substantial glacier advances began after 1970 (Fig. 5). Summer temperatures had fallen gradually after 1940, yearly mean temperatures following in the mid-1960s. Average monthly temperatures for 1961-1990 were lower than the corresponding averages for 1931-1961 in all months except February (Jonsson 1991).

The consequences of this cooling, and the associated reorganisation of water masses and ocean currents, were severe for the Icelandic population. After 1960 hay yields dropped in both north and south of the island (Fig. 6, Fridriksson 1969). Livestock production between 1962 and 1982 dropped by 27% per °C below the 1901-1930 mean of 3.2°C. Purchasing and transporting hay in 1965 to make up for the poor harvest cost 14.2 million Icelandic kroner. In 1968 15.1 million kroner were paid in compensation loans, and 4.7 million kroner in hay transport subsidies.

Introduction of modern technology had allowed fish catches to be five times greater in 1960 than in 1920. But reduction of sea temperatures affected the fisheries fundamentally by changing migration patterns, especially of herring, and forcing the fishing fleet to sail much further from home waters, risking the danger of icing of the super-structures of the ships. In 1965 and 1966 the herring catch was over 750,000 tonnes; by 1967 it had fallen to less than 500,000 tonnes, and by 1968 to 150,000 tonnes. This was one of the main factors which the led to devaluation of the Icelandic currency, and was the cause of political friction with Britain over fishing rights in the Icelandic sea area. In a maritime position, vulnerable to climatic shifts, Iceland felt the impact of adverse weather especially keenly because of her dependence on farming and fishing to support her modern urban population. No evidence of social factors which might have caused the difficulties afflicting the farming and fishing communities is to be found.

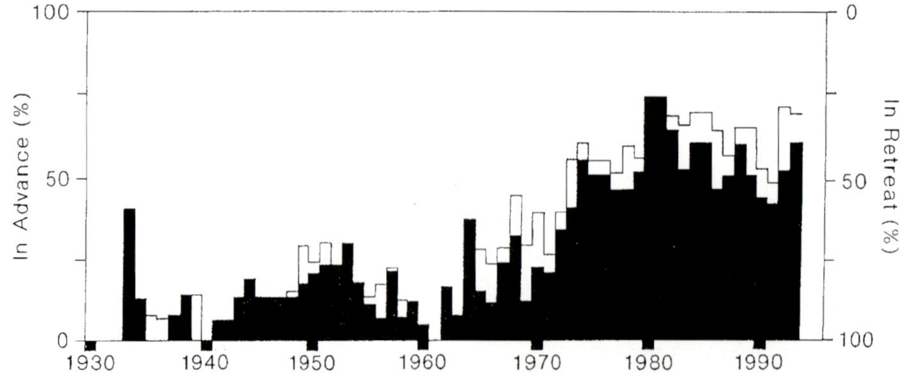

Fig. 5. The percentage of advancing (shown in black), retreating and stationary non-surging glaciers in Iceland 1930-1993. (from Sigurdsson and Jonsson 1995).

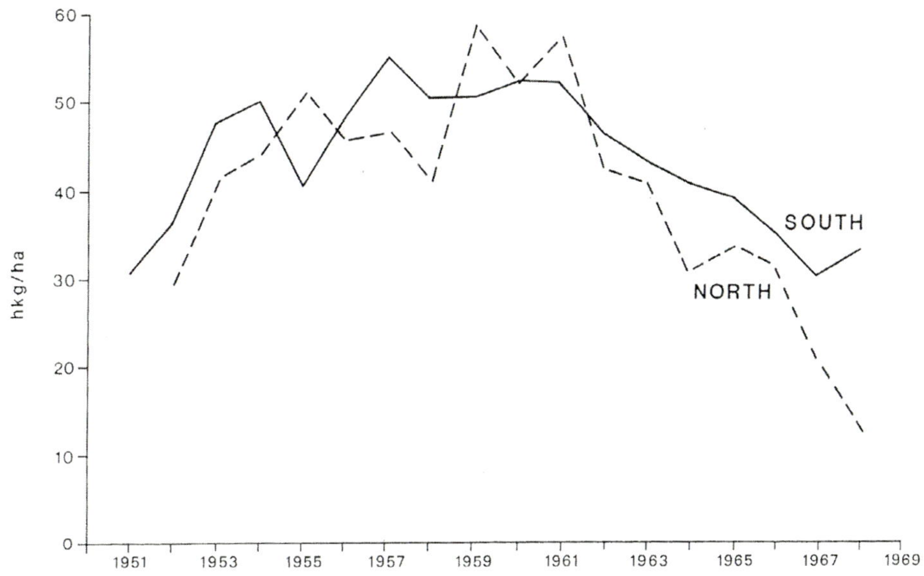

Fig. 6. Average yield of hay during 1951-68 in districts in the north and south of Iceland. From Fridriksson (1969).

The Icelandic glacial advances between the early 1960s and 1990s occurred while the Grosser Aletsch was retreating, but the percentage of the glaciers regularly monitored in Switzerland and Austria found to be advancing each year increased from near zero to over 50% by the early 1980s, though their expansion was on a lesser scale (Grove 1988, p.185, Fig. 6.7). Despite the marked global warming of the twentieth century, oscillations of climate have not ceased. Significantly between the early 1870s and 1980 twelve-month running means of northern hemisphere monthly mean snow cover increased (Matson and Wiesnet 1981), as the Icelandic glaciers extended.

Conclusions

Climatic changes on the century timescale involve changes in mean climatic conditions and in the incidence of extreme events. During the greater part of the last two thousand years it is Little Ice Age type conditions that have prevailed, the Medieval Warm Period having been preceded by two earlier periods of glacial advance. The incidence of extreme events was especially marked during the transition from the Medieval Warm Period to the Little Ice Age and also at times of ice advance within the Little Ice Age. Extreme events, especially if closely spaced, have effects on plant and animal life which inevitably affect human society and economies. Pre-industrial societies are, and were, more susceptible to the deleterious effects of such changes than the post-industrial societies in which we live. Nevertheless, we cannot ignore the possible harmful consequences of greater storminess, more intense rainfall, or longer droughts which could accompany climatic shifts in the future, although the likelihood of further episodes on the millennial scale comparable with the Little Ice Age and its forerunners is now low.

References

Barlow LK, Sadler JP, Ogilvie AEJ,Buckland PC, Amorosi T, Ingmundarson JH, Skidmore P, Dugmore AJ, McGovern,TH (1997) Interdisciplinary investigations of the end of the Norse Western Settlement in Greenland. The Holocene 7: 489-9

Bergthorsson P (1985) Sensitivity of Icelandic agriculture to climatic variations. Climatic Change 7:111-27

Boulton GS, Mason P, Ballantyne CK, Karlen W, Matthews JA, Nesje A (1997) Holocene Glacier Fluctuations in Scandinavia. Paläoklimaforschung, Palaeoclimate Research 24:5-33

Bourke A (1984) The impact of Climatic Fluctuations on European Agriculture. In: Flohn H, Fantechi R (eds) The Climate of Europe: Past, Present and Future.

Dordrecht, Boston, Lancaster, pp 269-314

Dahl SO, Nesje A (1994) Holocene glacier fluctuations at Hardangerjökulen, central southern Norway: a high resolution composite chronology from lake sediments and terrestrial deposits. The Holocene 4:269-277

Fridriksson S (1969) The effects of sea ice on flora, fauna and agriculture. Jökull 19:146-57

Grove JM (1996) The Century Timescale. In: Driver TS, Chapman GP (eds) Time-scales and Environmental Change. London & New York. Routledge, pp 39-87

Grove JM (1988) The Little Ice Age. Methuen, New York, pp 1-498

Grove JM, Battagel A (1983) Tax records from western Norway as an index of Little Ice Age environmental and economic deterioration. Climatic Change 5: 265-82

Harvey BF (1993) Introduction: The crisis of the early fourteenth century. In: Campbell BMS (ed) Before the Black Death: Studies in the „Crisis" of the Early Fourteenth Century. Manchester University Press, Manchester, pp 1-24

Holzhauser H (1984) 'Zur Geschichte der Aletschgletscher und Fieschgletscher', Physische Geographie 13, Geographisches Institüt der Universität Zürich , pp 1-448

Holzhauser P (1997) Fluctuations of the Grosser Aletsch Glacier and Gorner glacier during the last 3200 years: new results. Paläoklimaforschung, Palaeoclimate Research 24: 35-58

Houghton JT, Meira Fililho LG, Callender BA, Harris N, Kattenberg A, Maskell A (eds) Intergovenmental Panel of Climatic Change (IPCC) (1996) Climate Change 1995; The Science of Climate Change. Cambridge University Press, Cambridge pp 1-572

Jones PD (1994) Hemispherical surface air temperature variations: a reanalysis and an update to 1993. Journal of Climate 7:1794-1802

Jonsson T (1991) N'y medaltoel vedurthatta, 1961-1990. Jökull 41:81-87

Karlén W (1990) Glacier fluctuations in Scandinavia during the last 9000 years. In: Starkel L, Gregory KJ, Thornes JB (eds) Temperate Paleohydrology. Wiley, Chichester, pp 394-412

Lamb HH (1977) Climate: Present, Past and Future. II, Climatic History and the Future. Methuen, London, pp 1-835

Le Roy Ladurie E (1971) Times of Feast, Times of Famine. Doubleday, New York, pp 426

Lucas HS (1962) The great European famine. In: Carus Wilson EM (ed) Essays in Economic History. Vol.2. Edward Arnold, London, pp 49-72

Lyons MC (1989) Weather, famine pestilence and plague in Ireland, 900-1500. In: Crawford EM (ed) Famine: The Irish Experience 900-1900. John Donald, Edinburgh, pp 31-74

Manley G (1945) The effective rate of altitudinal change in temperateAtlantic climates. Geographical Review 35:408-17

Manley G (1974) Central England temperatures: monthly means 1659 to 1973. Quarterly Journal of the Royal Meteorological Society 100:389-405

Matson M, Wiesnet DR (1981) New data base for climatic studies. Nature 289:451-5

Messerli B, Messerli P, Pfister C, Zumbuhl HJ (1978) Fluctuations of climate and their geoecological significance, 1600 to 1976. Arctic and Alpine Research 10:247-60

Nesje A, Kvamme M, Rye N, Lovlie R (1991) Holocene glacial and climatic history of the Jostedalsbreen region, western Norway; evidence from lake sediments and terrestrial deposits. Quaternary Science Reviews 10:87-114

Nesje A, Rye R (1993) Late Holocene glacier activity at Sandeskardfonna, Jostedalsbreen area, western Norway. Norges Geografiske Tidsskrift 47:21-30

O'Brien SR, Mayewski PA, Meeker LD, Meese DA,Twickler MS, Whitlow SI (1995) Complexity of Holocene climate as reconstructed from a Greenland Ice Core. Science 270:1962-4

Ogilvie AEJ (1981) Climate and Economy in Eighteenth century Iceland. In: Delano Smith C, Parry M (eds) Consequences of Climate Change. University of Nottingham, pp 54-69

Parry ML (1978) Climatic change, Agriculture and Settlement. Dawson, Folkstone, pp 1-214

Pfister C (1978) Climate and economy in eighteenth century Switzerland. Journal of Interdisciplinary History 9:223-43

Pfister C, Schwartz-Zanetti G, Hochstrasser F, Wegman M (1996) Winter severity in Europe: the fourteenth century. Climatic Change 34:91-108

Pfister C, Luterbacher L, Schwarz-Zanetti G, Wegman M (1998) Winter temperature in western Europe during the Early and High Middle Ages (AD 750-1300). The Holocene 8:535-52

Sigurdsson O, Jonsson T (1995) Relation of glacier variations to climatic change in Iceland. Annals of Glaciology 21:263-70

Utterström G (1955) Climatic fluctuations and population problems in early modern Europe. The Scandinavian Economic History Review 3:3-47

Human Stature and Climate: The Impact of Past Climate on Living Standards

Joerg Baten

Department of Economics, Economic History, University of Munich, Ludwigstr. 33/IV, D-80539 München, Germany
e-mail: Joerg.Baten@econhist.vwl.uni-muenchen.de

Abstract: This contribution reviews the literature on anthropometric history that employs human stature as an indicator for nutritional status. The possibilities and problems of using this variable to measure the climatic impact on human history are considered. I argue that European climate in the 18th century significantly influenced nutritional status as measured with anthroprometric indices on a quinquennial or decadal basis. There was a causal chain leading from warmer or colder winters to higher or lower grain and protein production, and these factors determined real wages and finally human stature, only slightly modified in central Europe by the new production technology of potato growing.

Introduction

What is the significance of past climate for human history? I would think that the main influence on history consists of the impact via agricultural production on the living standard of the population. During most of human history, agricultural production represented a vast proportion of total production. Even in 19th century central Europe, some two third of median household expenditures was spent on food. In the time period before the mid-20th century, and in many respects even today, climatic development was an exogenous factor that influenced living standards.

But how can we measure historical living standards? One possibility is to extrapolate backwards national income accounts. But even if much progress has been made in this area, the lack of production data poses enormous problems. Many economic historians concentrated on the estimation of real wages, i.e. the income of the dependently employed part of the population. As wage rates can often be reconstructed for the time back to the middle ages and prices - at least for grain - were recorded on the market places for the same time period, this method yields interesting results. However, it gives little information about the self-employed - farmers, noblemen, master craftsmen etc. - and special groups such as women heads of households, slaves and many others. In addition, information about unemployment and particularly underemployment is not available, so the wage rate is only a rough approximation of real earnings.

These techniques of measuring human welfare have been supplemented during the last two decades by anthropometric research. As biologists and anthropologists found that the average human stature is to a large extent determined by the quality of nutrition and by the disease environment, hundreds of thousands of individual height measurements have been collected in large data bases to provide evidence about these aspects of human living standards. The aim of this article is to report the possibilities and problems of this measurement approach and to review the literature about the climatic impact on human heights. It should be noted, though, that the direct discussion of the climatic variables is not yet wide-spread among anthropometric historians, the main focus in this literature has been on (a) establishing reliable time series and cross-

From WEFER G, BERGER W, BEHRE K-E, JANSEN E (eds), 2002, *Climate Development and History of the North Atlantic Realm.* Springer-Verlag Berlin Heidelberg, pp 327-338

sections of humans and heights and (b) comparing those with income estimates and (c) putting them into the context of economic history debates, such as the discussion about the positive or negative impact of the industrial revolution on 19th century worker's living standards or the living standards of American slaves. But as this line of research is relatively new and has important advantages for measuring the impact of climate on human history in the long run, a review of the existing evidence could be an important contribution to this interdisciplinary volume. In addition, there are reasons why the impact of climatic variables on height should be explored in greater detail. The most critical component of the quality of nutrition, the animal protein content, cannot be measured until the late 19th century. As the literature reported below demonstrates, the climatic impact of this part of agricultural production was particularly strong. Therefore, a direct comparison of weather conditions and height reveals interesting findings that are otherwise impossible. I will argue that European climate in the 18th century significantly influenced nutritional status as measured with anthroprometric indices on a quinquennial or decadal basis. There was a causal chain leading from warmer (or colder) winters to higher (or lower) grain and protein production, and these factors determined real wages and finally human stature.

In the following, I will start with a discussion of available sources for estimating historical heights, then proceed in part 2 with reviewing some aspects of the literature on the climate-food-production relationship that relate to anthropometric history. Part 3 presents some of the results of anthropometric research and climatic conditions and the fourth section discusses the possibility of a direct climatic influence on human stature, without the linkage via agricultural production. Part 5 contains some considerations about possible future lines of research.

Height Data and its Estimation Problems

The majority of height data are of military origin. Military officers were always interested in the height of their soldiers or recruits, because some military abilities correlate with heights (e.g. shooting from a better position), and they wanted to catch their soldiers when they deserted. Additional valuable anthropometric information was collected in schools, prisons, hospitals, and in some special institutions (e.g. the slaves shipped between the American regions were all measured). Given the predominance of military sources in anthropometric research, the possibilities and problems of this type of information should receive the most attention.

The great advantage of military sources is their availability. Back into the 17th century, many soldiers were measured in a more or less accurate way. Moreover, their age or year of birth was always recorded. When large numbers of individual measurements are collected, problems such as heaping on even inches and similar phenomena average out. One important problem, however, is the minimum height requirement applied by many armies. The other main problem is selectivity in collecting the recruits from the population, as the researcher is not interested in the average height of the army, but the stature of the underlying population. We discuss the height requirement problem first.

The armies of Bavaria and Palatinate in the 18th century, e.g. published extremely high minimum height requirements (167.7 cm for both infantries). To make things more difficult, other army categories had separate height requirements: 175.1 cm for the elite groups, the so-called Grenadiers. Statistical methods ("kernel density estimates") have been developed to test whether this extremely high minimum height requirement was actually applied. It turned out in this example that the actual truncation point was estimated as 165.2 cm (instead of 167.7 cm). Later, the height requirement was substantially lowered, therefore the simple average does not mirror the true height trend. For the estimation of height trends and means, specialized techniques were developed: The Komlos and Kim (1990) method and the 'Reduced sample maximum likelihood estimator' (RSMLE). With the former method, the height distribution of every birth cohort is truncated at the highest minimum height requirement ever applied during the period under consideration. Then, the simple mean of the reduced sample is calculated, which is a constant function of the mean of the underlying 'true' distribution over time and across regions, assuming that the variance does not change dramatically. This characteristic was mathematically proven by Heintel et al. (1998). The resulting values are higher than the true mean and should be interpreted as an index. The advantage of the K&K method is its robustness against dis-tortions of the sample distribution. The second method, the RSMLE, enables regression of height on exogenous variables (e.g. dummy variables for age), and reports unbiased means, but it only yields reliable results if the truncation point is far enough below the mean (see Heintel et al. 1998). This method also cuts off all observations below the truncation point, but uses iterative maximum likelihood methods to find the most likely constants and coef-ficients for socioeconomic groups within the sample, taking into consideration the truncated distribution.

The second, and even more difficult problem concerns selectivity among soldiers, if volunteer armies provide the material. This occured in most European countries during the 17th and 18th centuries. Great Britain and the U.S. kept the system until the 19th and 20th centuries. If no general military duty existed, conditions on the labor market have to be discussed and the social composition of the soldiers (or, similarly, of prison inmates) have to be compared carefully with the underlying population.

For example it has been possible to ascertain that the downward trend in Bavarian heights between 1750-54 and 1770-74 was not caused by labor market conditions, because one would expect the falling real wages to increase the willingness of the better nourished part of the population to join the army, as their opportunity costs would decrease. Real wages declined during this period, therefore the selection bias arising from this economic development would have caused heights to increase, not to decline. Another biasing influence on volunteer armies might be the incidence of war. During major wars, people who have the choice prefer not to volunteer for armies, and the recruitment officers pay higher wages and premia and accept lower "quality" recruits. In the Bavarian example, the country was not directly involved in military conflicts during 1760-87 but, to give another example, the soldiers of the East India Company became suddenly shorter in 1857/58, when the Great Sepoy Revolt in India started. In sum, these data problems limit the possibilities of anthropometric research particularly in the 18th century.

Another problem are the weights that should be assigned to the environmental (i.e. here: climatic) impact on each year of the individual's growth period (from age 0 to 25 at that time). Baten (2000) argued that for adults the strongest impact is given during the very first years of life, whereas the year before measurement plays a crucial role for children and adolescents who are still growing.

These problems have been mostly solved during the last few years, but they make an annual comparison with climatic indices for the 18th century problematic: annual aggregation of the height data does not average out the random genetic variation to a sufficient extent. Instead, investigation of the climatic impact of this period should focus on the medium and long run. For the 19th century by contrast, the general service duty of all young males (except small groups of less than 3 percent) makes comparisons at the annual level more feasible.

Climate and Nutritional Status

Having introduced aspects of the dependent variable, its relationship with climatic factors can be considered. The relationship between climate and food production is naturally complicated. Pfister (1988a) demonstrated how problematic climatic extremes are for agriculture, both in terms of temperature and humidity. The monthly timing is very important; how warm and wet was it during the months of winter grain harvest, summer grain sowing and hay cutting? If, e.g. late snowfall was combined with a rise in temperature in May over 8°C, the threshold and interaction effects are also important.

Baten (1999) employed a simplified model of the relationship between climate and agricultural production. He first considered the climatic determinants of animal protein production that played a role in the nutritional experience of early modern societies, which has often been underestimated. Winter temperature is an important control of milk and meat production in Central, Eastern and Northern Europe.

Pfister (1988b: 38-39; see also Glaser 1991) has isolated four criteria that influenced the production of animal protein significantly in early modern Europe. Those relate to milk production in particular, but similarly apply to meat production.

1. The length of the period during which the pastures were snow covered.
2. When did temperature reach 8°C so that meadow grass began to grow, and when did growth stop in the autumn?
3. When was grain harvesting completed in the autumn?
4. How humid was it during hay cutting in summer?

The duration of snow cover - a product of both temperature and humidity - is decisive for the mortality of cows, because in favorable years the cows were taken out of the stables by March, and were quite healthy again in April and May. But in bad years the fodder stored for cattle feed did not last until the pasture grasses were growing again in April or May. In these circumstances the cows either died or their milk yields were diminished.

In Autumn the cattle were stabled as soon as the temperature fell to about 5°C. While the cattle were outside cut grass could be kept for winter. The last factor, the date of the grain harvest was related to customs in early modern European agriculture; after the harvest, the village community was allowed to bring its cattle on the fields to graze. If the grain harvest was later than normal - typically caused by low temperatures in April to June - this advantage was reduced. These four variables influencing animal protein production (both dairy products and beef) tended to be negatively correlated with winter temperature. The colder the winters and the longer a period of cold winters, the more difficult it was to maintain the cattle stock. Obviously, a single hard winter (as for example in 1740) by itself after a year favorable for roughage production did not affect the cattle stock as seriously as a series a cold years.

The fourth influence on the amount and quality of hay, the roughage for the winter period, was influenced by humidity in summer. If the summer was extremely rainy, the farmers await a dry period before cutting the hay. But if they waited too long the hay was of low quality, because too much protein was used for plant growth. If it started raining after the hay was cut, the water washed out the nutrients. Fortunately for the parsimony of the model, Glaser (1991) and Pfister (1988a) stressed the point that winter temperature and humidity are strongly correlated. The residual of regressing temperature on humidity must be left to the error term of the relationship between winter temperature and gross nutritional status for now.

Glaser (1991) reported that cold winters and short growing seasons also had a negative impact on grain harvests. Cold winters and short growing seasons also affected the manure produce, which was crucial for grain production. Moreover rye tended to develop a poisonous mold in cold winters as Matossian (1984) argued. Grain production was also affected by many other factors, including humidity during the harvest in July and August, and the date at which grain was sown in the autumn, and by drought as well as hail. As the last two factors do not correlate with winter temperature, the calories and vegetable protein available were not necessarily correlated with animal protein production. Thus substitution effects could buffer extreme cases of underproduction of either animal protein or grain.

Many sources recording grain production are available, but past production of animal protein, especially milk is poorly documented. Grain production was very important for urban populations, while meat was often regarded as 'luxury'. Milk, on the other hand, played an extremely important role in rural nutrition, but the amount produced was not recorded, because the peasants were not obliged to pay taxes on milk, and there was no other occasion when milk output could have been recorded. Insofar as climate influenced milk and meat production, using climatic data might give us a clue to animal protein production of early modern Europe.

The connection between climate and food production has been investigated for an earlier period (Pfister 1988b). Between the second and the last third of the sixteenth century, average spring temperature declined in Switzerland by 0.8°C, and average humidity in summer increased by 20 percent. Pfister has argued that milk production practically collapsed during this period. As the amount of roughage was reduced and the grazing season shortened, he estimated milk output to have been reduced by more than 50% (in the range of 50 to 200 million liters annually). For a population of 900,000 this meant a loss of 60 to 210 liters per capita. Grain production also declined by something between 10 and 33,000 thousand tons (10 to 33kg per capita per year, about 10-15 percent of output, Pfister 1988b:84-95). Even if these margins of error are quite large, the effect of climatic deterioration on milk production appears to have been more dramatic than that of grain. Although he was concerned with hay production for sheep and not with milk production, the results of Bergthorsson (1985: 113-123) are interesting in this context. He demonstrated for 20th century Iceland that hay yields were closely dependent on air temperatures both during the growing season and especially during the previous winter. Eckstein et al. (1982: 28) found a relationship between Swedish yields and winter temperatures 1756-1869, while autumn, summer and spring temperatures do not display this cor-relation. Galloway (1986) provided an excellent overview of the wider literature. Agriculture in Switzerland, Iceland and southern Bavaria were probably affected particularly strongly by climatic conditions on account of their dependence on animal products, but other European countries north of the Alps also suffered from deteriorating climate over a longer period of time. E.g. Galloway (1986:7) demonstrated the effects of temperatures on long run population growth, especially in the cold 14th and 17th centuries. It seems from this discussion that winter temperature may have had a crucial impact on protein production and human stature, particularly in the cold and mountainous parts of central, eastern and northern Europe.

Climate and Stature in the 18th Century

Southern Germany

Height estimates from several regions in central, eastern and northern Europe will now be compared with indices of winter cold. Baten (1999) found that there was an upward trend between the birth cohorts of 1725-29 and 1750-54 in Bavaria, Palatinate and Upper Palatinate, and a downward trend

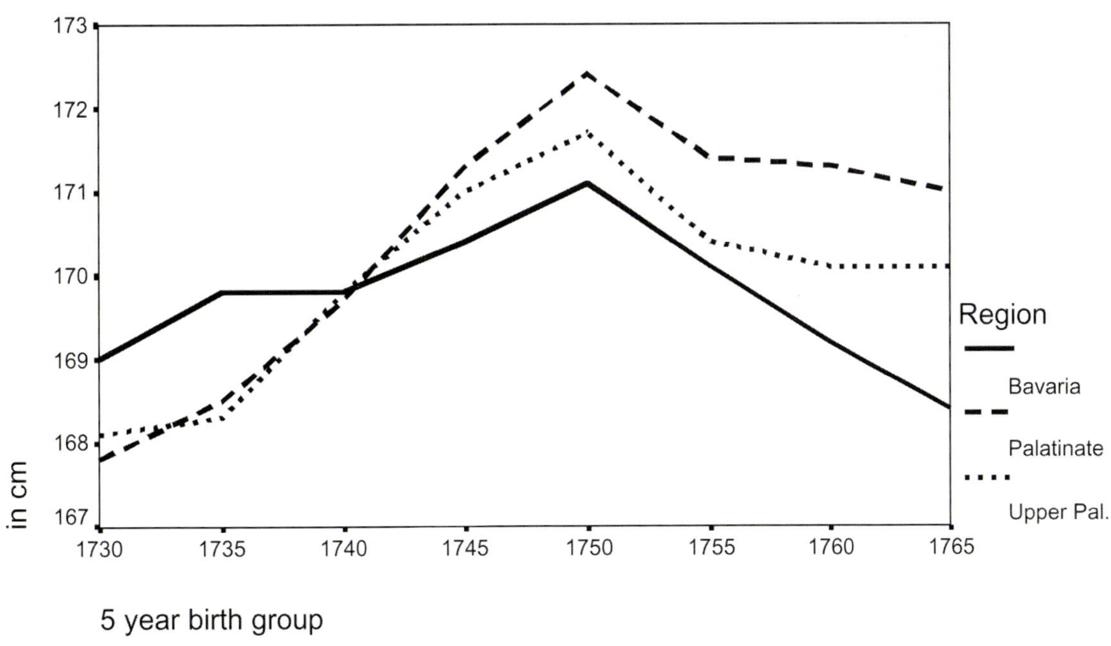

Fig. 1. Height trends in Southern Germany. Note: The level is higher than in Fig. 2, because adjustments for army categories were made. Fig. 2 represents only non-elite soldiers.

thereafter (Fig. 1). The region of Bavaria had the tallest soldiers during the 1730s, while in the last two birth cohorts soldiers from Palatinate rank at the top. The decline in Palatinate heights between 1750/54 and 1765/69 of about 1.4 cm is modest, compared to the decrease of Bavarian heights of 2.7 cm. Even if we take into account an error margin of 0.5 centimeter, the difference is considerable. After the 1770s, heights recovered somewhat in Bavaria (Fig. 2). The K&K method generally confirms these results.

How do these trends in heights compare with climatic data? As there are no temperature series for Bavaria before 1781, estimates from Pfister (1988a) for Switzerland serve as a proxy. In general, winter temperatures were rising up to 1745-49, followed by a dramatic downturn until 1755-59. Winters remained cold until the temperatures recovered in the late 1780s. If this time series is compared with the estimates for food production, real wages and heights from Baten (1999), the importance of lags becomes apparent. In favorable years, stocks of cattle and grain increased that enabled the population to maintain nutritional status through a series of adverse years. While the cattle stock was being reduced, additional protein (in meat) was made available for consumption. However, when climate became favorable again, it took some time to build up cattle stocks, especially in economies with low levels of interregional market integration. Therefore, Baten (1999) assumed that the average of winter temperatures in the current and the past five-year-period had an impact on nutritional status. This time-lag makes sense in terms of protein production, as cows typically live for five to ten years, depending on breed. While those are modern figures, this life-expectancy was broadly similar since prehistoric times. The same is true for oxen, the main draught animal in southern Germany, an important item in the capital stock of the grain producing farmer.

The trend in winter temperatures and heights is similar (Fig. 2). The warm winters between the 1730s and the early 1750s (even with the exceptionally cold year of 1740) enabled Bavarian farmers to increase their cattle stock and grain produc-

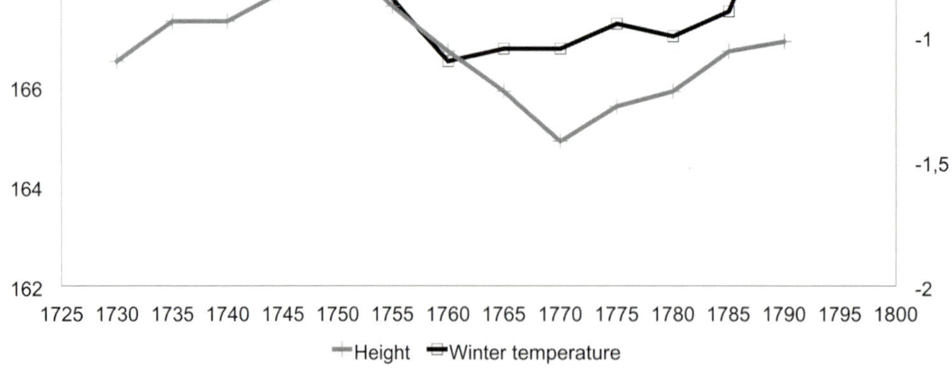

Fig. 2. Height in Bavaria and Swiss winter temperature. Note: Only non-elite-soldiers represented. Winter temperature is the average of the current and previous quinquennial.

tion, while the cold years after mid-century had a negative impact on heights. The extremely cold year of 1771/72 that paralleled that famine crisis of the same years is not evident, as the surrounding years were relatively warm. Thereafter, the relationship is less close. The famine crisis promoted the spread of the potato. This agricultural innovation shifted the food supply, even if the temperatures rose only slowly. This reminds us of the possibilities to solve shortage problems; possibilities that dampened the impact of climatic factors on the agricultural sector.

Austria-Hungary and Sweden

How do these trends in heights compare to other European countries? In brief, they are quite similar. Komlos (1989) estimated trends for five regions of the Habsburg monarchy. He found an upward trend in all five cases until the 1740s, and declining heights thereafter (Fig. 3).

In Sweden, heights increased until the 1740s and then declined until the 1780s, thereafter starting a recovery as Heintel et al. (1998) estimated (Fig. 4). Not shown in the figure is an especially strong increase in this cold northern country in the warming period between the 1800s and the 1820s. Winter temperature in Sweden (calculated, again, as the mean between current and previous five year period) increased until the early 1740s and started a strong decline thereafter. Temperatures fell from –1 to –5 degrees Celsius. The 1790s were somewhat warmer, partially explaining the recovery of heights.

England

For England, the downward trend after the middle of the 18th century is also visible (Fig. 5), although it seems to have started later. British height trends, especially those in the 18th century, have been the subject of debate. Komlos' (1998) re-estimation of Floud et al.'s (1990) army data set displays the same declining trend until 1800 as related estimates of a sample of Marine Society boys in London (Heintel and Baten 1998; Floud et al. 1990). The maritime

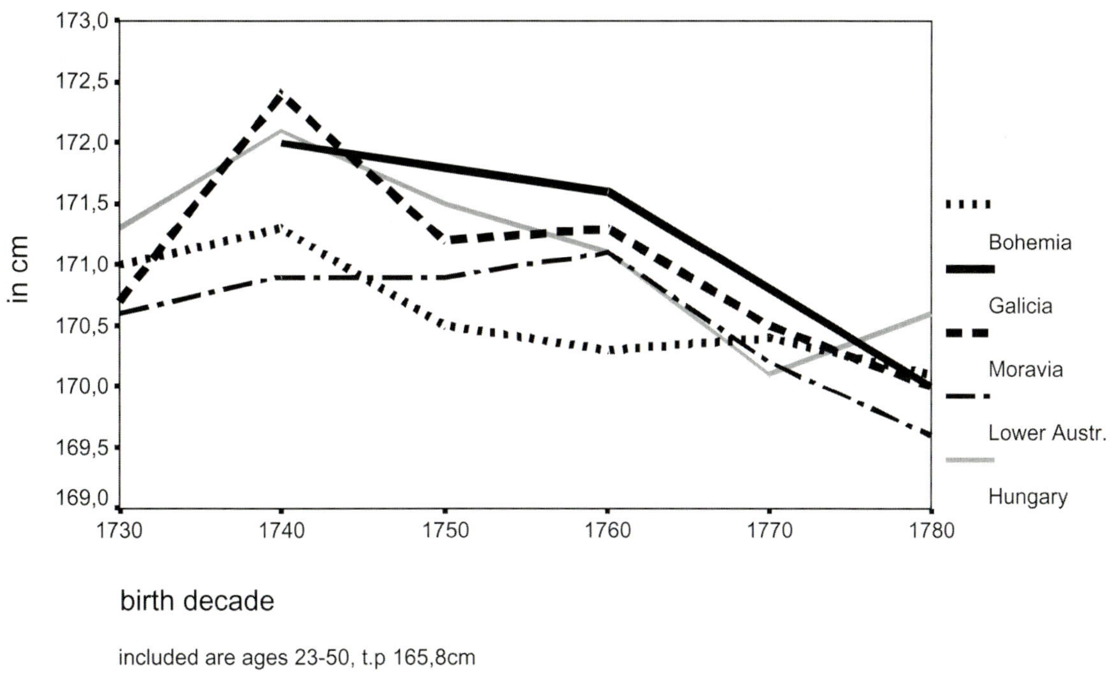

Fig. 3. Trends in height by region, Habsburg soldiers. K&K estimates (=index numbers, not true heights). Source: Komlos (1989).

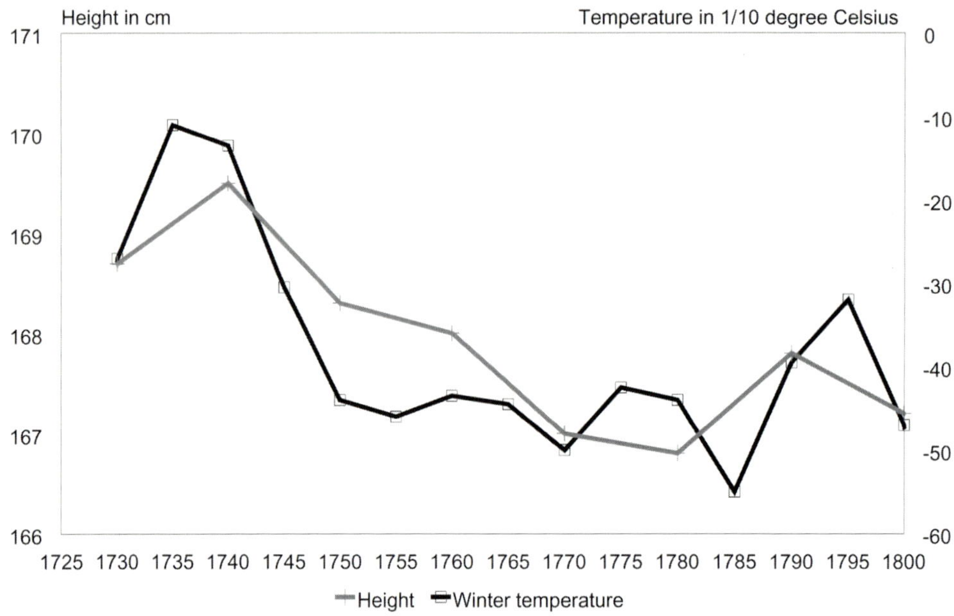

Fig. 4. Height and winter temperature in Sweden. Sources: Height: Heintel et al. (1998); Ages 21-49. Temperature: Moberg and Bergström (1997).

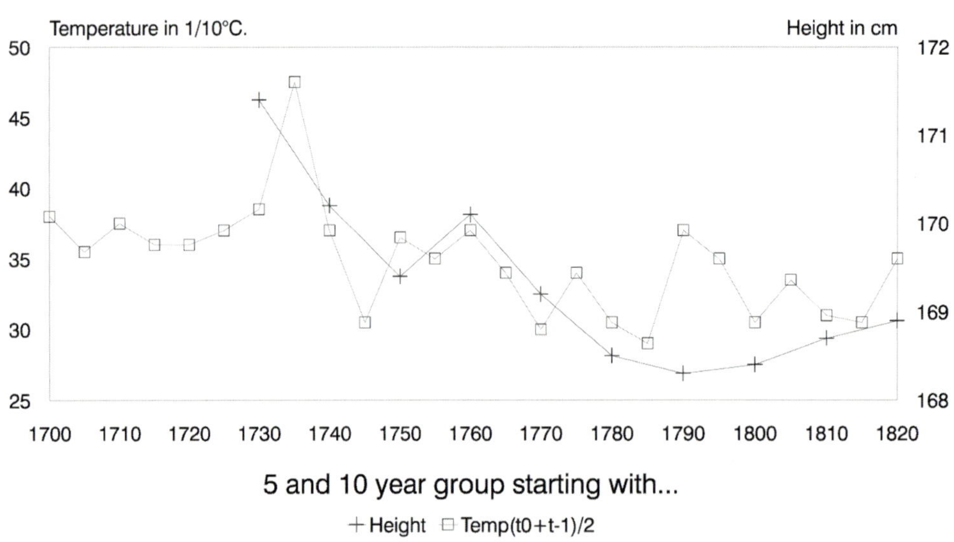

Fig. 5. Height trends and winter temperature in England, 18th century.

climate of the British Isles differs from the continental climate of Austria-Hungary, Switzerland and Bavaria. January temperatures in central England from 1781 to 1850 were about 4°C warmer than at De Bilt in the Netherlands. The variability of temperatures in England was also substantially smaller, as can be easily recognized by comparing maxima and minima. The warming trend until the 1730s, also favored better harvests in England and real wages clearly rose. It might also be the case that the Swiss climatic data is more accurate. The temperatures for England are contemporary instrumental measurements from a single source and the temperature data on Switzerland are estimated by a much more comprehensive proxy approach, see Pfister (1988b) for details. When, after the 1730s, temperatures fell once again, real wages began to fall, and continued with minor interruption until the turn of the century. Declining real wages also led to a decline in heights in 18th century England. In contrast to Southern Germany, English heights did not recover at the century's end, but fell again until the decade of 1800-09.

In summary, the following can be concluded with regard to England:

1. An influence of winter cold on real wages is evident, but it is much weaker than in southern Germany or Austria-Hungary.

2. There existed in England - for the seven decades after the 1740s for which data are available - a relatively close connection between real wages and heights.

3. A direct influence of climate on the decline in heights after the 1740s in England is far less obvious. The 1790s, especially, were relatively warm; heights, however, nearly reached a nadir. The French wars probably contributed to low real wages around the turn of the century.

Additional factors explaining the weak relationship between winter temperature and height come to mind. Within the climatic variable set, precipitation might have been more important than temperature. Another possible factor is the foreign trade in food. The foodstuff markets of the north European coastal countries were integrated by the 18th century (see Baten 1999 for a detailed discussion). After the climate had once more become unfavorable from the 1740s onward, England changed from foodstuff exporter to net importer (additional, especially demographic variables were also very important). From the mid-18th century onwards, England paid its food imports with the income from her industrial exports. Because of favorable sea transportation possibilities, the country could be supplied with grain from the Baltic Sea area and with livestock products from Ireland. In an investigation of the connection of climate and nutritional status in Great Britain, therefore, the climate in the Baltic area, as well as the climates of England and Ireland are important. We would expect that this would be reflected in English real wages, because when the climate worsened in grain-producing regions, the cost of living in England went up. Actually, data on the freezing of the Baltic Sea show that from the beginning of the 1750s to the turn of the century, the Baltic area was becoming cooler (Koslowski and Glaser 1995:84-85). The most extensive freezing of the Baltic Sea took place around 1800, when both heights and real wages in England reached their historic nadir. After the catastrophic 1790s and 1800s, the freezing of the Baltic decreased until the 1820s, and real wages and heights both recovered in England. If one includes for England not only the effect of domestic climate, but also the winter temperatures in the grain supplying countries, then a negative influence of winter cold on heights becomes evident even in the country that was the most industrially developed at the time.

Is There a Direct Influence of Temperature on Height?

A direct influence of climatic factors on the growth of children (independent of its effect on food production) cannot be ruled out fully, as the investigation of this issue would necessarily use the same variables as those in the climate - food production - height causal chain described above. Yet the direct influence of climate on physical stature is not revealed by recent anthropological literature and cross-sectional evidence (see, e.g. J.M. Tanner 1990: 144-146, who only discusses very long-term effects of climate on body-shape). Given similar nutrition, people in northern Europe or in the Bavarian Alps would have been much shorter than people living in warmer areas according to this argument, but, this was certainly not true in the 18th or 19th centuries. At the current stage of research, one cannot reject the hypothesis that the disease environment or physical exertion contributed to the cycles in height, but there is no evidence supporting their (independent) role for the 18th century.

Possibilites for Future Research

I would like to draft some lines of further work in this field, because the discussion of research possibilities on an interdisciplinary conversation such as this Hanse volume might prove very valuable. I reviewed above the anthropometric literature on the 18th century and compared the quinquennial or decadal height estimates with winter temperature. However, the relationship between climate and height on an *annual* level has not yet been clarified in the existing literature to the best of my knowledge.

Especially for the 19th century, large data bases have been collected that allow height estimates on an annual level. The problem of selectivity among conscripts does not exist in many countries and minimum height requirements often play a small or negligible role. Age was always given.

Even if many of these studies are not without data and estimation problems, there is a lot of material available for comparison with climatic indices. But how could the impact of climate be modelled on an annual level? A good starting point would certainly be studies that estimated the weather impact on annual agricultural production and food prices.

The classic contribution in this field is certainly Pfister (1988a). He simulated the impact of temperature and precipitation on the price of grain. The climatic data in this study consisted of an index series (coded as -5-,-3,-1,0,1,3 and 5) from 1550 to 1980. Each index value represented a fixed part of the temperature and precipation distribution of 1901-60 (Pfister 1988b). Larger weights were given to particular months.

Higher temperatures typically lowered the grain price, particularly in

- September (weight 3) and October (weight 1)
- March and April (both weight 1)
- May to August (each month has weight 1)

Precipitation on the other hand increased the grain price, again in some special months:

- September/October (each w. 1)
- December (w. 2) and January (w. 3)
- March (w. 2), April (w. 2), May (w. 1)
- July (w. 4) and August (w. 1)

In effect, all monthly temperatures were included except November and December, September did get a higher weight. Precipitation is weighted in a

Author (year)	Region	Time	Sample	Remarks
Twarog (1997)	Württemberg	1852-1893[a]	soldiers	truncated, 6-yr-mov. avg.
Drukker/Tassenaar (1997)	Netherlands	1818-1940	conscripts	
Weir (1997)	France	1817-1922	conscripts	
Baten (2000)	Bavaria	1834-63	conscripts	15 regions in database
Baten (1999)	Bavaria	1829-51	conscripts	200 regions in database
Baten (1999)	Bavaria	1820-80[a]	convicts	male and female
Mironov (1997)	Russia	1866-1914	soldiers	
Mironov (1997)	Centr. Russia	1875-1980	n.a.	
Komlos/Huiber (unpubl.)	Aus.-Hungary	1870-1913	Conscripts	yet unpublished (new project)
Quiroga (1998)	Spain	1893-1954	conscripts	17 regions
Baten et al. (unpubl.)	Germany	1920-60	children	yet unpublished (new project)

[a] birth years (years of measurement otherwise)

Table 1. Anthropometric studies on 19th and 20th century NW Europe that report annual heights for 25 or more years slightly more complicated manner. In addition, extreme temperatures and precipitations were weighted more strongly than milder deviations from the mean, because the original -3/+3 index was translated to a -5/+5 index. Comparing this model with price changes - expressed as deviations from a 35 year moving average - Pfister achieved a high explanatory power with his model, particularly if the military conflicts that increased grain prices exogenously are taken into consideration. The large advantage of this model is its monthly structure that allows greater weights for some crucial months. Also, climatic extremes get higher weights than smaller deviations from the climatic mean.

Another approach was chosen by Khatri et al. (1998). The authors seek to explain the annual agricultural output in Britain between 1867 and 1913 with weather variables and economic inputs, such as labor, capital and land, achieved from a multitude of sources. The latter turned out as mostly insignificant, perhaps because those variables had to enter the model to avoid spurious regression effects from non-stationary data. Nevertheless, in their best model the authors were able to explain 55% of variation in agricultural output by considering annual rainfall and temperature alone. One might criticize that annual data do not reflect the different monthly weights that Pfister used in his model. In particular the last months of the year (after the harvest) have more influence on next year's grain production. The strength of this paper is its sophisticated econometric technique that uses semiparametric estimation techniques. The model of Khatri, Solomou and Wu does not assume sym-metric effects on both sides of the distribution. The authors find the positive rainfall extremes to be more influential than extreme drought years. Very low temperatures (below 8.5 degree Celsius) are making agricultural production difficult, but also temperatures around 9-9.5 degrees Celsius were suboptimal. In contrast, very warm weather (>10 degrees) was beneficial and temperatures around 8.5 to 9 degrees turned out to have positive effects, too.

Comparing the results of these different models with the anthropometric series would be a worthy undertaking for future research. However, before this attempt can be made, a more detailed discussion of the highly non-linear and monthly nature of the short-term effects of climate on food production and heights seems to be necessary.

Acknowledgments

I wish to thank John Komlos, Christian Pfister, one referee, and members of the Munich ECONHIST research group for invaluable help in starting this line of research. I am indebted to Peter Waugh and Ursula Gütt for helping with the data collection and to Rüdiger Glaser for providing data on European climate. Research support by the DFG (German Science Foundation) is gratefully acknowledged.

References

Baten J (1999) Ernährung und wirtschaftliche Entwicklung in Bayern, 1730-1880. Steiner Stuttgart

Baten J (2000) Heights and Real Wages in the 18th and 19th Centuries: An International Overview. Jahrbuch für Wirtschaftsgeschichte

Bergthorsson P (1985) Sensitivity of Icelandic agriculture to climatic variations. Climatic Change 7: 111-127

Drukker JW, Tassenaar A (1997) Paradoxes of Modernization and Material Well-Being in the Netherlands during the Nineteenth Century. In: Steckel R, Floud R (eds) Health and Welfare during Industrialisation. Chicago and London 1997, pp 331-377

Eckstein Z, Schultz TP, Wolpin KI (1982) Short Run Fluctuations in Fertility and Mortality in Preindustrial Sweden. Center Discussion Paper No. 410, Yale University Economic Growth Center

Floud R, Wachter K, Gregory A (1990) Height, Health and History. Nutritional Status in the United Kingdom, 1750-1980. Cambridge Univ Press, Cambridge

Galloway PR (1986) Long-Term Fluctuations in Climate and Population in the Preindustrial Era. Population and Development Review 12(1):1-24

Glaser R (1991) Klimarekonstruktion für Mainfranken, Bauland und Odenwald anhand direkter und indirekter Witterungsdaten seit 1500. Fischer Verlag, Stuttgart/NY

Heintel M, Baten J (1998) Smallpox and Nutritional Status in England, 1770-1873: On the Difficulties of Estimating Historical Heights. Economic History Review 51(2):360-371

Heintel M, Sandberg LG, Steckel RH (1998) Swedish Heights Revisited. In: Komlos J, Baten J (eds) The Biological Standard of Living in Comparative Perspective. Stuttgart, pp 449-458

Khatri Y, Solomou S, Wu W (1998) Weather and Fluctuation in Agricultural Output: 1867-1913. Research in Economic History 18:83-102

Komlos J (1989) Nutrition and Economic Development in the Eighteenth-Century Habsburg Monarchy: An Anthropometric History. Princeton

Komlos J (1998) Shrinking in a Growing Economy? The Mystery of Physical Stature during the Industrial Revolution. Journal of Economic History 58(3):779-802

Komlos J, Kim JH (1990) Estimating Trends in Historical Heights. Historical Methods 23(3):116-120

Koslowski G, Glaser R (1995) Reconstruction of the Ice Winter Severity since 1701 in the Western Baltic. Climatic Change 31:79-98

Matossian MK (1984) Mold Poisoning and Population Growth in England and France, 1750-1850. Journal of Economic History 44:669-686

Mironov B (1997) Welfare and Economic Growth: Height of New-Born Children and Adults in Russia in 1855-1980. Contribution to the Munich Conference on Biological Standard of Living, Jan. 1997

Moberg A, Bergstroem H (1997) Homogenization of Swedish Temperature Data. Part III: The Long temperature Records from Uppsala and Stockholm. Intern J of Climatology 17:667-699

Pfister C (1988a) Fluctuations Climatiques et Prix Céréaliers en Europe du XVIe Siècle. Annales ESC 1:25-53

Pfister C (1988b) Klimageschichte der Schweiz 1525-1860. Das Klima der Schweiz von 1525-1860 und seine Bedeutung in der Geschichte von Bevoelkerung und Landwirtschaft. 3rd edn, Bern/Stuttgart

Quiroga GV (1998) Height Evolution in Spain, 1893-1954. An Analysis by Regions and Professions. In: Komlos J, Baten J (eds.) The Biological Standard of Living in Comparative Perspective. Stuttgart, pp 359-383

Tanner JM (1990) Foetus into Man: Physical Growth from conception to Maturity. Cambridge

Twarog S (1997) Heights and Living Standards in Germany, 1850-1939: The Case of Württemberg. In: Steckel R, Floud R (eds) Health and Welfare during Industrialisation. Chicago and London, pp 285-330

Weir D (1997) Economic Welfare and Physical Well-Being in France, 1750-1990. In: Steckel R, Floud R (eds) Health and Welfare during Industrialisation. Chicago and London, pp 161-200

Malaria Around the North Sea: A Survey

Otto S. Knottnerus

Uiterburen 47, NL-9636 EC Zuidbroek, The Netherlands
e-mail: ottoknot@xs4all.nl

Abstract: Malaria may have been introduced into the North Sea Basin in late Antiquity. It has been endemic at least since the 7th century, but its high-days were the Little Ice Age. After 1750 the disease steadily declined until it disappeared in the 1950s. The hot beds of malaria were largely restricted to brackish coastal zones, where the mosquito *Anopheles atroparvus* could thrive. In these zones death-rates were 25-50 pct higher than in inland areas. This was not caused by tropical malaria, but by the prolonged debilitating effect of tertian and quartan fevers. High summer temperatures as well as storm surges were usually followed by an outbreak of malaria. The impacts of ecological changes are discussed, as well as the effects of widespread malaria on the population's health and local culture. In spite of the high death-rates, popular concern with malaria fevers diminished during the 17th and 18th centuries. This was due to a medical fallacy according to which the chronic effects of malaria were diagnosed as 'land-scurvy'. The eradication of malaria in North-Western Europe had more to do with agricultural changes, hydrological measures and rising standards of living than with medical progress. The rise and fall of malaria took place largely independent of long-term climatic change.

Local environment, epidemiology and climate

During many centuries Northwest Europe's coastal populations have been better off than their inland neighbours. The marshland environment with its diversity of fish, shellfish, fowl and wild plants has been exploited since the 5th millennium BC by the sedentary Ertebølle and Swifterbant peoples. Subsequently, Neolithic, Bronze and Iron Age settlers learned to use the fertile salt marshes and riparian thickets for pasturage and agriculture. Extensive cattle-holding, as well as the cultivation of grains, broad beans, oil-seeds and vegetables provided the coastal population with a nutritious diet which fostered physical strength and body size. Hence, disorders caused by malnutrition or vitamin deficiencies, such as scurvy or rickets, must have been relatively scarce.

Medieval and early modern sources suggest that disease patterns in the coastal marshes were rather different from the inland districts (Dobson 1997; Haeser 1875-82; Hirsch 1883-86). Inland, we find many indications of ergotism and chorea, goitre, brucellosis, typhus, tuberculosis, relapsing fevers and respiratory infections, particularly after crop failures. In coastal areas, on the other hand, hunger crises hardly occurred, at least since the mid-16th century, when coastal wealth warranted regular grain supplies from the Baltic. Soaking terrains, damp atmospheres and chilling winds, however, reinforced many diseases. Hence, arthritic complaints (gout), asthma and hay fever were common. During hot summers intestinal infections could spread easily through canals and ditches, through which drinking water reservoirs were contaminated. Low-lying fields harboured fascioliasis, leptospirosis, tetanus and probably anthrax. Intestinal worms were common. Abundant clothing and insufficient hygiene may have reinforced cutaneous infections, including endemic syphilis and lepra. Regular commercial contacts made coastal areas more vulner-

From WEFER G, BERGER W, BEHRE K-E, JANSEN E (eds), 2002, *Climate Development and History of the North Atlantic Realm.* Springer-Verlag Berlin Heidelberg, pp 339-353

able to pandemics, such as influenza, smallpox, sweating disease and bubonic pest, as well as to sexually transmitted diseases. Additionally many cases of alleged scurvy have been reported, but this may be attributed to other causes, as we will see below. Although death rates were dropping on the whole, malnutrition and poverty related diseases, such as endemic syphilis, typhus and cholera, became more common after 1750, due to increasing social tensions.

The most important disease of the coastal marshes, however, must have been malaria. Though nowadays associated with the (sub)tropics, indigenous malaria has been present in Northwest Europe from the Early Middle Ages until the 1950s (Bruce-Chwatt and De Zulueta 1980; Winkle 1997; Bruce-Chwatts in Wernsdorfer and McGregor 1988; Schimitschek and Werner 1985; Jantsch 1948; Maisch 1938). As its occurrence is largely dependent on mean summer temperatures above 15°C, we might expect it to be fairly susceptible to long term climatic change.

The impact of climatic change on coastal epidemiology has not yet been determined. We are inclined to think that the medieval climatic optimum fostered the spread of malaria as well as intestinal infections. Moreover, rising sea levels associated with higher temperatures must have boosted hu-midity and salt water infiltration, thereby creating a more favourable habitat for both types of diseases. On the other hand, the drop in temperatures during the 14th century and more pronounced in the 16th and 17th centuries should have intensified the effects of diseases associated with coldness and humidity, whereas malaria and intestinal infections should have been in decline.

However, the rise and fall of malaria took place largely independent of global climatic fluctuations. This raises questions about the effectiveness of global models for explaining regional climatic change.

Malaria as the 'Great Debilitator'

The coastal marshes of Northwest Europe have been invaded by malaria at least since the first millennium AD. Few places with a temperate climate have been more afflicted by the disease than the embanked flood plains, fens and river estuaries surrounding the North Sea basin.

For many centuries, the main vector *Anopheles atroparvus* has occupied the coastal areas in considerable numbers, as it was undisturbed by competing mosquito species which could not stand the regular influx of seawater. Its larvae were able to grow up in stagnant pools and ditches with a salinity of 500 to 2,500 mg/l. In experiments some individuals are known to have endured a mixture containing almost a third seawater. In inland areas, on the other hand, the malaria mosquito was pushed aside by *A. messae*, which prefers bovine blood, as well as by other species which rarely feed on humans. Only in marshland areas where cattle were scarce, *A. messae* could become an important vector on its own. *A. messae* is able to transfer malaria at extremely low temperatures, down to 6°C (though by then it takes at least 44 days for the mosquito to become infectious). This may explain why cases of malaria have been reported from Scotland and Norway, where competing mosquito species were virtually absent. In contrast, *A. atroparvus* is largely restricted to the 15°C July-isotherm. In the Southern Baltic *A. atroparvus* and *messae* are both made responsible for malaria outbreaks, but current literature on the subject is obsolete. Another common species, *A. maculipennis s.s.* (*A. maculipennis typicus*), which prefers hilly areas, is only considered to be a risk in Southeast Europe. Finally, during prehistoric and medieval times *A. plumbeus*, which breeds in hollow trees, may have been a vector too (Jetten and Takken 1994; Jaenson et al. 1986; Bayliss 1985; Swellengrebel and De Buck 1938; Wesenberg-Lund 1920-21).

Malaria is caused by a parasite that incubates in the human liver, from where the victim's blood is infected. Female mosquitoes feeding on humans convey the disease, as the ingested gametocytes develop into sporozoites, which are then again injected into the next host. The patient is struck by intermittent fever attacks, which are regulated by the parasite's reproductive cycle. As soon as the attack starts, the victim is suffering from a shivering cold, which led to the popular expression that he or she has caught a 'cold'. Hence, intermittent fever and 'cold fever' were synonymous. The attack proceeds with high temperatures, lots of sweating and a deep sleep. Before the rise of modern medicine, the disease was known to become latent after several weeks or months, although the attacks often returned during the next spring or autumn, or even after a pause of several years.

In temperate zones 'benign' tertian fevers caused by *Plasmodium vivax* were most common. But particularly in areas where malaria was rampant also quartan fevers induced by *P. malariae* occurred. The former were identified by a two-day reproductive cycle, the latter by a three-day cycle. As a consequence, they were often referred to as two-day or three-day fever. These shifting terminologies are to blame for much confusion regarding the interpretation of historical sources. In addition, quotidian regular fever attacks could be observed, which used to be considered as a separate disease. Not before the second half of the nineteenth century was it recognised that quotidian fevers resulted from multiple infections in the patient's body.

P. vivax spreads far more easily than *P. malariae*. The latter is a rather ineffective agent, as it produces fewer gametocytes that might infect mosquitoes. It can only become epidemic when vector incidence is high and mean summer temperatures are above 16.5°C. Even then, *P. malariae* is a rather slow agent: it requires 35 to 45 days of favourable weather to spread, followed by an incubation period of another 20 to 40 days or more (Jetten and Takken 1994). The spread of *P. vivax* takes less time and lower temperatures. Nevertheless, quartan infections could be very persistent as they were able to remain dormant for decades. While tertian fevers mainly attacked children and adolescents, adults are known to have suffered from quartan fevers for many years. As soon as mosquitoes were abundant during a warm summer, the disease might spread again. On the other hand, the conditions favouring an outbreak of quartan fever would allow simultaneous outbreaks of tertian fever as well.

Apparently, tropical or 'malignant tertian malaria' with its oscillating fever attacks rarely occurred in Northwest Europe. It had to be imported from overseas areas, as its agent (*P. falciparum*) was poorly adapted to the available local mosquito species and was not able to survive in the patient for more than a year. To spread effectively mean temperatures had to be at least 18 or 19°C for several months (Jetten and Takken 1994; Harinasuta and Molineaux in Wernsdorfer and McGregor 1988). Nevertheless, subsequent hot summers may have been sufficient to cause incidental outbreaks. Apparently a series of epidemic fevers, which struck the southern North Sea coast in the years 1826 and 1827 was partly caused by tropical malaria (Martini 1938). In the city of Groningen 10 percent of the urban population died, while the surrounding countryside showed a death rate of 5 percent. Diseases were able to spread very rapidly, because most of the canals and ditches had been salted up during a severe storm surge in February 1825. The impacts of the outbreak, which was mixed up with local strains of tertian malaria and typhoid, were to be felt in the entire coastal area from Flanders to Denmark, as well as in the Baltic and in Britain (Ramaker 1998; Bruce-Chwatt and De Zulueta 1980; Andersen 1980; Haeser 1875-82). Even then, the tropical component of the outbreak must have been brought in from Southern Europe, since local mosquitoes were not susceptible to African or Asian malaria strains (Jetten and Takken 1994). Similar outbreaks have been observed in 1581/82, 1638 and 1859 (Knottnerus 1999; Bruce-Chwatt 1976; Fraatz 1929).

Compared to tropical malaria, the effects of tertian or quartan fevers seemed to be insignificant. People hardly ever perished as the direct consequence of an uncomplicated intermittent fever. Often people felt that the recurrent fever attacks were, though inconvenient, rather harmless. In the case of young children and pregnant women, who were most severely afflicted, the virulent manifestations of the disease were recklessly put down to other causes. In 1889 the medical doctor Wilhelm Olbers Focke from Bremen conducted an inquiry into the former occurrence of malaria in Northern Germany. From his own experience he recalled almost laconically: "Most marshland dwellers used to pass two fever periods during their life: First as children between the age of 2 and 10, later as adolescents from 15 to 24. Each fever period lasted one to two years" (1889). Therefore, it seems that local people merely put up with the disease, whereas they blamed the high death rates on other illnesses.

On this point, contemporaries must have been wrong: intermittent fevers were only a transitory phase during the course of the malaria infection. Much more dangerous were the chronic consequences, such as headaches, painful limbs, anaemia, hydrops, damage on spleen, liver and kidneys (in case of quartan malaria), and general exhaustion or cachexia. Chronic anaemia is known to cause many other disorders, from miscarriages to osteoporosis and porotic hyperostosis. Moreover, in the long run relatively mild quartan infections could do as much damage as tertian ones, because they were rather long-lived.

Malaria acted, as Mary Dobson states, "as a great debilitator. It was a disease which the people of the marshes permanently had to live with until they succumbed to its frequent attacks or died of secondary causes". Dobson was able to prove from demographic records that 16th and 17th century death rates in the Kent, Essex and Sussex marshes more than doubled the inland figures. The least healthy districts were those in which *A. atroparvus* was prevalent (Dobson 1980, 1997).

It was not only due to unfavourable ecological conditions that former marshland dwellers led an ailing life. Though the mosquito can be blamed for passing malaria from one individual to the next, humans sustain the disease. As soon as substantial parts of the human population are infected, the chance that mosquitoes transmit the disease increases greatly. On the other hand, malaria will gradually disappear as soon as the infection is contained to a few cases. In areas where a particular strain of malaria is endemic, almost everybody is infected. The local population will develop a certain immunity, which, however, can only been acquired through a brutal selection process which kills many children under five years old. Moreover, the immunity acquired is specific only for a distinctive strain. Any one who has endured a tertian infection, can always catch another quartan or tropical infection. Foreign immigrants, seasonal workers and travellers who lack immunity will be heavily affected too. Whenever tropical malaria is endemic, the selection process can be so effective, that immunity is genetically passed down to the next generation.

Our literature suggests that until the 19th century the North Sea coastal marshes were one of the so-called 'stable malaria regions', where the disease was hyperendemic. Nowadays this is a phenomenon restricted to tropical and subtropical regions. Therefore, several medical historians were doubtful if the mild strains of malaria of northern latitudes could ever have caused such strong demographic effects, as has been observed in tropical regions where *P. falciparum* is present. They may have been mislead, however, either by contemporary records which ignored the risk while immunity was still widespread, or by successive medical reports which failed to appreciate the impact of the disease because it was in decline by then (cf. Swellengrebel and De Buck 1938). Already during the 19th century many virulent malaria outbreaks were considered as incidents, which were to be contained by medical progress. Rather, one should consider the possibility that malaria only became epidemic (instead of endemic) after the chain of infection had been broken and the coming generations were without immunity. We had better heed the judgement of malaria-expert Peter Mühlens, who stated in 1936 that the epidemiology of the German coastal marshes originally "had a certain resemblance with the malaria of many tropical and subtropical beaches and flood plains" (cited in Knottnerus 1999).

Though they are normally associated with tropical malaria (De Zulueta in Bynum and Fantini 1994), it may be possible that some genetically based patterns of immunity, well known to the Mediterranean, were extant in Northwest Europe too. Several findings indicate that severe types of anaemia, such as thalassemia and dehydrogenase(G6PD)-deficiency or favism, which are specific for stable malaria regions, might have been present in our area (Knottnerus 1999). The latter harms individual male children, when they eat raw or insufficiently heated broad beans (*Vicia faba*), or when they inhale the corresponding pollen. As noted before, broad beans have been one of the most important foodstuffs in the coastal marshes since the Bronze Age. Moreover, their toxicity may have offered some protection against malaria to other individuals. A certain ambiguity towards beans, which can be found in popular folklore, may have been caused by these phenomena. Precisely because only specific individuals were affected by the deficiency disease, beans were ascribed with magical functions (Green and Danubio 1997; Burke 1996; Katz in Harris and Ross 1987). Another indication that selection processes were at work might be that blood group A, which enhances the suscep-tibility to malaria, is less frequent in certain coastal areas (Vonderach 1999; Bayliss 1985).

Three types of evidence can establish that it was primarily malaria that caused the raging fevers and enhanced death rates of the coastal marshes. Firstly, the manifest symptoms of the disease, such as intermittent fevers, anaemia and splenomegaly, which has been described as 'ague cake', 'side-stitch' or 'tightening of the heart'. Secondly, the geography of the disease, which shows that enhanced death rates were largely restricted to the brackish coastal zones in which *A. atroparvus* mosquitoes were indigenous. Thirdly, the seasonal pattern of disease, since it escalated in dry and warm summers when salt water infiltration intensified and when mosquitoes as well as plasmodiae could multiply more rapidly because of the high temperatures at hand. At 16°C *P. vivax* requires up to 30 days to develop within the mosquito, followed by a two week incubation period within the human body, to complete its reproduction cycle. With mean temperatures of 23-24°C the first period is reduced to 10-12 days or less. Only with temperatures above 32°C *P. vivax* does not survive (Jetten and Takken 1994).

Once malaria had exhausted the patient, other diseases often made an end to his or her life. Epidemics of (para)typhoid, dysentery, viral hepatitis A and leptospirosis raging during hot summers, as well as the common cold virus, influenza, pneumonia and typhus during wintertime, completed the destructive work which malaria had started. Measles and smallpox killed many children too. Intestinal diseases were fostered by the un-favourable drinking-water conditions in the coastal marshes, whereas leptospirosis could easily be caught during hard labour in the soggy fields. The effects of the frequent bubonic plague epidemics that harassed Holland until 1667 and Northern Germany up to 1714 must have been aggravated by malaria outbreaks. Several of these so-called plagues might be explained by malaria altogether. In 1617 the English ambassador Sir Dudley Carleton wrote about the Dutch Republic that "the pest is general ... and in some places hot". Obviously, he did not refer to the plague, but to common diseases such as malaria fevers and diarrhoea (cited in Knottnerus 1999:31).

Indeed, contemporary aetiologies were hardly fit to discriminate between the pestiferous 'hot fevers' observed and the seemingly 'cold fevers' of malaria. Intermittent, remittent, bilious, catarrhic and putrid fevers were easily combined into a general notion of miasmic disease. Moreover, Galenian medical doctrine tended to classify tertian malaria as a bilious fever, thereby reckoning it among the intestinal diseases. Quartan fevers, on the other hand, were categorised as a chronic disease caused by obstipation of the spleen, due to unhealthy food and lack of exercise (Swellengrebel and Honig 1925-28).

Introduction and Expansion (500-1500)

The history of malaria is intrinsically connected with the occupational history of the area concerned. Just as clearing the African jungle preceded the spreading of tropical malaria, so the occupation of the coastal marshes in the later Bronze and early Iron Age paved the way for the spreading of temperate malaria strains. In both cases a mosquito species occupied a recently created niche in the human habitat. *A. atroparvus* was only able to survive in the unfavourable coastal environment because of the hibernation chances it found in byres, pig-sties and human living quarters. Artificial ponds, natural pools and stagnant river branches provided the necessary sites for breeding.

Probably malaria was already present in Europe during the Neolithic, but it was hardly ever effectively transmitted. The existing malaria strains did not become hyperendemic in the most heavily populated parts of the central Mediterranean until the 5th century BC. The recent advance of the mosquito species *A. labranchiae* and *A. sacharovi* may have intensified this. As a consequence, men, mosquitoes and plasmodia were able to familiarise at large, so that little by little *A. atroparvus* became a significant vector as well. In addition, tropical malaria entered the scene, taking a high death toll on the indigenous population (Burke 1996; Bruce-Chwatt and De Zulueta 1980).

To what extent *A. atroparvus* populations were susceptible to different strains of malaria remains unknown. It seems, however, that the advance of tropical malaria into the western Mediterranean was delayed for some time. The *P. vivax* infection spread northward, maybe through Spain and south-western France, where competing mosquito species must have been absent. Here it gradually accommodated to colder climatic conditions by improving its ability to hibernate in the human liver as a hypnozoite (Molineaux in Wernsdorfer and McGregor 1988).

Exactly when malaria became endemic in the North Sea basin is an open question. It is very plausible, however, that the decimation of the coastal population during the 4th and 5th centuries was not only caused by rising sea-levels, large-scale migration and political turmoil, but also by epidemiological factors (Knottnerus 1999). As a rule, populations that are newly exposed to malaria suffer great losses. Although skeletal remains showing porotic hyperostosis or osteoporosis, resulting from childhood anaemias, have not yet been reported, future archaeological findings might provide us with a more precise date for the genesis of malaria endemicity (Burke 1996; Stuart-Macadam 1992).

Once a new strain of malaria had arrived, it could also be transmitted to inland mosquito species. At the same time, other strains may have spread northward through the river valleys, swamps and flood plains of Central Europe, where summer temperatures were high enough to sustain the disease. Though the mosquito might be a poor vector, quartan infections could hold out for many years. In either case *A. messae*, which had been a minor vector in the Mediterranean for a long time, must be held responsible for its further expansion.

By the 8th century, Britain's coastal plains had a bad reputation altogether. Literary sources such as the Vitae of Saint Cuthbert and Saint Guthlac claim that the "fens and marshes were haunted by the evil seed of Cain" (Newton 1993). The distinguished medievalist Hilda Ellis-Davidson actually assumes that "the monster Grendel in the poem Beowulf symbolised the dangerous climate of the fens, bringing plague to the king's hall" (1964). During the last years of Cuthbert's life (he died 687) the Lincolnshire marshes were depopulated by an unidentified epidemic. The abbess Hilda of Whitby died in 680 after six years of fever. An 8th-century translator identified her disease as 'lenctenadl' (spring ailment), therefore we might be sure that spring relapses from tertian fever were known by then. Hence, the year 673 must be seen as a minimum date for the arrival of *P. vivax* in the North Sea basin (MacArthur 1951).

Anglo-Saxon medical prescription books largely confirm that malaria was a main cause of disease. Bald's Leachbook (9th century) is very specific about the symptoms of malaria. Here the distinction is made between tertian fever, quartan fever and the frequent relapses in spring mentioned before. But even more attention is being paid to the chronic consequences of malaria, such as splenomegaly, side-stitches and severe anaemia, described as 'the halfdead disease'. Other medical books report about the cold fevers ('þa colan feforas') and the shiverings ('þe riþaþ') which accompanied the attack. Apparently, possession by evil spirits or 'dweorgs' (little people) was to be blamed for it (Cameron 1993). Subsequent Anglo-Norman sources describe the 'febris frigidas' or 'freide fevre'. The modern word 'ague' came into use in the 13th century. According to the malariologist Bruce-Chwatt, it has been derived from the French 'fièvre aigue' or Latin 'febris acuta', referring originally "to any acute febrile disease, and espe-cially to a fever accompanied by a shaking or shivering fit" (1976).

On the continent references to malaria are not found until much later. Nevertheless, the 9th-century Old High German expression 'ritto, rito' (Middle Dutch 'ridde, rijde', Old Norse 'riða, riðusótt') closely parallels the Anglo-Saxon terminology, whereas it seems to indicate that the shivering patient is being ridden by a demon. Interestingly enough, a late medieval saint's legend from

Scandinavia identifies tertian fevers as possession by a demonic snake ('vindormr'). Individual cases of possible malaria have been noted more than once. The founder of the archdiocese of Bremen, Saint Willehad, died of a severe fever in 789, while he was visiting the coastal marshes. According to the chronicler Saxo Grammaticus, king Sven Estridsøn of Denmark passed away in Southern Jutland, probably autumn 1076, when his heart felt tightened during a pestiferous fever attack. An early report about a possible malaria epidemic dates from 988, when diseases were rampant in the Southern Netherlands during a warm summer after widespread inundations. The Weser and Elbe river-banks are said to have been hit by a severe 'pestilence' after a storm surge in 1020 (Meineke and Schier 1995; Knottnerus 1999; Buisman 1995-98).

Though the coastal inhabitants had created a niche in the wetlands protecting them against human predators, their defence against natural enemies was largely insufficient. From the 11th century onward they even became more vulnerable, as they began to build up embankments against the sea, thereby creating a brackish habitat that generated additional risks. Storm surges were to become more destructive, whereas the fields were filled with slugs, mice, leather-jackets (*Tipula paludosa*), fluke-eggs, mosses and horsetail (*Equisetum arvense*). Furthermore the canals and ditches, being isolated from the sea, supplied new breeding-places to all kinds of insects, including malaria mosquitoes.

The inhabitants of the coastal marshes must have acquired a certain degree of immunity against malaria by then. Apparently, the Frisian warriors who - according to the sagas - captured Rome at about 1085 were able to endure the raging fevers that had wrecked earlier attempts. On the other hand, the crusaders and pilgrims coming back from Rome and Jerusalem must have brought many new infections with them (Knottnerus 1999; Maisch 1938; Martens and Hall 2000). A genuine epidemic of quartan fever is reported by the chronicle of the abbey of Gembloux, south of Brussels, during the hot summer of 1136 or 1137. The city of Cologne was struck by a quartan epidemic in 1192 (Lersch 1896; Keil 1993). A Danish cleric who suffered from a quartan fever, is said to have caught it while studying in Paris in about 1175. Indeed, it seems that *P. malariae* had become endemic along the moist and relatively warm river-banks of the Rhine, Danube and Rhone, from where it spread northward. A hundred years later the presence of quartan fevers or 'shivering disease' is attested as far north as Linköping in Sweden. Only Iceland remained free from malaria (Meineke and Schier 1995; Møller-Christensen 1959).

Reports about malaria abound in the Later Middle Ages. In 15th-century East Anglia the coastal parishes suffered high losses during warm summers. The famous chronicle of the abbey of Wittewierum (near Groningen) recounts an outbreak of various fevers during the warm summer of 1237, when the abbey's founder died of a quartan fever. The infirmaries had been overcrowded and all over Friesland hardly enough people had been left to nurse the sufferers (Gottfried 1978; Fraatz 1929). At the grave of the Frisian abbot Siardus of Mariengaarde (d. 1230) many sufferers of tertian fever were said to have been healed. According to a recent survey article, malaria may have had stronger restrictive effects on medieval population growth than the plague (Keil 1993).

Contemporary medical compendiums, such as Saint Hildegard of Bingen's Causae et curae, written in the 1150s, were quite familiar with tertian, quartan and quotidian fevers. The Danish physician Henrik Harpestræng (d. 1244) was acquainted with 25 remedies against different fevers, most of them tertian, quartan or quotidian ones, characterised as shiverings ('rithæ') or chilling sickness ('kaldæ sot') (Meineke and Schier 1995; Møller-Christensen 1959). In Bremen the patrician Arnoldus Doneldey differentiated in 1382 between cold fevers and burning fevers, the latter being identified as the newly arrived bubonic plague. Apparently, the doctors from whom he copied his prescriptions were very experienced in treating malaria. They were aware of duplicated malaria infections and spring relapses, which were considered not nearly so lethal as autumnal infections, and they described the rare blackwater fever (haemoglobinuria): a lethal complication among patients suffering from G6PD deficiency or severe tropical malaria, expressing itself in bloody urine (cited in Knottnerus 1999).

Although it is possible that these prescriptions merely reflected the lessons of Salernian medicine, it is more likely that its authors had learned about malaria from their own experience (Cameron 1993). Indeed, malaria was so common in the 14th century that a Hamburg Bible translator automatically imagined Christ not to expel an ordinary fever, but a cold fever ('dat kalde') from a patient (cited in Knottnerus 1999, after Luke 4:38).

High-days (1500-1750)

Still, the high-days of malaria were in the Early Modern Age (Reiter 2000; Molineaux in Wernsdorfer and McGregor 1988). Although the general drop in temperatures should have reduced disease frequencies, increasing human efforts to exploit the coastal niche had an adverse effect on living conditions. Intensified traffic increased the risks of contagion, whereas hydraulic innovations fostered waterways and ditches to salt up. Newly built floodgates, harbours and canals, meant to facilitate shipping-transport, enabled seawater to seep in more easily. Moreover, during summertime floodgates were often deliberately opened to prevent river-crafts from running aground, while allowing seawater to rush in driving fresh waters up into the scorched pastures.

Since the 14th century circumstances were deteriorating all over because of the salting up of bays, creeks and rivers, such as the Zuyder Sea, the Dollard and Jade estuaries, which affected the water qualities on the land side of the dikes. Although the existing dikes were strengthened and large tracts of salt-marsh were reclaimed from the sea, new embankments provided the mosquitoes with an even more favourable habitat. In addition, the artificial reclamation of vast lakes fostered the percolation of brackish ground-water from underlying salt-water domes. Frequent storm surges and wartime inundations had a detrimental effect on water qualities too.

Hence, mosquito populations must have been increasing, thereby boosting the risk of malaria. Acid waters leaking from the inland bogs might have curtailed the mosquito scourge, but for agricultural reasons these waters were kept at bay and - whenever possible - conveyed to the sea as soon as possible. Additionally, the relocation of settlements from the artificial mounds to the embanked polderlands must have intensified health risks too. Only the silting up of inland river-beds due to excessive grazing and forest clearance had an opposite effect, since the salt water border was driven back seaward. Already at the closure of the 18th century death rates on the Lower Weser river banks barely outstripped the inland figures (Hinrichs et al. 1988).

This is not the appropriate place for exhibiting a catalogue of malaria reports. Nevertheless, we may conclude that they were extant in almost every coastal district (De Baets 1998; Priester 1998; Dobson in Bynum and Fantini 1994; Norden 1984; Brouwer 1983; Schuberg 1927; Lemaire 1922; Wesenberg-Lund 1920-21; Schouten 1920; Trautman 1913; Nuttall et al. 1901; Focke 1889; Haeser 1875-82; Hirsch 1883-86). Everywhere we find abundant remarks in parish registers, regional monographs and medical reports, which might prove our point. Topical expressions, such as Fenland ague, Polder fever, Northern 'stier', Zealand fever, Dutch pip, Holstein marsh sickness or Eiderstedt stubble fever have terrified medical practitioners for centuries. Strangers were scared off, whereas newcomers had to fear for their lives.

Moreover, a 3-500 km zone in which malaria had a more incidental character surrounded the coastal districts. Here hyperendemicity was largely restricted to isolated pockets of marshland in which *A. messae* were abundant. Lack of cattle as an alternative host, as well as poverty and malnutrition were the main causes for maintaining secondary hotbeds of malaria. Particularly in the Netherlands, Northern Germany, Denmark and Scotland many inhabitants of scattered swamps, bogs and river valleys were infected. Peddlers and seasonal workers returning from the coastal districts as well as travellers from Southern Europe may have been responsible for spreading the disease. But eventually many secondary centres held out on their own. Malaria has also been endemic in many other wetland areas, from the Upper Rhine, Bavaria and Thuringia up to the Baltic. The salt mining towns, where *A. atroparvus* found appropriate breeding sites, were heavily inflicted.

Nevertheless, the coastal districts stood at the heart of every major malaria epidemic. According to an 18th-century physician from Friesland "the autumnal bilious fevers are our main national disease". Although few patients died, many fell seriously ill. One of his colleagues stated that the disease was so well-known that it needed no further description. A medical survey concluded in 1824 that probably no country in Europe, except for Italy, had been so badly afflicted by malicious intermittent fevers as Holland (Van Seventer 1969). Similar reports can be cited from almost every corner of the North Sea, particularly from low-lying islands and peninsulas that were surrounded by the sea. Regional monographs abounded in the 19th century until there was hardly a region, which did not have its own malaria literature. The reports were then collected and incorporated into medical surveys, whereupon they have been largely forgotten.

Repercussions on coastal society

The significance of malaria for the social history of the North Sea coastal marshes can be summarized into four statements:

Firstly, the geography of malaria is delineated by enhanced mortality rates. Hence, its spatial distribution can be largely determined with the help of mortality statistics. Unfortunately, only scattered figures have been published. As a reasoned guess, we would expect coastal death rates in the 16th, 17th and 18th centuries to have amounted to at least 30 to 50 per 1,000 as against 20 to 30 per 1,000 for inland districts (Table 1). Hence, every fourth or fifth death was indirectly caused by malaria or related diseases. To demarcate the importance of malaria as against other typical wetland diseases, subsequent data on the extent of salination are required. Incidentally, however, endemic malaria may have been conveyed to fresh water areas as well. Especially heathland villages bordering on the saltmarsh could have become infected by mosquitoes bred in neighbouring low-lying ponds.

For the second half of the 18th century enhanced mortality rates have been reported in a wide range of coastal districts from Furnes to Ribe and beyond (Knottnerus 1999). The excessive mortality was sufficient to restrict natural population growth, therefore causing several marsh regions to become dependent on immigration. Moreover, the step-by-step concentration of landholding into the hands of large farmers, which can be observed in many marsh districts, was hastened by the enhanced mortality rates.

Secondly, the chronology of malaria may coincide with storm surges, high summer temperatures and summer drought. Therefore, data on epidemics and death rates must be connected with climatic observations. We may assume that most epidemic outbreaks after storm surges or during prolonged periods of drought can be largely put down to malaria. Preliminary findings suggest that this is indeed the case. Moreover, there are many indications that epidemics in different coastal districts took a synchronous course. Hence, it seems justifiable to interpolate our scattered data into prolonged series for the entire southern North Sea coast, and maybe for the British marshlands as well (Dobson 1987; Fraatz 1929; Hanssen 1925; Lersch 1896; Creighton 1891-94).

Years	Pastoral sea marsh (Butjadingen) (n) = number of parishes	Arable sea marsh (Marne, Dithmarschen)	Heathlands (Bockhorn, Oldenburg)
1648-1652	42 (2)	-	-
1660-1664	34 (3)	-	-
1673-1677	55 (4)	35-45	-
1700-1704	48 (6)	30-40	21
1726-1730	101 (1)	40-45	-
1767-1771	44 (9)	42	22
1791-1795	45 (9)	36	22
1814-1818	21 (7)	20-25	25
1826-1830	51 (9)	40	32
1835-1839	30 (9)	21	29

Table 1. Crude Death Rates per 1,000 inhabitants in different districts (Sources: Norden 1984; Hinrichs et al. 1988; Lorenzen-Schmidt 1987).

Thirdly, the social topography of malaria implies that its impact on the poor was more pronounced than on the well-to-do. Malaria is known to be transmitted more effectively when people are insufficiently dressed, sheltered and nourished. Not only were members of poor families more frequently exposed to mosquito bites, also their general health condition made them more susceptible to the infection. Once fallen ill, their chances of recovery were lower because they could not afford medical treatment. Moreover, they were not able to compensate for the loss of income and working-days. There have even been sporadic reports of poor people who tried to avoid the risk by tarring their legs before working in the submersed fields. But such preventive measures seem to have been quite exceptional, at least in the 18th and 19th centuries when most adult males owned boots and shoes.

The tens of thousands of seasonal workers from inland districts who did much of the mowing, digging and reaping in the coastal marshes were extremely susceptible to malaria. Though they did not have the immunity of the indigenous population, they often slept in outdoor shelters or tents, drinking infected ditch-water and eating rancid bacon. Eighteenth-century estimations claim that a quarter to a third of them had caught an intermittent fever before returning to their villages. In several Westphalian districts special sick-transports were organized to take them home (Knottnerus 1999).

Fourthly, the cultural reflection upon malaria risks led to various preventive measures and curative remedies. But these did not always reduce actual exposure. In Britain as well as on the Continent most people were convinced that living near the sea-shore was hazardous. In accordance with current medical doctrine the medieval fear of the sea was transposed into a novel concern for miasmatic emanations. Early-modern man did not know about the role played by mosquitoes in transmitting malaria. Instead he believed that his disorders were caused by the hazes ascending from saltmarshes, swamps and ditches. Particularly in Germany and Denmark people thought they were better off in their heated parlors. It is questionable whether such a preventive strategy was effective at all. Mosquitoes may actually have preferred the stuffy living quarters, whereas they were scared off by the open fire places which were common in Holland and Britain (Knottnerus 1997).

Some traditional remedies may have been quite effective. Tobacco smoke for instance must have acted as a powerful mosquito repellent. In the coastal districts smoking a pipe was quite fashionable in the 17th and 18th century, not only among men, rich as well as poor, but also among married women (Knottnerus 1997). Several traditional drugs had antiseptic properties, such as pennyroyal (*Mentha pulegium*), already mentioned in Anglo-Saxon sources. Others might serve as a mosquito repellant (yarrow, *achillea millefolium*), they may have prevented renal failure (bogbean or 'fever herb', *Menyanthes trifoliata*) or they actually helped against diarrhoea (creeping cinquefoil, *Potentilla reptans*). The painter Albrecht Dürer, who caught a malaria fever in Zealand in the 1520s, took swallowwort (*Chelidoneum majus*) as a medicine. Distinct members of the *Artemisia* family, particularly wormwood (*A. absinthum*) and sea wormwood (*A. maritima*), have been used against malaria too. Possibly, they contain small doses of artemissinin as their Chinese counterpart *A. annua* does. Sea wormwood hass proved to be effective against liver damage (Dobson 1998; Tunón 1995; Janbaz and Gilani 1995; Cameron 1993; Winkle 1997).

Nonetheless, many remedies reverted to magic. Magical spells were common, as were the holy trees on which a fever could be tied. Some of these trees held out until the 20th century. Specific kill or cure remedies and bitter-tasting medicines were preferred because of their supposed capacities to expel evil. Expressions such as Flemish 'Noordsche stier' (Northern tribute), Frisian 'tjinst' (bond, bondage) and North Frisian 'thwung' (coercion) suggest that malaria fevers have been associated with possession by evil spirits until fairly recently. Obviously, people identified the haze as the agent of disease. Originally these expressions may have meant that the patient was paying tribute to the evil powers from the Northern seas (Knottnerus 1997; De Baets 1998).

Most of the remedies concerned were favoured because of their supposed ability to drive out the chills associated with malaria fevers. Liquor and wine were consumed in large quantities. Tropical products, such as tobacco, coffee and tea, were said to prevent illness. In the 19th-century Fenlands opium was used to comfort ailing children (Dobson 1980). Another drug which was said to avert fever was the sweet-flag (*Acorus calamus*), often used in liquor or a kind of tea. After being imported from Turkey in the sixteenth century, its proliferation in the coastal marshes may have been encouraged by medical beliefs. Purgatives were popular as well: in Holland the well-known Haarlemmer Oil, a forceful mixture of linseed-oil and turpentine, was considered as one of the most effective remedies against intermittent fevers and diarrhoea (Knottnerus 1999).

Medical progress did not always lead to appropriate answers. Though Peruvian bark or Cinchona has been widely used in coastal medicine since the end of the 17th century, the medical profession had many reservations. The drug was introduced by Spanish Jesuits in 1639 and became known in Britain, Belgium and Holland in the 1650s. Soon afterwards medical practitioners began to prescribe it against all kinds of intermittent fevers (Dobson 1998; Bruce-Chwatt in Wernsdorfer and McGregor 1988). In 18th-century Holland Cinchona was processed into several popular theriaca and sold in local shops as a panacea. Medical science, however, especially in Germany, turned away from Cinchona, as the usual doses were often too small to be effective, whereas many other diseases falsely diagnosed as malaria did not respond at all. Moreover, its bitter taste and its supposedly hot properties did not fit into the Galenian system. As an official drug, it only made a comeback after the 1820s, as the cheaper and more concentrated derivative quinine became available in drugstores and grocery shops. Since the 1860s many households kept their own stocks of quinine (Swellengrebel and De Buck 1938; Schouten 1920; Roth 1906; Focke 1889).

Indeed, it is striking to see how medical science slowly went astray, as it was searching for more adequate solutions. Since the mid-17th century many physicians began to downplay the risk of intermittent fevers. Instead, they became more concerned with a novel disease called 'land scurvy'. In spite of its name, land scurvy had hardly anything to do with the vitamin deficiencies among urban populations, which had been observed in the 16th century. Rather its symptoms, such as anaemia, hydrops and splenomegaly, largely coincided with the protracted effects of malaria. Case histories often began with a complicated quartan fever which triggered many other complaints not necessarily connected with malaria. As R. Elwyn Hughes rightly observed, "scurvy or 'the scorbute' suddenly became a convenient nosological safety net for the not inconsiderable diagnostic failures of the period" (Hughes 1990; Knottnerus 1997).

Thus, land scurvy came to be seen as the archetypical scourge of all coastal districts. It was said to be endemic in Britain, Holland and Northern Germany as well as in Scandinavia. According to Thomas Bartholinus' *De medicina Danorum domestica* (1666), scurvy and fever were the only two diseases which mattered in Denmark. Scurvy was above all a moral condition: its sufferers were not treated as victims but as individuals who had exposed themselves to the disease. The inhabitants of the prosperous coastal districts were said to be inflicted because of their immodest life-style. Moreover, nobody was entirely free from the affliction. In fact, the theory of scurvy was closely linked up with religious revival movements which gained ground in Britain and Holland. German physicians, on the other hand, argued that scurvy was an epidemic disease which affected any foreigner visiting the coastal districts. They suspected the Dutch had introduced it from abroad (Knottnerus 1997). Both views reflected the actual situation, in which the coastal population was accustomed to malaria, whereas newcomers fell ill.

Decline (1750-1950)

The decline of malaria has not yet been sufficiently explained. During the 18th century mortality rates in some regions were falling rapidly, whereas in others they remained largely the same. Particularly in Southeast England and the western districts bordering the Wadden Sea (Friesland, Groningen and East Friesland) population growth started early. But in Holland, Zealand and many German districts figures remained stagnant until the 19th century, whereas mortality rates were high up to the 1850s

(Dobson 1997; Knottnerus 1997; Norden 1984). Moreover, in the German Lower Rhineland as well as in the Baltic unprecedented outbreaks of malaria took place during the first half of the 19th century (Jaenson and al. 1986; Anderson 1980; Kortenhaus 1928; Wesenberg-Lund 1920-21). In general, tertian fevers got a more epidemic character instead of remaining endemic, whereas quartan fevers tended to become rare. As malaria outbreaks became more uncommon, seasonal peaks shifted from late summer to early spring (Swellengrebel and De Buck 1938; van Seventer 1969).

There are at least five possible explanations for the decline of malaria (Seventer 1969; Schuberg 1927; Focke 1889).

1. Medical progress: the massive use of Cinchona in Britain and the Netherlands may already have mitigated disease transmission in the 18th century. The subsequent introduction of quinine in the 1820s then made a further reduction possible.

2. Agricultural innovations: The introduction of the potato and the subsequent expansion of pig-breeding by cotters and agricultural labourers during the second half of the 18th century had a health-improving effect. Potato-growing not only im-proved general health by providing a nutritious diet to humans and their livestock. The supplement of vitamin C may also have been important to reduce the effects of anaemia (Green and Danubio 1997; Cameron 1993). Moreover, pig-breeding has diminished the risk of mosquito bites, as *A. atroparvus* prefers to feed on pigs (Jetten and Takken 1994). The 19th-century shift from traditional hairy swines to naked Chinese pigs must have reduced mosquito risks even more. Finally, the 20th-century introduction of modern byres and pig-sties led to a spectacular fall in mosquito populations.

3. Improved water management: Reduced water-levels and the replacement of field-drains by draining-pipes have reduced the number of breeding-places for mosquitoes (Schouten 1920). The introduction of American river-weed (*Elodea canadensis*) since the 1870s, the eutrofication of surface water and the subsequent spread of duck-weed (*Lemna gibba*) may have had a supplementary effect (van Seventer 1969; Focke 1889). Heightened dikes, larger canals and more effective sluices have reduced salt-water leakage and inundation risks, thereby destroying the natural habitat of *A. atroparvus*. The introduction of DTT has reduced the remaining populations even more.

4. Rising standards of living: The living conditions for cotters and farmhands have gradually improved during the 19th and 20th centuries: better and dryer housing, adequate clothing and sufficient fuel. These may have reduced the effectiveness of existing mosquito vectors. Traditional farm-buildings in which human living quarters, byres and pig-sties were found together under the same roof, have largely disappeared.

5. Parasitical evolution: Possible genetic variations in the virility of existing malaria-strains might have reduced malaria risks. Nowadays, even in heavily afflicted Indian communities, *P. vivax* is not a lethal disease. It may be, however, that its effects would be more detrimental, if it had not been accompanied by *P. falciparum* (cf. Harinasuta in Wernsdorfer and McGregor 1988).

We are inclined to come to the conclusion that the combined agricultural and hydrological innovations were decisive. In many districts where drainage efforts were delayed malaria was active for much longer. The Western provinces of the Netherlands, for instance, had higher death-rates than the North until the 1870s. The same holds true for Germany, where hydrological measures and housing improvements had to wait much longer. But other factors should not be ruled out, as they helped to change the balance between humans and mosquitoes to the latter's disadvantage. Apparently, all the causes combined have pushed the risk of contagion below a certain threshold, under which the disease could not be transmitted successfully. The existing degree of communal immunity may have acted as a supplementary warrant against reintroduction of the disease (Dobson 1980).

Conclusions

Is long-term climatic change relevant for explaining the history of malaria in Western Europe? As we have indicated, on the long range human interventions may have been more important than temperature fluctuations. Moreover, malaria incidence is largely dependent on summer temperatures. To a substantial extent the impact of global temperature fluctuations on Western European summer temperatures is muted by the shifting balance between continental and maritime climates (cf. Buisman 1995-98). Regional differences in mean summer temperatures, ranging from 14°C in Scotland to 17.5°C in the Rhine valley, were far more pronounced than long-term historical trends.

Indeed, the rise and fall of tertian malaria in Western Europe does not seem to correspond with temperature changes. Only the arrival of quartan malaria may have coincided with the medieval climatic optimum. Nevertheless, we might get another picture if we concentrate on 25-year periods or even decades. Here the peaks of malaria incidences seem to coincide with periods of high summer temperatures. Critical periods, such as the 1550s, the 1720s and the 1820s have all been characterized by warm summers, even when the winters were cold. Finally, we must consider the frequency of warmer summers: as soon as intermediate periods lasted long enough (at least four years), the chain of transmission could be interrupted. As a consequence, the next outbreak of tertian malaria might have become epidemic. This mechanism may have been responsible for the epidemics of the 1820s.

Will malaria return in Western Europe as a result of rising temperatures? Most epidemiologists agree that this is highly improbable. Even if *A. atroparvus* populations were to expand spectacularly, due to global warming and rising sea levels, the accessibility of medical services in 21st-century society are such, that infections would not remain unnoticed for long (Reiter 2000; Jetten and Takken 1994).

Acknowledgments

This paper has also been presented at the Third European Social Science History Conference, Amsterdam, April 12-15, 2000. I am thankful to Wiet Koren for his comments.

References

Andersen O (1980) A Malaria Epidemic in Denmark. In: Charbonneau H, Larose A (eds) The Great Mortalities: Methodological Studies of Demographical Crises in the Past. Ordina, Liège, pp 33-49

Bayliss JH (1985) Epidemiological Considerations on the History of Indigenous Malaria in Britain. Endeavour 9:191-94

Brouwer H (1983) Malaria in Nederland in de achttiende en negentiende eeuw. Tijdschrift voor Sociale Geschiedenis 9:141-59

Bruce-Chwatt LJ (1976) Ague as Malaria (An Essay on the History of Two Medical Terms). J Trop Med Hyg 79:168-176

Bruce-Chwatt LJ, de Zulueta J (1980) The Rise and Fall of Malaria in Europe: A Historico-Epidemiological Study. Oxford University Press, Oxford

Buisman J (1995-98) Duizend jaar weer, wind en water in de lage landen, 3 vols. Van Wijnen, Franeker

Burke PF 1996 Malaria in the Greco-Roman World: A Historical and Epidemiological Survey. In: Haase W (ed) Aufstieg und Niedergang der römischen Welt (ANRW): Geschichte und Kultur Roms im Spiegel der neueren Forschung, Part II: Principat. De Gruyter, Berlin New York, vol 37.3, pp 2252-81

Bynum WF, Fantini B (1994) Malaria and Ecosystems: Historical Aspects. Proceedings of a Rockefeller Foundation Conference. 18-22 Oct 1993. Societa Italiana di Parassitologia, Rome, Parassitologia, vol 36:1-2, pp 7-15, 35-60

Cameron ML (1993) Anglo-Saxon Medicine. Cambridge University Press, Cambridge, 17-18,117-29

Creighton C (1891-94) A History of Epidemics in Great Britain, 2 vols. Repr Cass, London 1965

De Baets P (1998) Malaria in West-Vlaanderen? Biekorf - West-Vlaams Archief 98:262-70

Dobson MJ (1980) Marsh Fever: The Geography of Malaria in England. Journal of Historical Geography 6:357-89

Dobson MJ (1987) A Chronology of Epidemic Disease and Mortality in Southeast England, 1600-1800. Institute of British Geographers, Cheltenham London (Historical Geography Research Group Series, vol 19)

Dobson MJ (1997) Contours of Death and Disease in Early Modern England. Cambridge University Press, Cambridge (Cambridge Studies in Population, Economy, and Society in Past Time, vol 29)

Dobson MJ (1998) Bitter-sweet Solutions for Malaria: Exploring Natural Remedies from the Past. Parassitologia 40:69-81

Ellis Davidson, HR (1964) Gods and Myths of Northern Europe. Penguin Books, Harmondsworth 1981, p 18

Focke WO (1889) Die frühere und jetzige Verbreitung der Malaria in Niedersachsen: Bericht über eine durch den Niedersächsischen Ärztevereinsbund veran-staltete Umfrage. Hannover

Fraatz P (1929) Beiträge zur Seuchengeschichte Westfalens und der holländischen Nordseeküste. Fischer, Jena (Arbeiten zur Kenntnis der Geschichte der westfälischen Medizin, vol 1, pp 35-68)

Gottfried RS (1978) Epidemic Disease in Fifteenth Century England: The Medical Response and the Demographic Consequences. Rutgers University Press, New Brunswick, pp 129-36

Green LS, Danubio ME (eds) (1997) Adaptation to Malaria. The Interaction of Biology and Culture. Gordon & Breach, Amsterdam

Haeser H (1875-82) Lehrbuch der Geschichte der Medicin und der epidemischen Krankheiten, 3rd ed. 3 vols. Repr Olms, Hildesheim 1971

Hanssen P (1925) Geschichte der Epidemien bei Menschen und Tiere im Norden: Nach Untersuchungen ausgehend von Schleswig-Holstein. Augustin, Glückstadt

Harris M, Ross EB (eds) (1987) Food and Evolution: Towards a Theory of Human Food Habits. Temple University Press, Philadelphia, pp 133-59

Hinrichs E, Krämer R, Reinders C (1988) Die Wirtschaft des Landes Oldenburg in vorindustrieller Zeit: Eine regionalgeschichtliche Dokumentation für die Zeit von 1700 bis 1850. Hülzberg, Oldenburg, pp 21-22

Hirsch A (1883-86) Handbook of Geographical and Historical Pathology. New Sydenham Society, London

Hughes RE (1990) The Rise and Fall of the "Antiscorbutics": Some Notes on the Traditional Cures for "Land Scurvy". Med Hist 34:52-64

Jaenson TGT, Lokki J, Saura A (1986) *Anopheles* (Diptera: Culicidae) and Malaria in Northern Europe, with Special Reference to Sweden. J Med Entomol 23:68-75

Janbaz KH, Gilani AH (1995) Evaluation of the protective potential of Artemisia maritima extract on acetaminophen- and CCl_4-induced liver damage. J Ethnopharmacol 47:43-47

Jantsch M (1948) Die Malaria. Ein geschichtlicher Überblick. Deutike, Vienna (Beiträge zur Geschichte der Medizin, vol 6)

Jetten TH, Takken W (1994) Anophelism without Malaria in Europe: A Review of the Ecology and Distribution of the Genus *Anopheles* in Europe. Wageningen (Wageningen Agricultural University Papers, vol 94,5)

Keil G (1993) Malaria. In: Auty R et al. (eds) Lexikon des Mittelalters. Artemis, Munich, vol 6, col 162-163

Knottnerus OS (1997) Die Angst vor dem Meer. Der Wandel kultureller Muster an der niederländischen und deutschen Nordseeküste (1500-1800). In: Fischer L (ed) Kulturlandschaft Nordseemarschen, Nordfriisk Instituut/Hever, Bredstedt Westerhever, pp 145-174 (Nordfriisk Instituut, vol 129)

Knottnerus OS (1999) Malaria in den Nordseemarschen: Gedanken über Mensch und Umwelt. In: Jakubowski-Tiessen M, Lorenzen-Schmidt KJ (eds) Dünger und Dynamit: Beiträge zur Umweltgeschichte Schleswig-Holsteins und Dänemarks. Wacholz, Neumünster, pp 25-39 (Studien zur Wirtschafts- und Sozialgeschichte Schleswig-Holsteins, vol 31)

Kortenhaus, F (1928) Das Wechselfieber in der Rheinprovinz und sein Verschwinden. Arch Gesch Med 20:120-36

Lemaire L (1922) Inondations et paludisme en Flandre maritime (1622-1922). Revue du Nord 8:173-209

Lersch BM (1896) Geschichte der Volksseuchen nach und mit den Berichten der Zeitgenossen mit Berücksichtigung der Thierseuchen. Karger, Berlin

Lorenzen-Schmidt KJ (1987) Ländliche Familienstrukturen in der nordwestdeutschen Küstenregion 1750-1870. Engelbrechtsche Wildnis (Archiv für Agrargeschichte der holsteinischen Elbmarschen, Beiheft, vol 3)

MacArthur WP (1951) A Brief History of English Malaria. Br Med Bull 8:76-79

Maisch M (1938) Über die Malaria und ihr Wirken in der deutschen Geschichte. Stroh, Backnang, pp 53-54

Martens P, Hall L (2000) Malaria on the Move: Human Population Movement and Malaria Transmission. Emerg Infect Dis 6,2, 103-107

Martini E (1938) Über die Malariaepidemie an der Nordseeküste 1826. Z Hyg InfektKrankh 120:36-43

Meineke E, Schier K (1995) Fieber. In: Hoops J, Jankuhn H (eds) Reallexikon der germanischen Altertumskunde, 2nd edn. De Gruyter, Berlin, vol 9, pp 4-11

Møller-Christensen V (1959) Feber. In: Kulturhistorisk leksikon for nordisk middelalder fra vikingetid til reformationstid. Rosenkilde & Bagger/Gyldendal/ Allhem, Kopenhagen Oslo Malmö, vol 4, col 208-10

Newton S (1993) The Origins of Beowulf and the Pre-Viking Kingdom of East-Anglia. Brewer, Cambridge, p 143

Norden W (1984) Eine Bevölkerung in der Krise:

Historisch-demographische Untersuchungen zur Biographie einer norddeutschen Küstenregion (Butjadingen 1600-1850). Lax, Hildesheim (Veröffentlichungen der Historischen Kommission für Niedersachsen und Bremen: Quellen und Untersuchungen zur Wirtschafts- und Sozialgeschichte Niedersachsens in der Neuzeit, vol 11), pp 85-95

Nuttall GHF, Cobbett L, Strangeways-Pigg T (1901) The Geographical Distribution of *Anopheles* in Relation tot the Former Distribution of Ague in England. J Hyg, Camb 1:4-44

Priester P (1998) De geschiedenis van de Zeeuwse landbouw circa 1600-1910. Wageningen Agricultural University, Wageningen (A.A.G. Bijdragen, vol 37), pp 37, 60-66

Ramaker M (1998) In de ban van de beet. De malaria-epidemie in de stad Groningen in 1826. Gronings Historisch Jaarboek 8:66-81

Reiter P (2000) From Shakespeare to Defoe: Malaria in England in the Little Ice Age. Emerg Infect Dis 6,1: 1-11

Roth M (1906) Die Geschichte des Wechselfiebers im Herzogthum Oldenburg. Oldenburger Jahrbuch 28:56-83

Schimitschek E, Werner GT (1985) Malaria, Fleckfieber, Pest: Auswirkungen auf Kultur und Geschichte: Medizinische Fortschritte. Hirzel, Stuttgart, pp 13-70

Schouten J (1920) Geschiedenis der malaria in ons land. In: Eerste jaarrapport van de Malaria-Commissie uit den Centralen Gezondheidsraad, Appendix XV. The Hague

Schuberg A (1927) Das gegenwärtige und frühere Vorkommen der Malaria und die Verbreitung der Anophelesmücken im Gebiete des Deutschen Reiches. Springer, Berlin, Arb R GesundhAmt, vol 59, pp 1-428

Stuart-Macadam P (1992) Porotic Hyperostosis: A New Perspective. Am J Phys Anthropol 87:39-48

Swellengrebel NH, Honig PJJ (1925-28) Bijdrage tot de geschiedenis der malaria in Nederland. Ned Tijdschr Geneeskd 69,1:2538-55; 70,2:1104-20; 1864-75; 72,1:38-47

Swellengrebel NH, de Buck A (1938) Malaria in the Netherlands. Scheltema & Holkema, Amsterdam

Trautmann A (1913) Die Verbreitung der einheimischen Malaria in Deutschland in Vergangenheit und Gegenwart. Arch Hyg Bakteriol 80:84-107

Tunón H (1995) Evaluation of Swedish Medicinal Plants as a Source of Anti-Inflammatory Drugs and Mosquito Repellents. Acta Universitatis Upsalienses, Uppsala (Comprehensive Summaries of Uppsala Dissertations from the Faculty of Pharmacy, vol 136)

van Seventer HA (1969) The Disappearance of Malaria in the Netherlands. Thesis, University of Amsterdam, Amsterdam

Vonderach A (1999), Serogenetische Bevölkerungsdifferenzierungen in Marsch und Geest des Elbe-Weser-Gebietes (Malariahypothese). Stader Jahrbuch 87/88 (1997/98):177-86

Wernsdorfer WJ, McGregor I (eds) (1988) Malaria. Principles and Practice of Malariology. Churchill Livingstone, Edinburgh, pp 1-59, 711-15, 913-98

Wesenberg-Lund, C (1920-21) Contributions to the Biology of the Danish Culicidae. Høst, Kopenhagen, K Danske Vidensk Selsk Sk, 8th Ser, vol 7, pp 1-210

Winkle S (1997) Kulturgeschichte der Seuchen. Komet, Düsseldorf Zurich, pp 707-81, 927, 1292

Patterns of Climate in Central Europe Since Viking Times

Rudolf Brázdil

Masaryk University, Department of Geography, Kotlářská 2,
CZ-611 37 Brno, Czech Republic
e-mail: brazdil@porthos.geogr.muni.cz

Abstract: Basic groups of historical-climatological data sets divided into man-made and natural sources are characterised. The general method of the historical-climatological analysis is described. Main results obtained of historical climatology during the last millennium in central Europe (climate fluctuations, circulation patterns and forcing, climatic anomalies and weather extremes, climate impacts and perceptions) are mentioned and the present state and prospects of future research are defined.

Introduction

Instrumental meteorological measurements, made systematically, provide the highest quality information source for climatic study. In central Europe regular instrumental measurements, especially of air temperature and pressure, generally began in the later half of the 18th century. Some earlier measurements, such as those made at several weather stations from 1718-1726 published in Polish Wroclaw, have yet to be processed. The foundation of the Societas Meteorologica Palatina (1780) was an important development preceding the establishment of national meteorological services, based on organisation of networks of meteorological stations in individual countries. Knowledge of historical climatology before the start of systematic measurements depends on the analysis of both of man-made and natural records. This knowledge is important for increasing understanding of the functioning of the climatic system when fluctuations were forced only by natural variability, and anthropogenic influence operated only on local, or at the most regional scales. This also applies to study of climatic impacts and weather extremes on human society. This paper is concerned with climatic fluctuations during the last millennium in central Europe, some of which also affected other regions.

Data

Data for climatic reconstruction of the pre-instrumental period can be either man-made or based on natural proxy evidence (Fig. 1).

Man-made evidence includes early instrumental records, made before the beginning of systematic meteorological observations, in most cases covering only short periods, or having sporadic character. Lack of detailed knowledge of the instruments used and of their locations makes linkage to the modern reference period difficult.

Descriptive documentary evidence includes both direct information about the weather and about phenomena and events dependent on the weather (for instance freezing of rivers, harvest or vintage). This is an exceptionally rich source, including written reports from annals and chronicles, daily weather reports, personal and official correspondence, travel diaries, economic records, pictures, leaflet newspapers, early journalism and so on. Epigraphic records of extreme weather events include inscriptions and marks on stones, rocks, bridges or buildings. Such records may include either sporadic or continuously recorded data.

Archaeological data, including evidence of extinction and desertion of settlements, have to be taken as resulting from the long-term effects of cli-

From WEFER G, BERGER W, BEHRE K-E, JANSEN E (eds), 2002, *Climate Development and History of the North Atlantic Realm.* Springer-Verlag Berlin Heidelberg, pp 355-368

<table>
<tr><th>Information \ Archives</th><th colspan="2">Natural</th><th colspan="3">Man-made</th></tr>
<tr><td>direct observation of weather and climate or instrumental measurement of meteorological parameters</td><td colspan="2"></td><td rowspan="3">Documentary</td><td>observed
• anomalies
• natural hazards
• weather situations
• daily weather</td><td>measured
• barometric pressure
• temperature
• precipitation
• water-gauge, etc.</td></tr>
<tr><td rowspan="3">indirect references: (Proxy data) indication of controlled or affected processes through meteorological parameters</td><td rowspan="3">organic
• tree rings
• fossil pollen
• animal and plant remains
• fossil wood (trees), etc.</td><td rowspan="3">non-organic
• ice-cores
• varves
• terrestrial sediments
• temperature of boreholes
• moraines, etc.</td><td>organic
• plant phenology
• yield of vine
• time of grain and vine harvest
• sugar content of wine, etc.</td><td>non-organic
• water levels
• snow fall
• freezing of water bodies
• snow cover, etc.</td></tr>
<tr><td colspan="2">• cultural: rogations
• pictorial • epigraphical</td></tr>
<tr><td colspan="3">material: • archeological remains</td></tr>
</table>

Fig. 1. A survey of evidence for reconstructing past weather and climate (after Pfister 1999).

mate and/or climatic extremes, the connection with climate sometimes being ambiguous.

With natural records, it is necessary to extract the climatic signal from a proxy series. As a rule the climatic information obtained relates to a certain type of data, such as air temperature, with a time resolution of months or longer. The most frequently used proxies are those relating to glaciers (changes in extent, ice cores), tree rings (width of rings and density of late wood), palaeobotany (pollen analysis), sediments (varves, erosion furrow fills) and geothermics (profile of ground temperatures).

Each of these data types has both advantages and drawbacks. They vary in degree of resolution, and sensitivity to climatic characteristics and have different geographical limitations. Cross-referencing and combination of such data are essential for achievement of the most complete reconstruction of past climates possible.

Descriptive documentary evidence provides the database for most historical-climatological studies; this has both advantages and limitations. Compared with natural proxy data, it has several outstanding advantages (Pfister et al. 1999):

1. Good dating control and high time resolution.
2. It makes possible disentanglement of meteorological elements, such as temperature, precipitation, snow cover, or wind.
3. It makes it possible to focus on anomalies and weather disasters.
4. It is sensitive to events throughout the year, with no seasonal restrictions.

Its drawbacks include (Pfister et al. 1999):

1. The structure is discontinuous and heterogeneous.
2. It is biased by the selective perceptions of observers.
3. It restricts analysis to simple and robust techniques of mathematical elaboration and interpretation.

For climatic reconstruction in central Europe it is important that descriptive documentary evidence is complemented by proxy data, particularly from tree rings and glacier fluctuations.

Methods

Whatever the character of the initial data, analysis involves the compilation of proxy chronologies which should eventually be interpreted to give a series of the meteorological element considered with respect to a given reference period. Analysis includes the following steps (Fig. 2).

Compilation of the Proxy Chronology

Compilation of the continuous proxy chronology should reach the period of instrumental measurements. Values from proxy evidence are usually obtained directly by measurement; those obtained from documentary evidence by quantitative interpretation of the basic data relating to temperature and precipitation indices. Compilation depends on the density and quality of the basic information, and is to some extent affected by the subjectivism of the researcher. A critical approach to the basic material and the method of processing and standardisation employed should guarantee elimination of inhomogeneities in the series (for instance the differing width of tree rings according to the age of a tree).

Calibration

The aim of calibration is identification of the relation (response function, transfer function) between the proxy characteristic and the meteorological element in the calibration period in which both values of the given proxy and the measured values of the meteorological element are available.

Verification

The relationship obtained is applied to the so-called verification period, for which are calculated the values of the meteorological element. These are then compared with the measured values. Various statistical techniques, including the correlation co-

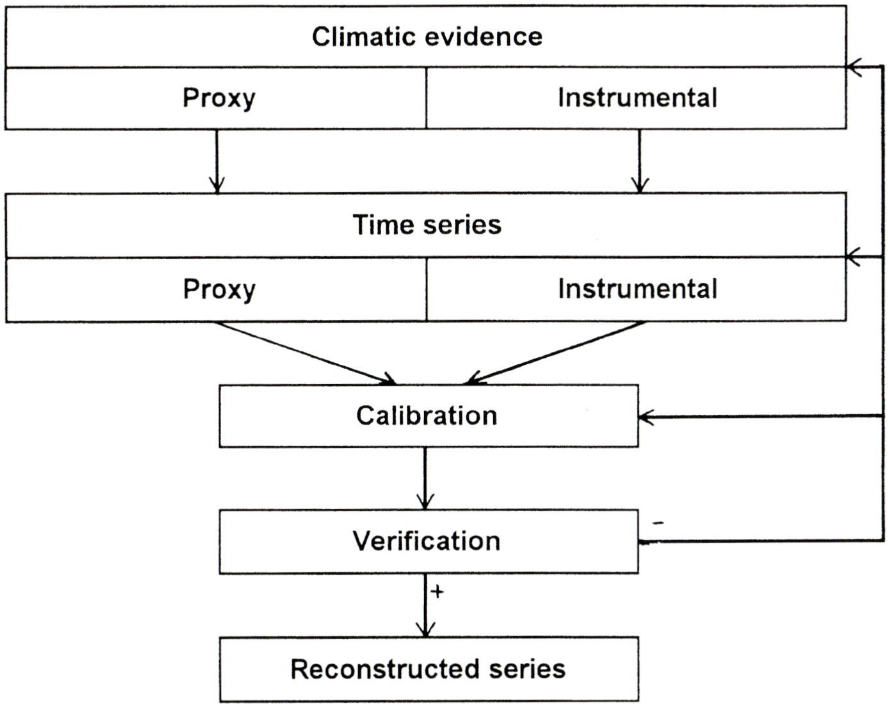

Fig. 2. The diagram of historical-climatological analysis.

efficient, the root-mean-square error and the t-test for the paired values, may be used to compare the measured and reconstructed values. It is also necessary to indicate what part of the variability of the series obtained can be expressed by the given response function. It is practical and useful in view of the normally shorter lengths of the calibration and verification periods, to exchange them and repeat the procedure during the next step.

Reconstruction

If the response function obtained expresses the variability of the series of the meteorological element under consideration with satisfactory precision, the chronology of the proxy can be used for climatic reconstruction. This procedure is based on the assumption that the response function is valid over the whole period of the reconstruction, which is not necessarily the case, if, for instance the phenological series has been affected by the introduction of a new variety, or by agrotechnology.

This methodological approach has been, with certain limitations (for instance with missing verification), applied in many climatic reconstructions in central Europe.

Traditional Division of the Climate of the Last Millennium

The climate of the past millennium has commonly been seen as including a Medieval Warm Period (MWP), a period of climatic deterioration, the Little Ice Age (LIA) and the current warming phase. Lamb (1977) placed the MWP in the period 950-1200 (but in 1150-1300 in the greater part of Europe), and the LIA in 1550-1850, with the most pronounced phase in the period 1550-1700. These terms have commonly been used not only by climatologists but also by those working in related sciences. The difficulties with this use of this terminology can be exemplified by tracing the history of the term LIA.

The concept of the "Little Ice Age" was first introduced by Matthes (1939) who stated that "We are living in an epoch of renewed but moderate glaciation - a "little ice age" that already has lasted about 4,000 years". In another paper he wrote that "... glacier oscillations of the last few centuries have been among the greatest that have occurred during the 4,000 year period ... the greatest since the end of the Pleistocene ice age" (Matthes 1940). But whereas Lamb (1977) places the start of the Little Ice Age in the mid-16th century, Porter (1986), on the basis of glacial studies in the Northern Hemisphere, considered that it had already begun around AD 1250. Pfister et al. (1996, 1999b) recognised that it was under way in central Europe shortly after 1300, because of the pronounced drop found in winter temperatures. Reconstruction of the climate of the Czech Lands (Brázdil and Kotyza 1995) indicates that the period of the 1340s-1380s can be considered a short temperature optimum, during which 60% of the total of extreme months (compared to the modern reference period) were warm. But Grove (1988) in her comprehensive treatise on the LIA left identification of its timing open in view of the distribution of data then available, although she was later convinced that the LIA was under way on a global scale by about AD 1250-1300 (see Pfister and Brázdil 1999; Grove in press). The whole matter was summarised by the conclusions of Jones and Bradley (1992) to the effect that:

a) The last 500 years have not experienced a monotonously cold Little Ice Age; certain intervals have been colder than others.
b) The coldest periods in one region are often not coincident with those in other regions. There is geographical variability in climatic anomalies.
c) Different seasons may show different anomaly patterns over time.

Similar difficulties exist with the use of the term Medieval Warm Period (Hughes and Diaz, eds. 1994). Sparsity of documentary data adds to the compilations in this case. But this sparsity is counterbalanced by the rapidly increasing quantity of data about the timing and extent of glacial fluctuations.

As it follows from above papers, there has been confusion about whether the terms LIA and MWP relate to glaciology or climatology. For instance, when Bradley and Jones (1993) came to combine their various series to give regional or hemispherical values, they produced a composite temperature anomaly series for the Northern Hemisphere which was very much what might have been expected from the glacial evidence. The term Little Ice Age relates to the behaviour of glaciers, not directly to climate. The Little Ice Age was the most recent period during which glaciers extended globally, their fronts oscillating about advanced positions (Grove in press). Their fluctuations were of course controlled by climate, but nonetheless the LIA is not a climatic term as such. Several combinations of meteorological factors may control an advance phase or a retreat phase. Consequently, the Medieval Warm Period can be defined as the period between the glacial advances of the LIA and the preceding phase of glacial expansion late in the penultimate millennium (Grove and Switsur 1994). In this period climatic conditions caused glacier mass balance to be negative, and volumes of glaciers to be reduced, so that they rectracted substantially, though their fronts no doubt fluctuated, as they have been observed to do during the warming of the 20th century. It means, that the LIA and MWP were not periods of unbroken cold and warmth respectively (Grove in press).

Central European Climate Fluctuations During the Last Millennium

Previous studies of central European climate fluctuations in the last millennium have been of the following types:

a) Papers presenting simple and weighted seasonal and annual temperatures or precipitation indices series with annual or decadal resolutions (for instance Glaser 1997; Rácz 1999).
b) Papers presenting seasonal and annual temperature and precipitation series calibrated for the present reference period with the resolution of years or decades (for instance Pfister 1988, 1999; Guiot 1992; Brázdil 1996).
c) Papers concerned with the special effect of a particular meteorological element (for instance Koslowski and Glaser 1999).

These papers do not, for the most part, cover the whole of the last millennium, and area usually concerned only with different parts of it. The extent of agreement between series relating to different parts of central Europe declines with increasing time resolution and increasing distance between the regions studied. This can be attributed not only to the differing quality of the basic data and the subjectivity of its interpretation, but also to the subjectivity of its interpretation, but also to the natural variability of the weather and climate of central Europe. In some cases quality of the conclusions is strongly affected by missing data; seasons and years for which there is no information usually being classed as average. Although there is normally satisfactory agreement concerning the most conspicuous anomalous periods, such as the cold period of the so-called Maunder Minimum in the 1680s and 1690s, or the period of cold, wet summers in the 1560s-1590s, it is difficult to present a comprehensive picture of climate fluctuations in central Europe covering the whole of the last millennium. The best way to achieve this seems to be to use the averaged series, as was the case when the 16th century temperature and precipitation series from Germany, Switzerland and the Czech Lands were averaged and expressed in the form of anomalies against the reference period of 1901-1960 (Pfister and Brázdil 1999). For compiling such series proxy data should be used on a larger scale (such as dendroclimatological or glacial data).

Several examples of reconstructed climatic series of the last millennium for central Europe are presented in Figs. 3-6. The questionable nature of the traditional delimitation of the LIA and MWP is evident from the variation in the annual temperature anomalies of selected grid points in central Europe (Fig. 3). The very cold period from the 1070s to the end of the 12th century, with the intensity of cooling dropping from west to east and from south to north, is particularly outstanding. However the temperature and precipitation anomaly series from the Czech Lands (Figs. 4-5) demonstrate differences in the fluctuations of individual seasons, as do the fluctuations in the 13th and 14th centuries, which are affected by the incomplete nature of the data. The temperature anomalies show that the degree of continentality, with colder winters and warmer summers, was greater over most of the last millennium than it has been in the 20th century. The Swiss series of winter and summer

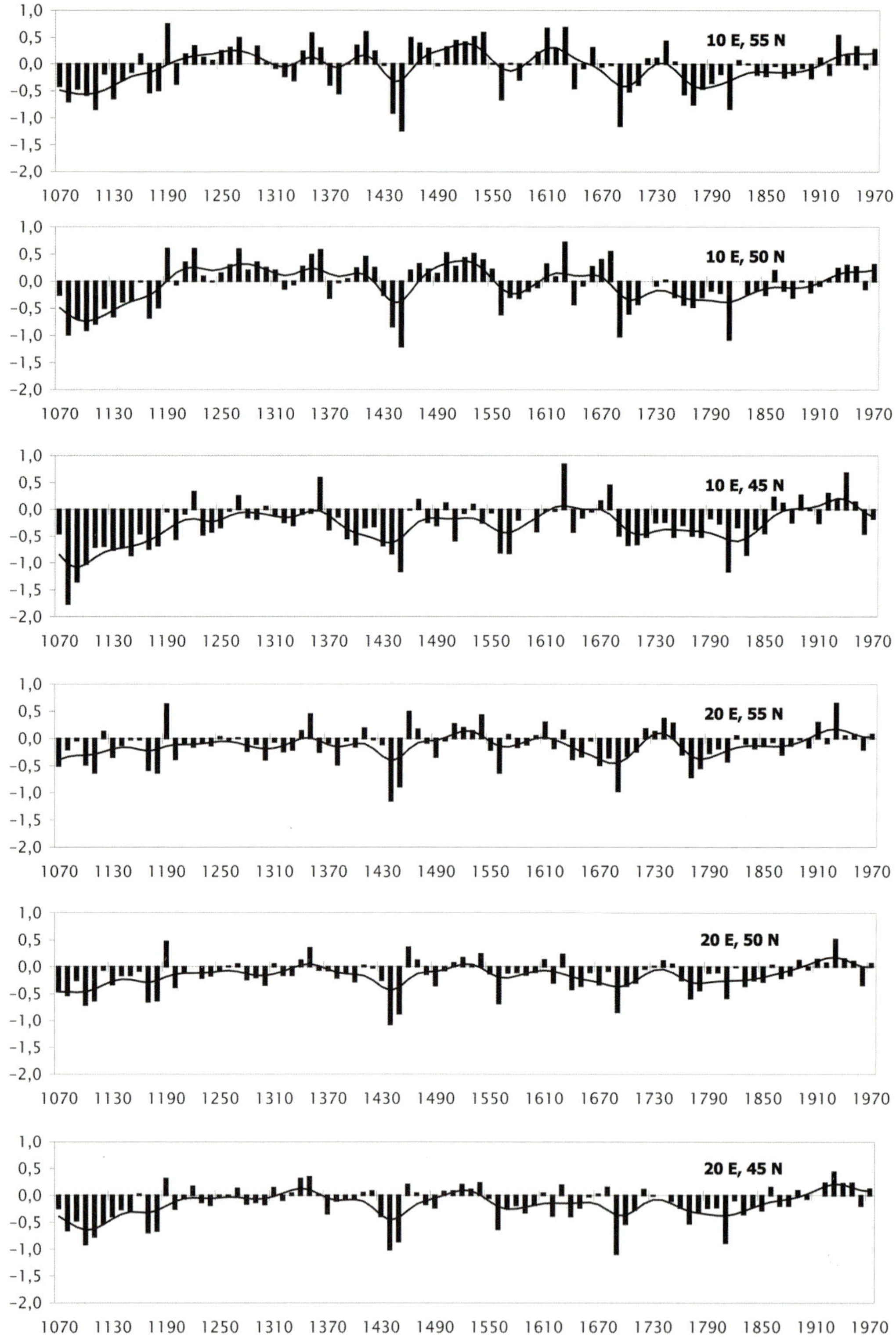

Fig. 3. The fluctuation of ten-year annual temperature anomalies (°C) in selected grid points in central Europe (reference period 1851-1979, data in Guiot 1992). Smoothed by the Gauss filter for 10 items. Years on the x-axis always mean the first year of the respective decade.

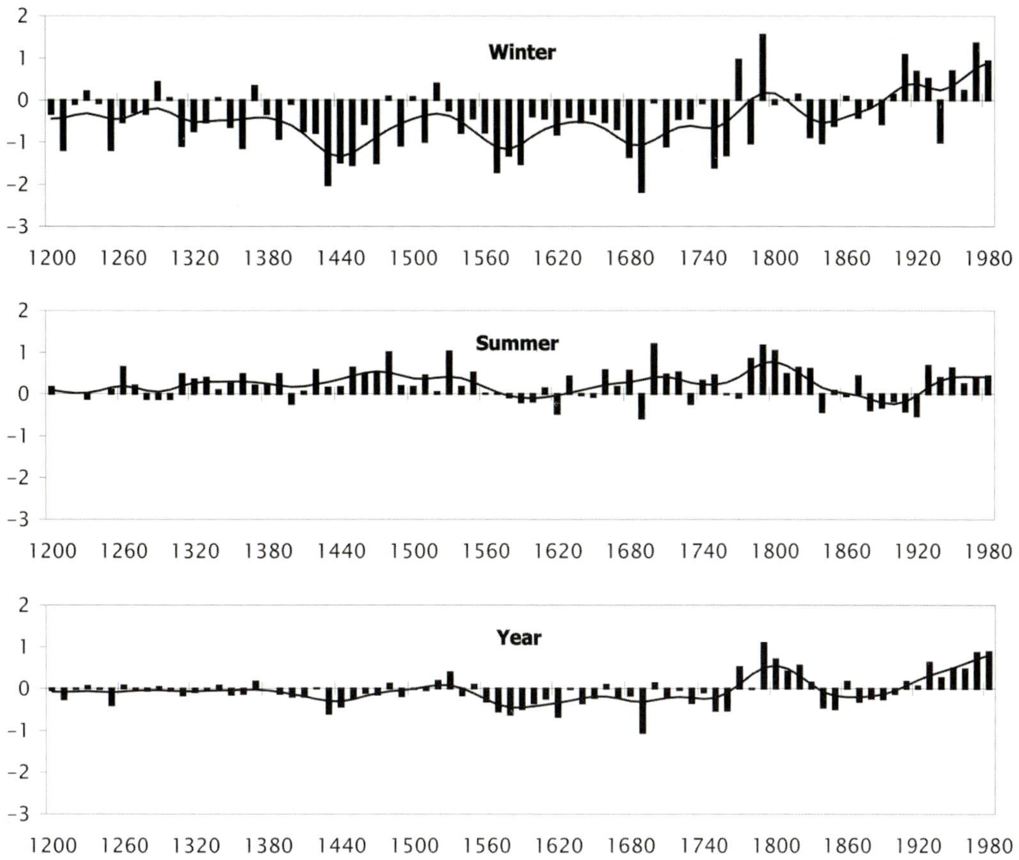

Fig. 4. Fluctuation of ten-year temperature anomalies (°C) of the winter, the summer and the year for Prague-Klementinum (reference period 1851-1950, measured values since 1771, according to Brázdil 1996). Smoothed by the Gauss filter for 10 items. Years on the x-axis always mean the first year of the respective decade.

temperature and precipitation anomalies since the 13th century (Fig. 6) show another view of central European climatic fluctuations at a higher resolution.

Circulation Patterns and Climate Forcing

The climatic characteristics of a given place can be understood as resulting from the interaction of radiation, circulation, geographical and anthropogenic factors. The extent of their influences will differ both in the course of a single year alone and over longer time intervals, such as decades and centuries. The nature of the circulation controls actual weather patterns, explaining, for instance, the occurrence of precipitation extremes and floods. Extension of the data base relating to the climatic history of Europe has made it possible to identify circulation changes on the basis of mean monthly sea-level pressure maps (Fig. 7). This work depends on the assumption that anomalies in temperature and precipitation must be related to circulation fluctuations caused by changes in the positions of the prevailing pressure systems. For the reconstruction of circulation patterns in the past it is possible to make use of situations characterised by the occurrence of anomalous patterns of temperature and precipitation during the period of instrumental measurements. Recently this approach has been used in the completion by more exact

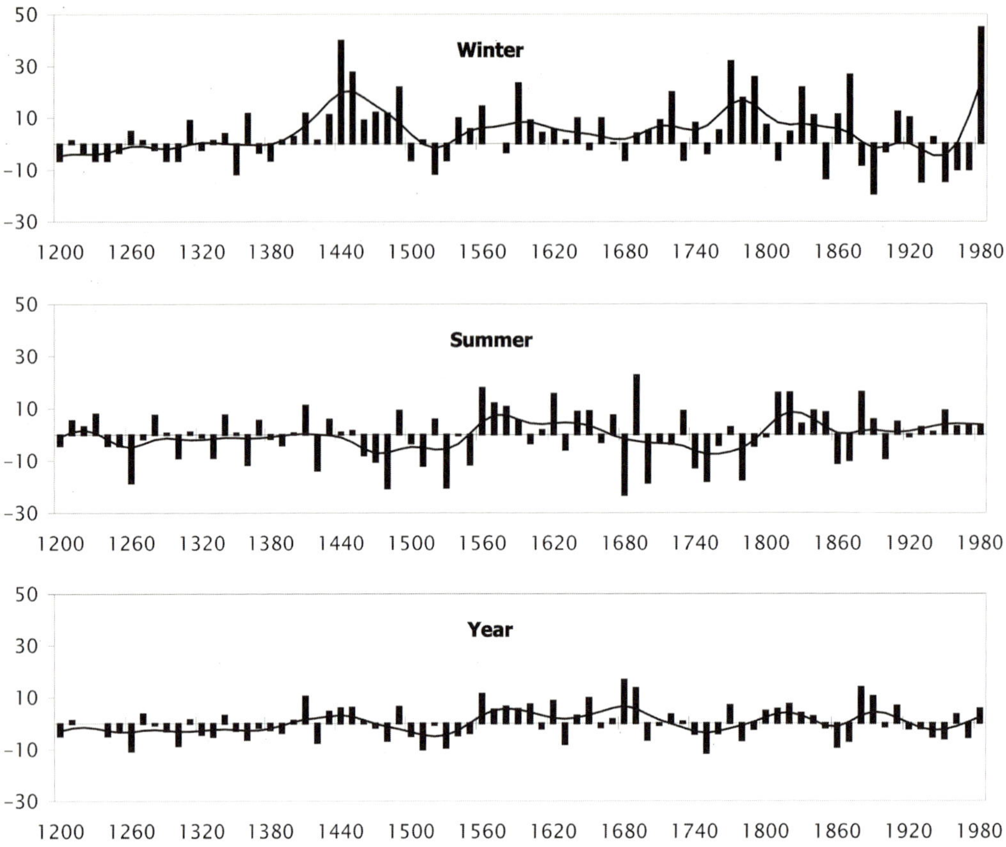

Fig. 5. Fluctuation of ten-year precipitation anomalies (%) of the winter, the summer and the year for Prague-Klementinum (reference period 1851-1950, measured values since 1804, according to Brázdil 1996). Smoothed by the Gauss filter for 10 items. Years on the x-axis always mean the first year of the respective decade.

statistical methods, including canonical correlation, of the analysis of circulation patterns during the Maunder Minimum period (1675-1704) originally carried out by Wanner et al. (1994).

Reconstruction of climate before the period of instrumental observation is important in order to improve knowledge of natural climate variability, when not affected by the activities of man. In central Europe anthropogenic influence occurred on a local as well as regional scale because of activities such as deforestation. As the concentration of greenhouse gases did not change sufficiently to affect climate on the pre-industrial part of the past millennium, the main forcing factors were evidently solar variation, volcanism, and changes in the ocean-atmosphere system, associated with the operation of factors such as ENSO and the conveyor belt. The traditional approach is to identify linkages between meteorological series and those of natural forcing factors such as ^{10}Be as an indicator of solar activity and various indicators of volcanic activity (see Pfister and Brázdil 1999). General circulation models (GCMs) make it possible to analyse past climatic forcing on a broad scale. Thus Rind and Overpeck (1994), Rind (1996) and Mann et al. (1998) showed the significant role of solar forcing, volcanic activity and internal variations in the ocean-atmosphere system. According to Hunt (1998), using a 500-year GCM run, climatic anomalies associated with the MWP and LIA

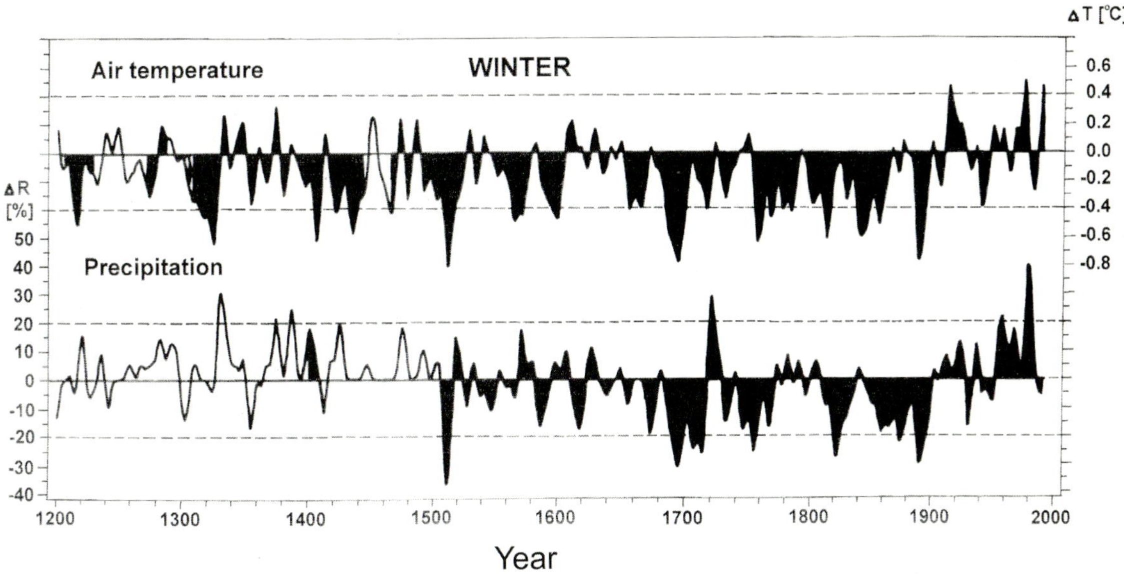

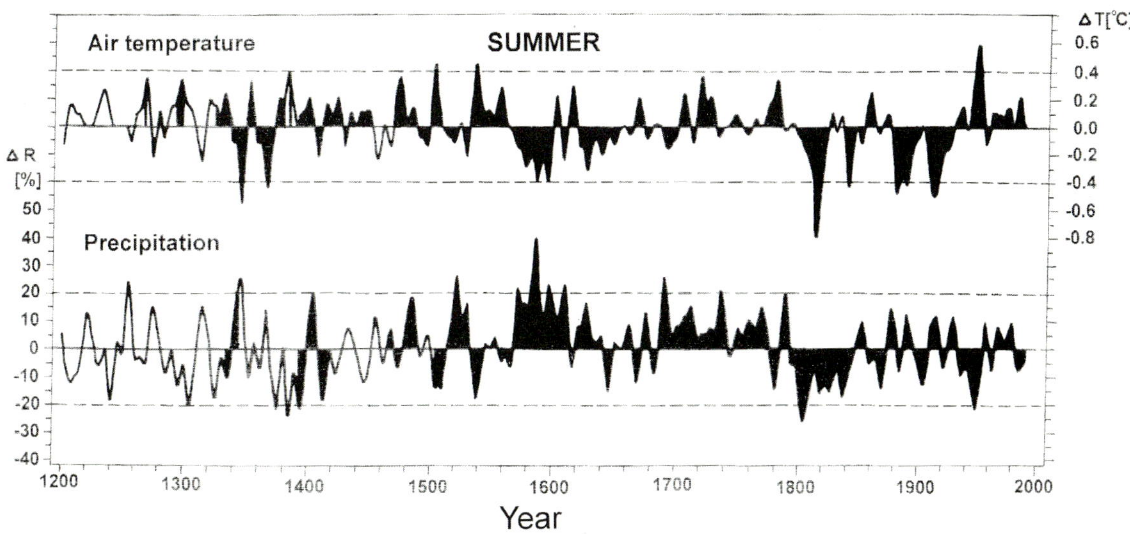

Fig. 6. Fluctuation of air temperature ΔT and precipitation ΔR anomalies smoothed by eleven-year running averages in Switzerland (reference period 1901-1960). For sectors not marked in black the information base is still insufficient (Pfister 1991).

can only be attributed to the internal mechanisms of the climatic system. His results do not exclude the possible involvement of external forcing agents, though these would enhance existing naturally induced climatic features rather than initiating them. Historical-climatological data are of essential importance for verification of the GCM.

Climatic Anomalies and Weather Extremes

Weather extremes and climatic anomalies have always affected the activities of man and human society, to some degree, as they do at the present time, when the material damage they cause is increasing. The increase is partly due to greater in-

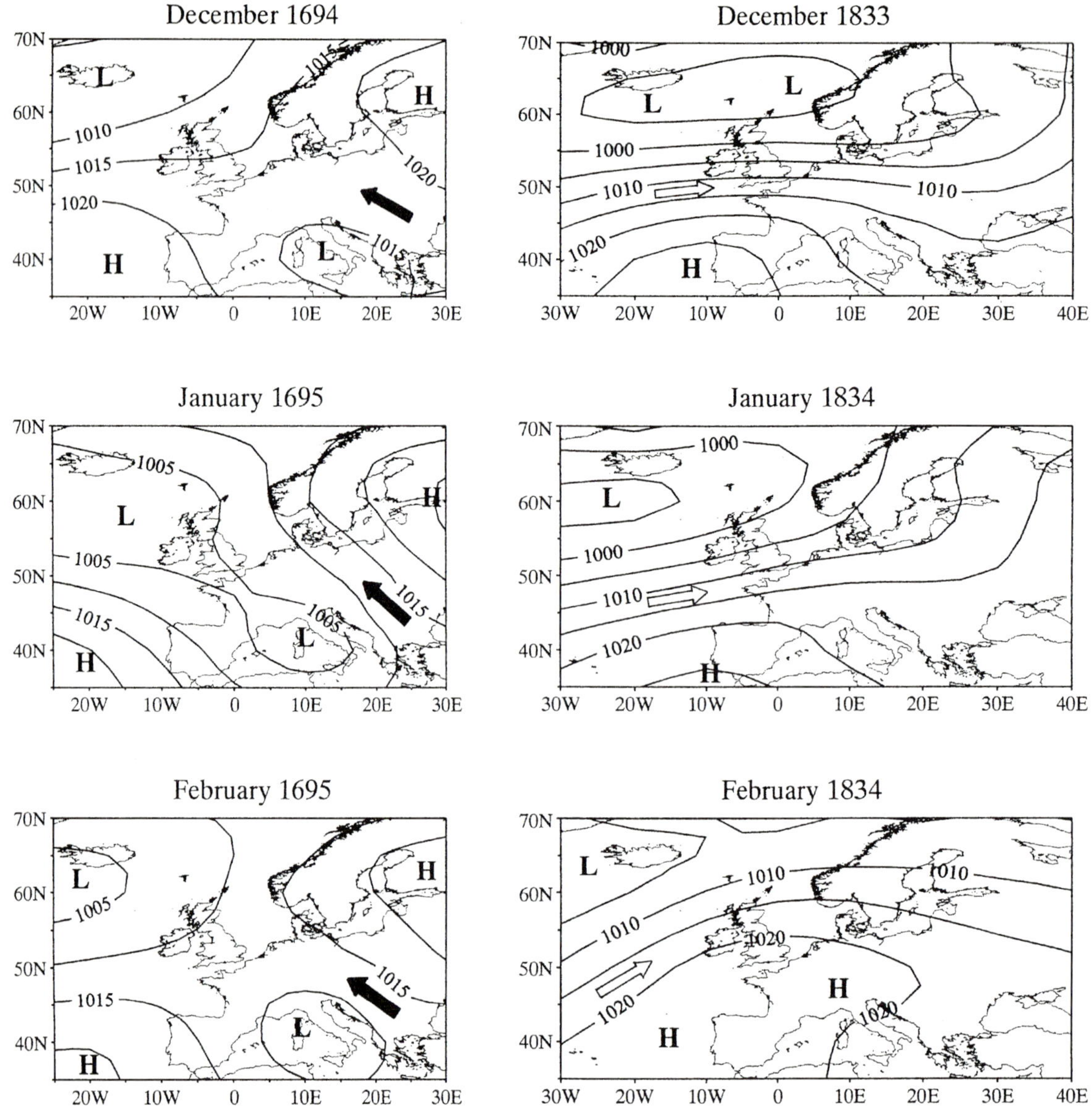

Fig. 7. Examples of reconstructed mean monthly sea-level pressure charts in the Atlantic-European region (maps on the left side according to Luterbacher et al. 2000, on the right side according to Jones et al. 1987): black arrows – cold air advection, white arrows – warm air advection.

tensity of settlement and construction in places which are liable to be affected by flooding, landslides etc. Even nowadays it is impossible to avoid considerable loss of human life, as with floods in the Czech Republic and in Poland in July 1997. Documentary descriptions makes possible to extend series of such extremes as floods, gales and hailstorms far into the past. Such evidence provides information about their intensity, seasonality, causes and impacts. This is all the more important because series of this sort from the instrumental period are relatively short. Contemporary extremes can also provide analogs for historical events.

Systematic records of the impact of extremes are useful for compilation of their chronologies. Thus De Kraker (1999) utilised data from accounts of dikes protecting a polder region in northern Flanders to trace the occurrence of high tides, storms and storm surges in the period 1488-1609.

Weather extremes have immediate consequences, but the effects of climatic anomalies may be time lagged. This was demonstrated by Pfister's (1999) study of anomalies in monthly temperature and precipitation in Switzerland, which took into account occurrence of warm or cold, dry or wet conditions, and their combinations. In the period 1560-1895, overlapping for the most part with the LIA, 71% of anomalous months were cold, of these 40% were rated cold, 17% cold-dry and 14% cold-wet. Between 1896 and 1988 the proportion of anomalously cold months dropped to 56%, while between 1988 and 1997, during the recent warming trend, cold anomalies completely disappeared. Figs. 8a, b show changes in the anomalies of individual months expressed in terms of their ten-year frequencies. A more noticeable increase in climate variability from the latter half of the 16th century to the mid-18th century is evident in Fig. 8c.

Climate Impacts and Perceptions

Climatic impacts on human society during the last millennium has received only rather random attention. This has resulted firstly from long standing unwillingness to admit the possibility of climate and weather affecting historical events and processes, and secondly from the insufficiency of the basis of versions of climatic history available for particular geographical regions, which have not been calibrated using the observed characteristics of climate during the period of instrumental measurements. On the other hand, the use of unverified documentary sources has also been important (see Parry 1978). The use of general terms such as the MWP and LIA in comparisons of climatic conditions with developments in human history and society did not lead to useful results and consequently led to the refusal to consider or the overlooking of the role of climate in historical analysis.

The influence of climatic factors is most commonly considered in relation to the need to safeguard the food supply of a population, as the degree of success of the harvest is fundamentally affected by variations in the weather though other factors such as wars, diseases, and the prevalence of pests, can also affect the outcome. Thus, for instance, Bauernfeind and Woitek (1999) analysed quantitatively the impact of seasonal temperature and precipitation on the price fluctuations of several commodities, including rye, butter and honey, at Nuremberg, Cologne, Augsburg and Munich during the period 1490-1620. In the last third of the 16th century climatic changes became the most significant factor affecting food prices.

In central Europe viticulture is at its northern limit and therefore the growth, yield and sugar content of grapes are closely dependent on climate. Wine was an important source of income for institutions and for the landed gentry on the basis of tithes, taxes or sharecropping. Landsteiner (1999) compared wine production in the Zurich area, Württemberg, Lower Austria and western Hungary in the period 1550-1630. The most striking feature in all four series was a slump in wine production in the 1580s initiating a consecutive series of crop failures continuing until the late 1590s or the early 1600s. These patterns fit with the long sequence of cold, wet summers. It had far-reaching consequences. In Lower Austria the collapse of wine production, leading to high prices, led the public to switch from wine to beer consumption. The Hapsburg Emperor, who drew a substantial part of his income from wine production, had to double the heavy export duty on wine in order to be able to meet the interests on the state debt. Moreover harvest failures put a severe strain on the budgets of urban wine producers, and deficient grain harvests reduced the real wages of the vine dressers by 50%.

The connections between large-scale witch-hunting and the distinctly worsening weather in central Europe between the 15th and the late 17th centuries demonstrate the vulnerability of society to meteorological shocks. Behringer (1999) found a dramatic increase in witch-burnings after the early 1560s, that is after the worsening of climate. Most witches were burned for weather-making, which was perceived as having caused crop failures, floods and cattle diseases related to the "unnatural weather", that is to weather sequences or single extreme events of a type which nobody had seen or experienced before.

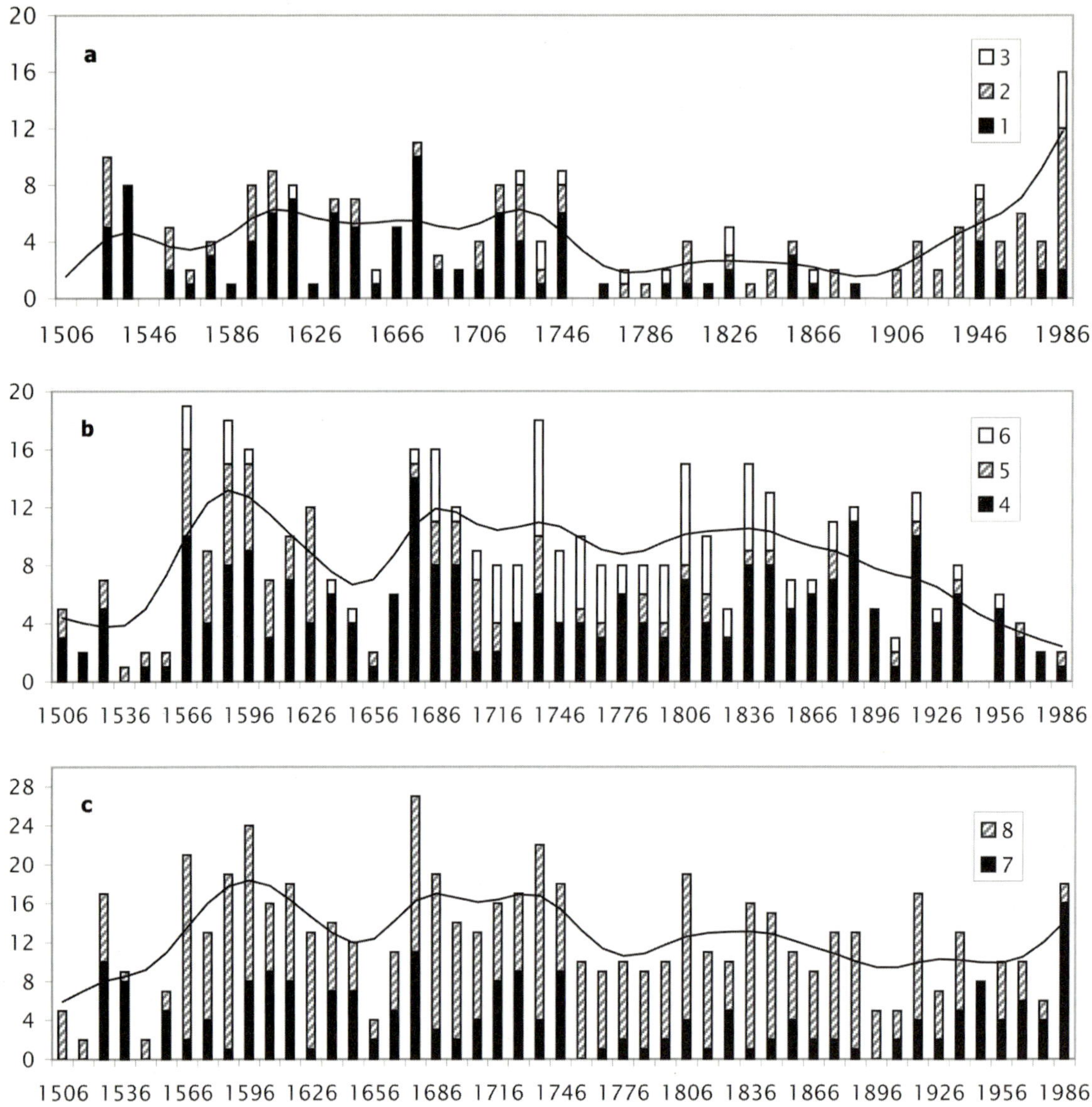

Fig. 8. Ten-year frequencies of warm **(a)**, cold **(b)** and warm+cold **(c)** monthly temperature anomalies in Switzerland in the period 1506-1995 (data according to Pfister 1999): 1 - warm, 2 - warm-dry, 3 - warm-wet, 4 - cold, 5 - cold-wet, 6 - cold-dry, 7 - warm, warm-dry, warm-wet, 8 - cold, cold-wet, cold-dry. Smoothed by the Gauss filter for 10 items. Years on the x-axis always mean the first year of the respective decade.

Prospects for Further Research

Investigations in historical climatology have made it possible to obtain a view of the main features of climatic fluctuations during the last millennium in central Europe. This has depended on the use of reference to the contemporary climate of the 20th century. Arising out of the knowledge obtained by the reconstruction of the 16th century climate (Pfister et al. 1999a) further research should be directed to:

a) Complementation of the existing data base, and compilation of series of temperature and precipitation indices for individual countries and regions and their cross-control (including identification of their common features and differences).

b) The reconstruction of temperature and precipitation in the form of deviations from the reference period, and compilation of averaged series for central Europe, as a basis for the study of their fluctuations and identification of climate forcing.
c) Reinforcement of the new mathematical-statistical methods used in the reconstruction of temperature and precipitation series on the basis of descriptive documentary evidence (for instance similar like in dendroclimatological analysis).
d) Compilation of series of weather extremes and their evaluation in terms of intensity, causes, seasonality, impacts and long term trends
e) Reconstruction of circulation patterns as a basis for understanding climatic fluctuations and the clarification of regional differences.
f) Closer engagement with climatic modellers and use of GCM outputs in the study of climatic forcing and for overcoming drawbacks in documentary evidence.
g) Use of high-resolution data on temperature, precipitation and weather extremes as inputs in impact studies intended to illuminate the effects of climate on history and society.

In future special attention must be paid to the advantages and drawbacks of the types of data used to further knowledge of the history of climate. The current degree of collaboration in historical climatology and the potential for these studies already revealed in central Europe give reason for optimism about the prospects of obtaining further and deeper knowledge of the climatic variability during the last millennium.

Acknowledgments

The author would like to thank Jean Grove, Cambridge, for style corrections and raising critical questions as well as Jürg Luterbacher and Elena Xoplaki, Bern, for preparation of Fig. 7. This study has been elaborated thanks to the financial support of the Grant Agency of the CR in awarding Grant No. 205/98/1542.

References

Bauernfeind W, Woitek U (1999) The influence of climatic change on price fluctuations in Germany during the 16th century price revolution. Climatic Change 43:303-321

Behringer W (1999) Climatic change and witch-hunting: The impact of the Little Ice Age on mentalities. Climatic Change 43:335-351

Bradley RS, Jones PD (1993) 'Little Ice Age' summer temperature variations: their nature and relevance to recent global warming trends. The Holocene 3:387-396

Brázdil R (1996) Reconstructions of past climate from historical sources in the Czech Lands. In: Jones PD, Bradley RS, Jouzel J (eds) Climatic Variations and Forcing Mechanisms of the Last 2000 Years. Springer-Verlag, Berlin, Heidelberg, New York, pp 409-431

Brázdil R, Kotyza O (1995) History of Weather and Climate in the Czech Lands I. Period 1000-1500. Zürcher Geographische Schriften 62, Zürich, pp 1-260

De Kraker AMJ (1999) A method to assess the impact of high tides, storms and storm surges as vital elements in climatic history: The case of stormy weather and dikes in the northern part of Flanders, 1488 to 1609. Climatic Change 43:287-302

Glaser R (1997) On the course of temperature in Central Europe since the year 1000 A.D. Historical Social Research 22:59-87

Grove J (1988) The Little Ice Age. Methuen, London, pp 1-498

Grove J (in press) The initiation of the Little Ice Age in regions round the North Atlantic. In: Jones PD, Davies TD, Ogilvie AEJ, Briffa KR (eds) Climate and Climatic Impacts Through the Last 1000 Years. University of East Anglia

Grove JM, Switsur R (1994) Glacial geological evidence for the Medieval Warm Period. Climatic Change 26:143-169

Guiot J (1992) The combination of historical documents and biological data in the reconstruction of climate variations in space and time. In: Frenzel B, Pfister C, Gläser B (eds) European Climate Reconstructed from Documentary Data: Methods and Results. Gustav Fischer, Stuttgart, Jena, New York, pp 93-104

Hughes MK, Diaz HF (eds) (1994) The Medieval Warm Period. Kluwer, Dordrecht, Boston, London, pp 109-342

Hunt BG (1998): Natural climatic variability as an explanation for historical climatic fluctuations. Climatic Change 38:133-157

Jones PD, Bradley RS (1992) Climatic variations over the last 500 years. In: Bradley RS, Jones PD (eds) Climate Since A.D. 1500. Routledge, London, New York, pp 649-665

Jones PD, Wigley TML, Briffa KR (1987) Monthly mean pressure reconstruction for Europe (back to 1780) and North America (back to 1858). DOE Technical Report No. TR37. U.S. Dept. of Energy, Washington, D.C., pp 1-99

Koslowski G, Glaser R (1999) Variations in reconstructed ice winter severity in the Western Baltic from 1501 to 1995, and their implications for the North Atlantic Oscillation. Climatic Change 41:175-191

Lamb HH (1977) Climate: Present, Past and Future. Vol. 2: Climatic history and the future. Methuen, London, pp 1-835

Landsteiner E (1999) The crisis of wine production in late sixteenth-century central Europe: Climatic causes and economic consequences. Climatic Change 43:323-334

Luterbacher J, Rickli R, Tinguely C, Xoplaki E, Schüpbach E, Dietrich D, Hüsler J, Ambühl M, Pfister C, Beeli P, Dietrich U, Dannecker A, Davies TD, Jones PD, Slonosky V, Ogilvie AEJ, Maheras P, Kolyva-Machera F, Martin-Vide J, Barriendos M, Alcoforado MJ, Nunes MF, Jónsson T, Glaser R, Jacobeit J, Beck C, Philipp A, Beyer U, Kaas E, Schmith T, Bärring L, Jönsson P, Racz L, Wanner H (2000) Monthly mean pressure reconstruction for the Late Maunder Minimum period (AD 1675-1715). Int J Climatol 20:1049-1066

Mann ME, Bradley RS, Hughes MK (1998) Global-scale temperature patterns and climate forcing over the past six centuries. Nature 392:779-787

Matthes F (1939) Report of Committee on Glaciers. Transactions American Geophysical Union 20:518-523

Matthes F (1940) Committee on Glaciers, 1939-40. Transactions American Geophysical Union 21:396-406

Parry ML (1978) Climatic Change, Agriculture and Settlement. Folkstone, Dawson

Pfister C (1988) Klimageschichte der Schweiz 1525-1860. Das Klima der Schweiz von 1525-1860 und seine Bedeutung in der Geschichte von Bevölkerung und Landwirtschaft. Paul Haupt, Bern, Stuttgart, Bd. I, pp 15-184 and Anhang, Bd. II, pp 15-163 and Anhang

Pfister C (1991) Belastung und Belastbarkeit. Klima und Energie in der Geschichte Europas seit dem Mittelalter. In: Die Menschen und das Klima. Ascom Holding AG, Bern, pp 12-19

Pfister C (1999) Wetternachhersage. 500 Jahre Klimavariationen und Naturkatastrophen (1496-1995). Paul Haupt, Bern, Stuttgart, Wien, pp 1-304

Pfister C, Brázdil R (1999) Climatic variability in sixteenth-century Europe and its social dimension: A synthesis. Climatic Change 43:5-53.

Pfister C, Schwarz-Zanetti G, Hochstrasser F, Wegmann M (1996) Winter severity in Europe: the fourteenth century. Climatic Change 34:91-108

Pfister C, Brázdil R, Glaser R (eds) (1999a) Climatic Variability in Sixteenth-Century Europe and Its Social Dimension. Kluwer, Dordrecht, Boston, London, pp 1-351

Pfister C, Brázdil R, Glaser R, Barriendos M, Camuffo D, Deutsch M, Dobrovolný P, Enzi S, Guidoboni E, Kotyza O, Militzer S, Rácz L, Rodrigo FS (1999b) Documentary evidence on climate in sixteenth-century Europe. Climatic Change 43:55-110

Porter SC (1986) Pattern and forcing of Northern Hemisphere glacier variations during the last millennium. Quarternary Research 26:27-48

Rácz L (1999) Climate History of Hungary Since 16th Century: Past, Present and Future. Centre for Regional studies of Hungarian Academy of Sciences, Pécs, pp 1-158

Rind D (1996) The potential for modeling the effects of different forcing factors on climate during the past 2000 years. In: Jones PD, Bradley RS, Jouzel J (eds) Climatic Variations and Forcing Mechanisms of the Last 2000 Years. Springer-Verlag, Berlin, Heidelberg, New York, pp 563-581

Rind D, Overpeck J (1994) Hypothesized causes of decade-to-century-scale climate variability: climate model results. Quarternary Science Reviews 12:357-374

Wanner H, Brázdil R, Frich P, Frydendahl K, Jonsson T, Kington J, Pfister C, Rosenorn S, Wishman E (1994) Synoptic interpretation of monthly weather maps for the late Maunder Minimum (1675-1704). In: Frenzel B, Pfister C, Gläser B (eds) Climatic Trends and Anomalies in Europe 1675-1715. Gustav Fischer, Stuttgart, Jena, New York, pp 401-424

On the Holocene Water Balance in Central Europe and Several Historical Consequences

Klaus-Dieter Jäger

University Halle-Wittenberg, Department of Art, Oriental and Antiquarian Sciences, Brandbergweg 23c, D-06120 Halle (Saale), Germany

Abstract: Climatic conditions have influenced the patterns of human settlement and activities for many millennia. Combining information from archeological and paleoclimatic studies, these patterns can be reconstructed. The nature and frequency of extreme events is of special significance in affecting human history, and water supply plays a crucial role. In Central Europe, sedimentary sequences with alternating fresh-water deposits and buried soil horizons testify to change in availability of water. Archeologic objects are quite common within the buried soils and can be used for age assignment. Soil formation and lake expansion correlate over large areas, emphasizing the climatic origin. During times of drought, settlements moved closer to water sources with shifts both toward lake shores and to higher altitudes. A prominent record of human response to drought cycles in central Europe is represented in the so-called "Burgenhorizonte" (fort horizons) representing times of construction of fortifications against sporadic immigration from drought-stricken regions, e.g. during the Bronze Age. At the end of the regional Bronze Age the climate became more continental with significant consequences for human settlement, culture, and history.

Introduction

An intelligent discussion of future climatic developments substantially depends on our knowledge of past climatic conditions and their changes. Unfortunately, the climate record based on direct measurements is restricted to the last 150 years or so. Trends and oscillations exceeding this time-range cannot be captured by this record. However, the climatic record can be extended back to several preceding centuries by applying man-made proxies from historical sources (Pfister 1997; Glaser 2001; Deutsch 1994). The shortcomings of proxy records can be diminished by the use of multiple indices for the same target information (that is, independent control) but they can never be eliminated entirely.

Using proxies from the geo- and biosciences (Fischer and Wefer 1999) likewise can greatly widen the foundation of climatic evidence comprising the centuries prior to the start of instrumental record. Furthermore, such proxies open the door to climate reconstruction for the millennia preceding any written record.

Even before the time of written records, of course, climatic conditions have influenced the patterns of human history as reflected by the traces of settlement and human activities. The discovery and examination of such patterns is a task of archaeology. By means of archaeological observations we should be able to find traces of climatic change and human activities for the periods before the appearance of written documents. Using such proxies in combination with climatic indicators from the findings of geo- and biosciences greatly enriches the palaeoclimatic record. The contributions from archaeology in this context are threefold:

From WEFER G, BERGER W, BEHRE K-E, JANSEN E (eds), 2002, *Climate Development and History of the North Atlantic Realm.* Springer-Verlag Berlin Heidelberg, pp 369-375

1) Archaeological objects can provide precise time-markers which can help to date past climatic events or conditions;
2) archaeological analysis of settlement patterns bring out not only the fact, but also social consequences of climatic change;
3) the interpretations of archaeological observations are commonly methodically independent from those of natural sciences.

Climate is a complex phenomenon and has many components with different consequences with respect to human beings (Freye 1978; Hentschel 1982). Therefore, as far as the course of climatic development during the last 10 (and more) millennia is concerned, it does not suffice to reconstruct the annual mean values and their seasonality. In ad-dition, the occurrence of extreme events and their frequency has to be considered, as they can affect human history in important ways.

A Focus on Water Supply

Commonly, climate reconstruction proceeds from a focus on palaeotemperature, as deduced from the palaeontological examination of fossils (such as pollen, molluscs etc.) from Holocene sedimentary sequences (Starkel 1977). In the context of human history precipitation and water supply are at least equally important.

Reconstruction of the past water supply and its changes throughout the last millennia is a rather complicated task. This is why it is necessary to gather all available proxies from palaeoecological and geological investigations related to sedimentary sequences as well as from archaeological examinations based on settlement patterns to obtain clues about the entire water supply of any landscape unit. The water supply reflects the regional water balance, but the balance integrates a number of factors besides average precipitation, such as temperature, evaporation and seasonality.

One approach to the detection of short- and long-term changes of the water balance in the interior of Central Europe has been a geological one – the comparative stratigraphical examination of Holocene calcareous fresh-water deposits. In such sedimentary sequences, there is commonly an alternation of calcareous layers and buried humus horizons (of rendzina type). The former originated under subaquatic conditions, the latter testify to dry terrestrial circumstances.

Remarkably, archaeological objects are quite common in the buried humus horizons. Thus, the various temporary surfaces can be dated by means of archaeology, and – independently from their respective local stratigraphic sequence – a chronological correlation is rendered possible. Available observations cover especially the southern part of Central Europe (Fig. 1), with evidence for an alternation of periods favouring the subaquatic deposition of carbonates with others favouring the occurrence of buried soils (Jäger 1982, 1999).

Periods dominated by soil formation and periods dominated by lime deposition, as seen in the local sequences, are synchronous over a vast area and suggest climatic factors covering an area of com-parably large extension.

There is an interesting relationship between the dry-wet cycles in the interior of Central Europe and apparent sea-level variations in the coastal regions of the Netherlands and adjacent countries as seen in the displacement of shorelines in the marginal zone of the North Sea (Pons et al. 1963). The chronology of these transgressions and regressions is based on many ^{14}C datings and supported by archaeological observations. Upon comparing the dates of shoreline displacements with those of the alternation of buried soils and calcareous layers in the stratigraphy of fresh-water lime deposits in the interior regions of Central Europe, it appears that periods favouring carbonate deposition at the inland sites correspond to those of sea transgression at the coast of the Netherlands (Fig. 2). Conversely, periods favouring soil formation at these sites (evidenced by buried soils) correspond to time slices of sea regression at the Dutch coast. The general agreement suggests a common climatic cause (Hageman and Jäger 1992).

The poor fit near 5000 BC probably is due to difficulties in dating; a 500 year-shift in boundaries would bring the stratigraphies into alignment.

Recent investigations of lake- level changes in Central and Northern Europe during the last decades (Joos 1982; Jäger 1987; Harrison, Frenzel et al. 1998) indicate that contemporaneous periods of

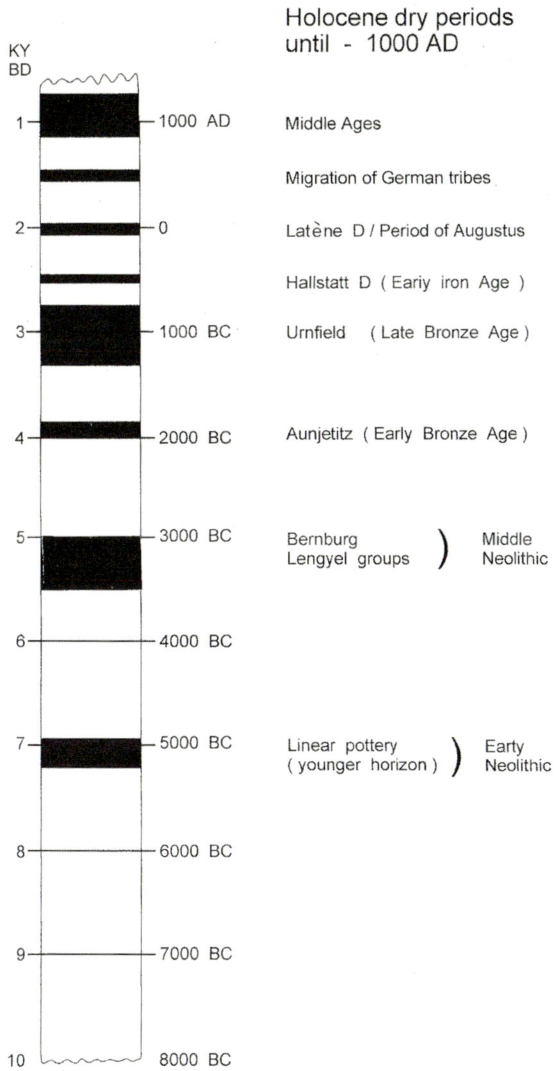

Fig. 1. Holocene dry periods in the Central European inland as derived from stratigraphical sequences in fresh-water lime deposits (from Jäger 1999).

high and low lake levels occurred throughout Europe.

Evidence for synchroneity in dry-wet cycles is further strengthened by a chronological coincidence of low level periods in many inland lakes with the periods of frequent soil formation in the Holocene tufa sequences of southern Central Europe and those of sea regression at the coast of the North Sea. Human response to alternating lake levels in the inland is well known since the middle of the 19th century. It is the displacement of ancient beach settlements, evidenced by the archaeological traces of previous sites and at present below actual lake level, which enhances the preservation of findings as well as precise archaeological dating.

In the southern part of Central Europe the so-called lake-dwellings ("Pfahlbauten" in the German archaeological literature) illustrate one of the reaction patterns of human beings to these changes. In the northern part of Central Europe similar observations are available from the lakes in the circum-Baltic young moraine regions. This may be exemplified by the Schwielochsee in eastern Brandenburg (Christl 1988) and the Mueritz in Mecklenburg (Prehn 1987), as well as by lakes in Poland (Niewiarowski 1976).

Other contemporaneous changes of human settlement imply the same duality. The periods of low lake levels correspond to times when sites moved closer to water sources in the dry hill areas, e.g. in Thuringia during the Late (Urnfield) Bronze Age. During the same periods the upper boundary of the settlement belt and of proven land use in the mountainous uplands of interior Central Europe shifted to higher altitudes (Christl and Simon 1995), partly to more than 1000 m a.s.l. (Liptovská Mara in Slovakia).

Clearly, the main reason for this shift to higher altitudes was the need to find water. The annual mean precipitation has increased – as presently the case – from the foothills to the mountains. An archaeological study related to the Thuringian Forest (Thüringer Wald) has demonstrated that during the Bronze Age this shift of the settlement area up the mountain slopes took place step by step (Donat 1969) in an apparent attempt to cope with increasing drought in the Late Bronze Age.

The utilization of caves by prehistoric people in the karstified regions of Central Europe is of some interest in the climatic context. The periods centred on ca. 3000 B.P. with low lake levels (i.e. low water supply) provides a lot of archaeological traces of human presence in the caves (Walter 1985). As a rule, increasing precipitation causes increasing moisture in the hollows covered by a porous limestone roof. Presumably, people did not like to live in a wet cave. However, an exception to the rule is a cave system near Bad Frankenhausen in north-

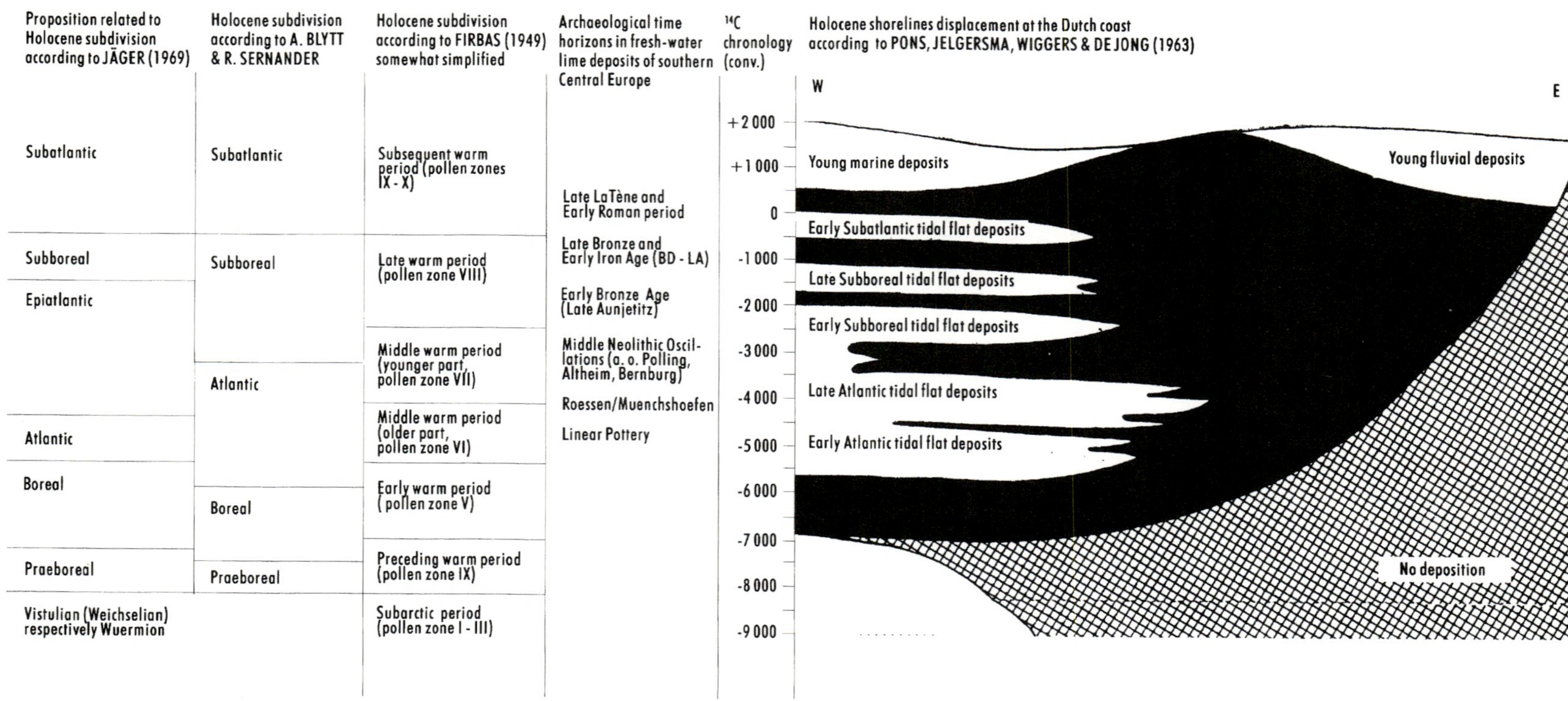

Fig. 2. Correlation of dry periods in the Central European inland and shorelines displacement at the Dutch coast (from Hageman and Jäger 1992).

ern Thuringia. It has yielded a rich collection of archaeological materials from the Middle Bronze Age, probably a wet period.

The caves of Bad Frankenhausen have originated in an anhydrite rock formation where water from precipitation causes its transformation to gypsum which expands and seals the roof. Therefore, even during periods of moisture the caves remained dry, and access was not affected. Factors of religious or social nature presumably can affect cave occupation as well, but the simplest explanation – whether caves changed from wet to dry or vice versa – should be considered first.

Climate, Migrations and Settlement Density

During the last millennia, just before the appearance of the first written records related to climatic history, a repeated alternation of relatively dry and comparatively moist periods took place in Central Europe.

Following a long dry period near 1000 BC there were several shorter periods, roughly once every ca. 400 or 500 years. Assuming that these dry periods are real (as suggested by both archaeological and palaeoecological findings) it should be asked what may have caused them. Several possibilities arise:

1) Decrease of precipitation,
2) increase of evaporation due to higher temperature and/or stronger winds.
3) a combination of both (1 and 2),
4) changes of the seasonality of precipitation and temperature,
5) human impact on water discharge and evaporation.

If the alternation of relatively dry and moist periods should be accepted, another chronological coincidence will likely gain interest. In the early fifties, the German archaeologist Neumann (1954) already introduced the concept of so-called "Burgenhorizonte", i.e. time slices favouring the foundation of numerous fortifications as shown by archaeological excavations, inventories, and maps throughout all parts of Central Europe. Neumann (1954) has exemplified this concept by means of a region in southern Thuringia, but subsequent research confirms the regional validity of the burgenhorizont concept on a broader scale in Central Europe.

The suggestion is that dry periods have caused migrations out of drought-stricken regions, as for example the Pannonian Basin or inner Thuringia where archaeological evidence suggest a decrease of the population accompanied by partial emigrations. The destination of emigrants, we should assume, were regions where the shortage of the water supply was of less concern – such regions have been occupied more or less everywhere. Thus, conflict could arise between previous residents and new immigrants. One can readily imagine that relatively dry periods became times of unrest, with an increased need for protection, giving rise to an increasing number of fortifications. In the same vein, the manufacture of arms flourished during such times of unrest, as exemplified by two time-slices of the Central European Bronze Age – the early part and the late one (Urnfiel Bronze Age). Both these periods may be taken as times of unusual dryness.

We postulate three types of population movement during the Urnfield period:

1) Redistribution of settlements within narrow bounds aiming at an approximation to water sources and resources, such as the beaches of lakes, rivers, and creeks.
2) Shifting of occupied areas to higher altitudes of mountainous regions where water is more plentiful.
3) Migrations from disadvantaged to favoured regions, with regard to access to water.

One of the chief impacted regions in this respect was the Pannonian Basin, practically as an outpost of the East European steppe zone. It was by no means coincidental that this was one of the main regions from where a most remarkable migration in prehistoric Europe had originated, the well-known Urnfield movement (in the German archaeological literature "Urnenfelderbewegung"). Its ramifications across the Balkan Peninsula include the so-called "Doric movement" of early Greeks (Kimmig 1964).

Last but not least, a more complicated and yet surprising relationship requires discussion. Repeatedly, detailed archaeological maps adjusted to time slices following each other in chronological sequence and covering time spans of comparable length, give the impression that precisely the comparatively dry periods in the younger prehistory of Central Europe are characterized by an extraordinarily high intensity of settlement activity and land use. This general impression is greatly strengthened by a set of maps related to the Neolithic millennia in Central Europe recently presented by Preuss (1996). These studies have inspired more detailed work on mapping the selected chronological horizons of comparable duration within the period of the Roman Empire, the so-called Germania Libera (Frenzel 1995; Faensen 1998a, 1998b). Likewise, two time slices of the Central European Bronze Age in the area influenced by the Lusatian culture were recently presented, along with adjacent archaeological groups in the eastern part of Central Europe (Neubert and Jäger 2000).

Increased land occupancy during dry periods seems to contradict our conclusions stated earlier regarding settlement density and water supply. How can this paradoxical finding be explained?

Extended land use means forest cleaning and vast open spaces typical of continental climate. Agrarian land use subsequent to deforestation has caused effects of "anthropogenic continentalisation" (Heinrich and Jäger 1987). Thus, one may ask if arid periods can be understood as resulting from intensified human impact on the landscape, due to an extended agrarian land use following the higher population density in comparatively moister regions. The reason why dry periods seem to favour a high density of settlements in these areas may be similar to explaining why people actually drain marshes and swamps: Agriculture depends on the availability of dry ground and a sufficiency of water. Thus the optimum for agricultural land use and settlement lies in periods of drier climates, as long as drought is avoided.

The chronological coincidence of inland climatic oscillations with the alternating displacement of shorelines at the coast of the North Sea is an argument for the prevalence of natural fluctuations (Hageman and Jäger 1992).

However, climatic fluctuations influencing the water balance in the interior regions of Central Europe may well have been modified by human impact on the environment.

Conclusions

The main cause for the fluctuating water balance in prehistoric Central Europe was the changing climate. At the end of the regional Bronze Age a trend toward a more continental climate took place, independent of any human impact on the landscape, but with significant consequences for human settlement, culture, and history.

References

Christl A, Simon Kl (1995) Nutzung und Besiedlung des sächsischen Erzgebirges und des Vogtlandes bis zur deutschen Ostkolonisation. In: Germania 73 (2), pp 441-462 (Frankfurt / Main and Mainz)

Christl G (1988) Ur- und frühgeschichtliche Fundplätze im Uferbereich des Schwielochsees sowie im nördlichen angrenzenden Spreetal. Befunde und Aussagen zur Besiedlungs- und Landschaftsentwicklung. In: Veröffentlichungen des Museums für Ur- und Frühgeschichte Potsdam 22, Berlin, pp 229-244

Deutsch M (1994) Die Hochwassermarken an der Unstrut. Unpublished manuscript (diploma thesis, Martin-Luther-Universität Halle, Institute of Prehistoric Archaeology)

Donat P (1969) Ein Beitrag zur relativen Chronologie der Hügelgräberbronzezeit Südwestthüringens. Zeitschrift für Archäologie 3(1):16-35

Faensen B (1998a) Karte der Besiedlung im freien Germanien in der Zeit um Christi Geburt (50v.-50n. Chr.). Unpublished manuscript (Martin-Luther-Universität Halle, Institute of Prehistoric Archaeology).

Faensen B (1998b) Karte der Besiedlung im freien Germanien zu Beginn der Völkerwanderungszeit. Unpublished manuscript (Martin-Luther-Universität Halle, Institute of Prehistoric Archaeology).

Fischer G, Wefer G (1999) Use of proxies in Palaeoceanography. Springer, Heidelberg Berlin, pp 1-735

Frenzel B (1995) Projektgruppe „Terrestrische Paläoklimatologie". In: Akademie der Wissenschaften und der Literatur Mainz, Jahrbuch 1995, Mainz und Stuttgart, pp 244-306

Freye H-A (1978) Kompendium der Humanökologie. Gustav Fischer, Jena, pp 1-420

Glaser R (2001) Klimageschichte Mitteleuropas. 1000 Jahre Wetter, Klima, Katastrophen. pp 1-227

Hageman BP, Jäger K-D (1992) Zur stratigraphischen Verknüpfung holozäner Meeresspiegelbewegungen im Küstenraum der Nordsee mit Klimaschwankungen im mitteleuropäischen Binnenland. In: Billwitz K, Jäger K-D, Janke W (eds) Jungquartäre Landschaftsräume – Aktuelle Forschungen zwischen Atlantik und Tienschan. Berlin, Heidelberg, New York, 157-169

Harrisson SP, Frenzel B, Huckriede U, Weiss MM (1998) Palaeohydrology as reflected in lake-level changes as climatic evidence for Holocene times (Palaeoclimate Research, 25) VIII., Gustav Fischer, Stuttgart, Jena, pp 1-180

Heinrich W-D, Jäger K-D, (1987) Paläoökologische und landschaftsgeschichtliche Interpretation einer klinalen Merkmalsvariation bei rezenten und fossilen Populationen von Crocidura leucodon (Mammalia: Soricidae). In: Petermannns Geographische Mitteilungen. Gotha, 131 (2) pp 93-98

Hentschel G (1982) Das Bioklima des Menschen. Berlin, pp 1-192

Jäger K-D (1982) Stratigraphische Belege für Klimawandlungen im mitteleuropäischen Holozän. Zeitschrift für Geologische Wissenschaften 10 (6): 799-809

Jäger K-D (1987) Zum gegenwärtigen Kenntnisstand über nacheiszeitliche Wasserstandsschwankungen an Binnenseen im Jungmoränengebiet des nörd-lichen Mitteleuropa. Wissenschaftliche Zeitschrift der Ernst-Moritz-Arndt-Universität Greifswald, Mathematisch-Naturwissenschaftliche Reihe 36 (1-2), Greifswald, pp 44-48

Jäger K-D (1999) Ur- und frühgeschichtliche Klimabeeinflussung durch Intensitätsunterschiede agrarischer Landnutzung? In: Cziesla E, Kersting Th, Pratsch ST (Eds) Den Bogen spannen ... – Festschrift für Bernhard Gramsch zum 65. Geburtstag (Beiträge zur Ur- und Frühgeschichte Mitteleuropas, 20), Weißbach, pp 515-522

Joos M (1982) Swiss Midland-lakes and climatic changes. In: Harding A (Ed) Climatic Changes in Later Prehistory. Edingburgh, pp 44-51

Kimmig W (1964) Seevölkerbewegung und Urnenfelderkultur. Studien aus Alteuropa (Beihefte der Bonner Jahrbücher, vol. 10), Bonn, part I, pp 220-283

Neubert A, Jäger K-D (2000) Archaeological mapping as an aid in elaboration of palaeoclimatological statements. In: Palaeoklimaforschung 29, Stuttgart, Jena etc.

Neumann G (1954) Sieben Gleichbergburgen nach dem Forschungsstand von 1952. In: Frühe Burgen und Städte (Wilhelm Unverzagt zum 60. Geburtstag). Akademie-Verlag, Berlin, pp 7-16

Niewiarowski Wk (1976) Wahania poziomu wod w jeziorze Pakoskim w swietle badan geomorfologicznych i archaeologicznch. In: Studia Societatis Scientiarum Torunensis VIII C (Geografia i Geologia) Nr. 4-6, Torun, pp 193-211

Pfister Chr (1997) Wetternachhersage. 500 Jahre Klimavariationen und Naturkatastrophen (1496-1995). Bern, Stuttgart, Vienna, pp 1-304

Pons LJ, Jelgersma S, Wiggers AJ, Dejong JD (1963) Evolution of the Netherlands coastal area during the Holocene. In: Verhandelingen van het Koninklijk Nederlands Geologisch Mijnbouwkundig Genootschap, Geol Serie 21-2, pp 197-208, 8 encl

Prehn B (1987) Zu Hinweisen auf Seespiegelschwankungen der Müritz. In: Wissenschaftliche Zeitschrift der Ernst-Moritz-Arndt-Universität Greifswald, Greifswald, Mathematisch-Naturwissen-schaftliche Reihe 36 (1-2), pp 49-51

Preuss J (Ed) (1996) Das Neolithikum in Mitteleuropa – Kulturen – Wirtschaft – Umwelt vom 6. Bis 3. Jahrtausend v.u.Z. – Übersichten zum Stand der Forschung. Vol. 3 (comprising 11 folded maps). (Wilkau-Haßlau: Beier and Beran)

Starkel L (1977) Paleogeografia holocenu. (Warszawa: Panstwowe Wydawnictwo Naukowe), pp 1-362

Walter D (1985) Thüringer Höhlen und ihre holozänen Bodenaltertümer. (Weimarer Monographien zur Ur- und Frühgeschichte, 14)

"Narrowest-Ring" Events in the Irish Oak Chronology: Uncertainties in Reconstructing Cause and Effect in Prehistory

Michael G.L. Baillie

The Queen's University of Belfast, School of Archaeology and Palaeoecology, University Road, UK-Belfast BT7 1NN, Northern Ireland
e-mail: m.baillie@qub.ac.uk

Abstract: A number of "Narrowest-Ring" events have been identified in Irish oaks, with close to calendrial precision, back to 4500 y ago. The nature of these events will be impossible to establish unless putative causes can be precisely dated, independently of the tree-ring chronology. But even if such precise dating were available, alternative causes can be envisaged. Thus, while we can test, in principle, whether an hypothesis regarding the nature of a "Narrowest-Ring" event is acceptable (through adequate dating), we can never establish that it is in fact correct. In addition, it will always be difficult to tie human activities to dated tree-ring events, because archeological dating tends to be imprecise even when using dendrochronology (which yields a date for when a tree was alive, or felled, but not for the date of its use in building). What dates are available for Irish prehistory suggest that major changes in population and culture are associated with the major calamitous events recorded in the tree-rings of Irish oaks.

Introduction

Archaeology provides many examples of abrupt change in the flow of history in Ireland (and elsewhere, for that matter). To what extent can we find the cause or causes of switching from one "period" to another? Cause and effect of course, are to do with timing; without a reliable chronology little or no progress can be made in this task. With dendrochronology, some exquisite dating, with calendrical precision, becomes possible. Archaeologists have readily embraced the results of dendrochronological dating and the method has slipped effortlessly into the role of a "routine dating method" with commercial dating widely available for oak timbers in much of northern Europe. There is no question that dendrochronology has provided much useful information regarding the timing of events in history and prehistory. However, there is another side to the story. Dendrochronological results are forcing us to face the fact that it is all but impossible to demonstrate cause and effect in the prehistoric past. In particular, the relationship between human activities and environmental forcing remain – unfortunately – quite obscure. Here I make use of several narrowest ring events from the Irish oak chronology; these are drawn from the following list as published in 1988 (Baillie and Munro), namely 4370 BC, 3195 BC, 2345 BC, 1628 BC, 1159 BC, 207 BC and AD 540.

(Note; the convention in this article is BC = calendar dates/archaeological dates; CalBC = calibrated radiocarbon age estimates; 'BC'= ice-core dates; "BC" = quasi historical dates/annals)

The So-Called Hekla-4 Event of 2354-2345 BC

Within the long tree-ring chronology for Irish oak there is what has been called a 'narrowest-ring event' at 2345 BC (Baillie and Munro 1988; Baillie 1995a). On examination this event is seen to start in 2354 BC, thereafter growth declines into the narrowest rings at 2345 BC. Examination of the samples which show the most pronounced effect indicates that the trees worst affected were those growing on the fenlands to the south of Lough Neagh. One tree in particular (Q1764) from close

From WEFER G, BERGER W, BEHRE K-E, JANSEN E (eds), 2002, *Climate Development and History of the North Atlantic Realm.* Springer-Verlag Berlin Heidelberg, pp 377-385

to Lough Neagh shows a growth anomaly which could easily be consistent with raised water levels during the 2354-2345 BC episode, while another (Q9172), from BallyMacombs More to the north of Lough Neagh shows a severe physical damage in 2354 BC (Baillie 1995b, 1999). It is fair to say that something unusual and detrimental to tree growth happened in the North of Ireland in 2354 BC and in the years immediately following.

When the narrowest ring events were first identified the most plausible explanation for their occurrence, given the presence of acid layers in the Greenland ice-core records at broadly similar dates, was environmental stress due to significant volcanic activity. It was therefore interesting when Pilcher et al. (1995) showed, using multiple stratified high-precision radiocarbon determinations - a so called 'wiggle match' - that the widespread Hekla 4 tephra dated closely to the range 2310+/-20 CalBC. The initial assumption had to be that it was the eruption of this Icelandic volcano which caused the effects in and after 2354 BC. However, on reflection it is clear that this simple correlation is not necessarily justified. Philosophically, to conclude that Hekla 4 erupted in 2354 BC is to force the evidence: the tree-ring event and the volcano may be in the same ballpark, but to force one to cause the other is both arrogant and foolhardy. Several relationships are possible, that is,

i) Hekla 4 did cause the 2354 environmental event by triggering increased rainfall after the eruption

ii) some tectonic activity associated with the eruption altered the drainage of Lough Neagh, or possibly the whole water table

iii) the eruption came before 2354 BC, at 2354 BC or after 2354 BC

iv) the Hekla 4 eruption had nothing to do with the 2354-2345 BC events etc.

So, how can one move the debate on the dating of Hekla 4 and its relationship to the environmental event forward? The answer is to introduce the other relevant environmental recorder, namely the ice-core records. As the volcano Hekla is relatively close to Greenland it seems reasonable that tephra from the eruption might be encapsulated in the annual snowfall records represented by the Greenland ice cores. The question then reduces itself to 'can the tephra be found?' and 'in which annual layer does it occur?'. Here we enter an interesting realm. There are four Greenland ice cores which might encapsulate Hekla 4 tephra. Of the four, the first (taking them in chronological order) was Camp Century. The Danish workers did not record any significant acidity in the 24th century 'BC' in the published record of the Camp Century core (Hammer et al. 1980). The Dye 3 and GRIP records have only ever been published back to 2100 'BC' so cannot be interrogated for this event (Clausen et al. 1997). The American GISP2 core is therefore the only one which offers target dates: these are 2398+/-2 (more realistically +/-40 given by Alley et al. 1997) 'BC' and 2310+/-2 (more realistically +/-40) (Zielinski et al. 1994). So, the question is whether Hekla 4 tephra occurs in one (or neither) of these layers.

The situation is complicated by the issue of the dating of the GISP2 core. Although the global error estimate for the core is of the order of +/-1% (suggesting the +/-40 quoted above by Alley et al.), it was believed by Zielinski that the dating accuracy is significantly better than this (Zielinski et al. 1994). However, this author has suggested that, on other evidence, the dates of the GISP2 core may be too old by circa 40 years in the second and third millennia BC (Baillie 1996). If that hypothesis were correct then we might expect to find Hekla 4 tephra in the 2398+/-40 'BC' acid layer. Indeed if that were correct it would suggest quite strongly that Hekla 4 does belong in close proximity to 2350 BC (currently it is believed that the American ice-core workers have investigated the 2310+/-40 'BC' acid layer without locating Hekla tephra but have not analysed the ice at 2398+/-40 'BC').

By archaeological standards, everything is indeed very closely dated. Yet this exposes us to the ultimate dilemma. We want to know if the effects in the bog oaks were caused by the eruption of Hekla 4. Were they? The simple answer is that we will probably never know with certainty. We can demonstrate this by looking at the best possible set of circumstances in reconstructing the course of events. First we need to set out some of the boundary conditions.

i) The Hekla 4 eruption can never be dated to the exact year of occurrence on Iceland itself. The only way to attempt that would be to use dendrochronology to date a tree killed by the eruption in Iceland and there are no appropriate buried trees. Radiocarbon can never provide a calendrical date for the eruption and no other method stands any chance of calendrical precision.

ii) The date of arrival of Hekla 4 tephra in bog deposits in Northern Europe can never be dated to the exact year, because no dating method will ever allow the dating of peat deposits to that level of precision. This also applies to the occurrence of Hekla 4 tephra in varve deposits; it is unlikely that prehistoric varves will ever be dated to calendrical precision.

iii) The exact relationship of the damage effects in the Irish trees to the arrival of the Hekla 4 tephra can never be known with certainty.

This leaves the location of Hekla 4 tephra in an ice layer in Greenland as the only route to dating the eruption of Hekla 4. So, let us imagine that the ice-core workers did succeed in locating Hekla 4 tephra in a single annual layer of ice in a Greenland core. The question would then reduce to how well that ice layer could be dated. We already know that the ice-cores are not precisely dated back in the third millennium BC. As noted, realistic estimates of the dating error is around +/-1% or +/-40 years.

If Baillie (1996) is correct, then the 1644+/-20 'BC' acid layer in Dye3 is the same as the 1636+/-7 'BC' acid layer in GRIP and is probably the same as the 1669 'BC' acid layer in GISP2 and in all probability all of these relate to the1628 BC Irish tree-ring event and the 1627 BC bristlecone pine frost ring event of LaMarche and Hirschboeck (1984); though it has to be pointed out that there are other acid layers in GISP2 with ice-core dates of 1623 'BC' and 1602 'BC'. It is possible that if spaced events could be found which coincide with the spacing between environmental effects in the precisely-dated tree-ring chronologies, then the ice-cores could be tied down somewhat more closely. The question is whether this procedure could ever give a date, for the layer containing the Hekla 4 tephra, to a single calendar year. Indeed even if it did and it turned out that the tephra occurred in a layer which in fact dated to the calendar year 2354 BC, cause and effect could not be proven. We would still be left with the possibility that the postulated rise in water level of Lough Neagh was due to a related tectonic or groundwater effect rather than directly due to the eruption of the volcano Hekla. Thus it is fair to say that we will never know the answer to our initial question.

Although we may not ever be able to prove cause and effect, we can establish possible links between seemingly unrelated events.

We do have some evidence for what appears to be a rise in Lough Neagh water level in 2354-2345 BC. But we also know that the Annals of the Four Masters record lakes 'overflowing' in 2341 "BC" while Ussher dates the Biblical Flood to 2349 "BC" (Baillie 1995b). Moreover, Newton and Halley, whose scientific credentials leave little to be desired, both believed that the biblical Flood was due to a close call with a comet in 2349 "BC" (Schechner Genuth 1997). But a circumstantial case can be made that all the narrowest-ring events in the Irish oaks are in some way related to comet activity (Baillie 1999). We also have the observation of what sounds awfully like impact debris from a cosmic explosion in Syria in the vicinity of 2350 BC reported by Courty (1998). The question is, are we observing a global event? As Peiser (1998; 117) has put it

"At some time around 2300+/-200 BC, a large number of the major civilizations of the world collapsed. The Old Kingdom in Egypt, the Akkadian Empire in Mesopotamia, the early Bronze Age societies in Anatolia, Greece and Israel, as well as the Indus Valley civilization in India, the Hilmand civilization in Afghanistan, and the Hongshan culture in China the first urban civilizations in the world all fell into ruin at more or less the same time".

Indeed, it has to be noted that James Legge (1879), a nineteenth century writer on Chinese history, mentions that the first emperor Yâo comes to the throne in, 2357 BC and that sometime during his reign there is this reaction: "Yâo ... appears suddenly startled by the ravages of a terrible inundation. The waters were overtopping the hills, and threatening the heavens in their surging fury". Hints of a global wet event are widespread, if poorly dated by any absolute calendar, though it should also be noted that independent work on pollen profiles from Corlea Bog, Co. Longford, has suggested a notable and well dated wet phase in the immediate vicinity of the Hekla 4 tephra (Caseldine et al. 1998). It seems that this strange clustering of calamitous happenings in the 24th century BC raises the question whether the Hekla 4 eruption could just be a symptom of some bigger global event, rather than the cause of anything.

A focus on events at circa 2350 BC brings to mind the close proximity of this date to the start of the use of metalwork in Ireland - the arrival of Beakers and possibly Beaker Folk. The dates given for the start of metal use vary from around 2500 BC (Mallory and McNeill 1991) to around 2400 BC (Waddell 1998) or even to around 2300 BC (Cooney and Grogan 1994). All these estimates are deduced from calibrated radiocarbon dates. Thus, it is quite possible that the first metal users arrived in Ireland in the aftermath of a global event around 2350 BC, say, on a Tuesday sometime in the 2340s BC. Any attempt to date such an arrival by radiocarbon will inevitably produce estimates between 2500 and 2200 CalBC (i.e. any radiocarbon age in the 3800s BP tends to calibrate to this range (Pearson et al. 1986)). Here we face the standard dilemma. Tree-rings and ice-cores give rise to discussions pertaining to the time frame of a single century while archaeology is seeing through a 200-300 year window; yet the 'event' – the arrival of metalworking - may have been quite abrupt. The question is, if abrupt, did it follow immediately upon the event seen in the tree-rings?

The So-Called Hekla 3 event of 1159-1141 BC

The second event to be discussed relates to a severe time of stress which spans the years 1159-1141 BC in Irish bog oaks. The effects are extreme in these trees, which mostly exhibit no summer growth for around two decades. The cause is still unknown although there is a reasonable circumstantial case that the event is somehow related to an episode of volcanic activity (Baillie and Munro 1988; Baillie 1995a), or just possibly impact activity involving cometary debris (Baillie 1999). It should be noted that the narrowest-ring event at 1159-1141 BC, while it shows up very clearly in oaks which grew rooted on the surface of peat bogs, does not show so clearly in archaeological oaks i.e. those oaks which span the 12th century BC, but which are presumed to originally have grown on mineral soils (dry land). However, this latter observation does not detract from the unusual severity of the effect in the long bog-oak chronologies.

A discussion of the relationship of the stress event to the Hekla 3 eruption would be a repeat of the issues raised above under the Hekla 4 example with the proviso that Hekla 3 is much less well dated. Let us, instead, focus on the influence of the stress event – whatever its cause – on human activity in Ireland, in the mid-12th century BC. The issue of the age of Haughey's Fort, Co. Armagh, may serve to illustrate, again, the limitations of our ability to reconstruct past events. When Mallory undertook preliminary excavations at Haughey's Fort in 1987 the likely date range of the original construction was not known in any meaningful way. The best estimate was that it was likely to be 'Late Bronze Age or Early Iron Age' (Warner 1986). A suite of radiocarbon determinations from the base deposits of the three ditches and from various pits and hearths on the site yielded radiocarbon ages between 2833 and 2923 BP. This finding suggested that the occupation was relatively short-lived (a century or so), a view also taken by the excavator. However, given the shape of the radiocarbon cali-bration curve, interpretation of this tight cluster of radiocarbon dates produced a large range of possible ages of construction. Mallory and Warner were willing to settle for an age somewhere in the range 1250 to 850 CalBC (Mallory and Warner 1988). In effect, the cluster of radiocarbon dates can be handled in either an optimistic or a pes-simistic manner. Conceivably all ten radiocarbon determinations could be consistent with construction and use of the fort in as little as half-a-century following on 1150 BC (Baillie 1995b). More pessimistically, the same range of radiocarbon dates could be interpreted as falling anywhere in the three century span 1250-950 CalBC. Warner's preferred option, based on modeling/simulating the radiocarbon date distribution is a "short occupation (of the Fort) around 1100 BC" (Warner 1994), but he was forced to conclude, given the date, that the occupation span could have been up to 150 years anywhere between about 1250 and 950 BC (with the proviso that the span should include 1100 BC).

Clearly, to solve the issue of the construction date, it was necessary to obtain wood samples that would be amenable to dendrochronological dating. In fact in the 1989 season Mallory did find a significant piece of oak in the base of the inner ditch, sample Q7971. This small piece of oak eventually yielded 161 growth rings showing a significant correlation with an oak chronology from the area south of Lough Neagh, at a last-ring date 1166 BC, (Baillie and Brown 1991). Unfortunately, there was no backup (no unequivocal tree-ring match) with the more northerly Garry Bog/BallyMacombs More chronology which spanned the same period of time. Thus, the date of 1166 BC remains tentative.

It was known in 1991 that there was another problem with the sample, namely that it was incomplete. Even if the tree-ring match was definitive, and the last existing ring was 1166 BC this only showed when the tree still lived, but not when it died, since the rings next to the bark were missing. After careful analysis of the sample, the date must be stated as follows: the tree from which the sample came was felled in the range 1134 +/-9 or later. At face value this suggested that it was possible that Haughey's Fort was constructed either just after the 1159-1141 BC environmental event, or some indeterminate time later. Eventually, the 1166 BC tentative date could be confirmed (Baillie and Brown 1998). Nevertheless, the precise year of the felling of the tree remains unknown. In any case, it does not support an hypothesis that the fort was constructed as a result of the environmental event at 1159-1141 BC. It should be pointed out that, even had there been no missing rings, the relationship between the time of death of the oak and the building of the fort would have been open to discussion. After all, there was nothing to imply any relationship between the digging of the ditch and the felling of the tree from which the sample came. The sample could (optimistically) have been a fragment of an oak tree which was felled as the ditch was being dug, or it could have come from a worked timber, from a pre-existing structure in the vicinity before the ditch was dug. Depending on how long the ditch was open, it could even have come from a tree felled after the ditch was dug. To be truly useful in the testing of a specific age-related hypothesis, a wood sample should be

(a) datable i.e. have sufficient rings and a well defined ring pattern,

(b) complete out to the underbark surface, and

(c) situated in a primary constructional context.

As soon as this list is defined it is clear that we will never know the date of construction of Haughey's Fort, or any other hill fort for that matter, except perhaps by sheer good luck.Thus, we will never know if Haughey's Fort, or any other hill fort, was constructed as a direct response to the mid-12th century BC environmental event. The irony is that there is every chance, on the available evidence (see for example Osgood 1998), that an initial flush of hill fort construction did take place immediately after 1150 BC in Britain and Ireland, and that it was stimulated by environmental pressure. But this story will remain invisible to archaeologists, despite all the digging.

Continuity or Dis-Continuity in Human Population

Observations such as those relating to the 2350 BC and 1150 BC events raise another major question. Was the population of Ireland relatively stable in the past, or was it subject to wild fluctuations? The simple answer is that we do not know. However, if we see radically different survival rates for everything from oak structures to cultural artifacts, as we do, it raises the possibility that human populations may also have fluctuated dramatically. This author has been vocal in pointing out the existence of the catastrophic events in the Irish tree-ring records (Baillie 1995a, 1999) and the possibilities that these events might have had radical effects on human populations which were dependent on agriculture.

Let us look at the extreme possibilities regarding the human presence in Ireland, to stimulate dis-cussion. The ultimate limiting cases with respect to the human population of Ireland in the BC era are (theoretically speaking)

(1) that a single boat load of Mesolithic settlers arrived on this island once, more than 9000 years ago, and that until the time of Patrick everyone

on the island was a descendant of this original Mesolithic shipment. In this scenario all additional material developments, such as agriculture, ceramics and metals, were thrown ashore from boats with shouted instructions for use; the deliverers never actually stepping ashore (interestingly several Polynesian islands were treated this way by the first European visitors who heaved into the bay, traded and departed). In this limiting case Ireland was effectively only colonized once in prehistory;

(2) that Ireland was totally depopulated on one or more occasions in the first seven millennia BC, possibly by famine, plague or extreme environmental event. In this limiting case, each new population, bearing new cultural baggage, would have arrived into an empty land and rapidly colonized without opposition. The hypothetical arrival of metal users on that Tuesday in the 2340s BC would be a case in point.

Now, it is possible to get to this point in the discussion solely with the knowledge that archaeologists have recognized different cultures through their artifact types and that there were environmental events which might have had real effects on humans. So the construction of limiting cases for discussion does not depend in any way on prior literary sources such as myth or tradition. It is then fascinating to see what some early literature says on the subject. For example, Geoffrey of Monmouth saw fit to relay the following story about the coming of Brutus, "the leader of those who survived the fall of Troy" to our neighboring island, Britain. Apparently Brutus (after whom Britain is supposedly named) addressed the goddess Diana on the subject of a safe dwelling place "Tell me which lands you wish us to inhabit". The goddess replied

"Brutus, beyond the setting sun, past the realms of Gaul, there lies an Island in the sea, once occupied by giants. *Now it is empty* and ready for your folk. Down the years this will prove an abode suited to you and to your people; and for your descendants it will be a second Troy. A race of kings will be born there from your stock and the round circle of the whole earth will be subject to them."

Not, as you might say, a bad guess for a book written in the 1130s.

Now I am not suggesting that either limiting case is correct. But where is the truth? Does reality tend towards the continuity model of limiting case (1) or towards the catastrophic replacement model of limiting case (2)? It is into a discussion of tendency that even evidence as tenuous as that of Geoffrey can be introduced. I don't believe that Britain was uninhabited sometime after the fall of Troy. How-ever, Geoffrey allows us to point out that the concept of a large island being depopulated is not, or was not, totally unthinkable. Indeed, it has to be interesting that such a tradition has people, who had been associated with the massive fortress of Troy, which fell in or around the twelfth century BC (traditionally 1184 BC) arriving in these islands just around the time when our best-dated triple banked/ditched hill fort was constructed. After all, in talking of limiting cases, the very earliest date that Haughey's Fort could have been constructed - if sample Q7971 was actually felled during the construction of the fort, and if only the absolute minimum number of sapwood rings was missing, is 1152 BC.

So, is our second limiting case a better approximation of reality than the first? Currently we don't know. But if we think of the potentially catastrophic (and potentially global) event at 2354-2345 BC, and how that date could easily separate the Late Neolithic and the Beaker periods, we are forced to ask if there really was population continuity across that episode. Did the use of metalwork filter into a stable population, or did boatloads of metal users, themselves under pressure from environmental catastrophe, find their way to an uninhabited group of islands in the extreme north west of Europe?

Now, that said, one is forced to go back to the Annals and, as usual, Warner has already set the scene. Warner (1993) pointed out that

The Neolithic was still strenuous until about 2800 BC, but virtually absent after that date. The beaker culture was flourishing by about 2300 BC. The implication is that the change-over took place within a couple of centuries of 2500 BC.

But, of course, Warner had previously pointed out that, in his view, the early Irish Annals might just be preserving a memory of the catastrophic happenings as recorded in the tree-rings (Warner

1990). When we return to the Annals of the Four Masters and to the 24th century "BC" we find that under the date AM 2820 i.e.

2380 "BC" Nine thousand of Parthalon's people died in one week .... Ireland was thirty years waste till Neimhidh's arrival.

While under the date AM 2850 i.e.

2350 "BC" Neimhidh came to Ireland ... with his people

So, we are now talking quite realistically about what may be a global environmental event around 2350 BC (with precisely-dated local aspects involving a possible inundation of Lough Neagh in 2354 to 2345 BC) which might possibly have reduced the pre-existing Irish population and aided invading newcomers (who themselves may have been on the move due to environmental stress). While ancient Irish literature refers to the island being 'waste' for thirty years before the arrival of a new people in 2350 "BC". All that in association with a reference to lakes springing forth in AM 2859 i.e. 2341 "BC". This would all carry little significance were it not for the surprising fact that the Annals, under 2341 "BC" go on to list

These were the forts that were erected, the plains that were cleared, and the lakes that sprang forth, in the time of Neimhidh ... (including) ... Magh-Lughadh, in Uituirtre...

O'Donovan (1848) then proceeds to interpret this place name as follows "MaghLughadh: i.e. Lughadh's Plain, a district near Lough Neagh". How do we explain, in the context of what we now know about 2354-2345 BC, a pre-existing reference to lakes springing forth in 2341 "BC" which includes a mention of the plain of Lough Neagh? As usual Warner's initial insight appears to be inspired to the point of being prophetic. However not even Warner could guess that a story would emerge suggesting that Lugh - normally noted as a solar deity - may, in fact, be a comet god (Baillie 1999). Yet this reference to Magh-Lughadh in 2341 "BC" has to be seen in the light of Newton's belief that the Flood of 2349 "BC" was caused by a comet and Courty's evidence for what sounds like an extraterrestrial impact event around 2350 BC. Thus one is forced to ask whether Newton and Ussher (who provided the date used by Newton) and O'Clery were all privy to some lost lore with a core of truth? Indeed, was Ireland totally depopulated more than once in prehistory?

Now, while some of these curious associations are intriguing and do hint that there were periods of extreme human stress in the Irish past, they serve mostly to allow broad associations between environmental information from the tree-ring chron-ologies and information from conventional archaeological chronology. It has to be interesting that totally independent dendrochronology gives rise to an extended list of dates as follows

5400 BC	start of the Irish oak chronology (and indeed the start date for pine chronologies in Finland and Fennoscandia)
4370 BC	narrowest ring event
4000-3800 BC	apparent wet event/chronology building problem (Baillie 1995a)
3195 BC	narrowest ring event
2345 BC	narrowest ring event - so called Hekla 4 event
1628 BC	narrowest ring event - so called Santorini event
1159 BC	narrowest ring event - so called Hekla 3 event
948 BC	single year break in Irish bog oak chronology (Baillie 1995a)
430 BC	tree-ring downturn (Baillie 1995b)
229 BC	end of prehistoric bog oak chronology
207 BC	narrowest ring event

If we turn to Cooney and Grogan's book on *The Prehistory of Ireland, a Social Perspective* it is possible to list the following dates cited by them.

Mesolithic microblade to large blade transition	c5500 BC
Well defined start of Neolithic	c4000 BC
Breton settlers arrive in Boyne Valley	c3200 BC
Neolithic end/ Bronze Age start	c2300 BC
Middle Bronze Age	c1700-c1200 BC
Bishoplands phase of Late Bronze Age	c1200-c1000 BC

Roscommon phase of LBA	c1000-c 900 BC
Dowris phase of LBA	c900 –c 600 BC
One of most enigmatic periods in Irish prehistory	c7th-3rd C. BC

The similarity between these two lists implies that the environmental factors which affected the trees/their survival had some direct bearing on the happenings, the changes, in the archaeological record. The question posed by these two essentially independent lists is just how badly human populations were affected at some of the key narrowest ring events. How close did the island come to being de-populated and were there, in fact, some wholesale re-colonization's by the very populations seen by Waddell (1995) as forming a 'web of communities ... of the Irish Sea region'?

Acknowledgments

This paper was originally delivered at the 1998 Dublin WARP conference on wetland archeology and is published seperately in the proceedings of that conference. This research was supported in part by the EC Environment Research Programme, Contracts EV5V-CT94-0500 and ENV4-CT95-0127 Climatology and Natural Hazards.

References

Alley RB, Shuman CA, Meese DA, Gow AJ, Taylor KC, Cuffey KM, Fitzpatrik JJ, Grootes PM, Zielinski GA, Ram M, Spinelli G, Elder B (1997) Visual stratigraphic dating of the GISP2 ice core: basis, reproducibility, and application. Journal of Geophysical Research 102, No. C12, 26, 411-26, 423

Baillie MGL (1995a) A Slice Through Time: dendrochronology and precision dating. Batsford, Londonm pp 1-176

Baillie MGL (1995b) Dendrochronology and the Chronology of the Irish Bronze Age. In: Waddell J, Shee Twohig E (eds) Ireland in the Bronze Age. The Stationery Office, Dublin, pp 30-37

Baillie MGL (1996) Extreme environmental events and the linking of the tree-ring and ice-core records. In: Dean JS, Meko DM, Swetnam TW (eds) Tree Rings, Environment and Humanity: Proceedings of the International Conference, Tucson, Arizona, 17-2 May 1994. Tucson, Radiocarbon, pp 703-711

Baillie MGL (1999) Exodus to Arthur: catastrophic encounters with comets. Batsford, London, p 272

Baillie MGL, Brown DM (1991) A dendro date from Haughey's Fort? Emania 8:39-40

Baillie MGL, Brown DM (1998) The Dendro-Date from Haughey's Fort. Emania 17:45-46

Baillie MGL, Munro MAR (1988) Irish tree-rings, Santorini and volcanic dust veils. Nature 332:344-6

Caseldine C, Hatton J, Huber U, Chiverrell R, Woolley N (1998) Assessing the impact of volcanic activity on mid-Holocene climate in Ireland: the need for replicate data. The Holocene 8, 1:105-111

Clausen HB, Hammer CU, Hvidberg CS, Dahl-Jensen D, Steffensen JP (1997) A comparison of the volcanic records over the past 4000 years from the Greenland Ice Core Project and Dye Greenland ice cores. Journal of Geophysical Research 102, No. C12, 26, 7073 -26, 723

Cooney G, Grogan E (1994) Irish Prehistory: a social perspective. Wordwell, Dublin, pp 1-251

Courty M-A (1998) The Soil Record of an Exceptional Event at 4000 B.P. in the Middle East. In: Peiser BJ, Palmer J, Bailey ME (eds) Natural Catastrophes During bronze Age Civilizations. BAR International Series 728, pp 93-108

Hammer CU, Clausen HB, Dansgaard W (1980) Greenland Ice Sheet Evidence of Post-Glacial Volcanism and its Climatic Impact. Nature 288:230-5

LaMarche VC Jr and Hirschboeck KK (1984) 'Frost Rings in Trees as Records of Major Volcanic Eruptions'. Nature 307:121-126

Legge J (1879) The Sacred Books of China. Clarendon Press, Oxford, pp 1-492

Mallory JP, McNeill TE (1991) The Archaeology of Ulster from Colonization to Plantation. The Institute of Irish Studies, Queen's University Belfast, pp 1-367

Mallory JP, Warner RB (1988) 'The Date of Haughey's Fort'. Emania 5:36-40

O'Donovan J (1848) Annals of the Kingdom of Ireland by the Four Masters. Hodges and Smith, Dublin, Vol 1, pp 1-565

Osgood R (1998) Warfare in the late Bronze Age of North Europe. BAR International Series 694

Pearson GW, Pilcher JR, Baillie MGL, Corbett DM, Qua F (1986) 'High-Precision 14-C Measurement of Irish Oaks to Show the Natural 14-C Variations from AD 1840 to 5210 BC'. Radiocarbon 28:911-34

Peiser BJ (1998) Comparative Analysis of late Holocene Environmental and Social Upheaval: Evidence for a Global Disaster around 4000 BP. In: Peiser BJ, Palmer J, Bailey ME (eds) Natural Catastrophes During bronze Age Civilizations. BAR International Series 728, pp 117-139

Pilcher JR, Hall VA, McCormac FG (1995) Dates of Holocene Icelandic volcanic eruptions from tephra layers in Irish peats. The Holocene, 5, 1:103-110

Schechner Genuth S (1997) Comets, Popular Culture, and the Birth of Modern Cosmology. Princeton University Press, pp 1-365

Waddell J (1995) Celts, Celticisation and the Irish Bronze Age. In: Waddell J, Shee Twohig E (eds) Ireland in the Bronze Age. The Stationery Office, Dublin, pp 158-169

Waddell J (1998) The Prehistoric Archaeology of Ireland. Galway University Press, pp 1-433

Warner RB (1986) Preliminary Schedules of Sites and Stray Finds in the Navan. Complex. Emania 1:5-9

Warner RB (1990) The 'Prehistoric' Irish Annals: Fable or History. Archaeology Ireland 4, 1:30-3

Warner RB (1993) Tree-Rings, Catastrophes and Culture in Early Ireland: Some Comments. Emania 11:13-19

Warner RB (1994) Emania Varia 1. Emania 12:66-72

Zielinski GA, Mayewski PA, Meeker LD, Whitlow S, Twickler MS, Morrison M, Meese DA, Gow AJ, Alley RB (1994) Record of volcanism since 7000 B.C. from the GISP2 Greenland ice core and implications for the volcano-climate system. Science 264:948-952

The Pleistocene and Holocene Development of the Southeastern North Sea Basin and Adjacent Coastal Areas

Hansjörg Streif

Geological Survey of Lower Saxony, Stilleweg 2, D-30655 Hannover, Germany
e-mail: H. Streif@nlfb.de

Abstract: The development of the present-day lowland area along the southern North Sea coast has resulted from climatic processes and the related sea-level rise in the course of the last 7500 years. From a geological point of view, therefore, the richly populated landscape of barrier islands, tidal flats, and coastal marshes, has to be regarded as a temporary feature. Relics of comparable landscapes are known from several stages of the Quaternary period. These stages of development had a duration of only 4000 to 5000 years. Major climatic cycles of the Quaternary period caused sea-level changes with vertical amplitudes of about 100 m and shoreline displacements of 500 to 650 km. However, the geological record shows characteristic sea-level dynamics with subordinate phases of transgressions and regressions. For most of the sea-level rise of the last 10,000 years, human beings were fully at the mercy of changes in the natural regime. Only in the course of the last 2000 years have technical measures been used to defend the population against the effects of sea-level rise. This started with the construction of dwelling mounds and small dikes and eventually led to the modern, highly effective coastal protection measures. With regard to the cyclicity of the climate changes in the Quaternary, it has to be assumed that we are approaching the end of the present warm period and high sea-level stand, and cooling and sea-level lowering would normally be expected within the next 1000 years. This is in contradiction to the general belief in a continuous and accelerated sea-level rise due to the anthropogenic greenhouse effect. Given these opposing and poorly understood factors, questions regarding the interactions of human activities with natural processes, and the stability of the natural system, become ever more pressing.

Introduction

The geological evolution of the Quaternary period is characterised by cyclic climatic oscillations comprising a series of significant cold (glacial) and temperate (interglacial) phases. These changing climatic conditions had a major influence on the ice/water balance of the Earth. During the cold stages, precipitation was trapped in the vast ice sheets. This caused a growth of the ice sheets of Antarctica and Greenland and the formation of additional ice sheets locally up to 3000 m thick which temporarily covered extensive areas of North America as well as northern and central Europe. At times during the last 0.3 Million years BP, the British and Scandinavian ice sheets covered parts of the North Sea and the adjacent lowland areas and deposited till, glaciofluvial and glaciolacustrine sediments. As an indirect response to the climatic deterioration and the formation of ice sheets, the sea level fell to about 100 m below its present-day level. In contrast, phases of temperate climate were characterised by the melting of the ice sheets, marine transgressions and high stands of the sea level.

Hence, a full climatic cycle from interglacial conditions through a cold stage to the following interglacial period is reflected in the sinuous curve of sea-level changes. However, in most cases the sedimentary record of these cycles is fragmentary.

On the basis of geological observations, it is possible to look at the present-day conditions in the southern North Sea region in the context of the long-term processes of Quaternary landscape development. Comparisons can be made with respect to the

From WEFER G, BERGER W, BEHRE K-E, JANSEN E (eds), 2002, *Climate Development and History of the North Atlantic Realm.* Springer-Verlag Berlin Heidelberg, pp 387-397

amplitudes and rates of sea-level change as well as the range of shoreline displacements connected with marine transgressions or regressions. Such studies give a reliable picture of the great variability of the natural cycles that occurred during the last 0.3 Million years BP and permit comparison with the sea-level changes documented by tide gauges during the last 150 years. With regard to the issue of anthropogenic influences on sea-level rise, evaluation of both geological data and tide-gauge measurements is necessary to develop realistic scenarios of future sea-level movements.

Outline of the Pleistocene Sea-Level Changes

A series of marine transgressions and regressions appears to have occurred during the *Early* and *early Middle Pleistocene*, i.e. the time span between 2.6 Ma BP and about 0.3 Ma BP (Long et al. 1988). In a pioneer study, Zagwijn (1979) reconstructed shorelines of the marine transgressions that took place during the temperate phases of the Tiglian (2.1 to 1.7 Ma BP), the Waalian (1.3 to 0.95 Ma BP) and the Cromerian (approx. 0.7 to 0.4 Ma BP). However, as no exact figures are avail-able for the amplitudes of the sea-level changes, and only rough estimates exist of the duration of the warm and cold phases, the dynamics of the Early and early Middle Pleistocene transgressions and regressions are practically unknown.

During the *late Middle and Late Pleistocene periods*, sedimentation was dominated by glacial erosion and deposition. Three regional glaciations affected northwestern Europe to different extents, temporarily covering parts of Denmark, northern Germany, the Netherlands, Britain and parts of the present-day North Sea area. Marine beds belonging to the Holsteinian and Eemian interglacial periods occur intercalated in these glacigenic, fluvial, and terrestrial sedimentary sequences and marine to coastal deposits of Holocene age rest on top of them (Long et al. 1988; Streif 1990a).

The earliest indicators of glacial conditions on the present-day European mainland and in the southern North Sea date from the *Elsterian (Anglian) glacial period*. These deposits rest on a regional unconformity that generally lies at a depth of less than 40 to 50 m below the land surface or sea floor. Pleistocene sediments do occur in places down to depths of 500 m below the present sea level, but these thick sediments are only found in a complex system of deeply eroded, anastomosing palaeochannels formed underneath the ice cover. Morainic and glaciofluvial sediments, ice-push structures, and the system of subglacial palaeochannels demonstrate that the ice extended southward to at least 52°20' N. They also indicate that the Scandinavian and English-Scottish ice sheets temporarily formed a coherent ice mass covering most of the southern North Sea area during this glacial period (Cameron et al. 1993).

During the *Holsteinian (Hoxnian) interglacial period*, a rise in sea level caused marine sedimentation to extend across most of the present-day North Sea and parts of the adjacent land areas (Streif 1990a, Niedersächsisches Landesamt für Bodenforschung 1995). The age of this temperate period is still controversial, estimates ranging between 195,000 years BP and more than 370,000 years BP. Nevertheless, reliable dates are available for the duration of this period. Pollen analysis and counts of rhythmic laminations in fresh-water deposits in northwestern Germany demonstrated a duration of 11,000 to 12,000 years for the Holsteinian interglacial *sensu stricto* (Müller 1974a). If one includes the subsequent oscillations, this temperate phase did not exceed 15,000 to 16,000 years in duration.

A series of detailed studies have been carried out on the Holsteinian interglacial deposits in the Hamburg and lower Elbe area, which make this region a key region for demonstrating the interactions between sea-level changes and sedimentation processes. Shallow-marine sediments belonging to the Holsteinian interglacial occur in the Hamburg area as a 65-m thick infill of a förde-like palaeochannel system 10 to 75 m below the present sea level. The transgression started with fully marine conditions and a fauna indicating a water depth of 20 to 30 m. Evidence of warmer and shallower water and brackish conditions are found in the middle part of the sequence; a return to more saline and slightly deeper water conditions took place in the uppermost marine zone of the sequence. For Hamburg, Linke (1993) concluded that the

Holsteinian sea invaded the area in a single, uninterrupted transgression cycle with a vertical amplitude of more than 55 m and a rate of sea-level rise of at least 1 m per century. No data is available about when regression began or the amount of lowering of the sea level at the end of the Holsteinian interglacial.

At the maximum extent of the *Saalian (Wolstonian) glacial period*, the ice sheet covered Denmark, parts of northern Germany as well as of the northern Netherlands and extended about 40 km seaward of the present-day coast of the Netherlands and about 110 km off the Danish coast (Cameron et al. 1993). Major areas of the southern North Sea basin remained ice free, and on the western margin of the basin, the ice most probably did not extend across the present-day coastline of the British Isles. So far, no data is available about the low stand of the sea level during the Saalian glacial period.

The *Eemian (Ipswichian) interglacial period*, which was originally defined by palynological investigations of terrestrial deposits in the Netherlands and northern Germany, is widely accepted to correlate with stage 5e of the deep-sea record, which ranges between 128,000 and 117,000 years BP. Additional, independent results about the duration of this period have been obtained from counts of the annual layers in rhythmically laminated fresh-water deposits in northern Germany (Müller 1974b, Hahne et al. 1994). Together with pollen analysis they provide information on the duration of the individual Eemian pollen zones (Selle 1962; Menke and Tynni 1984) and they give the total duration of this period of temperate climate as 11,000 years.

Marine sediments belonging to the Eemian interglacial have been identified at many sites in the southern North Sea and along the adjacent coastal zone. They consist of shallow-marine and tidal-flat deposits. Based on the pollen zones of Selle (1962) and counts of annual laminations by Müller (1974b), the Eemian sea-level changes in the German North Sea coastal area can be documented in great detail. So far, no marine beds from the early Eemian period have been recorded. Marine deposition started with Eemian pollen Zone IIIa (period of mixed oak forest). An extremely rapid rise of the sea level at about 4 m per century is recorded during the 650-year period of Eemian pollen Zones IIIa and III b (mixed oak forest with increasing proportions of hazel). This rate slowed to 0.6 to 0.7 m per century in pollen Zone IIIc (mixed oak forest with increasing proportions of hazel and lime). Then a stagnation period of 4000 to 4500 years occurred during Eemian pollen Zone IV (hornbeam and spruce) and the beginning of Zone Va (pine and spruce). This was followed by a regression when the sea level fell about 25 m during the late Eemian pollen Zones Vb (pine, spruce and fir) and VI (pine). Thus, the sea-level curve for the Eemian interglacial period only shows a portion of the full transgression-regression cycle, starting with a steadily rising sea level, passing through a period of stagnation, and followed by a phase of steady lowering. In contrast to offshore studies in the North Atlantic (Fronval and Jansen 1997), the investigations in the coastal zone of the southern North Sea yielded no indications of minor oscillations within this cycle or distinct phases of retardation or acceleration of the sea-level rise during the Eemian interglacial (Streif 1990a).

During the *Weichselian (Devensian) glacial period*, the sea level was always more than 35 m below its present level in the southern North Sea area. The climatic shift to full glacial conditions (22,000 to 18,000 years BP) caused the regional sea level in the North Sea to fall to 110 to 130 m below its present position (Jelgersma 1979). This amount of lowering is in good agreement with observations from the Atlantic coast of the USA and Africa and provides strong evidence in favour of a world-wide eustatic sea-level lowering.

Ice temporarily covered smaller parts of the southern North Sea area during the maximum of this glaciation. On the west side of the basin, an elongated tongue of grounded ice extended from the English coast into the Dutch sector of the North Sea as far as the southern slope of the Dogger Bank and deposited a blanket of pebbly, clay-rich till. On the mainland, the western margin of the ice extended from the Hamburg area northward through Schleswig-Holstein and the southern part of Jutland, Denmark, forming a distinct end moraine. Just south of the Limfjord, the ice margin turned northwest into the North Sea. The continuation of this lobe of grounded ice is poorly defined offshore, but there was no connection across the North Sea between the UK and Scandinavian ice sheets during this glaciation (Cameron et al. 1993).

Between 22,000 and 18,000 BP, extreme climatic deterioration caused a lowering of the sea level to about -130 m and a migration of the North Sea coast to a position about 350 km north of the Dogger Bank (Jelgersma 1979). Extensive areas of the present sea floor were dry land and exposed to eolian and fluvioglacial depositional processes. Cryoturbation and frost-wedge structures indicate that periglacial conditions prevailed across the emerged areas. During this period, the rivers Ems, Weser, Elbe, and Eider were joined to a composite drainage system that discharged across the present sea floor in a 30 to 40 km wide valley to the central North Sea and the North Atlantic (Figge 1980). The River Rhine and its tributaries formed an independent system which flowed from the Netherlands to the southwest through the area of the present-day English Channel into the Atlantic.

The Marine Transgression of the Weichselian Late Glacial and Holocene Periods

Three phases of sea-level rise, schematically depicted in Fig. 1, can be distinguished in the landscape development of the southern North Sea and adjacent coastal areas during the Weichselian late glacial and Holocene.

The first phase comprises the sea-level rise from the low stand of about 130 m below the present-day sea level to 45 m below the present-day sea level, which occurred between 18,000 and 8600 BP. The sedimentary record of this phase is characterised by transgressive sequences of clastic, shallow-marine and tidal-flat deposits resting directly on terrestrial, glaciofluvial and morainic Pleistocene sediments. It indicates a general sea-level rise and a transgression that initially started in most places with erosion and proceeded with a unidirectional landward shift of the shoreline. Erosion processes might be responsible that no hints of the Younger Dryas event have been recorded in the southern North Sea area up to now.

Following the initial sea-level rise glaciomarine muds were deposited in relict palaeochannels and in the present-day the central North Sea. The earliest brackish-water incursion into low-lying areas between East Anglia and the Dogger Bank may have occurred as early as 10,000 BP, when the sea level was about 65 m below its present-day position (Jelgersma 1979; Eisma et al. 1981). With the continuing sea-level rise, tidal sand ridges formed west of the Dogger Bank. This elevated area became isolated by tidal erosion of the Outer Silver Pit to form a temporary island. Tidal-flat sedimentation became more widespread between 9000 and 8000 BP. The southern North Sea was connected with the English Channel via the Dover Strait at about 8300 BP, and fully marine conditions became widely established after 7000 BP (Eisma et al. 1981). As dated transgression levels are rare between 18,000 and 8600 BP, only rough mean values can be calculated for the rate of sea-level rise. A relatively low calculated average rate of about 1 m per century may be interpreted as an indication that the general rise of the sea level was interrupted by repeated phases of retardation during the colder phases of the Weichselian late glacial period.

Continuous transgression is also characteristic of the second phase of sea-level rise between 8600 and 7100 BP. However, the progress of inundation from about 46 to 25 m below present-day sea level can be reconstructed with considerable accuracy on the basis of dated transgression levels. Relics of submerged limnic muds and peat beds between the underlying Pleistocene strata and brackish marine Holocene sediments offer excellent conditions for pollen analysis and radiocarbon dating and yield precise time/depth data on the beginning of inundation. Streif (1990b) compiled all such data available from the southern North Sea and concluded that between 8600 and 7100 BP, the rate of sea-level rise amounted to as much as 2.1 m per 100 radiocarbon years. Shallow-marine and tidal-flat sediments deposited during this period are relatively thin in the southern North Sea. In general they are less than 5 m thick; only in tidal sand ridges or natural sediment traps like the submerged Elbe palaeovalley do they reach a thickness of about 15 m.

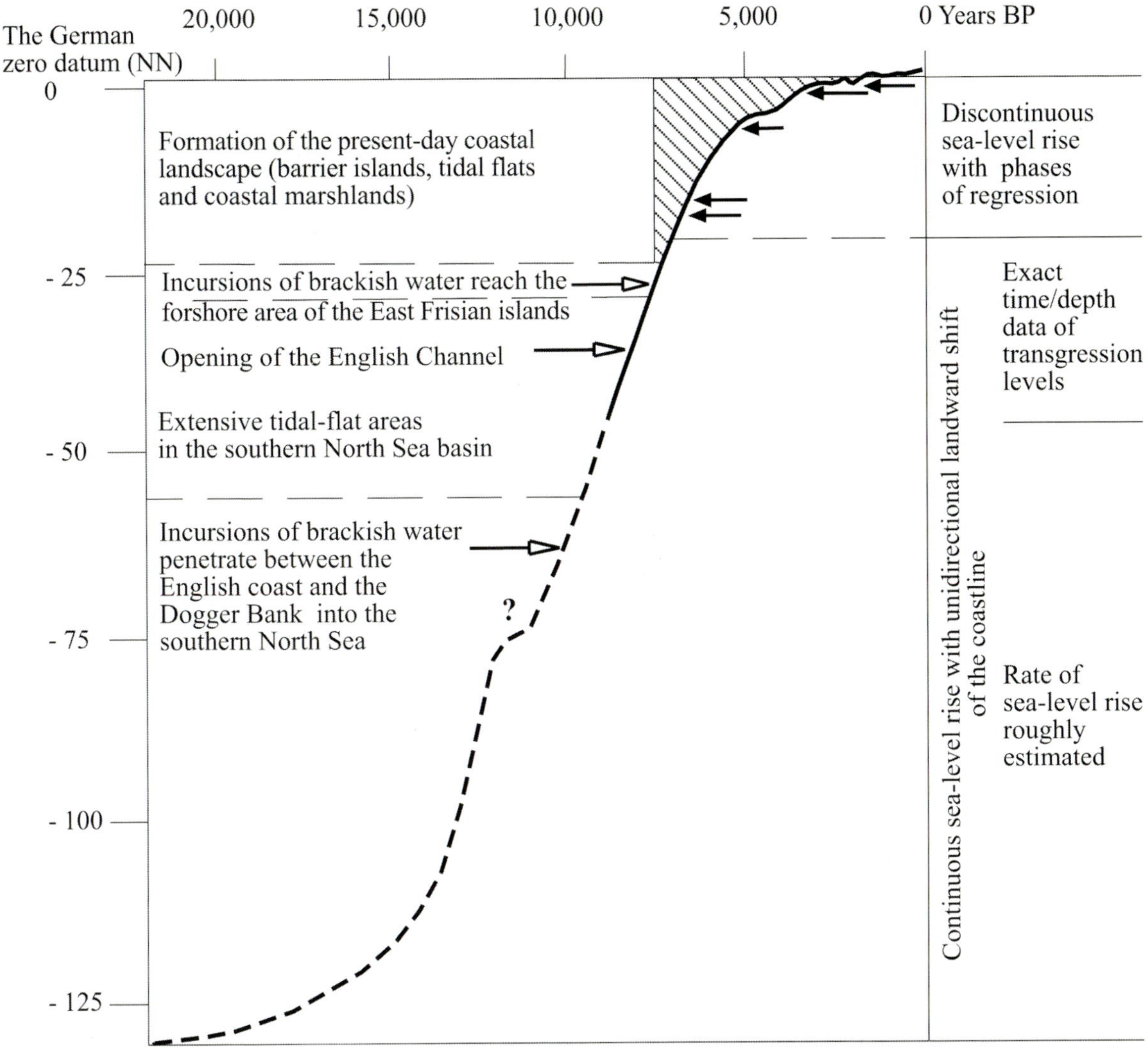

Fig. 1. Curve of the sea-level rise in the Weichselian late glacial and Holocene periods and the corresponding effects on coastal evolution. The hatched area marks the period of evolution of the present-day coastal area.

Finds of artefacts from the mesolithic period (10,000 to 6000 BP) made from bones and antlers indicate that human beings originally lived in the area which is nowadays inundated by the North Sea. The living area of these people was drastically constricted by the rapid sea-level rise and they were forced to escape on higher seated land areas.

The third phase of development began at about 7500 BP and has continued until today. It is characterised by a rise of the sea level from -25 m to its present position. This has led to the formation of the present-day coastal landscape (see also Flemming, this volume), which consists of three geomorphological units: the barrier islands, tidal flats, and coastal marshland. Generally, the coastal zone is 10 to 20 km wide; however, in the western Netherlands and in the funnel-shaped estuaries of the Weser and Elbe rivers it stretches 80 to 100 km inland. As shown in Fig. 2, the body of Holocene coastal sediments reaches its maximum thickness of about 20 to 25 m along the western coast of the Netherlands as well as in the region of the West

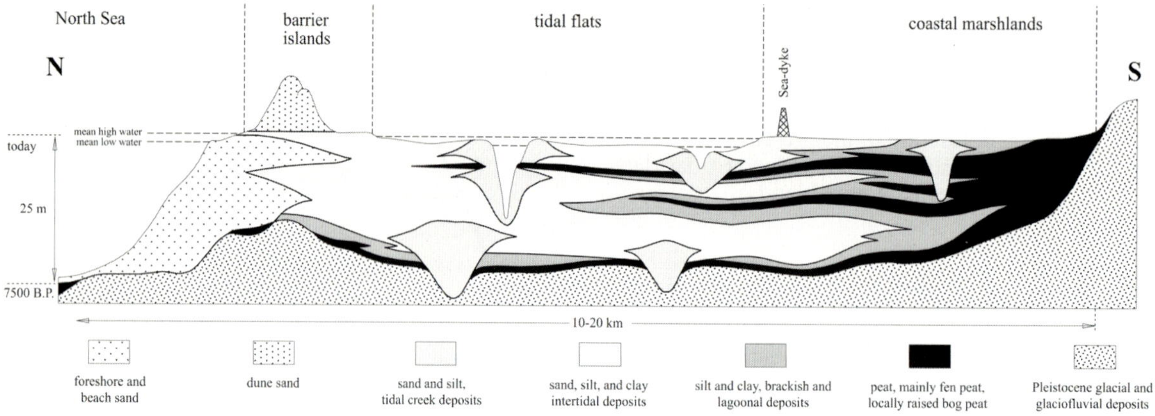

Fig. 2. Schematic cross section from the North Sea to the Pleistocene hinterland with the main lithologic units of the Holocene coastal accumulation wedge.

Friesian and East Friesian barrier islands (Streif 1990b; Van der Spek and Beets 1992) and on the seaward side of the open tidal-flat areas in the innermost part of the German Bight. On the islands, it is locally covered by eolian dune deposits up to 25 m thick. The coastal deposits become thinner in a landward direction and reach an average thickness of 10 m. On the landward margin they wedge out against the Pleistocene hinterland or merge into alluvial flood-plain deposits.

The boundary between the Pleistocene sediments and the Holocene coastal deposits is of multigenetic origin. In wide areas it is identical with the original land surface formed under fluvial and eolian conditions during the Late Pleistocene and early to mid-Holocene periods. Fossil horizons of podsolic soil can often be observed overlying the Pleistocene, indicating a vegetation cover and pedogenetic processes. The subsequent inundation started with a rising groundwater table, increasing swampiness and development of mires, forming what is known as "basal peat". This was followed by deposition of thick sequences of tidal-flat and brackish sediments with a series of intercalated peat layers formed in coastal mires. In places where relics of fossil soils, limnic muds and basal peat are missing and Holocene clastic sediments lie directly on Pleistocene deposits, tidal erosion has to be taken into consideration. An exception is areas which belong to former fluvial drainage systems where the original depositional environment prevented soil formation and deposition of limnic mud and peat. In general, tidal erosional features are dominant in the seaward parts of the coastal area, whereas especially good preservation conditions for the original Pleistocene landscape exist in the landward zone (Fig. 2).

Pleistocene sediments form the cores of the island of Texel, the Netherlands, as well as of the islands of Sylt, Amrum, and Föhr, Germany. In the subsoil of the West and East Friesian barrier islands, the boundary between Pleistocene and Holocene deposits generally varies between 20 and 25 m below the German zero datum (NN). However, high-standing erosional relics of Pleistocene deposits occur locally at elevations of about 10 to 5 m below NN, forming submerged cores to the present-day islands (Fig. 2). Characteristic of these relict features is a steep slope on the seaward side, which is interpreted as a buried cliff originally formed by marine erosion. The existence of these Pleistocene cores provides clear evidence that the barrier islands have maintained roughly the same positions during the last 2000 years. This however, is not compatible with the widely accepted opinion that the East Friesian barrier islands and the tidal inlets between them undergo continuous eastward migration (Gaye and Walther 1935; Luck and Witte 1974). The latter opinion is based on the present-day morphology, particularly the erosional features on the western end of the islands and accretion in the east. However, the presence of high seated Pleistocene relics provides clear evidence that continuous eastward migration of the islands and tidal inlets could not have taken place in the past. This would have led to levelling and complete removal of the Pleistocene cores.

The seaward part of the Holocene accumulation wedge (Fig. 2) consists of pure sand deposited in high-energy environments like the foreshore area, the breaker zone, and tidal inlets. The middle part of the accumulation wedge is built up of a cyclic sequence of fine sand, silt, and clay layers deposited in sheltered tidal-flat areas as well as in the brackish-water environment of coastal lagoons and estuaries. Fen peat and raised-bog peat locally occur at the base of these clastic sediments or as intercalated layers. In many places, these layers of basal and intercalated peat mentioned above merge into thicker units or a continuous peat sequence close to the Pleistocene hinterland.

The sedimentary sequence in the coastal zone shows major differences from that in the offshore area: the former is thicker and more complex. The cyclic alternation of transgressive overlaps (marine and brackish deposits overlying peat) and regressive overlaps (semiterrestrial peat overlying brackish and marine sediments) indicates repeated shoreline displacements, both landward and seaward, of several kilometres.

On the basis of a large number of radiocarbon dates of peat samples, Streif (1990b) constructed a band-like sea-level curve covering the last 8000 years and plotted it together with a histogram of the phases of the formation of intercalated peat layers (Fig. 3). According to this diagram, the older dates obtained in the coastal zone are compatible with a rapid and uninterrupted rise in sea level as demonstrated for the offshore area. However, from about 6500 BP onwards, temporary reversals are indicated by regressive overlap of peat layers over clastic sediments of marine origin. The oldest intercalated peats, dating from about 6500 and 6000 BP, have a very limited spatial distribution and are restricted to valley-like depressions in the pre-Holocene surface. Ideal conditions for the formation of intercalated peat apparently prevailed between 4800 and 4200 BP and between 3300 to 2300 BP. Intercalated peats from these periods occur throughout the coastal lowland area between the rivers Ems and Elbe. At a small number of locations, intercalated peat layers were found which yielded dates between 2000 to 1600 BP (Fig. 3).

Originally such regressive overlaps were explained as a result of a lowering of the sea level and the related groundwater level. However, more recent studies (Caspers et al. 1995) indicate that most of the intercalated peat layers have a more complicated genesis (Fig. 4). This view suggests

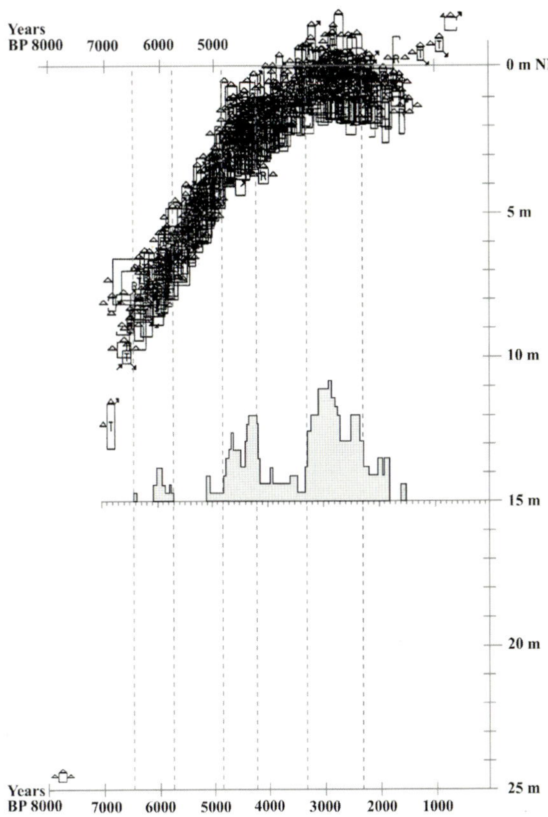

Fig. 3. Time/depth diagram with radiocarbon dates, plotted as error boxes, obtained from basal and intercalated peat layers in the coastal region between the rivers Ems and Elbe. A band-like sea-level curve of dates is shown in the upper part of the diagram, and a frequency distribution of samples from intercalated peat layers in the lower part. The dashed lines delineate time spans in which the conditions were favourable for formation of the intercalated peat layers. Water depth given in metres below German datum (NN).

that regressive overlaps occur during phases of slowly rising water table, when the rate of bog growth fully compensates or even exceeds the contemporary rise in sea level. A rising water table is inferred for two reasons. Firstly, it is essential for the accumulation of a sufficient amount of organic matter in a fen. Secondly, it is necessary to prevent the organic material from decomposing. In contrast, transgressive overlaps are apparently indicative of periods in which the rate of sea-level rise is higher than that of mire growth. In spite of the great variety of other local influences, the dates available from the coastal lowland areas indicate that the formation of intercalated peat layers requires low rates of sea-level rise not exceeding 35 cm per century.

Within the general trend of rising sea level, temporary falls in sea level can be demonstrated by several lithological indicators. Firstly, horizons of decomposed peat, which can be observed within sequences of relatively fresh peat, can be regarded as a type of fossil soil. Temporary lowering of the groundwater table reduces or interrupts organic production in a mire, and allows oxidation at the surface. Secondly, a rapid upward change from fen peat to raised-bog peat vegetation is also indicative of a lowering of the sea level and the related groundwater level. The third indicator is the presence of thin horizons of humic soil formed on clastic sediments of tidal flat and brackish environments. Such indicators of sea-level lowering can be demonstrated at about 2800 BP and 2000 BP. Additionally, there is strong evidence for another phase between 1395 and 1650 AD (Streif 1990b), which coincides with the first part of the Little Ice Age.

Phases of regression repeatedly offered human beings a chance to inhabit the coastal marshland

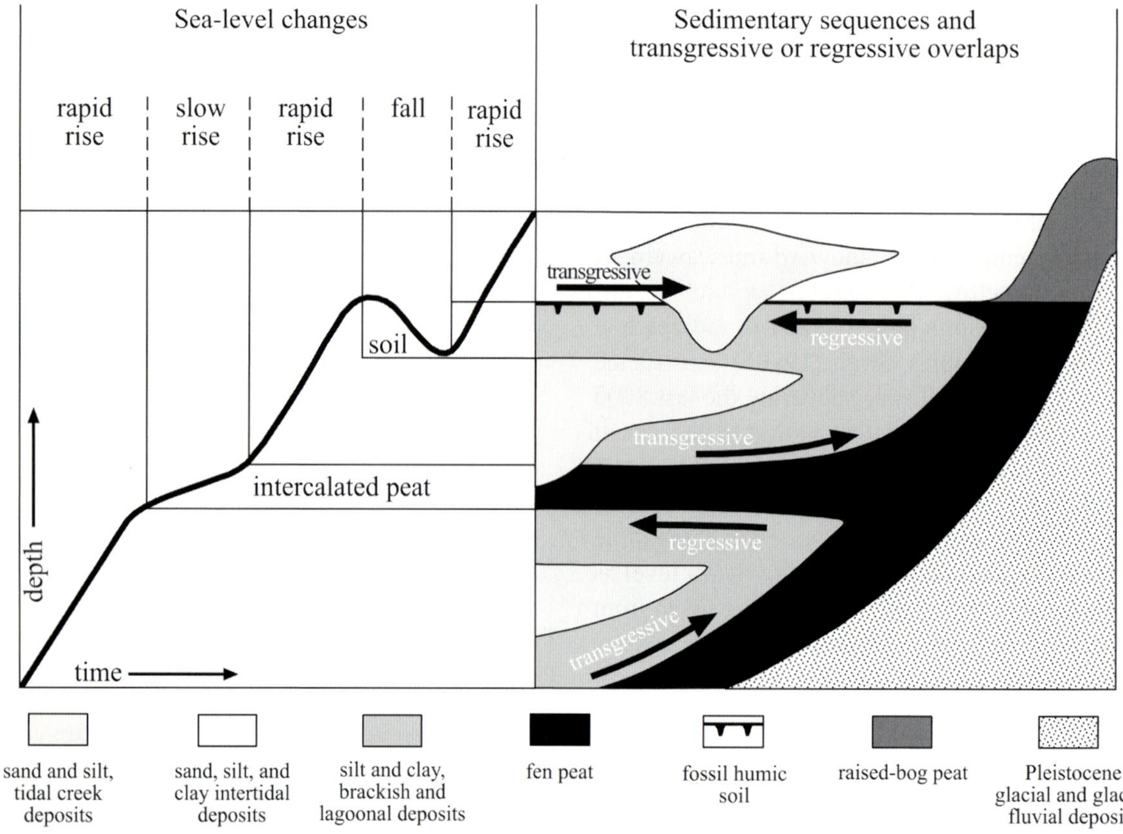

Fig. 4. Diagram showing the genetic relationship between sea-level rise and transgressive and regressive overlaps as might be found in the tidal flat and coastal marshland areas behind the barrier islands.

areas (see also Behre, this volume) and to built their settlements on the natural land surface. Along the German coast the oldest settlements date from the late Bronze Age to the early Pre-Roman Iron Age (900 to 300 BC). Subsequent inundation of the area destroyed the settlements and covered them by sand, silt, and clay deposits. A second colonisation period started at about 100 BC and ended in the 4th to 5th century AD. The last occupation phase began in the 7th century and has continued until now.

Altogether, the finds in the coastal zone reveal a complex history of transgressive and regressive cycles during the mid to late Holocene. It shows repeated phases of retardation and acceleration of the sea-level rise and, in addition, short phases of lowering of the sea level lasting 100 to 300 years (Freund and Streif 1999). Evaluations of tide gauge measurements along the German North Sea coast (Töppe 1994) which cover the time interval between 1881 and 1985 yielded average rates of 20 cm/century for the mean high tide level, of 12 cm/century for the mean tide level, and of only 4 cm/century for the low tide level. There is now evidence for an accelerated sea-level rise during the last decades.

Normally, sea-level rise and transgressions are looked upon as destructive and dangerous. This is expressed in legends about flooding and the Deluge, which exist nearly all over the world. These processes are indeed harmful for human beings, animals and plants in the inundation area. However, it has to be taken into consideration that same processes triggered also the accumulation of huge amounts of coastal deposits and created extended coastal lowland areas consisting of barrier islands, tidal flats and coastal marshland at high latitudes, and of reefs and mangrove swamps at low latitudes.

Under favourable circumstances, it is possible to quantify these accumulation processes. For instance, a mass balance study has been carried out for the coastal zone between the rivers Ems and Weser on the North Sea coast of Germany (Hoselmann and Streif 1997, 1998). This region comprises various sedimentary environments such as barrier islands, sheltered and open tidal flats, bay flats, and estuaries. Therefore, it is highly suitable for quantitative studies and can serve as a model for general mass transport and sediment accumulation connected with marine transgression. Based on an evaluation of data from about 23,000 boreholes sunk in a 4950 km^2 area, it was demonstrated that a total sediment volume of 41×10^9 m^3 has been deposited in the area since 7500 BP. This mass-balance study was based on the construction of structure contours on the base of the accumulation wedge of Holocene coastal deposits and by subdividing it into a series of horizontal slices 1 m thick. According to the study, about 90 vol.% of the wedge consists of sand, silt, and clay, i.e. material from marine or riverine sources that was deposited under tidal and fluvial conditions. Fen peat and raised bog peat, sedentary organic material formed in coastal mires, amount to 10% of the total volume. The dunes, which form impressive, 25 m high features on the East Friesian barrier islands, comprise only 1.3%.

The dynamics of the accumulation processes described above can be derived from a comparison of calibrated radiocarbon dates of the Holocene sea-level rise (Streif 1990b) with the corresponding volumes of the 1-m-thick sediment slices of the coastal accumulation wedge. This makes it possible to quantify the interaction between sea-level rise and the silting up of coastal lowland areas. For the investigated area, it was possible to demonstrate that the rate of accretion can keep pace with a contemporary sea-level rise up to about 1 m per century. This has been demonstrated for the early stage of coastal evolution between 8500 and 6500 calibrated years BP. Later, between 3900 and 1200 calibrated years BP, the rate slowed down to about 16 cm per century. Comparable figures of the accumulation processes in a tidal basin of the western Netherlands were published by Van der Spek and Beets (1992).

Conclusions

A series of cyclic climatic changes which influenced the balance between the volumes of the oceans and the inland-ice sheets caused repeated eustatic sea-level changes and regression-transgression cycles during the Quaternary. The cold phases of the Middle and Late Pleistocene had a duration of 80,000 to 100,000 years, whereas the interglacial periods were relatively short with a total length of 10,000 to 12,000 years.

Good evidence is available for the dynamics of the sea-level changes which occurred during the Holsteinian and Eemian warm periods. Each of these phases coincided with an uninterrupted transgression and high sea-level stand followed by a lowering of the sea level at the transition from temperate to cool climatic conditions. The phases of high sea-level stand had a duration of only 4000 to 5000 years.

From a palaeoclimatic point of view, the sea-level changes of the Weichselian late glacial and Holocene times are comparable with those of the preceding temperate periods described above. An obvious difference is the clear evidence that the Holocene transgression was interrupted by several phases of retardation of the sea-level rise, as well as by repeated phases of sea-level lowering. Concerning the long-term aspect of sea-level changes, it has to be taken into consideration that the thermal optimum of the present-day warm period occurred about 6500 to 6000 years ago. With regard to the cyclicity of the climatic changes in the Quaternary, it has to be assumed that we are approaching the end of the present warm period and high sea-level stand, and cooling as well as sea-level lowering may be expected within the next 1000 years. This is in contradiction to the general believe in continuous and accelerated sea-level rise due to the anthropogenic greenhouse effect. Unknown factors in these considerations, however, are the complicated interactions of the human impact with the natural processes and the stability of the natural system.

Geological knowledge about the formation of coastal landscapes and of the sedimentation processes is of great importance for discussions concerning the future evolution of the area which might be influenced to an unknown degree by the anthropogenic greenhouse effect. For instance, the second report of the International Panel on Climate Change (IPCC) presents scenarios of 30 to 65 cm of sea-level rise within the next century. These scenarios initiated considerable speculation about disastrous effects on coastal lowlands in the near future. This may be true for some extremely vulnerable coastal stretches of the globe. However, the results of the mass-balance study discussed above are strong evidence that natural accumulation processes along sandy to silty coasts can fully compensate a sea-level rise of up to 1 m per century. With regard to the southern North Sea region, an unknown factor in such considerations is, however, the extent to which coastal protection measures may have modified the hydrodynamic conditions and the availability of sediments. This implies some uncertainties for predictions of the future evolution of the area on the basis of the geological model of Holocene coastal accumulation processes.

References

Cameron D, Van Doorn D, Laban C, Streif H (1993) Geology of the southern North Sea basin. In: Hillen R, Verhagen H (eds) Coastlines of the southern North Sea. Amer Soc of Civil Engineers, New York, pp 14-26

Caspers G, Jordan H, Merkt J, Meyer K-D, Müller H, Streif H (1995) Niedersachsen. In: Benda L (ed) Das Quartär Deutschlands. Borntraeger, Berlin, Stuttgart, pp 23-58

Eisma D, Mook WG, Laban C (1981) An early Holocene tidal flat in the Southern Bight. Spec Publs int Ass Sediment, Oxford, 5:229-237

Figge K (1980) Das Elbe-Urstromtal im Bereich der Deutschen Bucht (Nordsee). Eiszeitalter u. Gegenwart, Hannover, 30:203-211

Freund H, Streif H (1999) Natürliche Pegelmarken für Meeresspiegelschwankungen der letzten 2000 Jahre im Bereich der Insel Juist. Petermanns Geogr Mitt, Gotha – Stuttgart (Perthes), 143 Jg:34-45

Fronval T, Jansen E (1997) Eemian and early Weichselian (140-60 ka) paleoceanography and paleoclimate in the Nordic seas with comparisons to Holocene conditions. Paleoceanography, Washington, 12, 3:443-462

Gaye J, Walther F (1935) Die Wanderungen der Sandriffe vor den ostfriesischen Inseln. Die Bautechnik, Berlin, 41:1-13

Hahne J, Kemle S, Merkt J, Meyer K-D (1994) Eem-, weichsel- und saalezeitliche Ablagerungen der Bohrung "Quakenbrück GE 2". Geol Jb, Hannover, A 134: 9-69

Hoselmann C, Streif H (1997) Bilanzierung der holozänen Sedimentakkumulation im niedersächsischen Küstenraum. Z dt geol Ges, Stuttgart, 148, 3-4:431-445

Hoselmann C, Streif H (1998) Methods used in a mass-balance study of Holocene sediment accumulation on the southern North Sea coast. In: Harff J, Lemke W, Stattegger K (eds) Computerised Modeling of sedimentary Systems, Springer-Verlag, Berlin, Heidelberg, New York, pp 361-374

Jelgersma S (1979) Sea-level changes in the North Sea basin. In: Oele et al. (eds) The Quaternary History of the North Sea. Acta Univ Ups Annum Quingentesimum Celebrantis, Uppsala, 2: 233-248

Linke G (1993) Zur Geologie und Petrographie der Forschungsbohrungen qho 1-5, der Bohrung Hamburg-Billbrook und des Vorkommens von marinem Holstein im Gebiet Neuwerk-Scharhörn. Geol Jb, Hannover, A 138: 35-76

Long D, Laban C, Streif H, Cameron TDJ, Schüttenhelm RTE (1988) The sedimentary record of climatic variation in the southern North Sea. Phil Trans R Sc London, London, 318 B: 523-537

Luck G, Witte HH (1974) Erfassung morphologischer Vorgänge der ostfriesischen Riffbögen in Luftbildern.- Forschungsstelle für Insel- und Küstenschutz, Jahresber 1982, Norderney, XXV: 33-54

Menke B, Tynni R (1984) Das Eeminterglazial und das Weichselfrühglazial von Rederstall / Dithmarschen und ihre Bedeutung für die mitteleuropäische Jungpleistozän-Gliederung. Geol Jb, Hannover, A 76: 3-120

Müller H (1974a) Pollenanalytische Untersuchungen und Jahresschichtenzählungen an der holsteinzeitlichen Kieselgur von Munster Breloh. Geol Jb, Hannover, A 21: 107-140

Müller H (1974b) Pollenanalytische Untersuchungen und Jahresschichtenzählungen an der eem-zeitlichen Kieselgur von Bispingen / Luhe. Geol Jb, Hannover, A 21: 149-169

Niedersächsisches Landesamt für Bodenforschung (1995) Quartärgeologische Übersichtskarte von Niedersachsen und Bremen 1:500,000. Hannover

Selle W (1962) Geologische und vegetationskundliche Untersuchungen an einigen wichtigen Vorkommen des letzten Interglazials in Nordwestdeutschland. Geol Jb, Hannover, 79: 295-352

Streif H (1990a) Quaternary sea-level changes in the North Sea, an analysis of amplitudes and velocities. In: Brosche, Sündermann (eds) Earth's rotation from eons to days. Springer-Verlag, Berlin, Heidelberg, New York, pp 201-204

Streif H (1990b) Das ostfriesische Küstengebiet - Nordsee, Inseln, Watten und Marschen. Samml Geol Führer, Borntraeger, Berlin - Stuttgart, 57, 2. Aufl.: pp 1-376

Töppe A (1994) Beschleunigter Meeresspiegelanstieg.- HANSA – Schiffahrt – Schiffbau – Hafen, Hamburg, 131 Jg.: 78-82

Van der Spek AJF, Beets DJ (1992) Mid-Holocene evolution of a tidal basin in the western Netherlands: a model for future changes in the northern Netherlands under conditions of accelerated sea-level rise. Sedimentary Geology, Elsevier, Amsterdam, 80: 185-197

Zagwijn W (1979) Early an Middle Pleistocene coastlines in the southern North Sea basin. In: Oele et al. (eds) The Quaternary History of the North Sea. Acta Univ. Ups Annum Quingentesimum Celebrantis, Uppsala, 2: 31-42

Effects of Climate and Human Interventions on the Evolution of the Wadden Sea Depositional System (Southern North Sea)

Burghard W. Flemming

Senckenberg Institute, Schleusenstr. 39A, D-26382 Wilhelmshaven, Germany
e-mail: bw.flemming@sam.terramare.de

Abstract: The most important constraints on barrier island genesis are a low coastal gradient, a low to moderate wave climate, low to intermediate tidal ranges, and a sufficient supply of sediment to construct and maintain the barrier and its shoreface. Sediment sources can be remote (river, updrift coast), local (upper shoreface) or a combination of both. With the exception of tidal range, the constraining factors are directly or indirectly controlled by climatic factors. Wave height and tidal range are loosely correlated and define the lower and the upper limits of barrier island stability within the mixed tide/wave energy zones. Since sea-level rise creates a sediment deficit in the tidal basin, the sediment budget ultimately controls the morphodynamic response and the resulting stratigraphy: progradational (remote supply > deficit), aggradational (remote supply = deficit), or transgressive (remote supply <-<< deficit). In the case of the southern North Sea, the western section from Den Helder (Netherlands) to the mouth of the Elbe river (Germany) the system is transgressive. From the Elbe river to St. Peter-Ording (Germany) the tidal flat system is aggradational-progradational. From St. Peter-Ording to the island of Sylt the tidal basins, which were originally created by coastal erosion (storm scouring) during the late Middel Ages, are aggradational-progradational. The island of Sylt itself is strongly transgressive. From Sylt to Skallingen (Denmark) the barrier system is again aggradational-progradational. Morphodynamic changes recorded in historical times (since about AD 1000) mostly reflect responses to human interventions in the western sector of the Wadden Sea (Den Helder to the Elbe river), whereas north of the Elbe river they are dominated by natural adjustment processes (erosion/accretion) which, in most places, incorporate a variable anthropogenic overprint. In the wake of a potentially accelerating sea-level rise, it is predicted that shore protection schemes will increasingly inhibit natural island dynamics and thereby threaten island stability, while storm-barriers will eventually have to be constructed in the lower reaches of all major rivers along the coast. In addition, the depletion of fine-grained sediments will continue, the system becoming increasingly more sand dominated and experiencing a further loss of mud flats and salt marshes.

Introduction

Barrier islands are found along many coasts of the world from the Tropics to the Arctic Ocean (e.g. Glaeser 1978). The genesis of barrier islands has been a topic of considerable debate in the past (cf. Schwartz 1973), but to this day no single mechanism has found general acceptance. Indeed, it would appear that barrier islands may evolve from a variety of antecedent conditions and nothing we know today would mitigate against the "multiple causality principle" suggested by Schwartz (1971). Once formed, barrier islands evolve in response to nu-merous factors, overriding constraints being defined by the coastal gradient, the tidal range, the rate of sea-level change, local sediment budgets, and the regional climate. In addition, human interventions have become increasingly more important in the course of the past millennium, mainly in the form of land reclamation and dike construction, thereby reducing tidal prisms and triggering large-scale morphodynamic adjustments. Various aspects of the Wadden Sea barrier island and tidal flat system, i.e. their evolution, stratigraphy, dynamics, and the effects of human interventions are dealt with by Dijkema et al. (1980), Ehlers (1988), Streif (1990), Flemming and Hertweck (1994), and Davis (1994).

The recent debate on climate change and the postulated acceleration of sea-level rise have stimulated government-sponsored investigations in many

From WEFER G, BERGER W, BEHRE K-E, JANSEN E (eds), 2002, *Climate Development and History of the North Atlantic Realm.* Springer-Verlag Berlin Heidelberg, pp 399-413

parts of the world to assess anticipated coastal responses and ways to mitigate potentially adverse effects. However, the development of coastal protection plans and/or other response measures require a sound understanding not only of coastal dynamics, but also of its natural evolution before human intervention. Of particular importance for the prediction of future coastal behaviour is the recognition and distinction between primary and secondary climate-driven effects.

This paper highlights the most important principles involved in barrier island evolution and morphodynamics, and the role played by overriding climatic factors. These principles are then used to reconstruct the Holocene evolution of the barrier island and tidal flat system of the Wadden Sea, including the impacts of man who has modified the size of many tidal basins by land reclamation and dike construction in the course of the past millennium. A prognosis is then made of the expected coastal response to the predicted acceleration in sea-level rise.

Physical Constraints on Barrier Island Development

Barrier island genesis is currently explained by three different mechanisms, circumstantial evidence being quoted in support of each case (cf. Schwartz 1971). Thus, one school favours vertical accretion and eventual emergence of offshore bars (e.g. King 1959). A second proposes the progressive breaching of prograding spits (e.g. Fisher 1968), whereas a third advocates the submergence of lowlands behind elevated coastal dune ridges (e.g. Hoyt 1967). While some authors accept the feasibility of at least two mechanisms (e.g. Shepard 1963), only one provides separate evidence for all three (Zenkovich 1967). Whatever the formative processes, the subsequent survival and further evolution of barrier islands is then constrained by specific interactions of a number of well documented physical processes (e.g. Hayes 1979; Davis 1994).

The most important constraints are considered to be a low coastal gradient, the availability of sediment, a low to moderate wave climate, and the tidal range (e.g. Hayes 1979). The former two are geological constraints which are indirectly linked to climatic factors on different time scales. The latter two are hydrodynamic constraints, of which mean wave heights are directly controlled by regional weather patterns and hence by climatic factors. A low coastal gradient is required to provide the space for the development and maintenance of a lagoonal water body or tidal flat system behind a barrier island. If the gradient is too steep, the tidal prism may become too small to maintain an inlet. As a consequence, the lagoon or tidal flat system would either rapidly evolve into a salt or freshwater marsh, or the barrier would amalgamate with the coast to form a dune or beach ridge. Of course, sufficient sediment must be available to construct and maintain the barrier, its shoreface, and the back-barrier tidal flat system. The sediment sources can be remote (river, updrift coast), local (adjacent shoreface) or a combination of both.

The two most important hydrodynamic factors, i.e. wave height and tidal range, can be correlated to define a variety of physical settings ranging from wave domination to tide domination. The two end-members are separated by three intermediate environments in which either the one or the other plays a more dominating role (Fig. 1; cf. Hayes 1979). This relationship adequately explains the occurrence or absence of modern barrier islands and tidal flat systems. In the case of barrier islands, the lower cut-off is controlled by wave action, the upper by the tidal range. In addition, there appears to be an upper limit controlled by the interaction of both waves and tides. It truncates the barrier island stability field along a line approximately defined by a wave height of H = 1.5 m and a tidal range of TR = 3.5 m at the upper margin, and a wave height of H = 2.2 m and a tidal range of TR = 1.8 m at the lower margin. However, this apparent high-energy limit for barrier island existence is entirely circumstantial, being based on the absence of such systems in the Present.

The relationship between wave height and tidal range highlights two important aspects: a) without waves, barrier islands would not develop at all, irrespective of the tidal range, and b) with waves, barrier islands are found neither along upper mesotidal and macrotidal coasts, nor along low microtidal and nontidal coasts. With increasing tidal range the tidal prism eventually becomes so large that the

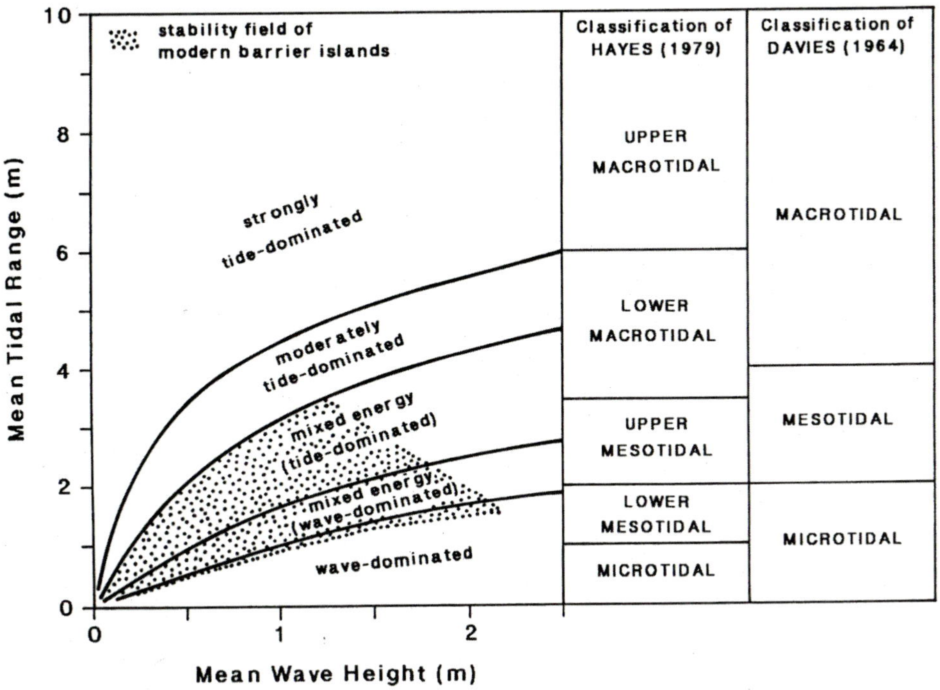

Fig. 1. Barrier island stability field as a function of mean wave height and tidal range (adapted from Hayes 1979). The barrier island systems of the southern North Sea would occupy a field bounded by tidal ranges of 1.4-3.0 m and mean wave heights of 1.0-1.5 m. Classification of tidal environments after Davies (1964) and Hayes (1979).

cross-sections occupied by adjacent flows leave no room for barrier islands between them. Whereas open or non-barred tidal flats are not found in wave-dominated environments below the barrier island stability field, they can occur in most tide-dominated environments above the barrier island stability field, provided other constraining factors are favourable. This interpretation is supported by the high correlation observed between tidal prisms and inlet cross-sections (e.g. O'Brien 1931; Van der Spek 1995). In the inner German Bight, for example, the upper limit of barrier island stability is reached at a mean tidal range of about 3.5 m. However, the slope in the line defining this upper limit in Fig. 1 suggests that the transition is to some extent compensated by wave action. Thus, with increasing wave action barrier islands can exist at increasingly higher tidal ranges until an upper limit defined by a mean wave height of about 1.5 m and a mean tidal range of about 3.5 m is reached. Conversely, along microtidal coasts, barrier islands will only survive if wave action is correspondingly low, the Gulf coast of the USA providing good examples (e.g. Davis 1994). Along nontidal coasts, wave action either prevents the formation of barriers altogether, or it produces prograding spits or barrier bars which protect lagoonal backwaters. Alternatively, very strong wave action may produce landlocked coastal lakes without direct connections to the sea. Good examples illustrating this morphological variability of nontidal sandy coasts are the partly landlocked lagoons found along the Nile Delta, the Gulf of Lions in the Western Mediterranean Sea, or the coastal lakes and lagoons lining parts of the Baltic and the Caspian Sea coasts.

Physical Controls on Barrier Island Dynamics

Once formed, barrier islands continue to evolve morphodynamically in response to numerous interacting factors, the most important being the rate of sea-level rise, the tidal prism, the seasonal wind and wave climate, and the annual sediment budget (e.g. Hayes 1979; Davis 1994). Aeolian processes and biological factors are also involved, in particular the marsh and dune vegetation which stabilizes the sediment. All of these are directly or indirectly linked to climatic factors. Together with the tidal range, the rate of sea-level rise modulates the size of the tidal prism and defines the sediment volume that is required to maintain a particular dynamic equi-librium. In addition, human interventions in the form of land reclamation (diking) and coastal protection schemes have had a strong impact on barrier island dynamics in many parts of the world over past decades and centuries. This is particularly well documented in the case of the Wadden Sea (e.g. Flemming and Davis 1994; Oost 1995).

A key factor in the morphodynamic response of barrier islands is the sediment budget. This is illustrated in a succession of four process-response models: a progradational, an aggradational and two transgressive ones (Figs. 2a-d; partly after Davis 1994). In budgetary terms, the progradational system implies a surplus of sediment in which the sediment supply from external sources exceeds the deficit created by sea-level rise (Fig. 2a). In aggradational systems the volume of imported sediment more or less balances the deficit (Fig. 2b), whereas in the two transgressive cases the deficit exceeds the imported volumes. The remaining deficit may be small (Fig. 2c) or large (Fig. 2d). Since in the latter two cases the barrier system must release sediment from its own depot on the upper shoreface, the island is displaced landwards in the course of back-barrier aggradation, the island migration rate being proportional to the deficit. The lower the import from remote sources, the faster the landward migration. If no sediment is imported then the entire deficit has to be replaced from within, and the whole system has eventually to be turned over in the course of continued sea-level rise (e.g. Kraft and John 1979).

The volume of the tidal prism is a function of the tidal range and the area of the tidal basin. As a result, for any given tidal range there is a high correlation between the other two parameters (e.g. O'Brien 1931). High correlations also exist among numerous other hydrological and morphological parameters, e.g. the area of tidal basins or volumes of tidal prisms and the cross-sectional area of inlets, the tidal prism

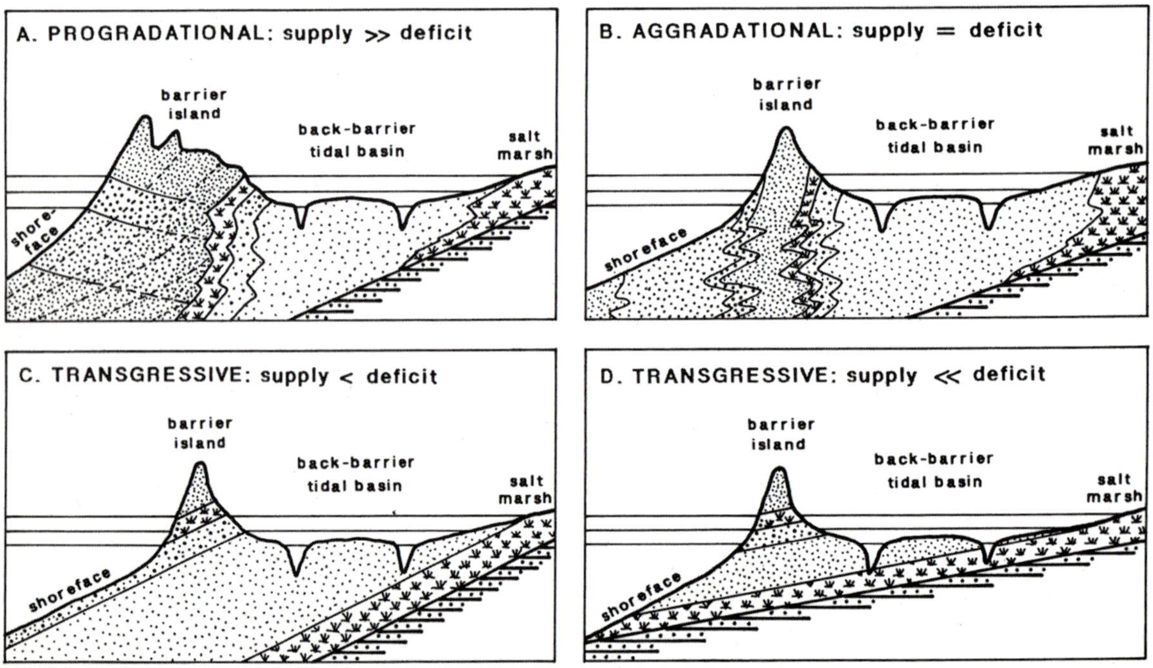

Fig. 2. Barrier island evolution in response to sea-level rise and a variety of different sediment budgets. **a)** Import >> Deficit. **b)** Import = Deficit. c) Import < Deficit. D: No Import or Import << Deficit.

and the size of ebb and flood deltas, or the tidal prisms and the depths of inlets and channels (e.g. Oost 1995; Van der Spek 1995). Two examples are illustrated in Fig. 3. Since the volume of the tidal prism is also modulated by wind-induced elevations of water levels, climatic factors are strongly involved in temporary morphodynamic adjustments on smaller temporal and spatial scales.

The Wadden Sea System

Physical Setting

At the time of the maximum Holocene transgression of the North Sea, the flooded lowlands extended as far south as Calais in France and as far north as the Jammerbugt in Jutland, Denmark, covering a distance of >1000 km (Fig. 4; cf. Ehlers 1988). The modern Wadden Sea, by contrast, stretches over a distance of about 500 km between Den Helder in the Netherlands and Skallingen in Denmark. Of this, the lower-upper mesotidal West and East Frisian barrier island chain, which commences with the island of Texel opposite Den Helder in the Netherlands, occupies about 250 km (cf. Fig. 1 for classification of tidal environments). The upper mesotidal to macrotidal section between the outer Jade channel and St. Peter-Ording on the Schleswig-Holstein coast, which straddles the mouths of the Weser and Elbe rivers, is about 100 km long. It is occupied by non-barred transgressive intertidal flats west of the Elbe river, and non-barred aggrading-prograding tidal flats to the north of the Elbe river. The final 150 km stretch up to Skallingen is upper-lower mesotidal, the southern section up to and including the island of Sylt being about 70 km long and comprising tidal flats and islands produced by storm scouring of the coastal lowlands during the late Middle Ages. The island of Sylt is strongly transgressive as documented by high coastal erosion rates. The Danish section, in turn, is characterised by aggrading-prograding barrier islands terminated in the north by the Skallingen barrier spit (Aagaard et al. 1995). The southern coastal sector between Calais and Den Helder was formerly also occupied by extensive intertidal flats fronted by barrier islands. Whereas the section south of the Rhine delta was totally reclaimed in historical times, the section north of the delta was infilled by aggradation and shoreline progradation in mid-Holocene times (e.g. Beets et al. 1992).

Fig. 4 also demonstrates that the occurrence and size of the barrier islands along the southern North Sea coast is evidently controlled by tidal range.

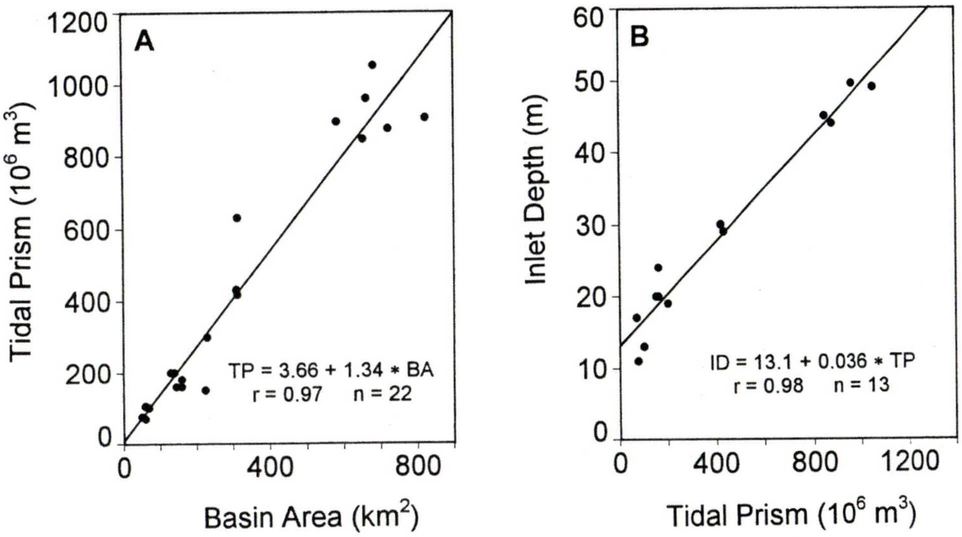

Fig. 3. The relationship between the area of tidal basins and the tidal prism (A), and the tidal prism and the maximum inlet depth (B) (modified after van der Spek 1995).

Commencing in the west, the mean tidal range at Den Helder is about 1.4 m, maximum island length reaching 30 km (e.g. Terschelling). Proceeding eastwards, the tidal range progressively increases to reach 2.4 m at the mouth of the river Ems, island length having reduced to 15 km (e.g. Juist). The last island of this chain (Wangerooge) is situated west of the outer Jade channel at a mean tidal

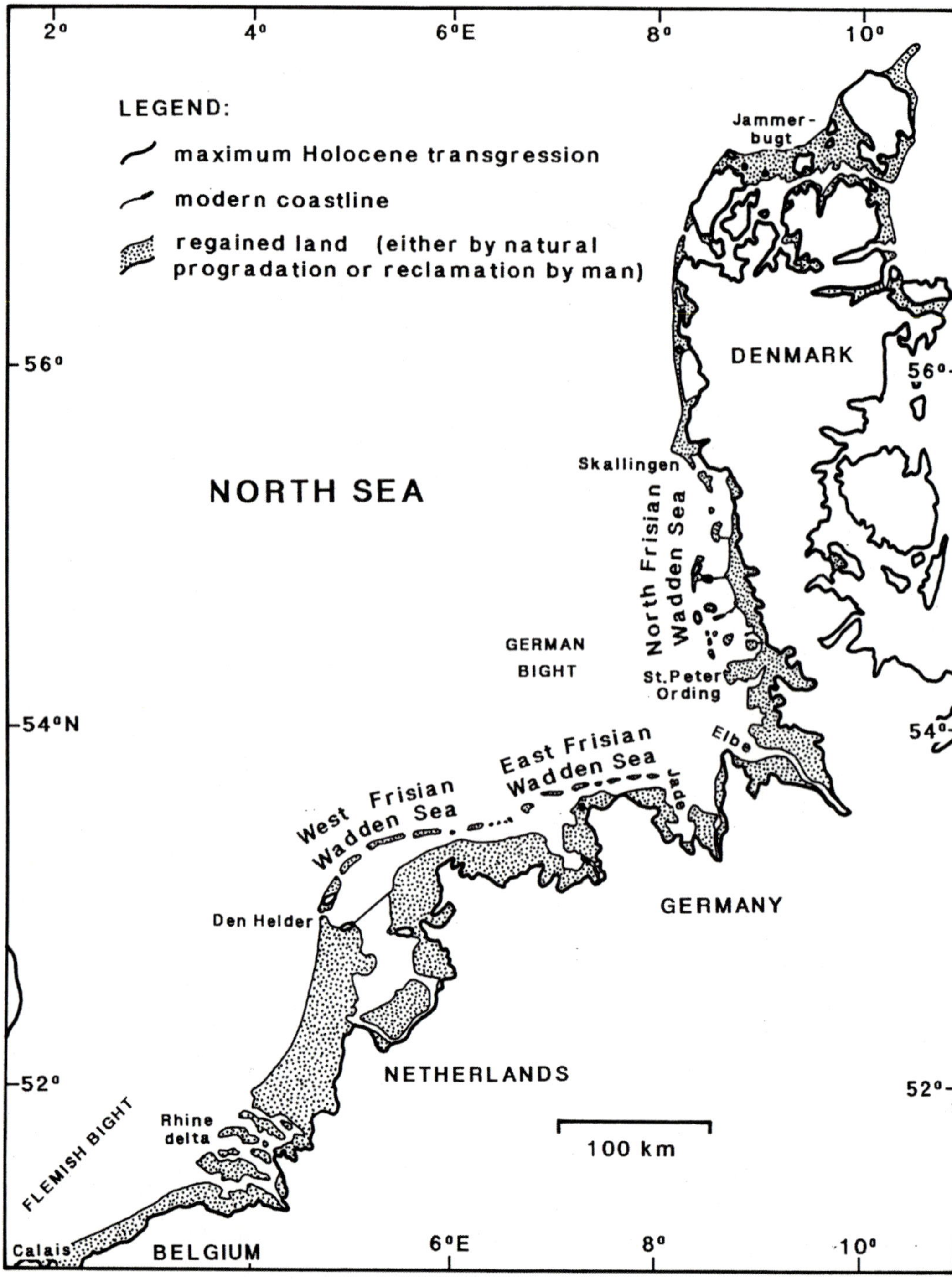

Fig. 4. Locality map of the barrier island system along the southern North Sea coast (modified after Ehlers 1988).

range of about 3.3 m, the length having reduced to <10 km. The central section is exposed to tidal ranges exceeding 3.5 m. Islands do not occur here, but ephemeral supratidal sand banks may develop from time to time. From the mouth of the river Elbe, where tidal ranges >4 m may occur, the trend reverses and tidal ranges progressively reduce to reach 1.5 m at Skallingen. Although not all the islands along this sector are true barrier islands, the Danish islands increase in length as would be expected. Thus, both Rømø and Fanø, situated south of the Skallingen spit bar, have a length of about 16 km.

Along the entire Wadden Sea coast the intertidal sediments are dominated by sandy sediments which display site-specific shoreward-fining gradients that are abruptly truncated at the foot of the mainland dike (cf. Fig. 7; Flemming and Davis 1994; Flemming and Bartholomä 1997). As demonstrated by Krögel and Flemming (1998), the mobility of non-cohesive sediment is higher at low water temperatures because of the high kinematic viscosities. Since the winter season is also the most energetic in terms of energy flux, the regional sediment distribution patterns are postulated to be adjusted to the fluid dynamic conditions of the cold season. During summer, when kinematic viscosities are lower, particle settling velocities increase, resulting in reduced sediment mobility and hence greater tidal flat stablility. Due to this temperature sensitivity, any climate change resulting in higher average water temperatures would clearly affect sediment dynamics, the overall tendency being towards greater sediment stability.

Holocene evolution

The evolution of the Wadden Sea coast commenced 7.5-8 ka BP when the recovery of the postglacial sea level reached the base of the modern shoreface at a depth of approximately 20-25 m below modern chart datum. In this situation the brackish waters of the shallow southern North Sea basin were able to penetrate into the numerous small river valleys draining the elevated Pleistocene sand bodies lining the coast. In this early stage of coastal evolution the tidal range was approximately 1.0-1.5 m (cf. Flemming and Davis 1994; Vos and Van Kesteren 2000). Under such conditions it is reasonable to assume that the river mouths were initially converted into estuaries and lagoons lined by lower mesotidal flats.

In this phase the seaward edges of the sand ridges separating adjacent river valleys were in all likelihood reworked by wave action and transformed into shore-parallel spits, in the shelter of which estuarine and lagoonal deposits could accumulate (e.g. Kosters et al. 1992). The sand spits will have accreted vertically by capturing windblown sand to form elevated topographic features, a process known to be strongly promoted by vegetation. In front of the estuary mouths first ebb-delta shoals began to develop, the remnants of which are still found locally along the coast at water depths be-tween 20 m and 25 m (Sha and De Boer 1991; Kosters et al. 1992). Although these early Holocene ebb-delta deposits have generally been interpreted as indicators for barrier island development, they are also compatible with the initial estuarine-lagoonal phase postulated above.

With continued sea-level rise local tidal ranges increased proportionally and the coast retreated gradually while aggrading vertically. This is documented by the lagoonal, intertidal and ebb-tidal delta sediments recovered in cores or mapped by sub-bottom profiling several kilometres north of the modern barrier islands (Hanisch 1980; Sha and De Boer 1991; Kosters et al. 1992). The stratigraphic evidence clearly suggests that even in the early stages of barrier island evolution, i.e. at a time when the nearshore water body was still rather shallow compared to today, and when the erosion of Pleistocene sands exposed at the seabed yielded much larger sediment volumes than today (e.g. Hoselmann and Streif 1997), the supply was clearly not high enough for in-place vertical accretion (aggradation), let alone seaward progradation. As a result, the entire coast appears to have had transgressive character in its early stages of evolution.

In the course of time, the increasing tidal range must eventually have led to a situation in which the waters of adjacent estuaries merged laterally at high tide behind the more elevated sand spits which were at first converted into temporary and later permanent barrier islands separated from the mainland coast by intertidal flats (Flemming and Davis

1994). Very similar models have been proposed by Zagwijn (1986) and Vos and Van Kesteren (2000).

The transformation of the barrier spits into barrier islands most probably did not occur simultaneously along the entire coast, but must have been completed by the time the rate of sea-level rise declined further around 6 ka BP when the tidal range off the central section of the East Frisian coast was of the order of 2.3 m. This transformation process marks the actual birth of the Wadden Sea. In Figs. 5a-c the Holocene coastal evolution is illustrated in a number of partly hypothetical re-

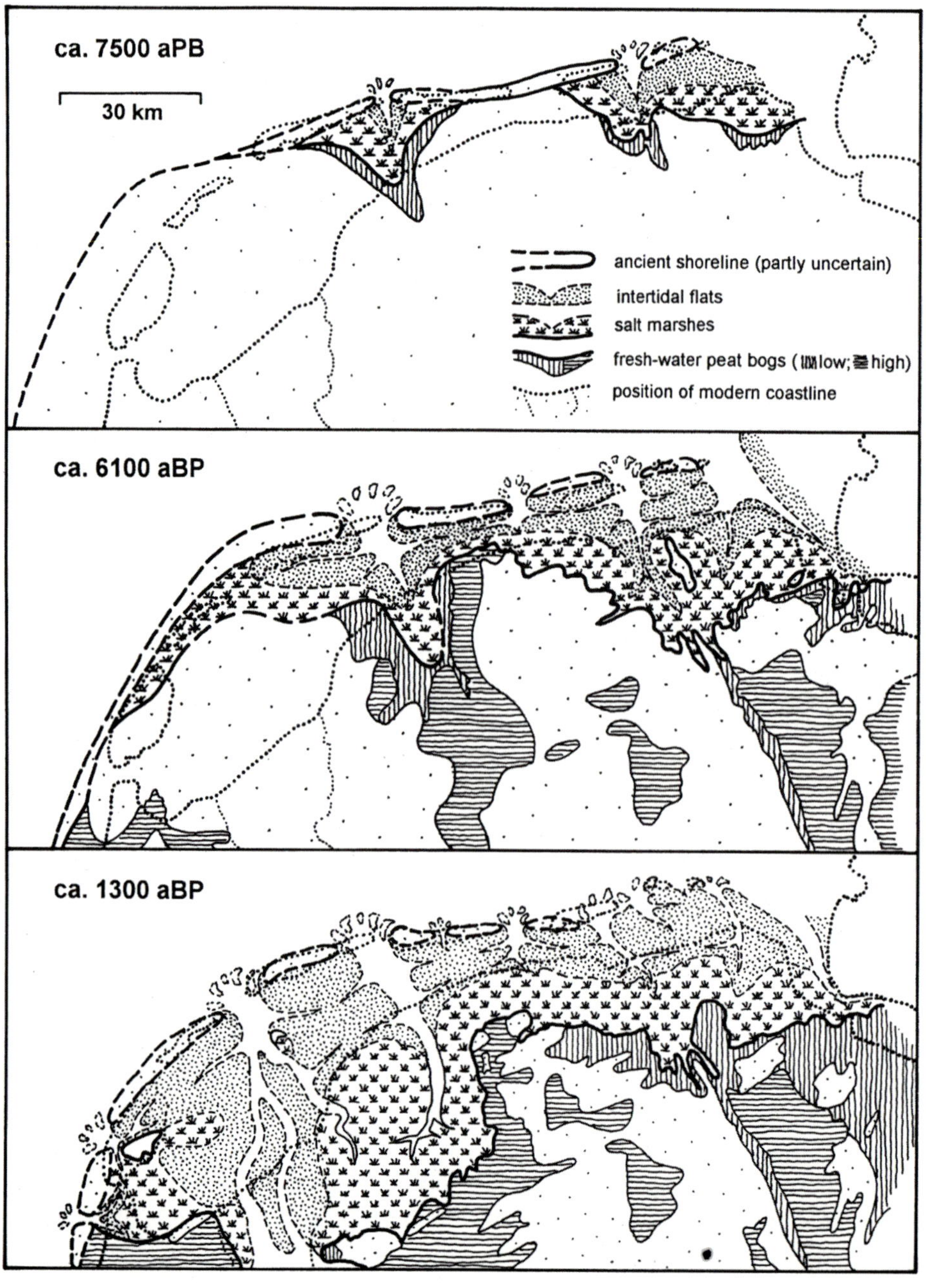

Fig. 5a. Hypothetical Holocene evolution of the West Frisian coast (Netherlands) over the past 7.5. ka (adapted from Zagwijn 1986).

constructions representing the West Frisian (Fig. 5a), the East Frisian (Fig. 5b), and the North Frisian Wadden Sea coasts (Fig. 5c).

Thus, from the very start much of the southern North Sea coast was initially characterized by transgressive barrier spit systems in which the sediment input from remote sources was smaller than the deficit created by sea-level rise. The same applied to the subsequent barrier island systems which survived by continually reworking their own

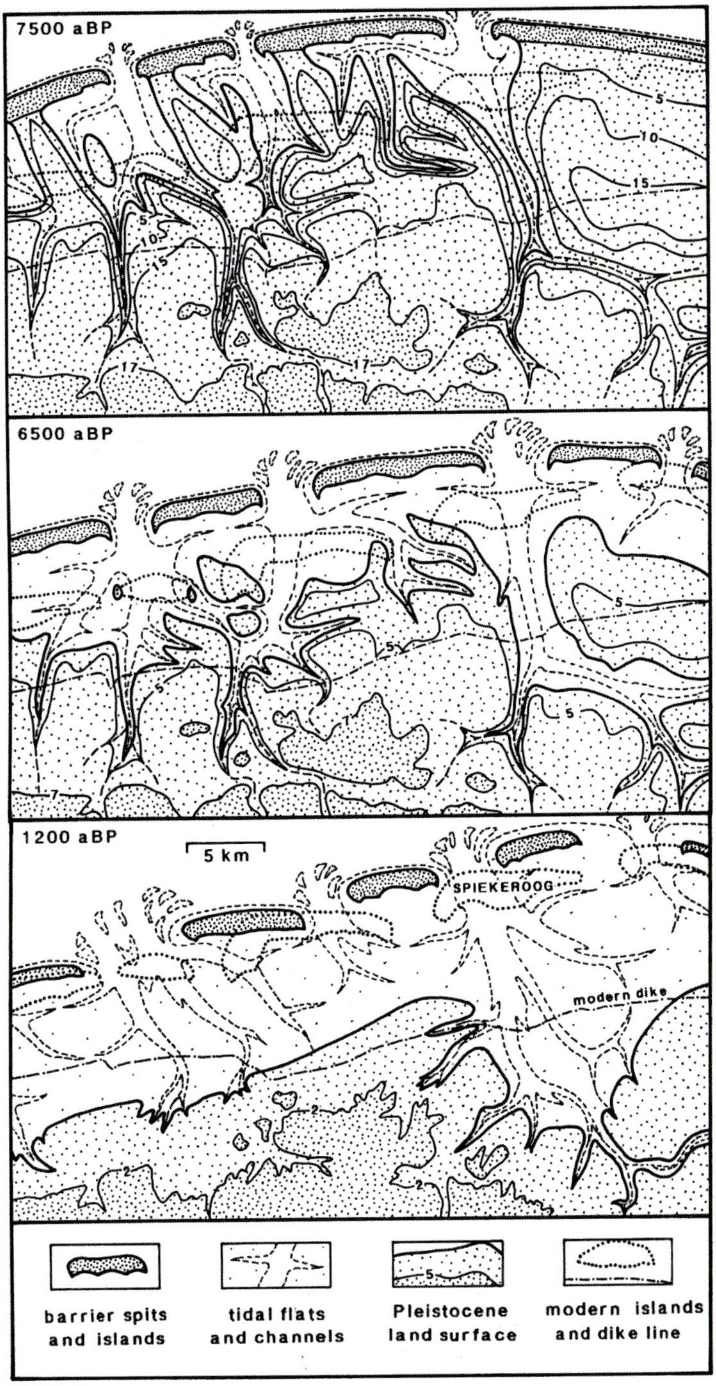

Fig. 5b. Hypothetical Holocene evolution of the East Frisian coast (Germany) over the past 7.5 ka (modified after Flemming and Davis 1994).

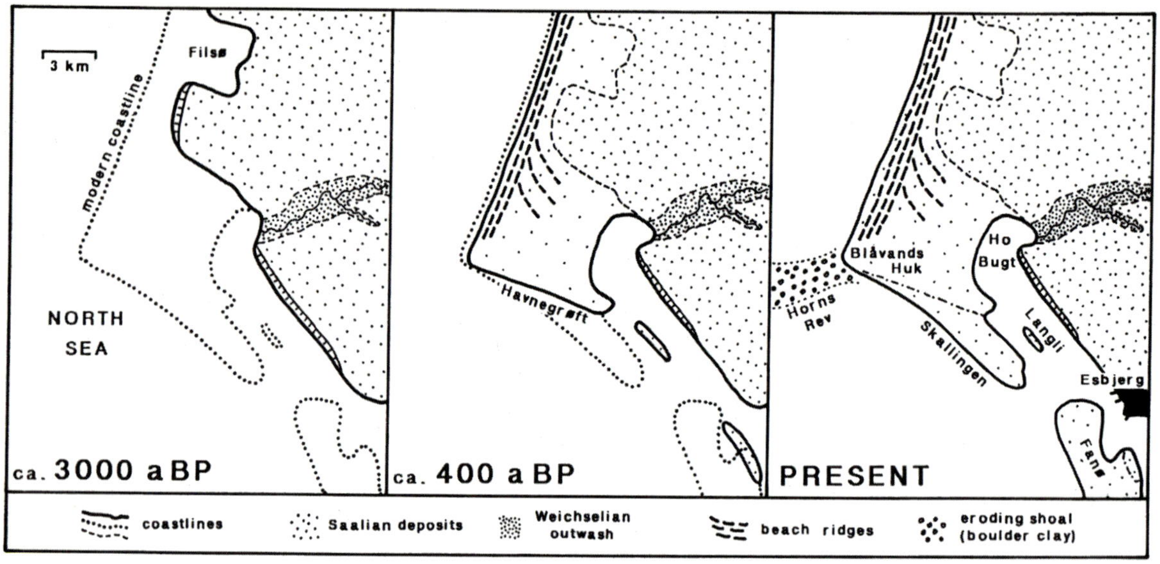

Fig. 5c. Hypothetical Holocene evolution of the North Frisian coast in the vicinity of Skallingen (Denmark) over the past 3 ka (modified after Aagaard et al. 1995).

sediment reservoir on the upper shoreface and transferring the material into the back-barrier basins to partly or fully compensate the deficit created by the rising sea. Locally, e.g. between the Rhine delta and Den Helder (e.g. Beets et al. 1992), between the Elbe river and St. Peter-Ording (e.g. Ehlers 1988), and along the Danish coast (e.g. Bartholdy and Pejrup 1994; Aagaard et al. 1995), the initial transgressive retreat gradually transformed into aggradation or regressive progradation when in mid-late Holocene times (around 6 ka BP) the rate of sea-level rise slowed down and former sediment deficits were locally converted into net surpluses.

Human impacts in historical times

An evaluation of historical charts of the West and East Frisian coastal sectors since 1300 AD revealed dramatic morphodynamic adjustments in the wake of land reclamation. In the case of the West Frisian islands this is illustrated by the infilling and subsequent reclamation of the Lauwers Sea in the Groningen District (Fig. 6a; after Oost 1995). Along the East Frisian coast it was observed that between 1650 and 1960 the length of the barrier islands progressively increased, while catchment areas and inlet widths decreased proportionally (cf. Flemming and Davis 1994). This morphodynamic adjustment has been interpreted to be a direct response to land reclamation along the mainland shore, in the course of which the areas of the back-barrier basins were substantially reduced in size (e.g. Fig. 6b). Since this is coupled with corresponding decreases in the volume of the associated tidal prisms (cf. Fig. 3), the reduced flow results in a reduction of inlet widths and hence growth of the islands. In some cases individual islands merely captured the sands of abandoned ebb-deltas, thus growing in length. In other cases, two smaller islands amalgamated into a single bigger one, the former inlet between them being filled by the influx of adjacent ebb-delta sands. A third type of adjustment involved the splitting up of larger islands, the individual sections amalgamating with adjacent smaller islands.

This large-scale re-arrangement of the morphology and drainage structure in historical times had in the past mainly been attributed to a general eastward migration of the barrier islands in response to a substantial longshore sediment drift. In more recent years this interpretation has been challenged

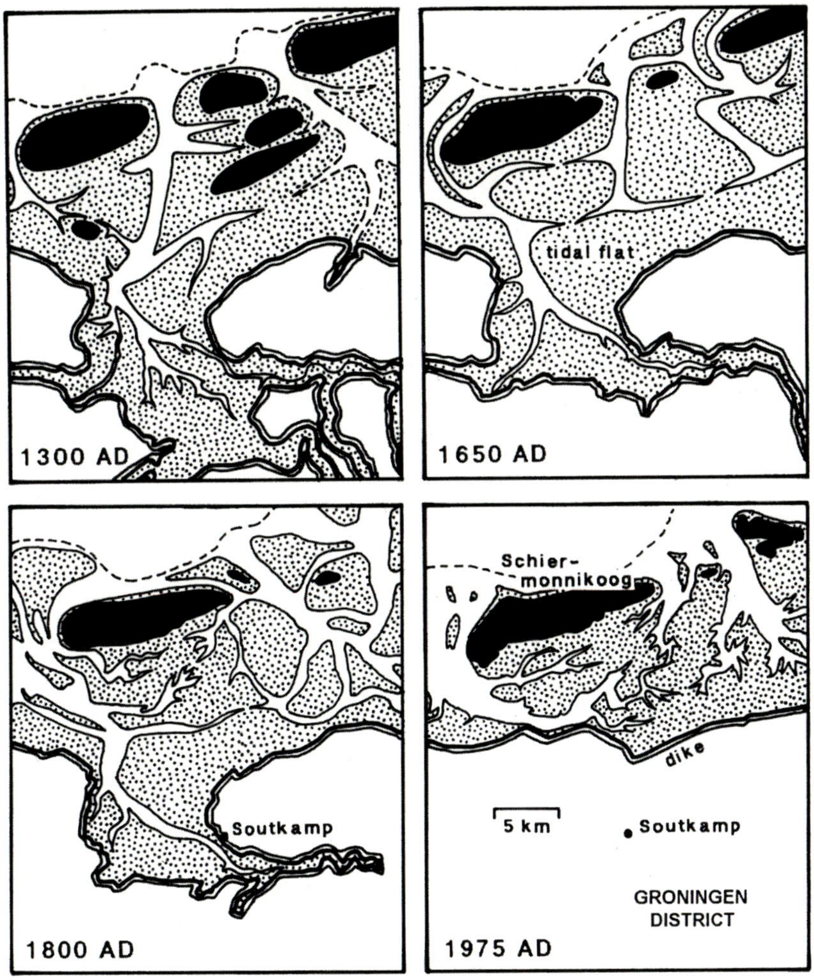

Fig. 6a. Changes in the morphology and drainage pat-terns triggered by the reclamation of the Lauwers Sea in the Netherlands (modified after Oost 1995).

in favour of changes attributed to a redistribution of local sediments triggered by land reclamation as outlined above (e.g. Streif 1990; Flemming and Davis 1994). The diking process was substantially influenced by the sometimes catastrophic damages caused by severe episodic storms. Thus, the North Frisian Wadden Sea between St. Peter-Ording and Sylt (Germany), was entirely scoured from the coastal lowlands during the late Middle Ages (Fig. 6c). The fate of the large volumes of sediment removed from this coastal sector is still unclear.

The progradational nature of the Danish Wadden Sea is well documented by the evolution of the barrier spit at Skallingen (Aagaard et al. 1995). On historical charts the former shoreline is shown to have been in line with the island of Langli which today is situated within the back-barrier tidal flats of Ho Bugt. At about 3000 BP the shoreline is thought to have been situated along the present mainland shore. Abundant sediment supply from the north led to strong headland growth which reached Blåvands Huk about 400 years ago. Since then the Skallingen barrier spit prograded south-eastwards. As a result, the shoreline rapidly shifted some 3 km to the west, thereby stranding the former barrier island of Langli in the middle of the newly created tidal flat area (cf. Fig. 5c).

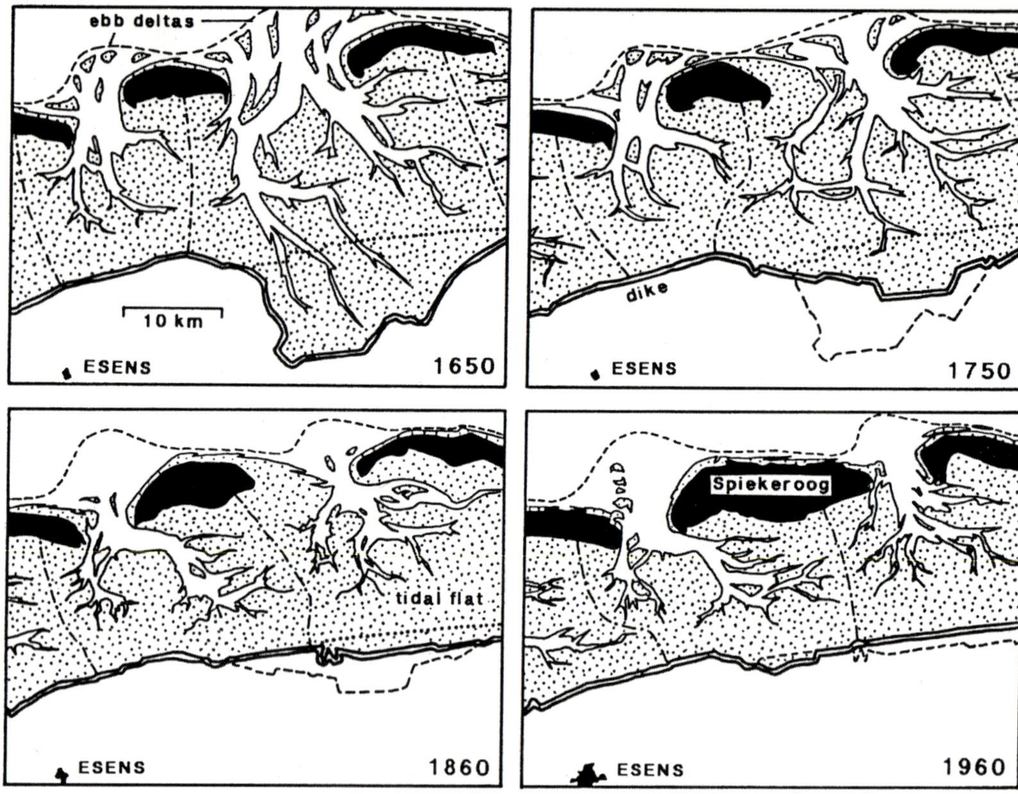

Fig. 6b. Changes in the morphology and drainage patterns triggered by land reclamation along the mainland shore of the East Frisian Wadden Sea (from Flemming and Davis 1994).

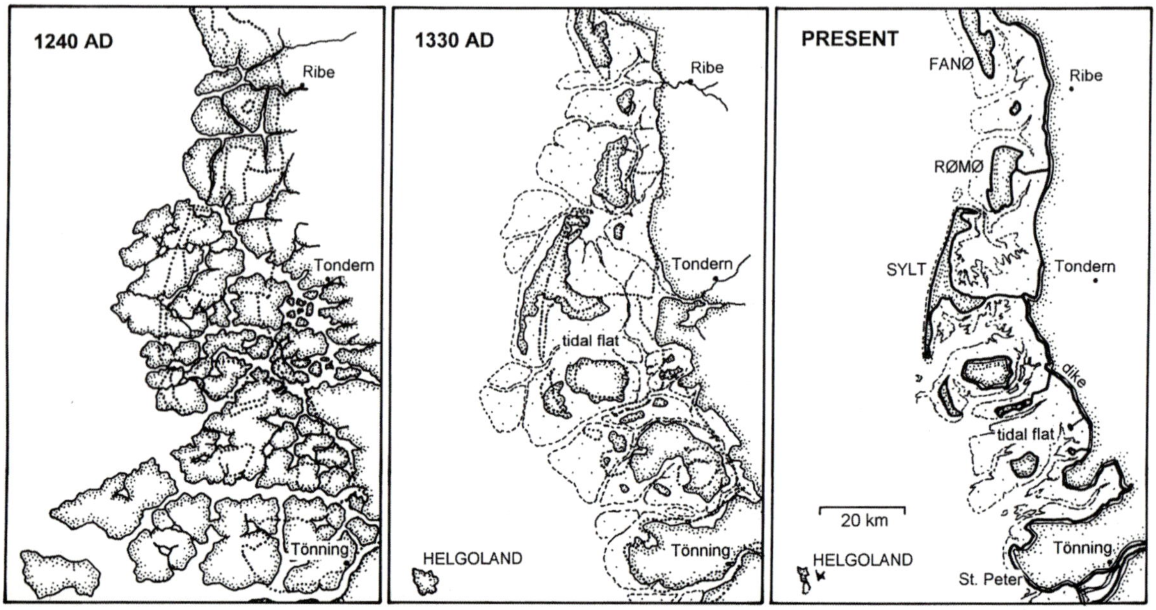

Fig. 6c. Evolution of the storm-scoured coast between St. Peter-Ording and the island of Sylt in historical times.

Predicting Future Coastal Evolution

The evidence and the inferences presented in this paper highlight the impact of both climatic and anthropogenic influences on the evolution of the Wadden Sea coast. The foremost factors controlling the dynamics of a transgressive barrier system is a rising sea level, the wave climate and the sediment budget. The higher the rate of sea-level rise and the lower the supply of sediment from remote sources, the greater is the shoreward displacement of the system per unit time. In addition, the high correlations between morphologic and hydrologic parameters of tidal basins imply rapid adjustment processes as soon as one of these parameters is disturbed. Under the present circumstances morphodynamic adjustments are likely to be triggered by changes in the regional weather patterns (climate change), anthropogenic interventions, or a combination of both.

In view of the existing and in future possibly accelerating sea-level rise, it must be anticipated that the barrier systems along the coast of the southern North Sea will respond with more vigorous morphodynamic adjustments than currently observed. At present, the relatively low rate of sea-level rise, as compared to earlier millennia, has expressed itself in the local loss of beach sand and minor displacements of some islands. In many places such sediment losses are being replenished by beach nourishment schemes. Existing coastal defence structures in the form of groins, revetments and sea walls are being maintained, but the erection of new structures are being avoided if at all possible. However, any acceleration in the current sea-level rise will rapidly aggravate the situation in existing problem areas, while creating new problems in currently stable areas. Coastal sectors, which in the past were characterized by aggradation or progradation, will probably continue to act as sediment traps, albeit with diminishing capacity, as long as their sources are not cut off.

The morphodynamic response of a transgressive barrier island system to sea-level rise in the absence of an external sediment supply is schematically illustrated in Fig. 7. Lacking any input from remote sources, the deficit has to be replaced from its own shoreface depot. In anticipation of a sea-level rise of 0.5 m within the next 100 years, the shoreward displacement of the island would in this case amount to about 0.5 km.

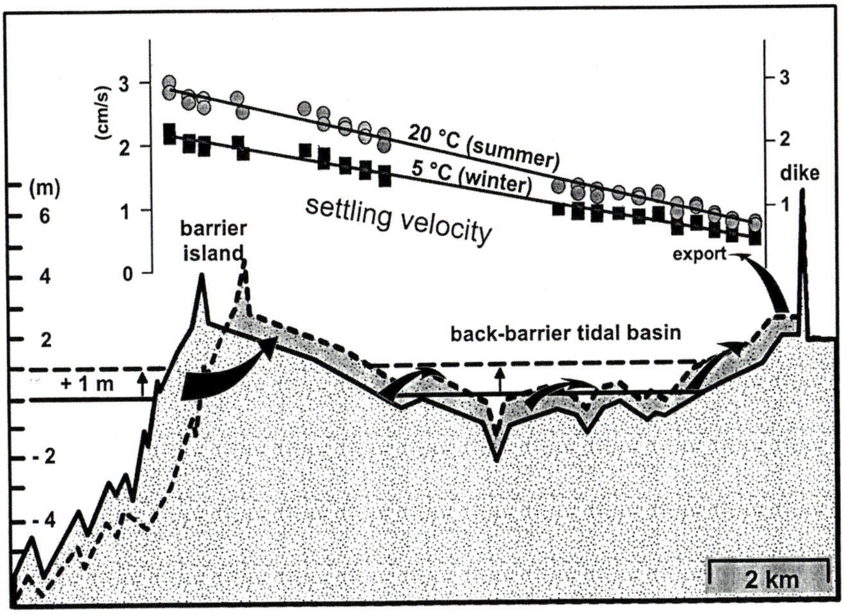

Fig. 7. Schematic diagram illustrating the morphodynamic response of a barrier island to a rise in sea level (modified after Flemming and Bartholomä 1997).

Lower rates of sea-level rise would reduce the migration rate proportionally. To counteract the Lower rates of sea-level rise would reduce the migration rate proportionally. To counteract the resulting shoreline erosion, the island beaches would have to be nourished on a long-term basis by as much as 0.5 * 10^6 m^3 per year in each case. While stabilizing the island beaches, this would not, however, compensate the sediment deficit created in the back-barrier region. As a result, the tidal basins would become increasingly more sediment starved, being gradually transformed from intertidal systems to tidal lagoons without intertidal flats, the Sylt-Rømø tidal basin being a good example. Furthermore, if storm frequency increases concurrently, then the actual volume of sediment required will also increase. Both sea-level rise and the size reduction of back-barrier tidal basins due to landward island migration will result in higher energy levels along the diked mainland shore. This would further deplete the Wadden Sea of fine-grained sediments, the system becoming progressively more dominated by sandy substrates. As a consequence, the gradual transformation of the ecosystem to a more psam-mophilic fauna will continue.

Finally, it is postulated that the lower estuarine reaches of all major rivers, in this case especially the Ems, Weser and Elbe, will eventually have to be protected by storm-surge barriers or large lock and sluice systems in order to avoid regular flooding of major cities and ports like Eemshaven in the Netherlands, Emden, Bremen, Bremerhaven, Cuxhaven, and Hamburg in Germany, and Esbjerg in Denmark. The more rapid the rise in sea level, the sooner such protection measures will become necessary.

References

Aagaard T, Nielsen N, Nielsen J (1995) Skallingen – origin and evolution of a barrier spit. Medd Skalling Lab XXXV, pp 1-85

Bartholdy J, Pejrup M (1994) Holocene evolution of the Danish Wadden Sea. In: Flemming BW and Hertweck G (eds). Tidal Flats and Barrier Systems of Continental Europe: A Selected Overview. Senckenbergiana marit 24:187-209

Beets DJ, Van der Valk L, Stive MJF (1992) Holocene evolution of the coast of Holland. Mar Geol 103:423-443

Davies JL (1964) A morphogenetic approach to world shorelines. Zeitschrift für Geomorphologie 8:127-142

Davis RA Jr (1994) Barrier island systems – geologic overview. In: Davis RA Jr (ed). Geology of Holocene Barrier Island Systems. Springer, Berlin, pp 1-46

Dijkema KS, Reineck H-E, Wolff WJ (eds) (1980) Geomorphology of the Wadden Sea area. Balkema, Rotterdam, pp 1-135

Ehlers, J (1988) The Morphodynamics of the Wadden Sea. Balkema, Rotterdam, pp 1-397

Fisher JJ (1968) Barrier island formation: discussion. Geol Soc Am Bull 79:1421-1426

Flemming BW, Hertweck G (eds) (1994) Tidal Flats and Barrier systems of Continental Europe: a Selected Overview. Senckenbergiana marit 24, pp 1-209

Flemming BW, Davis RA Jr (1994) Holocene evolution, morphodynamics and sedimentology of the Spiekeroog barrier Island system (southern North Sea). In: Flemming BW, Hertweck G (eds) Tidal Flats and Barrier systems of Continental Europe: a Selected Overview. Senckenbergiana marit 24:117-155

Flemming BW, Bartholomä A (1997) Response of the Wadden Sea to a rising sea level: a predictive empirical model. Ger J Hydrogr 49:343-353

Glaeser JD (1978) Global distribution of barrier islands in terms of tectonic setting. J Geol 86: 283-297

Hayes MO (1979) Barrier island morphology as a function of tidal and wave regime. In: Leatherman SP (ed). Barrier Islands from the Gulf of St. Lawrence to the Gulf of Mexico. Academic Press, New York, pp 1-27

Hanisch J (1980) Neue Meeresspiegeldaten aus dem Raum Wangerooge. Eiszeitalter und Gegenwart 30:221-228

Hoselmann C, Streif H (1997) Bilanzierung der holozänen Sedimentakkumulation im niedersächsischen Küstenraum. Zeitschr DGG 148:431-445

Hoyt JH (1967) Barrier island formation. Geol Soc Am Bull 78:1125-1136

King CAM (1959) Beaches and Coasts. Edward Arnold, London, pp 1-403

Kosters EC, VanMierlo BEJM, Verbeek NH, Posthumus BJ, McGee TM, Brouwer J (1992) Late Quaternary stratigraphic signature, offshore from the Dutch barrier shoreline. Sed Geol 80:199-212

Kraft JC, John CJ (1979) Lateral and vertical facies relations of transgressive barrier. AAPG Bull 63:2145-2163

Krögel F, Flemming BW (1998) Evidence for temperature-adjusted sediment distributions in the backbarrier tidal flats of the East Frisian Wadden Sea (southern North Sea). In: Alexander CR, Davis RA, Henry VJ (eds) Tidalites: Processes and Products. SEPM Spec Publ 61:31-41

O'Brien MP (1931) Estuary tidal prism related to entrance area. Trans Am Soc Civil Eng 1: 38-739

Oost AP (1995) Dynamics and sedimentary development of the Dutch Wadden Sea with emphasis on the Frisian Inlet. A study of barrier islands, ebb-tidal deltas, inlets and drainage basins. PhD Thesis, University of Utrecht, Geologica Ultraiectina 126, pp 1-400

Schwartz ML (1971) The multiple causality of barrier islands. J Geol 79:91-94

Schwartz ML (ed) (1973) Barrier Islands. Benchmark Papers in Geology 9. Dowden, Hutchinson & Ross, Stroudsburg, pp 1-451

Sha LP, De Boer PL (1991) Ebb-tidal delta deposits along the west Frisian Islands (the Netherlands) : processes, facies architecture and preservation. In: Smith DG, Reinson GE, Zaitlin BA, Rahmani RA (eds) Clastic Tidal Sedimentation. Can Soc Petrol Geol Mem 16:199-218

Shepard FP (1963) Submarine Geology. Harper & Row, New York, pp 1-557

Streif H (1990) Das Ostfriesische Küstengebiet. Nordsee, Inseln, Watten und Marschen. Sammlung geologischer Führer 57. Gebr. Borntraeger, Stuttgart, pp 1-376

Van der Spek AJF (1995) Reconstruction of tidal inlet and channel dimensions in the frisian Middelzee, a former tidal basin in the Dutch Wadden Sea. In: Flemming BW, Bartholomä A (eds) Tidal Signatures in Modern and Ancient Sediments. Blackwell Science, Oxford. IAS Spec Publ 24, pp 239-258

Vos PC, Van Kesteren W (2000) The long-term evolution of intertidal mudflats in the northern Netherlands during the Holocene: natural and anthropogenic processes. Continental Shelf Research 20:1687-1710

Zagwijn WH (1986) Nederland in het Holoceen. Rijks Geol Dienst, Haarlem, pp 1-46

Zenkovich VP (1967) Processes of Coastal Development. Oliver & Boyd, Edinburgh, pp 1-738

Historic Storms in the North Sea Area, an Assessment of the Storm Data, the Present Position of Research and the Prospects for Future Research

Adriaan M.J. de Kraker

Department of Human Geography, University of Amsterdam, Nieuwe Prinsengracht 130, NL-1018 VZ Amsterdam, The Netherlands e-mail: Krakeram@zeelandnet.nl

Abstract: A survey of historic storms, storm surges and sea floods in the North Sea area from about 1000 to the present day reveals that there is a rather complete and reliable time series of storm surges from the period of 1400 to 1700 for the Netherlands only. However, this time series does not include storms and high floods. Such a series has only become recently available for a small polder area near Antwerp from the period of 1488 to 1609. British, Danish and German time series of storm surges, storms and sea floods are very incomplete for the period earlier than the second quarter of the 18th century. Records from the Low Countries, owing to their special documentary sources starting as early as the 16th century in combination with additional data from Britain, Denmark and northern Germany, seem to offer the best opportunity of completing a useful time series of storms, storm surges and high floods for the period of AD 1500 to 2000.

Introduction

'A great, severe and tempest flood ..., never seen the like during our life time' an official of the Chapter of Our Lady at Kortrijk in Flanders told a government committee in 1513, when he was asked what he remembered of the storm surge of 14th of December 1511 (De Kraker 1997). Large scale flood on the coasts of the North Sea on Christmas Day of 1717, more than 11.500 casualties, 100,000 cattle drowned, 5000 buildings destroyed and about 6,000 square kilometres inundated (Gram-Jensen 1985).

These are just reports of two historic storms which battered the North Sea coast and which in turn led to large scale flooding. Two extreme weather events that hit a densely populated area that may be considered to be an economically very important region, with major harbours, such as Rotterdam, Hamburg, Antwerp, Le Havre and London. Moreover the coastal zone of the North Sea (from Dunkirk to Denmark) is very vulnerable with its broad estuaries and range of Wadden isles enabling floods to penetrate deeply into the hinterland of which parts are even far below sea level. Two examples of extreme weather events quite different in time and space, which set the scene for the North Sea region at a time of storm, storm surges and sea floods. This raises the first question in this paper, whether we have a complete overview of the history of storms of the recent onethousand years. And if the answer is no, is this due to the fact that there are no sufficient documentary sources left, or was there no tradition of observing wind in the coastal zone and no attention was paid to studying extreme weather events as mentioned before? Furthermore, this paper focuses on the prospects for future research on the subject of extreme weather events, e.g. historic storms and floods in order to be able to reconstruct wind force and to discover patterns in storms and storm frequency in historical times. Finally, it must be stressed that a complete overview of the historical storms on a time scale and a geographical scale can substantially

From WEFER G, BERGER W, BEHRE K-E, JANSEN E (eds), 2002, *Climate Development and History of the North Atlantic Realm.* Springer-Verlag Berlin Heidelberg, pp 415-434

contribute to our understanding of the transportation of air masses of a different moisture and temperature over Central Europe. Consequently this might also contribute to our knowledge of the boundaries of the Medieval Warm Epoch and the Little Ice Age, of the periodicity of fisheries and even of the changing shape of the Wadden isles.

Climate and Wind Patterns in the North Sea Area

In order to understand the pattern of wind, storms and storm surges in the North Sea region and its surrounding countries or parts of these countries, it is necessary to know how air masses flow within the climate system of the North Sea region.

The movement of air masses in the region of interest is largely determined by a general high pressure over Greenland with cold air masses and a low pressure region around Iceland, fed by warm air masses from the south. The front along which the warm and cold air masses collide is dominated by bad weather, rain and wind development of cyclonic storms. As wind speeds increase, such a system can develop into a severe gale, causing storm surges and sea floods, especially during winter seasons, when they have a far greater intensity than during summers.

The movement of a strong cyclonic storm from Iceland into the North Sea may be followed by means of the pressure charts of 29th January to 2nd February 1953 (Bijvoet et al. 1960). At 12.00 h on 29th of January (Fig. 1) there was a low pressure centre south-west of Iceland (1005 mb) which moved slowly south-east. At the same time there was a low at about 40°N in the middle of the Atlantic which hardly moved at all. Over France, a strong high is seen which moved slowly towards the north-west. At 12.00 h on 30th January the Icelandic low was now south-east of Iceland and greatly strengthened, while the high pressure area moved from southern France to the western Alps. The next day, on the 31st of January, a strong low lodged between Scotland and Norway, while a weak high south west of Iceland had strengthened and extended from south of Iceland over Iceland into Spain. The pressure gradients between this high pressure ridge and the low over the North Sea became extremely strong, generating forceful northerly winds. Finally a most severe storm surge developed at the place of the low, causing perhaps the largest natural disaster in Western Europe during the 20th century.

Air masses can flow from very different directions and these directions generally determine whether these air masses are relatively cold or warm, stable or unstable. Accordingly, Lamb (1972) made a classification of wind types. His classification of winds contains eight directional types: north (N), north-east (NE), east (E), south-east (SE), south (S), south-west (SW), west (W) and north-west (NW). There are two other categories: anticyclonic (A) and cyclonic (C) and still a third when no direction or distinctive type can be defined (Kelly et al. 1997). This additional unclassifiable type is called (U).

So whenever there is an air mass passing over from the north-west which is very unstable and consequently brings much rain, this type is called CNW. Travelling at a high speed this air mass can develop into a storm or storm surge overnight. An ANE type causes a stable air mass to be travelling from Scandinavia or northern Russia towards the North Sea. During winter such a high pressure system causes a drop in temperature below 0° Celsius. During summer the same pressure system causes warm and sunny weather.

However, there is more to it than only distinguishing wind directions and this is caused by the funnel shaped character of the shallow North Sea. Consequently north-westerly winds are piling up water into the southern funnel causing severe damage on the south-west Dutch coast, while westerly winds cause waters to be piled up in the German Bight. So in this respect the Lamb classification is not only useful in reconstructing wind direction and weather types as such, but his method enables us to determine a distribution of wind types on a longer time scale, especially the westerly and north-westerly winds.

Analysis of the variability of air pressure in the Atlantic has shown significant patterns on a geographical scale as well as on interannual and interdecadal time scales. In order to quantify this kind of variability the North Atlantic Oscillation index was introduced. The NAO-index shows how the Azores

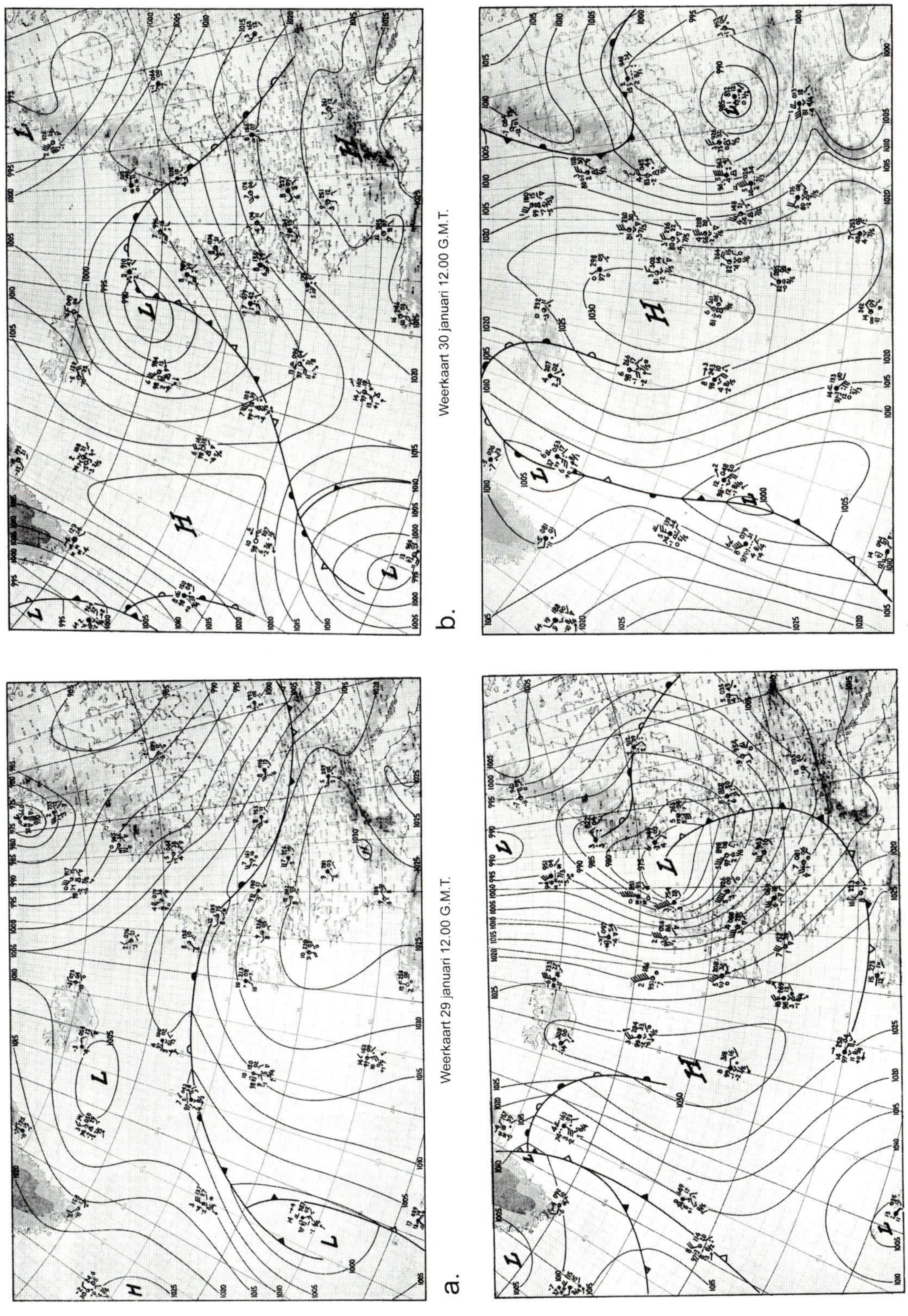

Fig. 1. Weather charts showing the development of the storm surge of 1953 in the North Atlantic area and above the North Sea: a.29 January 1953 at 12.00 G.M.T.; b. 30 January 1953 at 12.00 G.M.T.; c.31 January 1953 at 12.00 G.M.T. and d. 1 February 1953 at 12.00 G.M.T. (Bijvoet et al. 1960).

High and the Icelandic Low relate to the mean values of pressure between both air masses. If there is a positive NAO-index the differences in pressure between the Icelandic Low and the Azores High is above the average. And if the difference in pressure between the Icelandic Low and the Azores High is below average, there is, of course, a negative NAO-index. So a positive NAO-index correlates with more frequent eastward bound gales and must therefore be taken into consideration while reconstructing storms and storm surges on a longer time scale in the North Sea Region. Moreover, a positive NAO-index strongly correlates with the cyclonic W and NW-types from the Lamb classification of wind directions.

Survey of Historic Storms

In this chapter an attempt is made to see whether we have a complete overview of storms on a thousand years, time scale in the North Sea region as such and parts of it, such as the Dutch coast, the north western coast of Germany, the western coast of Denmark and the entire eastern coast of Britain. Therefore we need to know what the results are of the investigations on the subject under discussion carried out in these four countries and how these have contributed to the completing of the time series of storms.

The Medieval Period until 1800

Lamb and Frydendahl (1991) mention a storm that occurred around 120 BC, however his time series only starts in 1500. Gottschalk (1971) only considers the storms of AD 583 and AD 838 to have enough credibility to be considered as the oldest storms recorded in the Low Countries. For the German Bight Gram-Jensen (1985) found the first reliable storm for the period 1400 to 1449 (Table 1). For the medieval period a particularly great frequency of storm surges in the 14th century for the Netherlands can be distinguished. Gottschalk (1975) also noticed a marked increase of storm surge activity during the 1420s and from 1460 to 1483.

For the 16th century much more reliable documentary sources are available Consequently the reliability of the storm observations as mentioned by Gram-Jensen (1985) is now 73%. To the end of the century the number of storm observations for Denmark seems to be larger than for the Netherlands. This is partly explained by the fact that Gottschalk was interested in storm surges only, leaving out all other storms of minor importance. Her findings for the German part of the North Sea coast for the four quarters of the 16th century are: 6, 7, 8 and 13. Whereas this survey of the 16th century storms remains far from complete, no mayor conclusions may be deduced from the Danish, German and even British historical evidence, except for the fact that there seems to have been an increase in storm surges along the German and Danish part of the North Sea during the first decade and again starting from the middle of the 16th century. Recent research results of storm surges and the minor storms (De Kraker 1997, 1999) from the Wester Scheldt estuary, in-cluding all high tides, suggest a very unequal distribution of storms throughout the 16th century. It also reveals much more detail about every particular storm, storm surge sea flood that occurred in that small area.

For the 17th century storminess and storm surges tend to hit the northern part of the North Sea more frequently. But again the geographical scale is quite different. The storm surge that led to a large scale inundation in the southern Netherlands on 26th of January 1682, is not mentioned by Gram-Jensen (1985), nor by Lamb and Frydendahl (1991). Again it is impossible to have a complete picture of the 17th century storms, because the historical evidence shows some severe gaps and is by no means uniform and continuous.

On the other hand, Lamb has shown clearly that towards the end of the 17th century it becomes possible to use pressure maps, reconstructing historic storms and storminess for the period storms lasted. However, these maps can only be used as soon as a period is sufficiently covered by detailed storm observations (Douglas et al. 1978; Kington 1994; Pfister et al. 1994).

For the 18th century still more detailed information about storms and sea floods is available, especially from the instrumental recording of weather. Unfortunately, in the Netherlands this period has not been investigated yet. Only for the second half of the century data are available from the small

	Netherlands (Gottschalk 1971-1977)	Denmark (Gram-Jensen 1985)	Germany (Rohde 1977)	England (Lamb et al. 1991)	Vier Ambachten (De Kraker 1997)
< 1000	1	0			
1000-1099	2	0			
1100-1199	4	0			
1200-1299	6	0			
1300-1399	12 to 14	0			
1400-1499	17 to 20	0			
1400-1449	7 to 8	1			
1450-1499	11 to 13	0			
1500-1524	8	0		1	31 to 36
1525-1549	4	4		2	7
1550-1574	3	11		2	37 to 38
1575-1599	3	18		3	35 to 38
1600-1624	6	29	9	1	
1625-1649	1	30	11	2	
1650-1674	4	13	17	9	
1675-1699	3	11	13	9	
1700-1724		23	6	4	6
1725-1749		9	2	10	3
1750-1774		12	2	4	
1775-1799		10	22	9	

Table 1. Storm surges, storms and high tides in the North Sea Area, 838 to 1800.

coastal area along the Wester Scheldt, mentioned earlier. Evidence from Denmark is different again, especially for the first quarter of the 18th century, while storm and flood data from the Netherlands, the British Isles and Germany are very similar. Some of the storms and floods recorded between 1702 and 1704 were observed at Hamburg and on the British coast as well, e.g. the storm of 8th of December (Fig. 2).

The storm surge of 3rd and 4th March 1715 (Tables 1 and 2) seems to have caused disaster on the southern Dutch coast only, while the 1717 storm surge caused large scale flooding on the German and on the Danish coast, but was only mentioned in the southern Netherlands as just another severe storm. Recent study by Jakubowski-Tiessen (1992) has shown that the 1717 storm surge was by far the severest natural disaster that hit that part of the North sea coast, causing 'only ' 8374 casualties.

During the last quarter of the 18th century there seems to have been a strong increase in high floods on the German coast, which cannot be attributed to the fact that more measurements became available. For Hamburg alone values for the four quarters of the 18th century are respectively 5, 2, 2 and 22.

Thus in spite of the more detailed information on storms of the 18th century much remains to be investigated still, particularly for the early part of it. Questions such as, do we know all relevant storms and sea floods, do our data cover the entire region around the North Sea, and do we have

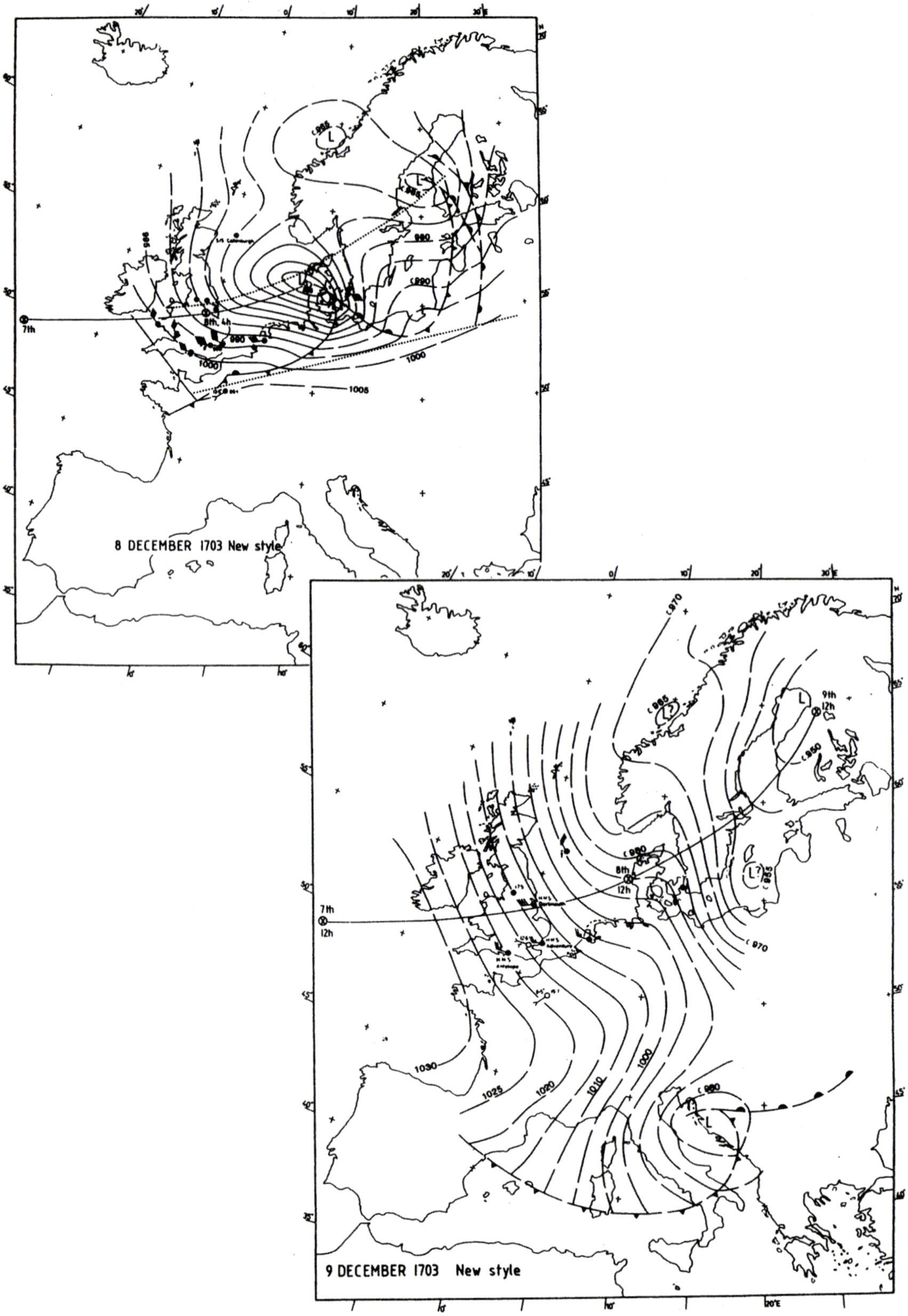

Fig. 2. Storm surges of 8th and 9th of December 1703 (Lamb and Frydendahl 1991).

date	*Hamburg*	*Embden*	*Damgast*	*Cuxhaven*	*Bremerhaven*
1701, 17 Oct.	*433 cm*				
1702, 28 Febr.	*444 cm*				
1703, 8 Dec.	*468 cm*				
1715, 4 March	*477 cm*				
1717, 24 Dec.	*562 cm*	*462 cm*	*490 cm*	*(425 cm)*	*(465 cm)*
1720, 31 Dec.		*398 cm*			

Table 2. Water levels and storm surges on the German coast, 1700-1725 (Source: Rohde 1977).

details of known storms necessary to be able to determine their importance in time and space, can only be answered following more research in different areas around the North Sea. Jakubowksi-Tiessen not only has corrected recent research results (Gram-Jensen 1985) considerably but makes us again aware of the need to be careful in using older studies for the reconstruction of 'historical' weather observations (Arends 1833).

Storms, Storm Surges and Sea Floods of the 19th and 20th Centuries

Data for storms, storm surges and sea floods of the 19th and 20th century are as complete as they can be. Unfortunately and in spite of the large quantities of accurate data of 'Rijkswaterstaat', the picture for the Netherlands remains far from complete.

Table 3 shows a large number of storm surges on the entire Dutch stretch of the North Sea coast. This large number is partly due to the way storm surges have been registered. When a severe northerly or westerly lasted longer than one day, of course it caused two or even three storm surges at each high tide as given in column A. Considering two or three successive storm surges as being one severe gale, the corrected figures for the Netherlands are given in column B. Still it can be concluded there was a considerable increase in the number of storm surges and consequently in storminess during the second half of the 20th century on the Dutch coast (Van Malde 1996; De Kraker 1999). The same is concluded by Rohde (1977), Gram-Jensen (1985) and Lamb and Frydendahl (1991) for the British Isles (Fig. 3). Higher frequency in the number of storm surges and storminess is closely related to the slow rise of the general sea level, at least as far as the North Sea is concerned (Jelgersma 1992).

It is clear that there is a general but, especially during the 19th century a very irregular sea level rise. This in turn calls for a careful comparison with time series of temperature and wind patterns and the reconstruct on maps of surface air pressure covering the period from 1850 to 2000 (see below).

Judging from the overview of storms on a thousand years, time scale and on a geographical scale for the North Sea region, it is clear that this does hardly exist. This makes it impossible to deduce general conclusions from the present state of research on storms and storminess in this area. In fact, this leaves us with the question, why does our knowledge show such serious gaps. Can it be that there is not enough historical evidence left or was there no tradition in observing wind in the coastal zone of the North Sea?

Studying Wind in and of the Past

People living on the shores of the North Sea were well aware of the dangers of storms and storm surges. This is not only sustained by the many reports and stories regarding such natural disasters. It may also be assumed that they developed a kind of system to record or measure these events for future reference long before the instrumental observations of wind direction and wind force was introduced in the course of the 18th century. As there is a marked difference between the recording of wind on land and at sea the two will be discussed separately.

period	*Netherlands*			*Germany*	*Denmark*	*Great Britain*
	number	*number*	*number*	*number*	*number*	
	A.	*B.*	*(Rohde)*	*(Jensen)*	*(Lamb)*	
1800-1824			*33*	*13*	*3*	
1825-1849	*9*	*8*	*30***	*17*	*12*	
1850-1874	*14*	*12*		*26*	*7*	
1875-1899	*31*	*23*		*10*	*11*	
1900-1924	*26*	*22*		*13*	*6*	
1925-1949	*25*	*19*		*13*	*17*	
1950-1974	*49*	*36*		*11*	*17*	
1975-1999	*44**	*39**			*18****	

A. storm surge
B. storm
** until 1991*
*** until 1840*
**** until 1989*

Table 3. Historic storms, storm surges and sea floods of the 19th and 20th century in the North Sea area.

Observing Wind on Land and at Sea before 1800

Wind enabled man to operate windmills, which were constructed with large sails so that they could catch enough wind to drive the millstones. Whenever the wind was changing direction, the entire mill could be turned around on its post or - as its construction evolved through time - only the rooftop had to be turned to the wind. At least in Holland mills remained the dominant factories until the second half of the 19th century.

Next to millers, still others had a profound knowledge of wind. One of them was Noppen (1706-1774), who was working for the 'Hoogheemraadschap van Rijnland' at Bilderdam, some miles south of Amsterdam, keeping a daily record of wind. Of course, there are some years and months missing, but his series begins in 1701 and continues until 1825 (Geurts and Van Engelen 1992). At the same time he recorded water levels at Amsterdam as well.

Starting in 1782 he observed winds twice a day as he did water levels. This leaves us with some 27,000 wind data before 1782 of the Amsterdam area and for the period 1782 to 1825 there are another 32,000 wind data. As it is shown in Fig. 4, the observer had only to fill in a form, indicating the year and particular month and, of course, his daily observations. Handling this kind of data statistically was rather easy. The contemporary example of observations at Amsterdam, which started in 1700 and continued right through until 1936 containing over a million observations, however shows that there are some serious calculation problems to be solved first, before these data can be fully utilised (Van Veen 1940).

During the 18th century others gradually began observing daily weather in the Dutch coastal zone using the first instruments to measure temperature and the amount of rain (Geurts and Van Engelen 1992). However, no satisfactory method of measuring wind force was applied yet. The observers on land had to restrict themselves to monitoring the direction of the wind and by using a kind of wording to express the wind force throughout the 18th century (Table 4).

Before the introduction of steamships, sailing ships were dominant. Life long experience led seamen to sail only during the 'summer' season when storminess was less than during 'winter' season. The calmer season was from April to October,

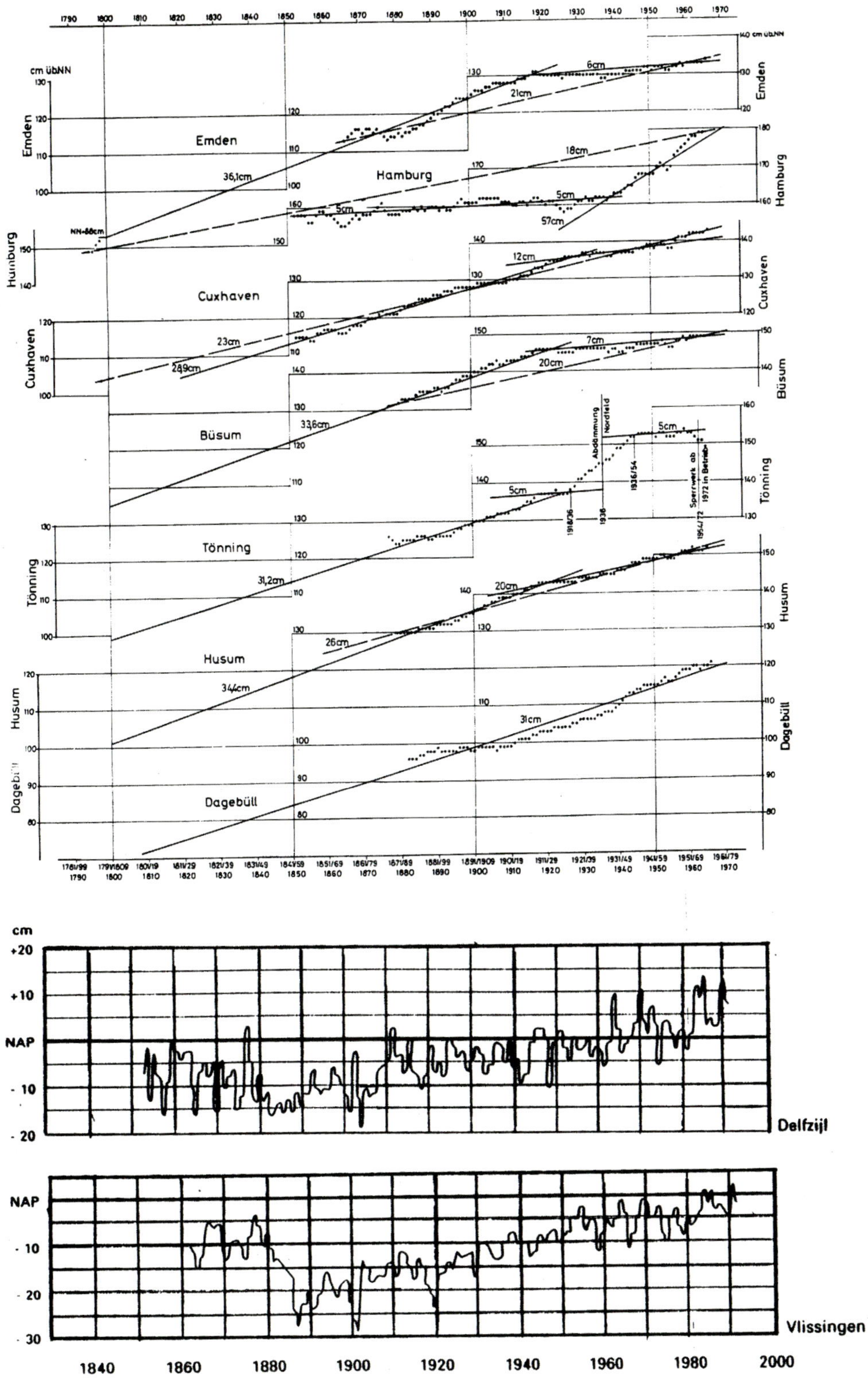

Fig. 3. Rising sea level Flushing, Delfzijl and Hamburg.

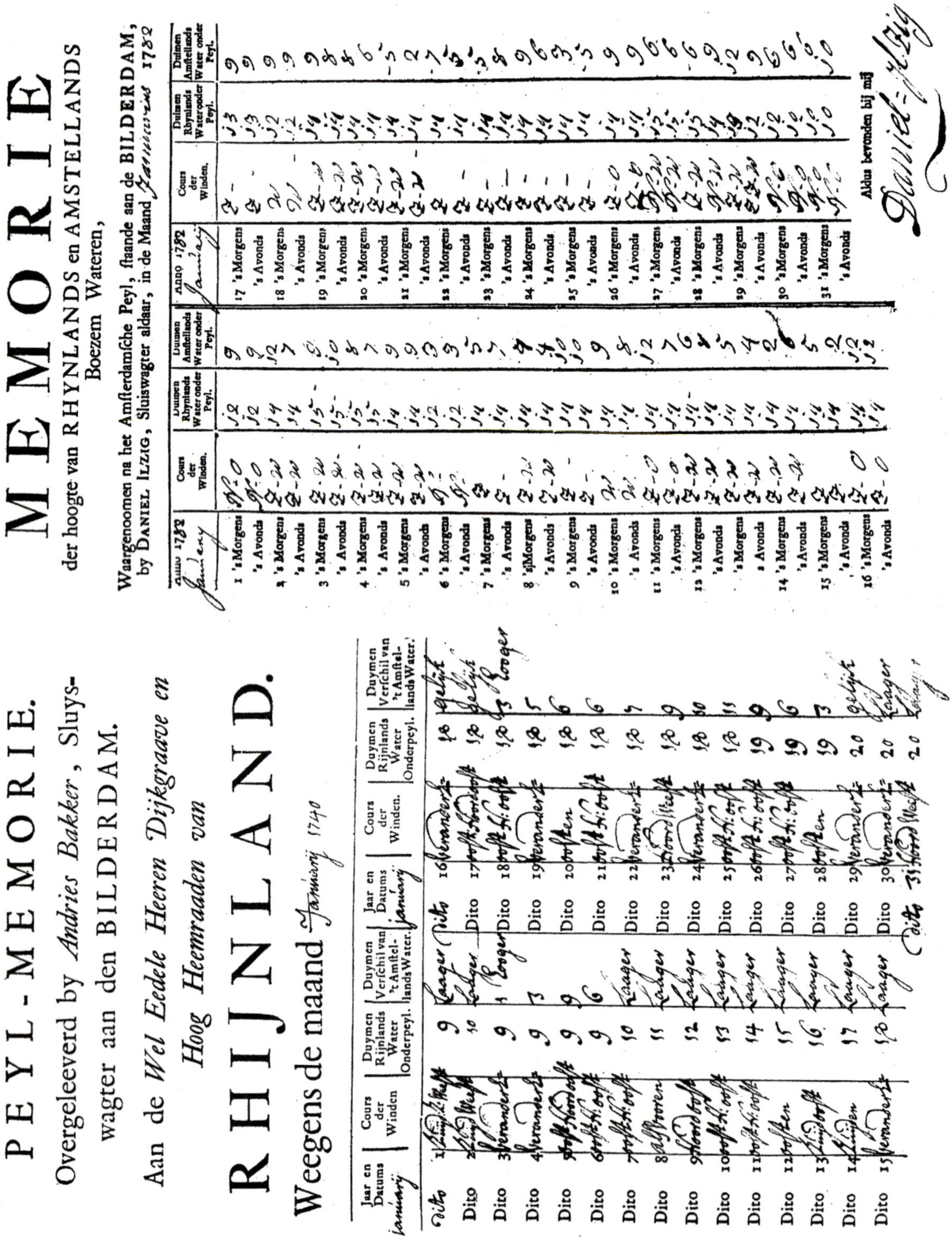
PEYL-MEMORIE. Overgeleeverd by *Andries Bakker*, Sluyswagter aan den BILDERDAM. Aan de *Wel Eedele Heeren Dijkgraave en Hoog Heemraaden van* RHIJNLAND. Weegens de maand Januarij 1740

MEMORIE der hoogte van RHYNLANDS en AMSTELLANDS Boezem Wateren, Waargenoomen na het Amsterdamsche Peyl, staande aan de BILDERDAM, by DANIEL ILZIG, Sluiswagter aldaar, in de Maand Januarius 1782

Aldus bevonden bij mij Daniel Ilzig

Fig. 4. Recording of wind directions at Bilderdam in January 1740 **(a)** and January 1782 **(b)** (Leiden, Hoogheemraadschap Rijnland, no 976-977).

windmill scale	*specification for the use of windmills*	*Beaufort scale of 1806*	
0	*calm*	*0*	*calm*
1	*some movement of the sails*	*1*	*light air*
02. Jan	*sails turn around gently*	*2*	*light breeze*
3 to 4	*staedy grinding*	*3*	*gentle breeze*
5 to 6	*steady grinding, perhaps striking one or more sails*	*4*	*moderate breeze*
7 to 8	*steady grinding with the striking of 1/4 or 1/3 of the sails*	*5*	*fresh breeze*
9 to 10	*steady grinding with the striking of 1/2 or 2/3 of the sails*	*6*	*strong breeze*
11 to 12	*steady grinding with the striking of 3/4 of the sails*	*7*	*moderate gale*
13 to 14	*steady grinding with the striking of all 4 sails*	*8*	*fresh gale*
15 to 16	*grinding too dangerous (stop the mill)*	*9 to 10*	*strong gale*
16 +	*grinding too dangerous (stop the mill)*	*11+*	*storm*

Table 4. The Dutch miller's wind scale of Jan Noppen, 18th century (Source: Geurts and Van Engelen 1992).

while from October to April more frequent storms were to be expected.

Whereas ship construction evolved rapidly, it became possible to sail even during the season of storminess. By the 16th century many ships sailed for the Americas and the Indies. So the seamen's knowledge of the wind and the various wind systems and patterns evolved as well. By the sixteenth century and even much earlier, seamen knew the best courses on the North Sea, in the Channel and even as far as Portugal and in the passage along the Sont into the Baltic. Unfortunately for us, few reports of such travels have been made and fewer still were kept. Those reports that have survived are Dutch ship logs of voyages to find the most suitable passage to the Indies. These documents are daily records of various aspect of synoptic weather. In Table 5 it is shown that wind direction, sailing direction, latitude, the speed of the ship and sometimes a description of the wind force are the most common features in ship's logs. However, some midshipmen even went so far as to distinguish ten columns, recording all kinds of data about the latitude, the number of day of the month, the name of the day, the part of the day and the hour, the direction the wind was blowing from, the change in the direction of the wind, and the number of miles sailed each part of the day, etc. (De Kraker 2000). Of course, such historical evidence is of the highest quality and rare!

In the course of the 18th century, the authorities of the Dutch East India Company gave instructions how official ship's logs had to be kept on their ships. In Great Britain such official naval instruction were given as early as 1731 (Oliver and Kington 1970). Dutch naval officers did not bother to develop an instrument for measuring wind force, but there seems to have existed a system of labels comprehensible and simple enough to be used at least until far in the 19th century. Whereas a descriptive wind force scale already existed at about 1700 in Britain and apparently one Danish example of about 1690 (Lamb and Frydendahl 1991), we must be confident that Dutch wording systems of wind force at sea must have been well developed. It is important that these are found in the immense archives of the Dutch East India Company at the Hague! From the wind direction's survey at the Bilderdam, which covers the period 1701 to 1770 it appears there has been some kind of wind scale in use ranking wind force from 1 to 16. Severe gale ('harde wind') was given 12/16 and severe storm or hurricane ('stormen') was given 16/16.

The Beaufort scale, introduced by the British naval officer, Sir Francis Beaufort at the beginning of the 19th century, was also introduced on Dutch

April 1595	wind	poistion of the ship	sailing direction, distance covered and other details
4	NE, ENE	Dover	
5	E		SSW and SW
6	E, ENE	Quissant	
7			SSW and SW
7 to 8	ENE, NE		SW, 32-33 miles
8 to 9	NE, N	Cape Fynistere	SW and SSW, 36 miles
9 to 10	N, NNW		
	morn. WNW		24 miles
10 to 11	unstable	40 degr. NL	SES, 9 miles, WSW 4 miles
11 to 12	calm		W, 6 miles
12 to 13	WN, NW, NE, EN		SSW, 14 miles (afternoon
13 to 14	N, NW, WNW		SSW, 28 miles
14 to 15		35 degr. 40 min NL.	SSW 28 miles
15 to 16	NW, NNW		SW, 30 miles
16 to 17			SSW, 24 miles
17 to 18			SSW-W, 30 miles
18 to 19			SSW, 4 miles until the morning
19 to 20	NE, ENE		W, 30 miles
20 to 21	NE, NE-E		SSW, 36 miles
21 to 22	NE, ENE	23 degr. 6 min. NL	SSW, 12 miles
22 to 23	NE, ENE		SW, 30 miles
23 to 24	NE, NNE	18 degr. 30 min. NL.	SSW, 36 miles
24 to 25			SW, 20 miles close to the coast
25 to 26		Isle of St. Jacob	sailing for 303 miles

Table 5. Monitoring wind on Dutch ships, April 1595 (Sources: National Archives at The Hague, Archives of the East India Companies, no 15, National Archives in Zealand (Middelburg), 'Staten van Zeeland' Archives, no 1210-II).

naval ships. According to a communication by Gunther Können the old system and the Beaufort scale were in use simultaneously for some decades. Further research will be necessary on this subject in order to reconstruct and calibrate the 17th and 18th century wind observations.

Studying Wind During the Recent Two Centuries

By the time the Beaufort scale was introduced national governments began to pay more attention to the impact of climate on society by founding various national meteorological institutes; measuring wind (direction and speed) reached a scientific level.

In the Netherlands the national institute 'Rijkswaterstaat' was founded in 1798, nowadays having its special department for the North Sea and its estuaries. As soon as the 'Rijkswaterstaat' had established itself firmly as a national government department, it gradually began to carry out observations all along the Dutch coast. In the course of the 19th century these statistics laid the foundation for all kinds of calculations and estimates of high and low tides, ranging from Flushing in the south as far as Bolsward in the Northeast.

And although 'Rijkswaterstaat' had to share its authority with the 'Provinciale Waterstaat', founded around 1851, many observations were carried out collectively, from which everyone involved in defending the dikes benefited. So starting about 1850s detailed information about storms, sea levels and all kind of particulars for the Dutch part of the North Sea coast are available.

Next to 'Rijkswaterstaat' the 'Royal Dutch Meteorological Institute' also began to carry out some investigations in this field and on wind observations in the Netherlands in general (see below).

From Germany observations of water levels dating from the period 1720 to 1753 have been kept (Rohde 1977). Although water levels have been observed at various places along the German part of the North Sea coast during the 18th century, most of these data must be considered lost. However, from seven places, some time series of observation of water levels have survived, such as those at Hamburg and Cuxhaven dating as far back as 1786 (Rohde 1977). During this century and perhaps even earlier, there must have been monitoring of wind as well. But this is not mentioned by Rohde (1977). Most of these early data are not uniform and have to be reworked.

It was only in the middle of the 19th century that a general and systematic recording of water level along the German North Sea coast began and consequently the recording of wind also. All these recordings have become the basis of the German research on sea levels and sea floods. Perhaps Arends (1833) was the first one to study the storm surges on the North Sea coast. Because it lacks a scientific basis, the results can hardly be considered reliable.

For Denmark the story seems to be very different. Here research on the history of sea floods was carried out by Gram-Jensen (1985). In his book he discusses some old chronicles and chroniclers in detail. He also elaborates on a variety of causes for the flooding of coastal lands. However, it does not become clear from which period the first Danish instrumental recordings of water level and floods date. Nevertheless, his time series of sea floods must be considered useful.

As for the recording of storms and sea floods in Great Britain, it is a story roughly similar to that of the Netherlands. Wind has been observed and consequently and systematically recorded at several observatories distributed over the British Isles. Much additional data came from ships' logs. All these data had to be collected and compiled in order to obtain a time series of historic storms. In this respect, much credit is due to Hubert Lamb, who has done a lot of investigations on historic storms in the North Sea region. Perhaps as much influenced by the work carried out by Gottschalk (1971, 1975, 1977) as by the eagerness to have a clear picture of the history of storms of the period from 1500 to the present day, Lamb finally succeeded in compiling the results of his research on the subject in a volume on historic storms of the North Sea (Lamb and Frydendahl 1991).

Present Position of Research and Prospects

Apart from the historical aspects of storm surges and sea floods, much research has been done on recent storms and storminess in general in all countries surrounding the North Sea

The British, Danish and German Situation

Judging from the British series of historic storms, it seems that there are no sufficient reliable observations until the second quarter of the 18th century. Recent research on wind seems to be much focusing on wind resource research (Palutikof et al. 1992) or individual storm surges only (Harding and Binding 1978).

Nevertheless, based on the Lamb classification of wind directions (Lamb 1972), the past record of circulation changes has been pushed back to 1861 (Kelly et al.1997). It would be a major step forward to push it much further back. It would be helpful indeed, to search for all relations between the time series of circulation changes 1861 to 2000 for the British Isles and the time series of sea floods and sea levels at various places along the Dutch, Danish and German coast for the same period as well. This can only be achieved by detailed information based on documentary sources which are reliable enough to reconstruct time series of wind. Such data have to be found in all kinds of record offices and local archives. In spite of the fact that the instrumental measurements of weather on a daily basis began in the late 17th century, including many wind observations, however, British documentary sources seem to be used predominantly for the reconstruction of temperature series (Jones and

Hulme 1996). So this leaves us with a great opportunity to work on the instrumental wind observations and still to investigate the remaining documentary sources, such as the ships' logs and e.g. the accounts of landed property situated in coastal regions of the North Sea (cf. below). Other recent British research (Hulme and Barrow 1997; Kelly et al. 1997) aiming at the reconstruction of surface air pressure patterns in Northern and Western Europe on a 10 or even on a 5 degree grid-scale going as far back as possible and searching for linkages with the NAO-index, has given some fine results, will also benefit from such a detailed research for historical weather observations.

As for the German and Danish wind research, much effort is spent on the more recent period. Important in this respect is the research carried out by Frydendahl et al. (1992) and Frich and Frydendahl (1994) on the Danish ships' logs (1675 to 1875) aiming at the transcription and the coding of over one million weather observations of which only a preliminary report exists. German research is focusing on finding causes of rising sea level (Von Storch et al. 1997). Still the outcome of this kind of research can only benefit from a complete and reliable time series of storms and sea floods over the last four or five centuries. Looking at the results published by Gram-Jensen (1985), the problem of reliability of storm data for the medieval period is about 0% and for the 16th to the 18th centuries it is about 73% to 90%. Thus the Danish time series earlier than 1700 and that of North Germany must remain far from complete as long as no final results of the research carried out by Frydendahl are known. On the other hand case studies on individual historical storm surges (Jakubowski-Tiessen 1992) supplying us with detailed information on the weather conditions leading to the natural disaster and its consequences remain important as well and should be carried out at the same time.

The Dutch situation

In the Netherlands storm patterns have been reconstructed for the period 1931 to 1960, from which it appears that the average wind force at Den Helder was stronger than at Flushing. Bouws (1978) studied winds during the period 1949 to 1976 and noticed an apparent increase in north-westerly gales. From recent research on wind and storm surges (Augustyn et al. 1990) ranging from at least Beaufort 8, it appeared that the dominant wind direction is SW (32%), except for the very heavy storms which tend to blow from the West (45%). Furthermore, sea level research in the Netherlands is very much restricted to the present and future condition of dikes and dunes along the North Sea (Jelgersma 1992). Because of increasing emphasis on environmentalist issues in politics and the 'Rijkswaterstaat' during recent years, research on historic storms seems to become neglected.

Other recent research on wind has been carried out within the framework of supplying parts of the Netherlands with cheap and clean wind energy (Wieringa 1981). The results of this kind of wind research have not been largely used for the research of historic storms.

As soon as Gottschalk finished her 'historical storm surge research' in 1977, only Dekker (1980) contributed to the discussion demonstrating that the periods of transgressions followed by periods of sedimentation must be clearly connected with the period proceeding the building of dikes. On the other hand he also demonstrated that as soon as the building of dikes began in the 11th and 12th centuries, flooding of large areas occurred at very irregular intervals and at different places. However, looking at the general picture we have of storms and storm surges, this is ready now for a fundamental review.

One of the shortcomings of Gottschalk's research proved to be the fact that she only has consulted chronicles, of which the larger part has been printed. Only for a few small areas additional documentary evidence kept in archives has been consulted. Because Gottschalk's sources were not uniform nor continuous, she could not construct a time series. Secondly, she ended her research with the year 1700, which leaves us with three centuries to be investigated still. Thirdly, Gottschalk covered a large area and has consulted a huge number of very different chronicles. That is why she was not able to discover the real cause of many a flooding of a particular area which in turn can be very helpful to assess the real character and force of a storm.

Recent research on storms in the Netherlands, covering only a small area of polders west of Antwerp, has revealed that human impact on the landscape during and after storm surges have been flooding an area is strongly conditioned by nature, economy and politics (De Kraker 1997). Of these economy and politics must be considered as being the dominant factors. Especially during wartime large areas were flooded. The same research has been carried out by consulting series of documentary sources which can be considered uniform and continuous. This has been very helpful in constructing a time series of storms. Moreover, it has enabled us to construct a method to measure the severity of storms (De Kraker 1999).

Still, other research on historic storms remains very restricted to smaller areas as well, such as the assessment of water levels reached by successive storms at Goeree and the number of storms observed at the Hondsbossche Zeewering between 1840 and 1879 reaching heights of over 1 m above the average (Aten 1997).

It appears that there are sufficient documentary sources still to be consulted, so that finally there will be a fairly complete picture of sea floods, storms and storminess on the North Sea coast of the Low countries starting as early as the 16th century. Several hundreds of independent polder boards ('waterschappen') have kept their archives fairly complete over the centuries. In the province of Zeeland and the northern part of Flanders, with its many former isles, dike accounts enable us to study wind from all directions. Furthermore the entire region around Amsterdam and as far north as Den Helder offers some fine opportunities to study winds from all directions. The Rijnland general dike account's series starts at 1509 and is almost complete. Finally the entire region surrounding the former Zuiderzee offers no fewer opportunities. In short, there are sufficient series of documentary sources which are uniform and continuous and which can enable us to construct a complete time series of storms, storm surges and sea floods over the last five centuries!

However, more is needed than a sufficient amount of documentary sources. Existing instrumental observations starting as early as around 1700 at e.g. the Bilderdam and Amsterdam should be made uniform in order that such a series can be use to reconstruct time series of winds (Table 6). The same applies to the existing series of observations on water levels.

Furthermore, constructing a time series of wind is one, assessing the intensity of wind is another. That is why some kind of uniform system must be applied, in which all details about storms (De Kraker 1997, 1999), such as time, space, cause (natural and human), consequences, etc., are of particular importance (Fig. 5).

For the Netherlands, a number of four test regions as mentioned (Fig. 6), perhaps to be extended with a fifth region somewhere on the Wadden coast will be enough to do the job. In addition to the storm data obtained from these four test regions, there is still much documentary source material from international and national shipping. Studying the busy overseas trade from the Netherlands to Denmark and the Baltic during the 16th to 18th century can supply us with many storm data to help us to complete our time series of storms and storm surges for the German Bight and western Denmark. Combining the efforts of economic historians and climatologists on the subject would be a major step.

Conclusions

In this paper the object has been to review the historic storms which have been recorded in the North Sea area. The main source of information consists of four distinctive time series of storm surges, sea floods from the Netherlands, Denmark and the British Isles and a time series of sea level recordings from Northern Germany. From this survey it has become clear that only as late as the second quarter of the 18th century there is a reliable series of storm surges and sea floods which seems to look rather complete. However, for the Netherlands no major research has been done for the period later than 1700. Moreover, storms of lesser importance have been included in neither time series on the subject . Except for the period 1861 to about 2000 for the British Isles, it is not possible yet to reconstruct a series of storms on e.g.

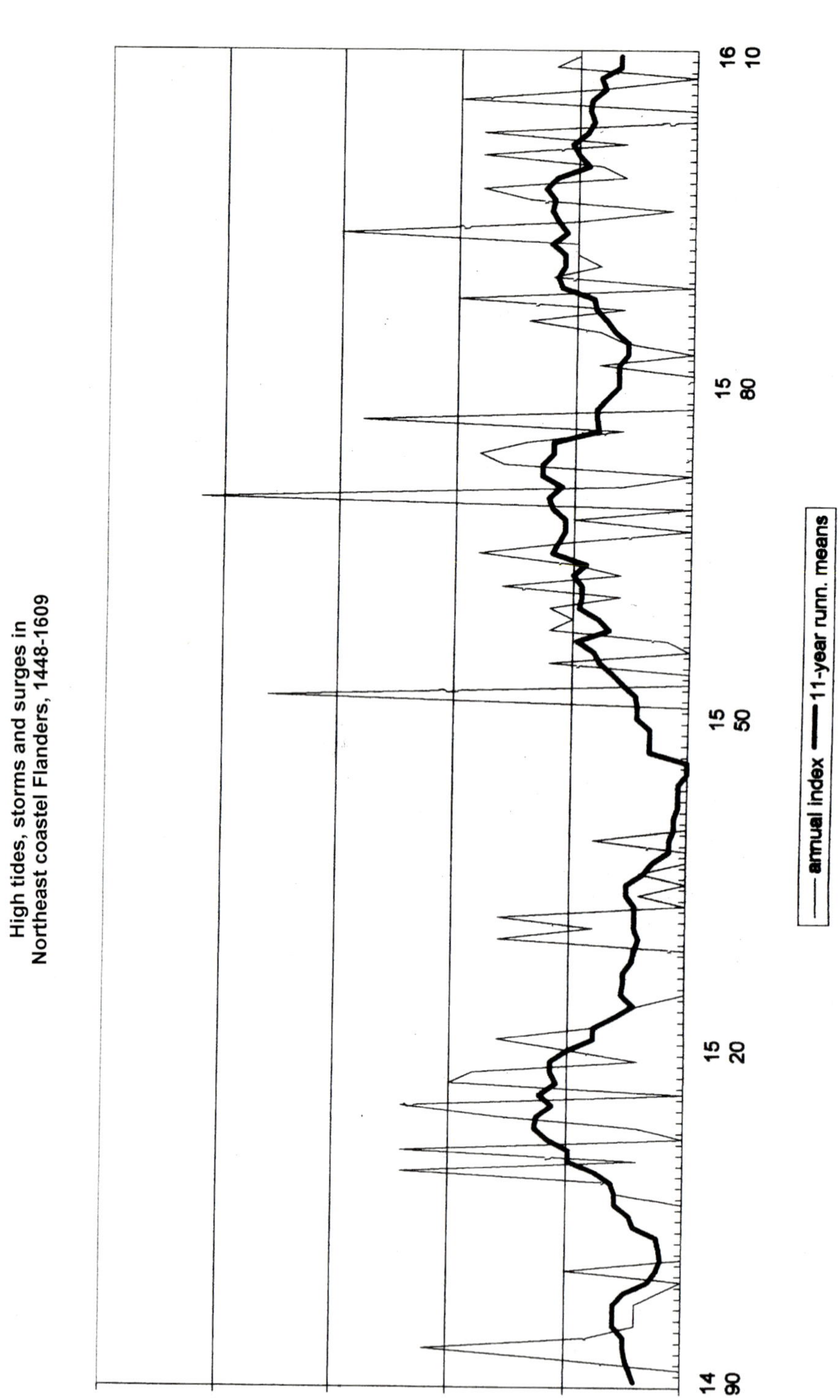

Fig. 5. Variability of storminess in the Wester Scheldt estuary during the 16th century (De Kraker 1997, 1999).

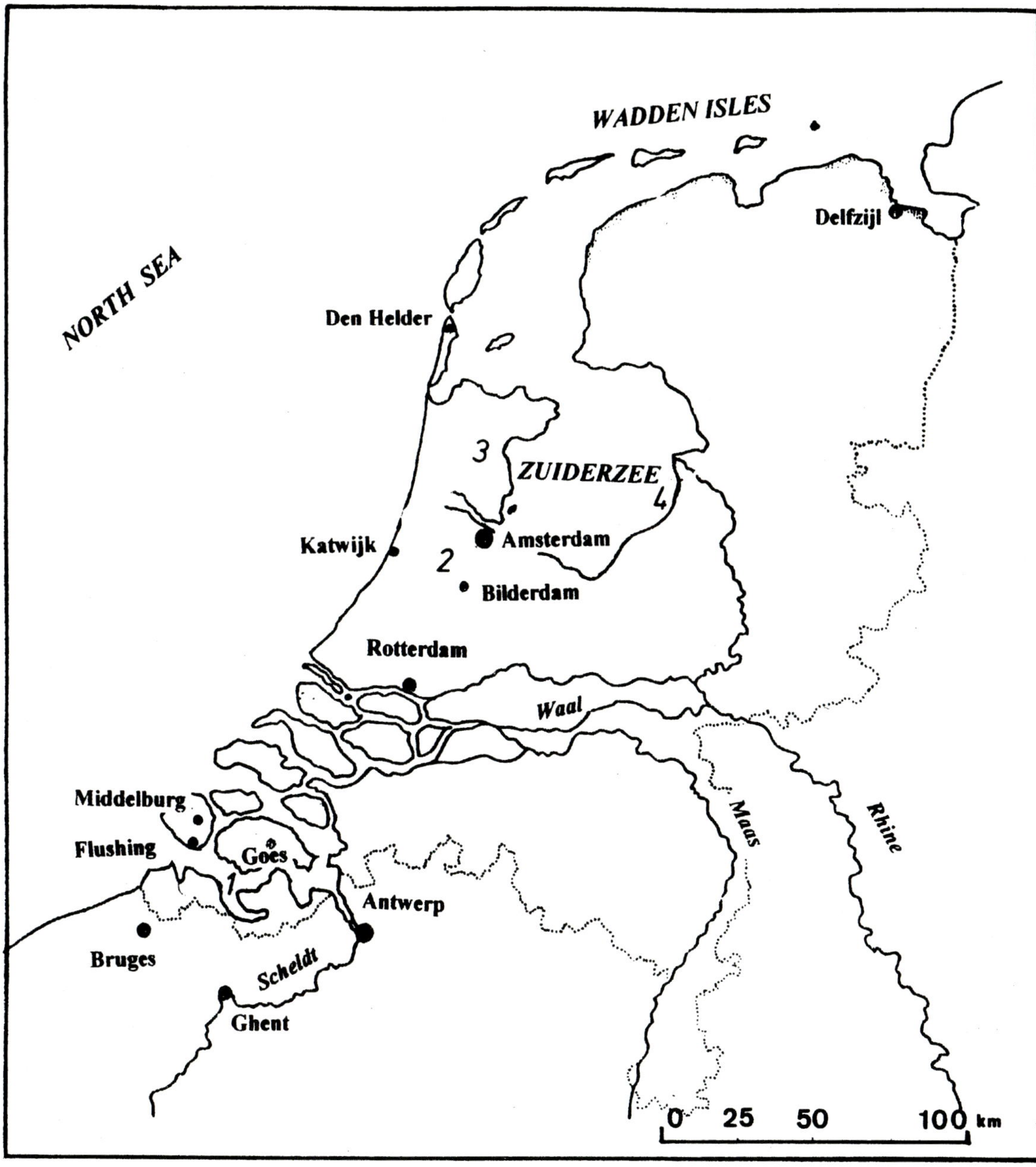

Fig. 6. Map of the Low Countries about 1700. Four areas for possible storm research in the future (author).

1. Delta area: North Flanders and the Dutch province of Zealand.

2. Area at Bilderdam-Amsterdam and the coastal region of the Haarlemmermeer.

3. The northern part of Holland and its coastal area: Noorder Kwartier.

3. Area surrounding the former Zuiderzee.

	Bilderdam, 1701 - 1770			Amsterdam, 1700-1750		Amsterdam, 1890-1937	
wind direction	days	%	wind direction	days	%	days	%
N	24.5	6.7	N	25	6.8	35	9.7
NNE	7.0	1.9					
NE	30.0	8.2	NE	42	11.5	39	10.8
ENE	6.5	1.8					
E	48.5	13.3	E	49	13.4	44	12.0
ESE	6.0	1.6					
SE	21.0	5.5	SE	25	6.8	49	13.4
SSE	7.0	1.9					
S	24.0	6.6	S	35	9.7	23	6.3
SSW	8.5	2.2					
SW	67.0	18.4	SW	87	23.8	84	23.0
WSW	12.0	3.3					
W	54.0	14.8	W	55	15.0	52	14.2
WNW	13.0	3.6					
NW	31.0	8.5	NW	47	13.0	39	10.6
NNW	5.0	1.4					
	365.0	100.0		365	100.0	365	100.0

Table 6. Wind directions in Holland, 1700 - 1937 (Sources: Leiden, Hoogheemraadschap Rijnland, no 973 (Bilderdam); Van Veen 1940 (Amsterdam)).

a thousand years' time scale and on a geographical scale for the entire North Sea area or parts of it.

Still there are very good prospects for research to continue. There are enough weather observations from instrumental measurements of the recent two centuries that can be included fairly easily into the reconstruction of time series of wind and wind patterns. If four or five distinctive areas in the Netherlands are chosen which are very sensible to wind and sufficient time is spent at record offices and in archives there is a fair chance of reconstructing a time series beginning in the 16th century. This kind of research includes, of course, all storms of lesser importance as well. And if documentary sources studied are continuous and uniform, new methods to assess storminess might be applied at the same time. Starting in the 17th century this also applies to the reconstruction of sea level recordings at various places along the Dutch, Flemish and (perhaps also) the British coast.

Finally, it would be very helpful to compare time series of sea level recordings with various other kinds of climatic time series for the North Sea area, starting with the 20th century and then going back from there. In the end there would be a fairly continuous thousand or 500 years' time scale of storms and sea level recordings on a geographical scale for the North Sea region. And linking it to the Lamb classification and the NAO-index, this time series of storms would be very helpful to understand the variability of air masses penetrating into the European hinterland as far as temperature and moisture are concerned.

Acknowledgments

I thank Dr. Meinte Blaas, Institute for Marine and Atmospheric research (IMAU), Utrecht University, Utrecht, the Netherlands; Prof. Dr. Chr. Pfister, Institute of History, University of Bern, Switzerland; Dr. David Prandle, Bidston Observatory,

Birkenhead, United Kingdom; Prof. Dr. Wolfgang H. Berger, University of California, USA; Prof. Dr. G.J. Borger, University of Amsterdam, Amsterdam, the Netherlands; Dr. Gaston R. Demarée, Royal Meteorological Institute, Brussels, Belgium; Otto Knottnerus, Uiterburen, Netherlands for their invaluable help.

References

Arends F (1833) Physische Geschichte der Nordsee-Küste und deren Veränderungen durch Sturm-fluthen seit der Enbrischen Fluth bis jezt. Verlag Schuster in Leer, Emden 1833. Reprinted in 1974, pp 1-355

Aten D (1997) Beslissende jaren voor de Hondsbossche. Dijkgraaf Foreest en de prijsvraag van 1864, in: Vrienden van de Hondsbossche, 13[e], pp 1-32

Augustyn B, Daan H, Van Mourik B, Messerschmidt D, Zwart B (1990) Stormenkalender 1964-1990. KNMI T.R. no. 2 revised, pp 1-138

Bijvoet HC, van der Ham CJ, Groen P, Rijkoort PJ, Schmidt FH, Verploegh G, Weenink MPH (1960) Meteorological and oceanographic aspects of storm surges on the Dutch coast. K.N.M.I. Bijdrage tot het rapport van de Deltacommissie, pp 1-200

Bouws E (1978) Wind wave climate in the Netherlands sector of the North Sea between 53 and 54 north latitude, KNMI Scientific Report W.R. 78-9, De Bilt, pp 1-54

Dekker C (1980) Resultaten van het historisch-geografisch onderzoek in Zeeland. In: Verhulst A, Gottschalk MKE (eds) Transgressions and the history of settlement in the coastal areas of Holland and Belgium. Proceedings Colloquium Gent 5-7 September 1978, pp 75-93

De Kraker AMJ (1997) Landschap uit balans. De invloed van de natuur, de economie en de politiek op de ontwikkeling van het landschapin de Vier Ambachten en het Land van Saeftinghe tussen 1488 en 1609. Uitgeverij Matrijs Utrecht. Thesis University of Amsterdam, Amsterdam, pp 1-464

De Kraker AMJ (1999) A method to assess the impact of high tides, storms and storm surges as vital elements in climatic history. The case of stormy weather and dikes in the northern part of Flanders, 1488 to 1609. Pfister Chr, Brázdil R, Glaser R (eds) Climatic variability in sixteenth-century Europe and its social dimensions. Special issue of 'Climatic Change', 43, no. 1: 287-302

De Kraker AMJ (2000) The oldest Dutch ships' logs (1595 - 1610), their interest for climate history in general and global climatic change in particular. In: Mikami T (ed) Proceedings of the International Conference on Climate Change and Variability - Past, Present und Future - held at Tokyo metropolitan University Tokyo, Japan, September 13-17, pp 95-101

Douglas KS, Lamb HH, Loader C (1978) A Meteorological Study of July to October 1588: The Spanish Armada Storms. Norwich CRU RP-6, pp 1-76

Frich P, Frydendahl K (1994) The summer climate in the Oresund region of Denmark A.D. 1675 tot 1715. In: Frenzel B, Pfister Chr, Gläser B (eds) Climatic trends and anomalies in Europe 1675-1715. High resolution spatio-temporal reconstructions from direct meteorological observations and proxy data. Methods and results. Gustav Fischer Verlag - Stuttgart - Jena - New York, pp 33-41

Frydendahl K, Frich P, Hansen C (1992) Danish Weather Observations 1615-1715. Danish Meteorological Institute, technical report 92/99, pp 1-22

Geurts HAM, Van Engelen AFV (1992) Beschrijving antieke meetreeksen, part V. KNMI publication, no. 165-V, pp 1-310

Gottschalk MKE (1971) Stormvloeden en rivieroverstromingen in Nederland. I. De periode voor 1400. Van Gorcum & Comp. N.V. Assen, pp 1-581

Gottschalk MKE (1975) Stormvloeden en rivieroverstromingen in Nederland. II. De periode 1400-1600. Van Gorcum & Comp. N.V. Assen, pp 1-896

Gottschalk MKE (1977) Stormvloeden en rivieroverstromingen in Nederland. III. De periode 1600-1700. Van Gorcum & Comp. N.V. Assen, pp 1-474

Gram-Jensen IB (1985) Sea Floods. Contributions to the climatic history of Denmark, Kopenhagen, pp 76

Harding J, Binding AA (1978) Windfields during the gales in the North Sea and the gale of January 1976 In: Met Mag 107:164-182

Hulme M, Barrow E (1997) Climates of the British Isles, present, past and future. London/NewYork

Jakubowski-Tiessen M (1992) Sturmflut 1717. Die Bewältigung einer Naturkatastrophe in der Frühen Neuzeit, R. Oldenbourg Verlag München

Jelgersma S (1992) Vulnerability of Coastal Lowlands of the Netherlands to a future sea-level rise. In: Tooley MJ, Jelgersma S (eds) Impacts of sea-level rise on European coastal lowlands, Blackwell, Oxford/Cambridge, pp 94-124

Jones PD, Hulme M (1996) Calculating regional climatic time series for temperature and precipitation: methods and illustrations, in: Internat Journ of Climat vol. 16:361-377

Kelly PM, Jones PD, Briffa K (1997) Classifying the winds and weather In: Hulme M, Barrow E (eds) Climates of the British Isles, present, past and future, London/New York, pp 153-173

Kington J (1994) Synoptic weather mapping, 1675 to 1715. In: Frenzel F, Pfister Chr, Gläser B (eds) Climatic trends and anomalies in Europe 1675-1715. High resolution spatio-temporal reconstructions from direct meteorological observations and proxy data. Methods and results. Gustav Fischer Verlag - Stuttgart - Jena - New York, pp 390-400

Lamb HH (1972) British Isles weather types and a register of the daily sequence of circulation patterns, 1861 to 1971. Geophysical Memoir, 116. H.M.S.O. London, pp 1-85

Lamb HH, Frydendahl K (1991) Historic Storms of the North Sea, British Isles and Northwest Europe, Cambridge University press, pp 1-20

Oliver J, Kington JA (1970) The usefulness of ships' logbooks in the synoptic analysis of past climates. Weather 25:520-528

Palutikof JP, Guo X, Halliday JA (1992) Climate variability and the UK wind resource. Journal of Wind Engineering and Industrial Aerodynamics 39:243-249

Pfister Chr, Kington J, Kleinlogel G, Schüle H, Siffert E (1994) High resolution spatio temporal reconstructions of past climate from direct meteorological observations and proxy data. In: Frenzel B, Pfister Chr, Gläser B (eds) Climatic trends and anomalies in Europe 1675-1715. High resolution spatio-temporal reconstructions from direct meteorological observations and proxy data. Methods and results. Gustav Fischer Verlag - Stuttgart - Jena - New York, pp 329-375

Rohde H (1977) Sturmfluthöhen und säkularer Wasserstandsanstieg an der deutschen Nordseeküste, Die Küste. Archiv für Forschung und Technik an der Nord- und Ostsee, Heft 30:53-143

Van Malde J (1996) Historical extraordinary water movements in the North Sea area, Mededeling Rijks Geologische Dienst 57:27-40

Van Veen J (1940) Is de heerschende windrichting te Amsterdam sedert 1700 gekrompen? Tijdschrift KNAG, 2nd series, 57e:686-707

Von Storch H, Reichardt H (1997) A scenario of storm surge statistics for the German Bight at the expected time of doubled atmospheric carbon dioxide concentration. Journal of Climate 10:2653-2662

Wieringa J (1981) Veranderlijkheid van de wind op korte en lange termijn, in: Nationale windenergie conferentie 1981. Den Haag/Veldhoven, B1 1-17

Climate Variability and Historical NW European Fisheries

Jürgen Alheit* and Eberhard Hagen

Baltic Sea Research Institute, Seestr. 15, D-18119 Warnemünde, Germany
**corresponding author (e-mail): juergen.alheit@io-warnemünde.de*

Abstract: Climate variability has affected marine ecosystems in NW Europe and their fisheries with concomitant implications for coastal populations. Long-term biological time series are essential for the retrospective analysis of a climate impact on marine ecosystems, however, they are scarce. Fish populations usually provide longer records than other biological components of marine ecosystems because of their economical importance. The dynamics of exploited fish populations are affected by environmental variability and man-made activities (fishing, habitat alteration) and retrospective studies will help to distinguish between the two. Records of the herring fishery off the Swedish coast of Bohuslän, in the Skagerrak, date back to the 10th century. This fishery is presented in this review as a case study for a climatically driven fishery the success and failure of which dramatically influenced the economy of an entire region throughout the last thousand years. A study of the climatological/ hydrographic scenario of the Bohuslän fishery periods and those of other European herring and sardine fisheries in Europe showed that they seem to be governed by decadal variability of the North Atlantic Oscillation. A comparison with the long-term dynamics of anchovy and sardine populations throughout the Pacific and in the South Atlantic points to a basin-wide and, perhaps even global, synchrony in fish population dynamics which might be driven by climate variability.

Introduction

Climate variability has affected marine ecosystems and their fisheries with concomitant implications for coastal populations and will do so in future. In order to improve our understanding of the influence of future climate variability on marine ecosystems, it is necessary to separate the effects of natural variability from anthropogenic impacts such as fisheries, habitat alteration and pollution. One way to solve this problem is to conduct retrospective studies using historical data series from a time when man-made influences did not exist, or were at a minimum. The longest data series we have from the marine realm, apart from information from sedimentary records, are based on fishery as fish have always been an important source of protein for coastal populations. The most useful long-term data are derived from small, schooling pelagic fishes. In the water column, pelagic fishes stay away from the sea floor in contrast to demersal or bottom-fish, such as cod (*Gadus morhua*). Small pelagic fishes such as sardine, anchovy, herring and others represent about 20–25 % of the total annual world fisheries catch. They are widespread and occur in all oceans. They support important fisheries all over the world and the economies of many countries depend on those fisheries. They do respond dramatically and quickly to changes in ocean climate. Most species are highly mobile; have short, plankton-based food chains and some even feed directly on phytoplankton. They are short-lived (3-7 years, except herring), highly fecund and some can spawn all the year round. These biological characteristics make them highly sensitive to environmental forcing and extremely variable in their abundance (Hunter and Alheit 1995). Thousandfold changes in abundance over a few decades are characteristic for small pelagics and well-known examples include the Japanese sardine, sardines in the California Current, anchovies in the Humboldt Current, sardines in the Benguela Current, or herring in European waters. Their drastic stock fluctuations often caused dramatic consequences for fishing communities, entire regions and even whole countries. Their dynamics have important economic and ecological consequences.

From WEFER G, BERGER W, BEHRE K-E, JANSEN E (eds), 2002, *Climate Development and History of the North Atlantic Realm.* Springer-Verlag Berlin Heidelberg, pp 435-445

They are the forage for larger fish, sea-birds and marine mammals. The collapse of small pelagic fish populations is often accompanied by sharp declines in marine bird and mammal pop-ulations that depend on them for food. Major changes in the abundance of small pelagic fishes may be accompanied by marked changes in the structure of the ecosystem. They are often accompanied by drastic changes in the abundance and species composition of zooplankton. The great plasticity of growth, survival and other life-history characteristics of small pelagic fishes is the key to their dynamics and makes them ideal targets for testing the impact of climate variability on marine ecosystems and fish populations. Consequently, this review will focus on the impact of climate variability on small pelagic fish stocks.

Herring Fishery in the Time of the Hanseatic League

The Hanseatic League was an organisation of merchants and trading communities, and herring (*Clupea harengus*) was one of the trading goods (the strength of) the Hanseatic League was built on. The growing population in Europe and the development of newly founded towns in the 12th and 13th century demanded more food and this demand was partly met by exploiting marine resources off northern Europe. The members of the Hanseatic League were instrumental in the development of commercial fisheries (12th-16th century) and fish trading. Its herring fishery was the first to reach the level of a real industry (Sahrhage and Lundbeck 1992). Most of this herring came from the small Öre Sound in the southern Kattegat between the Danish Island of Zealand and the (now) Swedish Peninsula of Scania, where sometimes enormous quantities of herring were encountered. Interestingly, this fishery developed into an international market with the rise of the League and faded again when the League deteriorated. This kind of fishery, which was combined with market activities and permitted every citizen to take part, provided he paid taxes and had his own net, is named "market fishery" in contrast to "urban fisheries" which could could be practised only by privileged groups (Jahnke pers comm). Merchants of the Hanseatic cities had moved their operations to Scania where the annual fishing season lasted from late July to late October. Most important was to cure the freshly caught herring, so it could be transported and stored for some time. This was achieved mostly by salting and, occasionally, by smoking the herring. Thus, the herring could be sent to many places in Europe, even to far away places like Italy. Besides, it is possible that salted herring was a convenient food for travellers, soldiers and beleaguered cities (Cushing 1982). There are estimates of annual catches of up to 10,000 and even 50,000 metric tons in the 15th and 16th century, which was a substantial amount for medieval times (Cushing 1982; Sahrhage and Lundbeck 1992). The "Scanian Fairs" became a major event, not only for trading herring, attracting merchants from quite a few European countries. The Hanse fishery declined in the 15th century and ceased to exist in the 16th century because of the frequent hostilities from the strengthened Danish kingdom and internal struggles among the cities of the North and Baltic Sea. Additionally, political reasons were responsible for the North Sea herring fishery to rise at the same time and become a strong competitor for the Baltic herring fishery. It is not clear whether the medieval Sound fishery had a climatic background. Early this century, the Swedish oceanographer Petterson noticed that cold winters characterised the climate of the Baltic region during the period of the Hanse fishery, lasting from the 12th to the 16th century. A less known, albeit more important herring fishery consisted in the Bohuslän fishery.

Bohuslän Herring Fishery in the Skagerrak

The Fishery

The Bohuslän region is a small coastal stretch on the Swedish Skagerrak coast to the north of Gothenburg. Already at the times of the Vikings, large amounts of herring migrated to the Bohuslän coast for certain periods to overwinter close to the shore. These periods lasted 20-50 years and were interrupted by even longer time spans of 50-70 years. Nine of these Bohuslän periods are known (Fig. 1) and information on the oldest stems from the Scandinavian Sagas which have often been shown to be relatively accurate (Höglund 1978; Cushing 1982; Lindquist 1983; Sahrhage and Lundbeck 1992; Alheit and Hagen 1997; Jahnke 1997):

end of 10th century-early 11th century (ca. 970-1020)
end of 11th century-early 12th century (ca. 1110-1130)
end of 12th century-mid 13th century
end of 13th century-mid 14th century (1307/1330)
mid 15th century
1556-1589
1660-1680
1747-1809
1877-1906.

Naturally, the further one goes back in time, the less reliable are the data, however, the data from the last four periods are very accurate. The catches were quite considerable, particularly when considering that the herring were not caught by fishing vessels, but mainly from the shore using beach seines. More than 200,000 MT (metric tonnes) were caught during the fishing season of 1895/96, and in the 18th century 270,000 MT were reported for one season. The reason why we know so much about this fishery is that there were no traditional fisheries, nor people fishing at the Bohuslän coast. All its rather poor inhabitants knew about was farming. So, whenever the herring came to Bohuslän again and started a new fisheries period, the whole economy of the region changed com-pletely and these remarkable changes were remembered for a long time and written down in sagas, chronicles, tax reports, etc. In fact, the Bohuslän herring were not Baltic herring, instead, invaders from the North Sea. This was proven by investigating the remnants (skeletons and vertebrae of herring) of train oil factories of the 18th century.

Socio-Economic Implications

The herring periods determined the economy and life in the Bohusläm region as described in a series of articles by A.V. Ljungman (1880, 1883). It is assumed that the flourishing trade in the early 12th century in Konungahella was connected to the second Bohuslän period. During the third Bohuslän period, in the first half of the 13th century, fisheries people and their families settled on the small islands

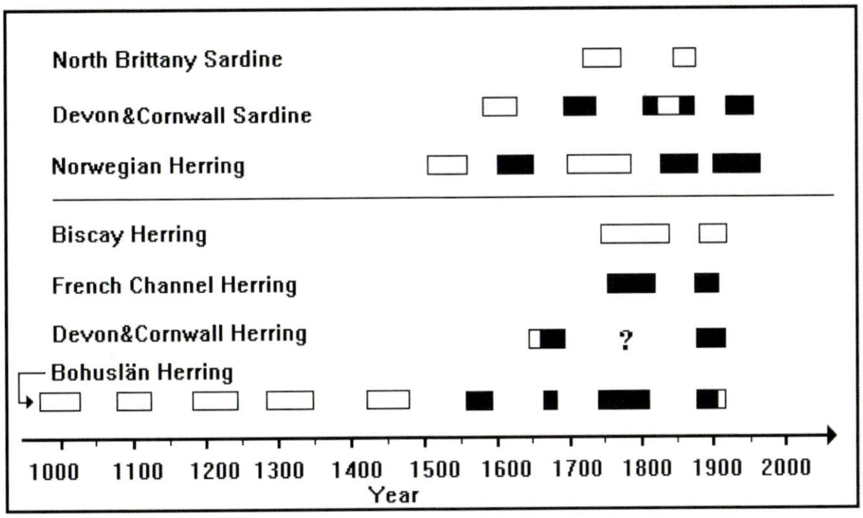

Fig. 1. Historical periods of European herring and sardine fisheries. Sources: Bohuslän herring (Höglund 1978, Lindquist 1983, Sahrhage and Lundbeck 1992); Devon and Cornwall herring (Southward et al. 1988); French catches of herring in the English Channel (Binet 1988); Bay of Biscay herring (Binet 1988); Norwegian spring spawning herring (Cushing 1982); Devon and Cornwall sardine (Southward et al. 1988); northern Brittany sardine (Binet 1988). Open rectangles depict periods where the extensions are not precisely known. Modified after Alheit and Hagen (1997).

off the Bohuslän coast and churches were built. In the course of the fourth period, the first half of the 14th century, Marstand experienced an economic boom and the Hanseatic cities were granted the privilege to participate in the fisheries. During the fifth period, in the mid 15th century, Marstrand became a city (1442) and its church was finished (1460). During the following period, 1556-1589, King Frederick II from Norway established custom stations to collect taxes from the fisheries. In 1561, the Marstrand Law-Books were published and included "Rules and Regulations for the Bohuslän Herring Fisheries". Marstrand, because of its wealth acquired through the herring fisheries, had to pay the same amount of war taxes as the large city of Bergen, very much to the dismay of its population. Foreigners were allowed to participate in the fish trading. It is reported that annually thousands of ships from Denmark, England, France, Germany, Holland and Scotland arrived at Bohuslän to transport salted herring to other European countries. Thousands of people from neighbouring countries settled with their families at the Bohuslän coast, built houses and "made a good living on the fisheries". The negative consequences of this economic boom was a strong increase in crime, prostitution and drinking. The herring schools migrating to the Bohuslän coast during the eighth period were so large the local population were unable to take full advantage of it immediately. When the herring rearrived after 70 years of absence, there was, as usual, a lack of experienced fishermen, nets and markets which had to be imported from abroad. The Swedish government subsidised the fishery in order to bring more people into the region and to fully exploit the fishery. It is estimated that up to 50 000 people annually came to the Bohuslän region during the peak of the period. A number of superintendents were appointed to oversee that all regulations were kept. Often, the season came to an end prematurely, not because of lacking fish, but because the markets could not absorb all the herring. In 1787, there was a count of 338 salting houses, 429 train oil factories, 358 large nets and 2100 herring boats at the Bohuslän coast. The towns of the Bohuslän region prospered considerably and Marstrand quadrupled its population during this period. However, crime, liquor and prostitution caused great problems again. These fisheries were very profitable and it was understood that the city of Gothenburg owed its flourishing condition during the latter part of the 18th century to the East India trade and the herring fisheries. The export of herring and train oil, derived from herring, gave a new impetus to commerce and had a beneficial influence on all branches of trade and industry. The production of train oil was an exceptionally good source of income, not only for the manufacturers, but also for the government. Export rates for fishery products were second to those from the iron industry. The benefit of the fishery was noticed particularly when the herring suddenly ceased to arrive. As a consequence, the whole economy collapsed and Bohuslän became once more a poverty stricken region.

Climate Impact

The last seven Bohuslän periods coincided with cold climatic periods as indicated by the winter mildness/severity index reconstructed by Lamb (1972) (see Alheit and Hagen 1997) (Fig. 2). The Bohuslän periods seem to have been triggered by large temperature declines and their termination usually coincided with temperature increases. The variability

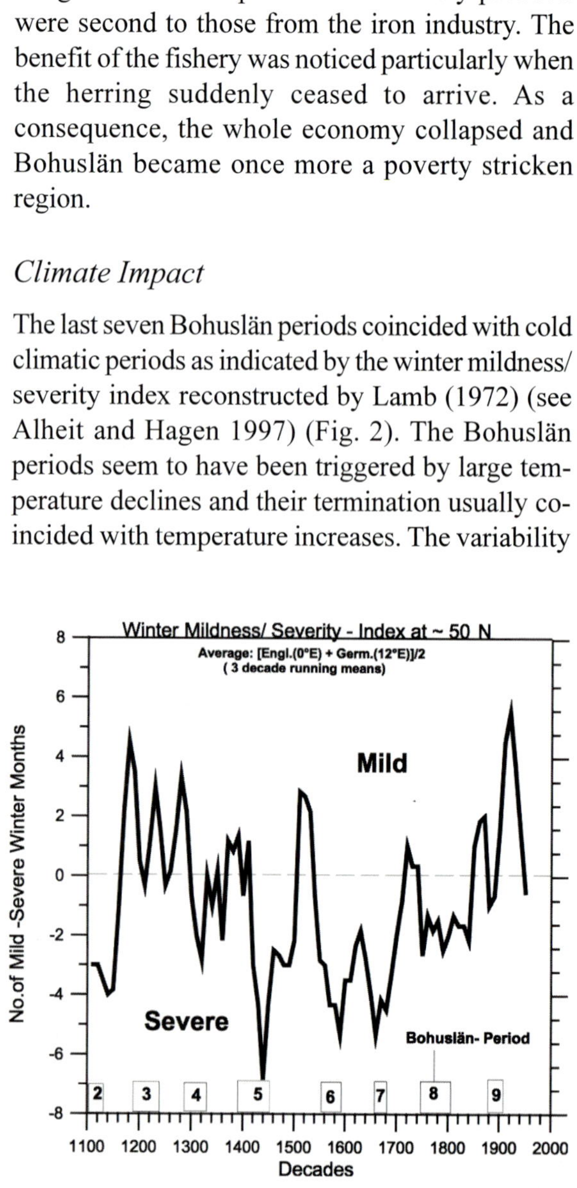

Fig. 2. "Winter mildness/severity index" (Lamb 1972) and Bohuslän periods which correspond to negative peak values (modified after Alheit and Hagen 1997).

of regional air temperature over NW Europe on decadal time scales is controlled by the North Atlantic Oscillation, the difference in air pressure between Iceland and the Azores. A high NAO index is associated with an anomalously low sea-level pressure in Iceland, strong meridional pressure gradients over the North Atlantic and intensified westerlies. Such a situation is climatically accompanied by the inflow of cold air masses from Alaska and Greenland which are warmed over the NW Atlantic ocean. Consequently, the westerlies transport air masses of moderate temperature towards Europe which results in mild winters over western Europe. The opposite situation is given by a negative NAO index, indicating a reduced westerly flow. Simultaneously, the inflow of cold air masses from Alaska and Greenland is reduced, whereas the inflow of cold Siberian air masses into eastern Europe is intensified, resulting in severe winters over Europe. Thus, the NAO exhibits two typical modes: one is connected with mild European winters while the other one coincides with severe winters. During the Bohuslän periods the following scenario was encountered: cold temperatures at mid- and higher latitudes of the eastern North Atlantic, sea surface temperature anomalies <0, increased ice cover off Iceland and the Baltic sea, NAO index negative with reduced westerly winds, minimum frequency of south-westerly winds over England and cold water in the North Sea, English Channel and in the Skagerrak. In spite of a number of hypotheses presented over the last 100 years (e.g. Corten 1999), the mechanism explaining how this climatic scenario induces the North Sea herring, directly by changes in currents or indirectly by food chain processes, to migrate for longer periods to the Bohuslän coast is not clear.

Long-Term Climate Forcing of European Herring and Sardine Fisheries

Not only the Bohuslän herring fishery was connected to the variability of the NAO, but also several other European herring and sardine fisheries give evidence of decadalscale variability either oscillating in phase with the Bohuslän periods or alternating with them (Fig. 1) (Alheit and Hagen 1997).

Norwegian Spring Herring Fishery

The Norwegian spring herring is the largest stock of the Atlanto-Scandian herring. Its main habitat is the Norwegian Sea and its total biomass may have ranged between 15-20 million MT (metric tonnes) when it was not exposed to fishing pressure. Historical records indicate that the Norwegian spring herring has undergone large fluctuations during the last 500 years when periods with large catches have alternated with periods of extreme scarcity. The well-being of the population of some coastal areas in Norway depended very much on these fluctuations. The periods of the Norwegian spring-spawners and the Bohuslän herring alternate with each other (Fig. 1).

Herring and Sardine Fishery off South-Western England

The distribution areas of the herring, preferring cold waters, and sardine which isadapted to warmer temperatures (also called pilchard, *Sardina pilchardus*) overlap in the English Channel. Fishing for both species off the south-western tip of England, Cornwall and Devon, has been reported since the 16th century (Southward et al. 1988). The geographical boundary between the two species seems to shift back and forth on a decadal scale and, consequently, periods of the herring fishery have alternated with those of the sardine fishery and were in phase with those of the Bohuslän herring (Fig. 1). Also, they seem to be linked to the Russell Cycle which describes the synchronous alternation of appearance and disappearance of a large number of pelagic species, zooplankton and fish, including fish eggs and larvae, in the western Channel, which has been recorded since 1924 (Cushing 1982). When the Russell Cycle changed from a warm-water to a cold-water system between 1965 and 1979, the herring failed to re-appear off the coast of Devon. Although the herring catches were relatively small as compared to the yields of the Norwegian spring-spawning herring or the Bohuslän herring, it was a major loss for the city of Plymouth when the herring fishery collapsed in 1930 (Southward et al. 1988).

French Herring Fishery in the English Channel

There are only few data on herring and sardine fisheries of French fishermen in the English Channel since the 18th century . The herring periods of this fishery were from about 1750 to 1810 and from 1880 to 1910 (Binet 1988; Fig. 1).

Bay of Biscay Herring Fishery

A small herring population exists in the Bay of Biscay off southern Brittany which was first mentioned in 1728. Due to its relative insignificance, reports are rather sporadic, but indicate, nevertheless, periods of presence and absence (Binet 1988; Fig.1).

French Sardine Fishery in the English Channel

Occasionally, sardine fishery developed north of Brest, off northern Brittany. There are various reports from the period between 1726 and 1764. A notable decrease of sardine abundance was reported in the first years of the 19th century, but the fishery resumed in the 1860s and 1870s (Binet 1988; Fig. 1).

All these fishery periods are reported historically in terms of persistent presence or absence of either fish or fisheries at certain locations. The drastic changes between rich and poor fishing periods ensured the maintenance of historical records because the regional economies and the well-being of the local population were heavily affected. The fluctuations of these fisheries could be the result of real fluctuations of biomass or of decadal changes in migration routes moving the fish to areas not accessible to the limited range of the fisheries methods employed. However, fluctuations in biomass and migration routes are likely to be concomitant phenomena. The herring and sardine fisheries described here are divisible in two groups with alternating periods of occurrence. Group 1 comprises the Bohuslän herring, the herring off south-western England, the herring caught by French boats in the eastern English Channel, and the Bay of Biscay herring. They seem to be forced by the climatic scenario described above for the Bohuslän herring. Group 2, consisting of the Norwegian spring-spawning herring, the sardines off south-western England and the sardines caught by the French fleet in the English Channel, appear to be favoured by opposing climatic conditions: warm temperatures at mid- and higher latitudes of the eastern North Atlantic, positive sea surface temperature anomalies, no strong ice cover off Iceland, NAO index positive with intensified westerly winds, increasing frequency of south-westerly winds over England and relatively warm water in the North Sea, English Channel and the Skagerrak (Alheit and Hagen 1997). The fluctuations in European herring and sardine fisheries are related to temperature regimes and fluctuate on a decadal scale. Consequently, when searching for factors forcing these fish populations, one has to take into account climatic changes in the oceanic current system (on the basin-scale?) of the North Atlantic. Such variations influence the atmospheric circulation of the entire hemisphere. The fish populations dealt with in this study are widely separated from each other. In spite of the distances of several thousand kilometres between these populations, they seem to react to the same basin-wide forcing.

World-Wide Decadal Variability of Anchovy and Sardine Populations

Large anchovy (*Engraulis spec.*) and sardine (*Sardinops spec.*) populations live in the fertile upwelling ecosystems of the large eastern boundary currents (California, Humboldt and Benguela Current) and in the waters surrounding Japan. They support important fisheries, mainly for fish meal production, and the economy of the riparian countries of upwelling systems depends on these fisheries (see: John Steinbeck´s "*Cannery Road*"). The population dynamics of these anchovies and sardines are characterised by their inverse relationship to each other. When one species is doing well and supports a large biomass and high production, the other species usually sustains a rather low biomass (Lluch-Belda et al. 1992; Alheit and Hagen 2000; Schwartzlose et al. 1999) (Fig. 3), very much like the antagonistic population fluctuations of Eu-

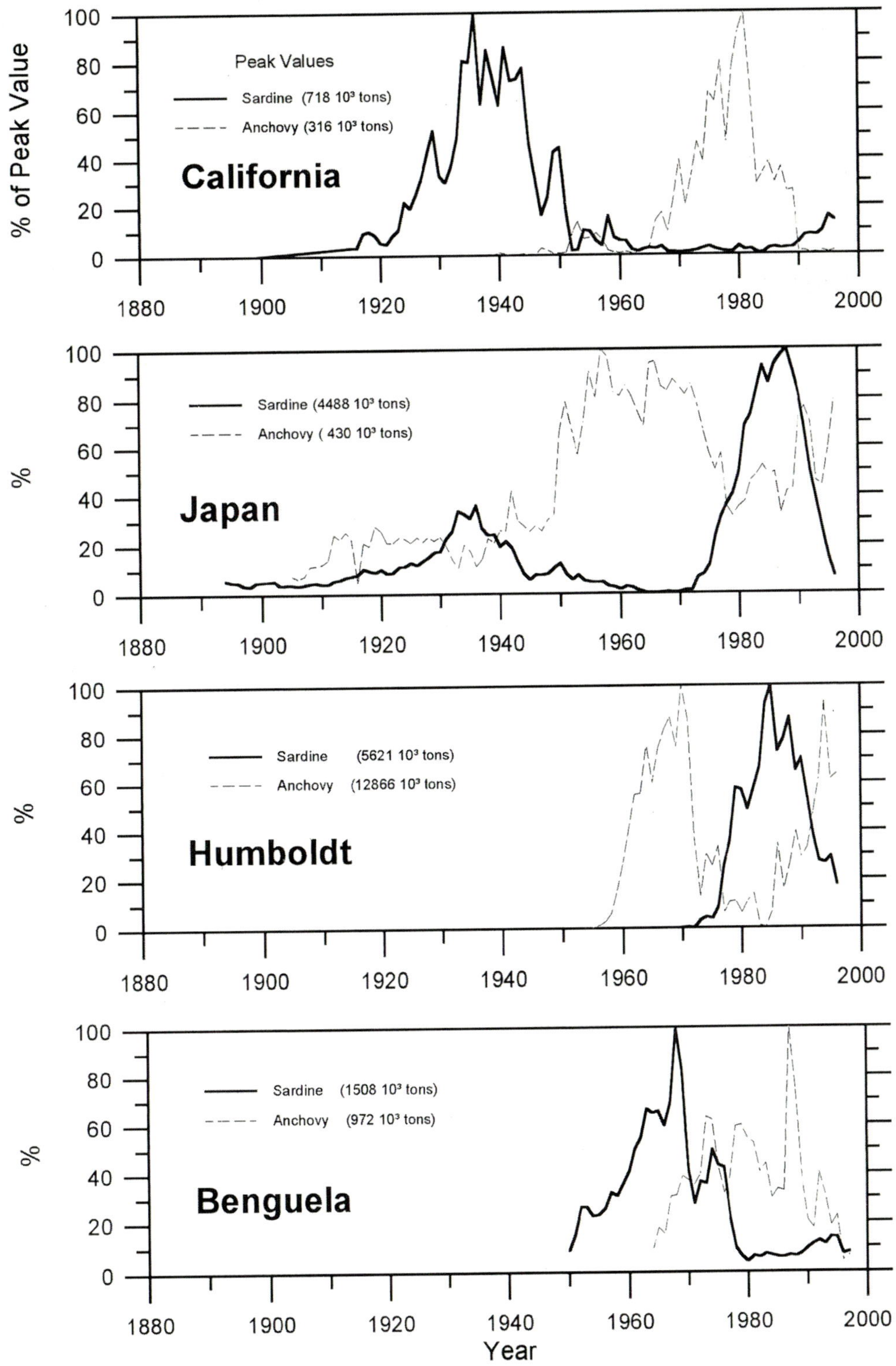

Fig. 3. Variation in the annual catch of sardine and anchovy in four major current systems during the 20th century. Variations in fish catches plotted as percentages of maximum historical annual values.

ropean herrings and sardines (Fig. 1). These shifts between sardine-dominated and anchovy-dominated states seem to restructure the entire ecosystem, because concomitant qualitative and quantitative changes in ecosystem components other than sardines and anchovy populations have been observed. Due to their dramatic and long-lasting nature, these switches have been termed "regime shifts". Regime shifts seem to happen synchronously all over the Pacific ocean. Thus, the sardine populations off Japan, in the Humboldt and in the California Current supported high biomasses during the 1930s and the 1940s. These stocks decreased dramatically during the 1950s and 1960s, rose again in the 1970s and 1980s (except California) and have reduced their biomass since the late 1980s (Fig. 3) (Alheit and Hagen 2000; Schwartzlose et al. 1999). Interestingly, North Pacific salmon populations exhibited the same population swings. In contrast, the sardine stocks in the Benguela Current in the South Atlantic showed alternating population fluctuations and had their maximum biomass when the Pacific sardines were going through low population levels.

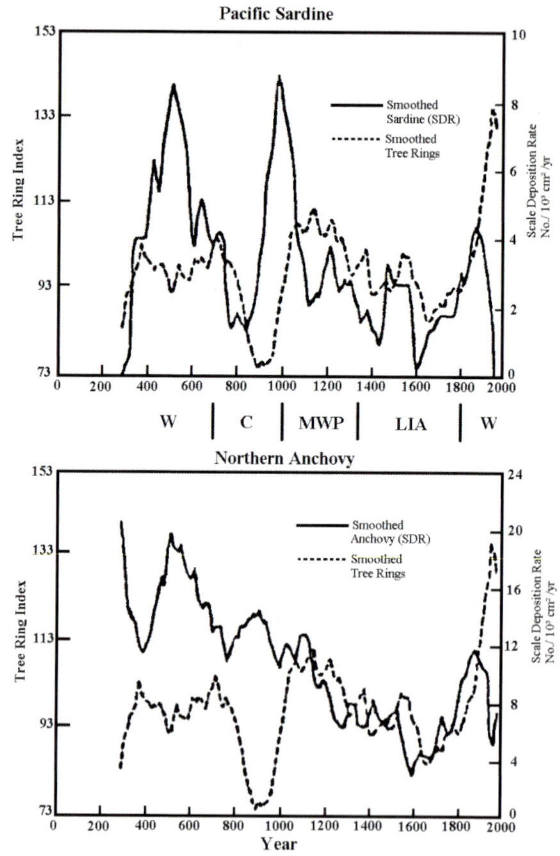

Fig. 4. Low frequency variability (periods > 150 year) of proxy time series of pelagic fish abundance in the California Current (Baumgartner et al. 1992) compared with bristlecone pine tree ring widths (after Francis et al. 1998). (W = warm; C = cold; MWP = Medieval Warming Period; LIA = Little Ice Age).

Centennial-Scale Fluctuations of Anchovies and Sardines

The upwelling systems of the California, Humboldt and Benguela Currents have varved anaerobic sediments which preserve fish scales. Accumulation rates of fish scales in aged sediment cores indicate relative historical fish abundances and can be calibrated to recent biomass values in order to infer historical biomass data. Pioneering work has been done by Soutar and Isaacs (1974) on sediment samples from the Santa Barbara Basin in the California Current. Baumgartner et al. (1992) extended the work of Soutar and Isaacs and reconstructed a 1750-year time series proxy of anchovy and sardine abundance from A.D. 270 to 1970 (Fig. 4). It clearly demonstrates that anchovy and sardine populations have undergone dramatic natural fluctuations over the last 2000 years in the absence of any fishery. Similar studies of anaerobic sediments in the Benguela and Humboldt Current are in progress.

Teleconnection Patterns and Global Synchrony of Fish Population Dynamics

Clearly, fish populations which are separated by thousands of kilometres and do not have any opportunity to interact with each other fluctuate synchronously (Bakun 1998a; Alheit and Hagen 1997, 2000). These teleconnection patterns are found within ocean basins (e.g., sardines in Pacific ocean) and between ocean basins (e.g., sardines in the North Pacific and in the Southern Atlantic). Even more striking is a comparison of long-term historical time- series data. There is a surprising alternation of the Bohuslän fishing periods and the sardine fishing periods off Japan (exception: sardine period 1790-1803) (Fig. 5). Or, in turn, good periods for European sardines match with those of Japanese sardines. Similarly, when the periods of the Swedish Bohuslän herring fishery are superimposed on the paleochronology of sardine and anchovy scales from the Santa Barbara Basin, California, they seem to coincide with the high abundances of anchovies and low abundances of sardines off California (Schwartzlose et al. 1999) (Fig. 5). These teleconnection patterns among fish populations seem to be widespread and might indicate a global synchrony of fluctuations involving many important fish populations around the world (Bakun 1998a). It is difficult to assume that the surprising match of these population fluctuations is just a pure coincidence and the question arises what forces fish populations to fluctuate synchronically on a global scale.

The Climate Connection

Although the synchronous population swings of anchovies and sardines in the Pacific and the South Atlantic have been reported for several years now, there is no satisfying theory or hypothesis explaining these teleconnection patterns, nor is there any clear explanation on the causes of regime shifts, the switch from an anchovy dominated ecosystem to one in which sardines dominate. It is highly unlikely that the forcing signal travels through the ocean as the fish populations are too far separated from each other for such a signal to initiate population changes at the same time in, for instance, the North Pacific and the South Pacific. Consequently, environmental forcing of the population dynamics of anchovies and sardines must be brought about by atmospheric processes which have been shown not to be linearly coupled with ocean dynamics. Because of the decadal nature of the fish population dynamics it is much likely that climate variability is the driving force.

According to Bakun (1998 b) the synchronous variations of fish populations seem to be generated by some type of very large external forcing, most probably through climatic teleconnections acting through the atmosphere. He excludes temperature as the key environmental variable as temperature trends have not been consistent in the various regions and suggests that the most likely external driving force must be mechanical. Wind stress acting on the sea surface is the predominant mechanism to transfer kinetic energy and momentum

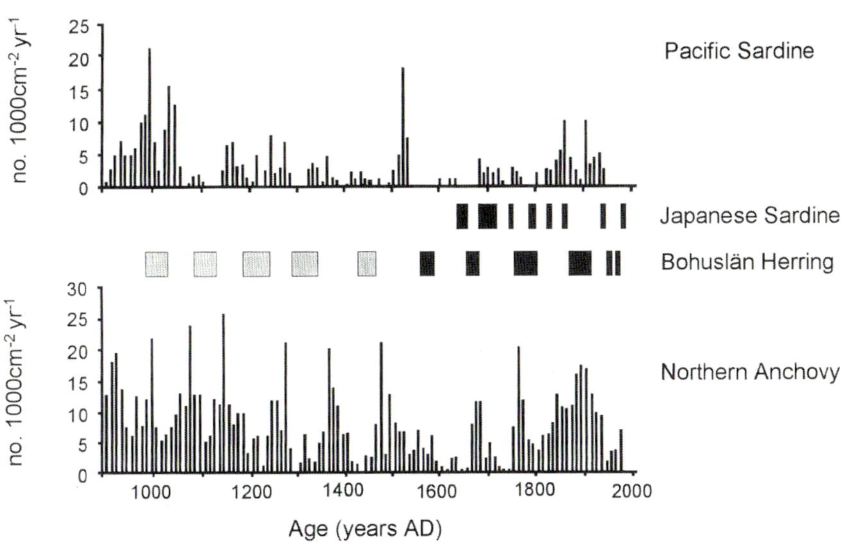

Fig. 5. Coincidence of Bohuslän herring periods (Alheit and Hagen 1997) and Japanese sardine periods (Yasuda et al. 1999) with fluctuations in paleosedimentary sardine and anchovy scale deposition in the Santa Barbara Basin, California (bar graphs, Baumgartner et al. 1992).

between the atmosphere and the ocean. This input of kinetic energy works against stratification of oceanic water. The oceanic stratification regionally depends not only on net heat fluxes between ocean and atmosphere but also on lateral advection processes acting on different oceanic scales not only in time, but also in space. Oceanic currents redristribute heat from low latitudes to high latitudes and influence atmospheric forcing fields around the globe. Consequently, one should expect that the causal mechanism for the observed regime changes applying to sardines and anchovies might be a sum of processes driven by the action of the wind on the sea surface and depending on regional anomalies in the heat transport of oceanic current systems.

Baumgartner et al. (1992) have analysed the frequencies of the paleoecological time series relating to Californian sardines and anchovies. The low -frequency variability (component) agrees well with tree ring width of bristlecone pine trees which is a proxy for prevailing temperature regimes, also reflecting major climate epochs of the last 2000 years, such as the Medieval Warming Period (950-1350 AD), the Little Ice Age (1400-1800 AD), and global warming which began at the end of the Little Ice Age) (Francis et al. 1998) (Fig. 5).

All these correlation studies indicate more or less clearly that fish population fluctuations and regime shifts are the consequence of climate variability, however, they do not prove it. We are aware of the physical forcing source (coupled ocean/atmosphere system) and the reaction of fish populations, but we do not know the processes leading to changes in fish population dynamics or as stated in the title of a recent paper by Baumann (1998): The fallacy of the missing middle: physics → ...→ fisheries.

A crucial question is whether effects of climate variability work their way right up through the food chain before fish stocks are affected, or whether there is a more direct impact on fish populations (Bakun 1998a). Francis et al. (1998) presented a conceptual model of physical-biological linkages to the food web to explain the impact of climate variability on marine ecosystems. Based on a modelling exercise of phytoplankton response to physical forcing, Polovina et al. (1995) suggested that changes in the mixed layer depth caused by the shift and intensification of the Aleutian Low have resulted in changes in biological production regimes from primary production to fish production. In contrast, Bakun (1998a) rejects the idea that climatic effects work their way through complex planktonic food webs as trophic interactions in marine ecosystems are characterised by chaotic behaviour. He concludes that the biological dynamics involved must be very simple such as a rather direct effect of the external forcing acting either on the fish themselves at some sensitive life stage in their complex life cycles, or directly on a primary food source.

Implications for Fisheries Management

The evidence that fluctuations of exploited fish populations are to a large extent governed by climate variability does not mean at all that no fishery management will be required. In addition to the impact of natural variability, the dynamics of exploited fish populations are influenced by fishing mortality. The new evidence gives us a valuable tool for a better fisheries management. For example, fishing activities could be adapted to whether fish populations are on the ascending or the descending slope in the course of their decadal population variability. The problem, of course, is to find out which regime and which phase of the regime are prevailing. This is the task of research projects such as the IGBP Global Change Core Programme "GLOBEC" (Global Ocean Ecosystem Dynamics Programme) which has the goal to improve our understanding of the structure and function of the global ocean ecosystem and its response to physical forcing.

Acknowledgments

We are indebted to Victor Smetacek and Han Lindeboom for critically reading the manuscript and to Carsten Jahnke for correcting some of the historical aspects of the medieval herring fisheries in the Baltic Sea.

References

Alheit J, Hagen E (1997) Long-term climate forcing of European herring and sardine populations. Fish Oceanogr 6:130-139

Alheit J, Hagen E (2000) The effect of Climate Variation on Pelagic Fish and Fisheries. In: Jones PD, Ogilvie AEJ, Davies TD, Briffa KR (eds) History and Climate: Memories of the Future. Kluwer Plenum, pp 247-266

Bakun A (1998a) Ocean Triads and Radical Interdecadal Stock Variability: Bane and Boon for Fishery Management Science. In: Pitcher TJ, Hart PJB, Pauly D (eds) Reinventing Fisheries Management. Kluwer Academic Publishers, Dordrecht, pp 331-358

Bakun A (1998b) Global climatic teleconnections and fisheries ecology. A perspective (with comments on the Benguela system). Paper distributed at International Symposium on Environmental Variability in the South-East Atlantic, Swakopmund, Namibia, 30 March – 1 April 1998, pp 1-10

Baumann M (1998) The fallacy of the missing middle: physics → .... → fisheries. Fish Oceanogr 7:63-65

Baumgartner TR, Soutar A, Ferreira-Bartrina V (1992) Reconstruction of the history of Pacific sardine and northern anchovy populations over the past two millennia from sediments of the Santa Barbara Basin. California Calif Coop Oceanic Fish Invest Rep 33:24-40

Binet D (1988) French sardine and herring fisheries: a tentative description of their fluctuations since the eighteenth century. In: Wyatt T, Larraneta MG (eds) Long Term Changes in Marine Fish Populations. Proc Int Symp, 18-21 November 1986, Vigo, Bayona, pp 253-272

Corten A (1999) A proposed mechanism for the Bohuslän herring periods. ICES J Mar Sci 56:207-220

Cushing D (1982) Climate and Fisheries. Academic Press, London, pp 1-364

Francis RC, Hare SR (1994) Decadal-scale regime shifts in the large marine ecosystems of the North-east Pacific: a case for historical science. Fish Oceanogr 3:279-291

Francis RC, Hare SR, Hollowed AB, Wooster WS (1998) Effects of interdecadal climate variability on the oceanic ecosystems of the NE Pacific. Fish. Oceanogr. 7:1-21

Höglund H (1978) Long-term variations in the Swedish herring fishery off Bohuslän and their relation to North Sea herring. Rapp Proc-verb Réun Cons Int Explor Mer 172:175-186

Hunter JR, Alheit J (1995) International GLOBEC Small Pelagic Fishes and Climate Change program. GLOBEC Report No. 8, pp 1-72

Jahnke C (1997) Heringsfang und –handel im Ostseeraum des Mittelalters. Thesis, Kiel University, Kiel

Lamb HH (1972) Climate: Past, Present and Future. I. Fundamentals and Climate Now. Methuen, London, pp 1-613

Lindquist A (1983) Herring and sprat: fishery independent variations in abundance. FAO Fish Rep 291:813-821

Ljungman AV (1880) The great Bohuslän herring fisheries. US Commission Fish Fisheries VI:221-239

Ljungman AV (1883) The Bohuslän sea fisheries and their future. US Commission Fish Fisheries VIII:89-98

Lluch-Belda D, Schwartzlose RA, Serra R, Parrish R, Kawasaki T, Hedgecock D, Crawford RJM (1992) Sardine and anchovy regime fluctuations of abundance in four regions of the world oceans: a workshop report. Fish Oceanogr 1:339-347

Polovina JJ, Mitchum GT, Evans GT (1995) Decadal and basin-scale variation in mixed layer depth and the impact on biological production in the Central and North Pacific, 1960-88. Deep Sea Res 42:1701-1716

Sahrhage D, Lundbeck J (1992). A History of Fishing. Springer-Verlag, Berlin, pp 1-348

Schwartzlose, RA, Alheit J, Bakun A, Baumgartner T, Cloete R, Crawford RJM, Fletcher WJ, Green-Ruiz Y, Hagen E, Kawasaki T, Lluch-Belda D, Lluch-Cota SE, MacCall AD, Matsuura Y, Nevarez-Martinez MO, Parrish RH, Roy C, Serra R, Shust KV, Ward NM and Zuzunaga JZ (1999) Worldwide large-scale fluctuations of sardine and achovy populations. S Afr J Mar Sci 21:289-347

Soutar A, Isaacs JD (1974) Abundance of pelagic fish during the 19th and 20th centuries as recorded in anaerobic sediment off the Californias. Fish Bull, US 72:257-273

Southward AJ, Boalch GT, Maddock L (1988) Fluctuations in the herring and pilchard fisheries of Devon and Cornwall linked to change in climate since the 16th century. J Mar Biol Ass 68:423-445

Yasuda I, Sugisaki H, Watanabe Y, Minobe SS, Oozeki Y (1999) Interdecadal variations in Japanese sardine and ocean/climate. Fish Oceanogr 8:18-24

Changes in Coastal Zone Ecosystems

Han J. Lindeboom

*Netherlands Institute for Sea Research (NIOZ), P.O.Box 59,
NL-1790 AB Den Burg, The Netherlands
e-mail: hanl@nioz.nl*

Abstract: Marine ecosystems are not in steady state, but exhibit continuous changes in production and species composition of different trophic levels. In this paper, examples of phenomena observed in different long-term data series, mainly collected in coastal seas, are given. Sudden changes in biomasses or species composition, a reversal of trends, increased seasonal variation and cyclic behaviour all seem to contribute to the interannual and decadal variability of the North Sea ecosystem. The indications that weather and climate play an important role in this variability are discussed and the possible effects of climate change are summarized. However, the complex nature of marine ecosystems hampers the analyses of clear cause-effect relationships. The need for more long-term data sets is stressed.

Observed phenomena

Several long-term data sets on phytoplankton, zooplankton, macrofauna, fish and birds have been collected in the Wadden Sea and North Sea. Up till now, these data sets have been mainly applied to demonstrate the effects of human (mis)use on the ecosystem. However, when the various data sets are combined it is striking that certain changes are very sudden and not gradual, as one would expect from a gradual increasing human impact. Data from Cadée, Beukema and Swennen (Fig. 1) indicate that the algal biomass in the western Wadden Sea doubled between 1976 and 1978, followed by the macrobenthos in 1980. And in 1978 the breeding success of Eider ducks (*Somateria mollissima*) increased by several orders of magni-tude (Cadée and Hegeman 1993; Beukema 1992; Swennen 1991). Apparently, large changes in the Wadden Sea ecosystem took place at the end of the seventies. An analysis of data collected in the North Sea indicates that large changes also occurred in this ecosystem in the same period. In the vicinity of Helgoland, changes in the phytoplankton species composition were observed and the bio-mass of flagellates <10 μm went from 3 μg/l in the period '62-'77 to 15 μg/l in the period '78-'93 (Hickel et al. 1993). And analyses of the Continuous Plankton Recorder (CPR) data indicated a minimum in zooplankton abundance and a simultaneous shift in species composition in the late 70s (Aebischer et al. 1990; Reid pers. comm.). Shifts in macrofauna assemblages (Josefson et al. 1993) and benthic respiration rates were reported, all indicative of a rapid major change of the marine ecosystem in the entire North Sea in the late seventies. Another rapid change in the ecosystem could be observed in the early 90s when, for example, the breeding success of eider ducks once again collapsed (see Fig. 1). This coincided with a dramatic decrease in standing stocks of mussels and cockles in the Dutch Wadden Sea. The collapse of the North Sea fish stocks at the end of the previous century and in the thirties were also rather sudden changes. In addition to these rapid changes, other striking phenomena have been observed. The CPR data indicate that the monthly variation of the net colour index was about twice as high in the 80s as compared to the 70s. A major shift in the marine ecosystem in the second half of the seventies has also been recorded in the North-Eastern Pacific, hinting at a large scale cause of the phenomena observed (Ebbesmeyer et al. 1990).

From WEFER G, BERGER W, BEHRE K-E, JANSEN E (eds), 2002, *Climate Development and History of the North Atlantic Realm.* Springer-Verlag Berlin Heidelberg, pp 447-455

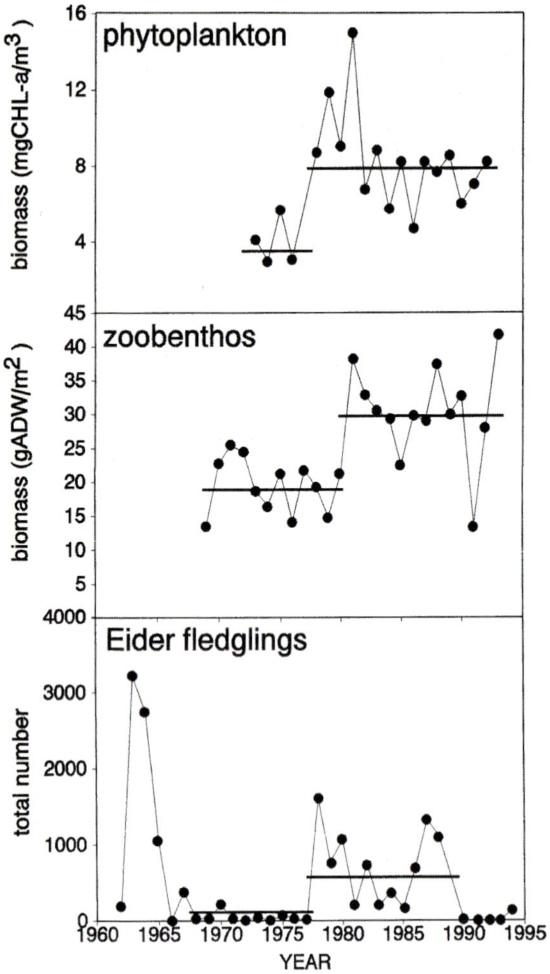

Fig. 1. Average chlorophyll-a concentration (after Cadée and Hegeman 1993), Macrofauna biomass (after Beukema 1992) and Eider fledglings (after Swennen 1991) in the western Wadden Sea. The different levels observed in the data set are indicated.

In several data sets, cyclic changes with different frequencies have been observed. A periodicity of high abundances of *Noctiluca scintillans* in three year intervals near Helgoland. Neudecker and Eurps (1996) found some indications for a 6-year cycle in the amount of shrimps landed around the North Sea. Between 1969 and 1996, the density of *Macoma balthica* recruits strikingly showed a maximum every 6 years, starting in 1973, with only two other maxima in other years (Dekker pers. com.). In the same period, cockles (*Cerastoderma edule*) also showed periodical recruitment patterns with optima every four years. However, their recruitment failed in the 90s.

The interannual variation of zooplankton biomass in the Gulf of Alaska seems to be dominated by the Pole Tide (~14 month period) and the quasi-biennial oscillation (QBO; (~24 month period), whereas 6 and 12 year periods were also significant (Conversi and Hameed 1997). Gray and Christie (1983) analysed plankton data from the North Atlantic and found evidence of 3-4, 6-7 and 10-11 years cycles, whereas benthic data suggested 6-7 and 10-11 years cycles. Witbaard (1996) found a 25-30 year cycle in the variation in annual shell growth of *Arctica islandica* collected at the Fladen grounds in the North Sea. Due to the fact that *A. islandica* grows older than 100 years he was able to create a data set long enough to detect this cyclic behaviour. For the CPR data a 30-year cycle has been suggested. The data on Eider fledglings hint at a possible 10-or 20-year cycle. However, up till now most data sets are too short for a reliable analyses of these cycles. Analyses of phenomena observed in marine sediments may help to solve this problem. Pike and Kemp (1997) found significant periodicities in the deposition of diatom mat laminae of ~11 years, 22-24 years and (~50 years in Gulf of California sediments deposited in the early Holocene. Cyclic deposition has been observed in many sedi-ments and their paper gives a selection of observations all indicating similar periodicities.

In conclusion, many longer data series indicate interannual and decadal changes among which the following phenomena may be observed: sudden rapid changes, gradual changes (e.g. in the direction of trends), changes in (seasonal) variability, changes in dominance of species and cyclic variation. What causes this behaviour of the marine ecosystem?

Possible Causes

Links with changes of short-term or large-scale weather patterns have been suggested in several papers. Wind, winter and/or summer temperatures or rainfall are mentioned as possible causes. A shift in storm frequencies or wind directions might cause changes in sediment water exchange in shallow areas or mixing of stratified waters in deeper areas (Lindeboom et al. 1995). Beukema (1990) found that the occurrence of cold winters strongly influenced the species composition of intertidal benthic communities. He demonstrated that the winter of 1979 affected the macrofauna in the Balgzand area. Other strong winters, but not all, led to similar effects. Comparing three different areas in the Wadden Sea, Beukema et al. (1996) showed several examples indicating that severe winters represent a major synchronizing factor among many of the zoobenthos species of tidal flats, both by immediately enhanced mortalitity rates in species sensitive to low temperatures and by enhanced recruitment in bivalve species some months later (Fig. 2). High temperatures in summertime may influence the distribution of species adapted to cooler waters. Cadée and Hegeman (1993) showed that there is a positive correlation between the freshwater input from Lake IJssel, and thus rainfall, and the length of algal blooms in springtime in the Marsdieparea. Therefore, specific differences in weather phenomena may explain local changes in the marine ecosystem. The long-term changes observed in the CPR data have been attributed to changes in westerly weather patterns (Aebisher et al. 1990). Well known is the El Niño-Southern Oscillation (ENSO) cycle resulting in the nearly complete failure of fisheries in South-American waters and many other ecological deviations worldwide every 4-7 years (e.g. Angel 1994).

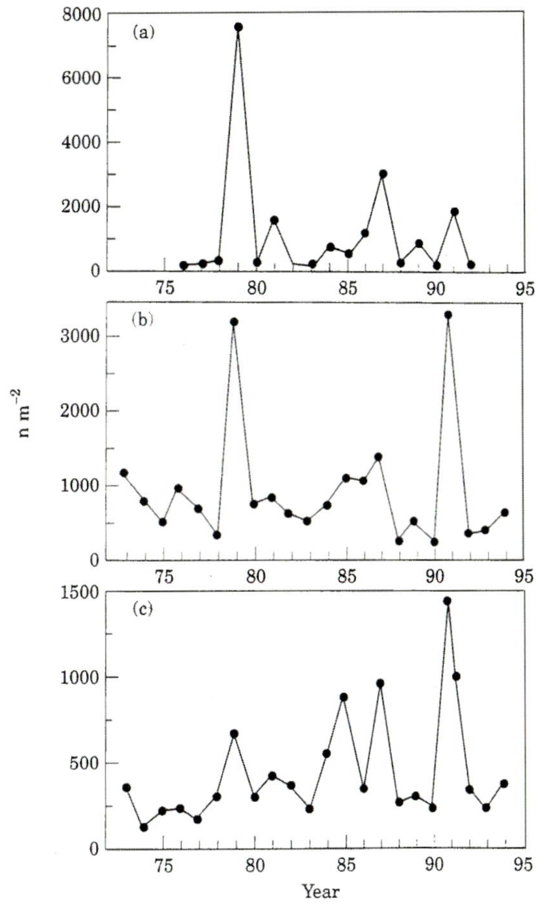

Fig. 2. Long-term changes in numerical densities (n, m-^{2}) of the bivalve *Macoma baltica* observed in late summer at **(a)** Norderney (German Wadden Sea; average of 4 stations), **(b)** Groningen (Eastern Dutch Wadden Sea; 5 stations) and **(c)** (Western Dutch Wadden Sea; 15 stations; Beukema et al. 1996).

Burroughs (1992) mentions cycles of 2, 3-4, 5-7, 10-12, 18.6, 22, 80-90 and 200 years observed with a relatively high frequency in meteorological records all of which might have a relationship with certain weather and astronomical features. Are these cycles reflected in the marine ecosystem? In the data set on phytoplankton (Fig. 1) the annual fluctuation shows an alternating pattern between years. Is this a coincidence or might there be a relationship with the quasi-biennial oscillation (QBO) often observed in weather patterns and other data series (e.g. Conversi and Hameed 1997). Although the data series is still short, the fledgling success of the eiderducks has periods with optima and minima of approximately 11 years. It has been argued that storm patterns in the temperate zone may be influenced by the occurrence of sunspots (Haigh 1996). Cycles of 11 or 22 years have been observed in several sedimentary records which hint at a possible influence of the 11 year sunspots cycle or the 22 year Hale cycle on marine systems. So far, many authors have criticized the possible relationships with sunspots, because the difference in solar irradiation is too small, others have pointed at the possibility of the effect of the changing magnetic field (Burroughs 1992; Haigh 1996). On the other hand, the 18.6-year period in the regression of the longitude of the node- the line joining the points where the Moon's orbit crosses the ecliptic (Burroughs 1992) may also have an influence on biota. Peaks in tidal forces due to this phenomenon occurred in 1950, 1969 and 1988. In the Wadden Sea a clear 18.6 year cycle was found in sedimentation rates (Oost et al. 1993). And organisms living in the intertidal area, such as the macrofauna shown in Fig. 1, may be subject to influences of this cycle. However, much longer data sets are needed before a relationship between a mix of different external cycles and possible phenomena observed in the marine environment can really be established.

For the sudden changes at the end of the 70s, a relationship with the great salinity anomaly which entered the North Sea in 1977/78 was proposed (Lindeboom et al. 1995). Not the salinity itself but a difference in macro- or micro-nutrient concentrations in this distinct water mass amount of water could have led to changes in the ecosystem. Taylor et al. (1992) indicated a correlation between latitudal displacements of the Gulf Stream and the abundance of plankton in the North-East Atlantic suggesting that changes in the hydrography of the westen Atlantic may influence the biota of the eastern part. The changes observed in the so-called Russell cycle described for the English Channel (Southwood 1980) were attributed to differences in the amounts of phosphate coming from the Atlantic Ocean into this area. Fromentin and Planque (1996) found a close relationship between the North Atlantic Oscillation (NAO) and the abundance of the copepods *Calanus finmarchicus* and *C. helgolandicus* in the eastern North Atlantic. In their analyses Gray and Christie (1983) showed evidence of clycles in hydrographic data for the North Atlantic Ocean with periods of 3-4, 6-7, 10-11, 18-20 and 100 years. They indicate a relationship between the hydrographic and biotic cycles. Witbaard et al. (1997) atributed the observed growth variation in *A. islandica* to hydrographic changes.

In many cases, human activities such as eutrophication, pollution or fisheries have been seen as a major driving force of observed changes. Along the Norwegian Skagerrak coast Johannessen and Dahl (1996) found a clear signal of ecosystem changes due to eutrophication. Beukema (1991) attributed the rise in the phyto-plankton and macrofauna biomass in the late seventies (see Fig.1) to the effects of eutrophication in the Dutch coastal waters. Indeed, the anthropogenic phosphate and nitrogen loadings of the river Rhine were high in that period. However, how does a gradually increasing nutrient concentration explain a sudden doubling of biomass? Eutrophication has certainly influenced the biota in the Dutch and German coastal zone (Gerlach 1990), and Norwegian waters (Johannessen and Dahl 1996), but other causes also seem to play a significant role.

Pollution may also influence the ecosystem. A dramatic decline in the Eider duck and Sandwich Tern (*Sterna sandvicensis)* populations in the Western Wadden Sea was observed in the early sixties. Later, the cause appeared to be the accidental pollution of the coastal zone with drins (Swennen 1991). Certainly, pollution can cause sudden changes, especially on a local level, but in general this leads to a deterioration of the biota and not to a sudden increase. It is unlikely that pollution with heavy metals or organic pollutants played a significant role in the large scale phenomena described above. An exception could be the pollution with TBT from antifouling paints on ships, which may have negatively influenced the occurrence of whelks (*Buccinum undatum* L.) in the Dutch coastal zone (Ten Hallers-Tjabbes et al. 1996).

Another important human impact factor is fisheries. The decline of herring stocks to very low levels in the mid-70s has been attributed to extensive fisheries. However, Corten (1990) has argued that in this case changes in hydrography may also have played a very significant role. Fishermen have been blamed for the tremendous decline of the mussel and cockle population in the Western Wadden Sea in the early 90s. But it is not unlikely that natural causes (e.g. weather changes) have played a role, whereby the fisheries simply caused an extra decline in already decreasing populations. The direct effects of beam trawl fisheries on biota has been clearly demonstrated (e.g. Lindeboom and de Groot 1998), and a gradual shift from long living, slowly reproducing species towards rapidly reproducing, short lived species may be the result. However, this cannot explain the sudden changes observed in the late 70s.

Other human induced causes of changes in coastal zone ecosystems include the building of structures like seawalls, dikes and jetties which may influence the local currents and morphology. The construction of polders, the mining of sand and gravel, and the subsidence caused by the extraction of water, oil or gas.

Up till now, I have looked for the possible cause of the observed variations outside the system, but it could also be caused by internal cycles in the marine ecosystem. High biomasses or high reproduction rates of one organism may alternate with similar phenomena in other organisms. These sorts of changes have been observed, e.g. in fish stocks (Hempel 1978), and using model calculations Silvert (1993) showed that sudden declines or increases may be explained by internal cycles. So until we have longer data series and more analytical and experimental evidence for certain cause-effect relationships, we should not rule out internal cycles as a possible cause of the phenomena observed.

Possible Effects of Climate Change

It is difficult to predict the ultimate effects of temperature changes on the marine ecosystem. In the literature many predictions have been made for different groups of organisms, but if we try to combine all these effects it becomes a very complex picture with many uncertainties.

Kawasaki (1991) states that global warming could accelerate coastal upwelling, and thus increase photosynthesis and the production of pelagic fish. However, changing current systems might change taxonomic composition of plankton leading to a movement of the distribution ranges of major fish stock.

Beukema et al. (1996) found evidence that cold winters enhance the breeding success of shellfish species (see also Fig. 2). Similar effects have been reported for flatfishes like sole and plaice.

For soft bottom macrobenthos communities on the Swedish West Coast Tunberg and Nelson (1998) proposed climatological linkages affecting primarily surface primary production resulting in bottom-up control of benthic population changes. They found a clear correlation with the NAO index. Kröncke et al. (1998) found the same for macro-faunal communities near Norderney (Germany), again suggesting that most of the interannual variability in macrozoobenthos can be explained by climate variability. Their data indicate that the benthic communities are severely affected by cold winters and not by storms or hot summers. However, this is not a general rule. Beukema (1988) compared changes in the faunal composition at three places in the Wadden Sea. The data on severe winter survival clearly showed a geographical trend of lower survival in the NE as opposed to the SW part of the sampling area, indicating important differences in reactions over a distance of approx. 200 km. De Vooys (1990) compared the macro-fauna communities of the Wadden Sea with those of the Seine estuary (mean temperatures 2°C higher) and the Gironde estuary (mean temperatures 4°C higher). He concluded that a rise of temperature of 2°C in the Wadden Sea would lead to an enrichment of ~20% of the number of species, whereas at a rise of 4% an enrichment of ~30% might be expected.

In the studies mentioned above, changes observed in the ecosystem were related to temperature changes. In other studies the changing faunal composition has been used as an argument that water temperatures are changing. Francour et al. (1994) used a variety of biological indicators to demonstrate that Mediterranean waters are becoming warmer. The same was done by Franke et al. (1999) who saw the recent arrival of the oceanic isopod *Idotea metallica* off Helgoland (Germany) as an indication of a warming trend in the North Sea.

For Gulf of Mexico oysters a significant correlation was found between length and reproduction stage and temperature and rainfall (Kim and Powell 1998). On the other hand the sex ratio and the oyster pathogen *Perkinsus marinus* prevalence were more influenced by variations in the local environment.

Changes in rainfall have also been mentioned in other areas as possible climate effects (Cadée and Hegeman 1993). These changes alter the river inputs and the salinities in estuaries and the coastal zone. Although many organisms in this area are adapted to highly variable salinities, shifts in species composition and dominance are to be expected, especially when the occurrence of extremes in salinities, either very low or high, increases.

Changes in wind velocities, directions or storm frequencies may have profound effects on the marine ecosystem. As a result, the water currents and water mixing may change. This in turn leads to changes in nutrient (re-)distribution, egg and larval transports, oxygenation and distribution of planktonic organisms. Lindeboom et al. (1995) have suggested that the shift from a pelagic dominated system towards a benthic dominated system, as has been observed in the North Sea could be the result of shifting wind phenomena (Fig. 3).

Global temperature rises are bound to be followed by rises in sea level, even from thermal expansion alone. The precise effects of sea-level rise in coastal environments will largely depend upon geomorphological processes, in particular the net result of changing rates of sedimentation and erosion (Beukema et al. 1990).

In conclusion, the coastal environment is and will be affected in a variety of ways by the possible climate changes. Consistent changes in temperature, windfields and/or rainfall will cause shifts in areas of distribution of plants and animals, changing the composition of ecosystems with risks of imbalance and extinction of rare and sensitive species (Beukema et al. 1990). For other species the

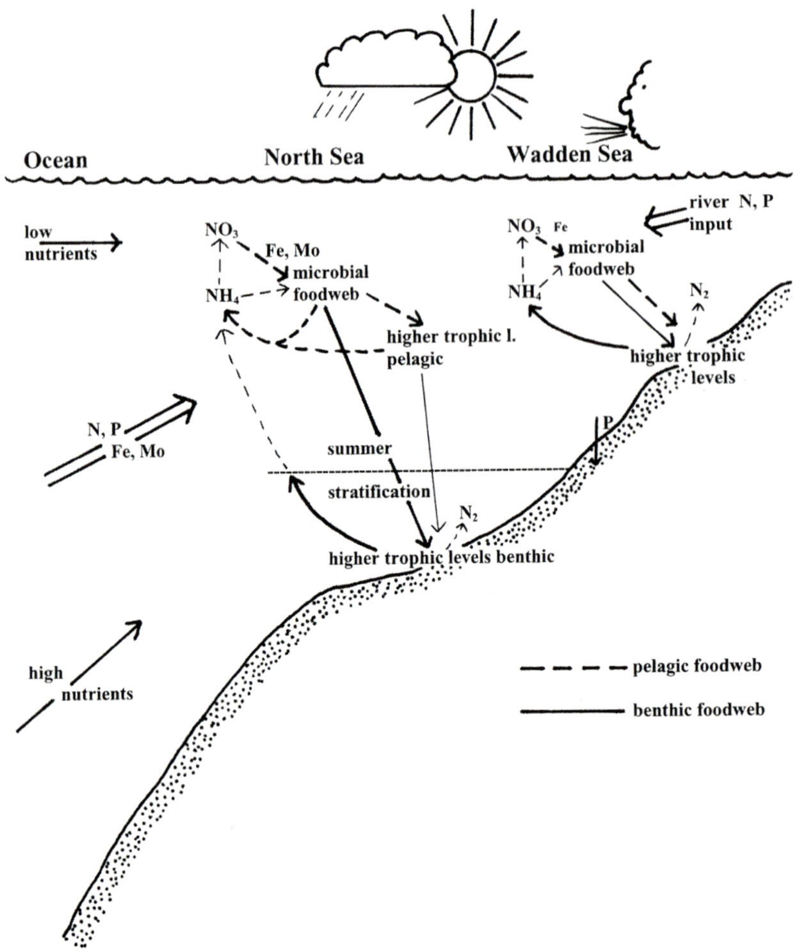

Fig. 3. The nitrogen cycle in a marine ecosystem switching from a 'benthic' to a 'pelagic' foodweb. In deeper areas, with summer stratification this migth lead to an extra sink for nitrogen.

breeding success may increase or decrease depending upon the occurring changes and the life strategies of the species involved. Since all these species also have very complex interrelationships, predictions of the actual effects of climate change are highly uncertain. Nevertheless, it is worthwhile to consider possible effects of climate changes on our highly variable coastal ecosystems, in particular to realise the gaps in our knowledge and to be prepared when climate changes become greater.

Discussion

Marine ecosystems are not in steady state but show large interannual and decadal variations. Periods with large algal blooms alternate with years or decades with relatively low algal biomasses. Biomasses or reproductive success may double or halve between two consecutive years. Periods with decreasing trends are followed by periods with increasing trends. Years with a high frequency of anaerobic events succeed years when no low oxygen concentrations are measured. Sometimes the ecosystem seems more or less stable for a certain period, but then the system seems to flip to another semi-steady state.

Can the interannual and decadal variation of the North Sea ecosystem be attributed to just one or a few causes or are we looking at a very complex combination of natural variability and human influences? For reasons of simplicity, or clarity towards financers or managers, it has been tempting to try to find the explanation in just one or a few causes (e.g. Lindeboom et al. 1995).

Anthropogenic activities certainly have an impact on the marine ecosystem. However, most often these are of a local and limited nature (eutrophication, pollution, mining) or lead to one way trends (fisheries). And considering the phenomena described above, it seems likely that a combination of factors induces the variations observed in the marine ecosystem.

When analysing the abundance of fish recruits in the western Wadden Sea, Philippart et al. (1996) found a significant correlation with the NAO index. However, they concluded that the actual cause-effect relationship might be very complex and related to transport rate of larvae across the North Sea and the quality of the nurseries, i.e. temperature, food availability and predation pressure. And although *Macoma balthica* recruitment in the western Wadden Sea was successful every six years for a period of over 25 years, there were also two other years that recruitment was above average. And although several maxima in recruitment occurred after a cold winter, the highest maximum occurred after a mild winter. So, just referring to cold winters as the cause does not explain all data observed. Lindeboom et al. (1995) introduced a set of hypotheses on how the changes might be linked to changes in weather patterns, hydrographical changes and changes in nutrient cycles, especially in the nitrogen cycle (see Fig. 3), all indicating that we are dealing with a very complex system. and that it will never be easy to explain all field observations. And similarly, it will be very difficult, if possible at all, to predict the effects of climate change at local or regional scales.

Long-term data series can help to answer questions on cause-effect relationships, and the continuous collection of data, often hampered by limited funding, should be strongly supported. These long-term data sets, in combination with the results of experimental laboratory and field studies, will be necessary if we are to answer the question whether we are looking at explainable phenomena with clear cause-effect relationships or at a more or less chaotic and unpredictable behaviour of the marine ecosystem.

References

Angel MV (1994) Spatial distribution of marine organisms: patterns and processes. In: Edwards PJ, May RM, Webb RN (eds) Large scale ecology and conservation biology. Blackwell, Oxford, pp 49-109

Aebischer NJ, Coulsen JC, Colebrook JM (1990) Parallel long-term trends across four marine trophic levels and weather. Nature 347:753-755

Beukema JJ (1988) An evaluation of the ABC-method (abundance / biomass comparison) as applied to macrozoobenthic communities living on tidal flats in the Dutch Wadden Sea. Mar Biol 99:425-433

Beukema JJ (1990) Expected effects on changes in winter temperatures on benthic animals living in soft sediments in coastal North Sea areas. In: Beukema JJ, Wolff WJ, Brouns JJWM (eds) Expected effects of climatic change on marine coastal ecosystems. Kluwer Academic Publishers, Dordrecht, The Netherlands, pp 83-92

Beukema JJ (1991) Changes in composition of bottom fauna of a tidal-flat area during a period of eutrophication. Mar Biol 111:293-301

Beukema JJ (1992) Expected changes in the Wadden Sea benthos in a warmer world: lessons from periods with mild winters. Neth J Sea Res 30:73-79

Beukema JJ, Wolff WJ, Brouns JJWM (eds) (1990) Expected effects of climatic change on marine coastal ecosystems. Kluwer Academic Publishers, Dordrecht, The Netherlands, pp 1-221

Beukema JJ, Essink K, Michaelis H (1996) The geographic scale of synchronized fluctuation patterns in zoobenthos populations as a key to underlying factors: climatic or man-induced. Ices Journal of Marine Science 53:964-971

Buchanan JB (1993) Evidence of benthic pelagic coupling at a station off the Northumberland coast. J exp mar Biol Ecol 172:1-10

Burroughs WJ (1992) Weather cycles Real or Imaginary? Cambridge University Press, Cambridge, UK, pp 1-207

Cadée GC, Hegeman J (1993) Persisting high levels of primary production at declining phosphate concentration in the Dutch coastal area (Marsdiep). Neth J Sea Res 31:147-152

Conversi A, Hameed S (1997) Evidence for Quasi Biennial Oscillations in zooplankton biomass in the subarctic Pacific. Journal of Geophysical Research 102 (C7): 15:659-15, 655

Corten A (1990) Long-term trends in pelagic fish stocks of the North Sea and adjacent waters and their possible connection to hydrographic changes. Neth J Sea Res 25(1/2):227-235

De Vooys CGN (1990) Expected biological effects of long-term changes in temperatures on benthic ecosystems in coastal waters around the Netherlands. In: Beukema JJ, Wolff WJ, Brouns JJWM (eds) Expected effects of climatic change on marine coastal ecosystems. Kluwer Academic Publishers, Dordrecht, The Netherlands, pp 77-82

Ebbesmeyer CC, Cayan DR, McLain DR, Nichols FH, Peterson HH, Redmond KT (1990) 1976 step in the pacific climate: forty environmental changes between 1968-1975 and 1977-1984. In: Betancourt JL, Tharp VL (eds.) Proc. 7th Annual Pacific Climate (PACLIM) Workshop. Calif. Dept. of Water Resources, Interagency Ecol Stud Prog Rep 26:115-126

Francour P, Boudouresque CF, Harmelin JG, Harmelin-Vivien ML, Quignard JP (1994) Are the Mediterranean waters becoming warmer? Information from biological indicators. Mar Poll Bull 28:523-526

Franke U, Hutter K, Jöhnk K (1999) A physical-biological coupled model for algal dynamics in lakes. Bulletin of Mathematical Biology 61:239-272

Fromentin J-M, Planque B (1996) *Calanus* and environment in the eastern North Atlantic. II. Influence of the North Atlantic Oscillation on *C. finmarchicus* and *C. helgolandicus*. Mar. Ecol Prog Ser 134:111-118

Gerlach SA (1990) Nitrogen, phosphorus, plankton and oxygen deficiency in the German Bight and in Kiel Bay. Kieler Meeresforschungen, Sonderheft Nr. 7, pp 1-341

Gray JS, Christie H (1983) Predicting long-term changes in marine benthic communities. Mar Ecol Prog Ser 13:87-94

Haigh JD (1996) The impact of solar variability on climate. Science 272:981-984

Hallers-Tjabbes CC ten, Everaarts JM, Mensink BP, Boon JP (1996) The decline of the North Sea Whelk (*Buccinum undatum* L.) between 1970 and 1990: A natural or a human-induced event? Mar Ecol 17 (1-3):333-343

Hempel G (1978) North Sea fisheries and fish stocks - A review of recent changes. Rapp. P.-v. Réun. Cons int Explor Mer 173:145-167

Hickel W, Mangelsdorf P, Berg J (1993) The human impact in the German Bight: Eutrophication during three decades (1962-1991). Helgol Meeresunters 47:243-263

Johannessen T, Dahl E (1996) Declines in oxygen concentrations along the Norwegian Skagerrak coast, 1927-1993: A signal of ecosystem changes due to eutrophication Limnol Oceanogr 41:766-778

Josefson AB, Jensen JN, Ærtebjerg G (1993) The benthos community structure anomaly in the late 1970s and early 1980s - a result of a major food pulse? J Exp Mar Biol Ecol 172:31-46

Kawasaki T (1991) Effects of global climatic change on marine ecosystems and fisheries. In: Jäger J, Ferguson HL. Climate change: science, impacts and policy. Proceedings of the Second World Climate Conference, pp 291-299

Kim Y, Powell EN (1998) Influence of climate change on interannual variation in population attributes of Gulf of Mexico oysters. Journal of Shellfish Research 17:265-274

Kröncke I, Dippner JW, Heyen H, Zeiss B (1998) Long-term changes in macrofaunal communities off Norderney (east Frisia, Germany) in relation to climate variability. Mar. Ecol Prog Ser 167:25-36

Lindeboom HJ, de Groot SJ (eds) (1998) The effects of different types of fisheries on the North Sea and Irish Sea benthic ecosystems. NIOZ Rapport 1998-1/RIVO-DLO Report COO3/98, pp 1-404

Lindeboom HJ, Van Raaphorst W, Beukema JJ, Cadée GC, Swennen C (1995) (Sudden) changes in the North Sea and Wadden Sea: Oceanic influences under-estimated? Dt Hydrogr Z Suppl 2:87-100

Oost AP, De Haas H, Ijnsen F, Van den Boogert JM, De Boer PL (1993) The 18.6 year nodal cycle and its impact on tidal sedimentation. Sedimentary Geology 87:1-11

Philippart CJM, Lindeboom HJ, Van der Meer J, Van der Veer HW, Witte JIJ (1996) Long-term fluctuations of fish recruit abundance in the western Wadden Sea in relation to variation in the marine environment. ICES Journal of Marine Science 53:1120-1129

Pike J, Kemp AES (1997) Early Holocene decadal-scale ocean variability recorded in Gulf of California laminated sediments. Paleoceanography 12:227-238

Silvert W (1993) Size-structured models of continental shelf food webs. In: Christensen V, Pauly D (eds) Trophic models of aquatic ecosystems. ICLARM Conf Proc 26:40-43

Swennen C (1991) Fledgling production of Eiders *Somateria mollissima* in The Netherlands. J Orn 132:427-437

Southwood DJ (1980) Low frequency pulsation generation by energetic particles. J Geomag Geoelectr 32, SII, 75

Taylor AH, Colebrook JM, Stephens JA, Baker NG (1992) Latitudinal displacements of the Gulf Stream and the abundance of plankton in the North-East Atlantic. J mar biol Ass UK 72:919-921

Tunberg B, Nelson WG (1998) Do climatic oscillations influence cyclicalpatterns of soft bottom macrobenthic communities on the Swedish west coast. Mar Ecol-Prog Ser 170:85-94

Witbaard R (1996) Growth variations in *Arctica islandica* L. (Mollusca): a reflection of hydrography-related food supply. ICES journ. Mar Sci 53:981-987

Witbaard R, Duineveld GCA, De Wilde PAWJ (1997) A long-term growth record derived from *Arctica islandica* (Mollusca, Bivalvia) from the Fladen Ground (northern North Sea). J Mar Biol Ass UK (in press)

The Impact of Harmful Algal Blooms in Natural and Human-Modified Systems of Northern Europe

Marina Montresor[1] and Victor Smetacek[2]

[1]*Stazione Zoologica 'A. Dohrn', Villa Comunale, I-80121 – Napoli, Italy*
e-mail: mmontr@alpha.szn.it
[2]*Alfred Wegener Institute for Polar and Marine Research, Am Handelshafen 12, D-27570 Bremerhaven, Germany*
e-mail: vsmetacek@awi-bremerhaven.de

Abstract: The discovery that certain phytoplankton species can render sea food highly toxic for humans was made only some 60 years ago. Information accumulating since then strongly suggests that harmful algae have had an impact on human diet, customs and also economy over many millennia. The current rapid advances in knowledge of algal toxins and their occurrence and role in ecosystems are described and the possible impacts speculated upon. Thus, we suggest that cyanobacterial toxins in drinking water have posed a serious threat to human health since the Neolithic and that the unusual drinking habits of the Europeans when compared to those of e.g. the Chinese might be explained in this light. In marine waters, toxins are transmitted to humans via shellfish that accumulate them by feeding on plankton containing toxic species. Since the toxins are highly potent, illness and deaths may have occurred often in the past. Regional differences in the acceptance of shellfish as human food might well be due to incidences of poisoning. Sedimentary records of toxic and related species suggest that marked fluctuations in abundance and distribution have occurred during the Holocene that provide additional information on regional climate fluctuation on century and perhaps decadal scales.

Introduction

Toxin production by land plants, sea weeds, fungi and various animals has been known for millennia but the discovery that some unicellular algae also contain potent toxins was made only in the 1930s (Sommer et al. 1937). Since then toxins are being found in increasing numbers of microalgae as a result of growing scientific and public awareness of the phenomenon (Hallegraeff 1993; Satake et al. 1997). It is now common practice to refer to these algae as 'harmful species', and their mass occurrence as 'harmful algal blooms (HAB)'. Included in the term are organisms which produce compounds toxic to humans, to organisms of economic interest to humans, i.e. fishes or bivalves, as well as organisms which can be a nuisance or cause economic loss when they build-up exceptionally large blooms. Toxicity is a natural property of the respective algae and their occurrence is part of the natural cycle in many regions. However, there is increasing evidence that human activities such as aquaculture and eutrophication promotes growth of toxic or at least undesirable species. Thus the term HAB is the marine equivalent of terrestrial 'weeds' and 'pests' and our growing confrontation with HABs is a reflection of increasing human use of aquatic environments (Smetacek 1998).

Although poorly documented, there is little doubt that harmful algae have had an impact on the human populations living along the coasts of northwestern Europe. Since harmful algae were not recognised and hence could not be specifically avoided as in the case of poisonous plants or mushrooms, their impact will have been more like that of pathogenic bacteria, that is, unpredictable. Whereas the occur-rence of disease and epidemics in the past tend to be well described and can often be diagnosed from accounts of the

From WEFER G, BERGER W, BEHRE K-E, JANSEN E (eds), 2002, *Climate Development and History of the North Atlantic Realm.* Springer-Verlag Berlin Heidelberg, pp 457-471

time, past human encounters with toxic algae are more difficult to pinpoint. Nevertheless, some diffuse knowledge on hazards due to toxic algae will have accumulated enabling interpretation of some peculiarities of regional diets and even resource-use patterns in this light.

In this essay we discuss algal toxicity against the background of phytoplankton ecology and evolution and contrast it with the situation on land. We next describe some of the major harmful species, their toxins and the various ways they impact human interests. This is followed by a section in which we speculate on how harmful algae might have affected people and their well-being in earlier times and left a mark on the collective memory, folk-lore and customs. Finally we present evidence from the sedimentary record indicating that the geographical distribution of some species of microalgae is influenced by local hydrography. Sedimentary records of such species can hence provide additional information on regional climate fluc-tuation on century and perhaps decadal scales.

Ecology of Phytoplankton and Toxins

The bulk of marine photosynthesis is generated by unicellular algae – the phytoplankton – that live in the surface layer of the sea. Because the algal cells are eaten almost as fast as they grow, they generally occur in low concentrations. Nevertheless, phytoplankton provide the food on which the bulk of marine animal life, including all commercially important fish stocks, is based. The organisms comprising the phytoplankton range across three orders of magnitude in size and are much more diverse in terms of pigmentation, physiology, mode of nutrition and life cycle pattern than the land plants. Thus, all land plants from mosses to trees and grasses belong to one evolutionary lineage whereas many phyla, widely separated from one another, contribute to the phytoplankton (Bhattacharya and Medlin 1998). However, the phylogenetic diversity of phytoplankton is not matched by their species diversity. Only about 4,000 species of marine phytoplankton have been described in contrast to 250,000 terrestrial plant species. About 100 microalgae are reported to be toxic (Sournia 1995) but this figure is difficult to compare with the number of toxic land plants as the majority of the latter contain metabolites that in some way or another are detrimental to animals that eat them. The fact that modern human nutrition is based on only a small number of plant species is testimony to this.

In the course of evolution, herbivores learn to cope with defences (chemical and mechanical) developed by plants and this ongoing process results in increasing specialisation between plants and their enemies. It was previously assumed that the situation would be very different in the plankton because the algae are diffusely distributed in the environment and much smaller than the bulk of the 'traditional' zooplankton, that were hence believed to feed indiscriminately on the phytoplankton. In the last decades, selective feeding has been shown to be much more widespread among planktonic grazers than previously assumed. For instance, high-speed cinematography indicates that suspension-feeding copepods, that comprise over 60% of marine zooplankton biomass, reject most of the particles conveyed towards their mouth by the feeding current they generate. Further, it is now accepted that the bulk of marine phytoplankton is eaten by heterotrophic protists of about the same size range as their food, also indicating that the degree of sprecies-specific selection is much greater than imagined. Thus, it would now appear that there is indeed ample scope for such "arms races" to develop in the plankton (Verity and Smetacek 1996) although the role of algal toxins in this light is not yet clear.

The phylogenetic heterogeneity of the phytoplankton arose because individual species of different lineages converged in the course of evolution on the same mode of nutrition: utilising ingested photosynthetic organisms as food-providing symbionts instead of digesting them (Raven 1997). This process of endosymbiotic evolution is ongoing amongst the eucaryotic unicellular organisms – the protists. Obviously, in terms of the 'selfish gene', mutual cooperation is one way for both sides to win the arms race and has been developed to perfection by land plants whose chloroplasts all originate from the same ancestor. The most plant-like of the phytoplankton are the diatoms which can only take up dissolved substances and are entirely dependent on their chloroplasts for nutrition. Most species from other groups are also entirely photosynthetic but there are many phytoplankters that can also collect, ingest and digest other particles such as bacteria and other planktonic organisms. Interestingly, toxins are most widespread amongst groups – in particular the dinoflagellates – that exhibit a wide range of feeding strategies. In contrast, only a few toxic diatoms have been recorded so far.

Because protists are recombinations of older, independent organisms, they will have retained metabolic pathways that contribute to fitness of the new unit. In keeping with the diversity of evolutionary lineages represented in the phytoplankton, the chemistry of their toxins are also very diverse and they also act in many different ways (Cembella et al. 1995). Some phytoplankton toxins are extremely potent, registering tens of thousands on a scale in which, weight for weight, cyanide is 1 and cobra poison 500. In comparison, the poisons of land plants appear harmless and are dominated by relatively few chemical types such as alkaloids. Some environmental bacteria have the most potent toxins. Tetanus toxin of the bacterium *Clostridium tetani*, that occurs naturally in enriched, preferably anaerobic soils, registers 100,000,000 on the above scale (Wyatt 1998). It appears unlikely that this toxin is part of the arms race between this bacterium and humans because infection only occurs from the environment and people infected with the tetanus bacterium often die. But what is the fate of the infecting tetanus bacteria in the corpse? They are not parasites residing within the human population like leprosy or syphilis bacteria that survive by infecting other individuals within the life time of their host. Instead, one could argue that the rotting corpse acts as a breeding ground for this anaerobic bacterium which subsequently retires into the enriched soil or forms cysts awaiting the next unwary person to infect. Indeed, it is well known that soil in contact with decaying corpses is full of dangerous bacteria. The role of ambush predator for a bacterium might seem far-fetched but has been suggested recently for a toxic dinoflagellate that kills and feeds on fish (Burkholder and Glasgow 1997). This case will be dealt with below, the point here is that toxins can play multifarious roles in the life cycle of the producers. Of course, toxins are not developed for the purpose but arise by chance mutations that, if they improve fitness of the mutant, enable the spread of the responsible gene within the population.

Animals that are regularly exposed to specific toxins in their food have evolved mechanisms to deal with them. Thus, carnivorous mammals are largely immune to the metabolic products secreted by anaerobic bacteria in putrefying meat. The same molecules are highly toxic to humans who suffer food poisoning as a result of eating rotting meat. Instead of developing an antidote to the toxins, we have developed a strong aversion to the smell of bad meat. Indeed, our sense of smell with regard to hydrogen sulphide (a by-product of rotting meat and itself toxic but only of the calibre of cyanide) is as acute as that of a dog; with the difference that dogs don´t find its smell disgusting. Similarly plant alkaloids are bitter and explains our natural aversion to this taste, in contrast to sweetness that generally signals palatability. The fact that humans are susceptible to a wide range of marine toxins and are unable to identify them by means of sense organs suggests that our history of dependence on sea food is not long enough to have led to evolution of protective measures. An alternative explanation is that occurrence of toxicity in sea food is too sporadic to have led to evolution of specific mechanisms. The spectacular incidences of mass mortality amongst marine animals (e.g. fish, birds and mammals) that have been related to occurrence of specific toxic algae (Geraci et al. 1989; Work et al. 1991; Scholin et al. 2000) indicates that even within marine ecosystems, organisms have not developed adequate responses to the entire gamut of toxins possibly due to the spatial and temporal variability in their occurrence.

There are various ways of dealing with toxins following ingestion. One is enzymatic break down or modification to a less toxic form in the gut. Another is to prevent the molecules from reaching their site of impact. Thus, neurotoxins can be sequestered where they do not encounter neurons and where they can be stored, broken down or excreted at convenience e.g. in the liver or comparable organs. The strategy is commonplace amongst insects that feed on toxic plants. Some even use the sequestered toxin to keep potential predators at bay. This is an elegant way out of the arms race. Humans do not normally eat insects and hence have little historical experience of this ecological phenomenon. However, there are many marine animals including bivalves (mussels, clams and oysters) as well as fish that can contain large doses of toxin ingested with their algal food that only marginally affect their well-being but can have a drastic effect on humans that eat them (Turner and Tester 1997). Of course this does not apply to all toxins and the same animals can be immune to one type of toxin but susceptible to another one. Again, it is difficult to formulate generalities that adequately cover the range of complex ecotoxicological relationships that have evolved in aquatic ecosystems.

Phytoplankton Distribution and Species Succession

In contrast to sessile land plants, phytoplankton cells are suspended in the water and their distribution pattern is strongly influenced by hydrography. Many phytoplankton species have a very wide distribution because they can thrive under a broad range of environmental conditions, whereas other species are very restricted in their occurrence. The notorious toxic species tend to belong to the latter category. Temperature and salinity but also nutrient availability and dependence on specific trace substances are invoked as determinants of species ranges. Phytoplankton species reach their peak abundance over periods of weeks at specific times of the year and are then replaced by other species. The vast majority of species belong to the rare or background flora of the sea, although they can be highly predictable in their occurrence, albeit at low concentrations. Each region has a characteristic pattern of species succession, with varying degrees of interannual variability with regard to bloom development, composition and biomass. The term 'bloom' is generally used to denote exceptionally high cell concentrations. Different species, harmful and non-harmful, can be in bloom state, i.e. in a phase of significant population increase, even if they do not reach high population densities (Smayda 1997).

In the North Sea, diatoms dominate the spring bloom and are replaced by haptophytes such as coccolithophorids and colonies of *Phaeocystis* in late spring. Medium-sized dinoflagellates and other diatom species are characteristic of the summer with large dinoflagellates – various species of *Ceratium* - appearing regularly in late summer and autumn. In some regions the annual cycle is terminated by an autumn diatom bloom. In contrast to diatoms which discreetly sink out of the surface layer after blooming, the colonies of the flagellate *Phaeocystis* tend to break up in the surface. Organic material released in the process is beaten to a stiff white foam that accumulates on the beaches of the Channel and the southern North Sea in early summer. This alga also releases a reduced sulphur compound (Dimethyl sulphide, DMS) with an unpleasant odour; however, at low concen-trations it is the 'tang of the sea'. Because foam is nowadays associated with pollution, the phenomenon is not popular with tourists, hence *Phaeocystis* is often classed as a HAB species.

Some dinoflagellates can also be quite conspicuous: the heterotrophic *Noctiluca* often accumulates to form stripes of coloured water in the German Bight in summer. This cosmopolitan organism is also responsible for intense bioluminescence but is not reported to have ever caused detrimental environmental effects. The late-summer/autumn *Ceratium* are also bioluminescent and can reach high concentrations but do not appear to be toxic. Dense blooms have caused anoxic conditions in bottom waters but this occurs rarely except for some restricted areas. The only toxic genus which occurs regularly during the summer is *Dinophysis* which never reaches high cell concentrations. However, because of its potent toxin, it is not advisable to eat mussels when this genus is present. Other toxic dinoflagellates can be responsible for harmful phenomena along the coasts of north-western Europe. They can form blooms in coastal bays and fjords from Portugal to Norway, with an apparently unpredictable interannual pattern, as in the case of the genus *Alexandrium*; they can form blooms related to upwelling patterns along the Atlantic coasts of the Iberian peninsula, as *Gymnodinium catenatum*, or they can be responsible for massive blooms both in coastal and offshore waters, as the fish-killing *Gymnodium mikimotoi* (= *Gyrodinium aureolum, Gymnodinium nagasakiense*). In the Baltic Sea blooms of cyanobacteria occur regularly in the summer (Wasmund 1997). Some of the species are toxic. Some common toxic phytoplankters have been depicted in Fig. 1.

The factors driving species succession are not well established and need to be differentiated in terms of those that initiate or terminate the growth phase of the various species. Different species have different environmental demands. Thus, diatoms grow rapidly and take advantage of a sudden improvement in growth conditions. Ceratia in contrast are slower growers but persist longer in the environment and hence accumulate biomass through the summer (Smetacek 1985). Physico-chemical factors such as a favourable light environment and nutrients are obviously very important in initiating but also terminating the growth phase but mortality induced by herbivores and pathogens (parasites, bacteria and viruses) as well as internal cycles can also determine disappearance of the species. Toxic algae can occur at any time of the year although their greatest incidence is in the summer and autumn. Since they represent a subjective grouping, no uniform theory can be developed to explain their occurrence.

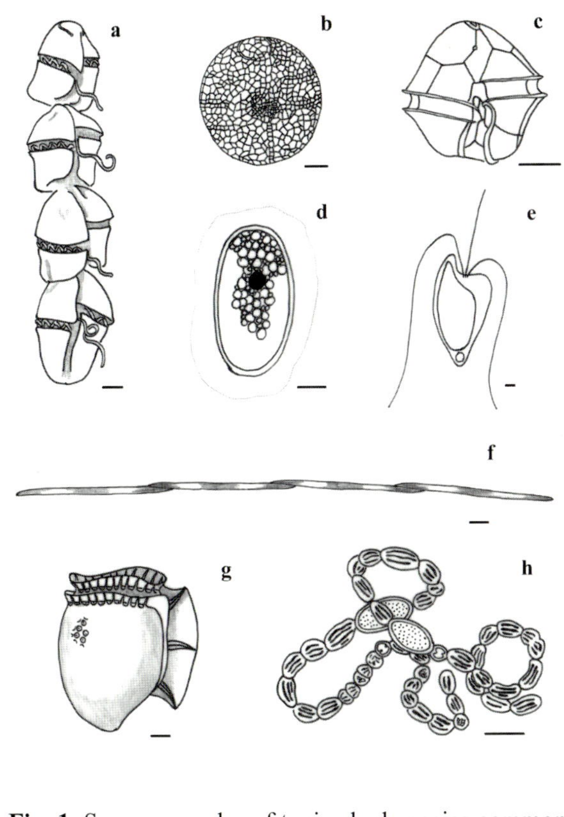

Fig. 1. Some examples of toxic algal species common in northern European waters: the dinoflagellate *Gymnodinium catenatum,* 4-cell colony **(a)**, and resting cyst **(b)**; the dinoflagellate *Alexandrium tamarense* motile cell **(c)** and resting cyst **(d)**; the haptophyte *Chrysochromulina polylepis* **(e)**; a chain of the diatom *Pseudo-nitzschia pseudodelicatissima* **(f)**; the dinoflagellate *Dinophysis norvegica* **(g)**; the cyanophyte *Anabaena flos aquae* **(h)**. Scale bars = 10 μm, with the exception of **(e)** where scale bar = 1 μm.

Many species of phytoplankton spend the inactive phase of their life cycle in the form of a resting phase which can differ considerably from the vegetative or growth phase. In some coastal species resting cysts are formed that settle out on the bottom and ensure presence of the species in a favourable locality (Montresor et al. 1998). It is from these resting-cyst forming species that historical information on local events can be derived. With increasing knowledge of the ecological preferences of various species we will eventually be able to reconstruct past climate regimes based on their presence in sediment layers the way terrestrial ecologists interpret the pollen record.

It should by now be clear that the science of harmful algal blooms is in its infancy. There is yet much to be learned and novel ecological interrelationships to be discovered. It is still common practice to group harmful algae according to their mode of impact on humans, i.e. whether they have a direct impact, are transferred to humans via sea food or whether they kill aquatic organisms but have no recorded influence on human health. Innocuous species that in some regions build up large biomass conspicuous to humans (discoloration of water, foam on the beach) are also grouped with the harmful algae. Such species are accorded nuisance status only when they occur on tourist beaches. For example, dense blooms of *Phaeocystis* that are common in the pristine Antarctic are, of course, not considered a HAB. In the following we shall deal first with algae that can poison drinking water, then with those that render food organisms toxic, followed finally by those that are responsible for mass kills of fish.

HABs and Drinking Water

Surprisingly few toxic algae have been reported from freshwater and of these, cyanobacteria (formerly termed blue-green algae) are by far the most important (Falconer 1996; Codd 1997). Only a few cosmopolitan genera are responsible that are all very common in northern Europe (Fig.1). Since cyanobacteria are autotrophs, they reach much higher concentrations than heterotrophic bacteria (photosynthetic magnification of the toxin), and can render drinking water fatally poisonous for humans. This is a major problem in many freshwater reservoirs around the world (Falconer 1996). Since lakes are generally not used as sources of drinking water in Europe, at least no longer, this potential hazard has lapsed from the collective memory. However, the authorities are well aware of this danger and some lakes in e.g. northern Germany, are closed to swimming when cyanobacterial blooms develop. So far only cow and dog deaths due to drinking water or licking off algal scum from the fur, respectively, have been recorded from northern Europe (e.g. Codd 1996; Nehring 1993). Based on the following considerations, we spec-ulate that cyanobacterial toxins have posed a serious threat to human life since the Neolithic and that the unusual drinking habits of the Europeans when compared to those of e.g. the Chinese might be explained in this light.

Cyanobacteria thrive best and reach high concentrations in eutrophic water bodies which are anoxic at depth (Paerl 1996). In such lakes the nitrate to phosphate ratio typical of oligotrophic, i.e. oxygen-rich conditions is upset in favour of phosphate. Because only cyanobacteria can fix nitrogen, they gain a competitive edge over the normal algae when phosphate is still abundant after reactive nitrogen (nitrate and ammonia) has been exhausted. Calm conditions are a prerequisite for blue-greens and small, wind-protected water bodies provide the longest windows of opportunity for growth. Many of the nitrogen-fixing cyanobacteria are toxic but although the toxins are highly potent (neuro- and hepatotoxins), the amount produced varies widely. Indeed there are some strains that never produce toxin and others that do it sometimes. The reasons remain unknown. Because the cyanobacterial cells contain minute gas vacuoles, their population ends up as a scum on the surface which contributes to the concentration effect. Indeed scum formation appears to be a necessary stage in their life cycle. When nutrient-poor, oxygen-rich lakes (oligotrophic) are fertilised by input of nutrients from external sources they become eutrophied and cyanobacteria thrive where they had been previously absent. In the seventies many large European lakes were eutrophied as a result of phosphate input from domestic and agricultural sources and summer scum formation occurred in sheltered bays. Heavily fertilised smaller water bodies become hypertrophic and carry scum longer than large lakes. Under certain conditions toxic cynaobacteria can occur even in winter (Jacoby et al. 1994). When nutrient input is curbed (e.g. Lake Constance) and/or the bottom waters artificially aerated, the lakes revert to their original clarity.

In many settlements and villages of the past, drinking water was collected from the surface of a central pond or small lake. This was also the recipient of runoff from the inhabited area, which, in the absence of sanitation practices, would be full of pathogens but also rich in phosphate due to organic pollution. The village and town ponds of presanitation Europe would have provided ideal environments for toxic cyanobacteria. Even today, village ponds tend to be hypertrophic but no one even thinks of drinking water from them. The collective memory is firm on these points: stagnant, smelling, turbid water is dangerous to drink and scum is simply disgusting. Although a natural phenomenon, most people today are not aware why scum has negative connotations. We are suggesting that the collective memory is based on experience with cyanobacterial poisoning that goes far back into human history. We argue next that these HABs posed a threat in northern Europe well into the Middle Ages.

One ancient method of improving the quality of drinking water was to dig wells and wall them off, even in areas where surface water was freely available. This would have curbed infection by pathogenic bacteria (e.g. cholera) but also the growth of photosynthetic cyanobacteria. However, wells were not present everywhere and particularly in the growing towns and villages, surface water had to suffice for many people and for livestock. Raised basins of water fed by an aqueduct were popular and convenient sources of drinking water in cities of the Middle Ages. Often a fountain was built in the middle of the basin and most people will assume that this served an aesthetic function. Today the words 'fountain' and 'well' have strongly positive connotations in all European languages. In German the word 'Brunnen' is used for both, which is strange because a fountain is very different from a well. Indeed, we suggest that the only thing they have in common is that they can both provide safe drinking water. But how does a fountain ensure safe drinking water? Because agitation prevents stagnation and growth of cyanobacterial scum. Although the cause and effect relationship will not have been known at the time, collective experience will have supported use of fountains to provide drinking water.

Even for people without access to a well or fountain and within besieged fortresses, it is reasonable to assume that incidences of cyanobacterial poisoning arose only sporadically because of the variability in toxin production. Indeed the obsession with 'Brunnenvergiftung' ('well and/or fountain poisoning') that commenced in some European countries during the Middle Ages might have been due to experience with outbreaks of toxic cyanobacteria, although other bacteria might have been equally responsible. At the time minorities, in particular the local Jewish population and people afflicted with leprosy, were accused of 'poisoning fountains, wells, or rivers, to transmit leprosy to healthy people, or to make them ill or die' (Ginzburg and Rosenthal 1993). The accused were persecuted as a result and pogroms authorised by local kings were carried out. The point is that even the most potent toxins available to humans at the time could not have rendered the drinking water supply of a village poisonous. But toxic bacteria and cyanobacteria could well have been the insidious agents.

It is reasonable to assume that people faced by the potential but unpredictable danger emanating from unprotected drinking water sources will have been looking for remedies. We suggest that the custom of drinking alcoholic beverages to quench thirst might well have its roots in this need. The practice of fermenting fruit juice probably arose for the sake of intoxication but it will have been noticed that drinking wine was much safer than drinking polluted water, because the alcohol content is sufficient to kill pathogenic bacteria and the wine would be free of cyanobacterial toxins. In northern Europe beer took the place of wine. The oldest evidence of beer-brewing comes from Egypt and Mesopotamia where barley and emmer were used. The grains were not dehusked and beer was strained or drunk through straws (Sherratt 1997). Beer was the common beverage in north Europe at the beginning of the historical period. Dyonisios of Alicarnassus (Roman Antiquities XII, 10) mentioned that "The Gauls at that time had not knowledge of wine … but used a foul-smelling liquor made from barley rotted in water". Greeks and Romans associated beer with the Celts, so that the latin word for beer 'cerevisia' (Spanish 'cerveza') is a compound of the Latin for cereal plus a Celtic element meaning 'water' (Sherratt 1997). Interestingly, it has been shown that adding bushels of barley straw to water suppresses growth of cyanobacteria (Pillenger et al. 1994) and is recommended today for use in reservoirs. Whether there is a connection between the customary use of barley for beer-brewing and cyanobacteria needs to be looked into. Perhaps this is merely a coincidence, but wheat straw is reported to have little effect on cyanobacteria.

Like wine, beer will have been a safe drink although it must have tasted and smelled horrible. Beer drinking became increasingly popular in the Middle Ages, and in northern Europe as much as 20-30% of the total cereal crop was used for the purpose (Sherratt 1997). It seems unlikely that intoxication was the sole reason for drinking beer, rather it was probably a form of drinking-water treatment albeit with a not too unwelcome side effect that helped tide over the long winter nights. The custom to quench thirst with alcoholic beverages arose in Europe but has spread to most of the globe only in the twentieth century. This is not to say that the Europeans were the only people who invented alcoholic drinks. But this is the only culture to have not only accepted drinking alcohol on a routine basis but to even claim, despite medical evidence to the contrary, that it is a healthy beverage. The empirical evidence for this is based on folk lore: the statements of ancient people claiming to have imbibed moderate amounts throughout their lives. Interestingly, it is customary in Europe to drink alcoholic beverages and to deny the effects of intoxication. So why drink an intoxicant and suppress its effect? Indeed, Europeans are better able to cope with alcohol and avoid addiction because, on a population basis, they have higher levels of the enzyme which breaks down alcohol than peoples of other continents. Perhaps this is a result of natural selection working through the drinking water. In China for instance, it is believed that cold water is bad for the stomach, so hot water is drunk to quench thirst. Since this tastes pretty bad, herbs and leaves are added to change the taste. Thus, the custom of drinking teas might well be the Asian way of coping with hazardous drinking water. We are not aware whether cyanobacterial toxins are destroyed by boiling, but certainly pathogenic bacteria are.

We conclude that experience of drinking water poisoned by HABs could well have been part of the collective memory of European populations for a long time – like predator fear - that more or less died out in this century following the widespread use of modern methods of water treatment. Whether drinking customs were indeed shaped in the past by hazardous natural water needs to be researched but it is highly unlikely that teas and alcoholic beverages will become less popular when it is shown that the danger has passed and thirst can now be safely quenched with tap water .

Toxic Sea Food

In coastal regions a variety of suspension-feeding bottom animals (benthos) including mussels, clams, scallops and oysters feed directly on phytoplankton by filtering sea water. These animals collect suspended particles and are generally not able to select against toxic algae when these are present in the water. Bivalves are apparently immune to most toxins produced by phytoplankton which are sequestered in their tissues for periods of many weeks after the toxic algae are no longer present. Bivalves can thus become vectors of very potent toxins to humans and all the attempts carried out up to now to detoxify them have proved to be of very little effect. Even the widespread belief that cooking would make bivalves safe for consumers is false: heat in fact does not eliminate the danger of poisoning.

Toxic species of the dinoflagellate genus *Alexandrium* and *Gymnodinium catenatum* (Fig.1) that also occur in coastal waters of north-western Europe produce potent neurotoxins that cause paralytic shellfish poisoning (PSP) by blocking the neuronal sodium channels. Incidents of poisoning leading to deaths have been recorded, but the majority of afflicted people recover after suffering partial paralysis. Apparently the degree of toxicity of the mussels varies considerably or there are differences in susceptibility between persons. The non-bloom forming dinoflagellates *Dinophysis* spp. (Fig. 1) produce toxins (DSP) which are also accumulated by bivalves and passed on to humans where they affect the gastrointestinal system causing vomiting and diarrhoea. New toxins are coming to light such as the case of amnesic shellfish poisoning (ASP) caused by domoic acid, an excitatory neurotransmitter which provokes the degeneration of neural cells. When neural cells in the hippocampus, (the part of the brain involved in memory), are affected, memory losses can occur. Amnesic shellfish poisoning caused by diatoms of the genus *Pseudo-nitzschia* (Fig.1) was first recorded in Canada in 1987 (Bates et al. 1998). Recently, ASP toxins have been detected in bivalves from European waters. It is likely that toxins in more microalgae will be found, the conclusion being that bivalves represent a hazardous food.

Sea food, in particular bivalves, have been a mainstay of human diet since prehistoric times and the remains of fossilized shells (called middens) along the coast are clearly due to human collectors. Middens are common along many shores in Europe and it is reasonable to assume that hunter-gatherer societies will have eaten shellfish whereever available. Based on the isotopic composition of the bones of mesolithic Danes, Tauber (1981) concludes that their diet must have been heavily based on sea food because it is comparable to that of modern Inuit. Dale et al. (1993) speculate that the shift from sea food to terrestrial food could have been related to the appearance of toxic blooms that led to the change in diet. However, there are alternate explanations such as the spread of agriculture and animal husbandry or migrations to the coast of populations from inland regions with no experience of sea food. An analogy between the geography of bivalve utilisation and that of mushrooms as a source of human food can perhaps be drawn. Thus, mushrooms are not utilised routinely by inhabitants of the British Isles in contrast to Eastern Europe where distinguishing edible from poisonous mushrooms has been developed to a fine art and passed down from generation to generation. Sporadic cases of poisoning have not led to discontinuation of the tradition because they are related to mistakes or lack of knowledge. However, because of the insiduous nature of marine toxins, a comparable body of knowledge regarding when and where to avoid bivalves could not develop. Not surprisingly, bivalves are shunned in many regions of the world such as the Middle East and the northwestern Indian Ocean although fish and crustacea are relished almost everywhere.

In the past centuries bivalves have not been particularly popular around the North Sea, Scandinavia and the British Isles possibly due to incidences of PSP and DSP. However, PSP will have been quite rare and restricted to a few regions in contrast to DSP caused by *Dinophysis* which is a regular component of summer plankton throughout the coastal regions of Europe. Before the advent of food quality control and hygiene, people often suffered gastro-intestinal problems which on occasion might have been due to DSP, but were not diagnosed as such. The fact that there exists a rule not to eat mussels in the months without the letter 'r' along the North Sea coast might well be derived from frequency of mussel related illnesses in the summer. The reason was not known at the time this rule was established and even today most people think it is due to mussels not being tasty or keeping badly in the summer.

In Europe the consumption of shellfish has increased dramatically in the past decades. This is partly due to increased production from shellfish farms but the main reason is the ability to monitor toxicity and guarantee safety for the consumers. Previously, shellfish were a regular part of the diet only in certain regions such as France and the Mediterranean. The first report of mussel farming in Europe comes from France, close to La Rochelle, where an Irishman started the cultivation of bivalves on poles in 1236 (Russo 1999). In Italy the cultivation of mussels (*Mytilus galloprovincialis*) only started around 1750. Romans were known to consume different bivalve species and they were apparently the first to start the cultivation of oysters close to Naples in the 2nd century BC and used terracotta collectors for the oyster larvae to settle. The cultivation of oysters spread to France around 1850 using the ancient Italian technique. Naturally occurring oyster beds along the Schleswig-Holstein North Sea coast were exploited till about the end of the nineteenth century after which they disappeared and have not returned. Currently mussel beds in the North Sea have deteriorated despite restrictions on overfishing. Possibly, environmental factors not yet understood are responsible for fluctuations in bivalve abundance and distribution (Lindeboom, this volume).

That fish can also be immune to potent neurotoxins is demonstrated by ciguatera poisoning which is transmitted to humans via tropical reef fish. A dinoflagellate (*Gambierdiscus toxicus*) living in epiphytic association with benthic macroalgae produces the potent neurotoxins gambiertoxin and ciguatoxin which are accumulated by both herbivorous and predatory fish (e.g. barracudas) without affecting them (Hallegraeff 1993). Since the fish are toxic only sporadically, cautionary measures such as testing the fish on a cat before eating is customary in areas where it occurs, and a folk lore for determining toxicity of fish and therapeutic practices has developed. Besides ciguatera-poisoning, little is known about the pathways through which different toxins can be passed up the food webs through zooplankton and fishes to birds and mammals. Since the effect of toxins on many zooplankers such as copepods appears to be minimal (Turner and Tester 1997), it is possible that algal toxins can be passed up planktivorous food chains as well. However, to our knowledge, there are no reported cases of algal toxins passed on to humans via fish outside the tropics. Possibly this is the reason why there are no restrictions on eating 'regular' fish in any region or religion outside the tropics that we are aware of.

Fish Kills

Some dinoflagellate species are toxic to fish and are responsible for mass fish kills, particularly salmon kept in cages for aquaculture although wild fish are also killed. Examples are *Gymnodinium mikimotoi* blooms along the North European coasts and in Japan and the neurotoxin-producer *Gymnodinium breve* in the Gulf of Mexico. An unusual case not reported from Europe is that of the dinoflagellate *Pfisteria piscicida* which occurs in several estuaries along the Atlantic US coast where it kills fish like an ambush predator. It has a large number of life history stages, including motile cells, non motile cysts and amoeboid stages and is reported to cause mass death of fish such as menhaden (a herring-like fish) and feed on cells from the fish surface (Burkholder and Glasgow 1997). Humans exposed to waters containing high concentrations of this species suffered from gastro-intestinal and respiratory symptoms and neurological disorders.

Tons of cage-farmed fishes have been killed by flagellate phytoplankters such as the haptophytes *Chrysochromulina polylepis, Prymnesium parvum*, and the raphidophytes *Chattonella subsalsa, Fibrocapsa japonica* and *Heterosigma akashiwo*. It was the unprecedented massive bloom of *Chrysochromulina polylepis* (Fig. 1) in the Kattegat and Skagerrak in May 1988 that brought the problem of HABs to the attention of a horrified European public. The media called it the 'killer alga' and treated it on a scale with the notorious 'killer bees'. Although humans were not harmed, fish but also many invertebrates and even macroalgae were killed along an extensive stretch of coastline and the aquaculture industry sustained huge economic losses. This species is widespread and a regular member of late spring phytoplankton but it only reached exceptionally high concentrations in 1988 due to an unusual combination of physico-chemical factors (weather, nutrients and hydro-graphy) (Granéli et al. 1993). Ichthyotoxins produced by Haptophyta and Raphidophyta are poorly known; they cause haemolytic and irritating effects, alter membrane permeability, but are apparently not transmitted through the food chain. It is likely that some other heterotrophic protists also contain potent toxins that are as yet unknown because the organisms, by virtue of their nutrition mode, cannot reach the high concentrations that photosynthesizing organisms can attain.

Are HAB Events Increasing? Eutrophication and Climatic Changes

A much discussed issue is whether the apparent increase in HAB events is actually occurring or whether it is a result of increasing awareness and intensified study (Hallegraeff 1993). If the former view is correct then it begs an explanation. Smayda (1990) argues that we are in the midst of a "global epidemic" brought about by human-driven changes. Northern Europe is one of the regions from which we have the best evidence, both because of the high level of anthropogenic impact in the area and because of the presence of nutrient and phytoplankton long-term records (Smayda 1990; Smetacek et al. 1991). Eutrophication due to enhanced nutrient loads (N and/or P, or other microelements) emanating from human activities (intensive agriculture, sewage discharge, acid rain) has caused an increase of phytoplankton blooms, both toxic and harmless, in many restricted coastal areas. There is evidence from the long-term records collected in Heligoland (Smetacek et al. 1991) that eutrophication has indeed occurred in the southern German Bight indicated not only by increasing nitrate loading but also a reduction in silica availability. The latter is due to several factors and results in a relative decline of the role of diatoms because they have an absolute silica requirement to make their cell walls. As a consequence, flagellates are favoured and since most HAB species belong to the latter heterogeneous group, an increase in their occurrence is to be expected. However, although it has been argued that *Phaeocystis* blooms have increased along the southern North Sea (Riegmann et al. 1992) blooms of toxic species have not increased concomitantly. This has happened in Tolo Harbour (Hongkong) and the Inland Sea of Japan where enforcement of measures to curb eutrophication resulted in a decline in toxic HAB events (Hallegraeff 1993). The geographic spread of HABs has also been related to transport of coastal species across oceans in the ballast water of bulk carriers (Hallegraeff 1993).

A shift in the species composition of coastal waters can also be brought about by changes in local circulation patterns. Thus, the residence time of coastal water can be either prolonged or shortened depending on whether exchange rates with the surrounding ocean decrease or increase respectively. Such changes can be brought about by cyclical phenomena such as the well-known El Niño but also the North Atlantic Oscillation (Strömberg 1997). For example, the extensive harmful algal blooms recorded along the southern Chinese coast in co-occurrence with the particularly strong El Niño event of 1997-1998 (Yin et al. 1999) would be the result of 'natural' climatic phenomena and not anthropogenically driven. Unless, of course, the natural cycles are accentuated by CO_2 induced global warming which is also under debate. Whatever the reason, the mechanisms resulting in increased HAB events need to be elucidated and require detailed studies of the life cycles and the ecology of the individual species populations in relation to shifts in hydrography.

Long-term monitoring of phytoplankton composition that are being carried out in different geographic areas provide valuable data for understanding relationships between specific environmental conditions/trends with changes in the phytoplanktonic community. Unfortunately, most series were commenced after pristine pre-eutrophication conditions had ceased to exist. One of the first long-term investigations of phytoplankton seasonality was commenced by Braarud in the Oslo Fjord in the early forties. He mentions that 'it is remarkable that dinoflagellates play such an important part in the plankton of the polluted regions' and mass occurrence of 'green flagellates' in enriched waters contrast with their reduced abundances in more pristine waters (Smayda 1993). Braarud was also one of the first to address 'red-tide' issues, although at that time HAB events were of little consequence for the Norwegian economy and shellfish were not a significant dietary item.

An alternative method to gain information on trends and patterns of species distribution over long temporal scales is to study the sediment record although most HAB species are not likely to be represented there. This approach based on shifts in diatom species composition has been successful in tracking eutrophication trends and the impact of acid rain in lakes. Unfortunately, most coastal marine diatoms are weakly silicified and HAB species such as *Pseudo-nitschia* are not preserved in oxic sediments. However, a number of HAB dinoflagellate species have fossilizable cysts that persist in marine sediments.

Cysts are non-motile resting stages produced by a number of dinoflagellate species. Many cysts have a cell wall consisting of a sporopollenin-like substance (dinosporine), which enables preservation in the sediments. Fossil cysts have been extensively employed as biostratigraphic markers and recent studies are supporting their use as paleoenvironmental and paleooceanographic markers for Quaternary strata. Cyst-forming, harmful dinoflagellates include many PSP toxin-producers such as several *Alexandrium* species, *Gymnodinium catenatum* (Fig. 1) and *Pyrodinium bahamense*. Cysts are also produced by *Protoceratium reticulatum*, a widely distributed species that has recently been found to produce yessotoxin (Satake et al. 1997) which causes DSP-like symptoms. *Alexandrium* species produce oval or rounded, featureless cysts, which are apparently not preserved in the sediments or, at least, are destroyed by the harsh palynologic methods used for the preparation of the samples. However, fossil cyst remnants of other species in marine sediment cores can serve as an indication of their abundance and distribution in the past.

A clear relationship between climatic changes and '*Gymnodinium catenatum*-like cyst' abundances is reported from Holocene core samples from the North Sea (Dale et al. 1993; Fjellsa and Nordberg 1996; Thorsen and Dale 1998). This species forms a microreticulate cyst which was thought to be 'unique' until very recently when two other species, closely related to *G. catenatum* but non-toxic, turned out to form very similar cyst morphotypes (Bolch et al. 1999). Due to their smaller size, the fossil microreticulate cysts recorded in the North Sea core samples should not be attributed to the toxic *G. catenatum*. We keep the notation '*Gymnodinium catenatum*-like cyst' when referring to the fossil cyst morphotype. These cysts first appeared in the Kattegat about 6200 yr BP corresponding to a warmer period. Peak abundances were reached at 4300-4500 BP, before disappearing at about 3600-3900 yr BP. At that time, there was a change in the oceanographic regime in the north east Atlantic/North Sea region and colder climatic conditions were established in the area about 3000 yr BP, as documented by the lowered timber-line in the Scandinavian mountains. '*G. catenatum*-like cysts' appeared again ca. 2000 yr BP and massive deposits were subsequently recorded between 700 and 800 yr BP, a period that corresponds to the Mediaeval warm period. Again, the subsequent disappearence of the species around 300 - 400 yr. BP is thought to be related to another cooling event: the coldest phase of the little Ice Age (1550-1600 AD) (Fjellsa and Nordberg 1996).

A study carried out on sediment cores from southern Norway over the last 2000 yrs (Thorsen and Dale 1998) confirmed similar distribution patterns of '*G. catenatum*-like cysts', supporting the use of these cysts as an indicator of warmer water intrusions on the southern Scandinavian shelf (Fig. 2). Although the fossil microreticulate cysts found in sediment cores may not belong to the toxic *G. catenatum*, these studies provide a good example of how changing climatic conditions can affect a component of the phytoplankton population. Small sized microreticulated cysts have been recently recorded in surface sediments of Denmark and the German Bight and they apparently co-occur with *G. catenatum* cysts along the Spanish Atlantic coast. Perhaps these findings might represent a signal of increased climatic warming of the last decades?

The distribution patterns of fossilizable dinoflagellate cysts in core samples has been studied also to test their possible value as indicators of anthropogenic eutrophication. Dale et al. (1999) examined sediment cores from the Oslo Fjord, whose inner basins are separated from the outer fjord by a shallow sill which restricts water exchange. Core samples date back to 1700, prior to the population increase in the area which started in the mid-1800s and in the early 1900s, associated with industrial development. The eutrophication signal comprises both an increase in total cyst numbers and a proportional increase of *Lingulodinium polyedrum* cysts, a species at times reported to cause dense blooms in coastal waters. Both total cyst numbers and the relative percentage of *L. polyedrum* cysts show a noticeable decrease by the mid 70s, concomitantly with improved sewage treatment. *L. polyedrum* cysts were previously considered as indicators of lower salinity, due to their occurrence in estuarine areas. However, their high abundances in shelf sediments

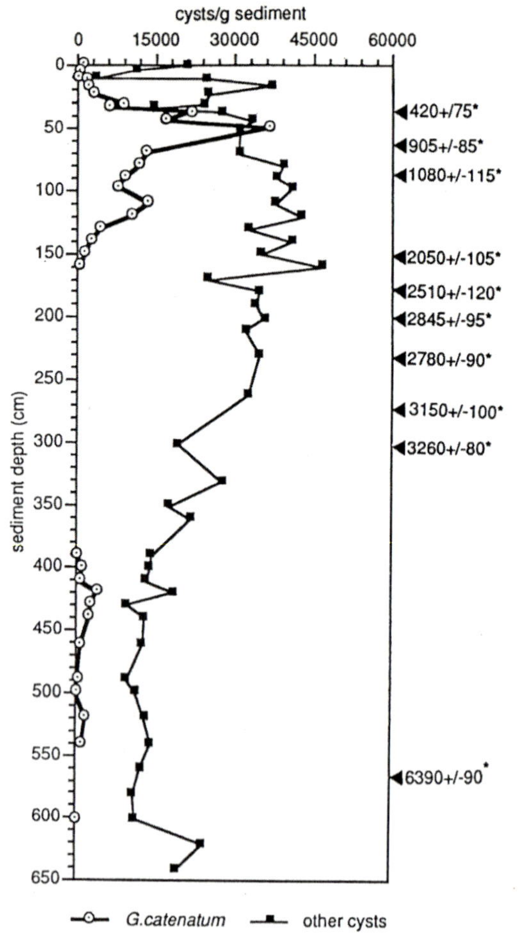

Fig. 2. Quantitative distribution of *Gymnodinium catenatum* and other cysts in Core 8902 from the Kattegatt *= ^{14}C datings (reprinted from Dale et al. 1993, p. 49).

associated with upwellings events and their positive relationship with eutrophication in the Oslo Fjord would suggest that these cysts are indeed good indicators of eutrophication and their presence in estuarine environments is related to the high nutrient loads generally associated with riverine outflows.

Conclusions

Without doubt, harmful algal blooms now represent a serious problem for aquaculture and we suggest that they have also played an important role in the history of northwestern Europe. With increasing pressure on the coastal environment and the growing demand for sea food the mitigation of HABs is a new challenge for marine scientists. Despite the large number of research programmes that have been launched in regions where HABs occur, little progress has been made in understanding, hence predicting their appearance. Judging from the spate of new information, this research is still in the information-gathering phase. The international programme 'Global Ecology and Oceanography of Harmful Algal Blooms' (GEOHAB) was launched in 1998 by the two international oceanographic organisations SCOR and IOC with an aim to co-ordinating research and co-operation to develop international capabilities for assessment, prediction and mitigation of harmful algal blooms. The scientific goal of GEOHAB is to: 'Determine ecological and oceanographic mechanisms underlying the population dynamics of harmful algae, by the integration of biological and ecological studies with chemical and physical oceanography, supported by improved observation systems.' This concerted effort will hopefully soon deliver the insights necessary to establish scientifically based programmes to cope with the problem. The ensuing knowledge will also shed new light on the history of human encounters with toxic phytoplankton.

Acknowledgments

We are grateful to Barrie Dale for comments on an earlier draft and to Federica Cerino for drawing Fig. 1.

References

Bates BB, Garrison DL, Horner RA (1998) Bloom dynamics and physiology of domoic-acid-producing *Pseudo-nitzschia* species. In: Anderson DM, Cembella AD, Hallegraeff GM (eds) Physiological Ecology of Harmful Algal Blooms. Springer-Verlag, Berlin, Heidelberg, pp 267-292

Bhattacharya D, Medlin L (1998) Algal phylogeny and the origin of land plants. Plant Physiol 116: 9-15

Bolch CJS, Negri AP, Hallegraeff GM (1999) *Gymnodinium microreticulatum* sp. nov. (Dinophyceae): a naked, microreticulate cyst-producing dinoflagellate, distinct from *Gymnodinium catenatum* and *Gymnodinium nolleri*. Phycologia 38: 301-313

Burkholder JM, Glasgow Jr HB (1997) *Pfisteria piscicida* and other *Pfisteria*-like dinoflagellates: behavior, impacts and environmental control. Limnol Oceanogr 42: 1052-1075

Cembella AD, Milenkovic L, Doucette G, Fernandez M (1995) Major phycotoxic syndromes and mode of action. In: Hallegraeff GM, Anderson DM, Cembella AD (eds) Manual of Marine Microalgae. IOC Manuals and Guides No. 33, UNESCO, pp 177-179

Codd GA (1996) Awareness of cyanobacterial or algal blooms in the Premonstratensian Monastery of the Green Loch, Soulseat Scotland, from the twelfth century, and cattle poisonings attributed to cyanobacterial hepatotoxins at this location eight hundred years later. Harmful Algae News, an IOC Newsletter on Toxic Algae and Algal Blooms, 15: 4-5

Codd GA (1997) Cyanobacterial blooms and toxins in fresh-, brackish and marine waters. In: Reguera B, Blanco J, Fernandez MªL, Wyatt T (eds) Harmful Algae. Xunta de Galicia and Intergovernmental Oceanographic Commission of UNESCO, pp 13-17

Dale B, Madsen A, Nordberg K, Thorsen TA (1993) Evidence for prehistoric and historic 'blooms' of the toxic dinoflagellate *Gymnodinium catenatum* in the Kattegat-Skagerrak region of Scandinavia. In: Smayda TJ Shimizu Y (eds) Toxic Phytoplankton Blooms in the Sea. Elsevier, Amsterdam, pp 47-52

Dale B, Thorsen TA, Fjellsa A (1999) Dinoflagellate cysts as indicators of cultural eutrophication in the Oslofjord, Norway. Estuarine Coast Shelf Sci 48: 371-382

Falconer IR (1996) Potential impact on human health of toxic cyanobacteria. Phycologia 35: 6-11

Fjellsa A, Nordberg K (1996) Toxic dinoflagellate

"blooms" in the Kattegat, North Sea, during the Holocene. Palaeogeogr Palaeoclimatol Palaeoecol 124: 87-105

Geraci JR, Anderson DM, Timperi RJ, Aubin DJS, Early GA, Prescott JH, Mayo CA (1989) Humpback whales (*Megaptera novaeangliae*) fatally poisoned by dinoflagellate toxin. Can J Fish Aquat Sci 46: 1895-1898

Ginzburg C, Rosenthal R (1993). Ecstasies: Deciphering the Witches' Sabbath. Penguin, USA

Granéli E, Paasche E, Maestrini S (1993) Three years after the *Chrysochromulina polylepis* bloom in Scandinavian waters in 1988: some conclusions of recent research and monitoring. In: Smayda TJ, Shimizu Y (eds) Toxic Phytoplankton Blooms in the Sea. Elsevier, Amsterdam, pp 23-32

Hallegraeff GM (1993) A review of harmful algal blooms and their apparent global increase. Phycologia 32: 79-99

Jacoby JM, Gibbons, HL, Hanowell R., Bouchard DD (1994) Wintertime blue-green-algal toxicity in a mesotrophic lake. J Freshw Ecol 9: 241-251

Montresor M, Zingone A, Sarno D (1998) Dinoflagellate cyst production at a coastal Mediterranean site. J Plankton Res 20: 2291-2312

Nehring S (1993) Mortality of dogs associated with a mass development of *Nodularia spumigena* (Cyanophyceae) in a brackish lake at the German North Sea coast. J Plankton Res 15: 867-872

Paerl HW (1996) A comparison of cyanobacterial bloom dynamics in freshwater, estuarine and marine environments. Phycologia 35: 25-35

Pillenger JM, Cooper JA, Ridge I (1994) Role of phenolic compounds in the antalgal activity of barley straw. J Chem Ecol 20: 1557-1569

Raven JA (1997) Phagotrophy in phototrophs. Limnol Oceanogr 42: 198-205

Riegman R, Noordeloos AAM, Cadée GC (1992) *Phaeocystis* blooms and eutrophication of the continental coastal zone of the North Sea. Mar Biol 112: 479-484

Russo GF (1999) Cambiamenti faunistici nel Mar Mediterraneo: un approccio zooarcheologico. Accademia Internazionale di Scienze e Tecniche Subacquee Ustica 22: 63-73

Satake M, MacKenzie L, Yasumoto T (1997) Identification of *Protoceratium reticulatum* as the biogenetic origin of yessotoxin. Natural Toxins 5: 164-167

Scholin CA, Gulland F, Doucette GJ, Benson S, Busman M, Chavez FP, Cordaro J, DeLong R, Vogetaere AD, Harvey J, Haulena M, Lefebvre K, Lipscomb T, Loscutoff S, Lowenstine LJ, Marin III RM, Miller PE, McLellan WA, Moeller PDR, Powell CL, Rowles T, Silvagni P, Silver M, Spraker T, Trainer V, Dolah FMV (2000) Mortality of sea lions along the central California coast linked to a toxic diatom bloom. Nature 403: 80-84

Sherratt A (1997) Economy and Society in Prehistoric Europe. Changing Perspectives. Edinburgh University Press, Edinburgh

Smayda TJ (1990) Novel and nuisance phytoplankton blooms in the sea: evidence for a global epidemic. In: Granéli E, Sundström B, Edler L, Anderson DM (eds) Toxic Marine Phytoplankton. Elsevier Science Publishing, New York, pp 29-40

Smayda TJ (1993) Flagellates and their blooms at sea: Trygve Braarud's contributions to our knowledge. In: Smayda TJ, Shimizu Y (eds) Toxic Phytoplankton Blooms in the Sea. Elsevier, Amsterdam, pp 3-12

Smayda TJ (1997) What is a bloom? A commentary. Limnol Oceanogr 42: 1132-1136

Smetacek V (1985) The annual cycle of Kiel Bight plankton: a long-term analysis. Estuaries 8: 145-157

Smetacek V (1998) How mainstream biological oceanography can profit from harmful-algal-studies and viceversa. In: Reguera B, Blanco J, Fernandez MªL, Wyatt T (eds) Harmful Algae. Xunta de Galicia and Intergovernmental Oceanographic Commission of UNESCO pp 109-113

Smetacek V, Bathmann U, Nöthig EM, Scharek R (1991) Coastal eutrophication: causes and consequences. In: Mantoura RFC, Martin JM, Wollast R (eds) Ocean Marine Processes in Global Change. Wiley and Sons, London, pp 251-279

Sommer H, Whedon WF, Kofoid CA, Stohler F (1937) Relation of paralytic shellfish poison to certain plankton organisms of the genus *Gonyaulax*. Am Med Assoc Arc Pathol 24: 537

Sournia A (1995) Red tide and toxic marine phytoplankton of the world ocean: an inquiry into biodiversity. In: Lassus P, Arzul G, Erard-Le Denn E, Gentien P, Marcaillou C (eds) Harmful Marine Algal Blooms. Lavoisier, Intercept Ltd, Paris, pp 103-112

Strömberg J-O (1997) Human influence or natural perturbation in oceanic and coastal waters - can we distinguish between them? Hydrobiologia 352: 181-193

Tauber H (1981) ^{13}C evidence for dietary habits of prehistoric man in Denmark. Nature, 292:332-333

Thorsen TA, Dale B (1998) Climatically influenced distribution of *Gymnodinium catenatum* during the past 2000 years in coastal sediments of southern Norway.

Palaeogeogr Palaeoclimatol Palaeoecol 143:159-177

Turner JT, Tester PA (1997) Toxic marine phytoplankton, zooplankton grazers, and pelagic food webs. Limnol Oceanogr 42: 1203-1214

Verity PG, Smetacek V (1996) Organism life cycles, predation, and the structure of marine pelagic ecosystems. Mar Ecol Prog Ser 130: 277-293

Wasmund N (1997) Occurrence of cyanobacterial blooms in the Baltic Sea in relation to environmental conditions. Int Rev Ges Hydrobiol 82: 169-184

Work TM, Barr B, Beale AM, Fritz L, Quilliam LA, Wright JLC (1991) Epidemiology of domoic acid poisoning in brown pelicans (*Pelecanus occidentalis*) and Brandt's cormorans (*Phalacrocorax penicillatus*) in California. J Zoo Wildlife Med 24: 54-62

Wyatt T (1998) Harmful algae, marine blooms, and simple population models. Nature & Resources 34: 40-51

Yin K, Harrison PJ, Chen J, Huang W, Qian P-Y (1999) Red tides during spring 1998 in Hong Kong: is El Niño responsible? Mar Ecol Prog Ser 187: 289-294

Climate and Human Induced Impacts on the Coastal Zone of the Southern North Sea

Victor Smetacek[1], Gerold Wefer[2], Jürgen Alheit[3], Frank Lamy[2], Adrian de Kraker[4], Burghard Flemming[5], Han Lindeboom[6], Hansjörg Streif[7], Tjeerd van Weering[6], Hans von Storch[9]*

[1]Alfred-Wegener-Institute for Polar and Marine Research, Am Handelshafen 12, D-27570 Bremerhaven, Germany
[2]University of Bremen, Department of Geosciences, Postfach 33 04 40, D-28334 Bremen, Germany
[3]Baltic Sea Research Institute, Seestraße 15, D-18119 Warnemünde, Germany
[4]University of Amsterdam, Department of Human Geography, Nieuwe Prinsengracht 130, NL-1018 VZ Amsterdam, The Netherlands
[5]Senckenberg Institute, Schleusenstraße. 39A, D-26382 Wilhelmshaven, Germany
[6]Netherlands Institute for Sea Research (NIOZ), P.O. Box 59, NL-1790 AB Den Burg, The Netherlands
[7] Niedersächsisches Landesamt für Bodenforschung, Stilleweg 2, 30655 Hannover, Germany
[8]Institute for Coastal Research, GKSS Research Center, Max-Planck-Strasse 1, D-21502 Geesthacht, Germany
** corresponding author (e-mail): vsmetacek@awi-bremerhaven.de*

Abstract: Ongoing changes in the morphology and ecology of the coastal zone and expected future scenarios with emphasis on the southern North Sea were discussed and research needs identified. The impact of sea level rise and the effects of resultant dyking activity need to be considered for the entire coastal zone as manipulation in one region can have unexpected effects in neighbouring areas. The extent to which the barrier islands can be maintained in their present state has also to be critically assessed. Anthropogenic impacts on coastal ecosystems can be direct (e.g. diking, fisheries) or subtle (pollutants). However, differentiating climate-induced from anthropogenic impacts is not always obvious. Long-term, integrated data sets provide the best context for evaluating observations of individual phenomena. Maintenance of monitoring programmes and introduction of new methodologies to extend coverage and detail is required for this purpose. The role of public perception was discussed at length as coastal management is not as much management of nature as it is management of human activity. Hence it is essential that social scientists be involved to a greater extent in decision-making processes.

Introduction

The coastal zone has been occupied by humans for many millennia but it is only since about the 16th century that anthropogenic impact has become a significant shaping force. Coastal plain environments such as the southern coast of the North Sea are most strongly impacted because dyke building fundamentally alters the land-sea interface, hence also the coastal sea and its ecosystems. Until recently coastal engineering was mainly concerned with protecting the land from ravages of the sea but experience gained so far has led to the realisation that the entire coastal system needs to be treated as a whole, as modifications in one area can lead to unexpected developments in adjoining regions.

From WEFER G, BERGER W, BEHRE K-E, JANSEN E (eds), 2002, *Climate Development and History of the North Atlantic Realm.* Springer-Verlag Berlin Heidelberg, pp 473-486

Besides, the impact on marine ecosystems is increasingly becoming a cause for concern.

This report summarises the discussion that were held by a multidisciplinary group at the Hanse-Wissenschaftskolleg. The intention of the discussions was to ascertain what has been learned so far and what can be expected in the future.

Coastline, Islands and Sediments

The southern North Sea is roughly triangular in shape. From the Rhine delta the coastline stretches in north-eastern to eastern directions as far as the Elbe estuary where it turns northward along the western coast of Schleswig-Holstein and Jutland, Denmark. A maximum water depth of the southern North Sea of about 70 m is encountered on the north-western flank of the Dogger Bank; however, the depth is generally less than 45 m in the inner part of the German Bight.

The following tidal zones can be distinguished along the southern North Sea coast. Low mesotides with an amplitude between 1 and 2 m occur between the Rhine delta and the island of Terschelling, the Netherlands. High mesotides with an amplitude of 2 to 3 m occur eastward from Terschelling as far as the island of Wangerooge. Low macrotides with an amplitude of more than 3 m occur in the innermost part of the German Bight (Siefert and Lassen 1985), between the tidal inlet of the Jade Bay and the peninsula of Eiderstedt, Schleswig-Holstein, reaching a maximum value of 4.2 m at Bremen on the Weser estuary. The tidal amplitude decreases in a northerly direction to that characteristic of high mesotides between Eiderstedt and Sylt, to that of low mesotides between Sylt and Blåvands-Huk, and to that of microtides on the north of Nymindegab, Denmark.

The present tidal conditions are partly mirrored by the coastal morphology. Continuous barrier systems partly with lagoons (Ringkøbingfjord and Nissumfjord) exist in the zone of small tidal amplitudes along the western coasts of the Netherlands and Denmark. Elongated barrier islands (West Friesian and East Friesian islands) and sheltered tidal flats occur in the zone which is affected by low to high mesotides. Characteristic for the macrotidal conditions in the innermost part of the German Bight are extended open tidal flats and rounded to sickle-shaped sandy shoals (Mellum, Scharhörn, Trischen, Tertius, and Blauort), which in part are mobile and ephemeral structures.

Additionally, the present-day morphology of the coastal landscape is strongly influenced by geological elements which were formed prior to the Weichselian late glacial and Holocene sea-level rise. Cores of Pliocene and Pleistocene sediments, surrounded by Holocene coastal deposits, crop out on the islands of Sylt, Schleswig-Holstein, as well as on Texel, in the Netherlands. Other structures were formed by the medieval storm surges which destroyed extended areas of former coastal marshlands and formed huge tidal bays like the former Zuider Zee (nowadays Ijsselmeer) and the Lauwers Zee in the Netherlands. Comparable structures are the bay flats existing along the coast of Niedersachsen, Germany, Dollart and Jade Bay. Severe destruction of former coastal marshland which occurred along the western coast of Schleswig-Holstein in the course of the storm surges of 1362 and 1634 AD created a special form of islands, the so-called Halligen. These marsh islands like Nordstrand, Pellworm and a group of smaller islands consist of Holocene tidal flat to brackisch sediments as well as of peat. Special forms of the Halligen are Föhr and Amrum which have cores of outcropping Pleistocene sediments. Additionally, Amrum is characterised by a spit-like sandy beach.

Late Holocene Sea-Level Evolution

First evidence for marine influences along the southern margin of the North Sea basin date at roughly 8000 years BP when the rising sea reached a level of about 25 m below the present and began to intrude into the lower valleys of local rivers and streams draining the more elevated Pleistocene sand bodies (e.g. Hanisch 1980; Zagwijn 1986). The postglacial sea-level rise is documented in a number of sea-level curves reconstructed for different sections of the coast. This is represented in Fig. 1 by the rise of mean high-water levels which are more relevant for coastal processes (modified after Flemming and Davis 1994). Besides having some features in common, the curves reveal quite distinct local trends. Thus, the curves rise more steeply up to about 6500 years BP (»10 mm/year), followed by a period of a more gradual rise (»1.5 mm/year) up to the onset of the "Little Ice Age" in the late Middle Ages. A distinct acceleration is again observed over the past 300 years or so, mean high-water levels currently rising by as much as 2.5 mm/year. This acceleration predates any anthropogenic influence and has evidently been overlooked in recent disputes over human impacts on climate change. The current estimates of global warming over the next century (up to 5.8°C) recently reported by the IPCC has led to a corresponding upward correction of the predicted acceleration in mean sea-level rise which is now estimated to reach as much as 8.8 mm/year over the next century. How this would affect the evolution of mean high-water levels is unknown, but a threefold increase in the current rate would not appear unrealistic.

Of interest in the context of this article are also the departures in the trends of the individual curves. Quite evidently, and irrespective of potential errors in the individual reconstructions, the relative vertical motions of the sea and the local land masses are not synchronised along the shoreline of the southern North Sea, even over distances as little as 100 km. This means that the effects of sea-level rise will have slightly different impacts depending on the location along the coast.

Any acceleration in the rate of sea-level rise will also result in larger tidal ranges. This is intimated in the model illustrated in Fig. 2 (modified after Flemming and Davis 1994; based on data reported in Franken 1987). The individual curves apply to different locations along the coast and show similar temporal trends as the evolution of mean high-water levels (curve 1: western-most Dutch Wadden Sea and northern-most Danish Wadden Sea; curve 2: north of Ameland, central Dutch Wadden Sea; curve 3: north of Spiekeroog, central East Frisian Wadden Sea, Germany). The increasing tidal ranges reflect the tidal response to a deepening water body.

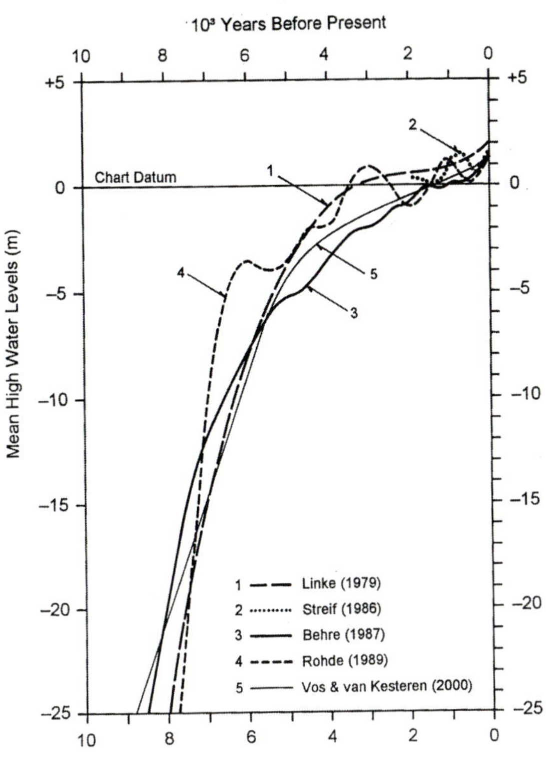

Fig. 1. Holocene evolution of mean high-water levels along different parts of the Wadden Sea coast as reported by different authors (modified after Flemming and Davis 1994). 1: Elbe Estuary near Cuxhaven, Germany (Linke 1979); 2: Island of Wangerooge, East Frisian Wadden Sea (Streif 1986); 3: East Frisian Coast (Behre 1987); 4: North Frisian Wadden Sea, Germany (Rohde1989); 5: Dutch Wadden Sea coast (Vos and van Kesteren 2000).

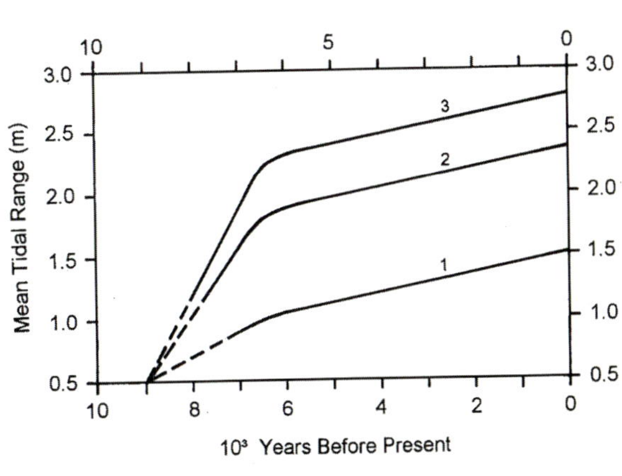

Fig. 2. Holocene evolution of mean tidal ranges at different locations along the southern North Sea coast (modified after Flemming and Davis 1994; based on numerical simulations of Franken 1987). 1: western-most and northern-most Wadden Sea; 2: central Dutch Wadden Sea; 3: central East Frisian Wadden Sea, Germany.

Although the amounts are small in the short-term (a few centimetres/century), they would nevertheless have to be added to the overall effects produced by the rising sea.

In summary, any impacts on the coast produced by an acceleration in sea-level rise will first be felt along the island shores. Beach erosion will increase at locations already experiencing erosion today, whereas currently stable shores will begin to show increasing signs of erosion in the future. Coastal management and protection will thus not only have to cope with locally increasing sand losses at increasingly shorter time intervals, but also with a rapid alongshore expansion of the affected zones.

Mean Sea Level Rise

The mean sea level has undergone an increase of the order of 100 m mainly due to growing volume of sea water in the world ocean in the past 10,000 years. This global increase of sea level is moderated by regional phenomena, mainly glacial rebound and other crustal movements (such as earthquakes).

On shorter time scales, changing circulation patterns affect sea level. On the local scale, anthropogenic effects, such as pumping gas and oil, building dykes and dredging rivers, may have a significant impact on the mean relative height of sea level and coastal land. Also the changing heat content in the ocean, reflecting atmspheric warming for natural or anthropogenic reasons, changes the volume of the ocean and thus the mean sea level. The latter effect is the only one simulated by present day atmosphere-ocean GCMs as a response to a changing composition of the atmosphere because of anthropogenic emissions.

The Intergovernmental Panel on Climate Change has found wide consensus that the global sea level has risen in the past 100 years by something of the order of 10 cm. Locally much larger increases have been found. An example is the record of Cuxhaven in the German Bight, where the mean high tide has increased by about 30 cm in the past century, reflecting probably mostly changing hydrographic conditions in the river Elbe (von Storch and Reichardt 1997). Also in Den Helder such an increase was recorded, whereas the increase in Esbjerg was much smaller (Figs. 3 and 4). In Smogen, at the Swedish coast in the Kattegat, sea level was sinking, likely due to the glacial rebound of land. In the past 40 years, mean sea level has also increased along the entire southern and eastern coast of the North Sea due to a change in the mean atmospheric circulation (Langenberg et al. 1999).

Increases in mean sea level lead to increases in levels of storm surges, even if the statistics of storms is unchanged. In fact, for Cuxhaven a significant increase in the frequency of severe storm surges was noticed, but it turned out that this increase was mostly due to the increase of the mean sea level, since the intensity of storms has not changed significantly in the past century. A similar finding resulted from a North Sea model simulation, where no significant upward trend for the intraseasonal percentiles since about 1960 was found (Langenberg et al. 1999; Kauker and Langenberg 2000). Seemingly, the increase of the intensity of storm activity since the 1960s after a decline since about 1920 (Alexandersson et al. 2000, Fig. 5) was not associated with a significant change in storm surge levels.

Changes in Tidal Range Due to Dyke Building and Channelling of Rivers

The combination of three factors influence the tidal conditions in the estuaries: the changing amount of fresh water discharge from the catchment areas, the tidal wave entering from the seaward side, and the morphology of the rivers. Whereas direct human influences on the water discharge and the tidal wave are of minor importance, anthropogenic modifications of the river morphology have a tremendous influence.

Various technical measures, aimed at different purposes, have been carried out in the estuaries of the rivers Ems, Weser, and Elbe as well as in the tidal inlet of the Jade Bay and have had strong influences on the natural tidal regimes. Of greatest importance were repeated dredging measures for the deepening of shipping channels and the construction of longitudinal training walls as well as groins. All these measures have been carried out to concentrate the natural currents in one main

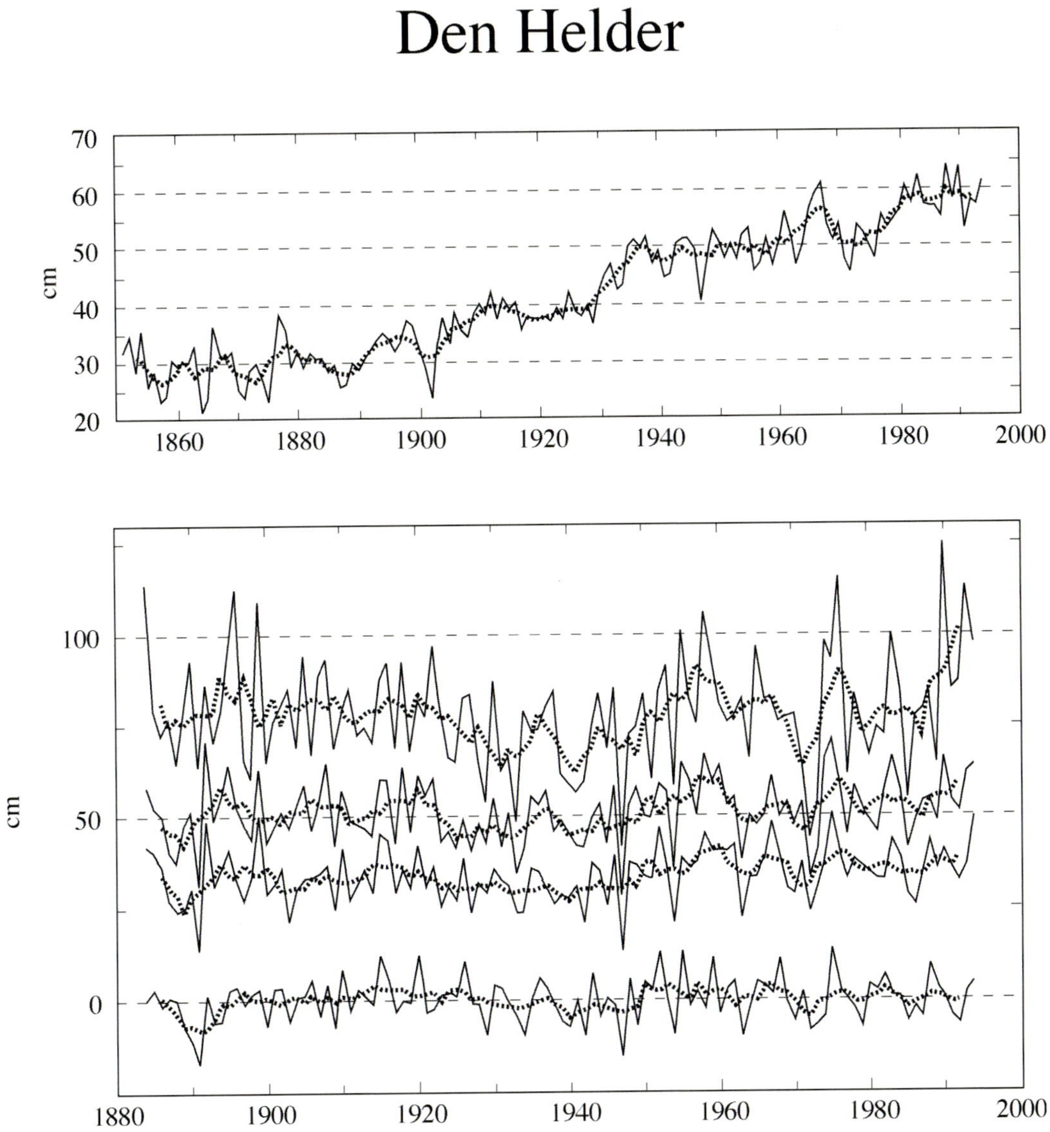

Fig. 3. Changing mean sea level (top) and of intraseasonal (DJF) percentiles after subtraction of trend in mean sea level (bottom) in Den Helder (Pfizenmayer, pers. comm.).

channel with great water depth. Additional influences have come from harbour constructions and extensive high water protection works, such as dykes and barrages against storm tides. At present artificial weirs set the upper limit of the estuaries. Such constructions exist on the river Elbe at Geesthacht, 48 km above Hamburg, on the Weser at Bremen-Hemelingen and at Herbrum on the river Ems.

The most impressive example of anthropogenic influences on the tidal regime can be observed at the lower course of the river Weser (Streif, this volume). The tidal amplitude was 13 cm at the tide gauge at Bremen in 1882. At present it reaches an amplitude of 4.18 m which is the highest tidal range along the entire German North Sea coast. This enormous change is the result of a series of technical measures which began with the first Weser

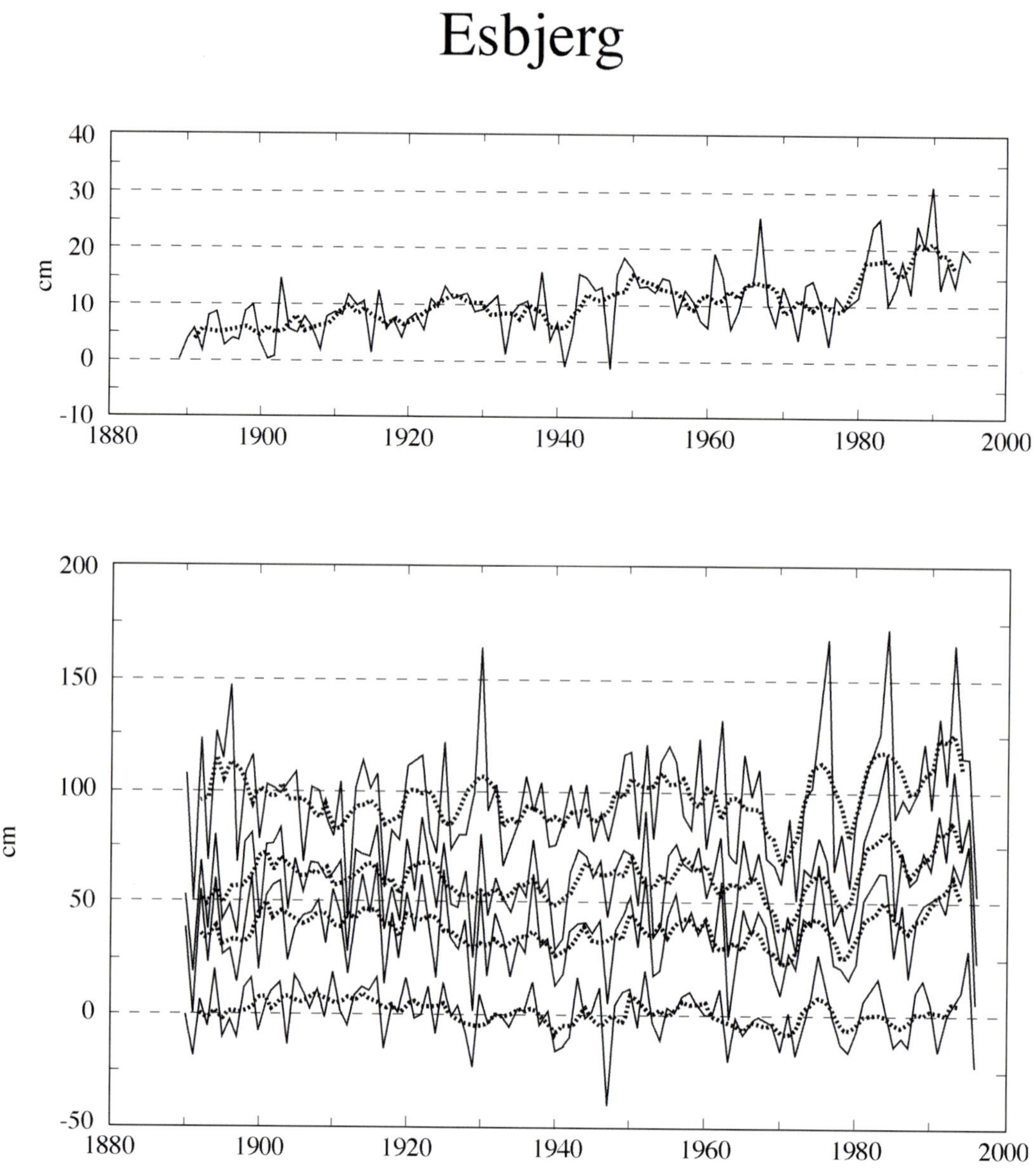

Fig. 4. Changing mean sea level (top) and of intraseasonal (DJF) percentiles after subtraction of trend in mean sea level (bottom) in Esbjerg (Pfizenmayer, pers. comm.).

correction in 1877 and led step by step to the present day conditions. The drastic increase in tidal range is mainly due to a lowering of the low water level and to a small rise of the high water level. Comparable trends, however with a lower amplitude, can be observed at the river Elbe. In the time span between 1887/96 and 1978 the tidal amplitude increased from 1.87 m to 3.35 m at Hamburg-St. Pauli and 0.07 m to 1.43 m at Geesthacht. In the period 1881/85 and 1980, the tidal amplitude of the river Ems increased from 2.18 m to 2.86 at Leer and from 0.2 m to 2.06 m at Herbrum.

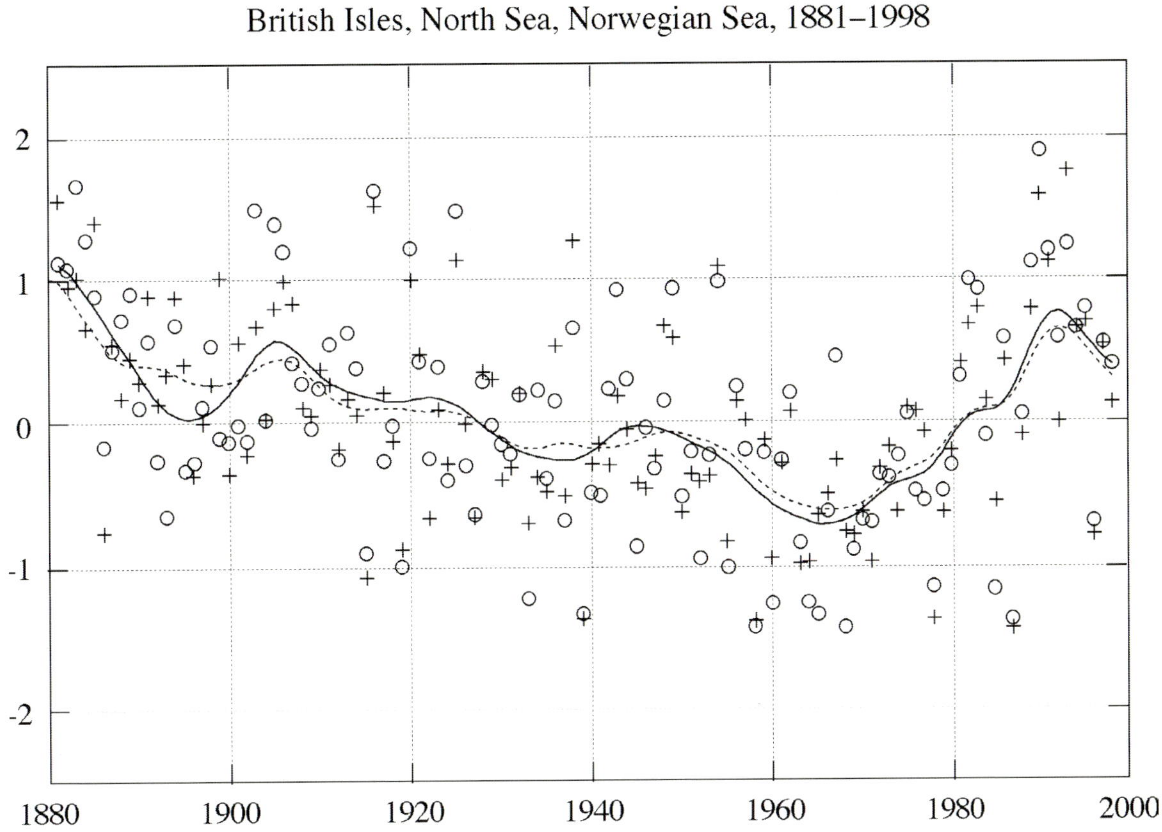

Fig. 5. WASA's index of storm intensity, derived from daily geostrophic winds calculated from various local pressure readings in the greater North Sea area. (Alexandersson et al. 2000).

Causes of Changes, Human Influence on Natural Processes

As already outlined sea level change is not a monocausal effect. Processes on geological time scales take part in changing sea level as well as processes on time scales of hours like storms and tides. Many of these processes are natural, but human impact is steadily increasing since a few centuries. Initially the human effect came about by the construction of dykes, later the dredging of rivers, the extraction of gas and oil from off-shore and on-shore deposits had a significant role in changing water levels on average and in extreme storm surge situations. Presently the main concern is related to Global Warming, i.e., the implication of global climate change due to human emissions of greenhouse gases and aerosols into the atmosphere. These emissions cause increased concentrations of the radiatively active substances in the air, which are expected to increase the amount of heat in the ocean and thus to an increase of sea level of the order of a few decimeters within the next hundred years. The ablation zones of small and large glaciers are expected to increase, changing the mass balance of these large stores of water. At the same time, global warming is also expected to intensify the global water cycle, so that more water may be deposited in the accumulation zones of the glaciers on Greenland and Antarctica. The net effect is currently unknown. Detailed research indicates that the Antarctic ice cap may actually grow, while that of Greenland is found in some analyses to shrink.

The different processes have different time scales and different forcing mechanisms. The robustness of their predictions vary accordingly. The geological processes may be considered on the managerial time scale as constant linear trends, while the prediction of changing morphodynamics and related hydrodynamics is hardly possible today. The global climate change scenarios are continuously improving, but they suffer notoriously from problems specifying regional and local features (the North Sea is a regional feature, and the Southern North Sea a local feature in this context) and precipitation statistics are needed for the assessment of the mass balance of Greenland and Antarctica.

Shifts in Perception, Polarisation of Position

The perception of the coast has seen significant changes in the past centuries and decades. Originally the coast was little more than the end of the world, the boundary to the unknown and threatening world of the ocean (Corbin 1990). People living at the coast were considered mostly backward; houses had no windows to the sea. In the late 18th century people's perception of the coast was changing. The coast (and the mountains) became the location of nature, of health and recreation. With the increasing importance of commerce and travel, the harbours became the centres of economic action – the coastal area gained an advantage over inland areas. This again changed, when shipping industry declined due to globalisation, air travel became a mass transportation system (with implications for tourism), when other means of bulk transportation (pipe lines) became available, and when the new economy emerged. Today, the coastal areas seem to have little advantages over inland areas and many coastal zones are in a depression or decay. What is left, is a potentially high environmental quality, but even this is seen as endangered by many people, because of Global Warming, pollution and over-use of resources.

Today, "the coast" is subject to various social conflicts, reflecting different uses and views of the coast (Döring et al. 2001). The perception of the coast as something natural and undistrurbed, with a healthy ecology and "soft" tourism, is in conflict with the view of a widely managed and manageable economic zone. The local population is demanding self determination (with respect to, e.g., fishery and tourism) and reject remote interference from the state government or the EU bureaucracy, whereas NGOs find further regulation needed to protect the environment against human degradation.

Recommendation for Future Research

Whereas the principles of island morphogenesis are reasonably clear, there is still considerable controversy regarding the behaviour of individual islands. Comparisons between the islands along the German Bight need to be carried out to ascertain similarities and differences. Thus, islands in front of a dyked coastline lose mud, because this material cannot be redeposited further inland. The fate of this mud is not clear and it is assumed that this material is being deposited in deeper areas of the North Sea. This needs to be verified by sediment coring.

Time series of indicators of the environmental state need to be recorded, continued and made available to the scientific community. Process based models of the ecology, the hydrodynamics, the morphodynamics and the flux and deposition of substances need to be improved. Data assimilation schemes, which combine limited dynamical knowledge (i.e. process based or empirical models) and limited observational evidence, need to be developed in order to allow for a synoptic monitoring of the coastal environment. With such data, the effect of natural variations and man-made forcing can be discriminated. The coastal module of the Global Ocean Observation Systems (GOOS) has been following goals similar to those outlined here.

Past development of environmental stress and the impact of regulation measures (e.g. related to lead in gasoline) need to be reconstructed and critically analysed, to what extent public control was efficiently exerted, or if the regulation had no significant or an unwanted impact.

Climate change scenarios need to be downscaled to the local scale of the coastal zone, and need to be combined with scenarios of future changes of other environmental factors (such as use of polders or shipping channels, emission of substances), of economy (e.g. role of agriculture and new economy) and public attitudes (e.g. with respect to flooding of polders in case of surges). The relative importance of the different factors need to be assessed, and strategies need to be defined which create conditions and options fa-vour-able for socially demanded developments.

The barrier islands need to be subjected to a cost/benefit analysis because it is not clear whether they can be maintained the way they are. The different views, and the dynamics of changing these views, need to be analysed by social scientists. In particular, the perceptions and attitudes of lay people need to be related to the process knowledge acquired by environmental scientists. To do so, case studies of past conflicts (for instance with respect to the status of the Nationalpark Nordfriesisches Wattenmeer, the reasons for the seal pandemic of 1988 or the effects of out-phasing of lead from gasoline) need to be made, and scenarios of future, changing perceptions should be attempted.

Effects of Climate and Humans on Ecosystems

Role of Climate and Natural Variability

Shifts in climate patterns occur at various scales of which the best documented are those that have the most drastic impact on the biosphere – the glacial/ interglacial cycles. Climatic changes oc-curing at scales of several years, such as El Nino, are also well documented, and their impact on upwelling ecosystems reasonably well understood. However, climate regime shifts occurring at decadal and century scales also occur but they are poorly documented and the mechanisms of their impact on ecosystems the subject of much speculation. Evidence is increasing that decadal-scale changes, such as those caused by the North Atlantic Oscillation (NAO), are accompanied by characteristic responses in marine ecosystems. It is imperative to gain a better understanding of the decadal variations because these appear to have substantial effects on ecosystem structure and functioning and also occur on the same time scales as the effects of human impact. Indeed, the areas where such cycles are believed to occur, and have been observed, coincide with regions where human presence and exploitation patterns have had strong effects on ecosystems. Thus, fluctuations in commercially exploited fish stocks have the longest documentation, whereas records of changes in the structure and productivity of the supporting ecosystem are meagre and only available from a few regions.

The fishing periods of the Bohuslän herring in the Skagerrak are known for the last millenium. During these periods lasting for several decades, the herring fishery was very strong. They alternated with periods also lasting for several decades when herring stayed away from the Bohuslän coast. These decadal fluctuations seem to be governed by the NAO (Alheit and Hagen, this volume). Similar ups and downs in small pelagic fish stocks in upwelling regions of the Pacific have been documented for the twentieth century. These stocks are dominated alternately by sardines or anchovies whereby the regime shifts last about three times those of El Nino episodes and do not seem to be related to them. Indeed, the shifts occur simultaneously in all three upwelling regions – off California, Japan and Peru – suggesting that the changes are not brought about by propagation of a particular hydrographical regime which would entail lag times in the different populations. Fisheries biologists attribute these shifts to teleconnections and are still searching for the underlying causes. Sediment records from the Santa Barbara basin off California indicate that the shifts from sardine to anchovy domination go back a very long time. Unfortunately, such sedimentary records are not available from other regions.

Another potential source of information on herring fisheries along the southern North Sea coast is documentation of the salt trade. Salt has been traded as a commodity since Roman times. In coastal areas it was extracted from peat and used on a large scale to conserve fish starting in the 14th century. When the salt-containing peat layers were exhausted, salt was shipped from the Gulf of Biscaye to the Dutch coast where it was refined and used for conserving herring. A lot of documentary sources exist on this salt trade from the 14th – 17th century which could provide interesting information on herring consumption and consequently on the issue of variability of the herring fisheries as a whole. Further, archaeological research in Belgium has provided time series of fish remains from various sites starting around 1300 and earlier in which herring and other species appear.

Other fish stocks such as cod and their relatives seem to undergo population variation at similar time scales to the pelagic fish although these large demersal fish are individually much longer lived. Their fluctuations have been explained by successful recruitment of a single year class. Thus a weather singularity which favours recruitment of a particular year class can have an impact for 20 years or more. Such an explanation cannot apply to e.g. anchovies as the periods of their domination include many generations. It is not yet clear how climatic shifts can cause fluctuations in so many different types of fish species.

Lindeboom (this volume) has presented a number of examples showing that marine ecosystems are not in steady state but undergo periodic shifts in biomass levels and abundance of species that have been difficult to explain so far. The shifts recorded from coastal systems of the North Sea have tended to be abrupt and cannot be related easily to long-term secular changes such as eutrophication or pollution. On the other hand, data collected by the Continuous Plankton Recorder surveys (CPR) from the eastern N. Atlantic and the North Sea indicate that the average phyto-plankton and zooplankton biomass fluctuated about twofold for periods of about a decade with transition periods of increasing or decreasing biomass over the entire study region. These large-scale shifts in biomass levels have been attributed to changes in the strength of westerly winds over the N. Atlantic but the mechanisms are not yet clear.

In recent years research on harmful algal blooms has intensified because of the economic effects on aquaculture, tourism and fisheries. Montresor and Smetacek (this volume) have discussed the role of these blooms on human history over the past millenium. Many toxic species belong to the dinoflagellates that tend to attain maximum population size in summer and autumn, in contrast to diatoms that appear earlier in the year when nutrients are more plentiful. The fact that dinoflagellates can build up equally high biomass levels in shelf regions in the second half of the year suggests that they have a very different strategy relative to seasonality of their environment. Thus, some species make benthic resting cysts that are preserved in sediments. Their sediment records indicate that the shifts are in the century scale and follow features such as the "medieval warm period" or the "little ice age". This could well be related to changing position of water masses bringing in populations from elsewhere that then establish themselves by means of benthic cysts. This type of periodic appearance and disappearance will apply particularly along the outer limits of the distributional range of a given species.

Can we Distinguish Human Impact from Natural Changes?

Human impacts on the coastal North Sea are considerable and range from subtle to obvious. Obvious impacts are caused by coastal engineering such as dykes, macropollutants, aquaculture, introduction of exotic species, beam trawling, cockle fishery and increasing tourism in breeding areas. Subtle changes can be due to micropollutants, shifts in nutrient ratios, the direct effects of ship traffic or the influence of local perturbation on the larger system. Some of these impacts are dealt with below.

Dyking results in the diversion of mud deposition, reduction of salt marshes, brackish water areas and muddy tidal flats. These affect the local benthos and, by divorcing the mud flat from its natural extension -the landward salt marshes - can exclude species with life-cycle stages that are dependent on the salt marsh.

The combination of excessive fertiliser use as well as the dumping of untreated sewage into coastal waters of the North Sea via the major rivers resulted in significant increases in the concentration of the macronutrients nitrate and phosphate. In addition, the ratio between nitrogen and phosphorus - the Redfield ratio of 16:1 - changed in the North Sea over time because it was technically easier to remove phosphate than nitrate. This eutrophication led to shifts in the dominant species of phytoplankton and increasing growth of green macroalgae along the shore. The latter have been held responsible for the "black spot" phenomena along some mud-flat coasts.

The silica concentration dropped with the increase in nitrogen - a trend since observed in all affected rivers - resulting in lower contribution of diatoms to total phytoplankton production. One of the shifts in phytoplankton species composition is reflected in the ratio of diatoms to non-diatoms which happen to be all flagellates. Their contribution to algal biomass is reported to have jumped at the turn of the eighties (Lindeboom, this volume). The species which increased most spectacularly is the colonial flagellate *Phaeocystis* which is often grouped with the harmful algae because its blooms lead to unsightly and odourous accumulation of foam on the beaches.

The apparent increase in flagellates in the North Sea plankton is of some concern because it encourages the incidence of toxic algal blooms during the summer months when nutrient levels used to be low. Whether toxic algae have indeed increased over the past years and, if so whether this is due to eutrophication, the spread of aquaculture or simply better monitoring is being debated (Montresor and Smetacek, this volume). It seems that the spread of toxic species is occurring at an alarming rate, partly due to transport and introduction of toxic species to formerly innocent waters.

The harmful effects of xenobiotics (man-made chemicals that are taken up by organisms) such as DDT and PCBs was a big issue in the seventies and eighties and legislation controlling their release to the environment has resulted in a significant reduction of their concentrations in natural waters. This also applies to heavy metals whose con-centrations in the rivers has gone down significantly. More recently, tributyl tin (TBT), an ingredient in antifouling paint, is reported to induce malformation and hence malfunction of the sexual organs of the whelk (*Buccinum*). Since this species has recently disappeared from the North Sea it is very likely that TBT is the sole cause. This large, carnivorous snail used to be very common and played a major role not only in the food web but its shells were occupied by other organisms and hence its presence increased biodiversity of the North Sea. If TBT is indeed the cause, then banning its use should result in return of the whelk.

From the previous section it will be clear that natural ecosystems are subject to considerable fluctuations, both in terms of total biomass as well as abundance of dominant organisms, which are very poorly understood. Hence, separating human impacts from natural variability is possible in only a few obvious cases. It should also be pointed out that both climate and human effects are external forces. Even though we have trouble separating them, it should be remembered that internal controls such as predator-prey and host-pathogen cycles also play an important role in controlling the popu-lations of wild organisms. Knowledge of these aspects is even more limited than environmental factors. In summary, much more research effort is needed before an understanding of the factors shaping structure and functioning of marine ecosystems and their major components can emerge.

Shifts in Perception of the Human Population

Before the middle of the last century marine ecosystems were perceived as providers of resources for humans to exploit or as convenient waste dumps. The environment at large was not a concern. The shift in perception was initiated in the 1960s by Rachel Carson´s book "The Silent Spring" which brought environmental pollution to the attention of the public and led to the current polarisation of positions between "exploiters" and "conservationists". The media exploited the new concern for their own ends with headlines such as "Is the North Sea dead?" in the seventies. Some years later, public concern over pollution peaked again when the "killer alga" *Chrysochromulina polylepis* spread in the Kattegat and Skagerrak followed by mass death of harbour seals along the Dutch, German and Danish coasts in May/June 1988. Later that same year, much publicity was given to masses of slime that washed up on the tourist beaches of the Adriatic. The slime was of algal origin and, like the happenings in the North Sea, it was attributed to anthropogenic impact. Since then, none of the phenomena that appeared the same year 1988 and that have since been shown to be entirely independent of each other, have recurred.

The seal population is now larger than before the pandemic and slime blooms in the Adriatic have been shown to have occurred also in the past centuries, albeit sporadically. So what is the state of public perception now? To find out, natural scientists need help from their colleagues in the social sciences.

Why is it necessary to know public perception at a given time? Because on the one hand phenomena that occur naturally but rarely, ought not to cause concern, whereas insidious changes of the kind reported by Carson need to be brought to the attention of the public. A current example is tributyl tin (TBT) and its effect on the whelk mentioned above. This is not the only species to have disappeared from the North Sea. Some such as the Atlantic gray whale became extinct altogether about 300 years ago, whereas many species of benthic invertebrates are no longer found in the Dutch part of the North Sea. Some of these species are reported to be extinct or on the verge of extinction. Although extinction of species is a perfectly natural process, the rate of modern extinction is unacceptably high. Since the preservation of biodiversity has become a major global concern, the state of marine biodiversity should be brought to the attention of the public. The policy to be pursued following public education cannot be decided by scientists; their duty is to inform the public. Bridging the gap to the social sciences and involving them in research projects should be pursued more actively. The current rapid development of scientific knowledge should not blind us to the fact that many areas touched on above are still poorly understood.

The growth of the nature conservation movement and the increasing popularity of ecotourism are clear indications of future trends. Nature is being regarded as much a part of our cultural heritage as the evolution of our own societies. Because of the general improvement in the standard of living in the last decades, particularly in Europe, one can well anticipate an increase in the value of unspoiled nature. Indeed the last semi-intact ecosystems of Europe lie in its seas and will well be regarded as equally worthy of protection as ancient houses and monuments. Hence the need to protect the remaining habitats such as the Ems-Dollard estuary that have not yet been "captured" by humans.

The movement to protect the Waddensea as a national park is a case in point. Currently it is not known to what extent shrimp fishing is at odds with the aims of the park. Such issues are currently being decided and it is the task of the scientists to provide guidance to policy decisions. Forecasts of the kind required need to be based on a better understanding of marine ecosystems. This can only arise on the basis of comprehensive data sets which need to be collected on a standardised, long-term basis. In addition intensive interdisciplinary studies of marine systems will have to be carried out in order to provide the data necessary to construct simulation models that, through continuous assimilation and sensitivity cycles, might some day achieve a reasonably realistic rendition of nature. Hope comes from nature herself: the fact that, despite the apparent complexity of habitat and species biologies, specific patterns arise in widely separated areas that bear sufficient resemblance to one another as to indicate the operation of some fairly strong and predictable forces shaping geomorphology and, in tandem, also ecosystem structure.

Research Needs

Currently, our understanding of the structure and functioning of marine ecosystems is hampered by lack of knowledge of decadal and century –scale variability, both in terms of secular trends as well as abrupt shifts in the state of the system. It is clear that only long-term, integrated data sets can provide the context for evaluating observations of individual phenomena. Hence there is an urgent need not only for maintaining regular monitoring using the original methods but also to enhance them with new methodology that will extend coverage and detail. For example, sedimentary records could be exploited more systematically to ascertain past variability. It will also be necessary to routinely measure parameters which are now suspected of being important but could not be measured earlier. One obvious path is to link biological measurements with the coastal module of the Global Ocean Observation Systems (GOOS) now being developed and which is following goals similar to those outlined here. New methods such as gene probes and other biosensors combined with flow cytometry hold much promise for following changes in plankton populations, as also remote sensing combined with ground truth measurements.

Such modernised monitoring programmes can also contribute to enhancing basic knowledge on the functioning of marine ecosystems. For instance the factors driving species succession in the plankton are still poorly understood although the phenomenon was first described a century ago for Kiel Bight. The sequence and dominance status of the species involved had barely changed between 1906 and 1972/73 (Smetacek 1985). However, a decade later a fundamental shift in species dominance was reported. The current status is unknown and it would be of interest to repeat the measurements carried out in the early seventies. The hypothesis to be tested is that, following the reduction in nutrient input during the nineties, the Kiel Bight system will be reverting back to the "pristine" situation that prevailed for at least 75 years before eutrophication had an impact. Alternatively, the system has reverted to a state characterised by other species. In either case, the results will shed light on the resilience of pelagic systems. Such studies have been carried out in many lakes that have regained their pristine pre-eutrophication status following reduction in nutrient input but there are only few equivalent studies from marine systems.

Monitoring benthic systems will profit from demarcation of areas protected from beam-trawling and other destructive fishing methods. Re-introduction of vanished species or ecosystems (such as sea-grass beds) will be worth pursuing. Networks established by the various coastal stations will promote assessment of shifts in geographical distribution patterns. It will also be necessary to specifically study the impact and fate of introduced species as the results will shed light on patterns of interaction between indigenous species. Careful, coordinated assessment of the biodiversity of characteristic areas will serve as the context for evaluating future trends. Needless to say, modelling provides a powerful tool to systematically combine the results of process-oriented research and descriptive monitoring.

The Waddensea coast is widely believed to be unsuitable for aquaculture of fish, hence fishing of wild stocks may well continue into the future. The lesson learned from history is that sustainable yields are not to be expected. Regardless of the direction in which utilisation of resources will go - the neces-sity for management will increase. The more solid the knowledge on which management is based - as is the case with the food and health services - the more effective it will be. In this context it should be pointed out that coastal management is not management of nature but rather management of human activity.

References

Alexandersson H, Tuomenvirta H, Schmith T, Iden K (2000) Trends of storms in NW Europe derived from an updated pressure data set. Clim Res 14:71-73

Behre K-E (1987) Meeresspiegelbewegungen und Siedlungsgeschichte in den Nordseemarschen. Vorträge der Oldenburgischen Landschaft 17:1-17

Corbin A (1990) Meereslust. Das Abendland und die Entdeckung der Küste. Fischer Taschenbuch Verlag, pp 1-414

Döring M, Fischer L, Settekorn W, von Storch H (eds) (2001) Küstenbilder, Bilder der Küste. Hamburg University Press (in press)

Flemming BW, Davis RA Jr (1994) Holocene evolution, morphodynamics and sedimentology of the Spiekeroog barrier island system (southern North Sea). Senckenbergiana maritime 24:117-155

Franken AF (1987) Rekonstruktie van het paleogetijklimaat in de Noordzee. Delft Hydraulics Laboratory, Report X0029-00, pp 1-74

Hanisch J (1980) Neue Meeresspiegeldaten aus dem Raum Wangerooge. Eiszeitalter und Gegenwart 30:221-228

Kauker F, Langenberg H (2000) Two models for the climate change related development of sea levels in the North Sea. A comparison. Clim Res 15:61-67

Langenberg. H, Pfizenmayer A, Von Storch H, Sündermann J (1999) Natural variability and anthropogenic change in storm related sea level variations along the North Sea coast. Cont Shelf Res 19:821-842

Linke G (1979) Ergebnisse geologischer Untersuchungen im Küstenbereich südlich Cuxhaven. Probleme der Küstenforschung 13:39-83

Rohde H (1989) Wasserstandsänderungen und technische Konsequenzen für Küstenschutzbauwerke. In: Wilhelm H-F (ed) Klimaveränderungen durch den

Menschen. CDU-Fraktion Niedersächsischer Landtag, pp 62-76
Smetacek V (1985) The annual cycle of Kiel Bight plankton: A long-term analysis. Estuaries 8:145-157
Streif HJ (1986) Zur Altersstellung und Entwicklung der Ostfriesischen Inseln. Offa 43:29-44
Von Storch H, Reichardt H (1997) A scenario of storm surge statistics for the German Bight at the expected time of doubled atmospheric carbon dioxide concentration. J Climate 10:2653-2662
Vos PC, van Kesteren WP (2000) The long-term evolution of intertidal mudflats in the northern Netherlands during the Holocene; natural and anthropogenic processes. Continental Shelf Research 20:1687-1710
Zagwijn WH (1986) Geologie van Nederland, Deel 1: Nederland in het Holoceen. Rijks Geologische Dienst, Haarlem, pp 1-46